P9-BYG-774
Rexburg, Idaho
83440
356-6957
15.00

Animal Agriculture

A Series of Books in Agricultural Science

Animal Science

Editors: G. W. Salisbury
E. W. Crampton (1957–1970)

Animal Agriculture

The Biology of Domestic Animals and Their Use by Man

Edited by

H. H. Cole   Magnar Ronning

University of California, Davis

W. H. Freeman and Company
San Francisco

COVER: FAO Photo

Library of Congress Cataloging in Publication Data

Cole, Harold Harrison, 1897–
Animal agriculture; the biology of domestic animals and their use by man.

Includes bibliographies.
1. Stock and stock-breeding. 2. Domestic animals.
I. Ronning, Magnar, joint author. II. Title.
SF61.C59 636 73–20360
ISBN 0–7167–0791–8

Printed in the United States of America

9 8 7 6 5 4 3 2 1

Preface

Our primary aims in *Animal Agriculture: The Biology of Domestic Animals and Their Use by Man* are to provide a description of the livestock industry, not only in the United States but also worldwide to the extent feasible, and to present the scientific basis for livestock-management practices. Although this book is addressed to college students without extensive backgrounds in biology, the discussions of the nutrition, inheritance, and physiology of domestic animals—knowledge of which is unparalleled except possibly for a few laboratory animals—should make this book of interest to all students in biology. It should be valuable also for the continuing education of those with careers in the animal industries and professions. Though the metric system will soon be universally adopted, the English system of weights and measures has been widely used in the research on which this book is based; so we have not converted the data to metric values. To facilitate conversion, a conversion table is inserted inside the front cover.

Although the primary emphasis here is on the livestock industry in the United States, we have included much discussion of the nutritional, physiological, genetic, and sociological factors that explain the presence of the llama in the Andes, the reindeer in the Arctic, the camel and the fat-tailed sheep in the Mediterranean area, the Shetland pony in the Shetland Islands, and the water buffalo in southeast Asia, as well as the wide distribution of the Merino breed of sheep and its offshoots in hot, arid areas with limited forage, and the presence in England, Australia, New Zealand, Canada, Argentina, and the United States of the conventional meat breeds of sheep and cattle, which not only gain rapidly but also rapidly lay down external as well as intramuscular fat. The extent to which livestock producers should go in developing animals with limited fat reservoirs is a current topic of great importance. On the one hand, the production of animals with excessive fat deposits is costly; on the other hand, intramuscular fat is important in imparting juiciness, and, in addition, fat covering of the carcass is a valuable aid in proper aging of meat. An interesting development is the current popularity of the mid-European breeds of cattle, some of which originally may have been developed for draft purposes, but which have become important as sources of meat and milk. Their rapid growth and muscularity are their most important virtues in the United States, attracting attention especially for crossbreeding.

Sheep throughout the centuries have been one of our most important domestic species, but suddenly their restricted reproductive capacity, most of them being seasonal breeders, has reduced their importance in the United States, as has the diminishing importance of wool relative to the synthetic fibers, the cost of herding on the open range, and the intrusion of row-crop agriculture in areas previously devoted to grain raising and to sheep raising as a subsidiary enterprise. Yet there are extensive areas of so-called marginal lands, both hot and cold, wet and dry, which are best-suited for sheep production. This has stimulated research aimed at developing new technology and management procedures for effective utilization of this resource.

The opening chapters deal with the origins of domestication and with the role of animals in the welfare of mankind; later chapters discuss the biology of domestic animals and the classification and marketing of animal products. We have attempted to paint only a very broad picture of animal agriculture throughout the world, and to point out the major differences from the picture in the United States, which we have described in much more detail, since the latter is the one that will likely concern the student directly.

In a lifetime, one may have seen tractors, trucks, and automobiles replacing cattle and horses as primary sources of power in much of the world, wool and cotton being partly replaced by synthetic fibers, vegetable oils challenging animal fats, the development of new sources of plant proteins which may partially replace meat, and, finally, a human population crisis which will increase competition between animals and man for food plants and which will endanger the maintenance of an acceptable environment.

But do these factors herald the doom of the livestock industry? We think not. There are tremendous areas of the earth where human food crops cannot be grown—mountains, forest, desert, Arctic—but which, by managed forage production, could become valuable resources for raising animals that provide food and fiber for mankind. Animals are invaluable for their ability to upgrade materials of low nutritional quality. The aesthetic value of animal foods cannot be overlooked; scientists have not yet produced any synthetic foods that can challenge the favor which meat currently enjoys in well-fed populations or meet the demands for it in poorly fed nations. Finally, the value of domestic animals for pleasure and as man's companions cannot be overestimated. In the United States, the horse has been largely transformed from an animal for power and transportation to one for pleasure, and dogs have become primarily animals for social purposes; the human mind finds serenity in the company of animals.

So, all in all, it is our considered opinion that domestic animals will continue in the foreseeable future to play a dominant role in the life of man. It is with this confidence that this book is written. As James Branch Cabell said in *The Silver Stallion* (1926), "The optimist proclaims that we live in the best of all possible worlds; and the pessimist fears this is true."

February 1974

H. H. Cole
Magnar Ronning

Contents

Section III Animal Species, Breeds, Strains, and Hybrids

Section IV Inheritance and Animal Improvement

Section VII Livestock Management

Section VIII Classification, Grading, and Marketing of Livestock and Their Products

Section One

Animal Agriculture: its scope and future

Water buffaloes being used to plow rice paddies in Indonesia.

One

The Beginnings of Animal Domestication

The domestication of animals may well be derived from the social relationships of animal species, man being one of them.

F. E. Zeuner,
A History of Domesticated Animals

The oldest known individual that we can assign with certainty to the human family died approximately 5.5 million years ago (Howell, 1972). The first evidence of a domesticated animal, a dog, is from an archeological site in northern Iraq that has been dated as about 14,000 years old (Turnbull and Reed, in press). The first domestic food animals were sheep, but they were apparently not domesticated until less than 11,000 years ago (Perkins, 1964). Thus, man survived 99.8 per cent of his known history without domestic animals, and also, incidentally, without cultivated crops.

Throughout the overwhelmingly greater part of human history, then, men and women maintained themselves successfully on what they could hunt and collect. (If they had not been successful, of course, we would not be here today.) Throughout most of this 5.5 million years, the food-stuffs they gathered—mostly vegetables, but probably including an occasional tasty insect larva or small reptile—must have been vastly more important to them than the meat they trapped or hunted. Large game, at least, seems not to have become an important factor in human hunting until roughly 0.75 million years ago: the stone-tipped spear is a relatively recent invention, of some 80,000 years ago, and the bow and arrow is hardly older than the beginning of animal domestication.

Every animal is a part of a food web: to live, it must eat. Man, as a protoplasmic being, shares this necessity with other animals; food yields energy, and he and she of whatever species who get sufficient food have a better chance of producing and rearing offspring than do malnourished members of the same species. Any new invention that channels energy through human protoplasm will probably lead to a better chance for human survival and an increase in the population. Such inventions may have to do with *things* (the early use of sticks and stones for prying, hammering, and throwing; improvements in twine-making, wood-working and stone-chipping; shaping of raw metals and the smelting of ores; building of windmills and irrigation works), or with chemical processes (the use of fire for warmth and cooking; tanning of skins; heating clay to produce pottery; making of alloys; synthesizing new molecules to produce fabrics or kill insects), or with the increasing complexity of social organization (food-sharing, and carrying of food to be shared from place of gathering or killing to the home-site; maintainance of the family group as a cooperative entity; recognition of kinds and grades of kindship and degrees of responsibility; organization of suprafamily groups into tribes and nations). Speech and writing, a very recent offspring of speech, by which the wisdom of experience can be transmitted, are also such energy-savers. Energy saved is energy gained: one need not make the same wasteful mistakes one's ancestors did.

The domestication of animals and the cultivation of plants were and are energy-diverting processes. You might, at first, think these to be energy-producing processes, but neither man nor other living entities can produce or destroy energy. If, in the process, less energy is left for passenger pigeons and Tasmanian natives, who then become extinct, the successful energy-diverters are, for the most part, not sorrowful. The domestication of animals, particularly of those used for food, diverts energy in the sense that the various species usually eat foods that man can't or won't eat (grass, leaves, straw). They are protected by man in different ways, and then usually killed for human consumption as needed.

Although we can analyze now, in terms of physics (energy) and biology, the benefits of domestication to our ancestors, those people who were first successful at domestication had no such fancy ideas. Indeed, considering that the whole cultural pattern for millions of years had been to hunt and kill animals, it is a wonder that domestication ever happened at all.

Many hypotheses about how domestication began have been proposed (Downs, 1960; Zeuner, 1963), but we of today's civilization find nearly impossible the task of putting ourselves into the mental framework of men and women of ten thousand years ago. Those people had a rich and functional tradition of nature lore interwoven with magical beliefs and rites, many of which focused on success in the hunt.*

What Is a Domestic Animal?

In the category of "domestic animals," we may include those under the control of man throughout their lives, those whose breeding is or can be controlled by man, and those that may be dependent on man for protection or for food. We must exclude most animals in zoos and many animals (such as various kinds of rodents and primates) in experimental research centers because they have not truly been brought "into the house." We must also exclude elephants because, although they may be used by man as beasts of burden, they are caught wild and tamed by man. In contrast, we should include reindeer because, although they are free-living for the most part, so that the people of northern Eurasia who herd them must follow the herds to keep alive, the people intervene by selective killing, by taking milk sometimes, by using them as sledge animals,† and by castrating the older males to diminish excessive fighting (Figure 1.1).

A tamed animal is not domestic—it is an animal that has been taken from its wild envi-

* No matter that we don't believe in such magic; those people did, just as many American farmers well into the twentieth century believed in the semimagical powers of the phases of the moon and scheduled their plowing, planting, and harvesting accordingly. [We have some idea of what people believed 10,000 or more years ago because we can compare the archaeological evidence (from cave paintings, figurines, etc.) with the activities and beliefs of peoples who were still at a similar stage of technological development in modern times. Ed.] Why did hunters cease hunting for the daily meat and how did they gain the wisdom needed to protect, then keep, game animals so that a constant supply of meat and hides would be assured? This particular transition in the human way of life was not a simple one; we shall return to the subject later.

† Snowmobiles are now rapidly supplanting reindeer for winter travel in the roadless areas of northern Eurasia.

Figure 1-1 Lapland reindeer rounded up to select animals for slaughter and branding of young. From Zeuner, 1963. Hutchinson and Co. (Publishers), Ltd., London.

ronment and has learned by experience that man is a source of food and shelter, rather than a source of harm. Perhaps every bilaterally symmetric animal with an anterior nerve net (brain) can be tamed and taught to come for food at a signal; but many of these—the worms, for example—are too small for man to profit economically from such taming, and others—such as piranhas or predaceous sharks—have such limited capabilities for modifying their instinctive patterns of behavior that the experiment could be dangerous.

Taming merges indistinguishably into domestication, if the original wild-caught animals breed in captivity and are selected for particular qualities, such as docility, color, horn size, meat production, and so forth. Every kind of animal now domestic had ancestors that we would necessarily say were only tamed.*

* Tamed animals, such as elephants, can do useful work for man, just as domestic animals do. Even baboons, for example, can be trained to be goat herds (Dart, 1965).

A domestic animal, or one descended from a domestic population, can never become a wild animal. Domesticated animals that return to nature to survive and breed are termed *feral*. The distinction is a nice one, and intellectually useful, but not necessarily satisfying to a man who has had lambs killed by "wild dogs." The wild ancestor of the dog was the wolf, but the dog has been changed sufficiently in characters of bone and brain and teeth that, when it returns to nature, it remains a dog—albeit a feral dog—for all of its wild behavior.

Why and How Did Man Domesticate Animals?

Much has been written on the subject of why and how man domesticated his animals, but all is supposition; the "natural" processes of domestication by a primitive people have never been observed by a modern student, for the simple reason that all such domestication occurred long before any records were kept. Recent successes at taming or domesticating a variety of animals—including Norway rats (Richter, 1952), eland (Skinner, 1967), musk oxen (Teal, 1970; Wilkenson, 1971), and wolves (Banks, 1967)—have been accomplished in the experimental spirit of modern science and with a purposeful expenditure of time and capital. Additionally, the domesticators could foresee with some clarity the pattern of their experiments and the expected benefits of successful domestication.

Primitive man had no such advantages—certainly not during the earlier part of the period of animal domestication. It is probable that taming, and then domestication, occurred without people being aware of what was happening. Certainly, gatherers and hunters—the people who first domesticated animals—could not foresee any uses for those animals other than those they knew already, for meat and skins. Only later, after long experience and the change to a more sedentary life style, and after the accumulation of random mutations in the domesticates, would secondary uses of animals—such as for milk, wool, motive power, war, sport, and prestige—be realized.

When one considers that men had been hunters and consumers of animals for millions of years, the behavioral change required for them to become keepers and conservers of animals was a major cultural revolution. What changes of culture (and possibly personality) were necessary before groups of humans began to protect animals instead of killing them whenever possible?

Pet keeping has been suggested as one channel toward domestication, particularly of the dog. Father, so runs the idea, killed a wolf bitch and brought home the young pups, which appealed to the children and possibly to mother, and so the pups were kept to be reared in camp. Thus, wolf pups, who are friendly in their own group, learned to substitute the human family for the pack and to be submissive to man, the substitute for the alpha wolf, the leader of the pack. The wilder ones ran away; the tamer stayed, moved with man, lived on the offal from the hunt, hunted small game of their own, raised pups of their own in camp, and became dogs. Is that the way it happened, this first conversion of wolf to dog? We don't know; we now know that dogs were derived from wolves (*Canis lupus*) and, we believe, from no other species. The two remain completely interfertile to this day and, indeed, the separation of them into two different species is an artificial one prompted by our own prejudices. Structurally and behaviorally, they remain the same animal, except as man has genetically modified the domesticate by artificial selection. The wildness is gone in hand-reared dogs, but it can return with sudden viciousness in dogs not so handled. However, wolves are tameable too, particularly if they are "taken into the house" and given food, affection, fondling, and training from the age of four or five weeks (Banks, 1967).

That we can undoubtedly now, by selective breeding, make dogs out of wolves with considerable care and expense proves the deed possible, but it does not show how an earlier people did the same job under primitive conditions. Why did they want to do it—or *did* they want to do it? Was there any intent to domesticate, or did it come about as the result of a process of unintentional association between these two social species? It is true that dogs might have been able to locate some game that even skillful human hunters would have overlooked, but then it may well have been necessary to tie up the dogs to prevent them, in their enthusiasm, from chasing the game away. And how would the hunters have kept the dogs from barking to the detriment of the hunt? Such living primitive hunters as the Australian aborigines keep their dogs, the dingoes, more often for bed covers on cold nights than as partners in the hunt (Meggitt, 1965).

Some students of domestication have thought pet keeping to have been unimportant or absent in the origins of domestication. They have suggested that wolves, for example, learned to follow human hunters in anticipation of a meal after the hunters had stripped what they wanted from the carcass (Downs,

1960). As evidence for the possibility of such an association, we know that, on the American plains in the nineteenth century, wolves were attracted by the sound of gunshots, as indicating probable kills of bison (Young and Goldman, 1964). As far as we know, however, no such associations between adult wolves and human hunters has resulted in any close, permanent, social bond; and such scavenging of the remains of large game killed by man probably occurred for several hundreds of thousands of years with no further intensification of human–canine relationships. Lastly, a wealth of recent experimental evidence indicates that only by constant, close, and affectionate contact, usually starting before the wolf pup is six weeks old, can a man create a tame wolf. Thus, the bulk of the evidence recommends to us that, for the transformation of wolf to dog, pet keeping was the path to domestication. If man could and did indeed tame the wolf some fourteen thousand years ago, why did he not do it earlier? I suggest that the behavior of wolves did not change; therefore, the behavior of men must have changed, but of this we have no evidence other than the appearance of the dog.

My own interpretation of the archeological evidence is that dogs were not domesticated for food, although Herre (1969) has disagreed; however, where the remains of dogs are found in the earliest archeological sites of Europe, the Near East, and North America, they are always rare, whereas the broken bones of the hunted food animals—mostly hoofed mammals—are numerous by contrast. Thus, I believe that man and tamed wolf pup—and then, generations later, man and dog—have always had a relationship other than that of consumer and consumed, which conclusion does not indicate that the dog was necessarily a pampered pet. Indeed, we know that different people in the past (Aztecs and Polynesians, particularly) have raised dogs for food, and that, far from having a favored status, dogs in Moslem lands are usually regarded as unclean animals whose main usefulness to man is as a consumer of garbage and human filth in the village.

We know even less than for the dog of the pattern whereby the hoofed food animals (pigs, goats, sheep, cattle, yak, water buffalo, reindeer, and others—perhaps the camel and the horse also originally belonged here) were "taken into the house." We can see that, of the earliest of these animals to be domesticated (pig, goat, sheep, cattle), the pig is an omnivore capable of surviving on a wide variety of plant and animal materials, and the others are ruminants with the typical four-chambered stomach, which allows them to utilize cellulose as a source of energy. Although a ruminant cannot survive on a diet of wood chips, the bacteria in the anterior part of its stomach do break down much cellulose into simpler compounds, which the ruminant can then digest (it also derives nourishment from the excess population of bacteria, which passes on into the abomasum and intestine). Additionally, ruminants recycle their urea via the circulatory system to the rumen, thus conserving nitrogen and allowing them to survive on a low-protein diet. Thus, all the ruminants are well designed, by a long evolutionary pattern of selective adaptation, for survival on diets on which other animals would starve (Reed, 1969); but such inadequate foods as straw may have been what early domesticated sheep, goats, and cattle were sometimes given to eat by their human captors. Those that could survive and reproduce on such a diet were those that became domesticated.

The archeological evidence suggests to me that the early domestication of pig, sheep, and goat occurred in southwestern Asia, beginning perhaps as early as 11,000 years ago, but always in association with settled villages (Figure 1.2). I suspect that plant cultivation may have accompanied or preceded the domestication of these hoofed food animals.*

* Others (Higgs and Jarman, 1969; Jarman, 1969) think this view too restrictive, and suggest instead that these food mammals may have been domesticated much earlier and over a much wider geographical range. The only way to resolve this problem will be by intensive archeological research throughout the Old World.

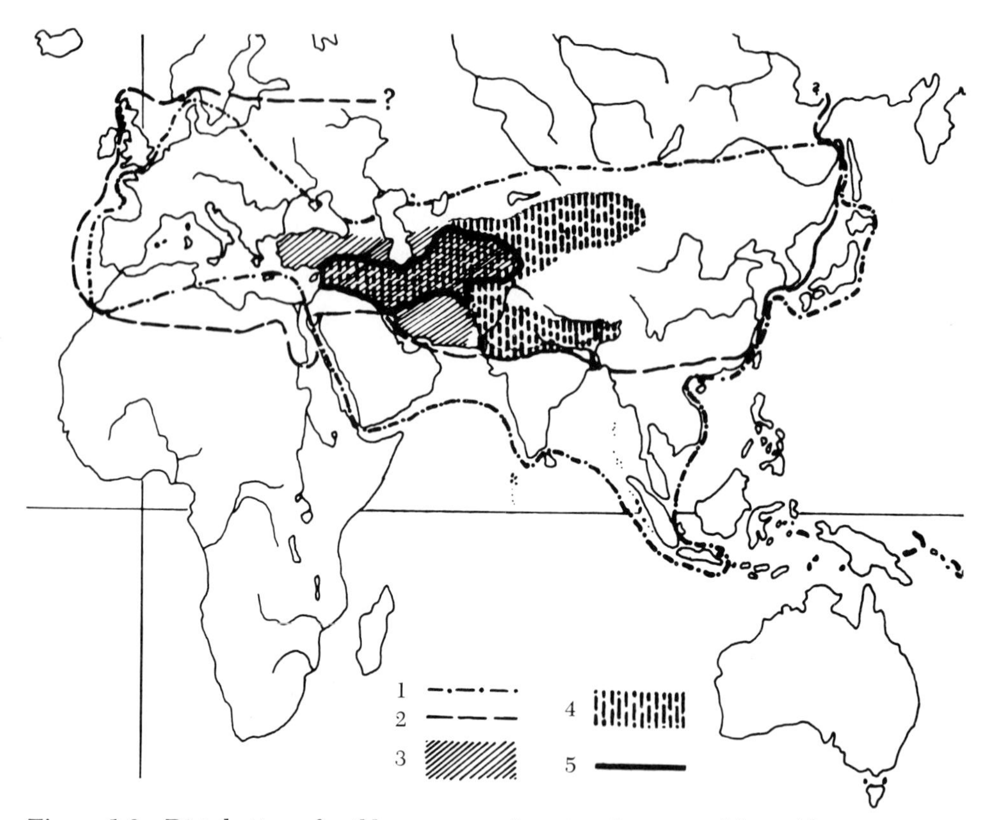

Figure 1-2 Distribution of wild ancestors of major domestic old world animals: 1, pig; 2, cattle; 3, goat; 4, sheep; 5, region of range overlap of sheep, goats, pigs, and cattle. From Isaac, *Geography of Domestication,* © 1970. Reprinted by permission of Prentice-Hall, Inc., Englewood Cliffs, N.J.

Even assuming such an association between early villages, plant cultivation, and the early domestication of pigs, sheep, and goats, we have no idea what behavioral paths of men and ungulates enabled the association to be established. Zeuner (1954, 1963) has suggested that, after plants began to be cultivated, several of the herbivores found the crops to be good grazing; even if the animals were driven off during the growing season, the stubble could have been eaten by them after the harvest. Such an association would not, as far as I can see, have led to domestication—unless the animals were, say, driven into holding pens or ravines, from which they could not leave except by a guarded entrance. Individual animals could then be killed for food as needed, and perhaps some people had the foresight to feed the pregnant females, so that they might survive to produce a new generation. From such practices, herding may have evolved. Such a path to domestication seems unreasonable, however, for the wild goats (*Capra hircus aegagrus*) that were ancestral to domestic goats: these wild goats dwell in such rough and rocky terrain that their range could hardly have overlapped early cultivated fields; instead, I suspect that hunters brought young kids into the villages, quite as hunters do today in the Zagros Mountains of southwestern Asia.

Sheep are more likely than goats to have entered into an association with primitive farmers by way of grazing on growing crops;

indeed, wild sheep (*Ovis ammon orientalis*) still descend from the mountains of Iran in the winter to graze on the wheat and barley growing in the fields of the foothills. Today, however—as I suspect was true in the past—such sheep are killed whenever possible, and it seems likely that this pattern of crop-robbing, as Zeuner has termed it, was not by itself an avenue to domestication. There had to come a time, I feel, when the animals had to be taken in hand, quite literally, and such handling is easiest done with the young.

The archeological evidence suggests that sheep, goats, and pigs (as well as the dog, of course) were all domesticated earlier than cattle. The wild cattle, or aurochs (*Bos primigenius*), of the Old World were powerful and formidable animals, with some males reaching a height of six feet at the shoulder; the males in particular had large horns, but both males and females were big and fast, and were undoubtedly capable of aggressive behavior as well as of competent defense. Even so, these animals, which were so magnificently portrayed in the Paleolithic cave art of southwestern Europe, were successfully hunted over most of their range. Once people had domesticated sheep and goats to supply them with the meat and skins they needed, why did they make the effort to domesticate an animal as large and strong as the aurochs?

Even though he hunted wild cattle, man seemingly regarded them with a respect verging on the religious. Thus, portrayals of these animals in the Paleolithic cave art often show them larger than life-size. Most revealing is a scene—seemingly not a hunting scene—drawn on the wall of a room in the early town of Çatal Hüyük, in south-central Asia Minor (Anatolia), dated at about 7800 B.P., in which an aurochs is portrayed larger than life-size surrounded by human figures that are minute by comparison (Mellaart, 1967). Also at Çatal Hüyük, definite evidences of a complex cattle cult have been excavated (Mellaart, 1967). Throughout the period of this cattle cult and the occupation of the town, which lasted roughly 700 years, cattle at Çatal Hüyük were becoming domesticated (Perkins, 1969). The earliest known remains of domestic cattle, however, are from northern Greece.

Cattle have continued to occupy a special place among man's domestic animals, by reason of veneration, affection, prestige of ownership, or simply economic value. Much of this special regard for cattle probably is derived from prehistoric cattle cults, in which, initially, the spirits of slaughtered wild cattle were worshipped and appeased and, later, domestic cattle were given a special place of honor in many religious ceremonies (Figure 1.3). We find this special attitude toward cattle still expressed in many ways today: cattle are sacred to the Hindus, and are not to be killed or eaten; the hill peoples of eastern India and adjacent areas breed a variety of cattle (the sacred mithan) primarily for sacrifice (Figure 1.4); the prestige of cattle-owners, among many East Africa herders, is related to the numbers of cattle they own and not to their actual economic value; in some Swiss villages, flower-bedecked cattle are important in ceremonial occasions; in early Crete, cattle were a part of special athletic spectacles (bull-leaping); in southwestern Europe (particularly in Spain), bulls are specially bred for formalized ritual sacrifice in public arenas; and, finally, in many parts of the world, great prestige has been attached to herders of cattle. Particularly on the Pampas of Argentina and in the Southwest of the United States, where the herders are mounted, these men have been given the deference they have demanded as being above the law and apart from other men. Thus, the gaucho and the cowboy, enshrined on the television and the silver screen, may be America's heirs to the traditions of an ancient and many-faceted Mediterranean cattle cult.

The pattern of the relationship between man and cattle does indeed suggest a religious origin for domestication, an idea put forward in the nineteenth century and recently reformulated by Isaac (1962, 1970). In correlation, Reed (1969) has introduced again the concept

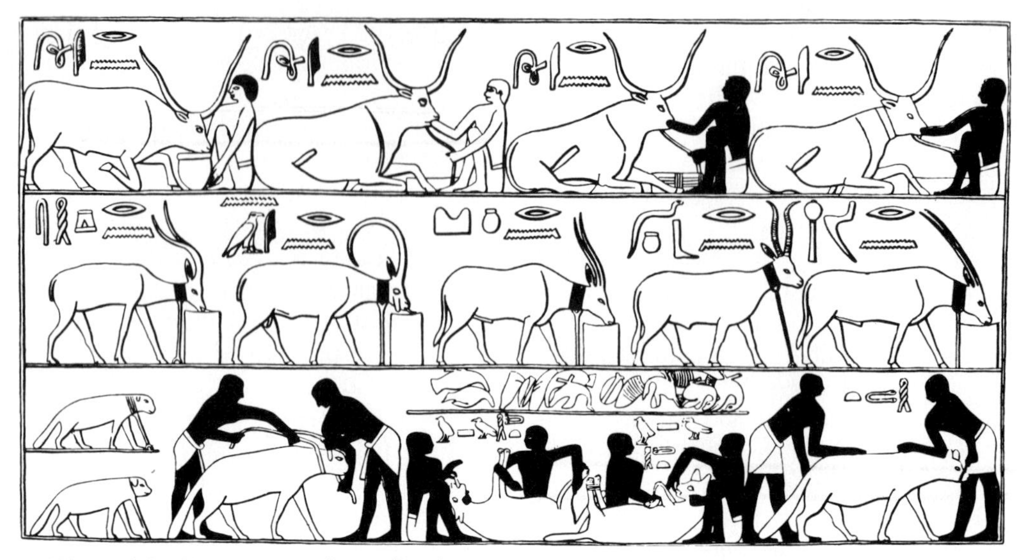

Figure 1-3 Drawings on the walls of tombs from the Old Kingdom, Egypt, c. 2500 B.C. In the upper panel, cattle seemingly are being fed by hand, as are hyenas in the lower panel. In the second group are a variety of antelope and gazelles, tethered and feeding in mangers. Adopted from Zeuner, 1963. Hutchinson and Co. (Publishers), Ltd., London.

of the holding pen, into which wild cattle could have been driven and held, the larger bulls then to be segregated for sacrifice and the other animals allowed to breed in semiwild captivity.

However, when all our ideas on the whys and hows of early domestication are summarized, we find that we simply don't know; all remains supposition, though some suppositions are more logical than others. We cannot now put ourselves back eight or ten or twelve thousand years into the minds of primitive peoples who left no written records of their cultures, nor can we as yet use other evidence to reconstruct their cultures in sufficient detail to understand the social and economic relations among men, and between men and other animals, that led to the earlier domestications.

When and Where Were Animals First Domesticated?

The earliest domestic animals, insofar as we can determine from study of the bones dug by archeologists from sites of prehistoric man, were all placental mammals. The domestication of some birds and other animals—bees and goldfish, for example—all occurred later than that of most of the placental mammals. Since the first record of a domestic animal, a dog, some 12,000 years ago, animals have continued to be domesticated, here and there around the world, and the process continues.

Domestications in the Old World

The dog was probably wide-spread across the northern hemisphere before any other domesticate existed. The earliest known site for domesticated dogs (dated 12,000 B.P.) is in the hills of northern Iraq in southwestern Asia. There were domesticated dogs in Japan, however, before the end of the Pleistocene, approximately 11,000 years ago (Shikama and Okafuji, 1958), in North America more than 10,000 years ago (Lawrence, 1968), and in England by 9000 B.P. (Degerbøl, 1961). The place of the earliest known occurrence is not necessarily the place of the first domestication,

Figure 1-4 The mithan, a type of cattle found in northeastern India whose primary role is as a sacrificial animal. From Simoons and Simoons, 1968. Univ. of Wisconsin Press, Madison.

however, because the former changes (sometimes almost year by year) with the happenstance of archeological discovery. Although we cannot, therefore, state the time and place of the first domestication of the dog with any certainty, current evidence indicates that it was probably in Eurasia. Lawrence thinks that the early dogs she has described from Idaho accompanied man into North America from Siberia.

Beginning almost 11,000 years ago, a number of domesticated hoofed animals were present around the northeastern corner of the Mediterranean, and this area, if any place, can be said to be the first center of domestication of "meat and hide on the hoof." These animals included sheep (domesticated from *Ovis ammon orientalis*), which were present in northern Iraq by 10,750 B.P. (Perkins, 1964); pigs and goats, which, along with sheep, were present in eastern Asia Minor (Anatolian Turkey) by about 9000 B.P. (Braidwood, *et al.*, 1971); and cattle, which, along with sheep, goats, and pigs, were present in Greece by 8500 B.P. (Protsch, 1970). By 7000 B.P., these four domestic food animals, plus the dog, were probably present over most of southwestern Asia (although not all of them have been found at every site that has been excavated), and sheep or goats, or both, had been taken into northern Africa as far as Libya (Higgs, 1967). At the same time, this quartet of food animals was being moved by herders and farmers up the Danube into central Europe, but they did not appear in Russia until more than a thousand years later (Murray, 1970, Chapter 3); domesticated pigs, however, may have had an early and independent center of origin in the Crimea (Krainov, 1960).

Beyond this primary center in the areas around the northeastern Mediterranean, evidence for such early domestication of sheep, goats, cattle, and pigs is lacking. I conclude, therefore, that the Near East (including Greece but excluding Egypt) was the primary area for the early domestication of these four basic food animals. The earliest known sites of settled villages with cultivated plants (wheat, barley, peas, and lentils) have been found in this area, too, and it is my opinion that these beginnings of settled villages, cultivated plants, and domestic animals are correlated parts of a single cultural evolution associated with relatively rapid, world-wide climatic change marking the end of the Ice Age (Wright, 1968; Reed, 1969).

Even if the Near East was the first area in which food animals were domesticated, it was not necessarily the only area. The idea of domestication could well have traveled faster than the movement of flocks or herds; or, quite independently, other people may have begun to accomplish what the people of the Near East had already done. Thus, pigs (*Sus scrofa*), which were wild over northern Africa and much of Eurasia, could have been independently domesticated in many areas, as they possibly were in the Crimea (Krainov, 1960) and probably were in northern China (Ho, 1970); it is possible that pigs were also independently domesticated in southeastern Asia, Egypt, and parts of Europe, but of this we have no firm evidence.

Animals not native to the Near East were obviously domesticated elsewhere, in areas in which their wild ancestor then lived. Several thousand years ago, wild horses (*Equus caballus* = *E. przewalskii*) were distributed over the grasslands of central Eurasia and westward into the forests of Europe. They were domesticated on the grasslands, somewhere between central Europe and the Gobi Desert (very possibly in southern Russia), perhaps by 3000 B.C. We know nothing of the pattern of human behavior that initiated the domestication of the horse—although other domesticated animals (perhaps including cattle) were present in southern Russia by that time, and their presence may have given rise to the idea that the horse, which had been hunted for millennia, could also be domesticated. At first, it is likely that domestic horses were eaten, but they were later harnessed to pull sledges and carts. After 2000 B.C., roughly, the heavy carts were modified to light war chariots, thus introducing an era of mobile warfare in which we still find ourselves today.

Two other species of equids have been domesticated—one only temporarily, but the other permanently. Before the general introduction of horses into the Near East, the people of the early civilizations in the lower reaches of the Tigris–Euphrates valley (Mesopotamia) had domesticated the native wild ass (*Equus hemionus*)—by all modern accounts, a most untractable animal. They used it, however, to pull clumsy carts on state occasions, and they even used such carts as war chariots. This animal was never of great importance as a domesticate, and its use disappeared after about 1800 B.C. Another equid, the Nubian wild ass (*Equus asinus*), was native to the drylands of northeastern Africa; it appeared as a domestic animal—sometimes ridden, but primarily, then as now, a bearer of burdens—by 3000 B.C. (Figure 1.5). We know

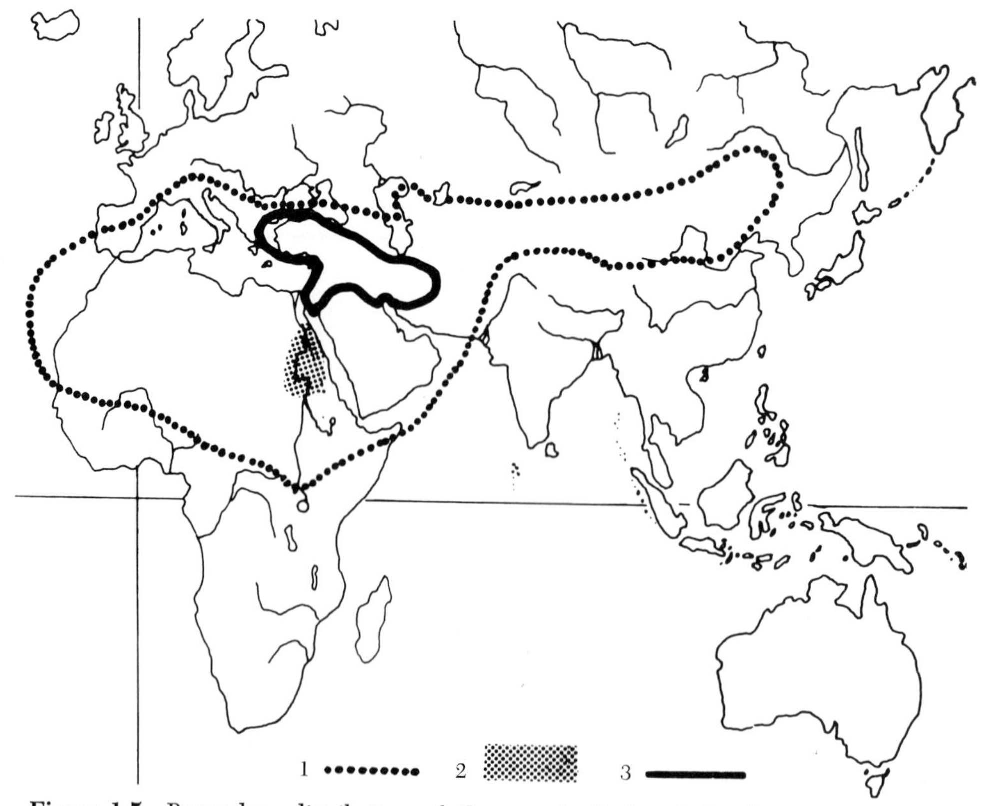

Figure 1-5 Premodern distribution of the ass: 1, limits of distribution of the ass; 2, presumed area of domestication of the ass; 3, oldest hearth of plant and animal domestication. From Isaac, *Geography of Domestication,* © 1970. Reprinted by permission of Prentice-Hall, Inc., Englewood Cliffs, N.J.

it as the donkey; it is one of the three animals (the others being the cat and the guinea fowl) that were domesticated in Africa.

Various kinds of cattle, other than those well-known in Europe and America, were domesticated in southern and central Asia from various wild ancestors. Zebus, or humped cattle, presumably were domesticated in India or an adjacent area from *Bos primigenius namadicus*, an eastern variety of the same species of wild cattle from which the western cattle were derived. Strangely, however, cattle with humps and looking very much like zebus were depicted accurately in carvings from Mesopotamia that have been dated at about 5000 B.C.; if domestication of this animal occurred in or near India, it must have been at a considerably earlier date, in order for there to have been sufficient time for the movement of the cattle to Mesopotamia, where, of course, more typical (to our western eyes) domestic cattle already existed.

We have no record of domestic zebus in India itself until about 2500 B.C.; if they were present there earlier than 5000 B.C., we would expect their bones to be found in appropriately dated archeological sites. It seems that the zebu was being moved eastward, as well as westward, at a respectably early date, because skeletons of cattle probably of this type that have been found in Thailand have been dated to 3500 B.C. or earlier (Higham and Leach, 1971). Because Thailand lies within the natural range of several types of wild cattle, but presumably not within that of *Bos primigenius*, the zebus whose skeletal remains have been found there must have been introduced from the west, presumably India.

If men domesticated cats, they did so only after cats had long been accustomed to men. Of all domestic mammals, cats remain the least changed morphologically, and probably psychologically, too. The close association of cat and man seems to have happened late in the history of domestication in the Old World, not until about 2000 B.C., in Egypt, which is a part of the natural habitat of the African wild tabby cat, *Felis catus libyca*. Cats of this variety are noted for their "tameness" when brought into zoos (Lorenz, 1955), and probably moved into Egyptian towns and villages, where stored foods attracted rats and mice. The path to domestication is not known, but the ancient Egyptians revered many animals, and may have lured kittens with food to watch their always interesting antics. Tamed, such cats could have led in a village a life not much different from that of their immediate "wild" ancestors.

Egyptians deified the cat, and prohibited its export, although some were smuggled out. After Egypt was conquered by the Romans, cats were spread throughout the Empire, and beyond, mostly as rodent catchers. In many societies cats have led a hard life, being feared or even hated; not everyone has responded to the social bond of affection which cats have offered to man.

Space limitations will not allow for the discussion of domestication of other species, but data concerning some of them are given in Table 1.1.

What Might Have Been, and What Might Yet Be

The world moves, unfortunately but seemingly irrevocably, toward ever more intense human occupation of all available surface, toward more steel placed vertically, more concrete laid horizontally. Room for wild and domestic animals is decreasing, the way our culture is going, but do we want that kind of cultural poverty? Before our animals are gone, let us think, for a moment, about the delightful diversity that could have been ours, had we and our ancestors arranged matters differently.

The culture of the ancient Egyptians is interesting in many respects; particularly fascinating to anyone interested in animals are the numerous experiments in taming and domestication that they accomplished (Boessneck, 1953; Zeuner, 1963). Fortunately for us, they left pictures of what they did. They kept and

Table 1.1.
Information on the domestication of some species not covered in text

Animal	*Probable area of domestication*	*Estimated time at domestication*	*Use of domesticate*	*Reference*
Donkey; Nubian wild ass, *Equus asinus*	Northeastern Africa	3000 B.C.	Beast of burden	Epstein, 1971; Isaac, 1970
Mithan, *Bos (Bibos) gaurus*	Northeastern India	Unknown	Sacrificial animal	Simoons, 1968
Banteng, *Bos (Bibos) javanicus*	Southeastern Asia	Unknown	Meat and draught	Zeuner, 1963
Yak, *Bos (Poëphagus) grunniens*	Tibet	Unknown	Beast of burden, milk, hides, hair	Zeuner, 1963
Water buffalo, *Bubalus bubalis*	Northern India?	Before 2500 B.C.	Meat, milk, draught	Epstein, 1971; Zeuner, 1963
European elk, *Alces alces*	Northern Europe	1500 B.C.?	Riding	Yazan and Knorre, 1964; Zeuner, 1963
Reindeer, *Rangifer tarandus*	Northern Eurasia	Incipient herding?, 14,000 B.P.?	Meat, hides, milk	Zeuner, 1963; Herre, 1955
Dromedary, *Camelus dromedarius*	Arabia	2000 B.C.	Beast of burden, meat	Epstein, 1971; Isaac, 1970
Bactrian camel, *Camelus bactrianus*	Northern Iran or Central Asia	1500 B.C.	Beast of burden	Isaac, 1970; Zeuner, 1963
Ferret, *Mustela putorius*	Circum-Mediterranean	400 B.C.?	Rabbit hunting, rodent control	Owen, 1969
Llama and alpaca, *Lama* sp.	Peru	1500 B.C.	Beast of burden, wool	Lanning, 1967; Herre, 1952
Guinea pig, *Cavia porcellus*	Peru	2000 B.C.	Meat	Lanning, 1967

tamed, even if they did not domesticate, a wide variety of bovids (gazelles, ibex, oryx, addax, and Barbary sheep).

The early Egyptians also seem to have tamed hyenas, which are shown in several paintings that span a period of 250 years. One such painting illustrates hand-feeding, with the hyenas on their backs and men stuffing food into their mouths. Hyenas have been maligned in our culture; if caught young, fondled happily, fed well on fresh food, and given an occasional bath, they might prove to be the best of man's friends, but no one has made the effort.

It is probable that a wide variety of other carnivores, in addition to cheetahs and ocelots, could be tamed and perhaps domesticated to join the cats and dogs and ferrets that we already have. All otters, and the sea otter in particular, seem psychologically preadapted for domestication, as do the family Bovidae, which has already contributed all the cattle, sheep, and goats to our list of domesticates. I suspect that almost all other bovids—including the bison and the Cape buffalo—could be domesticated, if we but made the effort. The effort *is* being made with the musk ox and the eland—with the latter, by a few interested individuals at personal expense and, so far as I have been able to determine, without government cooperation.

Seals and walruses seem eminently domesticable, and perhaps sea-lions would be domesticable, too, if we could learn how to handle the male during the breeding season. The sirenians (dugongs and manatees) are almost preconditioned to taming, and would not have to be "domesticated" so much as merely not frightened or disturbed. One could now, I think, domesticate the hippopotamus, probably easier than people in the past domesticated camels or cattle. Why not wart-hogs? Why not American antelope?

What a pity that, instead of killing them all or letting them become extinct, man did not domesticate ground sloths, mastodons, Irish elk, and a host of others. If early man in North America had only taken some baby saber-tooth tigers and raised them with tender loving care! We can say that our ancestors were not culturally advanced enough to understand extermination and extinction. Are we?

We, the people of the world, are killing the larger whales nearly as rapidly as possible. Of the whales, dolphins, and porpoises, probably all—in spite of size of some and the carnivorous behavior of others—are potential domesticates. Perhaps—some people think it's true—these animals have languages, and perhaps we can learn to talk with them. Certainly, they are doing more with those large and complex brains than feeding themselves and squealing noise at each other! We need absolute protection for them first, taming second, and some degree of domestication (perhaps only companionship) later. Of course, that's a big order, mainly because the whales and their allies are big animals.

Clearly, not all adventure and new research lies in outer space; biological adventures and new and fruitful discoveries remain here on earth for those who will but seek them.

FURTHER READINGS

Epstein, H. 1971. *The Origin of Domesticated Animals of Africa.* New York: Africana.

Isaac, E. 1970. *Geography of Domestication.* Englewood Cliffs, N.J.: Prentice-Hall.

Zeuner, F. E. 1963. *A History of Domesticated Animals.* New York: Harper and Row.

Two

Symbiotic Relations Between Plants and Animals and the Pollution Problem

And God blessed them, and God said unto them, Be fruitful, and multiply, and replenish the earth, and subdue it: and have dominion over the fish of the sea, and over the fowl of the air, and over every living thing that moveth upon the earth.

Genesis 1:28

The World Agricultural Ecosystem

Primitive Man's Relation to Nature

Man's emergence as a successful and resourceful member of the Earth's community of animal life lies far behind the mists that shroud his prehistoric beginnings, but we can be certain that, however varied his preoccupations may have been, obtaining food for survival was his constant concern. It still is. For unknown thousands of years, man was a wandering, food-gathering animal, precariously dependent on natural systems of animal and plant life for his needs. Hunger was often his lot, and insecurity and uncertainty were his constant companions. His role in the natural scheme of things was that of a consumer—in part a carnivore, in part a herbivore—and often that of a scavenger. Food scarcity and disease insured that his impact on the Earth's major ecosystems—whether forest, grassland, aquatic, or littoral—was minimal, because these limitations, and perhaps also warfare, kept his numbers very low in most parts of the earth.

Man remained dependent on intact, natural plant-and-animal systems for his food and other life needs until he began learning how to modify his environment. The historical events that lessened his direct dependence on Nature are not easily traced, but it is certain that, with the beginnings of agricultural innovation, the man–nature complex was profoundly altered. Although fresh-water and marine "farming" (which developed through the use of crude devices to capture fish for food) has a very long history, it was not until he learned something about the cultivation of plants and the domestication and care of animals, roughly 10 thousand years ago, that man became capable of altering the life-support systems of the biosphere. A greatly improved food supply favored increases in the rate of human population growth. In turn, increases in man's numbers, and the resultant stress on food-production practices, compelled flourishing human societies to make ever-increasing use of the natural resources available to them.

These two processes, population growth and increasingly more effective exploitation of food-production technology, have not always consistently reinforced each other, but they are now creating stresses on the earth's life-support systems that have incalculable consequences for the biosphere and, consequently, for man's own welfare. These stresses explain why many are now so singularly preoccupied with such matters as the use of natural resources, our relation to our environment, the population explosion, and man's arrogance toward nature, and why the term "ecological crisis" is fast becoming the rubric under which is classified all that is "wrong" in modern technology. Man's long-held belief that nature exists to serve him has become the object of much critical and damning re-examination.

Solar Energy and Life

Fundamental to all life is the process of photosynthesis, which "captures" solar energy and uses it for energy-requiring reactions that take place in the chlorophyll-bearing cells of plants. During these reactions, carbon dioxide, water, nitrogen, phosphorus, sulfur, and other inorganic elements are combined to form compounds found in the extremely varied structural and functional elements of a plant. Plants contain many kinds of carbohydrates, fats, proteins, vitamins, other organic substances, and minerals. These substances serve as nutrients for plant-eating animals. Certain types of microorganisms, termed chemoautotrophs, get their energy from inorganic compounds, but, aside from this remarkable exception, the energy that runs the life-support systems of the biosphere comes from the 150–200 million tons (dry weight) of organic matter that is produced annually by the major ecosystems of the earth: the forests, grasslands, oceans, lakes, rivers, deserts, and tundras (Woodwell, 1970).

Simply stated, plant life is the primary producer of organic matter, whereas organisms that eat plants are termed primary consumers. The terms *primary, secondary, tertiary,* and *quaternary* are used to designate ranks of organisms in the hierarchy of a food chain, starting with the producers of organic matter (plant life) at the base. Insects that feed on plants, for example, are primary consumers, whereas fish that feed on plant-eating insects are secondary consumers. If such fish become the prey of other carnivorous fish—a trout, for example—the food chain can be lengthened to include tertiary consumers. Man would be considered a quaternary consumer, if he caught, cooked, and ate the trout. The term *trophic level* is used to designate each of these successive or progressive transfers of energy from producers (plants) to plant eaters (herbivores) to carnivores that eat herbivores and, hence, to carnivores that eat other carnivores. This simple illustration of a food chain is not representative of complete food chains in nature, because many plant and animal organisms, when they die, become food substrates for a whole host of agents—microorganisms, primarily—that function as decomposers.

The flow of energy through echelons of consumers in natural food chains is inefficient. As little as 10 per cent of the energy stored in producers (green plants) is normally available for storage in the tissues of herbivores. Similarly, only about 10 per cent of the food energy stored in the tissues of herbivores is, in turn, stored in the tissues of herbivore-eating animals. Because the energy fixed by the photosynthetic activities of producers is limited, the total energy entering a food chain is also limited; thus, there is constant competition among consumers for food.

The cycle of chemical elements that are combined into organic matter by photosynthetic processes and the consumption of organic matter is not complete until much, if not all, of the organic matter is recycled and transformed to renew plant and animal populations. The complete cycle includes the recycling of chemical elements present in the fecal and urinary excretions of animals, those in the end products of the respiratory processes of all living organisms, and those in the products that are produced from the decomposition of dead organic matter by microorganisms (Figure 2.1). Balance—or, more accurately stated, approximate equilibrium—is attained when the production of organic matter and its consumption within a system of interacting plant, animal, and microbial life are nearly equal (at least under long term conditions), so that displacements in the cyclic flow of energy from

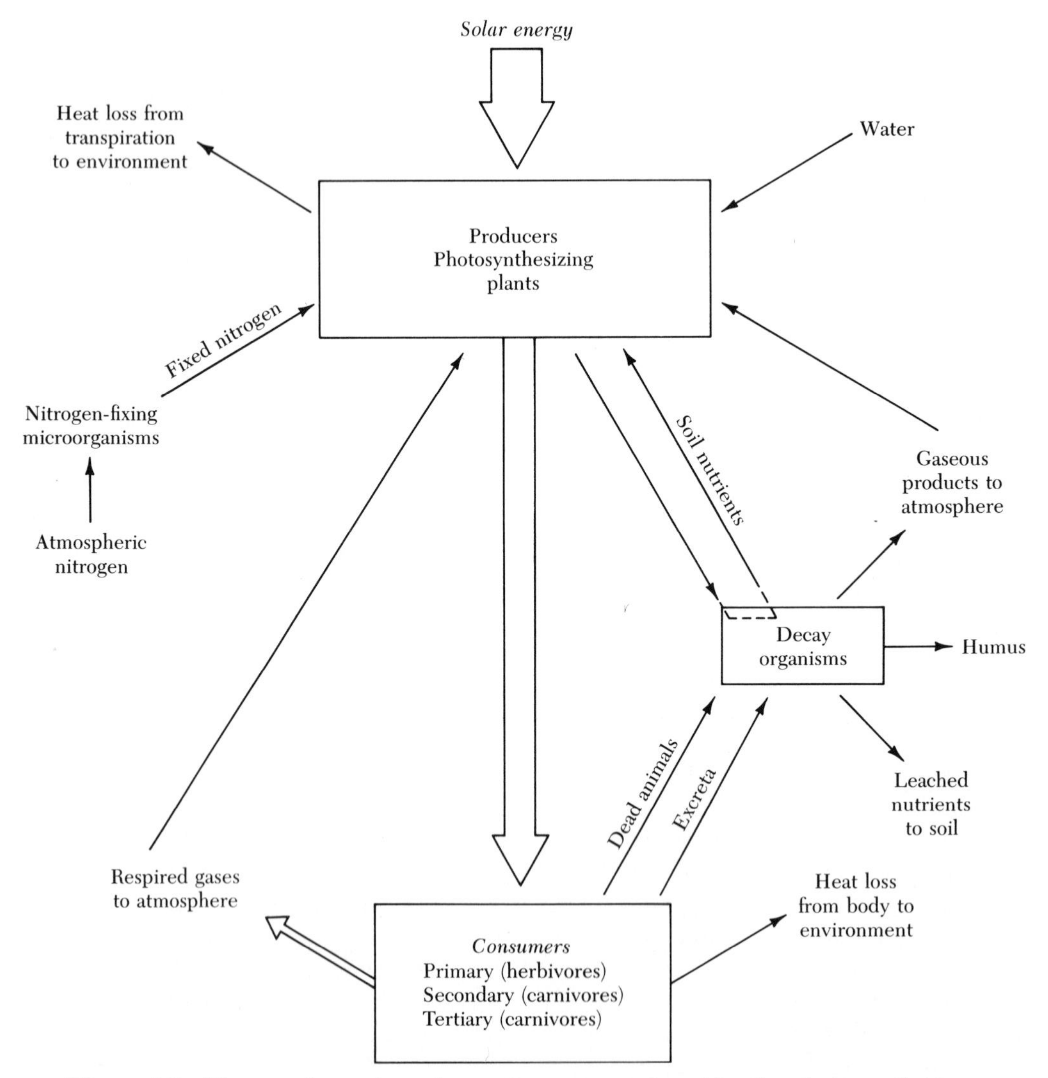

Figure 2-1 The transformations of organic matter produced by the photo-synthetic activities of producers (plants) by activities of consumers and decay organisms is presented in a very simplified form. Except for the food energy present in tissues of living plants, animals, and micro-organisms and decaying organic matter, energy fixed in organic matter by producers is in balance with energy "lost" to the environment as heat in the respiration of producers and consumers.

The chemical elements present in respired gases, excreta, and dead animals and plants are eventually recycled as shown to renew producer and hence consumer populations.

(Adapted from Woodwell, "The Energy Cycle of the Biosphere." Copyright © 1970 by Scientific American, Inc. All rights reserved.)

producers to consumers—and, subsequently, to decay agents—are minimized.

Obvious exceptions must be made for those chemical elements that are in dynamic exchange in the cycle of production and consumption of organic matter and in their transformation from one form to another in the present biosphere. During the earth's very long geological past, vast quantities of carbon, in the form of dead organic matter now represented by the fossil fuels (coal, oil shale, and petroleum), were accumulated beneath the earth's surface. Our increased utilization of fossil fuels during the past 150 years has added much carbon dioxide to the biospheric circulation of carbon, but the ultimate consequences of this human activity remain to be determined (Bolin, 1970).

A simplified model for a stabilized or mature ecosystem should not imply that ecosystems are closed systems. All ecosystems are open-ended, variably and delicately structured, and exceptionally fragile, as is now widely recognized. However, we are only now beginning to develop a fragmentary understanding of how the Earth's various ecosystems interrelate to maintain the integrity of the biosphere that allows life-support systems to function effectively. A few examples are sufficient to demonstrate the validity of this general concern about the man–nature complex.

Man living prior to the invention of agriculture could do little more than assume the role of an insignificant parasite on the intact natural energy-producing systems with which he was in contact. Animals killed in hunting, fish taken from the water, or plants and their fruits and seeds gathered for human use represented a minor diversion of the total energy fixed and transformed within the larger systems of the earth's plant and animal communities. Man's diversion of energy from the natural food chain, despite his best efforts to that end, was limited in two ways: he had not yet learned "how to tame nature," as it were, and the number of his kind was exceeded by that of every other kind of animal life. Primitive man, therefore, could divert for his own use only such food as he might obtain by hunting and gathering—that is, food wrested from his competitors, from plant-eating insects, herbivores, and carnivores. His control over plant growth was minimal and, for a very long time, limited to the use of slash-and-burn agriculture and to hand labor for controlling unwanted plants. The productivity of food plants was, in turn, limited to the natural fertility of the soil and the availability of water.

Modern man's energy (food) requirements and those of his domestic animals are derived almost exclusively from agricultural production and from the large-scale capture of aquatic animals. His dependence on the current contribution solar energy makes to agricultural production remains, but he is also dependent on the use of stored energy, the fossil fuels.

The present agricultural ecosystems are supporting, albeit in a rather spotty manner, 3.7 billion people, and they could provide food for an additional billion people, if current agricultural technologies could be widely adapted and put into practice. Fossil fuels are used in modern agriculture. This auxiliary energy is expended in endless ways to improve agricultural productivity: it is used for drainage and water control, clearing of forest land, seed-bed preparation, weed and pest control, fertilization of crop plants, and efficient harvesting, processing, and distribution of food products. Measured in units of kilocalories per square meter per year, the radiant energy of the sun is 1,500,000 units, and the work energy put out by one man is 6 units. The primitive farmer achieved an energy output in his crop of 20 units; whereas the modern farmer, by making an additional input of 135 units of fossil-fuel energy, achieves an output of 1,000 units. Only a very small fraction of the radiant energy of the sun is transformed by agriculture into organic matter as represented by crop yields (Odum, 1967). Analogous but equally valid considerations are applicable to man's developments in improving the efficiency of animal agriculture.

In these connections it matters little

whether the energy inputs are used to power farm or industrial machinery, grind feed, manufacture fertilizers and insecticides, or make paper in which to wrap foods. It is also important to appreciate that technology applied to uses of fossil fuel, and nuclear energy to bypass constraints on man's exploitation of natural ecosystems, has produced and continues to produce stresses on the human-environmental complex that could not have been predetermined or preguessed. Even so, it is ironic that to gain control over the biosphere and the capacity to alter it for better or worse, modern man is obliged to exploit the use of energy locked in fossil and nuclear fuels representing energy that reached this earth eons of time ago (Brown, 1970).

Resource Utilization for Plant and Animal Production

Primitive plant and animal production Field food and fiber crop plants and the many domesticated animals known to us today have, as their antecedents, wild counterparts domesticated by primitive man living in such parts of the Earth as South America, Africa, and Asia. Most of the main subsistance crops which are prominently associated with archeological records and folklore can be linked genetically to those that provide the bulk of the food eaten by man today. The extreme variability in cultivated strains of rice, wheats, and millets resulted in part from the deliberate efforts of farmers living in prehistoric time to select adaptable types of plants and animals, and in part from frequent gene exchange with wild germ plasm. These events have been most fortunate for us. The revolution made possible in man's current animal and plant breeding and selection programs would not have been as fruitful as it has been if it were not for the fortuitous preservation of the highly heterogenous populations of botanical and animal life throughout the agricultural historic past. While we can be critical of the haphazard occupations of primitive plant and animal agriculturists, we need to be reminded that our vaunted efforts to produce uniform crop varieties and strains of domestic animals sometimes end in disaster because of their vulnerability to disease.

The technological progress some early agricultural societies achieved remains to be marveled at to this day. The achievements of civilizations that once flourished in the valleys of the Nile, Tigris, Euphrates, and Indus provide ample testimony that early man possessed an uncanny capacity to exploit the advantages presented by well-watered fertile soils, mild climates, and the irrigation of crops. Food produced in these areas sustained millions of people and made possible the development of commercial, social, political, and economic institutions unparalleled in the history of mankind.

But in much of the rest of the world, man's mastery over the natural environment system of plants and animals was an exceptionally slow process until recently. While we may grant that the capacity for large-scale food production existed in the earliest systems of agriculture, and that primitive agriculturists used a surprising amount of ingenuity in making agricultural tools and implements and in using animal power to improve the efficiency of food-production practices, production of large food surpluses was not easily achieved.

Where land was plentiful, problems connected with exhaustion of soil, persistent weeds, and repeated crop failures caused by soil-borne organisms and plant pests were simply solved by employing cycles of burning, clearing, cropping, and abandonment of land, agricultural practices which prevail in many parts of the world today.

The main production input in such agricultural systems is the work of human beings aided in some instances by animal power. Actual tools and implements, while cheap and often made by the farmer himself, are inefficient and frequently not very effective. The low yield of food and fiber crops is often not unlike that of their wild counterparts. The main difference is that man had managed to grow more of one kind of plant in a smaller area. A similar analysis may be made of primitive animal-rearing systems. Many cattle-herding societies did little else than concentrate their animals on a grazing area until shortages of forage forced them to seek another. Animal productivity was often so low that net increases did not permit much more than the killing for food of very old and ailing animals. Since animal productivity was greatly limited by losses of animals to diseases and parasites and by lack of nutritious food, there was little incentive to improve animals by selective mating practices.

For analogous reasons, primitive agriculturists did not practice clean cultivation methods. The effort necessary to keep fields clean of weeds or wild food-plant species was seldom rewarded by offsetting improvements in yields. They may have observed, too, that clean cultivation left the land highly vulnerable to erosion by water and winds. Other soil-saving strategies were not often practiced by primitive farmers because they lacked the means to implement them.

Consideration of these types of agricultural systems, and there are many still in existence, leads one to accept that the food produced by primitive farmers was often only about enough to support the efforts of the agriculturist and the immediate members of his family. The natural energy circuits for such systems may reveal that an enormous quantity of solar radiant energy is available to be trapped for the production of plant and animal organic matter, but very little, whether measured as food calories produced per acre or as bulk weight of plant and animal life, can be produced without the addition of modern agricultural inputs, represented by the technologies that deal with plant and animal management, mechanization, commercial fertilizers, soil conservation and water control, and chemical control of weeds and plant and animal pests.

The best that can be said is that these agricultural systems did not create the problem of pollution that has been generated by highly industrialized societies, and that they can be made to function without the need for very much money. At worst they contributed to very serious declines in the quality of the natural environment through the erosion of soil and the deterioration of the value of forest and range lands, and they did not provide much opportunity for those caught up in them to raise their standard of living. Such circumstances were invariably exacerbated when the capacity of agricultural lands to support large numbers of poor people was outstripped. Despite the best efforts of primitive agriculturists in overpopulated, land-scarce countries of the world, the inevitable result was the production of fewer and fewer kinds of food plants, particularly of high-protein foods. Even today, throughout the densely populated parts of the world, high-yielding starchy food plants are favored since they provide the most calories per acre farmed, not because people necessarily like to eat them.

Modern plant and animal production We often attribute the slowness of man's progress in "taming nature" to a lack of scientific understanding of biology, soil fertility, pest control, and so forth. These were indeed major constraints, but we are only now realizing that the main obstacle was his complete dependence on hand labor and the work that could be done with animals.

Food-production practices were completely altered when man invented the arts and practices of plant and animal agriculture, and when he learned how to exercise controls over nature by the utilization of new sources of energy made possible by burning fossil fuels and by harnessing nuclear reactions for the production of power. Modern agricultural systems were developed out of a series of inventions in England, beginning about 1760, which completely revolutionized existing practices of industry and agriculture. But the most striking result of the industrial revolution was the tremendous increase in the productive power of man equipped with mechanical devices. The subsequent perfection of internal-combustion engines and the advances made in agricultural chemistry, plant breeding and production, and utilization of commercial fertilizers, and more recently in the synthesis of inorganic and organic materials for killing insects and weeds and for the control of microbial organisms, opened up possibilities for altering agricultural practices seemingly without limit. The impact of these revolutionary developments on and the consequences to human welfare have not been unmixed blessings. The enormous increase in the production of energy by the industrial use of coal, petroleum, water, and nuclear reactions as power sources enable us in one way or another to transform and to modify the environment, to produce useful materials not found in nature, and to alter natural ecosystems in many harmful ways.

Food-production capacities that can now be realized by the application of fossil-fuel-rich animal and plant production systems in the well-advanced, highly industrialized countries of the world have been measured in many ways, but it is best appreciated when we note that only 4 or 5 per cent of the total population in the United States is now needed on farms to produce food for the entire population, plus some for export.

The use of agricultural chemicals has unfortunately been accompanied by some serious problems. Fortunately, undesirable side effects resulting from large-scale use of toxic and persistent chemicals, particularly some of the synthetic organic insecticides, are now widely recognized.

If one lesson has been learned, it is that solutions to parts of a problem generate new problems. Furthermore, the immediate benefits gained from the application of science and technological innovations must be subject to question, close scrutiny, and reevaluation, particularly when evidence begins to accumulate that the benefits are outweighed by accompanying damages to the environment. This basic realization lies at the heart of man's continuing endeavors to minimize deterioration of the quality of the world's man-nature complex.

Agricultural Plant and Animal Production Systems

Modification of Practices: Direct Effects on Nontarget Organisms

Nature's system of checks and balances keeps the overall production, consumption, and decomposition of organic matter so highly stabilized that a large net production of organic matter is rarely possible. Obviously the cycling of organic matter and the energy-flow patterns in intact natural systems can be variable in the short run. Unfavorable weather, sudden emergence of a specific consumer organism, for example, an outbreak of locusts or the sudden drying up of a lake after an earthquake, are known to have caused serious localized disturbances in animal and plant communities, but in time the systems of energy production and utilization reach equilibrium conditions once more, albeit in some altered form. Thus, major functional and structural changes in an intact natural system occur over relatively long periods of time, often thousands of years.

This discussion so far leads to the conclusion that there is only one practical way for man to exploit, for his own needs, organic matter (food and fiber crops, meat and fish foods) produced by natural systems, and that is to short-circuit one or more links in the chain of producers, consumers, and/or decomposing organisms. Simply stated, he does this by the elimination of unwanted consumers (pests) and isolating and protecting producers (crop plants), so that the natural food chains are shortened to include himself and such consumers (farm animals) as he wishes to use for food or other purposes.

The best strategy in agricultural technology is to find vulnerable links, switch points, and energy-leakage pathways that can be controlled and manipulated so that net production of animal and plant products may be greatly increased. A convenient illustration of this point relates to man's effort to control insect pests. Losses (an energy leak) to insects in the production of the world's food crops are not easily estimated, but it is at least 10 per cent of what could be produced if they could be eliminated. One estimate suggests that, in the United States alone, an increase of 25 million metric tons of maize, wheat, rice grain, sorghums, soybeans, and potatoes over current levels of production could be achieved if losses to insect pests could be reduced by half. Control of intermediate insect hosts of parasites that cause diseases (malaria, for example) in man represent still another effort to solve an insect pest problem.

The use during World War II of a highly effective and cheap organic insecticide, DDT, followed by an astonishingly rapid acceptance of its use to control insect pests soon verified all the claims made for it. The health and economic benefits of its use in agriculture, home gardens, and mosquito control are well-known. Millions of people living in the malarial areas of the world are given respite from the misery and debilitation caused by malarial attacks, and plants and animals can be made productive at smaller cost.

During the initial stages of the widespread applications of DDT, little was known about its persistence in the environment and less of its cumulative buildup in food chains. Only a few of the organic insecticides, DDT being one of them, have caused injury to fish and wildlife. Because DDT is only slowly degraded in plants, animals, and soil, its use has been banned, except for special purposes. The widespread elimination of species because of its use that some had predicted has not come to pass.

The leakage of chemical fertilizers so lavishly used in intensive cropping practices or drainage from animal manure wastes that are accumulating in huge piles near large cattle feeding lots into fresh water courses and lakes likewise pose threats to nontarget organisms. Such problems, commonly lumped under the term "pollution," have arisen because excessive amounts of chemical elements and organic matter entering our rivers and lakes cause explosive growths of some types of aquatic life to the detriment of other types of aquatic life, including fish. All too often, these water-polluting agents and the far larger discharge of industrial and domestic wastes may so overburden the natural systems of chemical element reutilization and decomposition of organic matter that the entire system collapses and is terminated in a malodorous, poisonous, putrified "mess."

Systems of pest management, modifications in fertilization practices, and alterations in a host of other agricultural technologies based on the principles of applied ecology have been developed to cope with such problems. The classic example of screw-worm eradication from the southeastern United States through the use of the male-sterile technique represents one very promising, wholly specific approach to the control of pest species. The advantages of selection and breeding disease-resistant varieties of food and fiber crops as an alternative to the chemical control of pathogenic microorganisms are likewise well appreciated. It is equally obvious, however, that in using them we must find ways to avoid serious contamination of the environment with harmful chemical products and their residues.

Soil, Air, and Water Pollution

While technology has contributed immensely to our standard of living and is permitting more people to live in crowded urban communities than ever has been possible before, we are also exploiting our natural resources and producing wastes as never before.

In highly mechanized agricultural systems, the burning of fossil fuels contributes to air pollution, and intensive plant production practices require the use of large amounts of fertilizer and chemicals to control plant and animal diseases and pests. The massing of large numbers of meat and poultry animals on small areas of land create very serious animal-waste disposal problems. Irrigation of arid lands with water pumped from deep wells has resulted, in some instances, in very serious declines in depth of the natural water tables and intrusion of saline water. Impoundments of water behind dams are endangered by siltation caused by eroding soil entering streams. The runoff of chemical fertilizers, herbicides, and insecticides into rivers, lakes, and underground waters creates a source of chemical pollution of drinking water. Overfertilization of lakes and streams, resulting from entry of nutrients, whether from soil particles, fertilizer runoff, domestic sewage, or industrial effluents, often cause explosive growths of algae; massive dieoff of algae further depletes the oxygen content of water, resulting finally in death of fish and other aquatic life.

Agricultural Waste Disposal Problems

The problem of the handling and disposal of animal waste in large-scale total confinement systems for animal rearing and feeding is rapidly assuming major proportions in Canada and the United States. The animal-waste disposal problem is basically related to consequences that follow collections and movements of very large volumes of feed grains and commercially produced feed ingredients to small land areas where very large populations of farm animals are fed.

Prior to this revolutionary change in livestock production, farm animals were widely dispersed on large numbers of small farm units and for much of the year maintained on pastures. Manure disposal did not present much more than a routine chore, and it was a valuable source of crop nutrients when spread on fields where feed grains and forages were produced for the following year.

The developments that led to the manufacture of easily available, efficient, and cheap chemical fertilizers in the meantime greatly lowered the value of manure as a fertilizer, and made large-scale high-yield crop agriculture economically attractive. The subsequent production of large volumes of relatively cheap livestock feeds and economic pressures created opportunities for larger and larger production units. Feeding of large populations of farm animals under intensive-confinement livestock systems has been developed to take advantage of labor-saving devices for feeding and watering animals, feed preparation, economies associated with bulk purchases of feeds and feed ingredients, and other management and production cost-saving practices. Cattle-feeding operations capable of handling 50,000 animals annually are now common, as are very large pig-rearing and feeding installations. Hundreds of thousands of broilers and egg-laying birds are often located on areas of land that may not exceed 25 acres.

It is estimated that U.S. livestock produce about two billion tons of liquid and solid wastes per year, enough to fill a square mile ten feet high each day, or about ten times the amount of excrement produced annually by all the people in the U.S. today.

The magnitude of the problem is readily appreciated by a manager of a lot with 25,000 cattle who is faced with the task of handling 650 tons of manure daily. It has been estimated that a Colorado feedlot handling 120,000 cattle requires the labor of 12 to 15 men the year round just to clean lots. While most of the manure is eventually spread on 13,000 acres of land used for the production of corn silage, the manure-disposal expense per year is not offset by its value as a crop fertilizer. Since land is not readily available during much of the year for immediate spreading of animal wastes, manure must be stockpiled. Manure odors and the constant threat of drainage of liquid manures into surface and ground water supplies are therefore becoming one of agriculture's most vexing problems.

There have been numerous schemes for animal-waste disposal, but all have one or more limitations. Although land can be used as a disposal medium, there are problems to contend with. Too much manure applied to land can upset the soil's balance of nutrients, create odor problems, and result in stream pollution. Disposal of liquid manures in la-

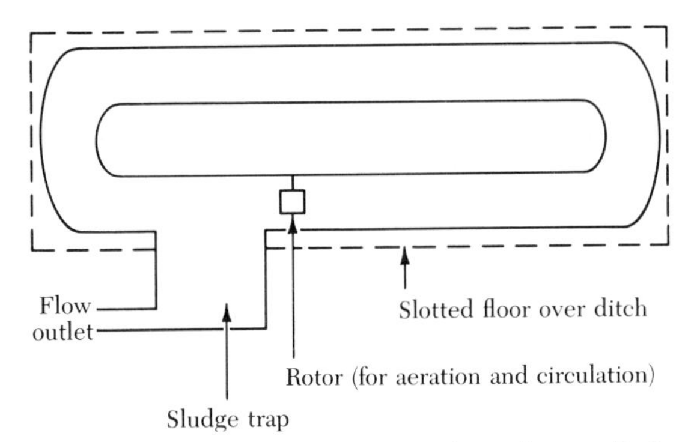

Figure 2-2 Diagram of livestock oxidation ditch installation. Cage rotor that aerates and moves liquid manure at a velocity that keeps solids in system in a closed loop system. Excess liquid and sludge are removed at intervals (Jones, Day and Dale, 1970).

goons, manure storage in silos until it can be pumped out and spread, manure drying, and septic tank systems are all being evaluated for performance and cost.

Problems associated with other agricultural wastes have been satisfactorily solved only in some instances. A wide variety of wastes and by-products of agricultural and industrial operations are processed into valuable animal feeds. Such materials include whey products, beet and citrus pulp, molasses, distillers dried grain, pea vines, beet tops, almond hulls, wheat bran, cottonseed hulls, and meals from a variety of seeds from which the oil has been extracted (cottonseed, linseed, and coconut meals, for example). The meat-animal processing industry also generates a host of products (meat scrap, blood meal and bone meal, tankage, animal fats, and hydrolized poultry offal) from salvaged animal wastes.

In some respects, manure disposal problems are not unlike those of urbanization, for whether wastes are produced by animals or man, their concentration in one location creates waste-disposal problems with the attendant hazards of water pollution and odors. Waste produced by the food-processing industry are not easily handled either, as so much of it is wet and cannot be economically segregated and collected. Satisfactory, practical designs for waste-handling facilities in agriculture and the food processing industry, which provide for cheap disposal without pollution, are therefore urgently needed (Figure 2.2).

Conclusions

Whatever the social, economic, political and legal costs are for changing the man-nature complex as we have known it for so long, there will be no easy escape from the realization that the options for decent human survival and welfare are few indeed. The current trend in the growth of human populations is likewise not acceptable.

The price to be paid for the correction of our ecological incompetence is a heavy one, but who can say man is not capable of paying it, given the reason that he cannot afford for very long to do otherwise?

FURTHER READINGS

Bolin, Bert. 1970. "The carbon cycle." *Scientific American,* 223, no. 3 (September 1970), 125. Available as Offprint no. 1193 from W. H. Freeman and Company.

Borgstrom, George. 1969. *Too Many: A Study of the Earth's Biological Limitations.* London: Collier-MacMillan.

Brown, Lester R. 1970. "Human food production as a process in the biosphere." *Scientific American* 223, no. 3 (September 1970), 160. Available as Offprint no. 1196.

Christensen, Raymond P., William E. Hindrix, and Robert D. Stevens. 1964. *How the United States Improved Its Agriculture*. Washington, D.C.: USDA.

Ehrlich, P. R., and A. H. Ehrlich. 1972. *Population, Resources Environment; Issues in Human Ecology*. 2d ed. San Francisco, Ca.: W. H. Freeman and Company.

Jones, D. D., D. L. Day, and A. C. Dale. 1970. "Aerobic treatment of livestock wastes." *Univ. of Illinois Agric. Exp. Stat. Bull.* 737.

Woodwell, George M. 1970. "The energy cycle of the biosphere." *Scientific American*, 223, no. 3 (September 1970), 64. Available as Offprint no. 1190.

Three

Animal Foods for Human Needs

"Man is a carnivorous production,
And must have meals, at least one meal a day;
He cannot live, like woodcocks, upon suction,
But, like the shark and tiger, must have prey;
Although his anatomical construction
Bears vegetables, in a grumbling way,
Your laboring people think beyond all question,
Beef, veal, and mutton better for digestion."

Byron,
Don Juan, *Canto* II, *St.* 67.

Animals as a Source of Human Food

Animals were used for food by man's ancestors long before modern man, *Homo sapiens,* evolved. Fossil remains of *Australopithecus,* which is regarded by anthropologists as a "near man," have been found in association with the bones of many animals, particularly members of the family *Bovidae,** that he used for food. This association may indicate that early man preferred the flesh of the *Bovidae* to that of other animals, as he does in most parts of the world today (Leakey, 1965), or may merely indicate that the *Bovidae* were the animals he could capture or whose carcasses he could find most easily. Man has eaten flesh long enough for the practice to have greatly influenced his evolution as a species, and to be a factor in determining his present-day nutrient requirements. The practice certainly was not acquired recently because of an increase in aggressive behavior, for research in Tanzania on the behavior of chimpanzees, who are among man's closest living relatives, has shown that they also eat flesh; they regularly hunt, kill, and eat a wide variety of other animals, including baboons, monkeys, young pigs, and bush bucks.

Man's ability to eat either flesh or vegetable foods, in combination with his manual dexterity, permitted him to sustain himself in a wide range of environments, and to become one of the most widely distributed animals on this earth.

* Living representatives of the family *Bovidae* include species of the genera *Bos* (cattle), *Ovis* (sheep), and *Capra* (goats). See Rice *et al.*, 1967, for a systematic classification.

Animals as Food in Primitive Societies

Primitive man ate the flesh from all animals in his environment, whereas modern man uses relatively few animal species for food. A compilation of all the animal species which have been used, in recent times, as food by various ethnic groups would be indeed extensive, and beyond the scope of this chapter. It would include practically all the members of the *vertebrata* (fish, amphibians, reptiles, and mammals), as well as many of the *arthropoda* (including *insecta*) and the *mollusca.* It appears that the two main factors which governed whether an animal was used as food by man were the ease with which he could capture it, and the existence of any cultural taboos, customs, or traditions which prohibited its consumption. Many of the more "unusual" animal foods of primitive man now constitute only gastronomical items in our diets, e.g.: snails, both land and marine; squid and other cephalopods; oysters, cockles, and mussels (mollusca); crayfish and crabs (crustaceans); and frog's legs (amphibians). These animals generally require high labor inputs for their harvest, which has precluded their extensive use by modern societies. It is interesting to note that primitive coast dwellers who lived on shellfish diets produced the greatest bulk of refuse for a given quantity of edible substance of any primitive people. Apparently pollution by man existed even in the paleolithic period! Primitive societies today provide outstanding examples of the essential role of animals in man's food chain. In environments where crops are not cultivated and the supply of edible native plant foods is limited, animal products generally constitute the major component of the diet.

Before the development of trade with Europeans, the diet of the Eskimo was almost wholly derived from the Arctic fauna. In summer, the migratory birds (e.g., ducks and geese) provided meat. Meat at other times came from the reindeer, which can maintain itself year round in the Arctic because of its ability to locate food under the winter snow, or from the hunting of seals and walrus, and from fishing. The latter animals derive their food from the food chain which ultimately relies on the photosynthetic organisms of the sea. Therefore, with the exception of a few berries and nuts which the Eskimo may have eaten in the summer, his diet was dependent on the plant-animal food chain.

In comparison with that of other continents, the native flora of Australia include very few edible plants, as is reflected in the fact that only one commercially cultivated food plant (the Macadamia nut) has originated from that continent. The diet of the aborigines in the higher-rainfall coastal areas included grass and other seeds and fruits, various roots, and game; but, in the interior of the continent, the vegetation comprises inedible xerophytic plants and grasses. Here tribes such as the Arunta subsist almost wholly on the game which graze either directly or indirectly upon these xerophytic plants. While the form of game is diverse, including kangaroos, wallabies, lizards, snakes, and emus and other birds and their eggs, these animals have the common role of being converters of otherwise inedible plant products into human food.

The Bedouins of the Arabian steppes also exploit an ecological niche similar to that of the Australian aborigines, but instead of hunting wild animals, they derive food from the domesticated camel. Frequently tribes may live for months on the milk of the camel, although this diet is supplemented generally with cereals purchased from the sale of camels or their by-products, such as hair.

Animals as Food in Modern Societies

We may legitimately inquire if it is necessary for modern man to use animals as a source of food and whether a complete diet could not be supplied totally from plants. This question may be approached from two aspects: first, nutritional necessity for the inclusion of animal products in the human diet; and second, whether there are any over-all benefits to be derived from the integration of animals with crop production.

Animals versus plant products in the human diet. For man to enjoy good health, his diet has to contain a number of specific chemical compounds which are referred to as the essential nutrients. These substances cannot be synthesized by the human body, or at least their rate of synthesis is suboptimal. These nutrients are classified under the four broad categories of energy (or calories), protein (or amino acids), vitamins, and minerals. All foods contain some of these nutrients, but some foods are much better sources than others. Animal products as a group are characterized by a high proportion of protein, and the amino-acid composition of these proteins more closely resembles man's dietary requirements than proteins of plant origin. For example, the proteins of the common cereals are deficient in the essential amino acid lysine, whereas milk and meat proteins contain a high proportion of this amino acid.

Our current knowledge of the nutrient requirements of man permits the formulation of nutritionally adequate diets based on plant and synthetic sources of these nutrients. Indeed, farm animals such as pigs and poultry which have qualitative nutritional requirements similar to those of man can be successfully reared on diets devoid of animal products. However, the preparation of these diets requires specific combinations of foods and frequently the addition of synthetic amino acids and/or vitamins before they become nutritionally adequate. A diet based solely on foods of plant origin, therefore, places constraints on which foods can be eaten. For example, no combination of basic plant foods derived from the cereal grains (e.g., wheat and maize) and the starch roots (e.g., cassava and potatoes) can give a balanced diet for man. All combinations will be deficient in some of the essential amino acids and probably certain of the vitamins, particularly those of the B complex (e.g., vitamins B_2 and B_{12}). These deficiencies can be corrected by the inclusion of relatively small amounts of animal products, such as meat, fish, or milk in the diet. Alternatively, the amino acid deficiency may be corrected by adding specific seed proteins (e.g., from soybean). However, this places a constraint on the selection of foods and in any event, soybeans cannot be cultivated in all parts of the world.

Therefore, one of the great virtues of animal protein in the human diet is that it forms an effective combination with a wide variety of other foods, particularly cereals, and allows flexibility in the selection of a balanced diet. Thus while a nutritionally adequate diet can be formulated from vegetable products, it is more readily achieved if animal products are a component of the diet.

There is good evidence to indicate that the mature height and physique of man can be affected markedly by dietary habits, in particular, protein nutrition. This may be illustrated by the comparison of the related Masai and Kikuya tribes of Kenya. The Masai, who lead a pastoral life, have a diet which includes substantial quantities of milk, blood, and meat, and their men are tall and vigorous. The Kikuya, who are agriculturalists, live almost exclusively on plant products, maize, millet, sweet potatoes, and yams. Their men are much shorter in stature than the Masai, and also have a lower standard of general health. Quite marked differences in mature height have been recorded between orientals who immigrated to the United States and their first-generation children, who enjoyed the better general nutrition of their new homeland. Similar differences in the height of children in Japan born before and after World War II have been observed. Better protein nutrition no doubt largely contributed to these differences in stature.

Besides the pure nutritional aspects of food, man derives psychic stimuli from eating. Foods of animal origin generally have a greater capacity to satisfy "appetite" and to produce a hedonic experience than foods of plant origin, even though they may have comparable nutritional value. Meat substitutes based on vegetable protein also require extensive use of artificial flavors, color, and stabilizers to make them resemble the natural product. Few of us would exchange a beef steak for a piece of modified soybean substitute even if we accepted that the nutrients in both were comparable.

Animals as an integral component of agricultural production. Only a small proportion of the Earth's surface is used for the production of crops. The areas which are not used have deficiencies in soil fertility, rainfall, topography, climate, or proximity to markets which preclude their participation in economic crop production. While the application of modern agricultural technology (e.g., land clearing, the use of fertilizers, irrigation) can expand the area of arable agriculture, often the economic costs of bringing this land into production are high, and in many places they are prohibitive. However, even modern technology is not capable of bringing areas of excessive slope into a stable system of crop production. Sole reliance on arable crops for human food would result in the utilization of only a very small proportion of the world's land surface and the non-arable areas of the world which produce grasses, herbs, and other plants would not contribute to man's food supply. However, grazing animals can utilize this plant material inedible to man as a source of food, harvest it at an economical cost, and bring rangelands into man's food chain. The exclusion of animals from agriculture would deprive man of this additional and valuable source of food. The productivity of rangelands is extremely variable, but can be very high. Milk yields in excess of 4,500 kg (or 150 kg of milk protein) per hectare per year are frequently obtained from cows solely grazing pastures in New Zealand. However, much of the world's rangelands are in low rainfall areas, suitable only for grazing at low intensities by meat or fiber-producing animals.

Because of man's digestive anatomy and physiology, he utilizes only selected portions of plants for food. For example, only the reproductive portion of the cereal-grain plants, which are the staple food of most of the world's human population, are eaten. The bulk of the plant, the photosynethetic and supportive tissues, cannot be used for human food. These plant products contain nutrients which can be utilized by animals, which can convert a proportion of the nutrients in these crop residues into food for man. While it is true that the efficiency of this conversion is frequently low, without animals the crop residues would make no contribution to the human food supply. Therefore, even in a system of intensive agriculture, it appears that at least for the foreseeable future there will be an essential role for animals as converters of plant products indigestible by man into human food. However, the direct use of land area capable of grain production for growing animal feed may decline as world food needs are met more adequately. The estimated efficiencies of livestock and crop enterprises in Table 3.1 support this contention.

A recent estimate of the total world production of animal products from agriculture and from fisheries is given in Table 3.2. In total weight, milk exceeds all other categories, but when all sources are expressed on an equivalent protein basis, the amounts of protein produced as milk and meat are similar, and, together with marine fish, account for 90 per cent of the total world animal-protein pro-

Table 3.1.

Estimated efficiencies of some livestock and crop enterprises (from Holmes, 1970). The animal enterprises are assumed to be self-sustaining populations producing at reasonable levels and on normal diets for British farms.

Animal enterprise or crop	*Energy yield in mcal/hectare*	*Protein yield in kg/hectare*
Dairy cows	2,500	115
Beef cattle	750	27
Pigs	1,900	50
Eggs	1,150	88
Wheat	14,000	350
Peas	3,000	280
Potatoes	24,000	420

duction. About 30 per cent of the world's supply of edible protein comes from animal products; of the remaining 70 per cent, about 50 per cent is derived from cereals. The relative contribution of cereals to the total protein intake varies from about 72 per cent for Pakistan to 16 per cent for the United States (Figure 3.1).

While the high concentration of protein of high biological value in animal products makes them of special value in man's diet, they also contribute significantly to the total energy, vitamin, and mineral nutrition of man. In general, countries with the highest caloric intakes also have the highest intakes of protein from all sources. Figure 3.2 shows this general relationship for 33 countries. Animal-protein intake is also related to the intake of calories, but there are several notable exceptions: Ireland, Yugoslavia, and the eastern block countries of Hungary, Poland, and Romania (Figure 3.3).

Studies conducted by FAO in 1964 indicate that there is a close relationship between total and animal protein intake and average income. Total protein intakes rise sharply up to a per-capita income level of about $1,000 per year (U.S. dollars), after which it tends to flatten out. However, the consumption of animal protein is linearly related to income and does not exhibit a plateau. Even within a country, the individual's consumption of meat products is related to his economic class. A recent survey from England indicates that there was a 75 per cent difference in consumption of beef, veal, and poultry meats between high and low economic classes. A similar change in the dietary habits of whole countries is found when we follow the rise in the purchasing power of individuals; the intake of meat increases at the expense of the cereal components of the diet. Food preferences in the United States over the last 60 years show either a relatively constant intake or an increase in the consumption of animal products, and a decrease of consumption of plant products such as potatoes and cereals (Table 3.3).

In European countries, the United States, Canada, and the middle-eastern countries bordering the Mediterranean Sea, the amount of protein provided to each person by meat and milk increases significantly as the countries become more highly developed and the proportion of the population in agriculture decreases. There is a highly significant between-country regression which indicates that developed countries with only about 5 per

Table 3.2.
World production of animal products, 1968. (From FAO, 1970b, p. 197.)

Source	*Weight (metric tons × 10⁶)*	*Approximate protein (per cent)*	*Total weight of protein (metric tons × 10⁶)*	*Percentage of total*
Milk	392.8	3.5	13.75	33.7
Meat	78.7	18	14.16	34.7
Eggs	13.5	13	1.75	4.4
Fresh-water fish	8.5	18	1.54	3.8
Marine fish	49.9	18	8.99	22.0
Crustacea, mollusca, and other invertebrates	4.8	12	0.58	1.4
Seals and misc. aquatic animals and residues	0.07	18		
Total	548.27		40.76	100.0

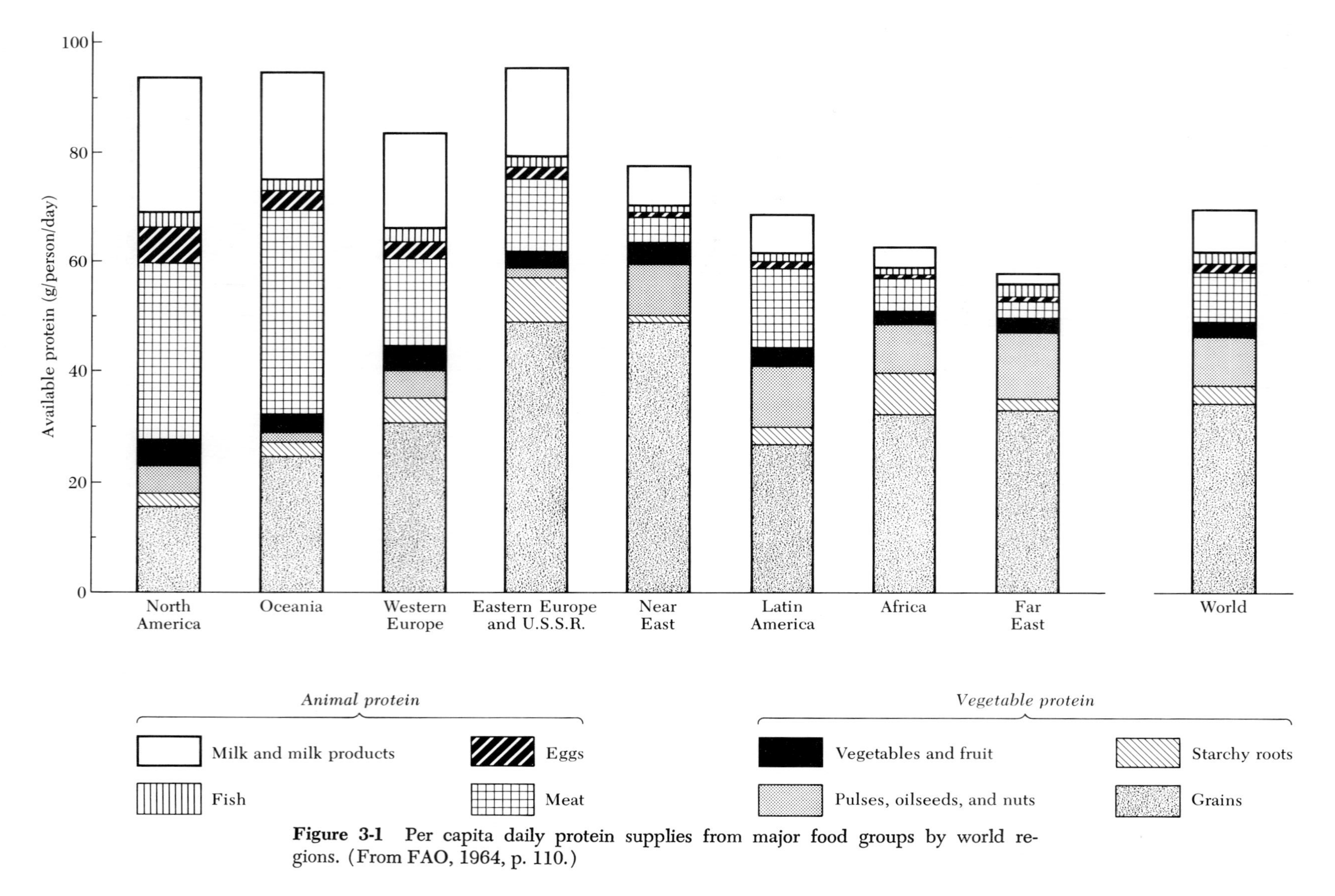

Figure 3-1 Per capita daily protein supplies from major food groups by world regions. (From FAO, 1964, p. 110.)

cent of their population engaged in agriculture (e.g., the United States) have the highest consumption of meat and milk products (Figure 3.4). This relationship applies to Western societies, or to those which aspire to Western standards, and does not include countries which are essentially exporters of primary products or Asian or Far Eastern countries. International trade in animal products is not accounted for in this relationship, and may affect the relative status of some countries; however, it illustrates that diets containing high levels of animal protein can be achieved with a very low work force engaged directly in agriculture. The advanced technology of present day societies could not have developed if modern agriculture had not released a large proportion of the community from the work of food production. This surely has been the remarkable achievement of modern agriculture.

Nutritive Value of Animal Products

The analyses of the foods of animal origin given in Table 3.4 show that within this group there is considerable variation in the chemical composition, particularly in respect to the ratio of protein to fat and the percentage of carbohydrate. Humans, when free to exercise their food preferences, select an array of foods, all containing different percentages of the es-

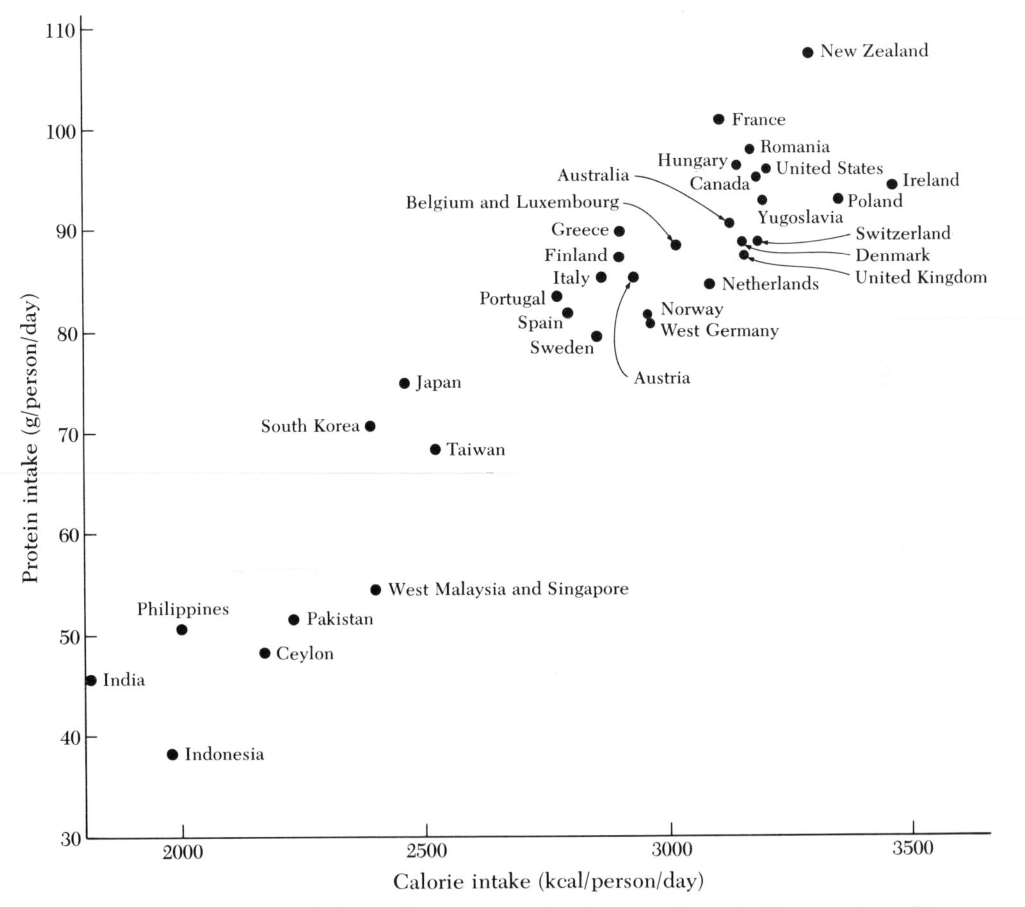

Figure 3-2 Per capita daily total protein and calorie intakes of 33 countries. (From FAO, 1969b.)

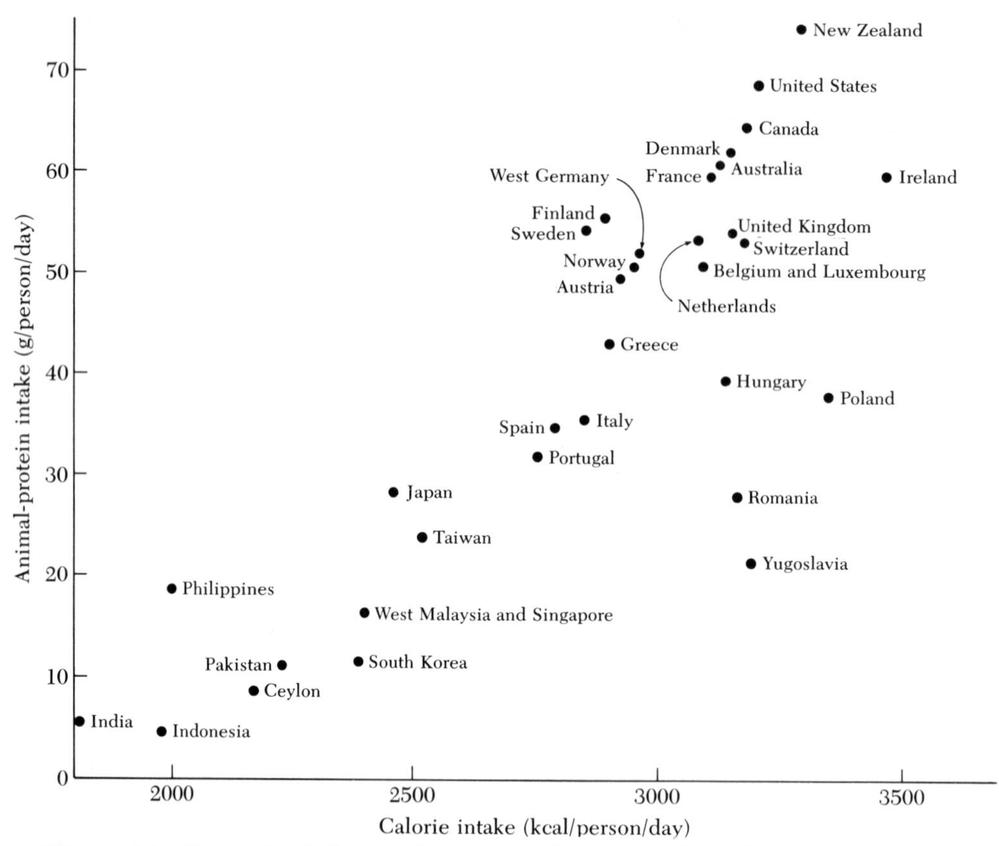

Figure 3-3 Per capita daily animal protein and calorie intakes of 33 countries. (From FAO, 1969b.)

sential nutrients, which we will discuss separately.

Protein

The nature of the structural components and the energy reserves in plant and animal cells are the primary determinants of their nutritional value. In plant cells polymers of glucose form both the main structural components (e.g., cellulose) and the energy reserves (e.g., starch), whereas in animal cells proteins are the main structural units and energy is stored primarily as fat. Therefore, animal tissues are always good sources of protein and may also be good sources of energy.

The protein content of foods is commonly expressed as crude protein, but this measurement gives no indication of the nutritive value of the protein. Nutritive value depends in the main on how close the proportions of amino acids in the protein molecules coincide with those required by the body for maintenance and the synthesis of new tissue. Since proteins of animal origin are derived primarily from animal tissues, their amino-acid composition more closely approximates that required by man than proteins of plant origin. There are various measures of the biological value of proteins which are related to the proportion of essential amino acids in them. One of these measurements is called the net protein utilization (NPU), which takes into account the digestibility and biological value of the protein. Table 3.5 gives the NPU of some common food proteins in relation to egg protein which has a high NPU value and is assigned the arbitrary figure of 100. Note that the animal proteins have higher NPU values than proteins of plant origin.

Meat Lean meat and muscle contains approximately 18 to 20 per cent protein, 1 to 10 per cent fat, and 1 per cent ash. The nutritive value of the various meats and cuts of meat are similar when they are compared on an equivalent fat basis. There is no sound evidence to suggest that white meats (poultry and fish) are inferior or superior to red meats (beef and lamb). The edible portion of fish is mainly muscle tissue, which has a similar gross composition to the muscle protein of warm-blooded animals. The fat content of fish is quite variable, which accounts for the range in caloric value. Cod, which is a low-fat fish, has a caloric value of less than 78 kcal/100 g, whereas salmon, a high-fat fish, has a caloric value of 222 kcal/100 g.

Milk Whole milk contains a lower percentage of protein than meat, but the greater quantity of milk and milk products in the diet compensate for its lower protein content. Milk and milk by-products contribute about 25 per cent of the total mean protein intake in the U.S. The protein of cheese, like that of milk, has a high biological value, but milk products in which the fat is concentrated, e.g., butter, ghee, and cream, contribute insignificant quantities of protein to the diet.

Carbohydrates

The animal products of flesh origin are low in carbohydrates, but milk and milk products contain appreciable quantities of the carbohydrate lactose. In nature, this carbohydrate is restricted to the milk of mammals. A number of milk products, e.g., cheese and yogurt, are produced by the fermentation of the lactose by microorganisms which in turn give a char-

Table 3.3.
U.S. civilian consumption of some animal products and of potatoes and cereals in lbs. per capita for the period 1910 to 1969.[a]

Product	*1910*	*1920*	*1930*	*1940*	*1950*	*1960*	*1969*[b]
Meat (retail cut equiv.)	125.9	117.9	112.8	124.4	125.4	134.0	147.7
Beef	55.6	46.7	38.6	43.4	50.1	64.2	81.8
Lamb and mutton	5.8	4.8	6.0	5.9	3.6	4.3	3.0
Pork (lean and fat cuts)	57.9	59.1	62.4	68.4	64.4	60.3	60.2
Total red meat[c]	136.2	128.1	121.7	134.1	135.5	144.1	158.5
Fish (fresh, frozen, and canned)	7.3	9.5	9.2	10.3	11.2	9.7	10.6
Poultry products							
Chicken	15.5	13.7	15.7	14.1	20.6	28.1	39.0
Turkey	1.0	1.3	1.5	2.9	4.1	6.2	8.4
Eggs	37.1	36.3	40.2	38.7	48.5	42.4	40.1
Dairy products							
Fluid whole milk	252	278	270	265	297	285	240
Low fat milks	62	53	48	47	34	27	54
Cheese	4.9	4.9	5.9	7.9	10.8	13.0	15.7
Ice cream, sherbet, etc.	1.9	7.6	9.8	12.3	19.4	24.5	27.0
Potatoes	184	130	123	114	99	100	100
Total flour and other cereal products	295	242	228	199	167	147	144

[a] From USDA (1965c), and USDA (1969d).
[b] Preliminary estimates.
[c] Excludes game, which accounts for about 2.5 lb/capita/year.

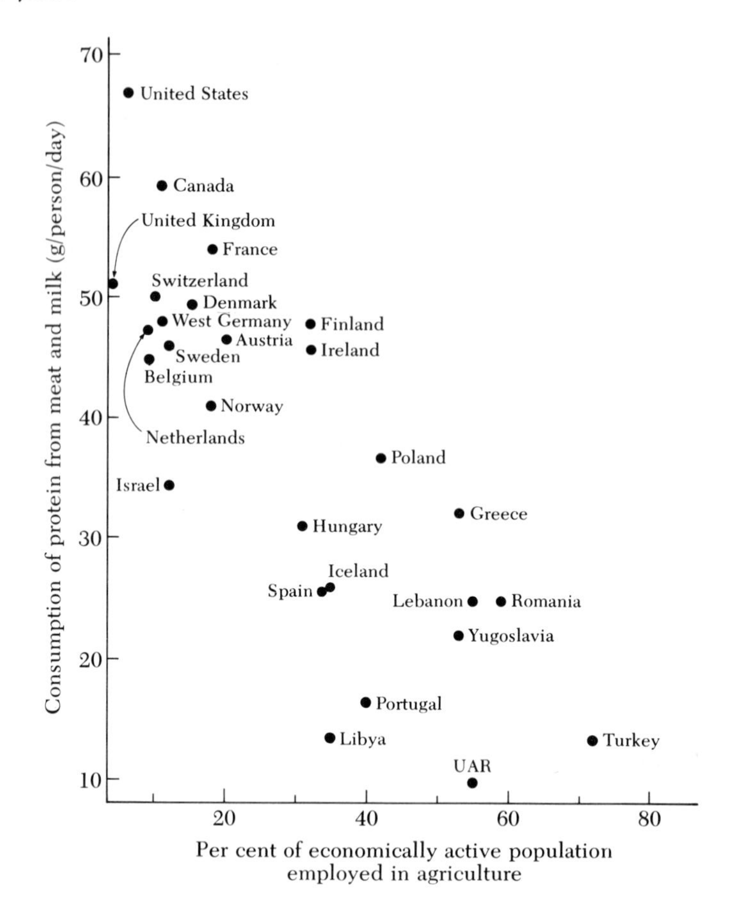

Figure 3-4 Consumption of protein from meat and milk in relation to the proportion of economically active population employed in agriculture. (Calculated from data in FAO, 1969b, 1970.)

acteristic flavor to these products. In some ethnic groups, such as the Negro and the American Indian, a higher proportion of the adult population is intolerant to lactose than in Caucasian populations. Products in which the lactose has been reduced by fermentation can be eaten by these individuals without incurring the side effects of nausea, diarrhea, and stomach pains characteristic of lactose intolerance.

Honey is a unique product of animal origin, in that it contains a very high percentage of carbohydrate and practically no protein or fat. The nutritive value of honey lies in its readily digestible simple sugars, fructose and glucose.

Fat

Fat provides a concentrated source of energy in the human diet, and contributes some 35 to 45 per cent of the total calories in prosperous countries. However, in poor countries it may contribute only 15 per cent or less of the total calories. Fat appears to be necessary in the preparation of foods of all civilizations, and affects the palatability of cooked foods.

Before the development of the oil-seed industries, western countries used animal fats, such as beef dripping and lard, as the main cooking fat, and butter as the main table spread. While the total consumption of fat in the U.S. has increased steadily this century, the proportion of animal fat has declined. In 1930 and 1969, the per-capita total annual consumption of animal fats was 28.3 and 14.19 lb., of vegetable fats 16.7 and 36.9 lb., respectively.

In recent years a high intake of saturated fat has been implicated in the etiology of atherosclerosis, but there is no unequivocable evidence to indicate that the consumption of animal fat *per se* will induce the disease. Diets high in fat are high in calories and may lead to obesity in people not engaged in heavy physical work. High-fat diets are essential for very active people expending over 4000 kcal/day, such as lumberjacks and people working in polar regions who have difficulty in obtaining this amount of energy from carbohydrate sources.

Vitamins

Experiments on the nutritive value of animal products early in this century led to the discovery of the fat-soluble vitamins A and D. Vitamin A occurs only in animal products, but the precursor, carotene, is present in all green plants. Animals store vitamin A in their livers and fat depots; so liver is a good source of this vitamin. Animal fats make a variable contribution to the human vitamin-D requirement, but contain little vitamin E. Animal products which have not undergone bacterial fermentation are poor sources of vitamin K.

Animal products are fair to good sources of the water-soluble B complex vitamins, but supply little vitamin C. Meat and milk products are fair sources of vitamin B_1 (thiamine); liver, milk, and eggs are good sources of vitamin B_2 (riboflavin); and the organ meats and fish are rich in nicotinic acid.

Studies on the anemias of man led to the discovery of the vitamins B_{12} and folic acid. Vitamin B_{12} was first isolated from liver, a rich source of this vitamin. Vegetarian diets which are free from fermented foods and contain no milk or eggs or other products of animal origin are devoid of vitamin B_{12}. People who voluntarily follow such diets occasionally develop anemias or nervous disorders due to vitamin

Table 3.4.
Composition of selected animal products (uncooked edible portion, choice grade).[a]

	Separable components (per cent)		*Percentage chemical composition (per cent)*					*Energy*
	Lean	*Fat*	*Water*	*Protein*	*Fat*	*Carbohydrate*	*Ash*	*(kcal/100g)*
Beef: Chuck	82	18	60.8	18.7	19.6	—	0.9	257
T-bone steak	62	38	47.5	14.7	37.1	—	0.7	397
Rump	75	25	56.5	17.4	25.3	—	0.8	303
Lamb: Leg	83	17	64.8	17.8	16.2	—	1.3	222
Pork: Ham	74	26	56.5	15.9	26.6	—	0.7	308
Poultry: Chicken, fryer (flesh only)			77.2	19.3	2.7	—	0.8	107
Turkey (flesh only)			64.2	20.1	14.7	—	1.0	218
Milk (cows) whole, liquid			87.2	3.5	3.7	4.9	0.7	66
Skim, liquid			90.5	3.6	0.1	5.1	0.7	36
Butter			15.5	0.6	81.0	0.4	2.5	716
Cheese, cheddar			37.0	25.0	32.2	2.1	3.7	398
Eggs: Chicken (whole)			73.7	12.9	11.5	0.9	1.0	163
Fish: Cod			81.2	17.6	0.3	—	1.2	78
Tuna (bluefin)			70.5	25.2	4.1	—	1.3	145
Trout (rainbow)			66.3	21.5	11.4	—	1.3	195
Salmon (king or chinook)			64.2	19.1	15.6	—	1.1	222
Honey			17.2	0.3	0	82.3	0.2	304

[a]From USDA, 1963.

B_{12} deficiency. Liver and green vegetables are rich sources of folic acid, but meat and milk are not good sources. Animal products in general are also good sources of three other B vitamins—pyridoxine, panthothenic acid, and biotin.

Minerals

Animal products contain a large number of essential minerals, and often make a significant contribution to man's dietary requirement for them. However, they are not essential for this reason; for example, meat is a good source of phosphorus, but this can be supplied from cereals. Milk and its by-products are excellent sources of calcium, and contribute some 75 per cent of the calcium in western diets. Cereals are poor sources of calcium, which is likely to be deficient in the diet of growing children and pregnant mothers that do not have milk in their diet. Generally a diet which is well-balanced for the major nutrients required by man contains adequate minerals.

Table 3.5.
The limiting amino acid and net protein utilization (NPU) of some common food problems.[a]

Food	*Limiting amino acid*	*Net protein utilization*
Egg	—	100
Pork	S[b]	84
Beef	S	80
Cow's milk	S	75
Fish	Tryptophan	83
Rice	Lysine	67
Millet	Lysine	56
Wheat flour	Lysine	52
Maize	Tryptophan	56
Beans	S	47

[a] From Davidson and Passmore (1969), p. 80.
[b] S indicates sulfur-containing amino acids (methionine and cysteine).

Cultures and Food Preferences: One Man's Meat

Food was early man's most precious possession and undoubtedly its procurement was his major preoccupation. He frequently attributed the availability of food to the humor of the spirits or deities; a lack of food was ascribed to the spirit being displeased or offended. The practices of offering food or of abstinence from particular foods have become in many cultures symbolic gestures of devotion or a desire to obtain the favor of some supernatural power. Food has assumed a symbolic significance for man, as well as being a material to supply his nutritional needs.

Although a great diversity of animals are eaten by man, when the animal species eaten by individual ethnic groups are compared, it is apparent that common species of animals eaten by one ethnic group are often disregarded or even actively avoided by another. Even within a culture there are often divisions based on age, sex, or social status as to whether it is proper for an individual to partake of the flesh of a particular animal. Many of these flesh avoidances are closely tied to religious practices, and the prophet or founder of a religion has often played a critical role in fostering rejection or acceptance of a particular food. It is interesting to note that it is foods of animal origin, particularly flesh foods, rather than plant foods, which occupy a predominant place in food-avoidance practices. Rejection of an animal as food seems to result from either its totemic relationship to the tribe, clan, or group, so that it becomes revered, or its being regarded as "unclean," so that its consumption defiles the individual. It appears that sometimes both factors work in combination: an animal which is initially revered may later be an object of revulsion. In this connection it has been suggested that domestication of animals may have resulted from their role in religious or ritual practices rather than from man's insight into the use of these animals as a source of food (Sauer, 1969).

The five major religions of the world, Christianity, Judaism, Islam, Hinduism, and Buddhism, present quite divergent views on the acceptability of various animals as food. For the Buddhist and certain groups in the Hindu society, no meat of any kind may be eaten, an observance which is in accordance with their reverence for all animal life. Hinduism currently prohibits the eating of beef or slaughter of cattle, practices which apparently evolved slowly over the last two thousand years. For millennia, cattle and cows in particular have been highly regarded in India, though writings indicate that they were sacrificed and eaten on ceremonial occasions from 500 to 100 B.C. The Buddhist-Brahmin struggle for supremacy in India may have been instrumental in the final abandonment of beef eating, for there is a tendency for castes in the attempt to elevate themselves to reject practices which are despised by other castes. The sacra surrounding Indian cattle can be, and have been, rationalized by anthropologists such as Harris (1970) on the grounds of being necessary to provide adequate numbers of animals for draught power and production of manure for cooking fuel. But this thesis does not necessarily preclude the achievement of greater productive efficiency by the application of controlled breeding and slaughtering practices.

While the practice of the avoidance of beef is almost totally restricted to the Indian subcontinent, the practice of pork avoidance is more extensive. The principal center of pork avoidance is the Middle East, where Moslems, Orthodox Jews, and Ethiopian Christians strongly reject pork as an item of food. Early Egyptian records (about 1400 B.C.) indicate that pigs were kept by the nobility at that time and presumably eaten, but by 400 B.C. the status of the pig had fallen. Although it is generally held that the Jews originated the prohibition on the eating of the flesh of the pig, and the Islamic practice arose from the Jewish prejudice, the basis for rejection by the Jews is not clear. It has been suggested at various times that this prohibition may have arisen for the following hygenic reasons: the pig is a scavanger; there is a possibility of infection of trichinella (a parasite of both pig and man); and pork decays easily in the high temperatures of the Middle East. While these are attractive theories, none of them stand up to critical study. Simoons (1961) suggests that the prejudice against the pig and pork may have developed first among pastoral people who held the settled farmers who kept pigs in contempt. It is fairly common for ethnic, religious, or other groups to be associated with the particular foods or animals they eat. For example during World War I, the American troops referred to the British, French and Germans respectively as "Limeys," "Frogs," and "Krauts." There is a generalized tendency of all civilized societies to associate the eating of foods strange to them as a mark of a less civilized or primitive state. A group which initially regards a certain food as being of "low class" could through continual reinforcement eventually consider it as totally unacceptable or even an item of revulsion.

While the rejection of beef and pork are the most publicized examples of active avoidance of the flesh of particular species, many others exist. Horse flesh is regarded as a preferred meat by the Mongol tribes and the world center of horsemeat eating stretches from Mongolia to Eastern Europe. The Buddha prohibited his followers from the consumption of horse flesh, and in Europe about 730 A.D. the Christian Church also tried to prohibit the eating of horse flesh. This appeared to have been an attempt to distinguish Christians from pagan tribes. Today a considerable amount of horse flesh is eaten in Europe, particularly in France, Belgium, The Netherlands, Switzerland, and Germany.

Camel flesh is highly regarded in most Moslem Middle-Eastern countries, but is rejected by non-Moslem groups. In this respect, it is interesting to note that, while the Moslems adopted many of the Jewish food observances, they claim that Jesus (whom they regard as a prophet) repealed the Jewish prohibition of camel flesh in the Levitican Code.

Although chickens and eggs are widely accepted food items in Western diets, centers of active avoidance of chicken flesh and eggs exist on both the African and Asian continents. In these areas avoidance appears to be based on the association of eggs with fertility rites, the use of chickens for divination, and the scavanging habit of the bird.

Christian societies are not without their flesh-avoidance practices, although these relate more to abstinence from flesh during certain periods of the year (e.g., Lent) than to total prohibition. However, the flesh from certain animals (e.g., dog flesh) is implicitly rejected by most Christians, either on the grounds that the "dog is man's best friend" or that dogs eat flesh so their own flesh is "unsuitable" or "unclean." It may shock some Westerners to learn that in parts of Africa and Asia, dog flesh is highly regarded and dogs are raised specifically for this purpose. More subtle flesh-avoidance practices which relate to the consumption of specific parts of the animal also exist in many societies. For example in the U.S., the offal meats are not highly regarded, and the spleen and lungs are rarely eaten. However, the pluck (heart, liver, and lungs) of the sheep is an essential ingredient in the Scottish dish haggis. Modern food processing (e.g., by the manufacture of small goods) often disguises some of these less-acceptable food items in forms which are well-accepted.

Despite the high nutritional value of all flesh of animal origin, food-avoidance prejudices and practices are the major determinants of whether an animal product will be accepted by a community. Many of the prejudicial barriers which are implicit can with time be overcome by education, but those which are derived from religious teachings provide an almost insuperable obstacle. It is clearly apparent that in international agriculture the study of food preferences and restrictions are of paramount importance before attempts are made to introduce new animal products into a culture.

FURTHER READINGS

Davidson, S., and R. Passmore. 1969. *Human Nutrition and Dietetics.* Baltimore, Md.: Williams and Wilkins.

FAO. 1964. *The State of Food and Agriculture.*

Harris, M. 1965. "The myth of the sacred cow." In A. Leeds and A. P. Vayda, eds., *Man, Culture, and Animals.* Washington, D.C.: American Association for the Advancement of Science.

Holmes, W. 1970. "Animals for food." *Proc. Nutr. Soc.,* 29:237.

Leakey, L. S. B. 1965. *Olduvai Gorge, 1951–1961.* Cambridge, Eng.: Cambridge Univ. Press.

Rice, V. A., F. N. Andrews, E. J. Warwick, and J. E. Legates. 1967. *Breeding and Improvement of Farm Animals.* 6th ed. New York: McGraw-Hill.

Simoons, F. J. 1961. *Eat Not This Flesh: Food Avoidances in the Old World.* Madison: Univ. of Wisconsin Press.

Four

Animal Agriculture and the Population Problem

"And I brought you into a plentiful country, to eat the fruit thereof, and the goodness thereof; but when ye entered, ye defiled my land, and made mine heritage an abomination."

Jeremiah 2:7

Two aspects of the human population problem are of immediate concern to the student of animal science: first, the availability of foods of animal origin for the present population, and the reasons for variations in their availability; second, trends in human populations and the probable effects of population growth on the availability of foods of animal origin for future generations.

There are, of course, other aspects with which everyone—including the animal scientist—should be concerned. For example, there are the problems of adequate land for food production generally, of adequate living space, of overcrowding in cities, of adverse effects on economic development, of overuse of limited natural resources, of space for recreation, and of environmental pollution. The further development of animal production is related in one way or another to all these problems.

Population Trends

The human population is now expanding very rapidly, a phenomenon that is often referred to as a population explosion. And it is an explosion, compared with the rate of increase that prevailed up to the nineteenth century (Figure 4.1). At the beginning of the first century A.D., the world's population was only about a quarter of a billion, and did not reach the half-billion mark until about 1650. By 1850 it had doubled, to slightly more than one billion. Within another century the number had more than doubled again, reaching approximately 2.5 billion in 1950, and is now expected to reach (or exceed) the 6 billion mark by the year 2000—that is, more than doubling again within 50 years.

It is clear that this upward thrust in population must slow down and level off. The questions are, when, and as the result of what forces? As the United Nations (1958) pointed out, although it required some 200,000 years from the time *Homo sapiens* emerged until his numbers reached 2.5 billion, at that time he was expected to add another 2 billion in a mere 30 years, and, if the then-existing rate

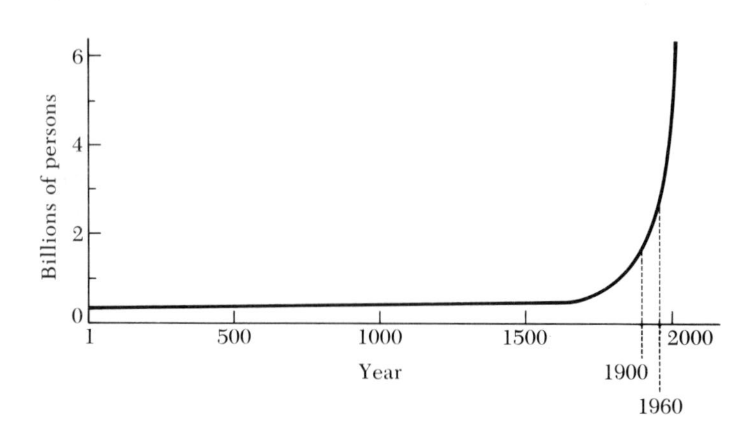

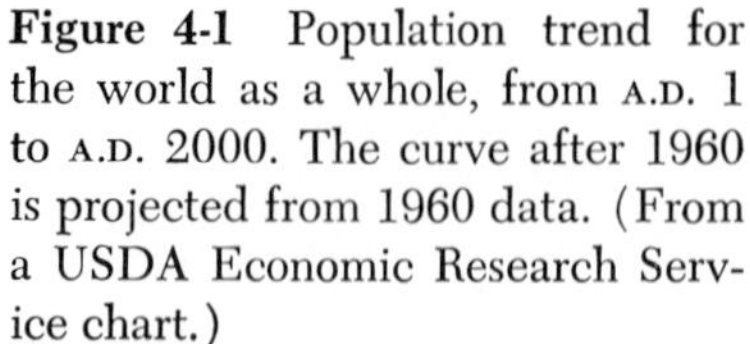
Figure 4-1 Population trend for the world as a whole, from A.D. 1 to A.D. 2000. The curve after 1960 is projected from 1960 data. (From a USDA Economic Research Service chart.)

of increase continued, within 600 years there would be only one square meter of land for each person to live on. Indeed, Bonner (1961) calculated that, if the same rate of increase continued, in 700 years people could only stand shoulder to shoulder, and in 7,000 years a solid mass of humanity not only would cover the whole of the earth's surface, but would be expanding outward from it at the speed of light!

The United Nations (1964) issued a provisional report on world population prospects, as assessed in 1963, which contained forecasts for the world and for regions and countries up to the end of the twentieth century. In round numbers, the world's population is expected to double—from 3 billion to 6 billion—in this 40-year period. Low and high estimates for 2000 A.D. were 5,297 million and 6,828 million, respectively, depending on the assumptions made. The medium variant is used for Table 4.1 as providing the most likely rate of increase, but it may prove to be too low. For example, at the time the above estimates were made, the United Nations had projected a world population of about 4,746 million (medium estimate) in 1985, but in a revised estimate (United Nations, 1969), the figure was raised by 187 million to a projected 4,933 million.

The projections in Table 4.1 show that the greatest increases will occur in the less-developed or developing regions, especially South Asia, Africa, and Latin America. As will be seen later in this chapter, the developing regions are the ones where even now there is not enough food, particularly foods of animal origin, to meet the needs of the population; hence population increases projected for these regions seem certain to aggravate an already difficult situation.

Differences Among Countries in Available Food Supplies

Countries differ greatly in the quantity and quality of their food supplies, and therefore in the degree to which their populations are well-fed or ill-fed. The differences for animal-protein supplies, and for the latter's availability to the countries' populations, are much greater than those for calories or total proteins, since countries differ much more in their ability to produce animal products (Figure 4.2), or to purchase such products from other countries, or both, than in their ability to produce food in general.

Variations in Supplies of Calories, Total Protein, and Animal Protein

Data are now available from FAO (1970b) on the amounts of calories, total protein, and animal protein in the national food supplies of 90 countries, and on the amounts of major

Table 4.1.

Population prospects for the year 2000, as compared with 1960, based on the United Nations (1964) medium-variant projections, for the world and for various regions. Figures are in millions.

Area	*1960*	*2000*
World	2,990	5,965
More-developed regions	976	1,441
Less-developed regions	2,014	4,524
South Asia	858	2,023
East Asia	793	1,284
Europe	425	527
Africa	273	768
Soviet Union	214	353
Latin America	212	624
North America	199	354
Oceania	15.7	31.9

animal products included in those food supplies. These data are summarized in Figure 4.3. Ireland ranks first, with 3,450 calories, followed by New Zealand and the United States with 3,290 and 3,240 calories, respectively. In general, supplies of total protein and animal protein decrease as caloric supplies decrease. Notice that calorie supplies in Ireland are nearly double those in Somalia, the country lowest in calories, i.e., 3,450 compared with 1,770 calories. Also, several countries that are relatively high in both caloric and total-protein supplies are quite low in animal-protein supplies.

Total-protein supplies vary from a high of 111.7 grams per person per day in Uruguay, to a low of 41.4 grams in Indonesia. In Yugoslavia (no. 4), Bulgaria (no. 16), Romania (no. 19), and Turkey (no. 33), the available supplies of total protein per person per day are relatively high, whereas those of animal protein are relatively low; obviously, these countries obtain a very high proportion of their protein from plants.

Since animal protein is generally much better for meeting human nutritional needs than plant protein, let us consider further these data on animal-protein supplies. Uruguay ranks first, with 77.4 grams of animal protein per person per day, whereas Rwanda is lowest, with only 3.6 grams. Thus, the best-fed country in this respect has over 21 times as much animal protein per person as the poorest-fed country. Of these 90 countries, 62 have less than 30 grams of animal protein per person per day available to their people, and 17 of them have 10 grams or less.

Sources of Animal Protein

The major sources of animal protein are, of course, milk, meat, eggs, and fish. Apart from the differences in animal-protein supplies already noted, there are also substantial differences in the degrees to which countries rely on these four sources for their animal products (Figure 4.4).

For milk, which includes milk products excluding butter, expressed as fresh milk, the range in grams available per person per day is from a high of 899 grams in Finland (no. 9) to a low of 2 grams in Indonesia (no. 81) and Korea (no. 84). In this respect, the United States (no. 3) shares seventh place with the Netherlands (no. 12).

For meat, which includes poultry and game, expressed as dressed carcass weight including offals, the range in supplies per person per day is from 310 grams in Uruguay to lows of 5 grams in Ceylon (no. 80) and 4 grams in India (no. 87). The United States (no. 3) ranks fourth, with 299 grams.

For eggs, expressed as the equivalent of fresh eggs, the range is from 58 grams per person per day in Israel (no. 21) to an amount too small to be recorded in Rwanda (no. 90). Ten countries have only one gram per person per day.

For fish, supplies are roughly the same as for eggs. Japan (no. 29) ranks highest, with 89 grams; and five countries (Bolivia, Iran,

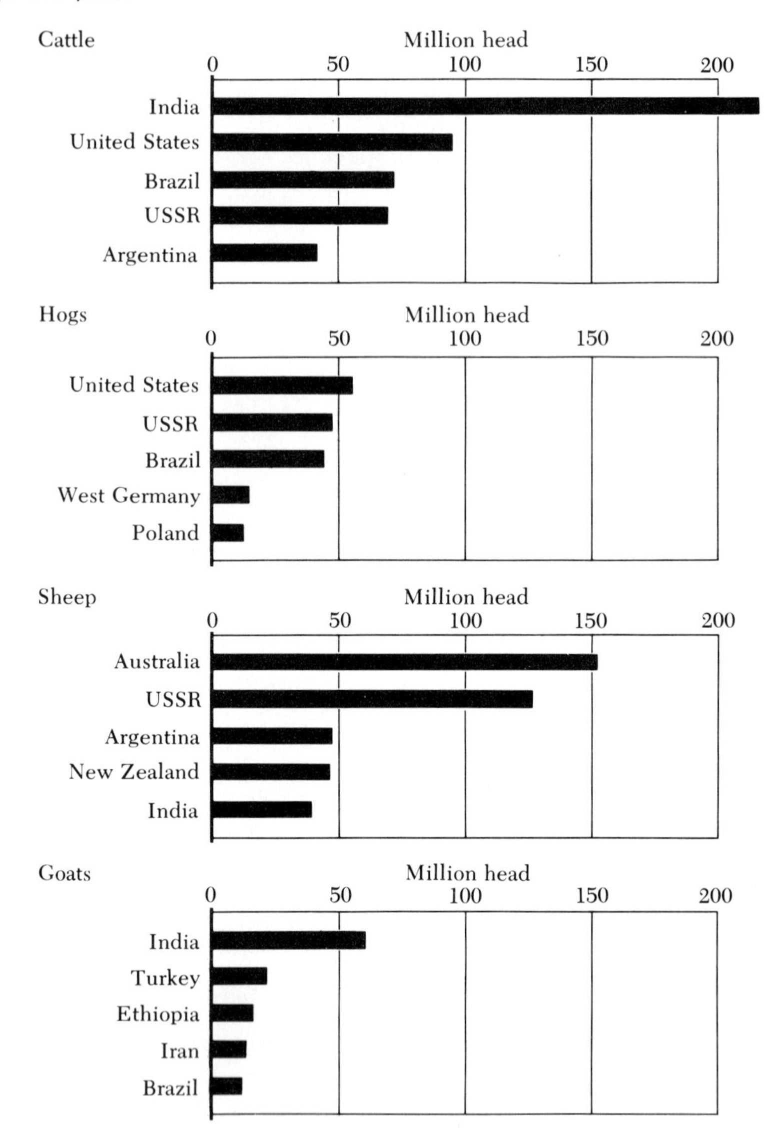

Figure 4-2 Some leading countries in livestock production. Data are averages for the years 1957–1961, except that the figure for Indian cattle is the average of 1957 and 1961 production, that for Indian goats is the production for 1961. The data for India include Jammu and Kashmir, and the term cattle here includes buffaloes.

Syria, Afghanistan, and Rwanda) have no measurable quantities.

Some Reasons for Differences in Animal-Protein Supplies

Many factors interact to determine how much and what kinds of food man has available to meet his needs in any particular country or area (Phillips, 1964). Among the more important factors that limit the supplies of animal products are the following:

Inadequate purchasing power for the acquiring of animal products, which generally are more expensive than foods of plant origin;

Lack of adequate systems for the integration of animal production into highly intensive systems of crop production;

Failure to integrate pastoral or nomadic systems of animal production with agricultural production on arable lands;

Lack of a tradition of animal production, in countries where man has, for centuries, been primarily dependent on plant foods;

Lack of a tradition among many pastoral

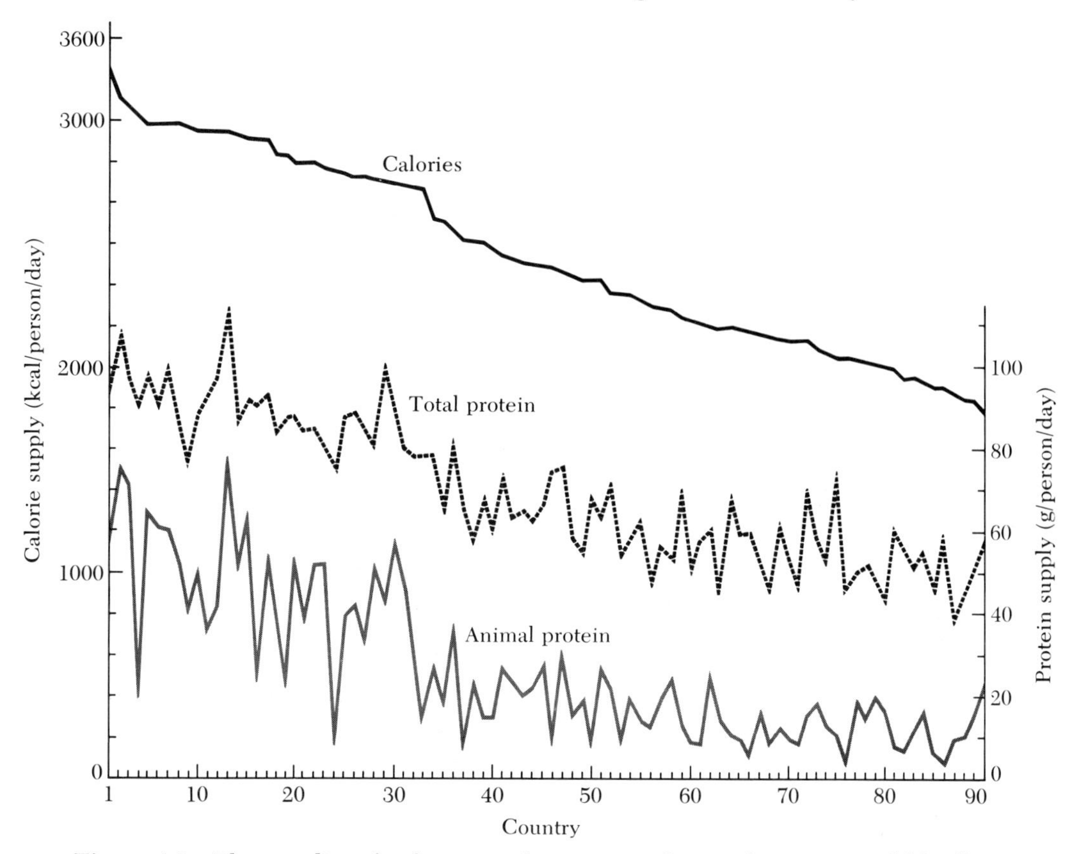

Figure 4-3 The supplies of calories, total protein, and animal protein available (but not necessarily consumed) per person per day in 90 countries. The countries are ordered from left to right in descending order of caloric supplies. (FAO data.)
Key to horizontal scale is:
1, Ireland; 2, New Zealand; 3, United States; 4, Yugoslavia; 5, Canada; 6, Denmark; 7, France; 8, United Kingdom; 9, East Germany; 10, Belgium and Luxembourg; 11, USSR; 12, Hungary; 13, Uruguay; 14, Argentina; 15, Australia; 16, Bulgaria; 17, Poland; 18, Netherlands; 19, Romania; 20, Austria; 21, Czechoslovakia; 22, Switzerland; 23, West Germany; 24, UAR; 25, Italy; 26, Israel; 27, Portugal; 28, Norway; 29, Greece; 30, Finland; 31, Sweden; 32, South Africa; 33, Turkey; 34, Chile; 35, Brazil; 36, Spain; 37, Jordan; 38, Costa Rica; 39, Mexico; 40, Libya; 41, Lebanon; 42, Paraguay; 43, Taiwan; 44, Surinam; 45, Venezuela; 46, Syria; 47, Japan; 48, Ivory Coast; 49, Jamaica; 50, Korea; 51, Panama; 52, Albania; 53, Madagascar; 54, Nicaragua; 55, Gambia; 56, Mauritius; 57, Peru; 58, Colombia; 59, Kenya; 60, Pakistan; 61, Guatemala; 62, Gabon; 63, Malaysia; 64, Tunisia; 65, Morocco; 66, Nigeria; 67, Uguanda; 68, Ceylon; 69, Tanzania; 70, Cameroon; 71, Ghana; 72, Mali; 73, Sudan; 74, Bolivia; 75, Ethiopia; 76, Mozambique; 77, Ecuador; 78, Honduras; 79, Philippines; 80, Dominican Republic; 81, Afghanistan; 82, Algeria; 83, Iran; 84, Iraq; 85, India; 86, Rwanda; 87, Indonesia; 88, El Salvador; 89, Saudi Arabia; 90, Somalia

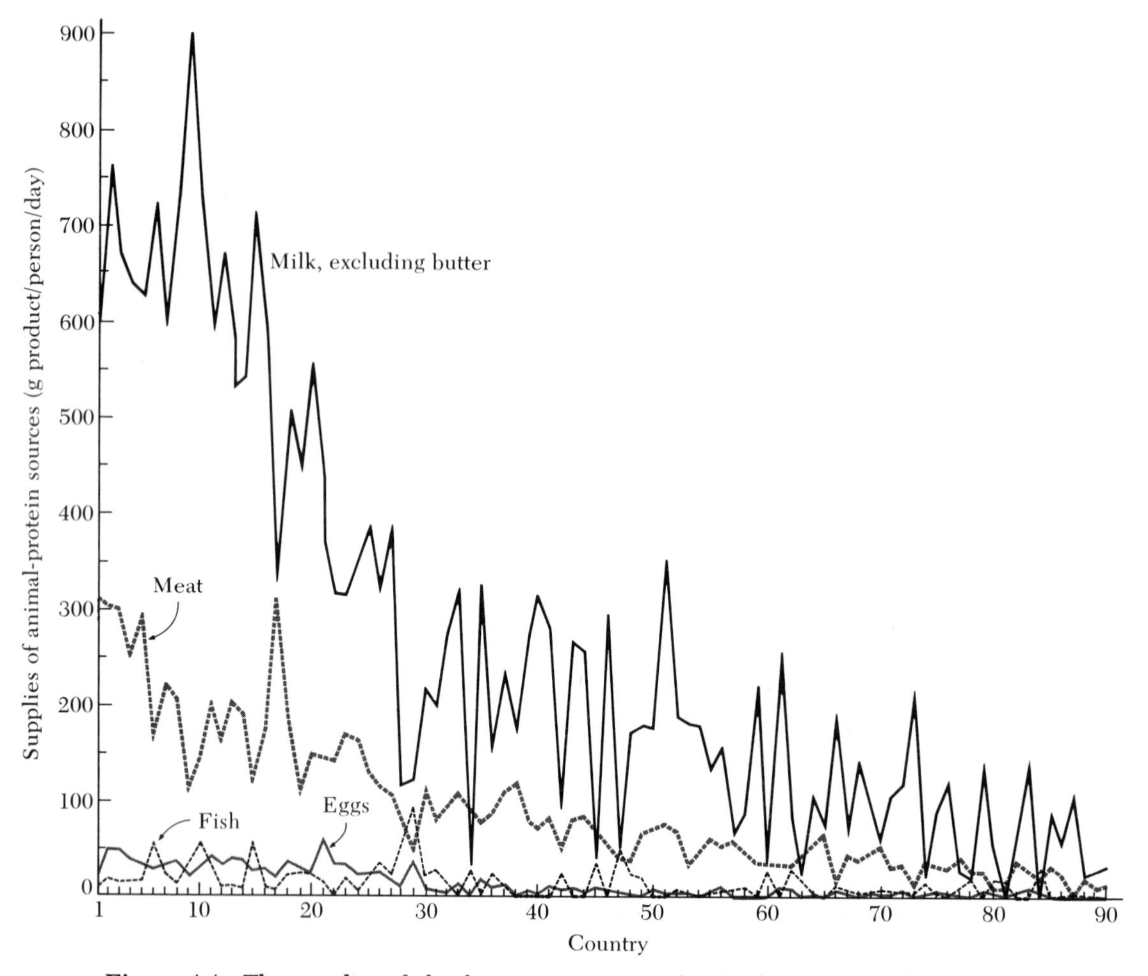

Figure 4-4 The supplies of the four main sources of animal protein in 90 countries, in grams per person per day.
Key to horizontal scale is:
1, Uruguay; 2, New Zealand; 3, United States; 4, Canada; 5, Australia; 6, Denmark; 7, France; 8, Ireland; 9, Finland; 10, Sweden; 11, United Kingdom; 12, Netherlands; 13, West Germany; 14, Austria; 15, Norway; 16, Switzerland; 17, Argentina; 18, Belgium and Luxembourg; 19, Greece; 20, Poland; 21, Israel; 22, Hungary; 23, East Germany; 24, Czechoslovakia; 25, Italy; 26, Spain; 27, USSR; 28, Portugal; 29, Japan; 30, South Africa; 31, Venezuela; 32, Chile; 33, Romania; 34, Gabon; 35, Lebanon; 36, Panama; 37, Bulgaria; 38, Paraguay; 39, Colombia; 40, Somalia; 41, Yugoslavia; 42, Surinam; 43, Costa Rica; 44, Albania; 45, Taiwan; 46, Nicaragua; 47, Philippines; 48, Jamaica; 49, Peru; 50, Brazil; 51, Sudan; 52, Ecuador; 53, Dominican Republic; 54, Iraq; 55, Libya; 56, Mexico; 57, Uganda; 58, Mali; 59, Turkey; 60, Gambia; 61, Honduras; 62, Malaysia; 63, Ivory Coast; 64, Kenya; 65, Bolivia; 66, Mauritius; 67, Tanzania; 68, Iran; 69, Saudi Arabia; 70, Ethiopia; 71, Tunisia; 72, UAR; 73, Pakistan; 74, Cameroon; 75, Morocco; 76, Syria; 77, Madagascar; 78, Ghana; 79, El Salvador; 80, Ceylon; 81, Indonesia; 82, Guatemala; 83, Jordan; 84, Korea; 85, Afghanistan; 86, Algeria; 87, India; 88, Nigeria; 89, Mozambique; 90, Rwanda

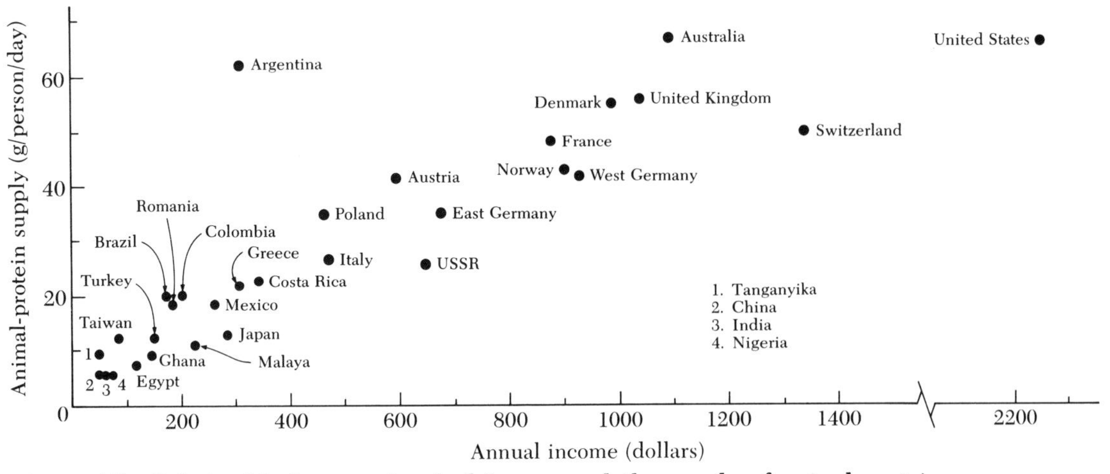

Figure 4-5 Relationship between level of income and the supply of animal protein per person per day for selected countries. (From Brown, 1963.)

peoples of harvesting and storing hay or other feed for livestock;

Religious beliefs and social customs which dictate the kinds of food to be eaten, and which, in particular, may prevent the consumption of meat and meat products;

Inadequate programs for the control of animal diseases and parasites;

Lack of adequate facilities for the marketing, processing, and distribution of animals and animal products;

Low productivity of meat, milk, or eggs among animals that have not been selected and bred for genetic superiority in these traits;

Lack of trained people competent to provide specialized training in animal science and production, and to provide leadership in animal research and extension programs;

The relatively high cost of animal research, particularly with the larger animals, which inhibits development of soundly based animal-improvement programs in the poorer countries;

The lack of readily available capital for financing animal-improvement programs; and

A tendency in many countries to orient their national agricultural-development policies primarily toward improvements in plant production, and to overlook some of the important potentials for improvement in animal production.

These factors are not listed in any order of importance or priority. What is important, and what should receive high priority in planning for improvements, will vary from circumstance to circumstance. Much more could be said about each of the factors. Some further comments follow on the two of them that are most closely related to the topic of this chapter.

Relation Between Diet and Income

It is evident from the information presented earlier on national food supplies that the poor countries are generally the poorly fed countries, in both the quantity and the quality of the diet. Brown (1963) pointed out that calories available per person in national food supplies tend to rise rapidly as annual income per person increases from around $60 up to $300. As income increases from $300 to around $1,000, the calories available increase much less rapidly, and beyond $1,000 per person, they hardly increase at all.

In contrast, the share of energy derived from starchy foods (grain products, roots, tubers) tends to decline as income increases. This share may be as high as 80 per cent in countries like the People's Republic of China and Nigeria, where people are largely dependent on foods such as rice or cassava, but is less than 25 per cent in the United States. Where income is low and people depend primarily on starchy foods, there is little incentive for the local animal industry to develop, and little possibility of meeting nutritional needs by importing animal products.

Protein intake also shows a rather close relationship to level of income. It increases quite rapidly until incomes reach about $300 per person annually, more slowly thereafter.

The consumption of animal protein is more closely related to income (Figure 4.5). For the 30 countries included, animal-protein intake increases quite steadily as income increases—except for Argentina, a country with a good supply of animal products and a traditionally high level of meat consumption.

In a study of 100 underdeveloped countries, Hoffman (1960) found that 52 had per capita annual incomes under $100, 39 had incomes ranging from $100 to $300, and income in the remaining nine ranged from $300 to $700. These countries encompassed 1.25 billion people. If the People's Republic of China is added, it is evident that over half the world's people simply do not have enough money to purchase substantial amounts of animal products.

Suppose, for example, that the average income in a country is $60 to $65, of which about $35 is spent for food. This is a food budget of about 10 cents per person per day. A day's ration of perhaps 30 grams (about 1/15th of a pound) of plant protein might be purchased for from 0.33 to 1.67 cents, or 30 grams of skim-milk powder for around 2 cents. However, for animal products such as cheese, beef, and eggs, the cost of 30 grams of protein would increase to between 7.5 and 13.5 cents per day. Obviously, a person with a food budget of 10 or even 20 cents a day will have to obtain much of his protein from the least expensive plant sources.

Increasing Pressures on the Land

Animals require space, for housing, for grazing, and/or for the production of harvested roughages and concentrated feeds. It follows that they must compete with many other uses of the land: with plant agriculture, other than for the production of animal feed; with forests, particularly plantation forestry; with factories, public buildings, streets, highways, and airports; and with housing for man himself. Such competition must necessarily become more intense as populations increase. In some parts of the world, the competition from food crops is already so great that there is little room for animal production, for example, among the intensively used rice terraces in Indonesia and in other countries in Asia and the Far East.

If we compare the amount of arable land in the world about 1962 to the numbers of people in 1920 (see Figure 4.1) and projected for 2000 A.D. (medium United Nations estimate), we find that in those eighty years the amount of arable land on which to grow each person's food will have shrunk from 1.92 acres to 0.6 acre. So as the population increases further, animals will face even stiffer competition from plants for direct food production, and from other users of the land. The competition from plants will take other forms besides immediate competition for each acre of land. For example, more attention will be given to plants that supply higher-quality plant proteins, such as oilseeds, pulses, and nuts. Crops bred for higher or better-quality protein, such as high-lysine corn, will also receive increasing attention. Special processing of present crop products, such as wheat turned into bulgur or enriched with lysine, may also contribute to the increasing competition. If this competition, whatever its form, is to be met, there must be substantial increases in the efficiency of animal production.

Future Possibilities and Trends

How, and how much, man will use animals in the future will depend on many factors, which more often than not will be complexly interwoven (Phillips, 1968b). These factors are many; only a few can be outlined briefly here.

Requirements of an Expanding Human Population

An increasing human population has an obvious and direct bearing on the animal industry. As the population increases, needs increase; insofar as people can afford to purchase animal products, producers will try to meet their demands. Applying simple arithmetic, if the human population doubles between 1960 and 2000, as is expected, and if the consumption of animal products per person remains unchanged, then the requirements for those products will have doubled in 40 years.

But, as has been seen earlier in this chapter, the level of nutrition is quite low in many countries. It is estimated at least 40 per cent of the world's people are malnourished. Much of this malnutrition results from lack of high-quality protein and of other nutritive elements which meat, milk, and eggs supply. FAO (1963) has estimated that, for the world as a whole, total food supplies would have to be increased by 174 per cent, and foods of animal origin by 208 per cent, by the year 2000 to bring about even reasonable improvement. Moreover, in the developing countries, where the nutritional problem is greatest and the population is increasing most rapidly, the output of animal products would have to be increased by 483 per cent even to achieve diets averaging only 2,450 calories and 21

Figure 4-6 Bedouins milking sheep in the Bekaa Valley, Lebanon. Sheep produce about 1.5 per cent of the world's milk supply. (FAO photo.)

grams of animal protein per person per day.

Certainly, this expanding need will provide an incentive for increases in animal production, but the achievement of such large increases in so short a time staggers the imagination, and it is by no means certain that they can be achieved.

Need for Animals as a Source of Power

Animals still provide much of the agricultural power in many parts of the world (see Chapter 9). Although the general trend in the world is toward mechanization, conditions in the developing countries—small farms, scattered holdings, low economic levels, abundance of manpower, lack of adequate capital, lack of servicing facilities—mitigate against the rapid, universal adoption of mechanical power. So as man's numbers increase, and as pressures to meet his needs increase, animals will likely continue to provide much of the power in many parts of the world.

Flexibility of Animals as Transformers of Feed into Usable Products

The many ways in which animals contribute to man's daily needs are often overlooked, or lightly considered. Things agricultural are regarded by many as being commonplace, even mundane. City dwellers may enjoy steaks, milk, and a host of other animal products with little thought of how they are produced and made available in supermarkets. Children raised in cities may be no more familiar with our common domestic animals than they are with wild animals they see in zoos.

There are those who argue that, as man increases in numbers, he will be forced to depend primarily on products from plants, and that animals must therefore gradually decrease in importance. Perhaps such a trend is inevitable, but before accepting it as the only alternative, those who plan man's destiny should not overlook the great flexibility of animals as converters of plant products into products man has found desirable and useful.

Animals range in size from the small breeds of birds, which produce meat and eggs, to the largest types of cattle and horses (or even elephants, for draft purposes); so they can be adapted to many circumstances. Milk is supplied not only by the large dairy animals, such as cows and water buffaloes, but also by the relatively small goats and sheep (Figure 4.6).

Challenges to the Animal Industry

Although the question of future possibilities and trends has been approached in the foregoing sections primarily, but not entirely, from the standpoint of factors that appear to favor increased animal production, it must be recognized that many factors could adversely affect such increases (Phillips, 1963a, 1963b; Morley, 1968).

There are circumstances, for example, where it is highly unlikely that, in the foreseeable future, people will be able to meet their needs for high-quality protein from meat, fresh milk or other animal products. In such circumstances, other approaches must be used, such as obtaining proteins from pulses, oilseeds, or nuts, or by improving the quality of an available plant protein, for example, by increasing the lysine content of wheat, either artificially or genetically.

On the other hand, there are many circumstances in both the developed and the developing countries where animals can make substantially greater contributions than they do now, and do so *efficiently*. Furthermore, there are many parts of the world where food for direct human consumption cannot be grown, but which can be adapted to supply food for animals; examples include the polar regions and certain desert and mountainous areas. It is these situations that the student of animal science should seek to identify, and to which he should direct his attention.

FURTHER READINGS

Borgstrom, Georg. 1969. *Too Many*. New York: Macmillan.

Brown, Lester R. 1963. *Man, Land, and Foods*. Washington, D.C.: USDA Foreign Agricultural Economic Report no. 11.

FAO. 1963. *Third World Food Study*. Rome, Italy: FAO.

FAO. 1970b. *The State of Food and Agriculture, 1970*. Rome, Italy: FAO.

Phillips, Ralph W. 1963b. "The necessity of defining needs and establishing priorities for the solution of animal-production problems, taking into consideration the needs of human nutrition." *Proc. World Conf. on Animal Production*, I, 7–45.

UN. 1965. *World Population Prospects Up to the Year 2000*. Population Commission Document no. E/CN.9/186. New York: UN.

Section Two

Animal Products

A hand-operated pressure lift picking up four dozen eggs at a time. (USDA photo.)

Five

Meats

"All flesh is not the same flesh: but there is one kind of flesh of men, another flesh of beasts, another of fishes, and another of birds."

I Corinthians, XV:39

In this chapter factors determining meat quality and criteria for measuring acceptability of meat will be considered. Though the method of cooking of meat is important in determining its acceptability, many other factors, such as species, breed, sex, and age of animal, its management (feeding and exercise) before slaughter, and aging of the meat after slaughter, all play major roles in determining meat quality. Criteria used in measuring meat acceptability include color, texture, aroma, flavor, juiciness, and tenderness. Following consideration of these items, current trends and future developments in the meat industry will be briefly discussed.

Changes in Meat After Slaughter

At the time of slaughter, an animal's muscles are soft and pliable. Following death, the most obvious change that occurs is the hardening of the muscles, also a shortening and loss of elasticity and transparency. This change is referred to as rigor mortis.

Rigor Mortis

The term *rigor mortis* is Latin for "the stiffness of death." Its use, therefore, should be restricted to stiffness, hardness, rigidity, or inflexibility. This stiffening is due to the formation of permanent bonds between actin and myosin filaments. According to Davis (1967), confusion exists in resolving rigor mortis, because two distinct and different aspects are involved: (1) a shortening or a contraction of the muscle fiber; and (2) a stiffening or ability to bear an increased tension without much stretching, a loss of muscle extensibility. More than 130 theories exist with respect to muscle contraction.

Goll (1968) divides this extensibility aspect of rigor into: (1) a macroscopic phase (directly observed and measured as the ability of a muscle fiber or a bundle of muscle fibers to stretch under the influence of a given weight or force); and (2) a molecular phase (the ability of actin and myosin protein filaments in a single sarcomere to slide past one another). Postmortem muscle always loses both macroscopic and molecular extensibility. Both are closely related to adenosine triphosphate (ATP) depletion in postmortem muscle. Postmortem shortening or contraction is a direct cause of rigor mortis (Figures 5.1 and 5.2).

There is a resolution of rigor, the loss of the ability of postmortem muscle to maintain a shortened or contracted condition. Goll (1968) reported that the following chemical changes are considered to be characteristic of postmortem muscle: (1) anaerobic breakdown of muscle glycogen (animal starch) to lactic acid, starting immediately after death; (2) a lowering of *p*H (from neutral, characteristic in the live animal) due primarily to the formation of lactic acid; (3) a fall in phosphocreatine content of muscle, this drop occurring very rapidly after muscle glycogen reserves have been exhausted; and (4) a decrease in ATP concentration, this decrease occurring very slowly until after the disappearance of phosphocreatine, after which time it proceeds rapidly, usually to a level less than 20 per cent of that found initially.

The resolution of rigor, which occurs in postmortem muscle, is actually the loss of the ability of postmortem muscle to maintain isometric tension, also the lengthening or relaxation of contracted sarcomeres in postmortem muscle. An increased tenderness also follows, probably primarily due to degradation at the level of the Z-line. Further studies are necessary to provide a clear understanding of exactly what causes this resolution of rigor.

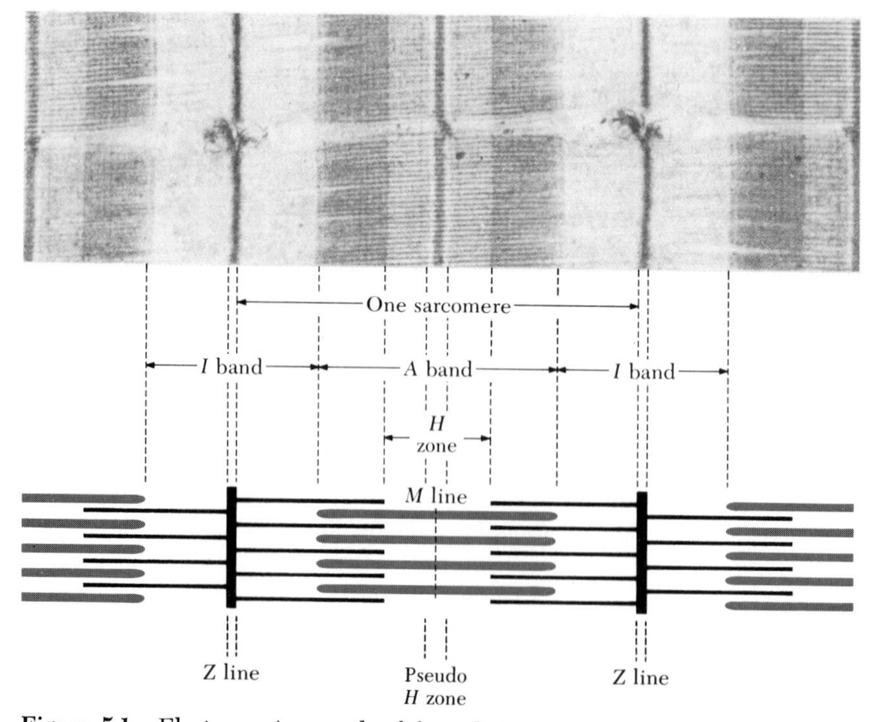

Figure 5-1 Electron micrograph of frog skeletal muscle with an accompanying schematic of the interdigitating thick and thin filament structure that is responsible for the striated appearance of skeletal muscle. (From "The Mechanism of Muscle Contraction" by H. E. Huxley. Copyright © 1965 by Scientific American, Inc. All rights reserved.)

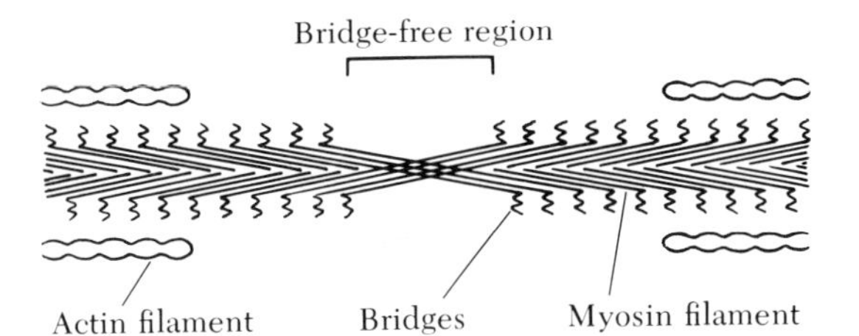

Figure 5-2 Schematic of the thick and thin filament structure of muscle showing how myosin cross-bridges can make contact with actin filaments. (From "The Mechanism of Muscle Contraction" by H. E. Huxley. Copyright © 1969 by Scientific American, Inc. All rights reserved.)

Chilling

Carcasses are normally refrigerated for 12 to 24 hours, immediately after slaughter, at a temperature of from 28°F. to 32°F., to remove animal heat. Carcasses should be quickly and thoroughly chilled, as soon as possible after slaughter, to insure keeping quality and to enable easier, more attractive cutting. Practices which hasten the loss of animal heat, cutting heavy carcasses, removing fat in the region of the crotch, facing off excess fat, etc., are employed. Chilling is done in rooms free of undesirable odors, insects, and rodents.

Aging or Ripening

Aging or ripening is the holding of certain kinds of meat, principally beef, after slaughter, usually under refrigeration, at temperatures ranging from 32°F. to 38°F., largely to increase tenderness and to develop flavor. Beef is usually aged for from 10 to 14 days, at a humidity from as low as 50 to 60 per cent to a maximum of 75 to 80 per cent in so-called dry aging, and at a relative humidity of 75 to 85 per cent in so-called wet aging. Proper air circulation is considered essential. Aging is actually brought about by enzymes indigenous to meat. Although bacterial spoilage will curtail aging, collagenases, produced by bacteria, aid in the breakdown of tissues. The most significant effects of aging take place in the first nine days after slaughter. Holding beef for periods longer than approximately 14 days, although resulting in additional tenderness and a more pronounced flavor, may be offset by excess trim and moisture losses. Veal and pork are not aged. Lamb and mutton are aged modestly.

The Tenderay process for tenderizing beef, introduced primarily through the efforts of the Mellon Institute (McCarthy and King, 1942), involves the use of ultraviolet lamps, in sufficient numbers, to curtail bacterial spoilage. At a temperature of 68°F., tenderization equal to that obtained in normal aging (as mentioned above), was accomplished in two days, at a relative humidity level of 85 to 90 per cent.

Mechanical tenderization has proven effective. The direct application of proteolytic enzymes to meat, now done in the home and in commercial food service, achieves tenderization by the addition of the enzymes papain, bromelin, and ficin. Since penetration is a limiting factor, such tenderization proves most effective on steaks. An even more unusual technique of meat tenderization involves the injection of a solution containing a combination of proteolytic enzymes, intravenously, in cattle prior to slaughter (Swift & Company's Proten beef).

Composition

Meat constituents may be classified in a number of ways. In the discussion which follows, composition will be described in terms of chemical and physical constituents.

Chemical

The composition of meat is outlined in Table 5.1. Water, the principal constituent in meat, varies from 70 to 75 per cent in most cuts. As chronological age increases, water content decreases.

Of the total solids, proteins rank first in amount and in importance. Proteins compose from 15 to 20 per cent of the edible portion of meat, the most abundant muscle protein being actomyosin, which consists of two myofibrillar proteins, actin and myosin, the globulins responsible for the contractile properties of muscle. Edible muscle tissue also contains: collagen, reticulin, and elastin proteins involved in skeletal functions; respiratory pigments, especially myoglobin (muscle pigment); nucleoproteins, the chief constituent of genetic material involved in cell heritability; enzymes (biological catalysts); and other protein components. The inedible portion of meat contains considerable amounts of the proteins, collagen (dermal portion of the skin, tendons, bone, and connective tissue), elastin

Table 5.1.
Chemical composition of typical adult mammalian muscle after rigor mortis but before degradative changes postmortem (per cent wet weight).[a]

Major constituents	*Subconstituents*	*Per cent*	
Water			75.5
Protein			18.0
Myofibrillar	myosin, tropomyosin, X protein	7.5	
	actin	2.5	
Sarcoplasmic	myogen, globulins	5.6	
	myoglobin	0.36	
	haemoglobin	0.04	
Mitochondrial, Sarcoplasmic reticulum, Sarcolemma, Connective tissue	cytochrome *C*	ca. 0.002	
	collagen, elastin, "reticulin", insoluble enzymes	2.0	
Fat			3.0
Soluble nonprotein substances			3.5
Nitrogenous	creatine	0.55	
	inosine monophosphate	0.30	
	di- and tri-phosphopyridine nucleotides	0.07	
	amino acids	0.35	
	carnosine, anserine	0.30	
Carbohydrate	lactic acid	0.90	
	glucose-6-phosphate	0.17	
	glycogen	0.10	
	glucose	0.01	
Inorganic	total soluble phosphorous	0.20	
	potassium	0.35	
	sodium	0.05	
	magnesium	0.02	
	calcium	0.007	
	zinc	0.005	
Traces of glycolytic intermediates, metals, vitamins, etc.		ca. 0.10	

[a] From Lawrie, 1966.

(connective tissue and ligaments), and keratin (the major constituent of hair, horns, hoofs, and the outermost layer of the epidermis), as well as blood protein (Giffee *et al.*, 1960).

Physical

The proportions of lean, fat, and bone at any stage in the development of meat animals are of interest and importance to the producer, packer, purveyor, and consumer. Body composition is influenced by genetic (heritable) factors, and by environmental factors, particularly feeding regimen, maturity, and sex.

The most accurate estimate of lean in a carcass is a chemical analysis of the whole carcass, a technique completely impractical for commercial applications and even for research purposes because of the time and cost involved. Powell and Huffman (1968) compared the accuracy of estimating beef-carcass composition by various methods with the chemical analysis of the fat-lean portion of the entire right side (Table 5.2). The Hankins and Howe method, although most accurately estimating carcass fat ($r = 0.94$) and carcass protein ($r = -.96$), was the least practical; hence yield grade, though slightly less accurate, was the best evaluation method when

Table 5.2.
Carcass evaluations.[a]

Method	*Correlation with* *Per cent of fat*[b]	*Correlation with* *Per cent of protein*[b]
Hankins and Howe[c]		
Per cent carcass fat (EE) = 2.82 + 0.77 [% fat (EE) 9–10–11 rib cut]	0.94	−.96
Tennessee[d]		
kg. carcass lean = 17.76 − 0.63 (fat th., mm.) + 0.2266 (carcass wt., kg.)	−.89	0.85
Wisconsin[e]		
Per cent retail yield = 16.64 + 1.67 (% trimmed round) − 1.94 (fat th., cm.)	−.88	0.83
USDA Yield Grade cutability[f]		
Per cent boneless retail cuts = 51.34 − 2.276 (fat th., cm.) − 0.462 (% kidney fat) − 0.0205 (hot carcass wt., kg.) + 0.115 (rib eye area, cm.^{2})	0.93	−.91
Illinois[g]		
Per cent retail yield = 67.99 − 0.0313 (carcass wt., kg.) − 2.52 (fat th., cm.) − 0.84 (kidney fat wt., kg.) + 0.06 (rib eye area, cm.^{2}) + 0.14 (conformation grade)	−.87	0.84
Oklahoma[h]		
total carcass fat, kg. $= \dfrac{\text{carcass wt., kg.} \times \text{SGf (SGlb} - \text{SGc)}}{\text{SGc (SGlb} - \text{SGf)}}$	0.92	−.89
Carcass specific gravity	−.92	0.89

[a] From Powell and Huffman, 1968.
[b] Determined by chemical analysis; for all correlations, $P < .01$.
[c] Hankins and Howe, 1946.
[d] Cole *et al.*, 1962.
[e] Brungardt and Bray, 1963.
[f] Murphey *et al.*, 1960.
[g] A.M.S.A., 1967.
[h] Guenther *et al.*, 1967. Here SGf is specific gravity fat, SGlb is specific gravity lean-bone mass, SGc is specific gravity carcass, EE is percent of fat, and th. is thickness.

calculated to the nearest 0.05; the Tennessee method ranked second.

Data available on the physical composition of meat indicates that, within limits, the fatter the animal, the lower the percentages of lean and bone. Studies on the physical composition of beef carcasses indicate that the percentages of lean, fat, and bone in the primal rib cut are very closely related to the percentages of these constituents in the entire beef carcass. The average percentages of separable lean, fat, and bone (including gristle), as obtained in a classical study, are shown in Table 5.3.

Table 5.3
Yields from different grades and weights of steer carcasses.[a]

	Prime	*High choice*	*Low choice*	*High good*
Slaughter weight, avg., lbs.	1,251	1,046	965	886
Dressed weight, chilled, avg.	777	613	548	513
Dressing percentage, avg.	62.1	58.6	56.8	57.9
Carcass:				
Lean, per cent	52.5	58.9	60.1	60.7
Fat, per cent	29.7	21.1	18.9	17.7
Bone, per cent	16.6	18.6	19.8	20.6
Wholesale (primal) rib:				
Lean, per cent	55.4	58.7	59.8	60.5
Fat, per cent	28.0	22.5	18.6	17.4
Bone, per cent	15.8	18.3	20.8	21.6

[a] Wilson and Co., 1943. Grade names changed here to accord with USDA meat grade-name changes since 1950.

Factors Determining Meat Quality

Consumer acceptability of meat depends on a number of considerations relating to the live animal, as its effect is ultimately discerned in the meat. The word quality, as it is used in general terms with meat, refers to its over-all eating desirability or appeal. In assessing quality, one must relate such live-animal influences as feeding and management (especially degree of finish), breeding, and age to such carcass characteristics as conformation (shape, meatiness, proportion of meat to bone, musculature), color, and texture, and fat (external and marbling, intramuscular fat, the interspersion of fat particles in the lean) to such palatability considerations as aroma and flavor, tenderness, and juiciness in cooked meat.

Live Animal

Feeding and management One of the major factors influencing the quality and palatability of meat is the method of feeding and management. Growth and fattening patterns differ considerably as a function of feeding. Cattle fed liberally on concentrates from birth to marketing tend to grow and fatten simultaneously, thus developing smaller and lighter carcasses over-all, reaching desirable finish characteristics at 16 to 18 months of age. This is the regimen for baby beef production. On restricted diets, young animals utilize available energy for growth, not for fat deposition. Such limitations are the typical feeding and management practices in beef-cattle husbandry, for example, where animals grow rather than fatten for several years, to attain large skeletal dimensions, and are then placed in dry lot (confined) for a concentrated fattening period, during which time large quantities of grain are fed, to place musculature on the large frame developed during the growing stage. Marbling is enhanced by maintaining a high plane of nutrition; it is the last factor to develop.

Breeding The influence of breeding on carcass characteristics is reflected in the history of livestock breed development. Selection from a heterogeneous cattle population resulted in the establishment of beef-type cattle,

cattle raised specifically and solely for meat purposes, and dairy-type cattle, cattle raised primarily for milk production, but used for meat when this function is terminated, or, in the case of bull calves, where demands for reproduction reduce the need for males so drastically that most male dairy calves are vealed or castrated and fattened. Most veal, therefore, is literally a byproduct of the dairy industry. Among the beef-type breeds are Angus, Hereford, and Shorthorn; among the dairy-type breeds are Holstein, Jersey, Guernsey, Brown Swiss, and Ayrshire; the Milking Shorthorn is the principal dual-purpose type. Type, those characteristics which make an animal useful for a specific purpose, would, therefore, indicate a superiority of beef type over dairy type for meat uses.

Selection has been able to achieve dramatic changes through the inheritance of desirable meat characteristics in a relatively short time. In swine, selection has resulted in the development of the intermediate meat-type hog, better designed to furnish the needs for pork than its predecessors, the rangy and chuffy types. In sheep, mutton-type and fine-wool types have been developed, the former primarily for meat with wool is secondary consideration, the latter principally for wool.

Ultrasonics, or high-frequency sound, has recently emerged as a promising nondestructive tool in live animal-carcass evaluation, from among other direct and indirect procedures, such as biopsy sampling, creatinine balance, the use of metal and electrical probes, X-rays, specific gravity, and visual appraisal. Stouffer and Wellington (1960) have not only measured fat thickness, through the use of ultrasonics, but defined the extent of underlying muscle, while simultaneously recording the results in a cross-sectional photograph. This method measures the differential efficiency with which ultrasonic signals of a known frequency are reflected by the muscle, fat, or other tissues against or through which they are transmitted. A comparison of an actual and an ultrasonically determined tracing appears in Figure 5.3.

Age The age of the animal at the time of slaughter has a marked influence on eating desirability. Consumers of meat have indicated a preference, based on a judgment of tenderness, flavor, juiciness, and suitable sizes of table cuts, for meat from animals in specific age categories: beef, from one to three years; veal, from 6 to 8 weeks; lamb, from 4 to 12 months; pork, from 6 to 10 months.

Palmer (1963) found that age accounted for only 6 per cent of the variation in tenderness in 538 carcasses from calves, steers, heifers, cows, and bulls, ranging in age from 5 to 99 months. Much higher correlations were obtained between tenderness and age of animals whose carcasses were devoid of marbling relative to more highly marbled carcasses, indicating that the interaction of age and marbling may be more important than age alone.

Exercise The increased use of pasture and its resulting exercise effect led Bull and Rusk (1942) to design an experiment to determine how the additional exercise affected gains and quality of beef. The data indicated that while exercise was expensive from the standpoint of feed cost (the exercised steers required considerably more feed per 100 pounds gain), the amount of walking required of cattle, even in scant pasture, did not have a detrimental effect on the quality of beef (carcass grade, physical composition of cuts, cutting percentages, or color of lean). On the contrary, it appeared that heavy exercise made the beef more tender. More recent studies of the effect of exercise during the growth of lambs have corroborated this observation; it was shown that the legs of exercised lambs were more tender than those of the unexercised controls (Spaeth *et al.*, 1967).

Sex A major source of variation affecting growth and development is sex. The recent consumer preference for beef cuts with a high proportion of lean to fat has prompted studies comparing meat from bulls with that from steers and heifers. Arthaud *et al.* (1969) reported that although bull carcasses were slightly lower in quality than steers, as determined by tenderness, grade, marbling, texture, and color, the average values of bulls were still acceptable. The fact that bulls grow faster than steers, combined with an increased yield of total retail product from carcasses of similar weight, suggest advantages in using young bulls for economical beef production.

Hedrick *et al.* (1969), comparing feedlot performance with carcass characteristics of half-sib bulls, steers, and heifers, observed that bulls were superior to steers and heifers, which were similar, in live-weight gain ($P < .05$) and feed conversion. Total weight and per cent of retail cuts were consistently greater for bulls than for steers and heifers, and in most instances greater for steers than for heifers. In all experiments, per cent of ether extractable constituents in the ribeye muscle were less and carcass grade lower for bulls than for steers or heifers, with no consistent differences noted between the latter.

A comparison of yield of trimmed cuts between steers and heifers usually shows that steers, on the average, have higher yields than heifers because heifer carcasses are generally fatter than steer carcasses of the same quality grade and weight. Because of this, heifers are usually marketed at lighter weights than steers. Heifers frequently have a higher proportion of carcass weight in the hindquarters than steers.

A study of the cutability of young (14.7 mo.) bull, steer, and heifer carcasses showed that bull carcasses had significantly larger ribeye muscle areas, less fat over this muscle, less marbling, less total fat trim, and a higher percentage of total retail boneless, closely trimmed cuts than did either steers or heifers. Heifer carcasses had significantly more fat than did either steers or bulls; bull and steer carcasses had a significantly higher yield of bone than did heifer carcasses, which were superior in the degree of marbling in the ribeye muscle and in carcass grade (King and Carpenter, 1967).

Increased interest in the production of meat from uncastrated males prompted a review of the recent literature on this subject by Field (1971).

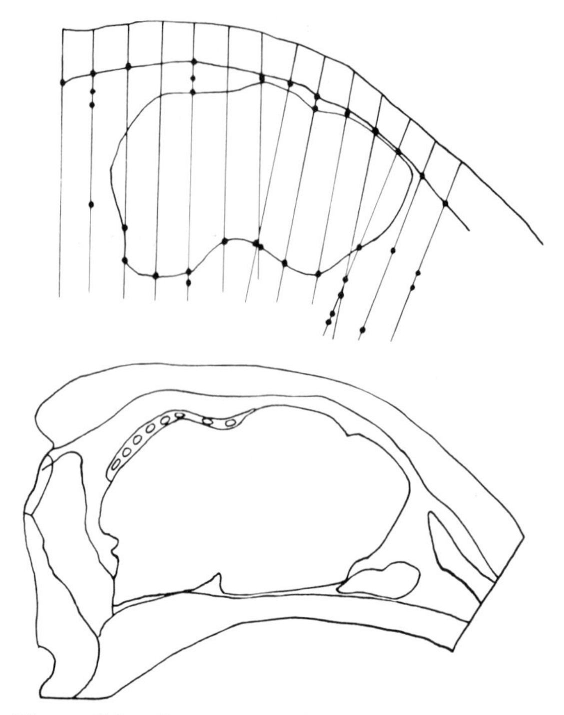

Figure 5-3 Comparison of an ultrasonically determined cross section at the thirteenth rib of a live beef animal with an actual tracing from the same location of its carcass. The predicted area was 9.0 in^2; the actual area proved to be 9.5 in^2 (Stouffer and Wellington, 1960.)

Criteria for Measuring Meat Acceptability

Appearance: Color and Texture

Color of lean is one of the more important factors determining the grade of meat. So-called "off colors" are discriminated against by the packer and retailer, and ultimately the consumer. The color of the lean of beef most desired is light (bright) cherry red; of pork, grayish pink; of lamb, light pink; of veal, pinkish brown. Dark beef, for example, is usually associated with advanced age—with tough, stringy, dry, and unpalatable meat. This normal beef color should not be confused with beef carcasses which cut dark red to purplish black in the lean, so-called "dark cutters," color usually accompanied by a sticky, gummy texture.

Color differences between muscles in the same carcass are quite common, notably in pork. This is probably an inherited characteristic. Two-toned effects in ham, created by pigment variations, have been associated with poor water-binding properties. A rapid rate of *p*H decline, resulting in acid conditions at a high temperature, causes the development of pale, soft, exudative (loss of moisture) muscle (PSE). In contrast to normal muscle, PSE muscle changes from dark red to a very pale color. Surface texture changes from a closely packed, sticky character to a smooth, moist, relatively loose structure (Briskey, 1964).

Variation in color does not exhibit a significant influence on the tenderness of fresh pork. Juiciness was significantly improved in darker muscles. Marbling appeared to improve palatability scores more in lighter-colored muscle than in darker muscle. The effect of marbling may be practically masked by darker muscles that retain more juice during cooking (Carpenter *et al.*, 1965).

Dark-cutting beef is caused by cattle being subjected to prolonged antemortem stress (Hedrick *et al.*, 1959), which results in depletion of glycogen (animal starch) in the tissues of the animal at the time of slaughter. The intensity and duration of stress, as well as the susceptibility of individual cattle to such conditions, particularly excitement, will determine the number of dark-cutting carcasses. Dark-cutting beef characteristically has a higher *p*H (lower acid content) than normal beef and is less permeable to oxygen from the air when cut. Since permeability to oxygen is necessary for color to brighten in freshly cut beef to achieve the desired bright cherry-red (oxygen from the air combines with the muscle pigment, myoglobin, forming oxymyoglobin, the process known as oxygenation), the color is similar to that of hot muscle immediately after slaughter. The dark-cutting condition occurs in all grades, breeds, sex conditions, and ages of cattle. Research has shown that there is no significant difference in eating qualities between dark and bright beef from animals of about the same age and degree of finish. Appearance and marketability are adversely affected by this condition, resulting in a great financial loss to the meat industry.

The color of fat varies with species, breed, and age of the animal, and with feeding regimen. Fresh pork and lamb fats are white. Beef fat varies from white to yellow, the latter usually indicative of older animals or dairy-type breeding, because of the presence of the pigment carotene, common in carcasses from grass-fed cattle. Color has no relation to tenderness, except that it may reflect some age or sex differences.

In studies conducted by Briedenstein *et al.* (1968), muscle firmness in heifer and cow carcasses was not influenced by maturity, but muscles which contained slightly abundant amounts of marbling were significantly firmer than those containing slight and modest amounts of marbling. Muscle texture increased in coarseness as maturity increased. Juiciness and flavor were significantly influenced by marbling, not by maturity.

Aroma and Flavor

Odor and aroma, important aesthetically and physiologically, is an important component of flavor, a complex characteristic also involving taste, texture, temperature, and *p*H. Fresh (raw) cold meat has a very slight odor; it is salty to the taste. Only on cooking is true flavor developed. Heating liberates odor from volatile chemical compounds and promotes taste by chemical changes and increased taste-bud sensitivity. The nature of the end products of the chemical reactions which occur is affected by the amount of heat used. Boiling meat in water, for example, results in a totally different over-all flavor than broiling or roasting. Odor and taste (oral sensations of sweet, sour, bitter, salt, hot and cold, and "mouth feel") in cooked meat results from water or fat-soluble precursors and by the liberation of volatile substances indigenous in meat. Crocker (1948) reported cooked beef flavor to be more odor than taste and indicated that hydrogen sulfide, amines of several kinds, and possibly indole were present.

Flavor differences exist from species to species, from animal to animal, even from muscle to muscle within the same animal. Lean-meat flavor precursors are the diffusable compounds present in meat extracts. The similarity in composition of the free amino acids and reducing sugars (most of the volatiles identified in cooked-beef studies can be accounted for by known reactions of the amino acids and sugars in beef extract) in pork, beef, and lamb, and the similarity of organoleptic qualities in water extracts of these meats, suggest a basic meaty flavor common to the lean portion of all meats, regardless of species. Fats can influence flavor; fats also serve as a depot for fat-soluble compounds that volatilize on heating and strongly affect flavor. The differences in the volatiles produced from the heated fats probably accounts for the differences in flavor of meat from the different species (Hornstein, 1967).

Herz and Chang (1970) extensively reviewed the literature on antemortem (age, breed, sex, feed) and postmortem (slaughtering, carcass handling, aging, storing, processing) factors relating to the flavor acceptability of cooked meat.

Flavor and tenderness are inversely related; the more tender the meat, the less flavorful; the less tender, the more flavorful; flavor is associated with amount of connective tissue. Older animals produce meat with more flavor than young animals, e.g., the distinctive taste of beef as compared with the absence of flavor in veal. Aging meat, under proper conditions, induces changes which give meat desirable flavor characteristics when cooked; overripening produces undesirable breakdown products resulting in unpleasant flavor.

Odor and flavor differences exist in pork. Warm cooked meat from mature boars often exhibits an objectionable odor. Patterson (1968) indicated that this odor is due to the presence of the steroid 5-α-androst-16-ene-3-one, isolated from boar fat.

Juiciness

Juiciness probably ranks second to tenderness in importance as a palatability factor in meat. It may be described initially as a manifestation of wetness produced by the rapid release of meat fluids, then as a sustained moistness, apparently a function of a slow release of serum and the stimulating effect of fat on salivation (Weir, 1960). The meat from young animals, e.g., veal, gives an initial impression of juiciness, but, due to its slight fat content, ultimately a dry sensation. In a study of the influence of marbling and carcass grade on the physical and chemical characteristics of beef, McBee and Wiles (1967) reported that, although variability was apparent within carcass grades, taste panel scores for tenderness, juiciness, and flavor increased as carcass grade increased from U.S. Standard to U.S. Good to U.S. Choice, to U.S. Prime, these differences being highly significant. Steaks from the older maturity group (19–30 mo.) were juicier and more flavorful than those from the younger group (8–19 mo.).

There is a great variation in juiciness scores for cooked meats from different species of animals, also for different cuts of meat. Cooking procedure has a great influence on juiciness. Those methods that result in the greatest fluid and fat retention yield the juiciest meat. Juiciness varies inversely with cooking losses (and degree of doneness); e.g., rare meat is juicier than well-done meat. Blumer (1963) in a review of the effect of marbling on the eating quality of beef, reported a reasonable association between marbling and sensory tenderness and juiciness, explaining this relationship as the effect of fat flavor on increased salivary flow.

Tenderness

Tenderness is the single most important factor affecting the consumer acceptability of meat, particularly beef; variations in tenderness between animals and between cuts of veal, lamb, and pork are less marked. Harrison *et al.* (1959) reviewed the literature related to factors affecting the tenderness of certain beef muscles. Szczesniak and Torgeson (1965) reported on methods of meat-texture measurement viewed from the background of factors affecting tenderness.

Connective tissue is the principal factor contributing to toughness in meat. Muscles containing much connective tissue, e.g., bottom round, are less tender and command a lower price than those containing little connective tissue, e.g., rib and loin. There is a species, breed, and muscle difference in connective-tissue content. Elastic and collagenous fibers are less abundant in tender muscle; both types increase in size as animals mature. Tenderness decreases as animals get older. Cartwright *et al.* (1958) reported that tenderness is heritable to an extent of more than 60 per cent (64.48 per cent over-all, 47.84 per cent within breeds and crosses).

Marbling (intramuscular fat), generally considered a desirable quality trait and long thought to be essential for tenderness, in recent years has been found to be insufficiently correlated with tenderness to be considered a real index of this attribute. Intramuscular fat has some influence on tenderness; most data indicate a low but positive correlation between marbling and tenderness. Slight amounts of marbling in beef seem desirable, but beyond a certain degree, marbling appears to be unrelated to tenderness (Bailey, 1964). Henrickson and Moore (1965) reported that the amount of fat within the muscle was relatively unimportant in steaks from animals under 20 months of age; it was more important in the tenderness, juiciness, and flavor of steaks from older cattle. Moody *et al.* (1970) reported that ribs with fine-textured marbling were significantly more tender (as measured by the Warner-Bratzler shear) and lighter in color than those with coarse texture. No significant differences were observed in flavor, juiciness, over-all satisfaction, or sensory (taste panel) tenderness.

Briskey (1963) reviewed the influence of antemortem and postmortem handling practices on properties of muscle related to tenderness, noting that such handling can influence the rate and extent of glycolysis, conditions of rigor mortis, ultimate *p*H, and structure and juice-retaining properties.

Muscles with small fibers are more tender than those with large fiber diameters. Muscle bundle size probably also plays a role in meat tenderness; tender muscles, e.g., longissimus dorsi, has more fibers of a similar diameter than do less-tender muscles, e.g., semitendinosus. Aging brings about a tenderization in meat; the rate and extent is a function of the temperature and length of the aging period. Treatment with proteolytic enzymes affects tenderness, as do weak acids and salts.

Cooking of Meat

Meat is cooked to create a more palatable product and to destroy those harmful organisms which may be present. Since the application of thermal energy produces changes in muscle, the eating quality of meat is greatly influenced by cookery method, temperature, and time. Level of tenderness of raw meat usually determines the method of cooking used. Tender cuts of meat become less tender on cooking. In meat cooked to a rare degree of doneness, muscle fiber proteins are tender and connective tissue is tough. Cuts with large quantities of connective tissue are usually less tender than those with little connective tissue. The time factor is important for collagen softening. Temperature is important for muscle fiber toughening. For meat high in connective-tissue content, cooking methods that involve long periods and low temperatures are best. Less tender cuts of meat are best cooked for long periods in moist heat; tender cuts in dry heat for a short time.

Current Trends and Future Developments

Major changes have and are taking place in the meat industry. Meat-packing establishments have proliferated away from Chicago, which for many years reigned as the center of the meat industry. Slaughter and processing plants now exist closer to the areas in which livestock are grown and fattened. Large conglomerates have replaced small operators. Companies are growing, fattening, slaughtering, processing, fabricating, and marketing direct to retail and institutional users of meat.

Changing eating habits of the consuming public, especially the move toward convenience, has revolutionized the industry. During the last 25 years, the use of processed (comminuted, e.g., sausage products) meats has increased tremendously. The sale of carcass meat, which saw a shift to wholesale or primal cuts, has been and is being ever more rapidly replaced by prefabricated and portion cuts. Portion-cut meat, originally used exclusively by hotels, restaurants, and related industries, is making inroads into retail outlets that cater to the homemaker. Meat fabrication, once done primarily in the retail outlet and the hotel or restaurant butcher shop, is now being done by the meat purveyor or in a central commissary. The constant improvements in preservation techniques, particularly freezing, portends an increasing usage and acceptability of meat so handled.

Precooked meats are being marketed for institutional users, e.g., rare, medium, and well-done roast beef (precooked and frozen), hamburger patties, corned beef, pork roasts, preseared steaks, and chops. The great demand for a select few cuts by fast food operators—also the airlines—has prompted the development of molded or formed steaks, combinations of cuts, encased, bound with natural meat salts or binders, molded, and cut, a trend which will increase. Roasts made from pieces of meat, handled in a similar manner, have also been developed, and are gaining acceptance.

Meat by-products continue to play a significant role in meat-industry profitability. Every part of the animal is still used for some purpose. Approximately two-thirds of an animal is represented by its carcass; the other third includes by-products, edible and inedible. Among the edible by-products are variety or extra edible meats (hearts, livers, kidneys, etc.), fats, blood, intestines, and gelatin. Among the inedible by-products are soap (from fats), glandular and other extracts (estrogens, insulin, thyroid hormone, cortisone, liver extracts, pepsin, etc.), hides and skins, glue, animal feeds, and fertilizer.

Meat analogs—synthetic, imitation, or artificial meat products—introduced some twenty years ago, and made from vegetable protein (primarily soybean protein, spun and textured, combined with artificial flavors, added nutrients, and coloring) are being accepted in the food industry (Wanderstock, 1968). The role of meat analogs is a complementary one—as "additives" to comminuted meats—in frankfurters, chilis, meat loaves, hamburger patties. Other major uses are in dietary and religious areas. These products should not compete with meat slices, patties, and large cuts. Meat analogs may well have a real place in satisfying the hunger needs of people throughout the world, but most certainly not as a replacement for animal meat.

FURTHER READINGS

American Meat Science Association. 1967. *Recommended Guides for Carcass Evaluation and Contents.* Chicago, Ill.: A.M.S.A.

Bendall, J. R. 1960. "Postmortem changes in muscle." In Bourne (1960), III, 227–274.

Breidenstein, B. B., C. C. Cooper, R. G. Cassens, G. Evans, and R. W. Bray. 1968. "Influence of marbling and maturity on the palatability of beef muscle, I: Chemical and organoleptic considerations." *J. Animal Sci.*, 27(6):1532.

Bull, S., and H. P. Rusk. 1942. "Effect of exercise on quality of beef." *Ill. Agr. Exp. Sta. Bull.* 488.

Cassens, R. G. 1966. "General aspects of postmortem changes." In Briskey *et al.* (1966), pp. 181–196.

Cory, L. 1950. *Meat and Man.* New York: Viking.

Goll, D. E., *et al.* 1966. "The chemistry of muscle proteins as a food." In Briskey *et al.* (1966), II, 755–800.

Herz, K. O., and S. S. Chang. 1970. "Meat flavor." *Advances Food Res.*, 18:2–83.

Hinman, R. B., and R. B. Harris, 1939. *The Story of Meat.* Chicago: Swift and Co.

Hornstein, I. 1967. "Flavor of red meats." In Schultz *et al.* (1967), pp. 228–250.

Lawrie, R. A. 1966. *Meat Science.* Oxford: Pergamon.

Newbold, R. P. 1966. "Changes associated with rigor mortis." In Briskey *et al.* (1966), pp. 213–224.

Weir, C. E. 1960. "Palatability characteristics of meat." In AMIF, 1960, pp. 212–221.

Six

Milk and Milk Products

Symbolics, as green cheese, curds and whey,
The milk of human kindness, and the milky way
Embody mystery, affection and a niche at the top
For the gift from nature's bosom, the cream of the crop.

Anonymous

According to the Grade "A" Pasteurized Milk Ordinance (Recommendations of the United States Public Health Service), milk is defined as "the lacteal secretion, practically free from colostrum, obtained by the complete milking of one or more healthy cows, which contains not less than 8.25 per cent of milk-solids-not-fat and not less than 3.25 per cent milkfat." Under this ordinance, only cow's milk may be marketed under the general name of milk; the source of milk from other animals must be identified. There is a relatively small market in this country for non-bovine milk. Goat's milk and some goat-milk products are produced domestically, and Roquefort cheese made from ewe's milk is imported from France. In various parts of the world, milk from other mammals is used for human consumption, including that from the water buffalo, carabao, camel, mare, ass, reindeer, and llama.

Regardless of source, milk contains water, lactose, fat, proteins, vitamins, and minerals (see Table 6.1), although how much of each will vary not only with the species, but also with different breeds of the same species, and, in fact, with individual animals of the same breed (Webb and Johnson, 1965; Jenness and Patton, 1959). The fat content varies the most. The composition of commercial milk does not vary substantially from day to day, because it is standardized at the plant to the same fat content, either by adding to or removing from the milk a calculated quantity of cream, or by blending milks of different compositions to yield the desired fat percentage. However, milk from Jersey and Guernsey cows standardized to 3.25 per cent fat content, would contain more nonfat solids than similarly standardized milk from Holsteins (see Table 6.1).

The animal scientist is interested in the factors, other than breed and individuality of the cow, which may influence yield and composition of milk. The significant ones are stage of lactation, composition of the ration, season of the year, level of nutrition, age of the cow, mastitis and other diseases, estrus, interval between milking, and exercise. These factors are also of interest to the dairy technologist,

Table 6.1.
Typical composition of milk from various species of mammals, given as percentages[a]

Species	*Fat*	*Lactose*	*Protein*	*Ash*	*Total non-fat solids*	*Total solids*	*Water*
Cow, Jersey	5.05	5.00	3.78	0.7	9.48	14.53	85.47
Cow, Guernsey	5.05	4.96	3.90	0.74	9.60	14.65	85.35
Cow, Ayrshire	4.03	4.81	3.51	0.68	9.00	13.03	86.97
Cow, Brown Swiss	3.85	5.08	3.48	0.72	9.28	13.13	86.87
Cow, Holstein	3.41	4.87	3.32	0.68	8.87	12.28	87.72
Goat	4.25	4.27	3.52	0.86	8.65	12.90	87.10
Ewe	7.90	4.81	5.23	0.90	10.94	18.84	81.16
Water buffalo	7.38	5.48	3.60	0.78	9.86	17.24	82.76
Mare	1.59	6.14	2.69	0.51	9.34	10.93	89.07
Reindeer	22.46	2.50	10.30	1.44	14.24	36.70	63.30
Human	3.75	6.98	1.63	0.21	8.82	12.57	87.43

[a] As cited in Webb and Johnson, 1965.

who must be aware of even very subtle changes in composition, since they may influence the behavior of milk during the manufacture of the various dairy products.

Our current knowledge of milk composition, though still incomplete, goes far beyond the observations of nearly 200 years ago, when milk was reported (Scheele, 1780) to consist of butterfat, casein, lactose, a small amount of extractive matter, a little salt, and water. Milk is a highly complex system of nutrients, other organic compounds, and salts. A list of the names of all of the compounds found in milk would number into the hundreds. Not all milk constituents are required nutrients for the suckling animal; some by-products of metabolism and certain hormones will pass across the walls of the alveoli of the mammary gland from the blood by simple diffusion. Examples of the major constituents of milk, and some of the minor ones, are given below.

Fats

The fat or, more precisely, lipid fraction of milk is very complex. Milk fat, often referred to as butterfat, is principally a mixture of triglycerides, which are compounds resulting from the reaction of glycerin with three molecules of fatty acids. Since milk may contain over 100 different fatty acids, the number of possible combinations is indeed great. Milk fat is unique, in that it contains many short-chained fatty acids with four to eight carbon atoms, whereas most other natural triglycerides contain mostly fatty acids which have 14 to 18 carbon atoms. Partly because of the short-chained fatty acids, the flavor of milk is also unique and difficult to duplicate. Approximately 1 to 2 per cent of the lipids are composed of several phospholipids, sterols, free fatty acids, waxes, and other materials. Milk lipids also act as solvents for vitamins A, D, E, and K.

Proteins

In milk, 19 different amino acids, which are the building blocks of proteins, have been identified. There are probably more than 30 different proteins in milk. For the sake of simplicity, it is commonly stated that milk proteins consist of several caseins and whey proteins, for which milk is, in fact, a unique source. However, numerous minor proteins are also present, including a number of enzymes, some of which are of considerable significance to the technologist.

Carbohydrates

By far the most prevalent carbohydrate in milk is lactose, also known as milk sugar. Chemically, it is a disaccharide which on hydrolysis yields glucose and galactose, sugars which are also found in small quantities in milk. The presence of additional carbohydrates has been reported, although their total number is not known.

Minerals

Some mineral constituents are present as salts; others are parts of more complex organic molecules. The inorganic and organic ions in milk include calcium, magnesium, potassium, sodium, phosphates, chlorides, citrates, bicarbonates, and sulfates. The trace elements include copper, iron, manganese, aluminum, zinc, cobalt, iodine, lithium, strontium, rubidium, and barium. Still others have been found, but some of these may be contaminants.

Vitamins

In addition to the fat-soluble vitamins A, D, E, and K, milk contains many water-soluble vitamins: ascorbic acid, thiamine, niacin, riboflavin, pyridoxine, folic acid, pantothenic acid, nicotinic acid, biotin, choline, B_{12}, and inositol.

Miscellaneous

To the many substances in this category, newly discovered ones are constantly being added. They include pigments, nitrogen-containing compounds other than proteins and amino acids, gases, aldehydes, ketones, acids, and sulfides. Additional compounds are formed as a result of processing, bacterial and enzymatic action, and oxidation.

Elementary Physical and Chemical Properties of Milk

An understanding of the physical and chemical properties of milk constituents is helpful in explaining the numerous phenomena encountered during the processing of milk and the manufacture of dairy products (Webb and Johnson, 1965; Jenness and Patton, 1959). A basic consideration is the manner in which the constituents are dispersed in the water portion of the milk. Two of the major classes of constituents, fats and proteins, are in colloidal dispersion, whereas most of the others are in true aqueous solution.

The colloidal dispersion of fat in water is termed an emulsion. The fat exists in the form of spherical globules, 0.1 to 10 microns in diameter, surrounded by a membrane which prevents the fat from coalescing or "oiling off." The membrane itself is a complex mixture of phospholipids, proteins, and enzymes, and carries a negative electrical charge. (See Figure 6.1.)

Fat globules are lighter in weight than the surrounding skimmilk, and therefore will rise under the influence of gravity. Eventually, they become quite concentrated at the top of their container and form a cream layer, which is essentially a high concentration of fat globules dispersed in skim milk: the higher the concentration, the richer the cream. Since very small fat globules rise quite slowly, homogenization, which drastically reduces the size of fat globules, produces a milk that does not form a cream layer during the normal shelf-life of market milk.

To break the fat emulsion and isolate the fat requires vigorous physical agitation, a chemical action, or a solvent action. The Babcock test for fat, on the basis of which most of the raw milk is purchased in this country, depends on the chemical digestion by sulfuric acid of

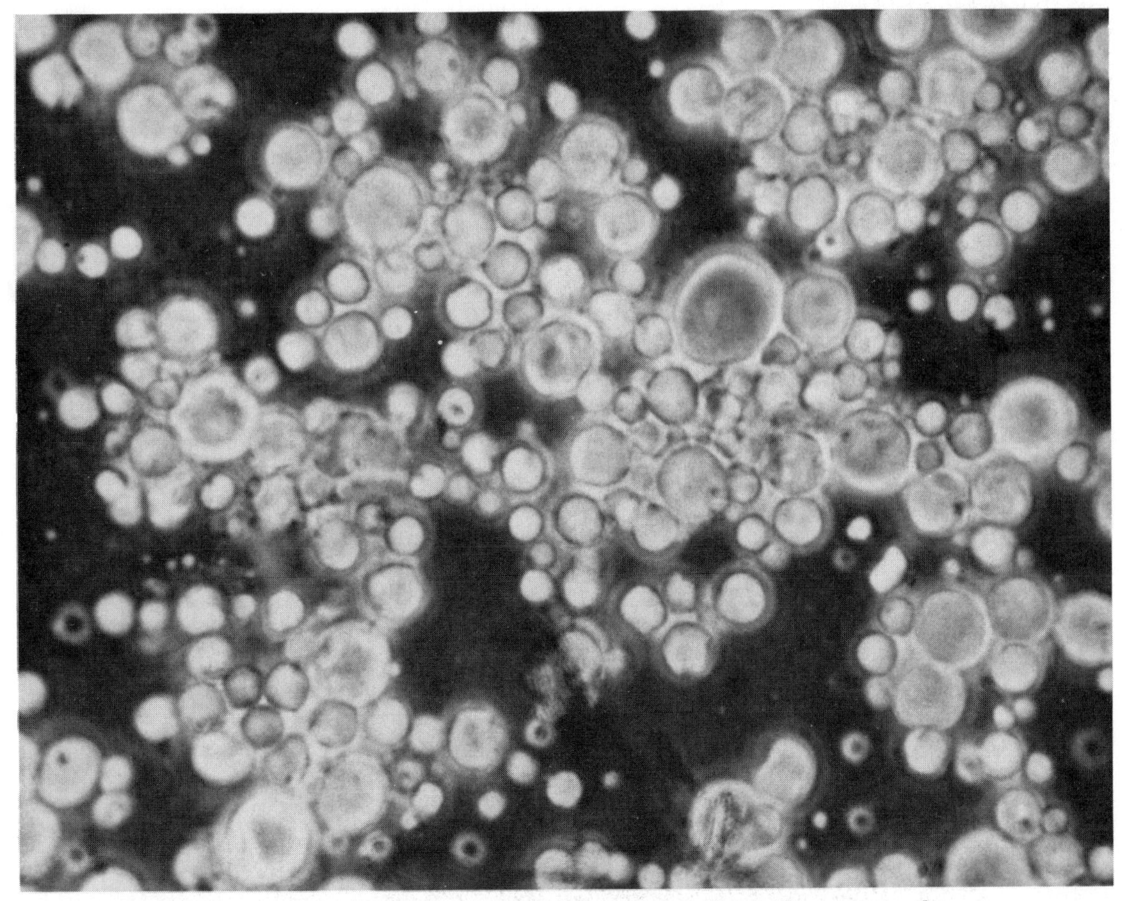

Figure 6-1 A photomicrograph of the fat globules in raw milk. Differences in the sizes of the globules can be seen.

all milk constituents except fat. The churning of butter is accomplished by physical action. (See Figure 6.2.)

Proteins are very large molecules whose molecular weight in milk ranges from 15,000 to 250,000, compared to approximately 800 for the molecular weight of triglyceride fat. The colloidal protein particles, which may consist of 100 to 100,000 molecules, are still (as much as 250 times) smaller than the particles of fat. Proteins are considerably heavier than water, i.e., their specific gravity is higher, but in normal milk the Brownian movement of the molecules offsets their tendency to settle out. Coagulation or further aggregation does not occur, because the colloidal particles are stabilized by a closely adhering film of water, and because their electrical charge causes them to repel each other.

Caseins are precipitated when their electrical charge is neutralized by the addition of a required quantity of acid or by the action of acid-producing bacteria. The other milk proteins, commonly categorized as whey proteins, are not precipitated by increased acidity alone, because their protective film of water stabilizes them. Heat treatment, however, alters the structure of the whey-protein molecules to render them coagulable, and a combination of heat and acid will, therefore, precipitate the whey proteins. It is significant that the heat treatment does not have to be applied simultaneously with the acidity increase for precipitation to happen. The acidity at which milk-protein coagulation takes place is best defined in terms of the hydrogen-ion activity, specifically, at *p*H 4.67. In its native state, as found in milk, casein is also coagulated by the action of such enzymes as rennin and pepsin.

Proteins provide the structural framework for many dairy products. They possess such properties as the ability to form a gel, or even a mat, and to bind water to a degree that may be controlled by the technologist in selecting his processing conditions. Cheese curd is an example of a protein mat during the formation of which fat globules are entrapped. Proteins may also react chemically with other milk constituents. For example, a reaction between proteins and sugars takes place when, on heating or drying, milk turns a brownish color.

Degradation or partial break-up of proteins into smaller molecules, a process called proteolysis, may yield both undesirable and desirable effects. This process is largely enzymatic, the enzymes originating from the milk or, more commonly, from bacteria. The manufacture of cheese is accomplished largely by a controlled proteolysis which gives rise to a mellow body on aging and contributes to the characteristic flavor.

The actual concentration of the proteins which qualify as enzymes in milk is quite small, but normally only a very small quantity of these remarkable biological catalysts is re-

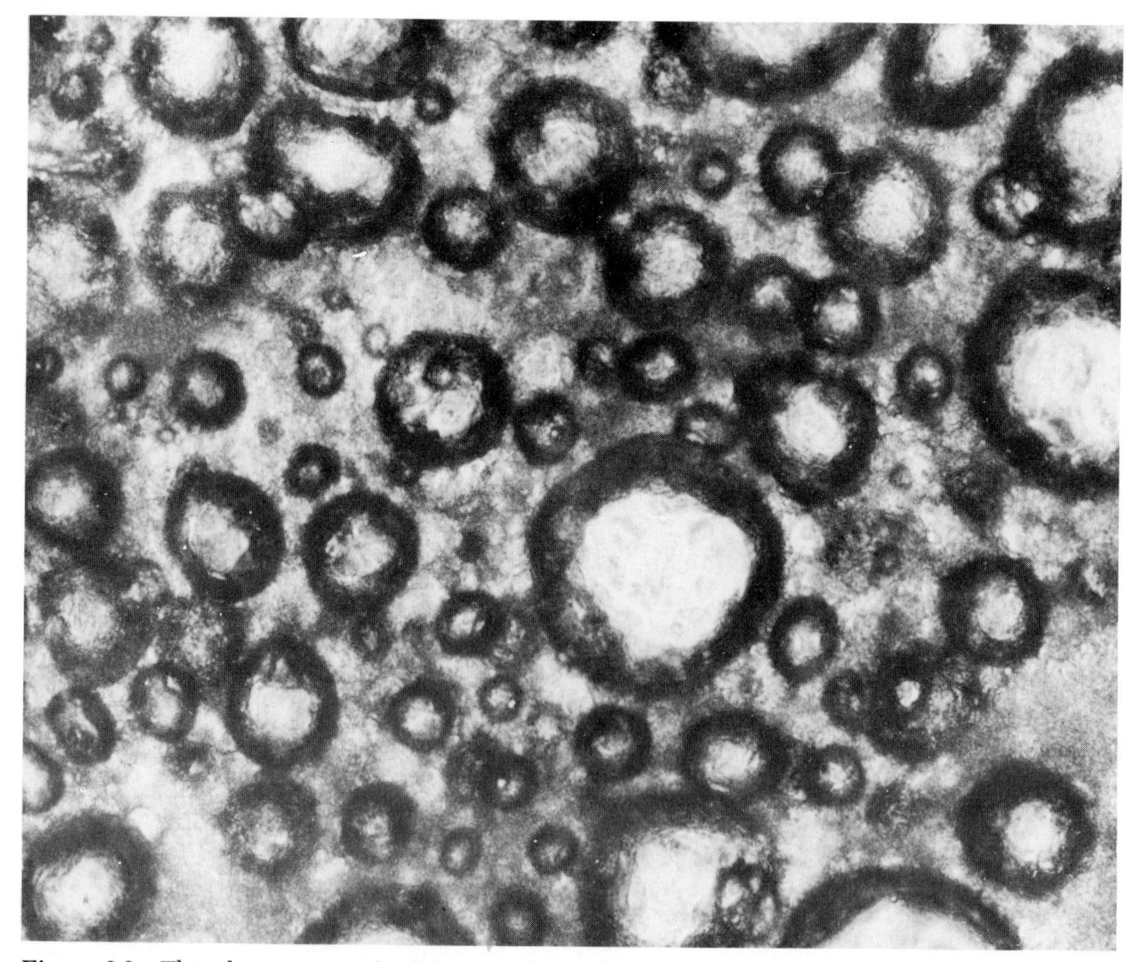

Figure 6-2 This photomicrograph of the air cells in whipped cream illustrates a stage in the process of butter churning. Fat globules are concentrating along the periphery of the air cell, and will form into butter when the air cells collapse as a result of the continued churning action.

quired to promote significant levels of chemical reactions. The enzyme lipase helps decompose fat by freeing one of its fatty acids; lactase catalyzes the decomposition of lactose into dextrose and galactose; phosphatase will split off phosphate groups; and each of the other enzymes in milk performs a similarly specific function.

The ability of enzymes to function in their normal manner is impaired or even inhibited by heat. The degree of inactivation of any given enzyme is predictable for any temperature and holding time at that temperature. Quite by accident, it turns out that the inactivation of the enzyme phosphatase occurs at the same time and temperature as is required for milk pasteurization; hence the presence of phosphatase activity in market milk is taken as evidence by Public Health enforcement officials that the milk was not adequately pasteurized.

In contrast to the proteins and fat, which are largely dispersed as colloidal particles consisting of many molecules, lactose exists in a true molecular solution. However, the solubility of lactose in water is not as great as that of the common table sugar, sucrose, or that of many other sugars. The sweetening power of lactose is only about one-fifth that of sucrose; so, even though lactose is the most prevalent solid in milk, the flavor of milk is not dominated by a sweet taste. It is interesting that the lactose concentration in human milk is higher than in cow's milk.

In certain concentrated dairy products, the low solubility of lactose may produce an undesirable texture defect because of the presence of lactose crystals. Whenever the size of the crystals exceeds 10 microns, there is a likelihood of a chalky or sandy texture. When milk is dried rapidly, as in making powdered milk, lactose cannot crystallize normally and dehydrates into an amorphous glass. In this state, it is highly hygroscopic until the glass absorbs sufficient moisture to crystallize.

The salts in milk exist largely in true solution, although some phosphates form colloidal dispersions. Both lactose and salts have a great influence on the osmotic pressure of milk, which remains essentially constant because any variation in lactose content is compensated for by a change in the salt content, and vice versa. The freezing point of milk, which is a phenomenon related to osmotic pressure, is so predictable that, when it is higher than normal, it signifies that water has been added to the milk.

The importance of milk salts must be understood in terms of their relationships to other milk constituents. Salts are responsible for stabilizing as well as destabilizing the colloidal particles, which is of practical concern in processes involving heating and freezing. Destabilization of the colloidal particles leads to a texture defect, and in severe cases to coagulation of the protein and coalescence of the fat. The fact that calcium salts are less soluble at higher temperatures than at lower, contrary to the behavior of most other salts, complicates milk processing, because calcium salts also affect colloidal stability. Even the slightest change in their distribution may have profound effects on the physical stability of a product.

Nutritional Properties of Milk

When a comparison is made of individual food items, little is found to contradict the statement that "milk is one of nature's most nearly perfect foods" (Rusoff, 1964; Jenness and Patton, 1959). Milk supplies energy as well as essential nutrients for man.

The milk constituents that are primarily sources of energy are fat and carbohydrates. Fat is the more concentrated source, in that it provides approximately 2.25 times as many calories per unit weight as carbohydrates (9 as opposed to 4 per gram). Cream, having a higher fat content than milk, provides more calories than an equivalent weight of milk. one pat of butter (½ tablespoon) supplies nearly 60 per cent of the calories supplied by an 8-oz. glass of skim milk, and a 1-oz. slice of

Cheddar cheese yields about 30 more calories than the same glass of skim milk.

The triglycerides of milk fat are largely composed of saturated and monounsaturated fatty acids, which are not essential in human nutrition and can be readily replaced in the diet by fats from other sources. Some polyunsaturated fatty acids also present in milk fat in small concentration, are considered to be essential nutrients. However, these fatty acids are more abundant in certain vegetable oils, such as corn, soybean, cottonseed, and safflower, so that the polyunsaturated fatty acids supplied by milk normally represent a small contribution to the total in the diet.

Lactose is the only carbohydrate present in sufficient concentration in milk to supply a significant amount of energy. However, there is some evidence that lactose may be important for other reasons. Galactose, a component of lactose, has been reported (White, 1959) to be required for the synthesis of galactosides in the brain and in the medullary sheaths of nerve tissues and for the myelin formation of cerebrosides in infants. Lactose is only slowly absorbed from the intestines (Rusoff, 1964), and while there it serves as a substrate for bacterial fermentation whose end-products include organic acids, a process believed to be responsible for improved absorption of such essential minerals as calcium, phosphorus, and magnesium from the intestine. On the negative side, some individuals cannot use lactose in their diet, because they lack the enzyme lactase, required to hydrolyze the sugar.

When considering the state of nutrition in the various parts of the world, one cannot escape the conclusion that the most critical shortage exists in proteins. Animal proteins are generally much better for human nutrition than those from a vegetable source. With protein accounting for 26 per cent of its solids, milk is an excellent source of one of the best proteins in nature. All the essential amino acids—lysine, leucine, isoleucine, methionine, phenylalanine, threonine, tryptophan, and valine—are present in generous concentration. The importance of adequate protein nutrition becomes apparent when it is visualized how a protein is synthesized in the body. All the amino acids must be available at the time they are needed; if only one of them is absent, the synthesis cannot proceed.

Although the different milk proteins have different physical and chemical properties, they do not differ markedly in their nutritional properties. Little is known of the physiological significance of the whey-protein components known as immune globulins. They are found in much higher concentration than all the other proteins in colostrum milk immediately following parturition, at which time they help protect the newborn calf against infection. Within several days, however, their concentration decreases rapidly in the milk, and eventually falls to about 1 per cent of the total protein. We don't know whether drinking the immune globulins in cow's milk provides any significant benefits to man.

The high biological quality of milk proteins is the main reason why some form of dairy farming should be encouraged even in countries that must rely largely on plant agriculture. By a carefully administered program of blending milk with vegetable extracts, such as soybeans, much could be accomplished in improving the existing state of nutrition.

Minerals function as components of skeletal units and of enzymes, as activators of enzymes, as acid-base regulators, and as osmotic pressure regulators. All the minerals believed to be essential in human nutrition are present in milk, although the concentration of some is more adequate than that of others. Most of the essential minerals can be obtained from a wide variety of foods, but without milk in the diet, not enough calcium may be available. Milk is also a reliable source of phosphorus, but at best only a fair source of iron, copper, and magnesium and a questionable source of iodine.

The detonation of nuclear devices has created concern over the appearance in milk of several radionuclides, particularly the thyroid-gland-bound iodine^{131} (I^{131}) and a companion of calcium, strontium^{90} (Sr^{90}). Milk is by no means the only food that contains these radioactive elements, but their presence in milk has been studied extensively because of the importance of milk in human nutrition. I^{131} has a relatively short half-life, but Sr^{90} has a half-life of 28 years. Fortunately, ion-exchange processes have been developed which can remove the radioisotopes from milk, and these may be put into operation if a resumption of aboveground nuclear detonations were to raise the radionuclide concentration above acceptable levels.

It is significant that only a small proportion of Sr^{90} consumed by the cow from feed is actually secreted into the milk. It also appears that the human body tends not to absorb Sr^{90} if it has an adequate intake of calcium. Since in milk the ratio of calcium to Sr^{90} is higher than in other foods, the consumption of milk actually safeguards against excessive Sr^{90} deposition in the bones.

In addition to such organic nutrients as carbohydrates, lipids, and proteins, the body requires minute quantities of the diverse organic materials called vitamins. Milk contains essentially all the required vitamins, but not in equal concentrations. It is an outstanding source of riboflavin, but a poor source of vitamin C. In commercial practice, it is customary to fortify milk with vitamins A and D up to the legally authorized levels that are printed on the outside of the container.

Much has been said in recent years about animal fats and their possible link to heart and circulatory diseases. Cholesterol, only found in animal fats, and saturated fatty acids have been implicated as the cause of the difficulty, whereas polyunsaturated fatty acids, prevalent in certain vegetable oils, have been reported to lower the blood-serum cholesterol level. The state of our present knowledge is insufficient to allow a clear interpretation of the very complex considerations which are involved here. On the one hand, it is difficult to believe that people would stop having heart attacks if they stopped consuming animal fats; on the other hand, it is common sense that overindulgence in any food is likely to be harmful. Certainly a person can best modify his diet on the advice and under the direction of a physician.

Organoleptic Properties of Milk

The term "organoleptic" refers to properties, such as taste, odor, consistency, and color, which are determined by one or more of the human senses. The flavor of milk, or that of any other food, is a composite sensation of taste and odor. Although taste may be easily classified in terms of the basic taste components of sweet, sour, bitter, and salty, it is odor that is usually the principal contributor to flavor, and odor is difficult to translate into language. It is much easier to refer to the fragrance of a rose, for instance, than to describe it in words. The organoleptic properties of milk and dairy products have been summarized by Nelson and Trout (1964).

The flavor of good milk is mild in both taste and odor. The milk salts and lactose are present in delicate balance, the resulting sensation being mildly sweet. The odor of milk comes from trace amounts of certain organic compounds whose identity is still not fully known. They may consist of fatty acids, methyl ketones, methyl sulfide, and others. The total odor is barely discernible, and thus milk has not only very little flavor but also very little aftertaste. This "bland" character of milk is largely responsible for its wide acceptance as a staple food—because one does not tire of it—and for the versatility of milk as an ingredient in the manufacture of other foods—because many other flavors may be imparted to it.

Much of the milk which reaches the marketplace contains varying levels of off-flavors, although the defects are almost always very slight and of little or no consequence. Feed flavors are very commonly encountered in milk, but at low concentration their presence is not objectionable. The flavor of feeds is transmitted to milk through either the digestive or the respiratory tract. The most common feed flavors are those imparted by ensilage, early spring grass, wild onion, and garlic. Effective precautions against feed flavors are to feed offending forages after milking and to remove cows from the pasture several hours before milking.

Feed and weed flavors are known as absorbed flavors, because the compounds which give rise to them seem to be only transferred by the cow from the feed to the milk. In parts of the country where feed and weed flavors are particularly serious, a vacuum treatment, either before or after pasteurization, may be applied to the milk to reduce their intensity. Absorbed flavors are more readily removed by vacuum treatment than any of the other off-flavors.

A rather serious group of off-flavors, classified as chemical, consists of defects described as rancid, oxidized, and light-activated. Milk containing these off-flavors is likely to be rejected by the processing plant because consumers would find it objectionable.

Rancid flavor develops in raw milk when, as a result of excessive agitation, alternate warming and cooling, or excessive foaming, the lipase is activated and liberates free fatty acids from the milk fat. There are indications that milk from cows in late lactation and from those affected by mastitis is more susceptible to the development of rancidity.

An oxidized flavor results from the attack of oxygen on the unsaturated fatty acids of the milk lipids. The fact that the reaction is strongly catalyzed by copper makes it imperative that milk not come in contact with any surfaces containing copper or copper alloys, and has been largely responsible for the almost universal adoption of stainless-steel equipment in dairy operations.

A light-activated flavor is produced when ultraviolet rays are allowed to penetrate into milk, where they catalyze certain chemical changes in the milk proteins. Exposure of milk to sunlight, even briefly, or to fluorescent lights for somewhat longer, will give rise to an off-flavor specifically described as "cabbagey" or "burnt feathers."

Microorganisms, such as bacteria, yeasts, and molds, introduced by poor sanitary practices, may cause undesirable changes in the flavor, consistency, and color of milk. If improper cooling or too high a storage temperature enables them to multiply in milk, different bacteria may produce such off-flavors as sour, malty, unclean, fruity, bitter, putrid, and rancid. Some organisms may cause coagulation of the milk proteins by acid or enzymatic action; others may produce pigments foreign to milk. It requires a relatively large number of organisms, approximately 5,000,000 per milliliter, to give rise to an off-flavor, and more than 100 times as many may be found in badly deteriorated milk.

Microorganisms are also largely responsible for the foul odors associated with ill-kept, poorly ventilated barns and stables. Cows need only inhale this foul air for the "barny" odors to be transferred into the milk. Improper sanitation, allowing organisms to grow on milking equipment and utensils, may be equally damaging to milk flavor.

There are certain miscellaneous flavors which may be encountered in raw milk coming from the farm. Milk from cows in late stages of lactation has a salty flavor because of an increased chlorine content, as does milk obtained from infected, mastitic udders. The inadvertent or willful addition of water to milk imparts a flat, watery taste which the skilled technician can distinguish from that of a milk naturally low in solids produced by certain animals.

The most serious of the miscellaneous off-flavors are those which are described as "foreign" and are due to the adulteration of milk with foreign substances. A good example is a "medicinal" or "disinfectant" flavor which arises when minute quantities of chlorine and phenol find their way into milk.

The discussion of organoleptic properties in this section has been confined largely to off-flavors which may be encountered in raw milk as it is produced on the farm. Some of these defects, as well as additional ones, may also develop because of faulty processing or manufacture of the milk or milk products, or because of improper or prolonged storage. The dairy technologist is concerned about all aspects of procurement, processing, packaging, and storage which affect the organoleptic properties of the finished product, because consumer acceptance is largely dependent on these properties.

Milk as a Bacteriological Medium

The field of dairy bacteriology has made great contributions to the dairy industry (Foster *et al.*, 1957). Bacteria are ubiquitous, microscopic, single-celled, living organisms that vary in size, shape, optimum growth temperature, resistance to heat, nutrient requirements, oxygen requirements, and pathogenicity to man, in their intra- and extracellular chemistry, and in their metabolic by-products. Some produce disease-causing or even life-threatening toxins; others bring about useful fermentations, such as the production of lactic acid from lactose. Under favorable conditions, bacteria rapidly multiply, by literally splitting in half, and thus doubling their number during every division. When all their growth requirements are met, some organisms may divide as frequently as every 20 minutes. If multiplication is allowed to proceed, the number of organisms may reach over 1,000,000,000 in a milliliter of milk. (See Figure 6.3.)

Milk provides favorable growing conditions for both desirable and undesirable bacteria. A healthy cow's udder contributes few bacteria, but milk may be easily contaminated by inadequately cleaned and sanitized surfaces. The Grade "A" Pasteurized Milk Ordinance

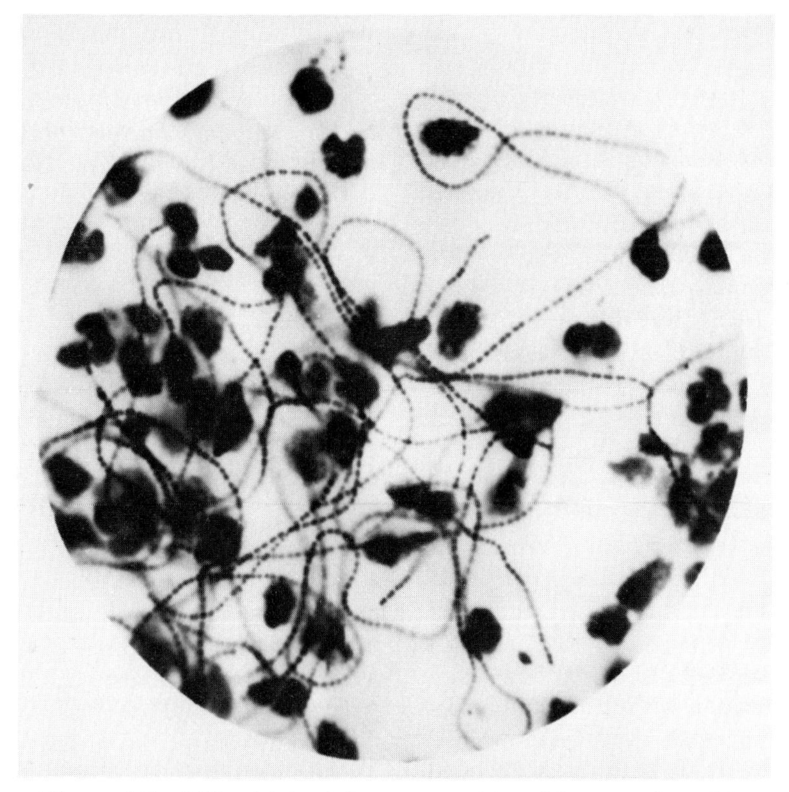

Figure 6-3 Milk obtained from a mastitic udder contains white blood cells and infectious bacteria. In this photomicrograph, the bacteria are seen as long chains of *Streptococcus agalactiae*.

(Recommendations of the United States Public Health Service) requires that prior to pasteurization raw milk must have a standard bacteriological plate count of no more than 100,000 organisms per milliliter. The count for pasteurized milk must not exceed 20,000 bacteria per milliliter. Under proper sanitary conditions and with good control over the storage temperature, much lower counts are possible, and are frequently achieved.

The famous French scientist, Louis Pasteur, was the first to discover that bacteria are destroyed by heat. The process of milk pasteurization, named in his honor, is designed to destroy any disease-producing bacteria which may be present without materially altering any other milk properties. Disease-producing organisms, such as those causing tuberculosis, undulant fever, septic sore throat, Q-fever, scarlet fever, typhoid fever, and food poisoning, may be transmitted to milk from infected humans or animals. Pasteurization standards were formulated by determining the time and temperature required to destroy *Mycobacterium tuberculosis*, and then increasing exposure to heat to provide an adequate safety margin. Later it was found that the temperature had to be increased more to insure the inactivation of the Q-fever virus, which led to the present minimum standard of 145°F for 30 minutes or 161°F for 15 seconds. More rigid standards apply to those dairy products that have a higher solids content than milk.

The most heat-resistant organisms in milk are not the disease-producing bacteria, which cannot form heat-resistant spores. The non-spore-forming bacteria that are capable of surviving legal pasteurization are, fortunately, not of Public Health significance, although they may be a factor in the keeping quality of milk. Some spore-forming bacteria may survive a heat treatment of 240°F for 15 minutes; ordinary pasteurization will obviously not destroy them. To prevent spoilage, therefore, pasturized milk must be held in refrigerated storage, preferably around 40°F, to prevent the growth of the surviving organisms. To permit storage at room temperature, milk must be either sterilized, dehydrated, or preserved by concentration and addition of sugar. Preservation of milk by chemical additives is not legally permitted.

Microorganisms other than bacteria which also may cause spoilage are yeasts and molds. Since they are capable of growing in the presence of considerable acidity, they are particularly likely to cause difficulties in fermented dairy products, such as cottage cheese, Cheddar cheese, and sour cream.

The dairy technologist is also concerned with desirable microorganisms, which are required in the manufacture of buttermilk and cheese. Molds of the *Pennicilium roqueforti* type produce the familiar veining in the Blue Mold varieties; a white mold is required for the rind of Camembert-type cheeses. A yeast of the genus *Debaromyces*, along with a bacterium, is required for Limburger cheese. Certain bacteria which produce lactic acid from lactose are indispensable in fermented milks and cheese (see Figure 6.4). The more important ones are *Streptococcus lactis, S. cremoris, S. thermophilus, Lactobacillus casei, L. acidophilus*, and *L. bulgaricus*. *Leuconostoc citrovorum* ferments citric acid to produce diacetyl, a highly pleasing flavor compound found in many fermented products.

Manufacture of Dairy Products

Before milk can be distributed and sold to the consumer, it must be processed to comply with legal requirements. It may be converted to a milk product, and it must always be suitably packaged.

Products which are commonly made in the milk plant fall into three classes: fluid milk and milk products, flavored milks, and fermented milks and milk products (for more detailed coverage, see Henderson, 1971, and Tuckey and Emmons, 1967). Fluid products include: homogenized milk, 3.25 per cent fat; low-fat milk, 1 or 2 per cent fat, skim milk, 0.5 per cent or less fat; half and half, 10.5 per

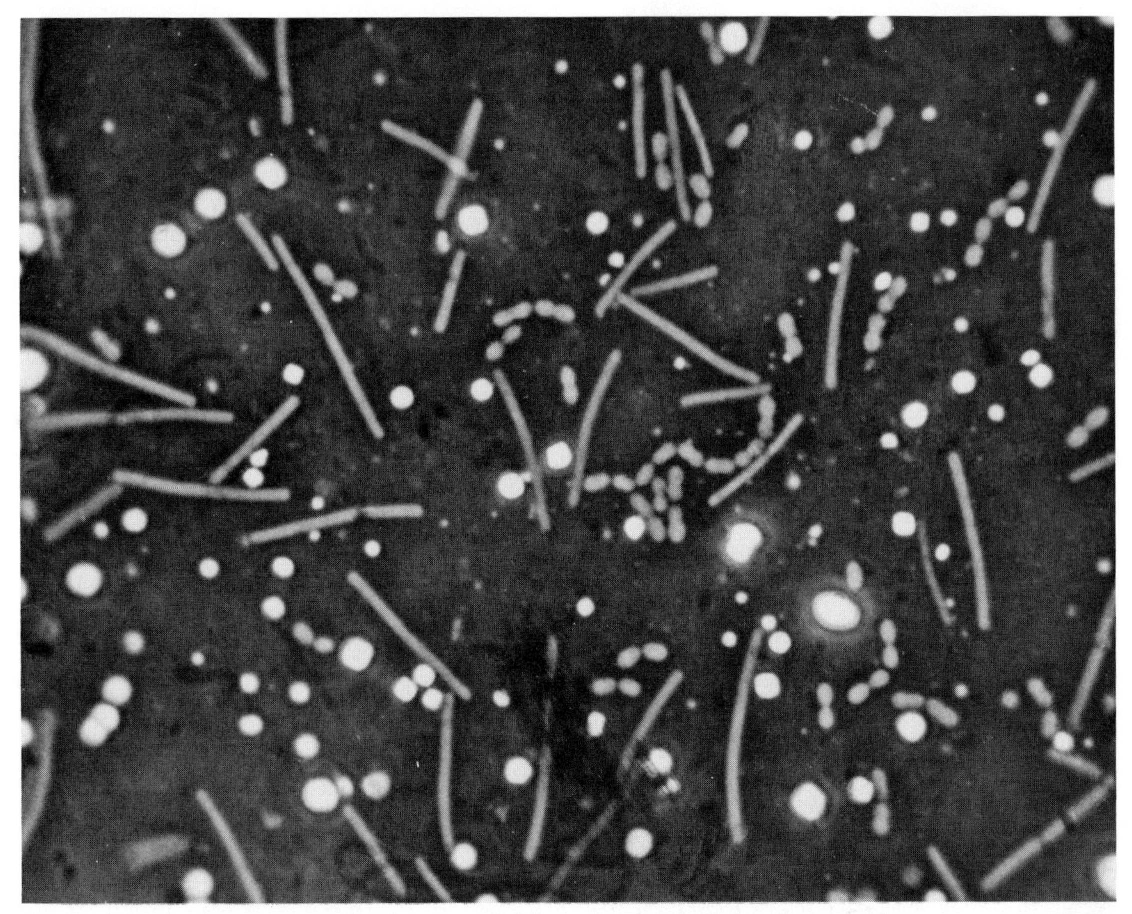

Figure 6-4 A photomicrograph of two useful bacteria required for the manufacture of yogurt: *Lactobacillus bulgaricus,* rod shaped; *Micrococcus thermophilus,* round shaped, in short chains.

cent fat; and whipping cream, 32 per cent fat. Flavored products are primarily chocolate milk (3.25 per cent fat), chocolate drink (less than 3.25 per cent fat), and eggnog (6 per cent fat). Both the flavored and unflavored products have a normal refrigerated (about 40°F) shelf-life of up to 15 days. However, special equipment is being introduced which uses flash heating within the sterilization range and packaging under aseptic conditions to yield a product with a refrigerated shelf-life of several months. Fermented milks may range in fat content between that of skim milk and whole milk. They are made by adding a culture of the desired lactic-acid-producing bacteria to a well heat-treated milk and incubating it at the proper temperature for optimal bacterial growth and flavor development. Each of the various fermented products, buttermilk, yogurt, sour cream, sour half and half, and sour cream dips, requires its own optimum processing for heat treatment, homogenization, fortification with additional milk solids, added flavoring materials if applicable, and packaging specifications.

Cottage cheese may also be made in the milk plant. Pasteurized skim milk is acidified by controlled bacterial action (a chemical acidification process has also been developed) to form a coagulum, which is cut and firmed by heat to yield the characteristic curd particles. Whey is removed, and salt and cream are added to the curd to season and raise the fat content to 4 per cent. Related to these products, but usually not made in a milk plant, are the unaged varieties of Cream, Neufchatel, and Baker's cheese.

Surplus milk is generally converted to dried skim milk, butter, and cheese. These products, however, are usually made in specialized plants, where certain economies may be realized by high-volume production (Webb and Whittier, 1970; Kosikowski, 1966; Hunziker, 1940; Hall and Hedrick, 1966).

Skim milk and cream are obtained by centrifugal separation. The skim milk is concentrated by boiling under vacuum to 30 to 35 per cent solids, and the concentrate is dehydrated by being forced under pressure through a spray nozzle, concurrently with a stream of hot air, into a chamber. Drying, under these conditions, is instantaneous, and the powder has good rehydrating properties. Instantizing, or making the powder even more easily dispersible, requires that a large, agglomerated dry particle be formed.

Butter is made by subjecting cream to a mechanical churning action. Sweet, fresh cream produces the highest grade of butter, based on flavor evaluation, and is a product which possesses excellent storage stability. Continuous butter-making machines are also commercially available. In one type, nearly pure fat is produced and is then reemulsified with skim milk. The principle employed in the other type is that of emulsion reversion of a high-fat cream. Federal standards demand that butter contain a minimum of 80 per cent fat.

Cheesemaking has been gradually transformed from an art to a science. The essential elements needed to make most of the varieties —which number into the hundreds around the world—are the enzyme rennin, obtained from the stomach of the calf, and the presence of specific microorganisms. Rennin causes a curd to form, which is cut, separated from the whey, and, after various treatments appropriate to the varieties involved, molded into its characteristic shape. Salt is incorporated either in dry form to the curd or by soaking the freshly formed cheese in a brine bath. Throughout the process, bacterial action continues, and bacterial enzymes remain active through the aging process, which varies with different varieties from a few weeks up to more than a year. Certain varieties like Limburger, Brick, and Camembert are surface-ripened, and require that the appropriate culture of microorganisms be introduced on the surface of the molded cheese.

Different varieties of cheese are created by variations in the following: milk from different animal species; the content of fat, moisture, and salt; type of microorganisms used; *p*H; temperature and humidity during aging; inclusion or exclusion of air during aging; and modifications in the manufacturing procedure. Cheese may be made from either raw or pasteurized milk, but when raw milk is used, the cheese must be aged for at least 60 days prior to sale. Some varieties are not palatable unless the aging process has been interrupted at the proper time, whereas others, like Cheddar cheese, may be consumed as either a young or an aged cheese. The most common varieties of cheese in this country are several variants of Cheddar (Colby, Monterey Jack), Swiss, Italian types, Brick, Limburger, and Blue.

Several varieties of cheese can be further modified into processed cheese, cheese foods, or dehydrated cheese products. Here the cheeses are blended, melted, combined with other approved additives, and packaged or dried, as the case may be.

Frozen desserts made with dairy ingredients include ice cream, ice milk, both hard and soft-served, and sherbets. Mellorine is a product which contains a vegetable fat, in place of milk fat, and non-fat milk solids (Arbuckle, 1966). The federal standards require that ice cream, before the addition of flavoring, contain a minimum of 10 per cent milk fat and 20 per cent total milk solids. The total solids content usually falls between 36 and 40 per cent. Typical ingredients for ice cream are cream, concentrated skim milk, milk, cane sugar, corn syrup, a stabilizer (usually a vegetable gum), a food emulsifier, and flavoring. The correct proportions of the ingredients are blended, pasteurized in a vat or by a high-temperature short-time process, homogenized by being forced through a small orifice at about 3,000 lbs per square inch pressure and cooled to 40°F. This ice-cream mix is flavored and partially frozen with agitation to a temperature of about 21°F. The ice-cream freezer is designed to lower the temperature of the mix rapidly while incorporating a desired amount of air. The partially frozen ice cream is packaged and hardened by lowering its temperature to −15°F.

Ice milk differs from ice cream in containing a lower fat content (2 to 7 per cent), but is otherwise manufactured like ice cream. Sherbets are largely fruit-flavored products containing between 2 and 5 per cent total milk solids.

The products described here are basic to the industry, and use up most of the milk produced. New products, however, are constantly being developed. By-products from cheesemaking (whey) and buttermaking (churned buttermilk) constitute a serious problem of disposal because of their high biological oxygen demand. The current emphasis is on their conversion to useful ingredients for animal as well as human food. The by-products may be dehydrated and used in that form, or fractionated into their constituents, of which lactose, casein, and whey proteins are particularly useful.

FURTHER READINGS

Arbuckle, W. S. *Ice Cream.* AVI, 1966.

Foster, E. M., F. E. Nelson, M. L. Speck, R. N. Doetsch, and J. C. Olson. *Diary Microbiology.* Prentice-Hall, 1957.

Henderson, J. L. *The Fluid Industry.* AVI, 1971.

Hunziker, O. F. *The Butter Industry.* Published by the author, LaGrange, Ill., 1940.

Jenness, R., and S. Patton. *Principles of Dairy Chemistry.* Wiley, 1959.

Kosikowski, F. *Cheese and Fermented Milk Products.* Published by the author, Ithaca, N.Y., 1966.

Tuckey, S. L., and D. B. Emmons. *Cottage Cheese and Other Cultured Milk Products.* Pfizer, 1967.

Webb, B. H., and A. H. Johnson. *Fundamentals of Diary Chemistry.* AVI, 1965.

Webb, B. H., and E. O. Whittier. *Byproducts from Milk.* AVI, 1970.

Seven

Wool and Other Animal Fibers as Raw Materials for Modern Use

"We must cut our coat according to our cloth, and adapt ourselves to changing circumstances."

Dean W. R. Inge, Anglican Prelate (1860–1954)

In selecting clothing for himself, man prefers materials that are supple, pleasing to look at, and comfortable to wear. No doubt such considerations led early man to use the hairs of animals for covering himself, and he found wool—the hair of sheep—especially suitable for this purpose. Moreover, he found sheep relatively easy to domesticate.

To a lesser extent, he adopted the hair of other animals for his use, including mohair from the angora goat, cashmere from the Cashmere goat, and various speciality hairs, such as camel hairs (hairs of the dromedary and bactrian animals) and the hairs of other members of the camel family, including the alpaca, vicuna, llama, and guanaco. Other animal fibers he uses, though they are of much less practical interest, include the hair of the Angora rabbit, of cows, pigs, horses, and reindeer, and of many small animals.

Although the hair from each species of animal and from each breed of the same species exhibits a characteristic range of properties, all hairs, including wool, are similar in origin, in chemical and physical structure and stability, and in mechanical properties. After centuries of selective sheep breeding, wool has become significantly more uniform in the properties desired in fibers to be used by man.

The apparel wools of today were developed from the soft undercoat of hairs in the ancestors of domestic sheep. The fleeces of many animals have coarse "guard" or "beard" hairs, along with an undercoat of soft, downy hairs. By means of selective breeding, the coarse guard hairs of the ancestors of domestic sheep have been almost entirely eliminated.

The balance of desired quantities in wool continues to be improved by selective breeding. Ideally, the fleeces should be as free as possible of contaminating materials and as uniform as possible in the qualities desired for specific uses. Among the properties of fibers that are important for textile use are length, size and shape of the cross section, density, luster, crimp, surface friction, moisture sorptivity, and color.

Hair and wool fibers possess these requirements. In certain uses, these fibers have advantages; for example, their relatively high moisture sorptivity contributes to the comfort in wearing apparel goods made of these fibers. On the other hand, the hairs and wool have disadvantages; for example, fabrics made of many of these fibers tend to shrink in laundering, as well as to degrade chemically in certain environments. By discovering the inner structure of the animal fibers, scientists are becoming better able to overcome the fibers' weaknesses and to add new and desired characteristics to them without sacrificing the advantages they offer.

Before we discuss some recent developments and prospects for the animal fibers' meeting requirements, we shall highlight the basic operations of wool processing. The processing of animal fibers other than wool is essentially the same as for wool; indeed, they are often blended with wool to make woven and knitted products.

A leading American textile engineer, Harold DeWitt Smith, once remarked "the essence of textile invention is the building up of large useful structures—threads, ropes, felts, nets, cloth—by combining tiny fibrous units in such ways that the individual strength and toughness of each fiber is accumulated, yet its individual freedom of movement, and hence its suppleness—although restrained—is largely retained."

Wood and hair fibers traditionally take a long and complex route from the animal to the finished textile product. In many places along this path, the fibers are subjected to adverse environments.

Traditional Operations in Wool Processing

The first step in processing wool after sorting and blending is scouring to remove grease and

Figure 7-1 Scoured (cleaned) wool is shown coming from the scouring machine en route to being dried preliminary to carding.

Figure 7-2 Wool fibers are being carded, the first step on processing after removing grease (lanolin) and other contaminants in the shorn wool.

dirt (Figure 7.1). The cleaned wool is then *carded* to separate and further blend the fibers and remove much of the remaining contaminants (Figure 7.2). Carded wool is used to make both woolen and worsted yarns. Woolen yarns are spun directly from the card sliver. The fibers used are generally short, and are not made parallel in the yarn. The softer and more open-structured woolen yarns go into making fabrics such as flannels, tweeds, and blankets.

In contrast, worsted yarns are hard and smooth, and go into fabrics such as men's suiting. Worsted yarns are made by several steps following carding. These include combing to remove short fibers (the so-called noils, used in woolens), drawing to align the remaining long fibers to make top, then further drawing and slight twisting to make roving, and finally spinning the roving into the worsted yarns (Figure 7.3). Some mills process wool only as far as top. These "top makers" sell to spinners, who in turn supply yarns to the fabric makers who supply finishers (Figure 7.4). From the finishers, the fabric goes to garment makers, then to wholesalers, and finally to retailers, who sell to the ultimate consumer. Usually about as much wool is used in the United States for woolens as for worsteds, though proportions will vary with the demands of style.

New Textiles

The remarkable growth of knitted outerwear, especially double knits, is an example of the adaptability of a specific textile structure to changing patterns in the textile trade. Double knits are characterized by a fine rib structure and full body. Both the back and the face of the fabric have a finished appearance, and are usable interchangeably. The flexibility of a

Figure 7-3 Spinning worsted yarns (at bottom) from wool roving wound on spools (at top). Roving is term given to the stage of worsted processing between top and yarn.

double knit is achieved by knitting two layers of fabric together with two or more yarns.

There are several reasons for the unprecedented movement to knit goods. Knitwear goods, because of their inherent high stretch, possess ease-of-fit along with less need for multiple sizes and garment adjustments. Moreover, knit goods have great flexibility of production, because the knitter is closer to the consumer than the weaver, is able to produce samples, and can shift quickly, when necessary, to high production. And, of course, knitting is less costly than weaving.

Another departure from traditional fabric usage is in bonded wools. A bonded fabric is produced by fastening two different cloths together with chemical agents. The scope of this development is wide, but cannot be discussed here.

Length and Fineness of Commercial Animal Hairs

The two broad commercial classifications of wools into apparel wools and carpet wools are based primarily on diameter and length. Wool fibers used in making woolen and worsted apparel goods are fine in diameter, whereas carpet-wool fibers are generally coarse and long. Carpet wools are derived from breeds of sheep that are not considered profitable to raise in the United States (although the Navajo raise limited numbers of one such breed).

In general, hair and wool fibers range in diameter from as low as 10 microns (25,400 microns = 1 inch) to over 100 microns. Apparel-wool fibers range from about 10 to 40 microns, carpet wools from 30 to 60 microns, kid mohair from 24 to 29, adult mohair 34 to 60, human hair 40 to 100, and horse hair 120 to 180. Various furs and cashmere can range from 10 to 16 microns in diameter.

The lengths of wool fibers after one year of growth can vary widely for different breeds

Figure 7-4 Wool yarns, on cones at top of a knitting machine, are knitted into a tubular fabric. Appropriate settings of the machine provide desired knit patterns.

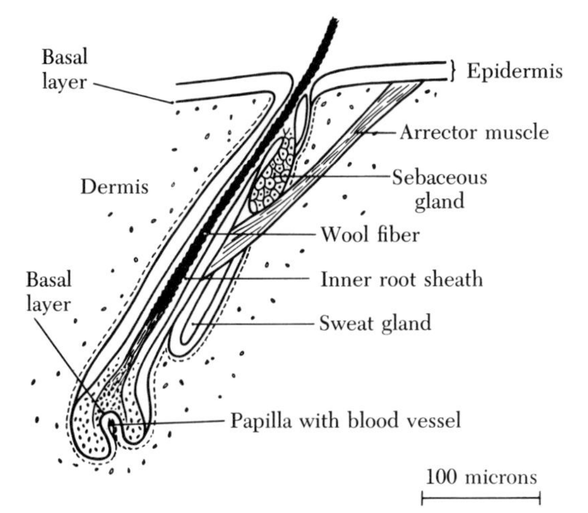

Figure 7-5 Illustrating the hair follicle in the skin. At the base of the follicle, living cells are produced. As they move up the narrow channel of the follicle, the living spherical cells transform into nonliving, fibrous, highly elongated, spindle-shaped cells surrounded by scales. (From Cole, *Introduction to Livestock Production,* 2 ed. W. H. Freeman and Co. Copyright © 1966.)

of sheep. The finest wools can be as short as 1 inch, the coarsest as long as 18 inches. In general, the lengths and diameters of wool fibers and the crimp (waviness) are associated properties. The finest wools can have as many as 30 crimps per inch of length, whereas the coarsest may have little, if any, crimp. Crimp is a desired textile property because it confers desired fullness of texture to a finished fabric.

Origin and Development of Animal Hairs

Our present conception of the formation and structure of hair and wool fibers comes from the application of many tools of research. Among these are microscopic techniques, some using visible light, others ultraviolet light or electrons. Another valuable technique involves analysis of pictures taken of the fiber with X-rays, from which details of the molecular structures in the fiber are deduced. Other important information is obtained from the absorption of light, visible, ultraviolet, or infrared. Information is also obtained from chemical studies of the protein and protein fragments obtained from wool and hair.

All this information is correlated with studies of the mechanics of normal and modified fibers, including their responses to stresses that cause stretching, bending, twisting, and breaking, such as are encountered in processing as well as in use. Such background information provides a necessary guide in adapting the animal hairs to meet modern requirements.

Wool and hair fibers are the end result of complex processes which originate at the base of tubular indentations in the animal skin called follicles (Figure 7.5). The processes of hair formation are similar and indeed related to those which produce the hornified and resistant outer layer of the epidermis of skin. Both the outer layer of skin and the outer layer of hair and wool fibers (called cuticle, or scales) consist of overlapping and flattened cells that are cemented together (Figures 7.6 to 7.8). In both cases the outer layers are composed of essentially pure keratin protein.

The follicles are themselves adaptations of the skin. The complex structural layers which make up skin extend well down into the follicle channel. At the bulk-like base of this channel is a projection from the underlying tissue called the papilla. The papilla is well

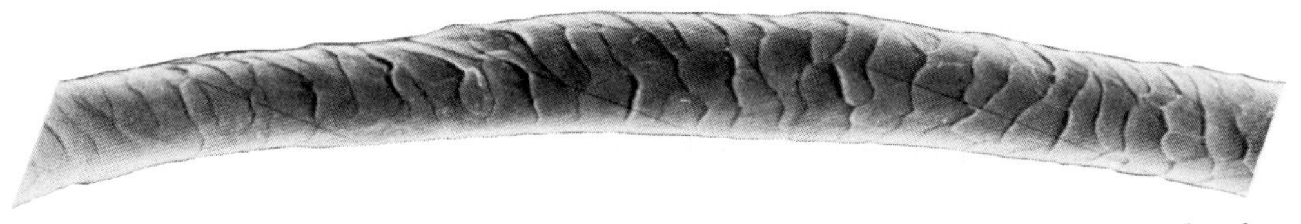

Figure 7-6 Enlarged photograph of a Rambouillet wool fiber showing its cuticle of overlapping scales. (Courtesy of W. H. Ward.)

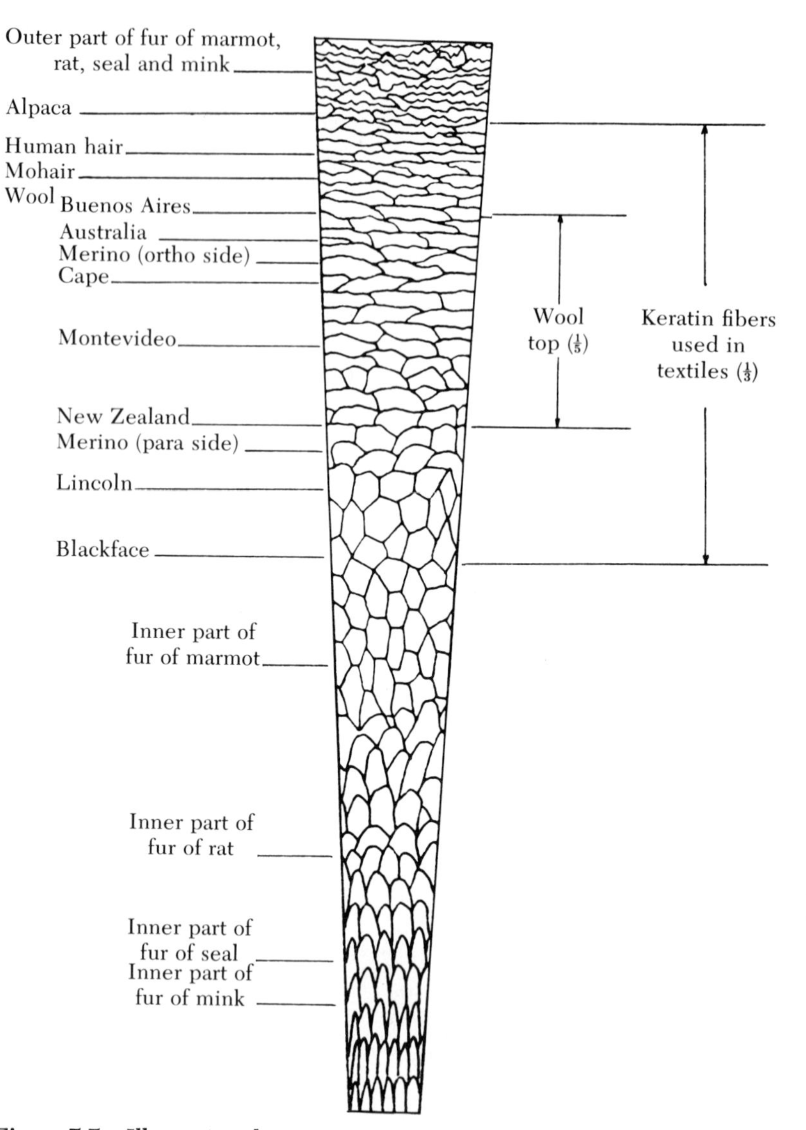

Figure 7-7 Illustrating the variation in cuticle structure and fiber diameter of various wools and fur fibers (Courtesy of P. Kassenbeck, Fraunhofer-Gesellschaft, Karsruhe, W. Germany.)

supplied with blood vessels. At its upper surface the precursor cells of hair are formed. The cells at this stage are similar to all living cells, in having a nucleus suspended in a fluid protoplasm. As soon as these cells have formed, they are displaced upward by other forming cells. It is during the passage of the cells out of the follicle that the complex transformation into hair takes place. This process involves solidification of the cell contents along with fibril formation.

The fibrils become fused together, and at the same time the cells change in shape and become cemented together. Some cells develop into the outer parts of the fiber, others the inside.

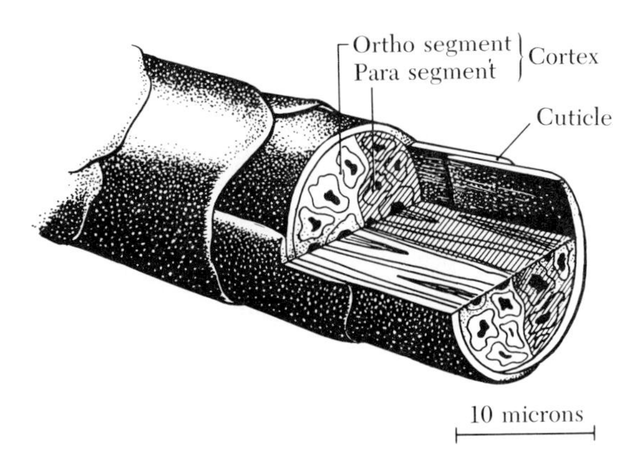

Figure 7-8 Illustrating internal morphology of wool fiber. The scales or cuticle surround the spindle-shaped cortical cells. One kind of cortical cell makes up the hard segment, called para-cortex; another kind makes up the soft segment, the orthocortex. The difference in elasticity of the adjoining segments is responsible for crimp in wool. The more elastic layer of ortho cells lies on the outside of the crimp wave. The overlapping scales are responsible for wool's ability to felt. (From Cole, *Introduction to Livestock Production,* 2 ed. W. H. Freeman and Co. Copyright © 1966.)

The cementing of cell structures involves chemical changes in the keratin protein, including the oxidation of sulfhydryl (—SH) groups (contributed by the amino-acid residues of cysteine) to form disulfide |—S—S—| linkages which join adjacent keratin molecules (Figure 7.9).*

Morphology

In gross and fine structures, hair fibers in general are considerably more complex than the epidermis of skin (Figure 7.10). The cuticle of hairs consists of at least three distinguishable components—an inner layer called endocuticle, a middle layer called exocuticle, covered by an ultra-thin membrane called epicuticle. An epicuticle is also present on the surface of epidermis. Although all three components are built of keratin proteins, their structure and properties differ. The epicuticle exhibits a significant water repellency. It is for this reason that undamaged wool can shed water to a considerable extent, even though the fiber as a whole absorbs a relatively large amount.

The entire cuticle complex surrounds the central fiber body, called the cortex. The cortex comprises as much as 90 per cent of the hair substance in fine wools. The cortex of many hairs and coarser wools contains a spongy structure called the medulla. Since its presence in fibers contributes to a dull appearance, in improved apparel wools the medulla has been more or less eliminated by selective breeding. The cortex of fibers may also contain pigment granules and remnants of the nuclei that were present in the precursor cells when they were formed at the base of the follicles.

The hair-fiber cortex is comprised of cemented-together, needle-shaped structures, the cortical cells, which originated from the single cells formed at the base of the follicle. They measure about 100 microns in length and 5 microns in width. They can be dissected into fibrillar bundles called macrofibrils, and these in turn into microfibrils.

In many wool fibers the cortex can be differentiated into two structures having different elastic and swelling behavior. These structures are called ortho-cortex and para-cortex. They are distinguishable, during microscopic examination of fiber cross section, by their difference in uptake of certain dyes and stains. The presence of ortho- and para-cortex, when laterally disposed in the cortex, determines the fiber crimp by virtue of their difference in elastic and swelling behavior. In some coarser wools,

* This natural process using sulfur has been copied by man in the process by which he transforms the sticky latex obtained from rubber trees into useful rubbers. In vulcanizing the latex, he similarly uses sulfur in order to tie together the rubber molecules. Depending on the degree of vulcanization, rubber materials are obtained which vary from highly elastic to stiff and hard.

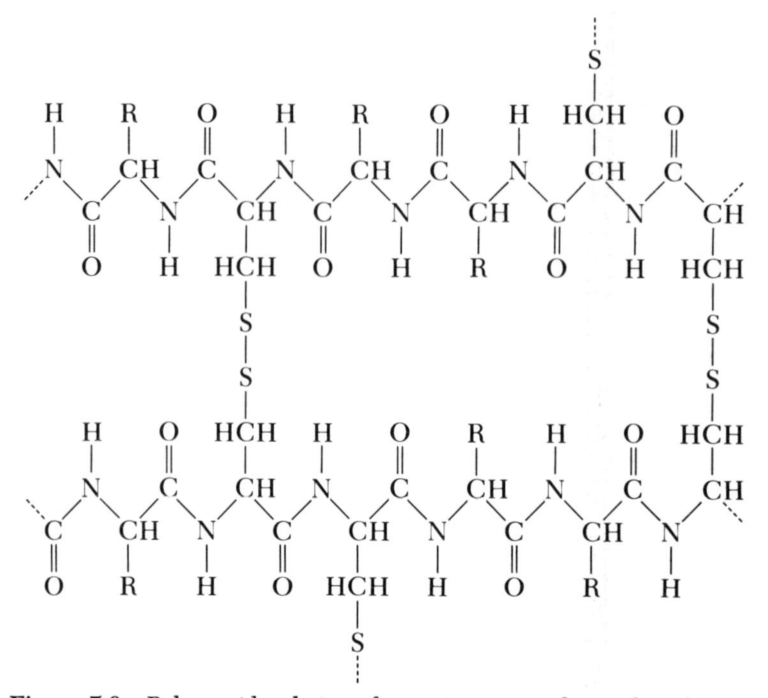

Figure 7-9 Polypeptide chains of proteins are tied together (cross-linked) through the sulfur bridges contributed by cystine residues in the protein chains.

the ortho-cortex may be present as a central part of the cortex surrounded by the para-cortex. The ortho- and para-cortex differ in structural packing of their component microfibrils. The microfibrils in the para-cortex appear more regularly packed than those in the ortho-cortex.

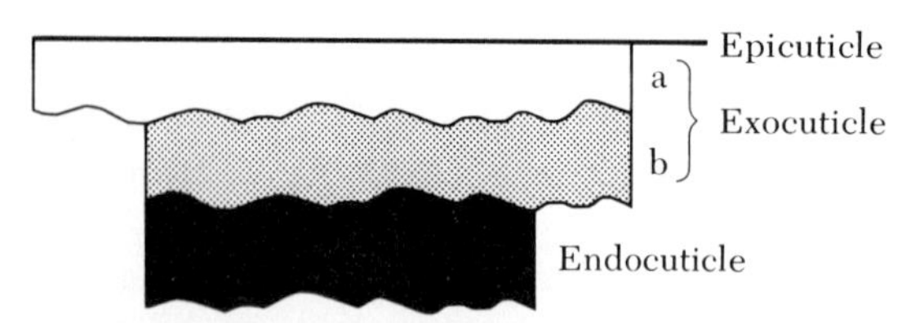

Figure 7-10 The relationship of different layers of the wool fiber cuticle. The three principal layers are respectively epicuticle, exocuticle, and endocuticle. These layers may have substructures, such as the *a* and *b* layers indicated for the exocuticle. (From Fraser and Rogers, *Australian J. Biol. Sci.*, 8:130, 1955.)

Keratin Protein Building Blocks

The ultimate building blocks of all components of the wool and hair fibers are the keratin protein molecules. Like most proteins, the keratins are long and flexible polymer chains (polypeptides) formed by chemical combination of amino acids. Depending on the number and kind of amino acids present and the manner in which they are tied together, these proteins have different properties. The highest degree of structural order of these molecules exists in the microfibrils, as is revealed by X-ray pictures and evidence from electron microscopy. The microfibrils have diameters close to 80 Ångstroms. They are built of clusters of eight or nine subunits called protofibrils, and each of these is built of three keratin molecules.

Much of our present conception of the structure of protein molecules in general comes from basic studies of hair and wool carried out in the 1930's by Astbury. He and his associates concluded from X-ray pictures taken of unstretched and stretched wool and hair that the keratin chain molecules of the unstretched fibers were "folded," and they became "unfolded" (extended) when the fibers were stretched (Figure 7.11). This process of unfolding and folding is reversible (provided the fibers are not stretched too highly).

Astbury called the folded structure "α-keratin" and the extended form "β-keratin." He classed the natural fibrous proteins in general into two groups: in the first the protein chain molecules are folded, as are the α-keratins, and in the other the chains are extended, as in the β-keratins. Examples of the first group other than the hairs are the myosin of muscle, epidermis of skin, and fibrinogen of blood. On the other hand, examples of fibrous proteins having extended chains in their natural state include the keratin protein of feathers, and the collagen of tendons and silk. The protein of feathers, hair, and wool contains from about 3 to 5 per cent sulfur. Collagen and silk are not keratins; they do not contain sulfur.

The folded configurations of the fibrous proteins conceived by Astbury and Street (1931) and Astbury and Wood (1933) have in recent years become clarified by Huggins (1943) and Pauling and Corey (1951, 1953) as helically coiled structures. It turns out that helical coiling is a natural tendency of all proteins. In fact, the non-fibrous globular proteins—for example, those present in eggs, milk, and seeds—exist in more or less helically coiled form, but they are not cemented into networks as are the α-keratins (Figure 7.12).

Knowledge of the inner workings of hairs and wool fibers is assisted by chemical studies in which the component protein structures are uncemented and then "fractionated" in order to isolate the individual units for study. On treating whole wool with chemical oxidizing or reducing agents, which cleave the disulfide cross links that tie these structures together, and then solubilizing and fractionating the uncemented proteins, one obtains proteins of differing amino-acid composition. Some of these are relatively higher in sulfur content than others. In no case yet has a protein been obtained having a single and reproducible molecular size. This makes the problem of determi-

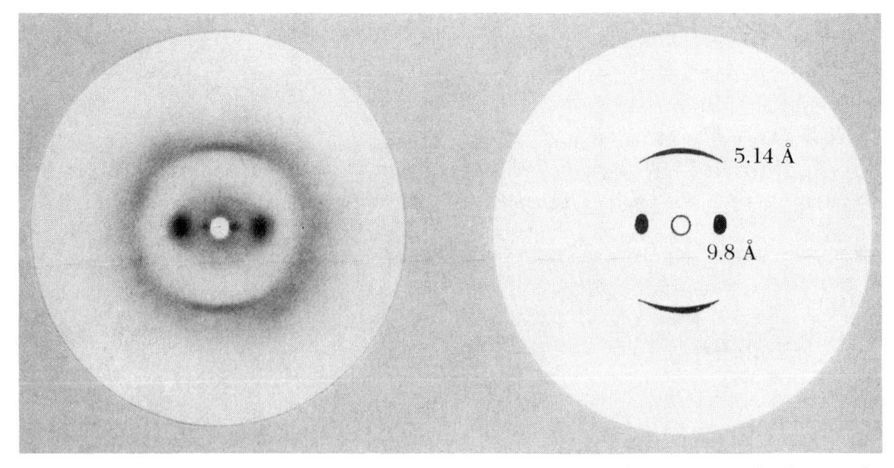

Figure 7-11 *On the left:* photograph taken by X-rays of an unstretched Lincoln wool fiber. A similar picture is obtained for all kinds of unstretched hair fibers. *On the right:* the essential features of the X-ray photograph; the 5.14 Ångstrom spacings indicate the pitch of the coiled constituent α-keratin molecules, and the 9.8 Ångstrom spacing reflect the width of the keratin molecule. (Courtesy of E. G. Carter, Int'l. Wool Secretariat, London.)

nation of the sequence of amino acids in these proteins difficult.

Chemical Stability and Instability

The keratin proteins are relatively inert to many conditions. They are remarkably tough and insoluble. They are adapted to serve higher animals as protective coverings. But we know that even with our own skin and hair, the keratin structures have limitations in the conditions they can withstand. They can undergo severe damage from excessive heat, alkali, sunlight, abrasive conditions, various oxidizing and reducing chemicals, and other adverse situations.

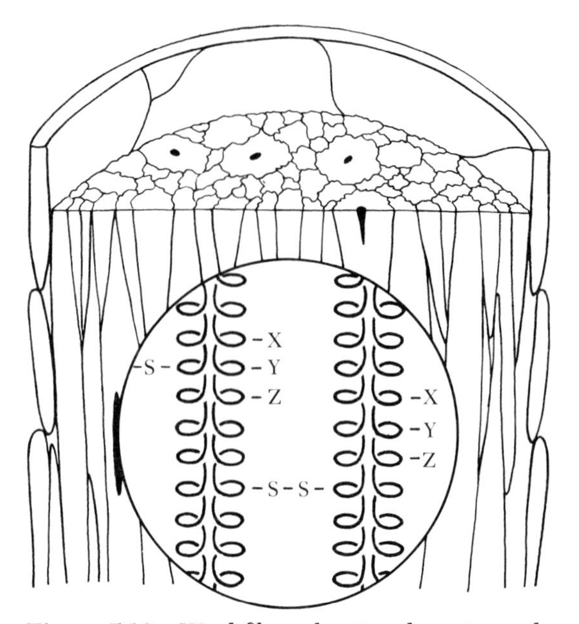

Figure 7-12 Wool fiber, showing the outer scale surrounding the cortex of needle-shaped cells. A schematically enlarged view of the coiled α-keratin molecules is shown in the circle diagram in the center. These molecules are cemented back to back as well as stabilized in the coiled state by relatively weak hydrogen bridges. Groups of coils are tied together by the sulfur bridges (disulfide cross links). The X, Y, Z letters symbolize chemically reactive centers along the keratin molecules at which moisture, dyes, and other modifying chemicals can attack. (From Cole, *Introduction to Livestock Production,* 2 ed. W. H. Freeman and Co. Copyright © 1966.)

Fiber-Tip Damage

Damage to wool can occur on the sheep in several ways. The condition called weathering involves the combined action of sunlight and moisture. This action at the ends of wool fibers on the exposed outer parts of the fleece results in "tip damage." Such damage is recognized by decreased fiber strength and elasticity, and modified dyeing properties. Extensive study of this problem found that unevenness of dyeings due to this cause can be minimized and often eliminated by careful selection of dyes and by use of chemicals which have special (surface-active) properties.

Yellowing

Damage to wools while still on the sheep can also occur from stains caused by pigments in urine and fecal matter and pigment produced by microorganisms in the fleece. Such stains are extremely difficult to remove even with the strongest bleaching conditions.

Yellowing of wools can result from complex influences during storage, during manufacture, and during use. These include yellowing caused by sunlight and by alkalies. With the relatively high demand in recent years for white and pastel-dyed wool products, the problem of discolored wools has become more serious.

Mechanical Properties

One reason why wool has received such widespread acceptance as a textile fiber is its characteristic and outstanding elastic behavior

(Figure 7.13). The elastic behavior of wool is also exhibited by other animal hairs. It is a reflection of their common unique structure.

When a slight force is applied to the hair, it responds first by uncrimping. When a higher stress is applied, the fiber, after uncrimping, responds like a semi-rigid spring, because of the helical structures in the cortex, which at first resist uncoiling by virtue of relatively weak stabilizing bonds (hydrogen bonds). When all of these bonds are severed by the applied stress (at about 3 to 5 per cent elongation), the fiber elongates relatively rapidly up to about 30 per cent elongation. During this extension, the helically coiled protein (α-keratin) structures uncoil into the extended (β-keratin) form (as revealed by the characteristic changes in the X-ray photographs taken of the stretched and unstretched fibers). When the stress is released and the fiber returns to its normal relaxed length, the coiled α-keratin state is recovered. The ability of wool to recover from stretching ("for wrinkles to hang out") is a desirable feature.

The wool fiber can also be bent sharply before breaking many times more than most other kinds of fibers. This property is probably one reason why wool rugs have lasting good appearance. This ability to take repeated flexings before breaking is attributed to the uncoiling and coiling of the helical structures around sharp bends. Such uncoiling would prevent undesirable localization of stresses that would promote breakage. Research is seeking means to further improve wool's ability to withstand mechanical abuse, especially to increase its resistance to abrasive wear.

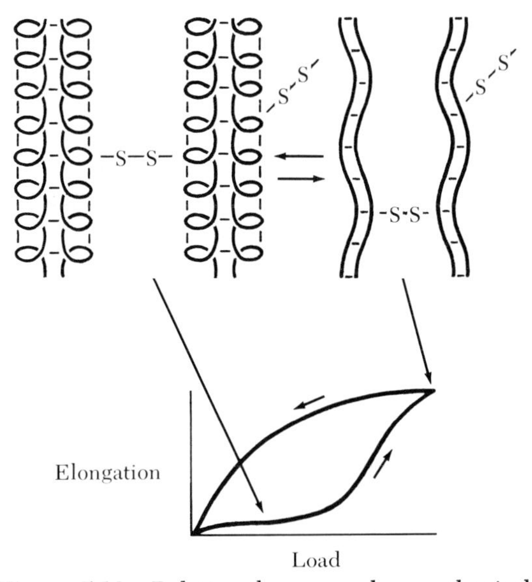

Figure 7-13 Relation between the mechanical behavior of hair and the ultimate keratin protein structure. The sulfur bridges help keep wool from being pulled apart when the molecules are uncoiled on stretching the fiber. The molecules recoil when stress is removed, providing they are not damaged. (From Cole, *Introduction to Livestock Production*, 2 ed. W. H. Freeman and Co. Copyright © 1966.)

Role of Moisture

An important feature of the animal fibers used for apparel textiles is their ability to absorb considerable amounts of moisture and still feel dry. This property is a principal reason why these fibers are comfortable to wear.

Most of the absorbed moisture is believed to be absorbed in the less oriented matrix of the fiber cortex in which the microfibrils are embedded. The more highly organized coiled structures of the microfibrils are preserved, and this helps maintain the fiber's elastic behavior as well as adequate wet strength.

The absorption of moisture occurs at various chemically active centers in the keratin molecules—for example, at the amino groups ($-NH_2$) located at the end of each keratin molecule and also on the side chains of lysine residues along the protein chains.

Wool fabrics can undergo relatively large changes in area with changes in moisture. This phenomenon has been recognized for some time and is sometimes referred to as hygral expansion. Hygral expansion can lead to undesirable garment appearance (bulging and seam puckering) especially if garment parts of different moisture content are sewn together during tailoring. Wrinkling and wrinkle recovery are also influenced by moisture (and humidity), and are important considerations affecting acceptance of lightweight wools (tropicals).

Setting of Wool and Hair

The fact that hair and wool will "set" has been used to advantage since the dawn of history. Women long ago discovered that when they wound their hair on some form of curler while it was wet, and then allowed the hair to dry, it would retain the curled state when the curling form was removed. The same principle is used in shaping, pressing, and pleating garments made of wool and other animal hairs.

Two kinds of chemical bonds in the fiber structure are involved, the disulfide bonds and the hydrogen bonds. When hair is set with ammonium thioglycolate (recognized by its skunklike odor), the disulfide bonds are opened by chemical reduction as follows:

$$|\!-\!S\!-\!S\!-\!| \longrightarrow |\!-\!SH + HS\!-\!|$$

With the hair in a desired configuration, these bonds are reformed by oxidation. The opening and closing of hydrogen bonds are promoted in a somewhat analogous manner:

$$\rangle NH - - O = C\langle \rightleftarrows \rangle NH + O = C\langle$$

For durable creasing and pleating, the wool garment is sprayed with an appropriate chemical reducing agent in solution, then pressed to give the desired pleat or crease, and finally dried under controlled conditions.

Setting is also used in fabric finishing. In traditional wool finishing it is the practice to treat the fabric with steam (decating) or with hot water (crabbing) in order to release stresses and strains from earlier processes, such as knitting or weaving. Unless the stresses and strains are released, they will contribute to shrinkage when the fabric is washed. This kind of shrinkage is called relaxation shrinkage. The setting of fibers which occurs on steaming contributes to the smoothness and luster of the fabric.

Felts and Felting

One of the oldest methods for producing fabrics without knitting or weaving is through felting of animal fibers. Felting is a characteristic ability of many of these fibers. It involves progressive entanglement of an assembly of fibers. It can occur in loose fibers, in yarns, and also in fabrics. It occurs with fibers which are moist or wet when suitably agitated. Shrinkage in a fabric or garment when it is machine washed is usually the result of felting. Felting shrinkage is distinct from relaxation shrinkage, which occurs in fabrics that have not had residual strains properly relaxed after previous handling.

The mechanism by which animal fibers undergo progressive entanglement on agitation is complex. The unique cuticle structure of the animal fibers plays an important role. By virtue of the overlapping structure of the cuticle (the scales), the hairs exhibit differential friction; that is, when rubbed the hairs tend to move more easily in one direction than in the other. Such differential movement favors progressive fiber entanglement into felt.

In the manufacture of wool felts, thin webs of wool fibers from the carding machine are set down in layers, then treated with heat, water, pressure, and rotating movement to induce felting into the tightly bonded felt structure.

Felts have many apparel and industrial uses. Angora rabbit hair has seen much use in manufacture of soft apparel felts. But because the cuticle structure of rabbit hair is not as rough as the cuticle of wool, the hair needs rather severe chemical treatment in order to induce felting. The chemical treatment, called carotting, promotes change in the cuticle structure ("raises the scales") to create a differential friction comparable to that of wool.

Toward Superior Machine Washables

Fortunately, research has come a long way toward overcoming the machine washability advantage which synthetic fibers have enjoyed over wool. Two general classes of processes are used. The first involves chemical (oxidative) treatments. This approach was developed first, and is being superseded by the second group of treatments, in which an ultra-thin insoluble sheath of synthetic polymer coats the fibers.

In order to achieve high shrinkage resistance by the oxidative treatments, the fibers are unavoidably damaged. This is reflected in loss of fiber substance (some of the fiber dissolves out in washing), a harsh feel to the treated goods, and poor appearance after repeated cleanings. Moreover, fiber strength is lowered, resistance to wear decreased, and dyeing properties are degraded.

In the second class of wool treatment, the deposited sheath of polymer masks the scale surface and changes the fiber's frictional properties so that the tendency to felting is minimized or eliminated. This process creates no loss in fiber substance, strength, or abrasion resistance. There is no change in moisture uptake by the treatments.

At first the polymer treatments of wool created an undesired stiffness in the fabric, and were relatively expensive. Polymer finishes are being refined and tailored specifically for effective modification of animal fibers. The WURLAN treatment developed by the U.S. Department of Agriculture opened wide interest in this approach. This treatment involves making and chemically anchoring a sheath of nylon or similar polymer around each fiber. This treatment has been used on millions of yards of fabric.

Permanent Press

Wool garments traditionally have been accepted as having a measure of lasting good appearance; for example, the heavier-weight wools easily shed wrinkles. Research is now providing to wools the additional features required of permanent press goods.

The concept of permanent press in garments is most simply described as the smooth-pressed condition which, together with sharp pleats and creases, lasts through repeated washings and tumble dryings so that the garments may be worn without ironing.

Permanent-press effects in fabrics and garments are being achieved in several ways. In some cases the shrink-resistant treated fabrics are treated with a chemical and then set, that is, cured, by heat treatment. In other cases the fabric is pretreated with a chemical, but curing is done after the garment is made. This is called the delayed-cure treatment. Still another method uses thermoplastic synthetic fibers blended with wool. These fibers can be stabilized as desired in the fabric or garment by heat treatment.

Because the permanent-press technology of textiles in general is relatively new, there are side problems in the treatment as well as in the product's use. Important progress is being made toward superior permanent-press products made of the animal fibers.

Among additional benefits sought are treatments that provide enhanced and durable resistance to wrinkling, to yellowing, to damage by insects, to soiling, to pilling, and to abrasive wear, as well as an ability to take the demands of increasingly high speeds of processing.

The Consumer Decides

Wool's position in the world market is not as vigorous as it once was. In 1969, wool accounted for 7.5 per cent of world fiber consumption, cotton 52.6, manmades 36.8, and silk and flax 4.1 per cent. Although wool's percentage of the market fell from around 10 per cent in 1960 to the present 7.5 per cent, volume rose slightly, from 1,541,000 metric tons to 1,657,000 metric tons.

As it has for the past decade, wool continues to hold between 10 and 11 per cent of the U.S. apparel fiber market. The per-capita mill consumption of apparel wool (scoured basis) in the United States in 1969 was 1.08 pounds. This figure includes imports of raw wool and manufactured products. This value is low compared with the years during World War II (maximum at 4.41 pounds per capita), but not too different from the 1.51 pounds per capita in 1932 and 1.22 pounds per capita in 1958.

The long-range markets for textiles are good. America is now spending over $45 billion annually for textile apparels. At the rate these markets are growing, the apparel market in 1980 may reach over $60 billion. As taste and fashion continue to grow in importance, and standards of living increase along with leisure time, the consumer will seek products with superior style, handling, appearance, comfort, and performance. He will gladly pay for such features, yet once he has acquired his apparels, he wants to "keep up appearances" with minimum effort and expense.

The textile industry is more conscious than ever of consumer likes and dislikes. By recognizing, defining, and computerizing consumer likes and dislikes along with changing styles, the wool, mohair, and other animal-fiber processors may come closer to predicting the qualities and amounts of specific textile products that will be in demand. Also, by having more precise specifications on consumer desires, the research and development laboratories can more scientifically design and produce textiles with characteristics that will be demanded. This whole approach is a far cry from simply making and offering to the consumer new products in the hope that a demand can be found to fit the product.

FURTHER READINGS

Alexander, P., and R. F. Hudson. 1963. *Wool: Its Chemistry and Physics.* London: Chapman and Hall.

Crewther, W. G., *et al.* 1965. "The chemistry of keratins." *Adv. Protein Chem.*, 20:191–346.

Dobb, M. G., *et al.* 1961. "Morphology of the cuiticle layer in wool fibres and other animal hairs." *J. Textile Inst.*, 52:T153–160.

Fraser, R. D. B., and G. E. Rogers. 1955. "The bromine allwörden reaction." *Biochem. Biophys. Acta*, 16:307–316.

Harris, M. 1954. *Handbook of Textile Fibers.* Washington, D.C.: Harris Research Laboratories.

Hearle, J. W. S., and R. H. Peters, eds. 1963. *Fibre Structure.* London: Butterworth.

Lundgren, H. P., and W. H. Ward. 1962. "Levels of molecular organization in α-keratins." *Arch. Biochem. Biophys.*, supp. 1, pp. 78–111.

———. 1963. "Keratins." In R. Borasky, ed., *Ultrastructure of Protein Fibers* (New York: Academic Press), pp. 39–122.

Lyne, A. G., and B. F. Short, eds. 1965. *Biology of the Skin and Hair Growth.* New York: Elsevier.

Mercer, E. H. 1961. *Keratin and Keratinization.* London: Pergamon.

Meredith, R., ed. 1956. *The Mechanical Properties of Textile Fibers.* New York: Interscience.

Meredith, R., and J. W. S. Hearle, eds. 1959. *Physical Methods of Investigating Textiles.* New York. Interscience.
Onions, W. J. 1962. *Wool: An Introduction to Its Properties, Varieties, Uses, and Production.* London: Ernest Benn.
Peters, R. H. 1962. *Textile Chemistry, Vol. I: The Chemistry of Fibers.* New York: Elsevier.
Proceedings of the Fourth International Textile Research Conference, Berkeley, California, 1971. New York: Interscience.
Swerdlow, M., and G. S. Seeman. 1948. "A method for the electron microscopy of wool." *Textile Res. J.,* 18:536–556.
Von Bergen, W., ed. 1963, 1969, 1970. *Wool Handbook.* Three vols. New York: Interscience.
Wool Science Review. Vols. 1–31. London: International Wool Secretariat.

Eight

Poultry Eggs and Meat as Food

"It's as full of good-nature as an egg's full of meat."

Richard B. Sheridan,
A Trip to Scarborough (1777)

The most important domesticated avian species are chickens, turkeys, ducks, geese, pigeons (squabs), and guinea fowl. Outside the United States, Canada, and a few other countries, poultry production and marketing statistics are scanty, nonexistent, or unreliable. For example, some people claim that ducks are commercially more important than chicken broilers in Asian countries; others say that geese are more important than turkeys in Europe. It is the considered judgment of the authors, however, that chickens and turkeys dominate the poultry industry worldwide.

The dominant position of chickens and turkeys derives from at least two factors: (1) ready availability of a large amount of research information developed in the past 30 or so years; and (2) development of economic integrated systems for the production and marketing of chicken eggs, and chicken and turkey meat. Newer knowledge of genetics, nutrition, management, disease control, processing, and marketing makes it possible to produce high-quality chicken eggs and chicken and turkey meat at very reasonable costs. New techniques of agricultural business management and marketing have made it possible to produce and market eggs and meat at relatively low prices to consumers.

Whether chickens, for eggs and meat, and turkeys, for meat, have higher theoretical potentials than ducks, geese, or the other domesticated species is not known. Duck strains and crosses have been bred which have excellent egg-laying capabilities, and it is known that the goose can utilize forage better than chickens or turkeys.

The composition and quality characteristics of duck eggs and duck and goose meat are somewhat different from chicken eggs and chicken and turkey meat. Some of these differences can be overcome by genetics, nutrition, and management. It is doubtful, however, that the high acceptability of chicken eggs and chicken and turkey meat can be matched by duck eggs, or by duck and goose meat.

Consumption of Poultry Meat and Eggs

Chicken and turkey meat and chicken eggs have achieved a substantial place among man's foods derived from animal sources. Figure 8.1 shows the per-capita consumption of chicken and turkey meat and chicken eggs in the United States during the period 1960–1970. These data show that currently the average person in this country consumes over 300 eggs per year, more than 40 pounds of chicken, and more than eight pounds of turkey. The consumption of poultry meat is exceeded only by that of pork and beef, which is still rising.

The Science of Processing and Marketing Poultry and Eggs

The application of scientific principles to the processing and marketing of poultry and eggs has been credited with a substantial role in achieving the high consumption of these products in the industrialized countries of the world. It will be the purpose of this chapter to show how science has been and can be applied to the processing of poultry and eggs as food. In this chapter three key topics will be discussed: (1) the structure and composition of the chicken, the chicken egg, and the turkey; (2) what quality in chicken eggs and chicken and turkey meat consists of, and how it is measured; and (3) the biological and environmental factors involved in the production of chickens and turkeys which influence composition and quality, and how they can be controlled.

Structure and Physical Properties of the Hen's Egg

Of the scientific treatises that have been written about the egg, at least three are now standard works for students in the field: Needham (1931); Romanoff and Romanoff (1949); and Carter (1968a).

Gross Structure and Composition of the Hen's Egg

The "average" hen's egg weighs about 58 grams, and has a three-dimensional geometrical form that mathematicians call a prolate spheroid. Figure 8.2 shows a cross section of the chicken egg, and Table 8.1 gives some typical measurements.

The shell is a somewhat brittle, highly mineralized structure with two closely adhering organic membranes. The outer membrane is firmly attached to the shell itself, while the inner membrane tightly encloses the egg contents. As the egg cools after being laid, its contents contract and the shell membranes separate, usually at the blunt end of the egg,

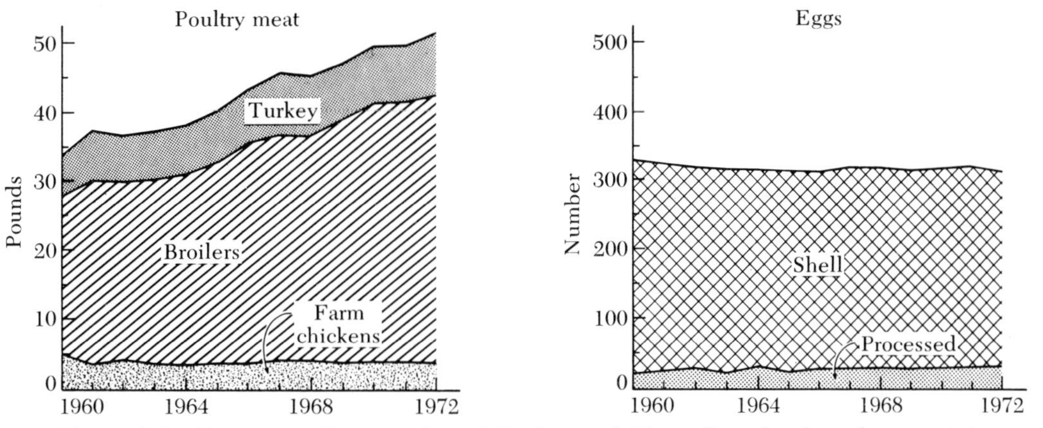

Figure 8-1 Per capita Consumption of Poultry and Eggs. Pounds of poultry meat is given in ready-to-cook weight, and processed eggs have been converted to the equivalent of eggs in shell. The figures for 1972 are estimates.

forming the aircell. Subsequent evaporation of moisture increases its size.

The albumen (commonly known as egg white) consists of four distinct layers. The outer layer, about 23 per cent of the white, is a somewhat viscous liquid. About 57 per cent of the albumen consists of a layer called the thick white. This soft, jelly-like layer has within it an additional liquid white, comprising about 17 per cent of the albumen. Finally, there is a layer of very gelatinous albumen adjacent to the yolk (about 2 per cent of the total). This layer is twisted at the poles of the yolk to form two structures called chalazae.

The sphere-shaped yolk occupies the approximate center of the egg, and is kept there by the surrounding thick white. The yolk contents are enclosed in the thin, transparent vitelline membrane.

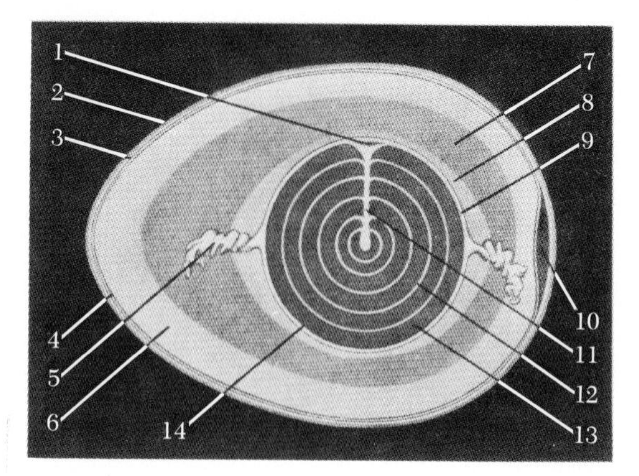

1. Germ cell
2. Shell
3. Outer shell membrane
4. Inner shell membrane
5. Chalaza
6. Outer thin albumen
7. Middle thick albumen
8. Inner thin albumen
9. Thick albumen (surrounding the yolk)
10. Air space
11. White yolk stalk
12. White yolk plate
13. Dark yolk plate
14. Vitelline membrane

Figure 8-2 Cross Section of Hen's Egg.

Chemical Composition, Including Nutrients of the Hen's Egg

The great complexity of the avian egg has made it difficult for the food scientist and the nutritionist to obtain reliable and complete information about egg chemistry as a basis for assessing its potential as a food.

Gross Chemical Composition

Table 8.2 presents general compositional data for three components of the hen's egg. The shell is a dry, highly mineralized structure. The small amount of protein it contains mainly makes up the cuticle layer on the outer surface, the organic cores of the mammilla layer, and, of course, the two shell membranes. By way of contrast, the albumen is almost entirely

Table 8.1.
Proportion of parts of the hen's egg.[a]

Portion	*Weight*	*Per cent*
Shell	6.1 gm	10.5
Albumen	33.9 gm	58.5
Yolk	18.0 gm	31.0
Total	58.0	100.0

[a] From Romanoff and Romanoff (1949).

Table 8.2.
Gross chemical composition of the hen's egg.[a]

Compound	*Shell (with membranes)*	*Albumen*	*Yolk*
Water	1.5	88.5	49.0
Protein	4.2	10.5	16.7
Lipid	0	0	31.6
Other organic constituents	0	0.5	1.1
Inorganic compounds (mainly $CaCO_3$)	94.3	0.5	1.6
Total	100.0	100.0	100.0

[a] These figures, which are percentages, are representative of what has been reported in various sources. They are *not* averages.

an aqueous solution of proteins. The yolk consists of about 50 per cent solid material, of which two-thirds are lipids and one-third proteins. All three parts of the egg contain small amounts of other constituents, which are extremely important to the quality of the egg as food.

Specific Composition of the Shell

The shell proper contains about 95 per cent calcium carbonate, as the mineral calcite, which is the same as that found in marble. The shape, size, and arrangement of the calcite crystals in the mammilla and the columnar part of the shell are such as to give the shell its strength and to some extent its well-controlled porosity. The shell also contains small amounts of magnesium, phosphorus, and other inorganic materials, but it is not known whether these are essential to its structure or merely chance contaminants.

The proteinaceous portion of the shell and its membranes has been shown, by some very elegant histochemistry, to consist of three different protein complexes. Its special composition no doubt gives the membrane its unique properties.

The pigments of the shell are complex compounds closely associated with the proteins of the cuticle and columnar layer. Brown eggs contain oöporphyrin, a reddish-brown pigment derived from hemoglobin.

Feeney (1964) has published an excellent review of the chemistry of the white proteins. In addition to amino acids, several of them contain substantial amounts of carbohydrates and sialic acid.

Albumin is the most plentiful protein in egg white. In cookery it imparts a firm structure to products such as cakes and custards as a result of heat coagulation. The globulin fraction is partially responsible for the excellent foaming and beating powers of egg white that make it so very useful in the manufacture of candies, angel food cake, meringues, etc. Mucin has the property of stabilizing egg-white foams and is, therefore, also important in cake and candy manufacture.

Interesting biological properties are possessed by lysozyme, conalbumin, avidin, and ovomucoid, all of which are nutritional inhibitors, compounds which interfere with the utilization of nutrients. Fortunately, these are counteracted when eggs are thoroughly cooked.

Yolk

Some protein chemists believe that most, if not all, of the lipid of the yolk is combined or complexed with protein (called lipoprotein). The protein-lipid complexes are important to the functional properties of the yolk, such as emulsifying power (e.g., in salad dressing) and foaming and coagulating powers (e.g., in sponge cakes and doughnuts).

The yolk contains a wide variety of fats, of which (1) almost two-thirds are ordinary triglycerides, (2) about 15 per cent are phospholipids, and (3) about 4 per cent is cholesterol.

The fat-soluble pigments are very important, in that they give the yolk its golden-yellow color. Yolk color is essential to the quality of many food products made from eggs (e.g., mayonnaise and egg noodles). These complex organic compounds, known as carotenoids, are derived exclusively from the feed.

Nutrient Composition

Eggs are high in food value, and are especially valuable for the health of infants and children, as well as for the infirm and the aged. They are an excellent source of high-quality protein, and contain substantial quantities of most vitamins, although, interestingly enough, vitamin C is absent.

Egg yolk is relatively high in cholesterol and saturated fatty acids, and consequently there is controversy about its use in diets of people prone to atherosclerosis (see Everson and Sanders, 1957).

Egg Quality

The food scientist generally defines quality in terms of those attributes of the product that determine marketability and consumer acceptability: (1) purity and safety, (2) sensory properties, (3) convenience, (4) shelf life, (5) functional performance, and (6) nutritive value.

Consumer Attitudes Toward Eggs

Jasper (1953) and Stewart (1967) have summarized consumer attitudes about eggs and egg quality. Both reviewers found that "freshness" and large size seem to be most frequently mentioned as important indicators of high quality. Why large size should be considered a quality factor in eggs is somewhat puzzling, but such confusing of quantity with quality seems not an uncommon trait among consumers.

Most Americans favor a white shell and a medium yellowish-orange-colored yolk, although many consumers in rural America and in England prefer a brown shell and a darker yolk color.

Some consumers prefer eggs with upstanding yolks and a large amount of thick albumen. There appears to be a negative reaction to prominent chalazae, possibly because consumers think they are foreign objects.

A most interesting survey conducted by Miller *et al.* (1960) revealed that consumers were not able to distinguish between the flavor of an egg that was one to two days old and of one held for 3 to 4 weeks at 50°F. This is significant, since most consumers think that there is a real difference in flavor between fresh and "not-so-fresh" eggs.

Consumers presumably react negatively to eggs with any obvious defects, such as large blood spots, dirty and/or cracked shells, or embryonic development.

Specific Quality Criteria

Soundness of the shell (freedom from cracks) is essential to prevent loss of contents and to minimize contamination and spoilage by microbes.

Shell strength is of importance in the handling and marketing of eggs to minimize breakage and loss of contents. The thickness of the shell is a good indirect indicator of shell strength, and is easily and accurately measured (Figure 8.3).

Shell color is generally measured by direct visual comparison using a set of arbitrary shell color standards. The eggs are classed as white, cream, light brown, or brown.

The porosity of the egg shell is an attribute of some importance, since it determines moisture loss from the egg and the penetration of spoilage microbes into the egg. The simplest and most reliable measure of porosity to moisture vapor is the rate of weight loss.

Albumen Quality

Wilhelm (1939) has published an excellent review of the early work on the quality of egg white. Figure 8.4 shows how the height of the thick albumen is measured with a special micrometer.

Figure 8-3 Gauge for Shell Thickness (Courtesy, U.S. Dept. of Agriculture).

Figure 8-4 Measurement of Albumen Height (Courtesy, U.S. Dept. of Agriculture).

Yolk Quality

Yolk index (height divided by width) has proven to be a useful method for measuring quality. Yolk color is generally measured by a scoring system based on comparison with a set of arbitrary color standards.

The functional properties of eggs have been studied extensively, especially the heat-coagulating ability, and the emulsifying, foaming, and beating powers. End-use tests (e.g. cake making) are commonly employed.

Eggs and egg products are subject to contamination by certain bacteria harmful to health and may be contaminated by pesticide residues. Testing for these contaminants involves highly complex and sophisticated analytical techniques and instrumentation (Riemann, 1969; Miller and Berg, 1969).

Structure and Physical Composition of Poultry

The structure and physical composition of birds have not fascinated scientists as much as has the avian egg. Consequently, our knowledge here is much more limited. Those aspects of this topic which are relevant to our interest, the use of poultry as food, are very different in kind from those we have just considered for the egg. For all practical purposes, the egg is essentially a nonliving system, whereas the bird is alive when marketed. Secondly, the conversion of a bird to food is a fairly complex process; it must be slaughtered, bled, defeathered, eviscerated, and further processed to remove inedible portions and to make the remainder suitable for use as food. Comparatively little has to be done to eggs to make them usable.

Structure

Only the following will be considered: (1) for chickens, only broiler/fryers, roasters, and cull hens; and (2) for turkeys, fryer/roasters, and young hen and tom turkeys. These types of birds constitute the majority of commercial poultry used for meat. The broiler/fryer is a young chicken of either sex, 7 to 9 weeks old and weighing about 2 or 3 lbs., which is ready to cook by broiling or frying (Figure 8.5). The roaster is an older chicken of either sex, 9 to 15 weeks old and weighing about 3.5 to 7 lbs., which is ready to cook by oven roasting. A cull hen is a mature female, usually 10 to 20 months of age and of variable weight, that requires cooking with moist heat (e.g., stewing).

A fryer/roaster turkey is a bird of either sex, usually less than 14 weeks old and weighing about 5 to 8 lbs., ready to cook by broiling, barbecuing, or frying. A young hen or tom turkey is a young bird, usually 5 to 7 months old and weighing 12 to 30 lbs., ready to cook by oven roasting (Figure 8.6).

Feathers *per se* are of little importance here, since they are removed in processing. However, feather color is important, since any noticeable pin-feathers in the ready-to-cook

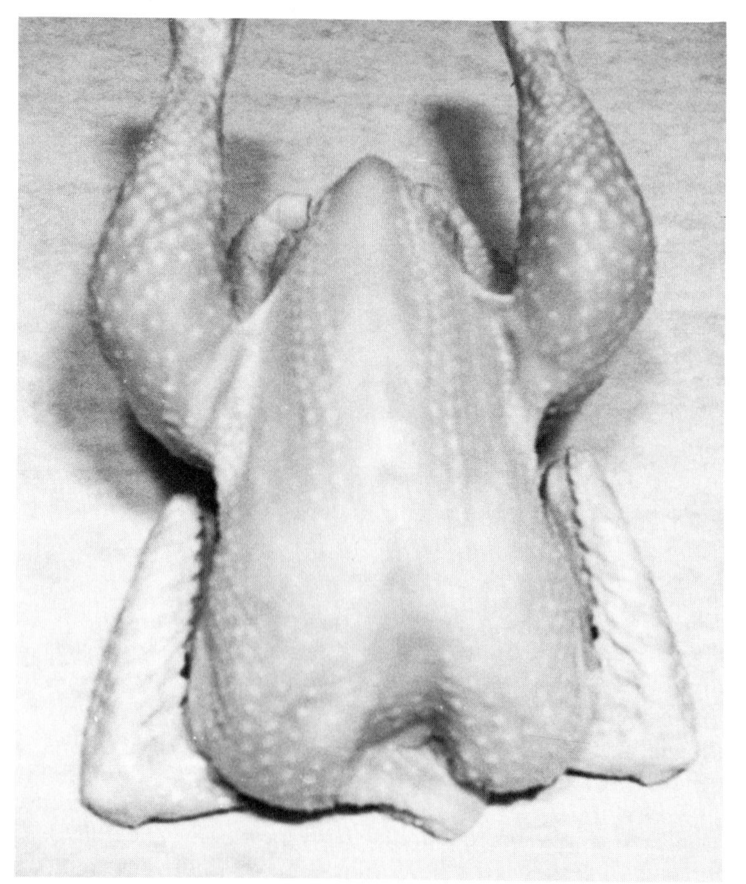

Figure 8-5 Dressed commercial broiler chicken (Courtesy of Foster Farms).

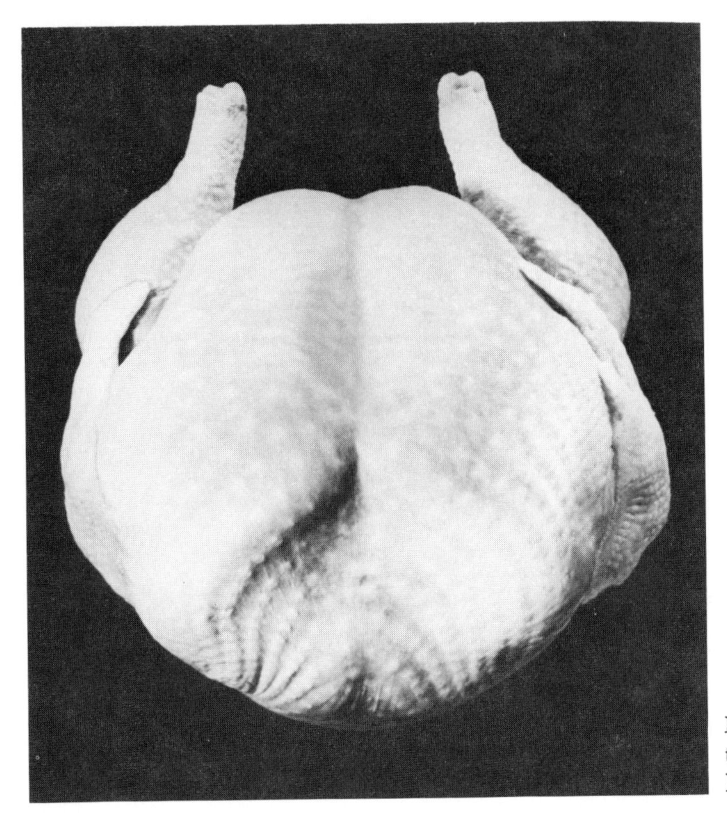

Figure 8-6 Dressed commercial turkey (Courtesy of Nicholas Turkey Breeding Farms).

carcass are considered a defect. For this reason, most commercial chicken and turkey strains and crosses are white feathered, since the pin-feathers of such birds are not very noticeable.

The gross anatomy of the chicken (Chamberlain, 1943) and turkey (Harvey *et al.*, 1948) has been reasonably well worked out. These two atlases provide a reasonably good guide to the structure of the muscles, bones, and joints of these two species, but little information about the feathers, blood, skin, and adipose tissue.

There are many different tissues in the chicken and turkey that make up the edible portion. The bulk of these are the skeletal muscles, which vary widely in size and shape. Skeletal muscle is made up of bundles of small fibers held together by connective tissue. A given muscle is classified as tender or not, depending on its content of connective tissue. The muscle fibers themselves are composed of still finer fibrils, which, in the chicken, are only about 1 micron in diameter. This structure is altered by rigor mortis, aging, and cooking, all of which influence the tenderness and texture of the muscle meat.

The connective tissue of muscle is also made up of fibers, but it does not possess fibrils. There are two types of connective tissues: (1) collagenous and (2) elastic. The latter tissues possess flexibility and are highly birefringent (i.e., they reflect light strongly and thus have a glistening appearance). On the other hand, collagenous tissue does not have much flexibility. This connective tissue is greatly altered by moist heat, which converts it to gelatin. This is an important aspect of the tenderizing effect of moist-heat cooking on meat.

Adipose tissues contain the depot fat of the bird. Most of this fat is under the skin, attached to the gizzard and heart, and in fat depots in and around the gastrointestinal tract.

Physical Composition

The proportion of parts in poultry varies tremendously, depending on species, age, sex, and environmental factors. Table 8.3 shows the proportion of parts for three types of chickens. These figures are of great economic importance. The proportion of carcass parts shows that, of the edible portion (flesh, fat, skin, and giblets), hens show the highest yield (57.9 per cent), roasters are intermediate (45.9), and broiler/fryers yield the least (43.0). Much of the edible portion of hens is fat.

The proportion of offals (inedible) in chickens (14.7 to 20.0 per cent) is a much larger proportion of live weight than in turkeys (13.01 to 13.84 per cent). The percentage of viscera in chickens and turkeys is about the same, but the percentage of carcass in chickens is considerably lower (62.2 to 66.6 per cent) than in turkeys (72.1 to 73.5 per cent). Likewise, the percentage of edible portion is likewise considerably higher for turkeys than for chickens, which is of great economic importance in producing products made up entirely of edible portions (e.g., TV dinners and canned poultry meat). Thus turkey meat is generally cheaper than chicken.

Chemical Composition (Including Nutrients) of Poultry

While scientists have been busy over the years analyzing all kinds of birds and their component parts, the job is far from complete. Two international symposia on the biochemistry and physiology of muscle have been held in the past few years (Briskey, Cassens, and Trautman, 1966; Briskey, Cassens, and Marsh, 1970).

Gross Composition

Detailed studies by Mitchell, Card, and Hamilton (1926; 1931) provide some basic compositional data (Table 8.4), which show that, for the edible portion, the percentage of total solids range from about 28 per cent for cockerels to about 45 per cent for hens. The high content of fat in the hens obviously contributes greatly to their higher total solids. Protein percentages are somewhat higher for the

Table 8.3.
Proportion of parts of chicken.[a]

Part	*White Plymouth Rock cockerels* 3-lb.[b] broiler/fryers	6-lb.[b] roasters	*White Leghorn hens* 4 lbs.[b]
Offals			
Feathers	6.4	8.1	5.8
Blood	4.1	4.2	3.5
Head	3.3	2.8	2.7
Shanks and feet	6.3	4.5	2.7
Subtotal	20.0	19.7	14.7
Viscera			
Heart	0.43	0.45	0.40
Liver	2.3	2.0	1.89
Kidneys	0.6	0.53	0.68
Pancreas	0.22	0.18	0.19
Spleen	0.21	0.16	0.13
Lungs	0.51	0.45	0.39
Testicles (or ovaries and oviduct)	0.02	0.26	2.52
Gastrointestinal tract (inc. gizzard)	10.0	7.9	9.72
Subtotal	14.3	12.0	15.9
Carcass			
Skin	7.0	7.6	7.69
Neck	3.7	3.4	2.34
Legs	21.5	22.2	18.1
Wings	6.7	5.9	5.72
Torso	23.4	26.4	32.7
Subtotal	62.2	65.6	66.6
Total bone in carcass	18.7	16.7	11.7
Total edible in flesh, skin, and fat (carcass only)	36.1	40.2	45.6
Total edible flesh, fat, skin, and giblets (heart, liver, gizzard, and kidney)	43.0	45.9	57.9

[a] Adapted from Mitchell, Card, and Hamilton (1926; 1931). All figures are percentages of live weight.
[b] Live weight.

Table 8.4.
Gross (proximate) chemical composition of total edible flesh, fat, and giblets (i.e., liver, heart, and gizzard) of chicken.[a]

Type and live weight	*Solids*	*Protein (N × 6.0)*	*Fat (Ether extract)*	*Mineral (Ash)*
Plymouth Rock cockerels				
3-lb. fryer/broilers	28.51	20.10	4.39	1.13
6-lb. roasters	28.51	20.22	4.46	1.40
Leghorn hens				
4 lb.	44.89	16.69	24.32	1.00

[a] Adapted from Mitchell, Card, and Hamilton (1926; 1931). All figures are percentages of live weight.

hens. The figures for turkeys are quite similar to those for the cockerels.

Detailed Composition

There are many different tissues and fluids in poultry carcass and the giblets, but here we should direct our attention to those portions which make up the important edible portions.

Muscle Skeletal muscle constitutes the major tissue of poultry that is used as food. Marsh (1970) aptly described skeletal muscle as "a machine for translating chemical energy into physical movement. . . . It is fitted with a supporting structure, an energy store, and a waste-exhaust system. . . . Finally, unique among machines, at the end of its working life it is both edible and nutritious."

The major chemical components of muscle are the myofibrillar proteins, myosin and actin, and two related proteins, troponin and tropomysin, all four of which, along with calcium, magnesium, and adenosine triphosphate (ATP), interact in a complex series of reactions to produce muscular activity and in a closely related post-mortem phenomenon called *rigor mortis*. The contraction and stiffening of the muscle after slaughter, and its resolution during the aging of meat, are directly associated with the initial toughening of meat following slaughter and the tenderization that follows.

Fat Some fat is found in the muscle tissues of poultry, but the great preponderance is located in the fat depots. Chicken and turkey fats are unique among animal fats in their high content (52 to 68 per cent) of unsaturated fatty acids. Chicken fat also contains appreciable quantities of tocopherol (vitamin E), which imparts stability to the fat (prevents oxidative rancidity) during processing and storage.

Pigments

There are two types of pigments of interest in poultry meat: the carotenoids, the same fat-soluble yellowish-orange pigments found in the yolk, and hemoglobin and myoglobin, the red pigments associated with muscle. The breast muscle of poultry is almost entirely free of pigment, thus causing it to appear essentially white. The red pigments change color during cooking, giving rise to the varying shades of pink to grey associated with the "rare" to "well-done" meat.

Nutrient Composition

Table 8.5 provides nutrient composition data on poultry, beef, pork, and lamb. These data show clearly the relatively high content of protein, as well as of two B-vitamins, riboflavin and niacin, in poultry muscle. Bird (undated) has reviewed studies on the nutritional value of chicken meat, and reports that there are significant amounts of thiamine present. He further reports that chicken liver is high in vitamin A, and also can be high in vitamin D, depending on the diet.

Energy content of poultry meat is low compared to that of red meat.

All these meats are good sources of the essential amino acids, although they are not as good as egg or milk.

The essential fatty-acid content (Mecchi *et al.*, 1956) of chicken and turkey are similar: linoleic acid, 13 to 20 per cent; linolenic acid, 0.7 to 1.3; and arachidonic acid, 0.2 to 0.7. These are substantial levels, and are somewhat surprising for animal fats. This, plus the relatively low level of saturated fatty acid in poultry fat, should make it attractive as a food, even for those suffering from heart disease.

Poultry Meat Quality

Pre- and post-slaughter inspections for poultry-meat quality are required in the United States for the production of safe, wholesome products. Carcass inspection is carried out by government veterinarians. Thus provision is made for the detection and condemnation of live animals and carcasses or parts which are considered to be dangerous or potentially dangerous to human health.

Attitudes About Quality

Besides being concerned with the wholesomeness and safety of poultry meat, consumer and trade attitudes about poultry-meat quality relate mainly to sensory properties. It is generally conceded that the following characteristics are important: carcass conformation, skin and fat color, flavor, tenderness, juiciness, and freedom from visual defects.

Tenderness is of particular concern, since consumers expect young birds to be tender when prepared by comparatively mild and rapid cooking methods. On the other hand, consumers understand that older birds (e.g., cull hens) must be cooked by using moist heat and require much more cooking time to produce a tender product.

Good conformation (plumpness of breast

Table 8.5.
Comparative nutrient composition of poultry and red meats.[a]

Type/portion	*Protein (per cent)*	*Energy (cal/100g)*	*Riboflavin (mg/100g)*	*Niacin (mg/100g)*
Chicken (roaster)				
Breast	31.5	138	0.3	10.5
Leg	25.4	168	0.6	5.0
Turkey				
Breast	34.2	204	0.4	10.5
Leg	30.4	227	0.88	4.1
Beef				
Round steak	27.0	233	0.22	5.5
Rump roast	21.0	378	0.15	3.1
Pork				
Loin chop	23.0	333	0.24	5.0
Lamb				
Rib chop	24.0	418	0.26	5.6

[a] Adapted from Scott (1956).

and compactness of carcass) is considered important for roaster-style chicken and turkey. It is also of some importance for frying chickens where the breast portion remains intact when displayed for purchase in the supermarket.

In the United States there is a strong preference for yellow skin and fat in chicken, whereas white skin and fat are preferred in the United Kingdom. On the other hand, consumers everywhere seem to prefer turkeys with a creamy white skin and fat.

Quality Measurement

Except for tenderness, quality in poultry is usually measured subjectively. Tenderness may be evaluated objectively, e.g., by measuring the force needed to shear the muscle. When properly carried out, shear force results correlate very well with sensory panel scores.

Skin and fat color is usually measured subjectively, using a set of arbitrary standards. Conformation is generally measured by viewing and handling the carcass, also comparing it with a set of arbitrary standards.

The odor of the ready-to-cook bird is probably the best indicator of incipient spoilage, although fishy and rancid-fat odors due to improper feeds are sometimes noted during or after cooking, especially in turkeys.

Safety of Poultry Meats

The veterinary inspection does not completely eliminate problems associated with eating poultry meat. The poultry industry, and incidentally also the red-meat industry, are continually faced with the problem of its products becoming contaminated to some degree with Salmonella and other hazardous organisms. Salmonella is especially a contaminant of egg products, where its presence is somewhat more serious than in poultry meat. Reliable and effective (but very tedious) methods for detecting and enumerating these organisms in poultry have been developed and are in use by some parts of the industry. Rigorous quality-control steps will have to be taken by the entire processing industry to keep this contamination to a minimum. In addition, it is necessary to impress on housewives and chefs the necessity for adequate cooking of poultry to destroy all such organisms in the meat before serving.

Biological and Environmental Factors Affecting Eggs and Poultry Meat as Food

Much work has been done to establish the biological and environmental bases for variability in the composition and quality of eggs and poultry meats; many factors have been shown to affect these traits: genetic background, nutrition, age, environmental factors, disease and medical treatment and general management procedures. These factors will be considered in Chapters 23 and 43.

FURTHER READINGS

Bearse, G. E. 1966. "Pesticide residues and poultry products: U.S.A. legislation and pertinent research." *World's Poultry Sci. J.*, 22:194–232.

Bird, H. R. (undated) "Nutritive value of eggs and poultry meat." *Nutrition Res. Bull. No. 5.* Chicago: Poultry and Egg National Board.

Brant, A. W., A. W. Otte, and K. H. Norris. 1951. "Recommended standards for scoring and measuring opened egg quality." *Food Technol.*, 9:356–361.

Briskey, E. J., R. G. Cassens, and B. B. Marsh, eds. 1970. *Physiology and biochemistry of muscle as food, Vol. II.* Madison: Univ. of Wisconsin Press.

Carter, T. C., ed. 1968a. *Egg Quality: A Study of the Hen's Egg*. Edinburgh: Oliver and Boyd.

Riemann, Hans, ed. 1970. *Food-borne Infections and Intoxications*. New York: Academic Press.

Romanoff, A. L., and A. J. Romanoff. 1949. *The Avian Egg*. New York: Wiley.

Stewart, G. F. 1967. "Fifty years of research in egg quality." *Commonwealth Sci. & Ind. Res. Org. Quarterly* (Australia), 27:73–82.

Nine

Contributions of Working Animals for Power, Herding, Protection, and Pleasure

A righteous man has regard for the life of his beast.
Proverbs XII:10.

Unlike the foregoing chapters of this section, which have dealt with products such as meat, milk, eggs, and wool, this chapter will focus on the use of animals based upon their ability to assist man in performance of work or in adding to his pleasure.

Technology has made tremendous strides in providing fast, convenient, and economic farm power and transportation. Extensive application of these inventions is admitted; however, it is still true that most of the power to do work on farms throughout the world is derived from humans* and livestock, nearly 90 per cent, in fact. Where cash is available and the economy permits, mechanical power takes over; otherwise, less sophisticated ways persist, and this holds for much of the food and fiber-producing areas of the world. Animals are used to carry man or freight, some to herd other animals. Some animals are primarily for the pleasure of man, as in games, racing, or showing, or even for their social value as pets or companions. In these various ways, animals work to contribute to mankind's welfare.

* So much of the farm produce in Nigeria is carried on the heads of men, women (Figure 9.1), and children that the crop statistics report "head loads"; hence, the inclusion here of the human along with animals as a significant source of power. In many parts of the world, people are accustomed to pulling or pushing carts or tricycles loaded with other people and/or produce. Humans draw water, cultivate fields, pound grain, and otherwise provide much physical power.

The Horse

"Wherever man has left his footprint in the long ascent from barbarism to civilization, we will find the footprint of the horse beside it," wrote John Trotwood Moore. The speed and agility of the horse have long been highly prized. The horse is found in a variety of climates, and he has been adapted to many needs of man, being highly trainable. There is anecdotal evidence that he can instinctively find his way home, sense danger, and detect the presence of water where it is difficult for man to find, but unfortunately well-controlled experiments to test the validity of this evidence have not been made. If properly managed, the horse demonstrates remarkable endurance.

Can we relate the endurance of the horse to his physical makeup? His feet are peculiar in that they are normally adapted to absorb terrific concussion shocks, especially at the canter and gallop. The bone structure, the cartilaginous plates, the angle of these parts to the ground, the tendinous plantar cushion with its sweat glands, the extremely vascular and nerve-filled laminae tissue, the hoof wall-secreting body situated above the laminae, and the hoof itself all go to make up the very important foot of the horse. This mechanism provides elasticity and toughness that is essential to the horse's usefulness.

The horse is found in cold, temperate, and subtropical climates. Those best adapted to cold environment have broad, blocky bodies, with the least surface area proportionate to the body weight (Figure 9.2). The Shetland pony of the Shetland Islands (60°N. latitude) is a good example of this conformation, with its heavy coat of hair, all making for efficiency in minimizing heat loss.

When the horse works, he tends to get hot, and somehow this heat has to be dissipated, else his body temperature (38.3°C or 102°F) will go up dangerously. When the horse becomes warm or hot, he sweats, and the evaporation of this perspiration cools the skin, and thus aids effectively in maintaining the animal's normal body temperature.

The hair coat affords a means of retaining body heat, or of slowing its escape. It is chiefly the air that is trapped in the hair that provides the insulation. Wind reduces the effectiveness of the hair in keeping the animal warm. As the length of day increases, shedding of the hair occurs, at latitudes far from the equator. Near the equator short hair prevails, and shedding and replacement go on most of the time. Thus, a minimum insulation or barrier to heat loss is assured, and at the same time some protection exists against solar radiation and environmental heat. Horses that are best adapted in the cold regions do shed their coats as spring advances, and later grow thick and long hair coats as the days shorten and weather becomes colder; the primary stimulus is photoperiodicity (changes in the length of day).

Figure 9-1 Women carry head loads of eggs. Fashola, Nigeria.

In Arabia and other hot parts of the world, the lightweight, fast horse has proven most suitable. His surface area is greater in proportion to his body volume, compared to those of the horse in the colder areas, thus improving the sweating process as a means of heat loss and body-temperature control, an essential factor in adaptation.

There are some limitations on the horse's adaptability. He has difficulty working at altitudes above 3,050 meters (10,000 feet). Too often at high altitudes the heart is overworked and becomes very big, and the animal eventually develops a critical condition. The Peruvian army maintained horses at different altitudes in the Andes Mountains, conditioned them to work at these levels, and thus avoided moving any one animal from one level to another. This made for increased efficiency in Peru's use of horse power. Reproduction was not possible at the higher levels (above 3,000 m). Horses cannot go long without drinking water, even though the vegetation they graze has a high water content. In this respect they differ from the sheep, camel, and llama.

Figure 9-2 A team of ten draft horses clearing land for farming in British Columbia before mechanization in the early part of the twentieth century. (Courtesy, Vancouver Public Library.)

Horses are expected to carry about one-quarter of their body weight, after being conditioned. In the Sudan, the pack horse of reputation has been the Kordofani because of his great endurance under hard conditions (Bennett *et al.*, 1948). He measures up to 140 cm. (14 hands).

Pleasure Horses

Many horses are used solely for pleasure. In a recent study in Lane County, Oregon, only 4 per cent of the horses in the county were being used as work horses on sheep and cattle ranches; the remainder were used for the most part in various recreational activities.

Of the sundry uses of horses for pleasure, riding is by far the most popular. The type of horse used for pleasure riding varies from ponies for children to Arabians, Quarter Horses, Thoroughbreds, Tennessee Walking Horses, American Saddle Horses, and crossbred animals carrying various degrees of these breeds. The type of horse preferred will depend to a large measure on the proficiency of the individual in horsemanship. Of prime concern to the vast majority is an animal with a gentle disposition, because most people involved in pleasure riding are not expert riders. Following disposition come features such as physical appearance, quality of gaits, and, for the sophisticated rider, style, speed, and a high-spirited disposition, in addition to other performance characteristics and conformation. Attractive and unique coat colors have added to the popularity of "coat-color breeds," such as the Palamino and Appaloosa.

Interest in pleasure horses has been stimulated by the 4-H horse programs under the supervision of the Agricultural Extension Service. These programs range from gaining primary knowledge of horsemanship up through breeding programs, the training of horses, and showmanship. Pleasure riding has many diversities. It may involve trail riding in urban, suburban, rural, or mountain areas; it may involve participation in polo or the fox chase; or it may involve competition riding in horse shows or rodeos.

Throughbreds and Standardbreds raised for running and harness racing, respectively, are an important facet of the pleasure-horse industry. Thoroughbreds which do not measure up for the race track may be used for pleasure riding, as jumpers, or as stock horses. Standardbred horses not adapted for the track are frequently used for competition in roadster classes at horse shows. In the early part of this century, many Standardbreds which did not qualify for harness racing were used as buggy horses.

Tests and measurements that might objectively evaluate horses used for pleasure purposes are scarce. Sound feet and legs are of paramount importance in horses regardless of their use. Equally important, irrespective of use, is sound wind (freedom from defects, such as roaring or heaves).

For certain cases, endurance is a highly valued trait. Dawson *et al.* (1950) have set standards for measuring endurance and other qualities in horses under saddle. For endurance tested on a 18.26 km (11.35 mile) course, they found that horses varied from year to year, stallions showed less fatigue then geldings, and mares were most susceptible to fatigue. Ease of riding was associated with length of stride, action at the walk, and temperament. Horses given the best scores for ease of handling at the walk, walked significantly faster than those given poor scores.

To many, there is satisfaction in caring for and handling horses, an element of companionship. In this sense, the horse may be said to have social value.

The Donkey

Hafez (1968) reports the world population of the donkey or ass as 42 million, with the largest numbers in Ethiopia, Mexico, Brazil, Turkey, Afghanistan, and Morocco. Tothill (1948) maintains that "the donkey is the transport animal of choice in the Sudan." It "has astounding strength and endurance." The long ears, relatively small body, long legs, and short hair all fit the donkey to hot climates. He is suited to cold climates, too, for his hair coat can be long and protective. The donkey browses from trees and bushes. He gives the impression of moving with the least expended energy, so smooth is his gait.

The donkey is extensively used as a pack animal. Government regulations in the Sudan limit the pack load to 45–67 kg (100–150 pounds), but many nongovernment donkeys carry 91–135 kg (200–300 pounds); their height is 90–110 cm, 9–11 hands at withers.

The donkey has been long used in mines for draft and pack, and has served long and well, carrying loads that otherwise would have moved on the backs or heads of men and women. Mechanization has displaced the donkey in the mines of the United States and Canada.

The riding donkey, "a well-nigh tireless mount" (Tothill, 1948), is a familiar scene in Sudan and other parts of the world. He is usually taller than the pack animal by a hand or two (10–20 cm.) and may walk straighter and faster. He is seen often loaded with canvas bags, water skins, or milk cans, with the rider nestled atop or amongst such freight. Although the donkey is not credited with long journeys commercially, I have seen many a camel train led by a man mounted on a donkey in Turkey.

The Mule

The mule has been maligned as stupid and falsely accused as obstinate when, in fact, he has merely sought to exercise judgment and limit himself to his capabilities. Unlike the horse, the mule can be self-fed with safety. Many a mule has been turned loose after his day's work into a corral with hay, concentrate, and water free-choice, refreshed himself, rested, and been fit for the next day's work, whereas horses typically overeat or overdrink, and suffer severely thereby (colic, founder). Hence a good reputation has built up for this offspring of the mare (horse) and the jack (male donkey or ass).

This beast of burden is found in many parts of the world where the jeep and other power transport have not displaced him. He can withstand severe stress and still keep going. In addition to packing farm products, he finds use in mines, forests, and sundry other places.

There are some 15 million mules in the world today, mostly in Spain, Ethiopia, and the Near East. There are some working on the farms of North America and in the forests. They have long found favor in the southern United States.

The mule can grow a thick coat of hair and adapt himself to the colder climates, but his forte is more often in the hot areas of the world, for he is heat-tolerant. The shape and texture of his feet minimize lameness. His speed and nimbleness gain him preference over cattle in mountain terrain.

Some traits lending to heat-tolerance are dominant, and thus the mule inherits these traits from the donkey. The hinny is also heat-tolerant. The hinny is smaller than the mule, and therefore is often the pack animal of choice when a man on foot must lead it, because its size and length of stride are more compatible with the man's.

Mules used for draft purposes measure 130 cm (13 hands) or more in height at the withers and weigh 340 kg (750 pounds) or more. They fit their task uniquely for a variety of reasons, some of which have been alluded to in the opening of this section. Inexpert drivers can handle mules more readily than horses. Mules are less likely to be injured or tangled up than horses.

Figure 9-3 The mule pack train in the Yukon. Photograph by Adams and Larkin. (Courtesy, Vancouver Public Library.)

Figure 9-4 European-type oxen drawing a steel plow in Mapuc (Photograph by Professor Wm. Furtick.)

Pack mules (Figure 9.3) are often smaller than draft mules, but may stand from 120–160 cm (12–16 hands) and weigh as little as 270 kgs (600 lbs.).

The Ox

For ages the ox has drawn the plow (Figure 9.4), tramped out the grain on the threshing floor, drawn man's carts and wagons, dragged logs (Figure 9.5), and in other ways done his bit, even to affording a carcass of meat after a life's work, and in the case of the cows (which are usually worked less than bulls and bullocks) providing some milk. The ox can pull about the same load as a horse of his weight, but walks only two-thirds as fast.

The European-type cattle (*Bos taurus*) are found in great numbers in Europe, the Americas, Australia, New Zealand, Turkey, and Russia. Humped cattle (*Bos indicus*), referred to as Indian cattle or Zebus, indigenous to India, Pakistan, and much of southeast Asia, are found generally in hot regions.

Throughout the tropical areas of the world, the various kinds of humped cattle, because of their capacity to withstand heat, are used for drawing farm implements, turning the water wheels for irrigation, and transporting produce on the farms and to the market places.

The Yak (*Bos grunniens*)

What moves in Tibet and Mongolia does so on the backs of yaks (Figure 9.6). Here in these high rugged ranges and plateaus, these heavy-bodied, long-thick-haired (with fine under-coat), sure-footed animals are the mainstay of the people. The yak is probably the closest living relative of the American bison. Yaks do admirably well at altitudes well above 3,100 m (10,000 feet) and often prefer to range at 5,180 m (17,000 feet). They are ridden and used as pack animals, and also are a source of meat, milk, and leather (Phillips, 1948).

Yaks stand 109 cm, some higher, and weigh 180–235 kg (400–520 lbs.) depending on the area. Hybrids (yaks crossed with cattle) weigh 100 kg. more than the parent yak, are well-liked for their higher milk and meat production, but show less stamina as work animals. The feet of the yak are especially

Figure 9-5 A 14-ox team (European-type) pulls logs, clearing land for farming in western Canada. (Courtesy, Vancouver Public Library.)

Figure 9-6 A yak train in Tibet. (Courtesy, Ilia A. Tolstoy.)

adapted to its task in the difficult terrain. The big hoof is somewhat cup-shaped, is very hard, and wears better than that of the horse or mule in mountain gravel, in marshes, and for plowing through snow.

Their gait is faster than that of most oxen. They walk about 2.4–3.2 km/hr. (1.5–2 miles/hr.). They regularly carry 55–73 kg (120–160 lbs.) as packs, for 13–16 km/day (8–10 miles/day) grazing as they go and without losing weight. Their short legs may sink in the marsh, but they propel the body as it slithers over the soggy surface, and on they go (horses would become excited and be lost).

The yak's testicles seem small, and are held close to the body in a small, tight scrotum. This is not surprising, since we recognize the thermoregulating function of the scrotum, and in the cold weather and thin air its job is to keep those testicles just cool enough to allow spermatogenesis and warm enough to be safe from freezing.

The Water Buffalo (*Bos bubalus bubalis*)

God's ugliest creature! But what a worker! The water buffalo finds its place in the flooded rice paddies, plowing and cultivating (Figure 9.7), and in the forest. He plods along no faster than 3.2 km/hr. (2 miles/hr.), and can keep this up for ten hours a day, provided he is working in the shade or, if working in the sunshine, is allowed to bathe every two or three hours; a mud bath will do, or he can be hosed down with water.

The water buffalo's unique characteristics that suit him to his work are his big feet, much larger than those of cattle; thus he does not mire down. His powerful legs and bulky body seem adapted to the muddy fields. His slow pace allows the driver to keep up with him although the going is hard. A pair can pull two tons on a good road, but even here the pace is slow; hence the water buffalo is not used extensively for drawing wagons. Ox

Figure 9-7 Water buffalos pull the harrow in the flooded rice paddy in West Java, Indonesia. (Photograph by Dr. Budi S. Nara.)

neck yokes fit well, especially on castrated males.

The water buffalo is semiaquatic in the wild state, a strong swimmer. He has 48 chromosomes (domestic cattle have 60). His digestive system handles coarse fodder better than cattle. Yeates (1965) reports this draft animal has only 1/6 the number of sweat glands found in European cattle, and describes the skin as physiologically inert, pigmented, and with very little hair; thus the necessity of frequently wetting the skin in wallows or by other means, especially where there is lack of shade.

Water buffalos (some 75–80 million) are used in the tropics and subtropics of Asia, Trinidad, the Philippines, the Mediterranean basin excluding France, northern Australia, and Indonesia. Often in Indonesia, people invest their funds in water buffalos, using them as banks, as it were.

The Deer

The Reindeer (*Rangifer tarandus*)

The reindeer of the cold Arctic and sub-arctic regions is easily broken to the halter and to work, whether to pack in summer or to pull a sled in winter. He can carry 27–32 kg (60–70 lbs.) on his back or 45–135 kg (100–300 lbs.) on a sled, and can travel 32–40 km/day (20–25 mi./day). In height he is 108–113 cm (42–44 in.), and from nose to tail measures 1.7–1.8 m (67–70 in.). At three years of age, the males average 118–136 kg (260–300 lbs.), the females some 36 kg (80 lbs.) lighter (Phillips, 1948).

Adaptation to the severe cold is recognized by the toleration of 9°C temperature in the lower legs in −30°C weather. This is accomplished by a special blood circulation in these distal parts. The reindeer hoof surface is expanded to prevent sinking in the snow. The dense, compact coat and subcutaneous fat help this cold-tolerant animal.

The Moose (*Alces alces*)

The moose (Figure 9.8) is a work animal, and the U.S.S.R. has set up experiment stations to improve on it as a draft, pack, riding and milk- and meat-producing member of the livestock world (Heptner and Nasimowitsch, 1967). In most places outside the United States, the moose is known as an elk.

To help this largest member of the deer family survive the rugged cold, we find a compact body and sturdy limbs affording the least surface area proportionate to his body weight, 680–820 kg (1,500–1,600 lbs.). He has a flabby and pendulous muzzle that overhangs the lower jaw by 7–10 cm. and probably facilitates browsing on twigs and moss. The thick coat is designed to entrap air with its coarse hollow hair. Like those of the reindeer, the legs and muzzle have special blood circulation that help tolerate the subzero (−50°C) cold.

Figure 9-8 A moose ("elk" in Europe) pulls the travois in northern Canada. (Courtesy, Vancouver Public Library.)

The Camel

It is a picturesque sight to watch a camel caravan sauntering over the desert sands. Mechanized transport is replacing the camel, but nonperishable freight, such as minerals, salt, and wool, is still moved great distances in the Near East and in central Asia by this animal (Figure 9.9).

Two camel species are common:

1. *Camelus bactrianus* of central Asia, eastward from Iraq, two-humped, with its thick winter coat (shed in the spring), and surefooted.

2. *Camelus dromedarius* of Arabia and the Near East, single-humped, suited to the hot desert climates, but not to areas where the ground may be wet because here it slips and falls, especially when under load.

George McLeroy in a private communication tells of his many conversations with camel breeders and drivers in Iraq. In olden days as many as 5,000 camels would make up a transport caravan moving at 4.8 km/hr. (3 miles/hr.). Riding camels travel at 8 km/hr. (5 m.p.h.), and can maintain this pace 10–11 hours a day for 12 or more days. Camel herders were encountered in the desert who remain away from water 5–7 days in a stretch. The average load of a mature camel is 200–300 kg, but varies with the season, distance, and nature of the load. The ordinary saddle in Iraq is straw padded with a central hole to fit over the hump; a wooden fork anchors the load equally divided to either side; a girth may or may not be used. The 1.5–1.8 meter stride of the walking camel is facilitated by the long legs and loose attachment to the thighs. Almost the entire length of the digital bones lies flat on the ground. The hoof is reduced to nail-like structures, and the whole yielding foot with its silent tread is admirably designed to support the animal on the shifting desert sands, or snow of the north country.

The camel's rumen has a number of small cavities or diverticula which are guarded by sphincter muscles, supposed to contribute to water storage. While the working camel can endure without water for long intervals, it must have feed each day. The upper lip is cleft, the mouth lining is tough, and the camel consumes all manner of thorny plants. While the camel does not sweat visibly, under extended exertion the hairless parts of the body may show signs of copious sweating. The respiration rate is 10–12 per minute during the cooler part of the day, 20–24/min. in the heat. The nostrils of the camel can be closed like eyelids. Camels are gifted with keen eyesight and a well-developed sense of smell. Coat colors vary from all white to dark brown or even black, but yellow is the most popular.

The hump consists mainly of fat with connective tissue (no vertebral spine protrudes into it), and varies in size, but may be 20 per cent of the body weight. This fatty tissue, when metabolised, yields an amount of water greater than the original mass of fat.

The camel excels in water conservation, low use, and low water loss by the kidneys and the gastrointestinal tract. Much water is reabsorbed into the blood from the colon.

In addition to its work as an overland carrier of freight and passengers, the camel pulls plows and operates irrigation pumps, besides providing milk and meat for local consumption.

The Llama (*Lama glama glama*) and Alpaca (*Lama glama pacos*)

The llama, guanaco, alpaca, and vicuna are part of the camel family, here listed in de-

Figure 9-9 The camel caravan brings figs to Izmir, Turkey. The "driver" leads on his donkey.

Figure 9-10 Llamas with packs on the Alto Plano in Bolivia. (Photograph by Professor Warren C. Foote.)

creasing order of size. The llama is 1.2–1.4 m (4–4.5 feet) at the shoulders and averages in weight about 115 kg (250 lbs.). The guanaco females that I saw were larger than llamas, 1.5 m (5 feet); these were on Tierra del Fuego (Chile) on fairly good pasture, which may account for this extra size over what is reported in the literature.

The llama has long been used as a pack animal carrying 35 kg, seldom as much as 50 kg. When the burden becomes uncomfortably heavy, the llama simply lies down and refuses to get up until relieved. When it is irritated or molested, it spits out a regurgitated watery mass with startling force and sure aim. Llamas are commonly used in herds or trains, and travel 20 km a day (Figure 9.10).

Male alpacas have been used as pack animals, but their load limits are less, so that wherever llamas have been available, the alpacas have been relieved of this task and left to their fleece bearing, which commercially is their forte. White alpaca wool is now in great demand and limited supply.

The other two members of this camel family have remained in the wild. All have ranged in the high Andes, where 3,600 m is low for the llamas and usually the alpacas range above 4,200–5,000 m. The vicunas are found above the alpacas and the guanacos at any of these levels, and even to the sea.

Several features about the llama fit it for work at extreme altitudes. The blood has extra oxygen-carrying capacity because the number of red blood cells is unusually high. Only limited amounts of oxygen are available at high elevations, where the atmosphere is rarified. The big hoof is cleft and has two hardened spurs that must be useful in the rocky, mountainous paths that are negotiated. The fleece is coarse, but affords a thick cover in the cold hours of the day, yet suits the warm hours too. Most llamas live and work in the high elevations near the equator, so the seasonal changes are largely limited to the 24-hour cycle. These are from below freezing to 15–25°C. The llama can go for days without drinking water, because it gets moisture from a wide variety of plants, including moss, lichens, and the hardy Andean grass "uchu" (*Stypa ichu*). Its endurance is remarkable considering the extreme altitude, the scarcity of feed, and the daily range of temperature changes.

Figure 9-11 Elephants in teak logging operation in northern Thailand (Courtesy, FAO.)

Figure 9-12 Dog team with sled and load in the Klondike. (Courtesy, Vancouver Public Library.)

The Elephant

The elephant possesses great strength and intelligence. Those in India, Burma, and Thailand, used largely for moving logs in the forests (Figure 9.11) and for transport on farms, are 2.4–3 m (8–10 feet) in height. Different strains are found in Africa, somewhat taller (3.3 m) and weighing 6–8 tons.

Because of the huge amount of feed needed to maintain the elephant and his delicate constitution, his usefulness has definite limits. The elephant attempts to avoid the bright sun. When he is urged to exert himself to his upper limit, he lets off a high-pitched squeal.

Because of the difficulty of reproducing in captivity, elephants are usually allowed to run wild. Individuals are selected from the herd, tamed, and trained to work. Typically, an elephant works under the command of only one driver.

The Dog

Many breeds of dogs have been developed by selecting for specific traits which fit them to perform useful tasks, such as herding or hunting. Others have been developed by selecting primarily for traits which make them suitable as companions to man.

Dogs have long been employed by Eskimos for drawing sleds in the arctic region (Figure 9.12). Dog teams travel up to 800 km (500 miles) in 10 days. They work in the tundra of the Arctic and in Antarctic explorations. Rescue dogs are on duty in the European Alps and Newfoundland in Canada. Police and army dogs are trained to protect and to attack on command.

After long training, dogs serve the blind as guides. There are 420,000 blind people in the United States, with 35,000 being added annually. The State of California is the first in the United States to have laws governing the operation and licensing of a guide-dog school. This institution trains dogs to guide and also the blind people in the use of the dog. Golden Retrievers, Labrador Retrievers, and German Shepherds are used there.

On the farm or the range, dogs are used to herd livestock. A dog that excels here is an invaluable aid, especially in herding sheep on the range (Figure 9.13).

The dog's capacity for smell and sight distinguishes him in tracking humans, game, and farm animals. The trait of pointing when in close proximity to game is a unique characteristic of some breeds of dogs. Retrieving of game is also a trait in which certain breeds excel.

Dogs come in great variety; so it is important to select those that can be expected to

Figure 9-13 Lady, Border Collie, five times champion of the Oregon State Dog Trials and owner, R. V. Hogg, work the sheep. (Photograph by R. W. Hogg and Sons.)

possess the traits most needed for the goal to be accomplished, in terms of whether it is at high altitudes, in the cold or the tropics, in the great outdoors, or in the shelter of the house or kennel, and whether he is to satisfy the commands of one master or of a group.

FURTHER READINGS

Brody, Samuel. 1945. *Bioenergetics and Growth.* New York: Reinhold.

Brody, Samuel. 1956. "Climatic physiology of cattle." *J. Dairy Sci.,* 39 (no. 6), 715–725.

Campbell, John R., and John F. Lasley. 1969. *The Science of Animals that Serve Mankind.* New York: McGraw-Hill.

Leeds, A., and A. P. Vayda, eds. 1965. *Man, Culture, and Animals in Human Ecological Adjustments.* Washington, D.C.: AAAS.

Mason, I. L., and J. P. Maule. 1960. *The Indigenous Livestock of Eastern and Southern Africa.* Edinburgh: Commonwealth Bureau of Animal Breeding and Genetics.

Rhoad, A. O., ed. 1955. *Breeding Beef Cattle for Unfavorable Environments.* Austin: University of Texas Press.

Stroock, S. I. 1937. *Llamas and Llamaland.* New York: Stroock.

Williamson, G., and W. J. A. Payne. 1965. *An Introduction to Animal Husbandry in the Tropics.* London: Longmans, Green.

Section Three

Description of Animal Species, Breeds, Strains, and Hybrids and their Distribution

A Sherpa woman milking a yak in summer pasture in Nepal. (FAO photo.)

Ten

Dairy Cattle and Other Animals Used for Dairy Purposes

The unique composition of milk is associated with its unique nutritional properties. Milk was evolved especially for the nutrition of young mammals and has a matchless combination of protein of high biologic value, of calcium, phosphorus, and vitamins. It is poor, however, in iron, copper, manganese, and apparently magnesium. Milk happens to supplement almost perfectly whole-cereal diets for adult humans.

Samuel Brody,
Bioenergetics and Growth

Dairy animals will continue to occupy an important position in the world's economy of food production because of the excellent and unique nutritional properties of milk, the high efficiency of the mammary gland, and the high gross efficiency of improved dairy animals, and because most dairy animals are ruminants and as such are capable of relatively high production utilizing forage and other food sources which man cannot use directly. Milk and dairy products also are one of the most economical sources of food energy and protein.

From the standpoint of milk yield per animal, dairy cattle, dairy buffalo, goats, and to a lesser extent sheep have been selected and bred specifically for milk yield. The recorded production of individual dairy animals and herds is indicative of the immense food-production potential of dairy animals. For example, the 1970 Annual Report of the Stanislaus County Dairy Herd Improvement Association of California lists 69 herds of 200 cows or more. Of these herds, 22, ranging in size from 215 to 679 cows each, had production averages of over 15,000 lb. of milk per cow. The top herd, with 402 cows, average 19,511 lb. of milk and 697 lb. of milk fat.

The above herds are almost exclusively composed of large Holstein cows. Under past and present economic conditions in most dairy areas of the world, the high total production per animal is emphasized. This is most readily met by the larger breeds of dairy cattle. To produce 1,000 lb. of FCM (fat-corrected milk, or, more precisely, milk of constant energy content), 26 large dairy cows or 200 small dairy goats would be required if all animals were of equal gross energetic efficiency. They would use the same amount of feed but the overhead, especially the labor, would be greater for the smaller animals (see Chapter 20). Under special conditions of environment and labor costs, or for other reasons such as milk composition, flavor, and allergy problems, dairy animals such as the goat and sheep play important roles in dairy production.

The Production of the World's Milk Supply

Commercial Dairy Animals and Their Distribution

Data gathered by the FAO (1969) indicate that most of the world's milk is produced by cattle. Table 10.1 presents recent estimates for regions of the world and a few of the countries contributing large portions to the supply. This summary indicates that over 90 per cent of the milk is produced by cattle. It would be difficult to further divide this into milk of European and Zebu breeds of cattle or into milk produced by specialized dairy, dual, or triple-purpose breeds. However, examination of the areas listed would indicate the largest part of the cows' milk is produced by dairy or dual dairy-beef breeds of European origin. Breeds of dairy buffalo (*Bubalus bubalis*) are important dairy animals of Asia, accounting for nearly 40 per cent of the milk production of that region of the world. Large quantities of both goat and ewes' milk, used in certain areas throughout the world, escape formal production reports. Table 10.7 shows that both goats and sheep are very important dairy animals, particularly in Asia, Europe, and Africa.

Dairy Cattle

Dairy cattle are widely distributed throughout the world from the subarctic regions of North America, Europe, and Asia to the tropics. With special attention to management, they are productive at these extremes, yet they are most adapted to the temperate zones of the world, and it is in these regions that dairy cows produce over 90 per cent of the world's milk supply (Table 10.1). Dairy cattle are classed in the genus *Bos*, and most authorities further divide them into two species or types: European cattle (*Bos taurus*), and the zebu (*Bos indicus*), the humped Indian cattle. These two species are perfectly interfertile. Mason (1969) lists some 274 major breeds of cattle, with about 44 of these being European cattle primarily for dairy purposes and five Zebu breeds

Table 10.1.
World production of milk.[a]

Region	*Cow*	*Buffalo*	*Sheep*	*Goat*	*Total*
North America	61,193	—	—	—	61,193
Central America	5,726	—	—	199	5,925
South America	18,030	—	22	146	18,198
Europe	146,185	82	2,736	1,608	150,611
U.S.S.R.	82,100[b]	—	100[b]	700[b]	82,900
Asia	26,365	18,971	2,756	2,664	50,756
China	3,200[b]	—	—	—	3,200
Africa	9,504	990	640	1,374	12,508
Oceania	14,077	—	—	—	14,077
World total	366,380	20,043	6,254	6,691	399,368
Per cent of total	91.7	5.0	1.6	1.7	100.0

[a] FAO, 1970. All figures are in units of 1,000 metric tons.
[b] Estimated.

primarily for dairy. The vast majority of breeds of cattle are classed by uses as some combination of milk, meat, and draft.

Bos Taurus as a Dairy Animal

United States dairy cattle Some important trends in the U.S. dairy industry are illustrated in Figure 10.1. Milk cow numbers have decreased steadily, yet because of increasing production per cow, total milk production has declined only slightly. Constant improvement in breeding, feeding, milking, and other management factors accounts for a large part of the upward trend in production per cow.

Dairy cattle breeds of the U.S. Most of the milk of the United States is produced by five breeds of cattle, Holstein, Brown Swiss, Ayrshire, Guernsey, and Jersey.

Cattle that eventually formed the base for the American dairy breeds were imported from 1812 to 1869. The Channel Island breeds have had small importations, more or less continually to date. Disease-control restrictions have limited importation from continental Europe since 1906. Breed associations were formed in the U.S. between 1868 and 1912. Details of breed history, importation records, productive characteristics, breed societies, herdbooks, and improvement programs may be found in the publications of Briggs (1969), Saunders (1925), and Cole (1966). Some major physical characteristics of our dairy breeds are presented in Table 10.2.

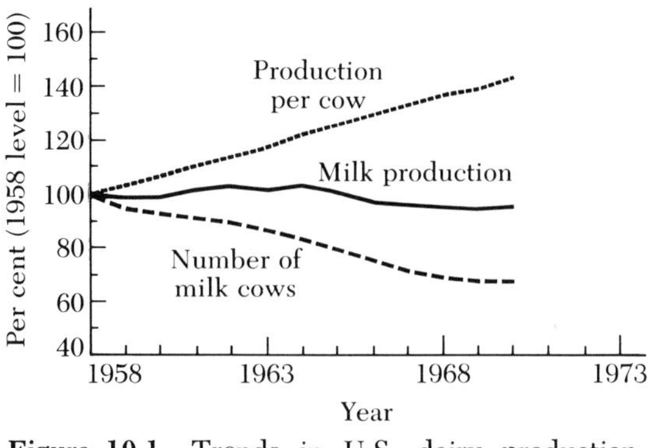

Figure 10-1 Trends in U.S. dairy production. (From USDA, 1970b.)

Livestock breed associations are formed for the purpose of guarding the "purity" of a breed and of promoting and improving the breed. The guarding of "purity" reverts to registry systems which assure that, after some initial period, all animals to qualify as purebreds of the breed are from registered parents. Open registry of dairy cattle will be discussed below. Breed associations depend largely on registration and transfer fees to exist and carry out their programs. Breed improvement is recognized as a vital function, but it cannot take the form of highly selective registration without seriously reducing an association's income and hence its programs. Thus while breed associations are sometimes criticized for not imposing severe qualifications for registry in order to insure that purebred stock remains distinctly superior to grade or non-registered stock, this is not always feasible. In recognition of this problem the Holstein-Friesian Association of America initiated the concept of Advanced Registry (AR) shortly after it was founded in 1885. While any purebred cow could be registered by the association, only cows meeting certain production requirements could be entered in Advanced Registry. This plan, to assure breed improvement through selection and breeding, was adopted with appropriate modifications by each of the associations. With the establishment of type-classification programs, similar plans to promote improvement of body form were initiated.

DHIA testing was initiated by the USDA in 1905 to include all cows in a herd, both grades and purebreds (see Chapter 20). Between 1925 and 1932 the breed associations established herd testing in their Herd Improvement Registry (HIR) programs. The DHIA and HIR programs had much in common, and by 1965 the associations modified their HIR testing and accepted, with appropriate supervision, DHIA records of purebred herds in what is termed the DHIR program. The AR program of testing selected individuals has now been dropped by all the breed associations, but in pedigrees one will see reference to HIR and occasionally AR records.

Dairy type-evaluation programs have been

Table 10.2.
Origin and characteristics of the major breeds of American dairy cattle.

	Ayrshire	*Brown Swiss*	*Guernsey*	*Holstein*	*Jersey*	*Milking Shorthorn*
Country of origin	Scotland	Switzerland	Island of Guernsey	Holland	Island of Jersey	England
Wt. of mature cow (lbs.)	1,200	1,400	1,100	1,500	1,000	1,500
Wt. of mature bull (lbs.)	1,850	2,000	1,700	2,200	1,500	2,200
Av. gestation (days)	277.9	289.7	283.9	278.8	278.8	—
Desirable color	Red, brown, mahogany, with variable areas of white.	Light to dark brown, udder often white.	Fawn, with white markings, yellow skin.	Black and white, and red and white.	Light gray to dark fawn, solid color or broken with white.	Red, roan, and white

an important function of dairy breed associations. Score cards for body form predate breed associations in several instances. Jersey breeders on the Island of Jersey adopted a score card for visual appraisal of that breed in 1834. Cattle judging and show-ring competition have played an important role in the development of our present cattle. In the United States the Purebred Dairy Cattle Association, composed of the five major breed associations, adopted a Unified Score Card for dairy cows and bulls in 1943 and revised this card in 1957. This has been the basic guide in dairy-type evaluation in this country. Type-classification programs were initiated by each of the breed associations: Holstein, 1928; Jersey, 1932; Ayrshire, 1941; Brown Swiss, 1942; Milking Shorthorn, 1943; and Guernsey, 1946. In these programs official evaluators classify breeders' animals using the unified score card or a modification of it. The programs are voluntary and supported by the breeders utilizing them. The objective has been to improve the physical strength, beauty, and longevity of dairy cattle. While often criticized for undue attention to minor points of physical form, these programs have set standards and provided data for scientific analysis by which future programs may be improved. The Holstein breeders initiated, in 1971, the provision that animals rescored up to five years of age will be scored without regard to previous scores. Scores can be raised, be lowered, or remain unchanged. This may lend important strength to a program that has been criticized as not fully recognizing animals that score well while relatively young but fail to "wear well" with increasing age. Breed improvement for type and milk production progresses with a growing volume of production testing data and a limited amount of body conformation data. However, most studies have shown a positive, though very low, genetic correlation between over-all type rating and yield. Details of the programs implemented by the breed associations have been recently reviewed by Porter *et al.* (1964). These are frequently updated and improved, and current information is best obtained from these associations.

With the practical elimination of any distinct gap between registered and improved grades of the breeds as mentioned above, the question of "opening" the herdbooks to admit certain outstanding grade animals has been debated for some time by most of the breed associations. Currently there are four well-regulated open-herdbook plans in operation. The objective is to recover outstanding unregistered females for the registered population. It is an opportunity to introduce new genes and gene combinations for production and type into the population without changing the major breed characteristics. These programs were adopted in 1969 by the Brown Swiss breeders and in 1970 by the Guernsey and Ayrshire breeders. The first phase in the Brown Swiss program is the identification of a desirable female, sired by a registered bull. If this animal is certified as having the type and color characteristics of the breed she is permanently identified; her female offspring, by registered sires, are then eligible for registry in an Identity Enrollment herdbook. The second phase consists of performance certification. If the identified female produces above breed average, and scores at least 82.5 in type, the identity-enrolled offspring become eligible for registry in the regular herd book. The Ayrshire and Guernsey programs are similar.

Holstein breeders, while having a much larger population of cattle, have recognized that their registered population may have lost valuable genetic material due to restrictive color requirements. The gene for red coat color exists in the breed at a low gene frequency. By 1968 there were over 1,000 red and white Holsteins registered in a new association for Red and White Cattle which had been formed in 1964. The color subject was extensively debated, and in 1969 the Holstein-Friesian Association formed a separate herd book for their red and white cattle and provided for the registration of "off color" black and white animals. In 1971 the association moved to accept red and white as well as black and white animals in the same herd book, simply designating red animals as such with their registration number.

Each of the dairy breeds has characteristic milk composition traits, but individual cows overlap breed ranges in most components. One composition trait that seems quite fixed in the Channel Island breeds, Jersey and Guernsey, is the form of the vitamin-A activity of the milk. In these breeds larger amounts of carotene are found, and the milk is characteristically more yellow in color than that of the Holstein, Ayrshire, Brown Swiss, or Milking Shorthorn. The vitamin-A value of the milk of these latter breeds may be equally high when on the same feed, but its form does not lend the yellow color to the milk fat. The Jersey and Guernsey breeds produce milk that averages distinctly higher in percentage of milk fat, protein, and solids-not-fat than the other breeds, the Ayrshire and Brown Swiss are intermediate, and the Holstein, as a breed, has the lowest percentage composition (Table 6.1). There is wide variability within the breeds between cows in milk composition.

Summaries of the National DHI cow testing program provide one of the good sources of information on U.S. dairying and dairy breeds. This material is released several times yearly from the USDA Agricultural Research Service in the form of DHI Letters. The present production status of cows on official DHI test that are identified by breed (grade as well as registered) is shown in Table 10.3. The righthand column of this table gives the records converted to a constant-energy basis. This permits a comparison of the total food energy that is represented in the milks of varying milk-fat content. Note that differences between breeds in energy production are less than weight of milk or of milk fat alone.

Figures 10.2 through 10.6 present selected individual cows of our major breeds that illustrate the genetic potential for high yield, strong, well-proportioned body conformation, and long productive life span. American dairymen have shown outstanding success in their application of science and husbandry skills in improving the productivity of their cattle.

Other parts of the world There are other parts of the world that have advanced dairy industries as well. From the standpoint of estimated average milk yield per cow per year, the ten leading countries listed in FAO (1970)

Table 10.3.
Lactation averages of cows coded by breed in official dairy-herd-improvement testing.[a]

		305-day, 2X, ME basis			
Breed	*No. of records*	*Milk (lb.)*	*Fat (%)*	*Fat (lb.)*	*FCM (lb.)*
Ayrshire	23,803	11,219	3.90	437	11,043
Brown Swiss	28,768	12,232	4.02	491	12,258
Guernsey	112,146	9,712	4.63	450	10,635
Holstein	1,387,737	14,032	3.61	506	13,203
Jersey	106,976	8,914	5.00	445	10,241
Milking Shorthorn	2,159	10,066	3.71	374	9,636

[a] From USDA (1971a).

are: Israel (11,000 lbs.) followed by Japan, the Netherlands, and the United States (9,000 lbs.), with the United Kingdom, Korean Republic, Sweden, West Germany, Switzerland, and Taiwan (7,000 lbs.) following in order. European breeds of dairy cattle, principally the Friesian (the Holstein of the U.S.), account for most of the milk of these dairy countries.

The breeds used for commercial milk production in England and Wales are much the same as those of the United States with the exception of the Brown Swiss.

Reports indicate that the Friesian, and what are probably related black and white breeds of various countries and regions, are among the very important dairy breeds of Europe. Other breeds that account for a high percentage of the dairy cattle of individual countries include: Red Danish, Jersey, Finncattle, Norwegion Red, Ayrshire, Swedish Red and White, Red and White Lowland, Simmental, Normandy, and Brown Swiss. The total number of breeds of cattle is very large. Many of these breeds are described in some detail by French *et al.* (1966).

Bos Indicus as a Dairy Animal

Cattle of the *Bos Indicus* species are known as Zebu or Indian cattle. They are characterized by a very marked hump over the top of the neck and withers, and in most instances by loose, excess skin, particularly at the throat and dewlap in contrast to European breeds. These cattle originated in India, Pakistan, and nearby regions, and have been of great importance to the economy of these countries for centuries. In recent decades dairy-breeding stations in Pakistan and India have developed high milk-producing strains of such breeds as the Red Sindhi, Sahiwal, and Tharparkar. The available feeds for cattle and dairy-feeding practices of the regions of the world where the Zebu breeds have developed have been generally quite restricted in comparison to European and American cattle feeding. The studies of Joshi and Phillips (1953) and of Mason (1969) show that draft qualities have been a strong consideration in selection and development of most of the zebu breeds of cattle.

Zebu cattle breeds Mason (1969) lists over 65 breeds of Zebu cattle, and draft is coded as the primary use of some 45 of these. Dairy is the primary use of some five breeds and the secondary use, after draft, of a fairly large number.

The Zebu breeds most frequently reported in dairy studies of the Indian scientific journals and in Dairy Science Abstracts are the Sahiwal, Red Sindhi, Tharparkar, Gir, and Hariana. Zebu dairy cattle are relatively late maturing, but it is not clear what effect higher nutrient intake during the growth period would have on this trait. There are numerous studies on gestation length of the Zebu breeds, and these suggest a range of between 284 to 287 days comparable to the various European breeds.

Production potential Compared to European dairy cattle, the Zebu breeds are often mentioned as having problems with short lactations, long dry periods, and long intervals between calving. Because of elevated temperatures and restricted feed, high-producing European breeds often fail where Zebu breeds make a reasonable production record. However, some of the inherent differences in productivity are illustrated by a recent report summarized in Table 10.4. These cattle were under the same feeding and management at a station in a desert environment. The authors noted that the Friesian crosses survived and produced without difficulty, whereas straight Friesians had failed to do well.

Zebu type cattle of Africa Numerous additional breeds of Zebu-type cattle are found in Africa. Joshi *et al.* (1967) describe some 38 breeds or types of African cattle, most of which possess Zebu characteristics and are of Indian or indigenous Zebu origin. Many of the breeds are used for milk and meat. Their draft use is also important. Osman (1966) points out that individual cows of the Kenana and Butana breeds of Northern Sudan compare

Figure 10-2 Ayrshire cow, Crusader's Joyce of Windy Top, pictured at 19 years of age. At age 21, she had dropped 15 living calves, and produced 206,888 lbs. of milk and 8,725 lbs. of milkfat. (Photo courtesy of the Ayrshire Breeders' Association.)

Figure 10-3 Brown Swiss cow, Ivetta, at age 5. All-breed champion lifetime butterfat producer, with 13,607 lbs. of milkfat in 12 lactations. Scored for type six times between ages 2 and 12, and rated as "5 excellent." (Photo courtesy of the Brown Swiss Cattle Breeders' Association.)

Figure 10-4 Guernsey cow. Fox Run AFC Fay represents the current ideal of her breed. She is present breed milk production champion and milkfat champion, producing 31,040 lbs. of milk and 1,736 lbs. of milkfat in 12 months as a 6-year-old milked twice daily. (Photo courtesy of the American Guernsey Cattle Club.)

Figure 10-5 Holstein cow. Reinharts Arthur-Farms Ballad, a world champion milk producer, with 40,980 lbs. of milk and 1,311 lbs. of milkfat in 12 months as a 6-year-old. (Photo courtesy of the Holstein-Friesian Association of America.)

Figure 10-6 Jersey cow. Pinnacle Jester Vol Janice, at 15 years of age. Total lifetime production was 183,573 lbs. of milk and 8,172 lbs. of milkfat, all on twice-daily milking. She was classified "Excellent," with a score card rating of 93 per cent. (Photo courtesy of the American Jersey Cattle Club.)

Figure 10-7 Danish Red cow, owned by Mr. Arne Karlby, Denmark. Production: 1960–61, 6,921 kg. of milk, 4.73% milkfat; 1961–62, 6,495 kg., 5.01%; 1962–63, 7,009 kg., 3.85%; 1963–64, 8,927 kg., 4.26%.

favorably in milking ability with Zebu cattle elsewhere in the world. Most of the dairying of the equatorial region of Africa is carried out at elevations of 5,000 to 10,000 ft above sea level. In a survey of world dairying, the International Dairy Federation reports that European breeds were introduced to this region of Africa, to cross with indigenous Zebu-type cattle and that with disease control and improved forage and management programs, there has been rapid progress in dairying since 1930. They report that there are presently herds of European Friesians, Guernseys, Jerseys, and Ayrshires that would be of acceptable productivity in their country of origin and that the crossing of European dairy bulls on local zebu cows is a continuing program.

Other Dairy Animals

The Domestic Water Buffalo (*Bubalus bubalis*)

The domestic buffalo is an important dairy animal in such countries as India, Pakistan, and Egypt. Over half of the milk production reported for these countries is from dairy buffaloes. Several European countries also have an important buffalo dairy industry, including Bulgaria, Roumania, Greece, Italy, Yugoslavia, and the U.S.S.R. The buffalo is important in many other countries of Asia and the Philippine Islands as well. Mason (1969) lists some ten major breeds of buffaloes, seven of which are stated to be developed and used primarily for dairy, three for dairy plus draft and meat purposes. The breed most frequently referred to in recent scientific literature is the Murrah and several authors refer to it as one of the best of the dairy breeds.

Body weights reported for buffalo cows range from 800 to 1,600 lbs., and a characteristic body size would appear to be near 1,200 lbs. They are noted by several authors to be able to consume relatively large quantities of coarse forage and produce a reasonable amount of milk from such feed. Sebastian *et al.* (1970) report the Murrah buffalo was significantly superior to Sahiwal cows in the digestibility of crude fiber, 79.8 vs 64.7 per cent.

Buffaloes' milk compared to cows' milk is richer in its content of fat, solids not fat, mineral, calcium, and phosphorous. The protein content and amino acid composition of the milk of the two species are not much different, however (Table 10.5). A wealth of detail has been published on the physical and chemical properties of buffaloes' milk and its products. This has been reviewed recently by Laxminarayana and Dastur (1968).

Buffaloes are most frequently reported to be less able than Zebu cattle to withstand high temperature unless they have access to water ponds or sprinklers. With proper management, however, they withstand high dry heat as well as humid climates (Laxminarayana and Dastur, 1968), and yield more milk per animal than Zebu cows. They are characteristically late maturing, with ages at first calving ranging from 35 to 48 months.

The buffalo is currently receiving a great deal of attention as a dairy animal. Presently it accounts for nearly 5 per cent of the world milk supply (Table 10.1), and it is a very promising animal for future improvement of dairying, particularly in tropical and semitropical regions.

Table 10.4.
Comparative performance of three zebu breeds and their first crosses with the European Friesian breed.[a]

Breed	*Number*	*Milk (lbs.)*	*Days of lactation*
Sahiwal	24	4,018	302
Red Sindhi	18	2,897	255
Hariana	28	3,814	274
Friesian × Sahiwal	10	7,322	327
Friesian × Red Sindhi	10	6,157	328
Friesian × Hariana	10	7,170	328

[a] From Bhasin and Desai (1967)

The Domestic Goat (*Capra hircus*)

The domestic goat accounts for 1.6 per cent of the world milk production (Table 10.1), ranking behind the cow and buffalo and just ahead of the sheep. Goats are distributed throughout the world. Countries of Asia, Europe, and Africa account for most of the reported production of goats' milk. There are probably about a million dairy goats in the United States, and although many of these are on DHI testing as are dairy cattle, very few production statistics can be found to characterize the industry and the major breeds. A significant amount of goats' milk is consumed

Table 10.5.
The yield and composition of milk of dairy buffaloes.

	Yield (lbs.)	*Milk fat (%)*	*Protein (%)*	*Lactose (%)*	*Ash (%)*
Range	2,000–7,000	5.1–9.3	3.6–4.7	4.5–5.0	.73–.80
Estimated characterization	5,000	8.0	3.8	4.8	.75

Table 10.6.
Some characteristics of the major breeds of dairy goats of the U.S.[a]

Characteristic	*Saanen*	*Toggenburg*	*Nubian*	*Alpine*
Origin	Switzerland (Saanen Valley)	Switzerland (Toggenburg Valley)	Upper Egypt and Ethiopia	France and Switzerland
Color	White to cream	Brown, white stripes on face, light legs	Black or brown with or without white marks	White to black with spotting
Hair	Short except back and flanks	2 types; short and long	Short, fine	Short except back and flanks
Body wt. of doe, lbs.	120	100	100	125
Lactation length, mo.	8–10	7–10	7–10	8–10
Ave. milk, lbs.[b]	2,150	2,140	1,690	2,130
fat, per cent	3.5	3.5	5.0	3.5
High milk, lbs.[c]	4,200	4,400	4,200	4,600

[a] These breeds are characteristically polled, but with occasional horned individuals, the Alpine having the highest frequency.
[b] Combined estimates of testing reports of 305-day breed averages.
[c] High 305-day records.

as fresh whole milk, but estimates of this are not available.

Breeds of dairy goats Mason (1969) lists some 62 breeds of goats, and 30 of these are tabulated as principally used for dairy purposes. Other breeds are used for dairy and fleece, and all are a source of meat. Most of the dairy goats of the U.S. are of the four major breeds shown in Table 10.6. Lactating does of two breeds are shown in Figure 10.8.

The goat may have been one of the earliest ruminants domesticated by man, and it may further have been domesticated primarily for its milk-producing ability. The modern dairy goat is truly a remarkable animal, since mature does of under 200 pounds body weight can produce upwards of 4,000 lbs. of milk during a 10-month lactation.

Distribution of goats The dairy goat is found under three general circumstances. One is in areas of marginal or developing agriculture, or in time of war or economic distress, where food supplies are short and total forage resources must be pressed into use. Here the goat can utilize poor, rough forage, and browse and negotiate terrain that cattle and sheep cannot graze productively. Under these severe conditions, goats will produce some food for man. When forage and feed become abundant, dairy cattle replace goats, since more food can be produced with fewer animals and hence less labor.

A second area in which the goat is important is for family milk production on small farms or in suburban areas inhabited by industrial or other urban workers.

The third area of dairy-goat husbandry is the commercial dairy producing goats' milk for a specialized market demand. These are the efficient, well-managed goat dairies of this country and many others that operate on productive land under conditions that readily support dairy cattle. The goat dairy in competition with high producing cows faces a situation nicely summarized by Mackenzie (1967) as follows: "A well-managed herd of goats will produce 200 gallons per head. Management and cropping for 30 goats is as laborious as the work of a dairy farm with 20 cows yielding 600 gallons per head, so the labor cost per gallon

Figure 10-8 Milking does of two of the dairy goat breeds, the Nubian (broken color), Laurelwood Acres Variety, and the Saanen, Laurelwood Acres Sabine. (Photo courtesy of Laurelwood Acres Farm, Ripon, California.)

of goats' milk is roughly twice the cost per gallon of cows' milk." The goat may have an unrealized dairy potential in the tropics, where it is one of man's most important domestic animals. It probably adapts better to harsh tropical environments than other productive ruminants. In areas where labor costs are relatively low, economic studies have indicated dairy goats may be competitive with zebu dairy cows and dairy buffaloes. Their economic and food production possibilities deserve serious consideration.

The Domestic Sheep (*Ovis aries*)

Few, if any, sheep are milked in the United States, yet the estimated world production of ewes' milk is very close to that of the goat (Table 10.1). Sheep are important dairy animals of many European, North African, and Middle Eastern countries. Countries of the Mediterranean region in particular account for most of the commercial production. Cheese is the primary use of ewes' milk, with yogurt and other cultured products and butter or clarified butter being next. A limited amount is used as fresh milk. Roquefort cheese made in Southeastern France and Pecarino Romano cheese from Italy are two of the main varieties of cheese made from ewes' milk.

Dairy breeds of sheep Mason (1969) lists some 320 breeds of sheep. Scanning this list, one finds a single breed, the East Friesian of Germany, used practically exclusively as a dairy animal.

The production and milk composition characteristics of five selected dairy breeds of sheep are presented in Table 10.7. Jenness and Sloan (1970) have reviewed the world literature on mammalian milk composition, and give the composition of ewes' milk as 7.4 per cent fat, 5.5 per cent protein, 4.8 per cent lactose.

Table 10.7.
Estimated milk yield and composition of some breeds of sheep used primarily for dairy.

Country	*Breed*	*Yield*[a] (*kg*)	*Fat* (%)	*Protein* (%)
Italy	Langhe	260	6.5	5.5
Israel	Awassi[b]	250+	7.5	—
Spain	Mancha	124	—	—
France	Lacaune	160	—	—
Germany	East Friesian	450	6.8	—
Czechoslovakia	Valachian	74	7.3	5.6

[a] The average lactation length is about 200 days.
[b] The "improved" Awassi. This breed is also found in several Arab countries.

Particular attention to selection for milk production has been given to the Friesian and the "improved" Awassi breeds. The Awassi are a fat-tailed breed found in Jordon, Syria, Lebanon, Iraq, and Southern Turkey. Finci (1958) noted that milk and wool production in this breed were not incompatible traits. One of the high milk-yielding flocks of this breed had an average fleece weight of 2.96 kg, while the breed average was 1.76 kg. The coarse-wooled dairy breeds, such as the Lacaune and other of the Roquefort and adjoining regions of southern France, account for most of the ewes' milk production. France reported for 1969 an estimated production of 30,031,000 metric tons of cow's milk, 729,000 of ewe's milk, and 301,000 of goat's milk. The milk production of Italy for 1968 is reported as 76.4, 3.1, and 1.0 million metric tons for cows, ewes, and goats, respectively.

East Friesian sheep have been imported to Israel from Germany to cross with the Awassi in an attempt to increase lambing percentage, eliminate the fat tail of the Awassi, and improve wool production without reducing milk yield. Final results of this work have not been reported.

There are a number of countries with large sheep populations where the primary income from the flocks is derived from milk. In Spain, for example (Mason 1967), some 4 million ewes were milked in 1964, with milk accounting for 65 per cent and wool 10 per cent of the income from the flocks. Yugoslavia had some 9.7 million sheep in 1964, all breeds of which were used for milk, meat, and wool. Milk is the principal product of the sheep industry of Greece. Meat, milk, and wool in that order are the products of the sheep flocks of Albania, Turkey, and Syria. The sheep of Israel are among the highest producing in the world, and they are kept primarily for milk, which is used in cheese and yogurt production. Some 70 per cent of the milking ewes of Israel are tested and recorded for milk production.

The Camel (*Camelus* sp.)

Veisseyre (1969) described the dairy industry of Tunisia and referred to a total production of 141 million liters of milk for 1964; 1.9 per cent of this was estimated to be camels' milk. Certainly from a world-wide standpoint camels' milk is of minor importance as a food. However, the camel is an important animal in the lives of many people of local regions of Africa and Asia, who depend upon it as a source of transport, meat, milk, wool, and leather. Most of the world's camels are found in India, Mongolia, Pakistan, Ethiopia, Somaliland, Sudan, Northern Kenya, and Tunisia.

Table 10.8 presents some recently published summaries of camels' milk composition. The dromedary of Pakistan is reported to yield from 2,700 to 3,600 kg of milk, testing 2.9 to 4.0 per cent fat under good feeding conditions during a lactation period of 9 to 18 months. The bactrian, in comparison, may average 1,800 kg of milk of 5 per cent fat in similar situations, according to Leupold (1968). Bhimasena Rao *et al.* (1970) point out the Bactrian camel is found in the deserts and colder, dry steppes of Mongolia and Turkestan, whereas the dromedary is found in the hotter, dry subtropics.

The Mare (*Equus equus*)

The milk of the mare has long been used by man, particularly by nomadic peoples, as a food source. It remains important in certain local areas of the world, although facts about the extent of its uses are difficult to obtain. The principal use of mares' milk in the U.S.S.R. is for the production of kumiss, a beverage of combined lactic acid and alcoholic fermentation. Many references dealing with mares'

Table 10.8.
Some mammals of potential but minor dairying importance.[a]

	Milk composition					
Species or type	*Fat*	*Protein*	*Lactose*	*Ash*	*Total solids*	*References and notes*
Camel, Bactrian	5.4	3.9	5.1	0.7	15.0	Jenness and Sloan (1970)
Camel, Dromedary	4.5	2.7	5.0	0.7	13.6	" " " "
Mare (U.S. light horses)	1.3	2.0	6.5	0.3	—	*J. Anim. Sci.*, 25:217 (1966) (10 samples)
Mare (several reports)	1.9	2.5	6.2	0.5	—	Jenness and Sloan (1970) (231 samples)
Yak	6.5	5.4	4.6	0.9	—	Schley (1967)
Chowrie, Urang	5.0–8.0	—	—	—	—	Schulthess (1967)
Chowrie, Dimschu	5.5–9.0	—	—	—	—	" "
Chowrie	4.5–6.2	—	—	—	—	Schley (1967)
Eland	9.8	6.8	3.9	1.1	—	Treus and Kravchenko (1968) (51 samples)
Musk Ox	11.2	5.3	3.6	1.8	—	Tener (1965) 2 samples, 1 animal
Musk Ox	5.4	5.2	4.2	1.1	—	" " 1 sample, 1 animal
Reindeer	10.4	18.6	2.8	1.6	—	Luick (1971) 150 samples, 5 animals

[a] See Chapter 29 for data on the composition of milk in some other species.

milk production and kumiss manufacture are to be found in *Dairy Science Abstracts* and *Animal Breeding Abstracts,* which are English abstracting journals.

The characteristic composition of mares' milk is presented in Table 10.8. Compared to cows' milk, mares' milk is low in fat and protein and high in lactose. Mares' milk has a fairly close resemblance to human milk. It has been evaluated as a food for human infants (Neuhaus 1959), and there are reports from Germany on its successful use for this purpose.

From the recent abstracts of Russian literature, the milking characteristics and management of mares for dairy purposes appear as follows. Milking as frequently as six times a day is referred to, and lactation lengths of 150 to 200 days are most common. Machine milking of mares is reported to be superior to hand milking, and the development of special milking machines is described. Yields of mares are reported frequently. One average of 400 mares ranging from 6 to 17 years of age is given as 1,883 liters (about 4,100 lbs.) in a five-month lactation.

The Yak (*Bos grunniens*) and Yak-Cattle Hybrids

The yak is distributed through the high mountainous regions of central Asia, where it and its crosses with cattle are important to the livelihood of the local peoples. The yak and its cattle hybrids are used for milk, meat, draft, and packing; their hair is used in making rope and tent cloth. There are numerous terms with various spellings applied to yak-cattle hybrids. The female is often termed chowrie. The male is called zopkio. A dimschu chowrie is the F_1 female of a cattle bull × yak cow, and the reciprocal cross female is termed urang chowrie. Numbers of yak and chowrie in Mongolia are estimated at up to 700,000, and it is of major importance to the Sherpas of Nepal. Yields of milk reported range from 227 to over 7,000 lbs. for lactations of 150 to over 300 days in length; milk composition is shown in Table 10.8. Further reference to this interesting animal can be found in Schulthess (1967), Upadhaya (1969), Rouse (1970), and Schley (1967). The apparent heterosis for milk yield presents an intriguing genetic study, and these animals represent another species that may be of additional use in future food production.

Other Mammals of Use, or Potential Use, for Milk Production

There are additional species of large mammals that are used for milk production in limited local areas, and others that have shown some potential for future development as dairy animals under special environmental conditions. They are all ruminants. One is the reindeer or caribou, *Rangifer tarandus,* of the deer family, *Cervidae.* Three others are of the family *Bovidae;* the eland, *Taurotragus oryx;* the gayal or mithan, *Bos frontalis;* and the musk ox, *Ovibos moschatus.* Crawford (1968) notes that the reindeer industry of Norway is of increasing importance, with some 500 Lapp families depending on it. Even by these people the milking of reindeer is limited, and production data are almost non-existent. Luick (1971), Table 10.8, presents some of the best milk-composition data available to date, since his animals were trained to hand milking and milked completely. The reindeer is a relatively small animal; the average adult weighs from 175 to 200 lb. Additional references to reindeer include the papers of Kelsall (1968), Epstein (1966), and Phillips (1948).

The eland, an antelope of central and southern Africa, was noted as early as 1848 to be a species that was particularly docile and tractable and thus suitable for domestication. Russian zoologists imported a small number into the Southern Ukraine in the early 1890's and have worked with them since. Recent reports pertaining to eland include those of Posselt (1963) and Treus and Kravschenko (1968). Eland's milk is particularly high in its protein content (Table 10.8). Yields of up to 1,400 lbs. in a 207-day period are reported.

The musk ox has evidently never been used as a dairy animal, and although some very interesting current domestication projects are underway in Alaska and Canada (Teal, 1970), the purpose is to investigate wool production rather than meat or milk. Milk-composition data are given in Table 10.8. Tener (1965) points out these are post-mortem milk samples, and the analysis with 11 per cent fat may have been largely colostrum, whereas the lower (5.4 per cent) testing sample was from a cow with a 4-day-old calf.

FURTHER READINGS

Briggs, H. M. 1969. *Modern Breeds of Livestock,* 3d ed. New York: Macmillan.

Crawford, M. A., ed. 1968a. *Comparative Nutrition of Wild Animals.* London: Academic Press.

Epstein, H. 1966. *Domestic animals of China.* Farnham Royal Bucks, England: Commonwealth Agric. Bureaux.

French, M. H., I. Johansson, N. R. Joshi, and E. A. McLaughlin. 1966. *European Breeds of Cattle.* Rome, Italy: FAO.

Jenness, R., and R. E. Sloan. 1970. "The Composition of milks of various species: A review." *Dairy Sci. Abst.,* 32:599–612.

Kelsall, J. P. 1968. *The Caribou.* Ottawa, Canada: Queen's Printer.

Leupold, J. 1968. "The camel—an important domestic animal of the subtropics." *Blue Book of the Vet. Prof.,* 15:1–6.

Mason, I. L. 1967. *Sheep breeds of the Mediterranean.* Farnham Royal Bucks, England: Commonwealth Agric. Bureaux.

Mason, I. L. 1969. *A World Dictionary of Breeds, Types, and Varieties of Livestock.* Edinburgh, Scotland: Commonwealth Bureau of Animal Breeding and Genetics.

Neuhaus, Von U. 1959. "Milch und milchgewinnung von Pferdestuten." *Zeits. für Tierzuchlung und Zuchtingsbiol.,* 73:370–392.

Porter, A. R., J. A. Sims, and D. F. Foreman. 1964. *Dairy Cattle in American Agriculture.* Ames: Iowa State Univ. Press.

Rouse, John E. 1970. *World Cattle.* Norman: Univ. of Oklahoma Press.

Simoons, F. J., and E. S. Simoons. 1968. *A Ceremonial Ox of India.* Madison: Univ. of Wisconsin Press.

Tener, J. S. 1965. *Musk Oxen in Canada: A Biological and Taxonomic Review.* Ottawa, Canada: Queen's Printer.

Treus, V., and D. V. Kravchenko. 1968. "Methods of rearing and economic utilization of eland in the Askaniya-Nova Zoological Park." In Crawford 1968a, pp. 395–411.

Zhigunov, P. S., ed. 1968. *Reindeer Husbandry.* 2d ed. Washington, D.C.: USDI and NSF.

Eleven

Cattle Used for Meat Production

And God made the beast of the earth after his kind, and cattle after their kind.

Genesis II:25

Number and Distribution

The world population of cattle and buffalo is estimated at approximately 1.1 billion head. Figure 11.1 depicts the distribution of cattle throughout the world. Many of these cattle are utilized primarily for work or milk, with meat only a by-product of their major intended usefulness. Most of the cattle produced strictly for beef are found in North and South America, Australia, and New Zealand. Table 11.1 identifies the countries with the most cattle and those with the highest per capita consumption of beef. Twenty countries account for over 70 per cent of the cattle, but several of these are not among the leading countries in per-capita beef consumption. Reasons for this apparent anomaly include the following. (1) Religious and other customs prevent the eating of the flesh of cattle. In India, with twice as many cattle as any other country, 227,000,000 head of cattle and water buffalo, it is illegal to kill any of the cattle and the slaughter of buffalo is restricted. In much of Asia and on some of the Pacific Islands, there are strict laws against killing a cow or a female buffalo capable of reproduction and often against killing a bull under a specified age. For countless ages the native African, often well-endowed with cattle, has slaughtered an animal only for a ceremonial feast. (2) Some countries have a large geographical area with a high ratio of number of people to the number of cattle, e.g., China. (3) Some countries have a small geographical area, but have a high ratio of numbers of cattle to numbers of people, e.g., Finland. (4) Some countries import a considerable amount of the beef they eat, e.g., Switzerland.

Types and Breeds

Nearly all cattle are classified into either the humped or humpless types. Most of the humpless cattle include the ancestors of the European cattle and the majority of cattle found

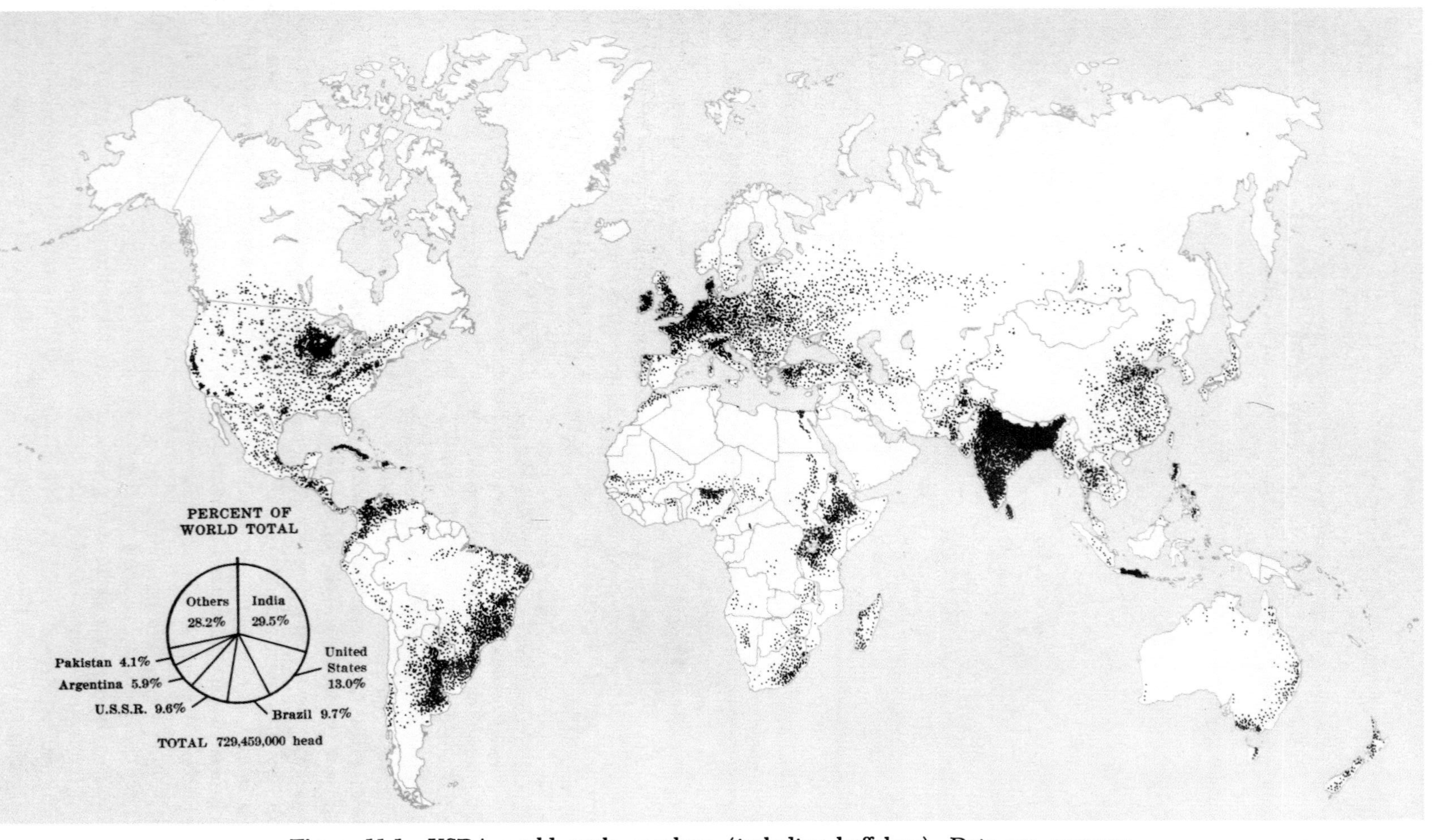

Figure 11-1 USDA world cattle numbers (including buffaloes). Data are averages for the years 1957–61. Each dot represents 100,000 head. (From USDA 1964, p. 58.)

Table 11.1.
Leading countries in cattle numbers and per capita beef consumption.

Cattle and buffalo numbers (1969)[a]		*Beef and veal consumption (1969)*[b]	
Country	*Number (mil. head)*	*Country*	*lbs. per capita*
1. India	227.0[c]	1. Argentina	182
2. United States	109.8	2. United States	114
3. Soviet Union	95.7	3. New Zealand	106
4. Brazil	92.3	4. Uruguay	98
5. China	86.3[c]	5. Canada	96
6. Argentina	52.0	6. Australia	93
7. Pakistan	41.9[c]	7. France	66
8. Mexico	27.5	8. Belgium	59
9. Ethiopia	25.3[c]	9. Switzerland	56
10. France	22.1	10. United Kingdom	54
11. Australia	20.6	11. Germany (West)	52
12. Colombia	19.6	12. Austria	49
13. Turkey	15.0	13. Italy	45
14. Germany (West)	14.1	14. Colombia	44
15. United Kingdom	12.1	15. South Africa	44
16. South Africa	11.8	16. Venezuela	43
17. Canada	11.5	17. Netherlands	42
18. Poland	11.0	18. Finland	42
19. Italy	10.1	19. Sweden	42
20. Tanzania	10.0[c]	20. Soviet Union	41
		21. Denmark	41

[a] From Foreign Agriculture Circular, U.S.D.A., May 1970.
[b] From Foreign Agriculture Circular, U.S.D.A., Jan. 1971.
[c] From Rouse (1970), vol. II.

in the United States. The humped cattle, usually called Zebu, had their ancient origin in Pakistan and India. They have spread both east and west into the warmer areas of Asia, Africa, South America, and the southern United States, where they seem to be best adapted. Figure 11.2 shows the world distribution of these two basic types of cattle.

The bison, musk ox, yak, and some others belong to the same family, *Bovidae,* as cattle. However, the water, swamp, and river buffalo belong to different families, and thus are distinguished from cattle. Of the 1.1 billion head of domesticated cattle and buffalo, approximately 100,000,000 are buffalo. They are used primarily for work, but also furnish milk. It has been estimated that slightly over 500,000,000 of the cattle are raised either for beef, milk, or both.

A breed of cattle has basically been defined as a race or variety, related by descent and similar in certain distinguishable characteristics. Mason (1969) mentioned 274 important and recognized cattle breeds throughout the world. He also listed several hundred other varieties and types of cattle that have not dis-

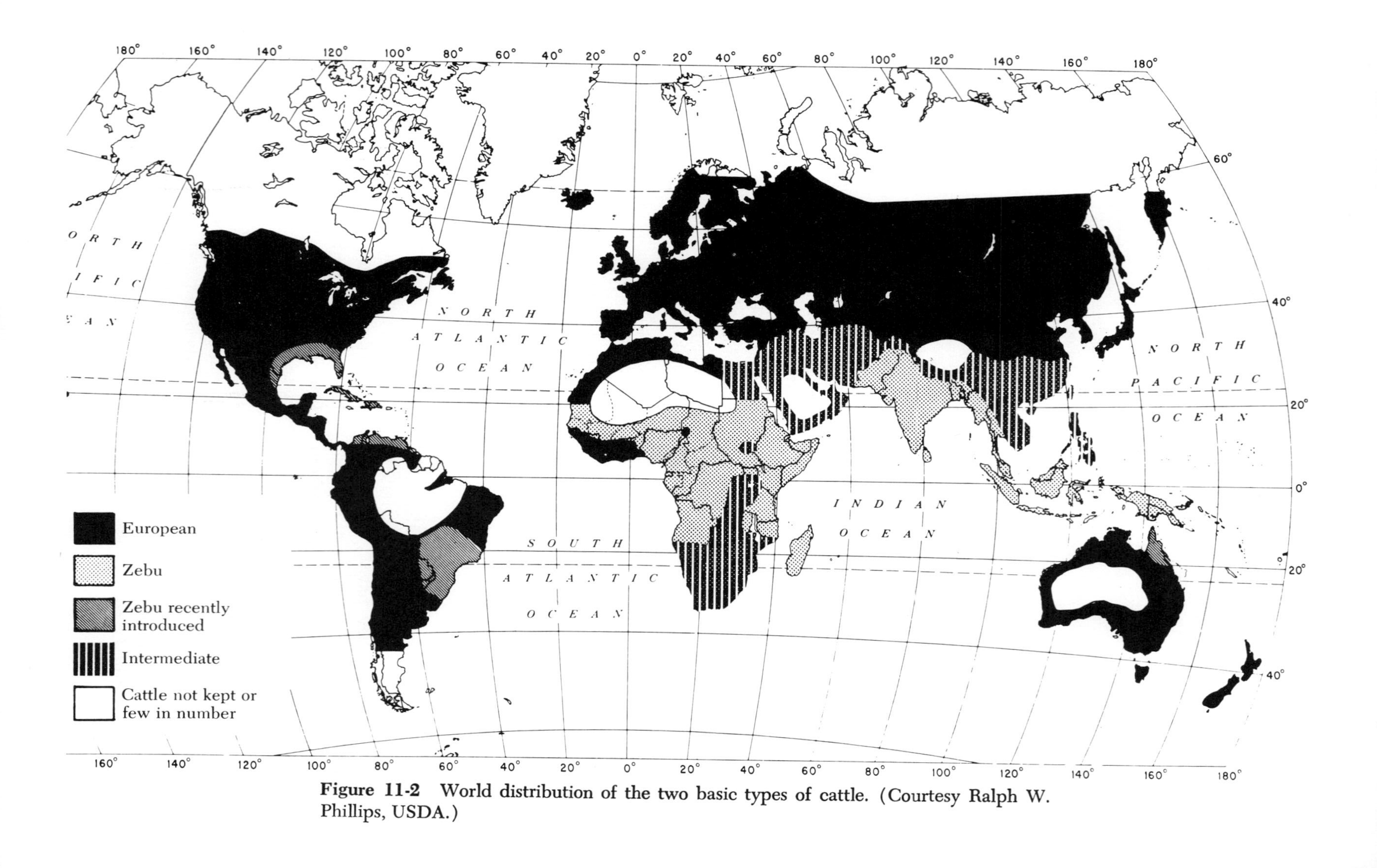

Figure 11-2 World distribution of the two basic types of cattle. (Courtesy Ralph W. Phillips, USDA.)

tinguished themselves as breeds. Some of the 274 breeds listed are basically the same, but have been given different names in different countries. The Holstein, for example, is called Friesan in Europe.

Some of the older, recognized breeds in the United States came into existence, as breeds, during the middle to late 1800's. Most of these breeds originated from the crossing and combining of existing strains of cattle.

When a breeder or group of breeders decided to establish a breed, distinguishing that breed from other breeds was of paramount importance; thus major emphasis was placed on readily distinguishable visual characteristics, such as color, color pattern, polled or horned condition, or rather extreme differences in form and shape.

New cattle breeds have come into existence in recent times, such as Brangus and Santa Gertrudis. These have been developed by combining desirable characteristics of several breeds. However, the same identifying characteristics, as mentioned above, have been considered in order to give the breed identity.

After some of the breeds were developed, it was not long before the word "purebred" was attached to some of these animals. Herd books or Registry Associations were established to assure the "purity" of the breed. Purebred refers to purity of ancestry, established by the pedigree, which shows that only animals recorded in that particular breed have been used to produce the animal in question.

When a person views a herd of purebred Angus or a herd of purebred Herefords, he is generally impressed with uniformity, particularly uniformity of color or color pattern. Because of this impression of uniformity of one or two characteristics, the word purebred has come to imply genetic uniformity of all characteristics. This is not the case, and if it were, breeds themselves could not be improved or changed where change is desired.

Some of the predominant breeds and types of cattle, in the top twenty countries in cattle numbers, are shown in Figures 11.3 to 11.17. Table 11.2 shows the registration numbers for the major breeds of beef cattle in the United States. Not reflected in this table is the influence of the so-called "exotic" breeds, primarily the Limousin and Simmental, that have been introduced into North America. These introductions have been primarily through a few bulls, with their use extended through artificial insemination. Attempts are now being made to use the several exotic breeds together with the British breeds in developing crossbreds with considerable uniformity as concerns specific economic traits. The Santa Gertrudis serves as an excellent example in which heat tolerance of the Brahman has been successfully combined with some desirable traits of the Shorthorn (see Table 11.3).

Breed Differences for Economically Important Traits

Traits such as fertility, milking ability, growth rate, and carcass traits are controlled, not by one or two pairs of genes, but by many pairs, and thus are more complex in their inheritance. Calves at weaning vary continuously from low to high weights; Figure 11.18 shows

Table 11.2.
Registration numbers for the major breeds of beef cattle in the United States.[a]

Breed	*Registration numbers (1970)*
Angus	352,471
Brahman	18,219
Brangus	6,540
Charolais	45,328
Hereford	236,617
Polled Hereford	160,374
Red Angus	5,096
Santa Gertrudis	19,025
Shorthorn	35,653

[a] From National Society of Livestock Record Association, Annual Report, 1970.

Figure 11-3 The Hereford breed, red body with white face and markings. The most numerous breed in the United States and a major breed in Argentina. (Courtesy American Hereford Association.)

Figure 11-4 The Aberdeen Angus breed, black and polled. The second most numerous breed in the United States and a major breed in Australia. (Courtesy *Aberdeen Angus Journal.*)

Figure 11-5 The yellow-tan Chinese Yellow breed. This is the local name for the indigenous cattle of China, which are also prevalent on Taiwan. (From *World Cattle*, by John E. Rouse. Copyright 1970 by the University of Oklahoma Press.)

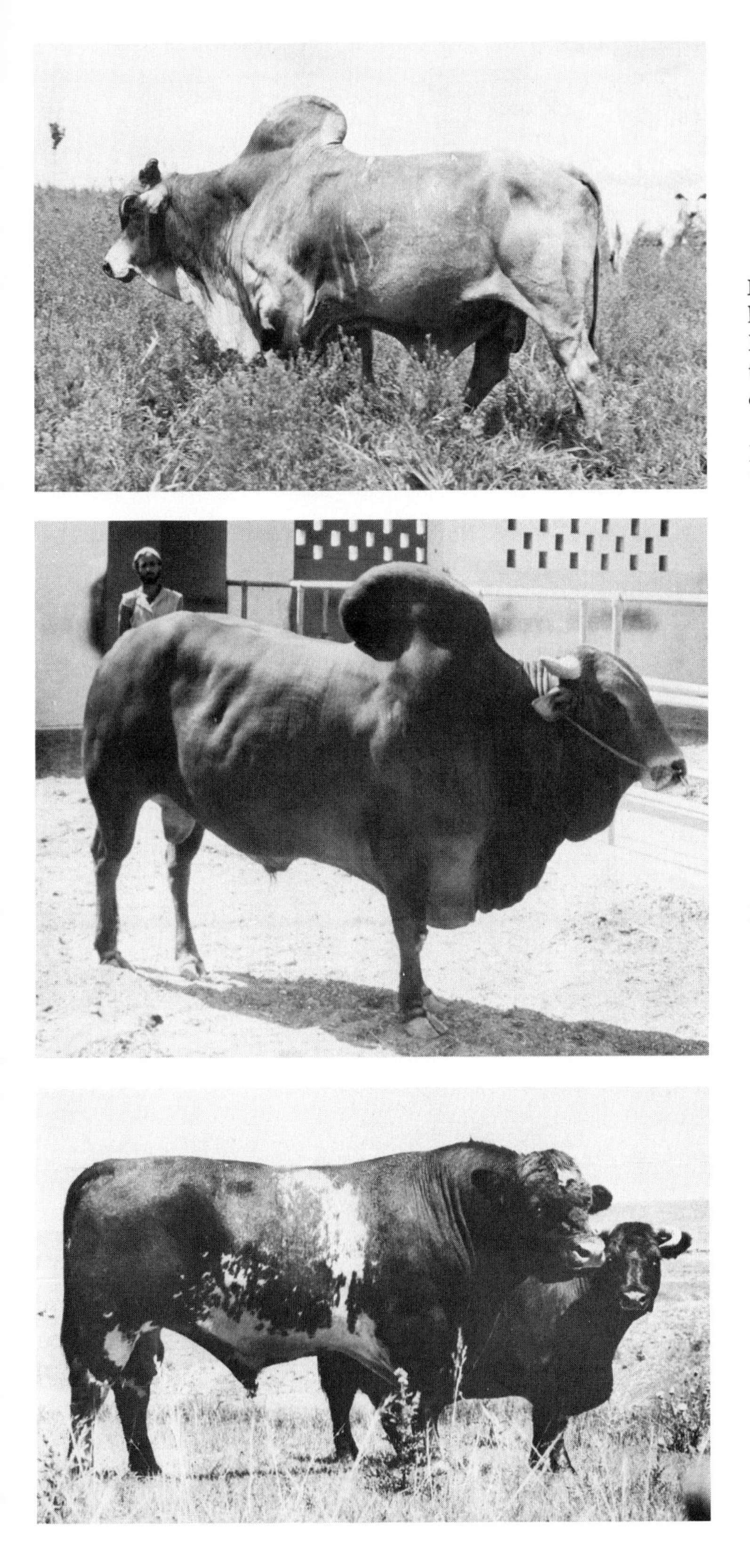

Figure 11-6 The Nellore breed, light grey to near white. Cattle in Brazil are of the Zebu type, and this breed is one of the well-defined breeds of that country. (From *World Cattle,* by John E. Rouse. Copyright 1970 by the University of Oklahoma Press.)

Figure 11-7 The Red Sindhi breed, light to very deep red. A prominent breed in Pakistan; however, most of their cattle are nondescript Zebus. (From *World Cattle,* by John E. Rouse. Copyright 1970 by the University of Oklahoma Press.)

Figure 11-8 The Shorthorn breed, red, roan, or white. This breed, along with the Hereford and Angus, is prevalent in Australia, and also has been influential in the background of commercial cattle in the United States. (Courtesy American Shorthorn Association.)

Figure 11-9 The Charolais breed, cream to pure white. A popular breed in France, and a preeminent breed of continental Europe. (From *World Cattle,* by John E. Rouse. Copyright 1970 by the University of Oklahoma Press.)

Figure 11-10 The Limousin breed, red to reddish yellow. The second most important breed of continental Europe. (From *World Cattle,* by John E. Rouse. Copyright 1970 by the University of Oklahoma Press.)

Figure 11-11 The Simmental breed, red with white markings. An influential breed in West Germany, Austria, Switzerland, Yugoslavia, the Soviet Union, and other parts of continental Europe. (From *World Cattle,* by John E. Rouse. Copyright 1970 by the University of Oklahoma Press.)

Figure 11-12 The Polish Black and White breed. Similar to the Holstein (Friesan) breed, and other black and white cattle with different breed names, this is one of the top four breeds in continental Europe. (From *World Cattle,* by John E. Rouse. Copyright 1970 by the University of Oklahoma Press.)

Figure 11-13 The Native Black breed, the most prevalent breed in Turkey. (From *World Cattle,* by John E. Rouse. Copyright 1970 by the University of Oklahoma Press.)

Figure 11-14 The Africander breed, from light to dark red, with yellows and greys occurring. The most important commercial breed in South Africa; also influential in other African countries. (From *World Cattle,* by John E. Rouse. Copyright 1970 by the University of Oklahoma Press.)

Table 11.3.
Comparative ratings on economic traits of 29 breeds of cattle now available to North American producers.[a]

	Cow traits				*Calf traits*			*Carcass*			*Bull traits*			*Breed's place in crossbreeding*		
Breed	*Age at puberty*	*Conception rate*	*Milking ability*	*Mature size*[b]	*Preweaning growth*	*Postweaning growth*	*Optimum slaughter weight (lbs.)*	*Cutability*	*Marbling*	*Tenderness*	*Fertility*	*Freedom from genital defects*	*Calving ease (sire effect)*	*Maternal*	*Terminal*	*Rotational*
Angus	1	2	3	A	3	4	950	4	1	2	2	2	2	X		X
Ayrshire	2	3	1	A	3	3	900	2	3	2	2	2	2	X		
Beefmaster	3	2	3	L	2	2	1,150	2	3	3	3	2	1	X		X
Braford	3	2	3	A	2	2	1,150	2	3	3	3	2	1	X		X
Brahman	5	5	3	A	3	4	1,150	2	4	5	4	4	1	X		X
Brangus	3	2	3	A	2	3	1,050	2	2	3	3	2	1	X		X
Brown Swiss	3	4	1	L	1	1	1,200	2	3	2	2	2	3	X		X
Charbray	4	4	3	L	2	2	1,250	1	4	3	3	3	4		X	X
Charolais	4	4	3	L	1	1	1,250	1	4	2	3	2	5		X	X
Devon	3	2	2	A	2	3	1,050	3	2	2	2	1	3	X		X
Galloway	2	2	3	A	4	4	950	3	2	2	2	1	2	X		X
Guernsey	2	3	1	S	3	4	900	2	2	1	2	3	1	X		
Hays Converter	2	2	2	L	2	2	1,150	2	3	2	2	1	3	X		X
Hereford	4	2	5	A	4	2	1,050	3	3	2	2	1	2	X		X
Holstein	3	2	1	L	1	1	1,200	2	3	2	2	1	3	X		X
Jersey	1	1	1	S	4	5	850	2	2	1	1	3	1	X		
Limousin	3	3	3	L	2	1	1,200	1	3	2	3	2	4		X	X
Maine-Anjou	3	3	2	L	1	1	1,250	2	3	2	3	2	5	X	X	X
Milking Shorthorn	2	4	2	A	3	3	1,000	5	2	2	3	1	2	X		X
Murray Grey	2	3	3	A	2	3	1,050	3	2	2	3	2	3	X		X
Polled Hereford	4	2	5	A	4	2	1,050	3	3	2	2	2	2	X		X
Polled Shorthorn	2	4	3	A	3	4	950	5	2	2	3	2	2	X		X
Red Angus	1	2	3	A	3	4	950	4	1	2	2	2	2	X		X
Red Poll	2	2	2	S	3	3	950	2	2	2	3	2	2	X		X
Santa Gertrudis	4	4	3	L	2	2	1,150	2	3	4	4	3	2	X	X	X
Scotch Highland	2	3	4	S	4	4	900	3	2	2	3	1	2	X		X
Simmental	3	2	1	L	1	1	1,250	2	3	2	2	1	5	X	X	X
Shorthorn	2	4	3	A	3	4	950	5	2	2	3	1	2	X		X
South Devon	3	3	2	L	2	2	1,150	3	2	2	3	2	3	X		X

[a] From Sumption, *et al.*, 1970. Numerical grade of 1 is high (desirable), 5 is low.
[b] S is small, A average, L large.

Figure 11-15 The Criollo-type cattle, many different colors. Prevalent in Colombia and several other South American countries. (From *World Cattle* by John E. Rouse. Copyright 1970 by the University of Oklahoma Press.)

the comparison of three breeds of beef cattle for weaning weight. The breed averages are noticeably different; however, the variation in this trait, for each breed, is rather extensive. It is of greater importance to the individual breeders to have high performing animals of a particular breed than to be concerned about the average superiority of one breed over another.

The example demonstrated with weaning weight could be shown for each of the other economically important traits. It would portray not only the great amount of variation within a breed, but also the fact that no one breed is superior in all economically important characteristics. Table 11.3 shows the relative ranking of the various breeds at this point in time. This information is based on the accumulated breed comparisons that have been made. Some of these rankings will no doubt change as certain breeds improve and as more extensive breed comparisons are made.

Figure 11-16 The Chianina breed, white. Popular in certain areas of Italy. One of the largest breeds of cattle, with mature bulls weighing from 3,000–4,000 pounds. (Courtesy American Chianina Association.)

Figure 11-17 Fattened Pinzgauer bulls. The Pinzgau breed is centered in Austria and adjacent areas of Italy and Germany. It is a large, rugged dual-purpose breed. (Reprinted by permission of Better Beef Business, Shawnee Mission, Kansas.)

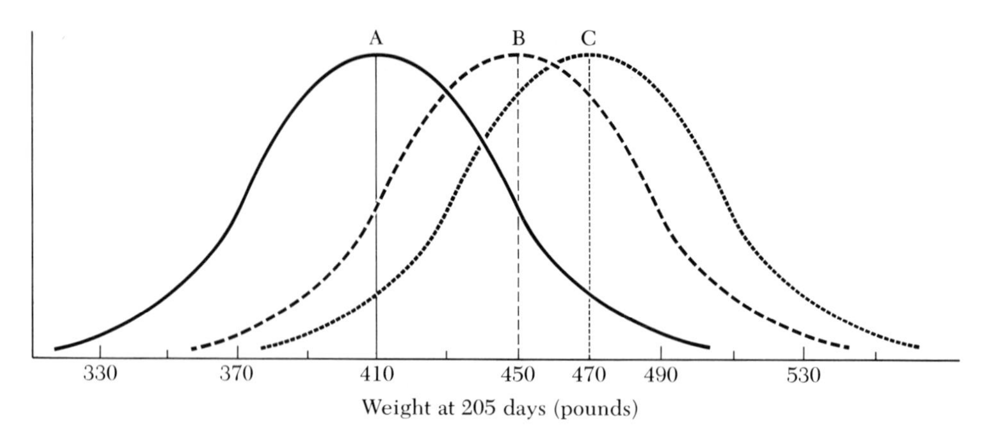

Figure 11-18 The comparison of the variations within three breeds of beef cattle for weight at weaning (pounds at 205 days).

The information we have today tells us that there are thousands and thousands of genetic combinations within a breed. The breeder's challenge is to be able to select out the more productive combinations and have them reproduce their kind. Breeds are not static and likely never will be. Man's needs and desires, which change from time to time, are imposed on the beef animals. Thus, new breeds come into existence, and the other breeders change selection emphasis on existing breeds to meet the desires of man.

FURTHER READINGS

Briggs, H. M. 1969. *Modern Breeds of Livestock.* New York: Macmillan.

Mason, I. L. 1969. *A Dictionary of Livestock Breeds.* London: Morrison and Gibb.

Rouse, J. E. 1970. *World Cattle.* 2 vols. Norman: Univ. of Oklahoma Press.

USDA. *Foreign Agric. Circular* for May 1970 and for Jan. 1971.

USDA. 1964. *A Graphic Summary of World Agriculture.* Misc. Pub. no. 705, rev. ed. Washington, D.C.: USDA.

Twelve

Swine: Plasticity in Meeting the Needs of Man

"Streamlining our swine production marks the most forward step in animal breeding in the last four generations. Fortunately, throughout human history the hog has proved the most plastic of all the domestic animals."

C. W. Towne and E. N. Wentworth, 1950

Swine adapt well to both temperate and tropical environments, and are found in all parts of the world. The pig, along with the dog, is said to have been the earliest domesticated animal in China (Epstein, 1969). Biblical reference to the pig (Leviticus 11:7) puts its domestication in the Middle East several centuries before Christ. According to Towne and Wentworth (1950), who have described the early evolution and domestication of the pig, ancestral forms of the pig existed more than 40 million years ago, and domestication of the wild hog began 7000 to 3000 B.C. in Europe. Apparently (Mellen, 1952), the pig was brought to the Western Hemisphere, to Haiti, by Columbus on his second voyage in 1493.

Except where religious or cultural taboos forbid the use of pork for human food, pork is in wide demand by people of all economic levels and provides an excellent source of nutrients for man. The pig, having a digestive system and nutrient requirements similar to those of man, competes directly with man for available food supplies. Although this competition for nutrients in a world with a rapidly expanding human population may spell extinction for the pig as a supplier of animal protein for man in the far-distant future, the conversion to pork of many feedstuffs and waste materials not suitable for or accepted by man insures the pig a useful and important place in the life of man for the foreseeable future.

Breeds of the World

Mason (1969) reports some 87 recognized breeds, and an additional 225 or more "varieties" of swine in the world (Figure 12.1). For the purposes of our discussion, a breed is a group of animals of similar characteristics and common genetic background whose individual parentage is registered by some private or governmental agency, whereas a "variety" is a group with distinguishable characteristics and a common geographic distribution, but

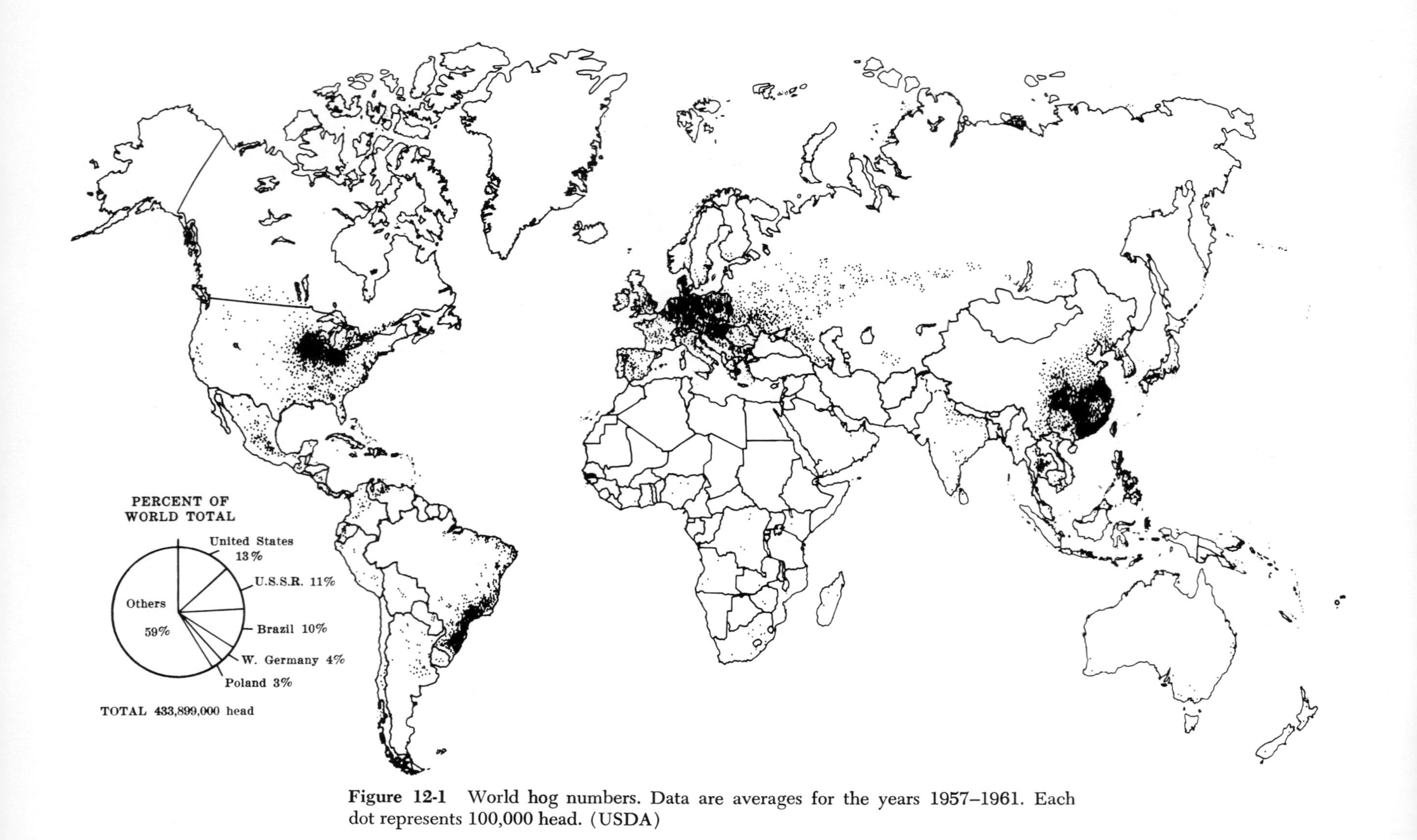

Figure 12-1 World hog numbers. Data are averages for the years 1957–1961. Each dot represents 100,000 head. (USDA)

Figure 12-2 A sow typical of the Ningsiang breed of China.

with no vehicle for registration or records of parentage.

All the domestic breeds and "varieties" of swine, despite differences in appearance, have the same number of chromosomes (forty) and are capable of interbreeding. This provides the opportunity for animal breeders and geneticists to utilize the existing germ plasm to change or improve the pig to meet the desires and needs of man.

Although most of the efforts in swine improvement have been made using the breeds of Europe and North America, there are opportunities to utilize traits (such as prolificacy in Chinese breeds) peculiar to swine in other parts of the world. For this reason, there is general recognition among animal scientists that care should be taken to preserve existing germ plasm.

Variability Among Breeds

The domestic pig occurs in many sizes, shapes, and appearances. The indigenous swine of China, consisting of some 40 breeds and varieties of economic value, are generally smaller and slower-growing, though more prolific, than European and North American breeds. For example, according to Epstein (1969), the adult sow of the Ningsiang breed of Hunan (Figure 12.2) a mature weight of about 100 kg. and a litter size of 10 to 12, and weans the pigs at two months of age at a weight of 12.5 kg.

In contrast, the Duroc (the most common breed of the United States) sow often reaches two to three times that mature weight and weans pigs that are twice as heavy at two months of age and that reach 90 kg. at five to six months of age. The indigenous breeds of Latin America represent yet other differences in over-all appearance (see Figure 12.3).

The Place of Purebred Swine in the United States

Of the more than 80 million swine produced each year in the United States, less than 1 per cent (approximately 200 thousand) are purebred. The vast majority are crossbred pigs, because there is systematic use of two or more breeds in a breeding program to produce slaughter pigs of high merit. The performance characteristics of crossbred swine are generally superior to those of the purebred parents. This increased productivity observed in the progeny when purebred parents of two differ-

Figure 12-3 An indigenous pig in a mountain village of northern Ecuador.

ent breeds are mated is commonly referred to as "heterosis" and forms the basis for the crossbreeding programs practiced by most commercial swine producers in the United States.

The key to continued improvement of commercial swine lies in the continued selection of superior purebred animals for use in crossbreeding programs. Each of the purebred swine associations in the United States has a systematic program of breed improvement through selection based on economically important inherited traits.

Table 12.1.
Major breeds of swine in the United States.

Breed	*Origin*	*Color markings and distinguishing characteristics*
Berkshire	England	Black with white face, feet, and tail; erect ears; short, dished face
Chester White	Pennsylvania	All white; drooping ears
Duroc	New York, New Jersey	All red (golden to mahogany); drooping ears
Hampshire	Kentucky	Black with white belt around shoulders including both forelegs; erect ears
Hereford	Missouri	Red and white (white face, at least two white feet; no more than one-third of body surface can be white); drooping ears
Landrace	Denmark	All white; large, drooping ears; long face
Poland China	Ohio	Black with white face, feet, and tail; drooping ears
Spotted	Indiana	Spotted black and white (up to 80 per cent white or black on body surface accepted)
Tamworth	England	All red (golden to dark); erect ears; long, narrow face and snout
Yorkshire	England	All white; erect ears

Table 12.2.
Purebred swine registrations in the United States in recent years.

	No. of animals registered			
Breeds	*1961*	*1965*	*1968*	*1970*
Berkshire	14,138	9,489	8,502	8,012
Chester White	17,112	15,624	17,115	19,934
Duroc	60,542	48,075	64,676	76,394
Hampshire	54,674	51,587	63,496	74,101
Hereford	—	—	697	—
Landrace	27,307	15,642	8,123	8,810
Poland China	26,409	24,821	19,512	16,102
Spotted	13,020	12,627	12,367	13,974
Tamworth	—	—	1,775	—
Yorkshire	26,817	27,663	40,033	56,506
Inbred registry[a]	—	562[b]	708	805

[a] This includes the Minnesota no. 1, no. 2, no. 3 and no. 4, Montana no. 1, Maryland no. 1, Beltsville no. 1, Palouse, San Pierre, and Conner Prairie Farms no. 1 and no. 2.
[b] 1966.

Breeds of Swine in the United States

The major breeds are listed in Table 12.1, along with their place of origin and some identifying characteristics of each breed. Most of the breeds of swine in the United States, including the two most common breeds, the Duroc and Hampshire, were developed in the United States.

The numbers of animals registered in the respective breed associations during recent years are summarized in Table 12.2. Each of the major breed associations has developed a system of on-the-farm testing whereby superior sires and dams are identified on the basis of the performance of their progeny. The criteria include number of pigs per litter, weaning weight of the litter, postweaning growth

Table 12.3.
Meat certification among major U.S. breeds.[a]

	No. of certified meat sires			*No. of certified litters*	
	Total to date	*1970*	*Per 1,000 registration*	*Total to date*	*1970*
Berkshire	155	9	1.12	1,920	138
Chester White	65	2	0.26	960	47
Duroc	695	57	0.75	5,280	411
Hampshire	1,156	52	0.70	12,571	603
Landrace	84	9	1.02	988	95
Poland China	456	12	0.74	2,811	115
Spotted	135	16	1.15	1,486	185
Yorkshire	361	21	0.37	4,966	482

[a] *National Hog Farmer*, March 1971, p. 59. Webb Publishing Co., 1999 Shepard Road, St. Paul, Minnesota 55116.

rate, and carcass measurements of representative pigs in the litter. Litters that meet all minimum standards are designated Certified Litters. Sires that produce some specified number of Certified Litters are designated as Certified Meat Sires, and repeat matings of sire and dam that have produced a Certified Litter are designated Certified Matings. Breed comparisons on this basis are shown in Table 12.3.

The Hampshire Registry Association has the longest history of carrying out Meat Certification programs, and at the present time has produced the largest number of Certified Meat Sires and the largest number of Certified Litters among the breeds of swine in the United States. It must be recognized, of course, that superior animals must be tested to be identified, so that the progress of any breed will depend on the soundness of its testing program.

Production Traits Among Breeds

No one breed has a monopoly on superior animals in any one productive trait or group of traits of economic importance, such as prolificacy (litter size), growth rate, efficiency of feed utilization, and carcass characteristics. It is possible to find individuals in all breeds that are outstanding in any one trait or group of traits. However, some breeds have developed a reputation for excelling in particular traits. For example, the Yorkshire and Landrace breeds are noted for their prolificacy, as is supported by the data in Table 12.4.

Table 12.4.
Litter size of nine major breeds of swine in the United States.[a]

Breed	*No. of litters*	*No. of pigs per litter*
American Landrace	192	10.5
Berkshire	177	8.1
Chester White	604	10.6
Duroc	367	9.8
Hampshire	447	9.6
Poland China	715	8.5
Spotted	95	10.0
Tamworth	29	9.0
Yorkshire	207	11.1
Total	2,833	9.6

[a] Self, 1959.

Table 12.5.
Carcass characteristics of some breeds of swine in the United States.[a]

	Duroc	*Yorkshire*	*Hampshire*	*Poland China*	*Spotted*
Back fat, cm.	3.7	4.0	3.3	3.6	3.8
Loin eye area (cm.^{2})	23.8	24.3	26.1	26.3	24.7
Per cent of lean cuts	52.9	52.5	54.5	52.2	51.4
Per cent of ether extract in lean	7.0	3.7	4.4	4.3	4.3

[a] Jensen *et al.*, 1967. Total of 585 pigs from 268 dams and 116 sires.

Hampshire pigs have a reputation for excelling in carcass leanness, based on their generally high placings in carcass contests in major all-breed swine shows and in the high rank often attained in testing-station performance tests. Jensen *et al.* (1967) found the lowest average backfat thickness and the highest average percentage of lean cuts for Hampshire carcasses among five breeds compared (Table 12.5). Breed comparisons made for several years at the National Barrow Show held annually at Austin, Minnesota, also show that Hampshire carcasses tend to be lean, but it is clear that by no means all the top carcasses are produced by Hampshires. The Grand Champion barrow over all breeds at the 1969 National Barrow Show is shown in Figure 12.4. The long, lean appearance of this pig represents the desired characteristics of present meat-type pigs. In other carcass traits, the Landrace and Yorkshire breeds are both noted for their length of carcass. The Landrace often has an extra pair of ribs, which is known to be associated with greater body length. Intact males (boars) excel females and castrated males (barrows) in growth rate and in carcass leanness. Because of these differences due to sex, it is important to consider the sex of the animal in making comparisons among or between animals in these traits regardless of the breed involved.

As emphasis on selection for meatiness has occurred among U.S. and European breeds, there has been an increased incidence of pale, soft, exudative (PSE) pork. Briskey and Lister (1968) have described this problem. It involves a loss in red meat color and in water-holding capacity of the muscle, making the pork less attractive at the meat counter and more prone to shrinkage losses. The syndrome seems to be more common in very lean pigs, and there is evidence that it is genetically related. Bray (1968) suggested that "PSE pork appears to be prevalent in Poland China, Hampshire, and Landrace breeds," and added that Danish researchers have found a high incidence in Danish Landrace swine and in the Pietrain, a Belgian breed noted for its large ham. Thus the problem is not confined to the United States. At the present time, there are not enough controlled observations on large numbers of animals to conclude that breed differences do, in fact, exist in susceptibility to this syndrome.

Again, it must be emphasized that making generalizations about strong or weak points of particular breeds is dangerous and can be fallacious, because of the wide variation among individuals within breeds. In fact, to have final proof of any assertion about the ranking of a particular breed in a particular trait, it would be necessary to measure that trait in every member of the breed concerned!

Plasticity of Swine

As one might guess from the earlier discussion, and from the wide differences in size and ap-

Figure 12-4 Grand Champion market barrow at the 1969 National Barrow Show, Austin, Minnesota, 1969. This purebred Hampshire had the Reserve Champion carcass with a live slaughter weight of 225 lb., carcass length of 79.0 cm. (31.6 in.), back fat of 2.65 cm. (1.06 in.), and a loin eye area of 44.0 sq. cm. (7.13 sq. in.). (Courtesy, Hampshire Swine Assoc.)

pearance of present-day swine around the world in the pictures in this chapter, swine respond rapidly to selection for particular traits. In 1950, scientists at the University of Minnesota, in cooperation with Hormel Institute, Austin, Minnesota, embarked on a program to develop a "miniature" pig for use in biomedical research. Starting with wild pigs obtained in the southern United States (Piney Woods), which are smaller than improved domestic breeds in the U.S., the program reduced the body weight by selection for small size from 39.2 kg. to an average of 17.5 kg. at 140 days of age after eleven generations. The average birth weight, but not number per litter, has decreased as well, although the only criterion of selection was body weight at 140 days of age (Table 12.6). Other "miniature" swine breeds have since been developed in other parts of the world by a similar procedure. One such project is underway at the University of Göttingen in Germany, using a Vietnamese pig and the Minnesota Miniature pig as the foundation stock (Figure 12.5).

With the increased demand for lean pork in the United States, most breeders are attempting to identify lean breeding stock. Sci-

Table 12.6.
Changes in body size of miniature pigs.[a]

Year	*No. of litters*	*Live pigs born per litter*	*Birth wt. in kg.*	*56-day wt. in kg.*	*140-day wt. in kg.*
1951	13	6.4	.89	8.8	39.2
1954	79	7.2	.85	7.1	35.8
1958	133	5.9	.72	6.0	22.6
1961	99	6.7	.77	5.8	17.5

[a] Dettmer *et al.*, 1965.

Figure 12-5 A Minnesota Miniature pig (left), a Vietnamese pig (right) and a cross between the two (center), being used in a project at the University of Göttingen, Germany, to produce a miniature pig for medical research.

entists at the U.S. Department of Agriculture, Beltsville, Maryland (Hetzer, 1967), undertook an experiment to determine the effects of selecting breeding stock on the basis of a single trait: backfat thickness as measured by a live-probe technique. Purebred Duroc and Yorkshires were used. Within each breed, a high-fat line and a low-fat line were selected. Within ten generations (ten years), three distinctly different populations were developed, within each breed: a randomly selected control group, a group with little back fat, and a group with much back fat. The rate of change obtained in each direction for each breed is shown in Figure 12.6. Although back fat was the only trait considered, it is interesting to note the other changes in carcass measurements that accompanied the changes in backfat thickness. Percentage of lean cuts and cross-sectional area of the loin eye at the tenth rib, both measures of carcass leanness, were both increased in the low-backfat groups of both breeds.

A classic example of the effects of a sustained progeny-testing program on carcass characteristics is that of the Danish Swine Testing program. Denmark depends on an export market for pork as an important part of its national agricultural income. To maintain a high-quality product that competes favorably on the foreign market, the first progeny-testing station was started more than 65 years ago in Denmark, and this program continues today with few changes. The Danish Landrace is essentially the only breed in Denmark; so the emphasis on improvement is centered on one breed. The Danish Landrace of today is perhaps the most uniformly lean breed in the world. By selecting boars for growth and carcass leanness, the average back fat has continued to decline and indices of lean have continued to improve over many years of testing (Table 12.7).

The small size of the Minnesota Miniature pig, the two extremes in fatness of the Yorkshire and Duroc lines in the USDA experiment, and the leanness of the Danish Land-

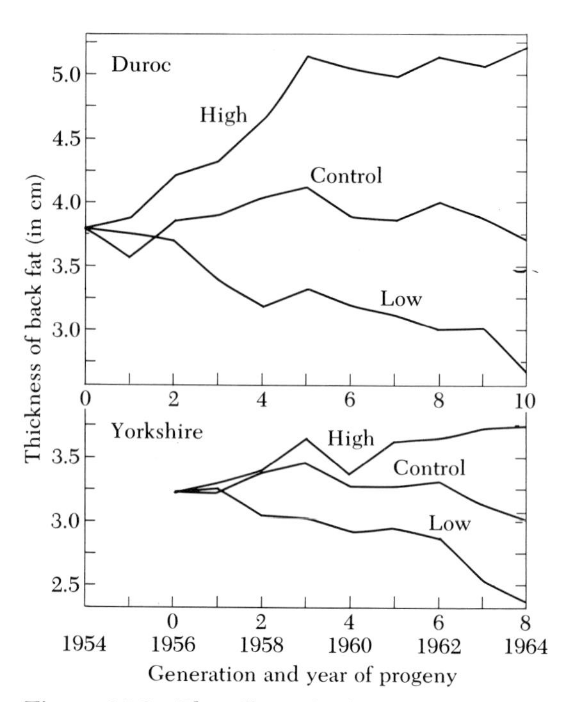

Figure 12-6 The effect of selection on back fat thickness in Duroc and Yorkshire pigs at USDA (reprinted by permission from *J. Animal Sci.* 26: 1244, 1967).

Table 12.7.
Changes in carcass traits of Danish Landrace pigs.[a]

Year	*Body length cm.*	*Back fat cm.*	*Area of loin eye sq. cm.*
1926/27	88.9	4.05	—[b]
1936/37	92.8	3.49	—[b]
1946/47	93.4	3.36	—[b]
1957/58	94.8	3.05	29.3[c]
1965/66	96.3	2.46	30.1
1966/67	96.3	2.41	31.1
1967/68	96.0	2.36	32.5
1968/69	96.4	2.34	31.8

[a] Clausen *et al.*, 1970, pp. 74–75.
[b] Not available.
[c] 1959/60.

race pig all illustrate the plasticity of the pig and its versatility in responding to new needs or new uses.

Breeds for the Future

The variability within breeds and between breeds in productive traits and the plasticity of breeds in response to selection pressure offer much hope for the future of the pig in contributing to man's needs. The preservation of the diversity of germ plasm existing among the many breeds and varieties of swine in the world is important so that unforeseen needs can be accommodated. This diversity of germ plasm can be used in forming new breeds and in improving existing ones.

FURTHER READINGS

Bray, R. W. 1968. "Variation of quality and quantity factors within and between breeds." In Topel (1968), pp. 136–144.

Hetzer, H. O., and W. R. Harvey. 1967. "Selection for high and low fatness in swine." *J. Animal Sci.*, 26:1244.

Mason, I. L. 1969. *A World Dictionary of Livestock Breeds, Types, and Varieties.* Edinburgh: Commonwealth Bureau of Animal Breeding and Genetics.

Self, H. L. 1959. *Traits of Major Breeds. National Hog Farmer* Swine Information Service Bulletin no. B3.

Thirteen

Sheep, Goats, and Other Fiber-Producing Species

"My job as president is to separate the sheep from the goats and to keep the goats from getting sheepskins." *

A remark credited to Bancroft Bentley, President of Simmons College, Boston

Domestication and Early History of Sheep

Sheep and goats were probably among the earliest animals to be domesticated—some six to eight thousand years ago, as shown by evidence reviewed by Reed (1959). The tractability of sheep, which can be observed in wild forms today, and the versatility of their products, including meat, milk, wool, and skins, were no doubt dominant factors leading to their early husbandry by man. Sheep are sometimes, though rarely, used even for work and sport.

Sheep and goats are the most versatile of all domesticated animals in the wide variety of foods they can survive on. Killebrew wrote of sheep in 1880:

> "No domestic or wild animals are capable of existing on more different sorts of food. Weeds, grasses, shrubs, roots, cereals, leaves, barks, and even in times of scarcity, fish and meats all furnish a subsistence to this wonderful animal. They will, in the great pine forests of Norway and Sweden, subsist upon the pungent resinous evergreens through a hard winter. The cultivated grasses of the temperate zone, clover and cereals, are as a matter of course the best food for them, but in the absence of these they will gnaw the barks and crop the leaves of the forests. Among the Laplanders, when all other kinds of food fail, they will eat the dried fish of the people, or the half-rotten flesh of the walrus; or in the cases of extreme destitution, they will eat the wool off each other's backs."

Breeds of sheep have been developed in relatively isolated geographic areas where adaptation to peculiar environmental factors has been dominant. Thus the Merino of Spanish origin thrives in hot, dry climates, and their herding or flocking instinct is well-developed. These sheep have been accustomed

* Seriously, there are some real problems in distinguishing sheep and goats taxonomically, and some of these are indicated in this chapter. See also Reed and Schaffer, 1972.

through many centuries to long trails, and those that failed to stay with the flock probably did not leave descendants. On the other hand, the British breeds—usually kept in enclosed pastures in a more northern climate—do not band together as tightly as the Merino, and also, like wild types, have retained the tendency to breed at such a time that lambing will occur in the spring, when temperatures and plant growth favor survival. The fat-tailed and fat-rumped sheep in the deserts of Asia store fat during the lush season that can be drawn on during the often long periods when plant growth is dormant. Fine-wool Merinos were developed in a dry climate, the coarse- and long-wool types, such as the Lincoln and Romney, in more humid climates.

Breeds of sheep have also resulted from selection for specialization to fill the need for meat or for a particular kind of wool. Some native breeds in Central and Southern Europe, such as the German East Friesian or the Italian Langhe, are kept primarily for milk; the world-famous Roquefort cheese is made from ewe's milk.

The wild Rocky Mountain bighorn sheep has long been present in North America, but it has never been domesticated; all our present domestic breeds of sheep or their ancestors have been imported. Domesticated sheep were first introduced on the American Continent by the Spanish conquerors (Carmen *et al.*, 1892); Columbus brought sheep on his second voyage in 1493. Sheep were brought into the English Colonies almost as soon as they were settled, beginning with Jamestown in 1609.

Figure 13-1 A Rambouillet ewe. Courtesy Ameri Rambouillet Sheep Breeder Association.

Important importations of Spanish Merino sheep into the United States began about 1801. Of particular interest are the Merinos that have been bred at Rambouillet, France, since 1786. Some of these were brought to the United States in 1840, but the large importations from both France and Germany were made in the 1890's (Figure 13.1). These Rambouillets and Merinos have been the foundation of the western sheep industry, and they and their crossbred descendants today constitute a large majority of this country's ewe flocks, especially those of the western range states and Texas (Figure 13.2). In many states of the Middle West and East also, the Western ewe is the chief producer of lamb meat and wool for market.

The early sheep imported from England were not of the quality of modern English breeds (Connor, 1918), because breeding improvements only began in England in the eighteenth century. Before then, the English sheep were relatively coarse, leggy, late-maturing animals, but with good foraging qualities. The increase of industrial development in the United States, with the influx of foreign labor accustomed to eating mutton and lamb, resulted in large imports of mutton sheep. Continued importation of other medium-wool mutton breeds, such as the Hampshire, Suffolk, Oxford, Shropshire, and Cheviot, now account for most of the meat-type breeds that are most common in the United States.

Wentworth, in *America's Sheep Trails* (1948), presents a comprehensive review of the history of sheep production from the early development in the eastern colonies, and the movements into the Middle West and Southwest in the early 1800's, to the completion of the westward movement at the close of the Civil War. In the following years, sheep numbers fluctuated from 35 million head to a peak

Figure 13-2 Columbia ewes on a ranch in Colorado. This breed was developed by crossing one of our largest-bodied breeds, the Lincoln, onto a fine-wooled breed, the Rambouillet. (Courtesy, Columbia Sheep Breeders Association of America.)

of more than 50 million at the beginning of World War II. Sheep numbers, as they are referred to in the United States, denote breeding sheep one year or older. Few mature wethers are maintained only for wool in this country, unlike many other regions of the world. In 1970 numbers of sheep in the Unted States had declined to slightly less than 20 million, although sheep numbers of the world were at an all-time high of over one billion head. There are more domesticated sheep in the world than any other domesticated species.

Sheep of the World

Few countries in the world have no sheep (Figure 13.3). They are found in tropical countries and in the arctic, in hot climates and in cold, on the desert and in humid areas.

There are over 800 breeds of sheep in the world (Mason, 1969), in a variety of sizes, shapes, types, and colors. Not all sheep have wool—the sheep of the Congo and the Fulani of West Africa are two breeds that have hair. The colors of sheep may be white, brown, black, red, or combinations. Wool-type breeds have become most important in the southern hemisphere, particularly in Australia, New Zealand, South Africa, and Argentina and Uruguay in South America, but both fine- and long-wool breeds are found all over the world. Fat-tailed and fat-rumped sheep are common in the desert and semidesert areas of the Near East, Asia, and Africa. They usually produce carpet wool, although wool yields are generally low and the sheep are valued for many purposes. Dairy breeds of sheep are more common in Central and Southern Europe. Northern short-tailed varieties, noted for their prolificness, are found in the Scandinavian countries. Meat breeds originating from Britain, which are common there as well as in the United States and Canada, are widely distributed throughout the world.

Sheep breeds specialized for products other than meat and wool are uncommon in the United States. Although milking of sheep is

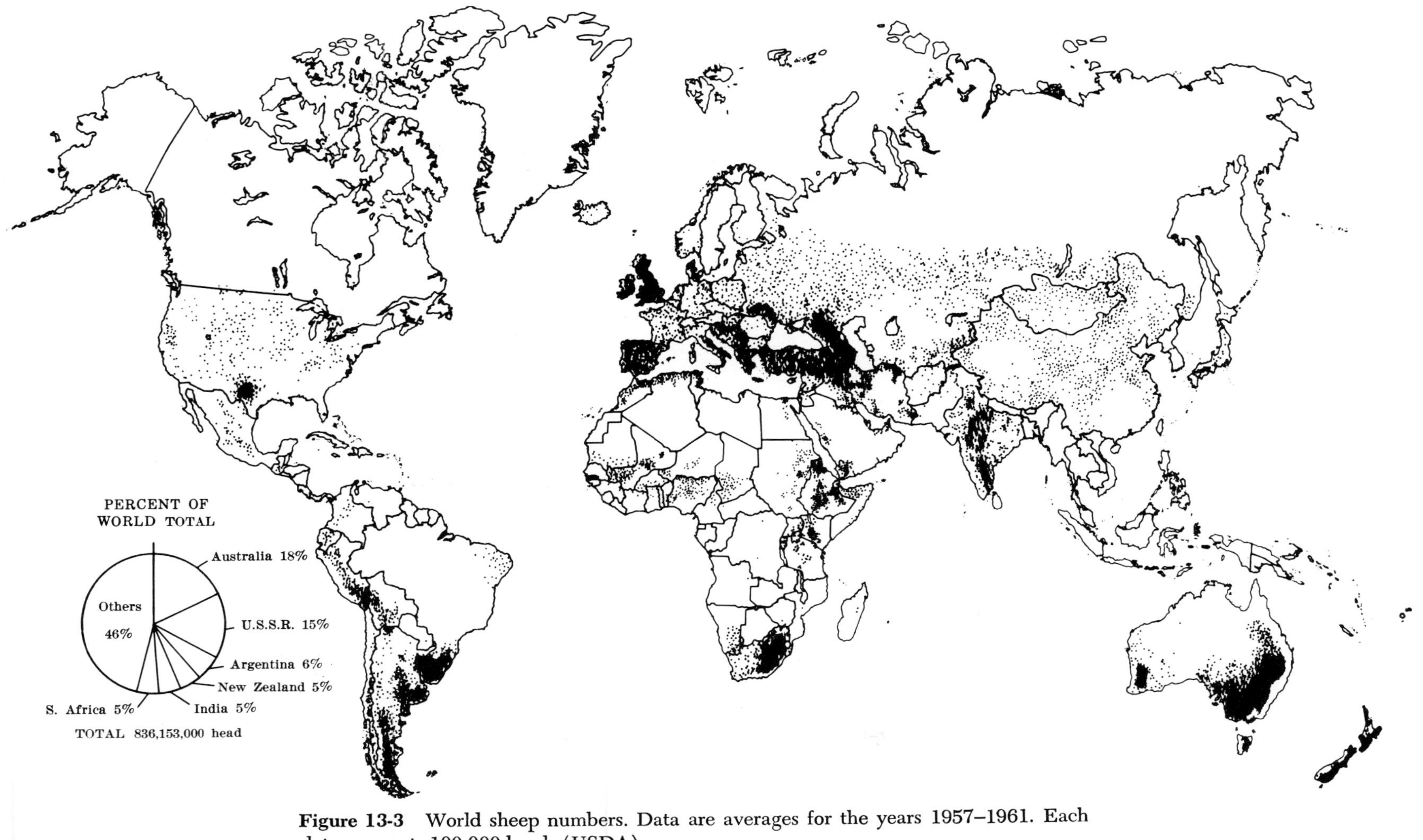

Figure 13-3 World sheep numbers. Data are averages for the years 1957–1961. Each dot represents 100,000 head. (USDA)

common in many countries of the world, such breeds are often triple-purpose, being also kept for meat and wool. The production of fur pelts and skins from sheep is important in many parts of the world, particularly in the U.S.S.R. and the Middle East.

World Breeds That May Contribute to the American Sheep Industry (Scott, 1971; Terrill, 1970)

The Border Leicester are large sheep (rams weigh 250 to 325 lbs.) with long, coarse wool, native to Scotland. They have white faces, bare of wool as are the legs. While they are not noted for prolificacy, they are good milkers and superior mothers. They are used in the United Kingdom, New Zealand, Canada, and the high rainfall areas of Australia and South Africa to produce crossbred ewes for fat lamb production. A few are being used experimentally in the United States.

Finnish Landrace or Finnsheep are also being used experimentally in this country (Figure 13.4). They are native to Finland and are small (rams weigh up to 150 lbs.). Their most important trait is high prolificacy, with lamb crops of 250 to 450 per cent. The gestation period is short, and they may breed at almost any time of the year. Results of work

Figure 13-4 Finnish Landrace ewe (Finnsheep) with triplets. This is a relatively small-bodied breed noted for its prolificacy. This breed should be useful in crossing with breeds with more desirable wool or meat characteristics. (Courtesy Edward Arnold, Publishers, Ltd., London.)

Figure 13-5 An Awassi ewe. This breed was developed in the Near East. Milk production records as high as 1,950 pounds for one lactation have been reported. (Finci, 1957.)

to date indicate the breed is capable of transmitting part of this high fertility to crossbred offspring.

The Texel is raised in the Netherlands. It has a white face, and rams weigh from 240 to 290 lbs. It has meat-type conformation, is prolific, and is noted for milk production. It produces 10 to 13 lbs. of wool that measures 48's to 50's in fineness.

The East Friesian in Germany is another breed that is noted for milk production. It is similar to the Texel in appearance. Still another breed used for dairying is the Awassi, found in much of Asia and parts of Africa (Figure 13.5). It has the ability to withstand severe heat and to survive on poor grazing conditions in low-rainfall areas. Its fat tail provides a reserve storage of energy for periods of scant feed supplies.

In Israel, the East Friesian is being crossed with the Awassi to increase milk production. The Awassi can produce 40 to 60 gallons during an eight-month lactation, whereas the East Friesian can produce 3 to 4 times this amount. The East Friesian does not have a normal size tail, and neither do the crossbreds, which may present a problem under the customary methods of raising sheep.

The Ile-de-France has been selected to produce lambs out of season. It has a white face, and is of medium size and wool yield.

The North Country Cheviot, native to Scotland, is found in Canada and is being tested experimentally in the United States. It is similar in appearance and other characteristics to the Cheviot, its ancestor, except that it is much larger. Ewes will weigh 150 to 170 lbs., with rams weighing 100 lbs. more.

The Romano of the U.S.S.R. is reported to have lambing rates comparable to the Finnish Landrace. Its body conformation makes it a poor source of meat, but fur from the lambs and carpet wool from the mature sheep are produced.

The fat-tailed Han-Yang of China is about the size of the Finnish Landrace, and is also reported to have litters. Estrus occurs throughout the year.

Sheep Breeds in the United States

There are 23 purebred breed registry associations in the United States representing 20 distinct breeds (see Tables 13.1 and 13.2 for data on 15 breeds). Many occur in such small numbers that they are of little significance to the industry. Briggs (1969) has discussed in detail the history, characteristics, distribution, and importance of sheep and goat breeds in the United States. In commercial sheep production, few sheep are maintained as straightbreds.

The traditional method of classifying breeds of sheep is according to the grade of wool they produce: fine wools, 60's and finer; medium wools, 50's to 58's; long wools, 48's and coarser. These numbers refer to the number of hanks of yarn that can be obtained from a pound of clean wool. A hank is 560 yards.

It is now more popular and meaningful, especially for commercial production, to group the breeds into the following classes: wool or ewe breeds; meat or ram breeds; other breeds.

The six wool or ewe breeds are shown in

Table 13.1.
Breeds of sheep in the United States.

					Fleece		
Classification	*Breed*	*Country of origin*	*Approximate date of origin*	*Approximate date of first importation*	*Fineness*[a]	*Length*[b]	*Weight*[c]
Wool or ewe breeds	Delaine Merino	Spain	Early	1801	Fine	Medium	Heavy
	Rambouillet	France	1786	1840	Fine	Medium	Heavy
	Debouillet	United States	1920	—	Fine	Medium	Heavy
	Corriedale	New Zealand	1880–1910	1914	Medium	Medium long	Heavy
	Columbia	United States	1912	—	Medium	Medium long	Heavy
	Targhee	United States	1927	—	Medium fine	Medium	Heavy
Meat or ram breeds	Suffolk	England	Early 1800's	1888	Medium	Short	Light
	Hampshire	England	Early 1800's	1881	Medium	Medium	Medium
	Shropshire	England	1860	1860	Medium	Medium	Medium
	Dorset	England	About 1815	1887	Medium coarse	Medium	Medium
Other breeds	Southdown	England	Late 1700's	1803	Medium	Short	Light
	Montadale	United States	1933 on	—	Medium	Medium	Medium
	Romney	England	Early	1904	Coarse	Long	Heavy
	Karakul	Asia	Ancient	1909	Coarse	Long	Light

[a] In general, fine wool averages 64's or finer, medium wool from 50's to 62's, and coarse wool 48's or coarser, in U.S. Numerical Grades.

[b] In approximate terms of one year's growth, long wool would be 6 in. and longer, medium wool from 2 to 6 in., and short wool under 2 in.

[c] Heavy fleeces generally average over 12 lbs., medium fleeces from 8 to 12 lbs., and light fleeces under 8 lbs., for one year's growth from mature ewes producing lambs.

Table 13.1. The color of the hair on the face, ears, and legs is white which is an advantage in quality wool production. They are selected for: (1) quality of wool production; (2) reproductive efficiency; (3) mothering ability; (4) milk production; and (5) adaptability to a variety of environmental conditions, including herding. It is estimated that these six breeds comprise 70 to 80 per cent of all the sheep in the United States.

The four meat or ram breeds are selected for: (1) rapid growth; (2) large mature size; and (3) a high meat-to-bone ratio.

These ram breeds are crossed on the ewe breeds to produce crossbred market lambs. The resulting heterosis of these offspring increases the vigor and livability of the newborn. The crossbred ewe also has several important attributes, such as prolificacy, milk production, and body size.

The remaining breeds are used in crossing under special conditions. The Southdown cross produces a high-quality carcass for special eastern markets, but is small. The Karakul breed is one of the oldest, dating back several thousand years, but is rare in the United States. Karakul lambs are sacrificed for their fur pelts a day or two after birth.

Ewe or Wool Breeds

All Merino sheep are so closely related and so similar in type that they may be considered together. Merino sheep, originating from

Spain, have had and still do have great influence on wool production of the world and in the formation of new breeds. The "Marvelous Merino" of Australia has proved that this breed has amazing adaptability to varied and harsh environments. The Merino was the most important breed in the early history of the sheep industry in this country, but because of small body size, slow growth rate, and poor carcass conformation, its numbers have rapidly declined except in the Southwest.

The Rambouillet was developed in France from the Spanish Merino through the encouragement of Louis XVI. It has met with great success in the United States, and today is the foundation of most of the western sheep flocks. It succeeded where the Merino failed because of its larger size and freedom from excessive skin wrinkles while still producing excellent wool. Like any descendants of the Merino, it has the great flocking instinct so important to range sheep production.

The Corriedale originated in New Zealand during the late nineteenth century (Figure 13.6). Breeds used as foundation were the Lincoln, Leicester, and Merino. Corriedales were imported to the United States in 1915, and have proved very satisfactory for range and farm flock production. They perhaps come nearer to being a truly dual-purpose sheep than any other breed.

The other three wool or ewe breeds originated in the United States. The Columbia and Targhee breeds were established at the U.S. Sheep Experiment Station, Dubois, Idaho, and released to the public about 1920 and 1940, respectively. No other breeds of sheep in this country were developed by research stations. Both of their breed associations require that the offspring of registered parents must pass inspection to be registered by an approved inspector of the association. This is unique in registering, compared to most associations. Lincoln-Rambouillet crosses were used to found the Columbia, whereas the Targhee carries approximately ¼ Lincoln and ¾ Rambouillet breeding. These breeds were developed as productive and efficient northern range sheep,

Table 13.2.
Body characteristics of sheep breeds in the United States.

Classification	*Breed*	*Body size*[a]	*Wool on face*[b]	*Hair color*	*Horns*	*Registrations in 1970*
Wool or ewe breeds	Delaine Merino	Medium	2	White	Rams only	109
	Rambouillet	Large	2	White	Rams only	7,683
	Debouillet	Medium	2	White	Rams only	1,275
	Corriedale	Medium-large	2	White	Polled	9,020
	Columbia	Large	2	White	Polled	7,809
	Targhee	Medium-large	2	White	Polled	1,570
Meat or ram breeds	Suffolk	Large	1	Black	Polled	36,491
	Hampshire	Large	2	Black	Polled	23,064
	Shropshire	Medium	2	Black	Polled	5,215
	Dorset	Medium	2	White	Rams and ewes: polled or horned	—
Other breeds	Southdown	Small	3	Gray	Polled	5,677
	Montadale	Medium	1	White	Polled	2,419
	Romney	Medium	2	White	Polled	1,105
	Karakul	Small	1	Black	Rams only	—

[a] Large mature rams in good condition would generally range from 250 to 350 lbs., medium rams from 200 to 275 lbs., and small rams 160 to 200 lbs. Ewes are 50 to 75 lbs. lighter.

[b] 1 = completely bare; 2 = wool down to eyes; 3 = wool extending below eyes.

Figure 13-6 Corriedale ewe. (Courtesy American Corriedale Association.)

and this objective has been accomplished. They also are well-adapted to farm flock commercial production for mating with any of the ram breeds. Lincoln and Rambouillet breeds were used to establish the Panama breed in Idaho about the same time as the Columbia breed was being established. The Panama breed was established by private breeders under range conditions.

The Debouillet was developed in New Mexico from crosses of Delaine-Merino and Rambouillet sheep in the late 1920's. Selection has been for adaptability to arid range conditions, and they are particularly well-adapted to range sheep production in New Mexico and Texas.

Ram or Meat Breeds

Unlike the ewe breeds, all five ram breeds originated in England. Suffolk sheep were first imported into the United States in 1888 (Figure 13.7). They excel in rapid growth and muscling, thus being most suitable for crossing on the wool breeds for commercial lamb production or for the production of crossbred ewes. These crossbred ewes are much preferred over straight Suffolks, which are one of the poorest sheep in weight and quality of wool produced. Although commercial flocks of straight Suffolks are therefore few, they are widely distributed throughout this country for crossbreeding.

Hampshire sheep, like the Suffolk, are popular for crossing with the ewe or wool breeds, since they too excel in growth and carcass merit (Figure 13.8). Hampshire sheep are found throughout the United States, where they are used for crossing under range conditions and crossing or straightbred in the farm-flock states. Since they produce slightly more wool than the Suffolk, there are more straightbred commercial flocks in the Midwest region. The majority of experiments comparing Hampshires and Suffolks as sires used in crossbreeding show no significant differences.

Shropshire sheep for many years were the most popular farm-flock breed (Figure 13.9). Theirs has been the tragic example of show-ring fancy nearly eliminating a breed. From the 1930's into the 1950's, the "ideal" Shropshire face was covered with wool. Immediately after World War II, importations of open-face sheep were made from England and this type of Shropshire has been encouraged to date. Shropshires today are open-face, large, prolific, and produce more wool than Suffolks or Hampshires. They are excellent for crossing or as a straightbred commercial sheep, but have not regained their earlier widespread popularity because other breeds such as the Suffolk expanded rapidly and proved their worth, while Shropshire breeders were selecting for "wool from the tip of the nose to the tip of the toes."

In 1948, a polled mutation occurred in the naturally horned Dorset flock at North Carolina State College. The inheritance of this polled characteristic is still not fully understood, since it is not a single dominant gene as in cattle. In 1958, the College made 15 polled rams available to the public, and polled Dorsets now far out-number those with horns. Dorset ewes have a shorter anestrous period than most breeds, and the rams are aggressive during hot weather. It is these traits that make them suitable for crossing on wool-breed ewes.

Breeds of Sheep Without Associations in the United States

In Tables 13.1 and 13.2, only the breeds that have formal Registry Associations were listed. Since Briggs (1969) defines a breed as "a group of animals that as a result of breeding and selection have certain distinguishable characteristics," there are others worthy of brief mention. Most are not common outside their area of origin.

The Navajo sheep of New Mexico have been used for many years to produce wool for weaving the famous Navajo blankets. The "native" sheep of the southeastern United States, while not truly native to this country, have developed, through natural selection, traits necessary for survival in a warm, humid climate.

Figure 13-7 Suffolk ram. The Suffolk is one of the most popular meat sire breeds in the U.S. (Courtesy, National Suffolk Sheep Association.)

There are at least four breeds of Minnesota sheep developed at the State Agricultural Experiment Station. They are mostly three-breed crosses, and are known as Minnesota 100, 102, 103 and 105.

Foreign or exotic breeds found in this country in somewhat larger numbers include the Finnish Landrace or Finnsheep, Border Leicester, and North Country Cheviot, which were discussed previously. Most of these are being imported from Canada, in order to avoid exotic diseases.

Figure 13-8 Hampshire ram. Hampshire rams are used extensively in crossing on white-faced ewes for lamb production in western U.S. (Courtesy American Hampshire Sheep Association.)

Research on Comparison of Breeds

Experimental comparisons of sheep breeds have been uncommon. Cooper and Stoehr (1934) under range conditions found Columbias to be heaviest in body weight, followed by Rambouillets and Corriedales. Columbias and Rambouillets produced heavier fleeces than Corriedales, and Columbias and Corriedales produced greater numbers and pounds of lamb per ewe than did Rambouillets. Losses from dead and missing sheep were lowest for Rambouillets and highest for Columbias. Terrill and Stoehr (1942), studying the same breeds under the same conditions, found that Corriedales produced more pounds of lamb per pound of ewe than did Columbias or Rambouillets, and Rambouillets produced more pounds of wool per pound of body weight than did Corriedales or Columbias. Bennett *et al.* (1963) found Targhees, Rambouillets, and Columbias ranked in that order in lambs weaned per ewe bred. The order for average weaning weight and pounds of clean wool was Columbia, Targhee, and Rambouillet.

Sidwell *et al.* (1962, 1964) found that Hampshires, Merinos, Shropshires, and Southdowns ranked in that order in lambs weaned per ewe bred, and in the same order, except that Shropshires preceded Merinos, in average weaning weight (lambs weaned per 100 ewes: 100.3, 98.0, 80.3 and 78.6, respectively).

Suffolk lambs were heavier at weaning than Corriedale lambs in a Wyoming study (Botkin and Paules, 1965). Corriedale ewes produced significantly more wool.

Rams of Minnesota 100, 102, 103 and 105 did not produce lambs at weaning as heavy as Hampshire or Suffolk rams (Singh *et al.*, 1967).

A comparison of Suffolk, Targhee, and Shropshire sheep at Wisconsin (Bradley, 1964) showed Suffolks to be superior in growth and Targhees superior in wool production. There were significantly fewer Targhee lambs lost from birth to weaning than with Suffolks and Shropshires. Since there was no significant difference in numbers of lambs born, the result was that Targhee ewes produced as many pounds of lamb at weaning as the Suffolk and, together with their superior wool weight, were more productive ewes.

There may be important differences in flavor of lamb meat from different breeds, although little is known about such a comparison. Cramer (1970) found Hampshires to have the least amount of mutton flavor, followed by Columbia, Targhee, and Rambouillet. Flavor appeared to improve with decreasing fineness of wool. Lamb consumption could be increased in this country if causes for objections of some consumers to flavor and cooking odors could be lessened or removed.

One must use caution in drawing conclusions from breed comparison research. At best they serve only as indicators, because the breeds used are not representative of the population of that breed over the entire country. Similarly, each experiment was conducted in only one particular environment.

Breeds of Goats

Goats are similar to sheep, and some are difficult to distinguish from sheep; the tail of the goat is said to turn always upward and that of the sheep downward, yet this may be unreliable. Horns, *per se,* do not distinguish between goats and sheep, nor does wool or hair. The absence of facial glands, of lachrymal pits in the skull, and of glands in the hind feet, and the presence of a beard and odiferous tail glands in the male, are all characteristic of goats (French, 1970). It is not certain that crosses of sheep and goats have produced viable offspring, but fertilization and early embryonic growth do occur (Gray, 1954). Goats, as well as sheep, produce a wide variety of products; milk, mohair, meat and skins are the most important.

Figure 13-9 Shropshire ram. Selection for excessive face covering reduced their popularity as a meat sire in the west, but the modern open-faced Shropshire is still popular in the midwest. (Courtesy American Shropshire Registry Association.)

Goat hair is used for a number of commercial purposes such as the manufacture of felts, velours, cords, carpets, and brushes. By far the most important type is that from the Angora goat, which is known in the trade as mohair.

Angora Goats

Angora goats appear to have originated before Biblical times, since references to the use of goats' hair are found in the Bible (Thompson, 1901). The area of origin surrounding Ankara (Angora) in Asiatic Turkey is a high plateau of about 3,000 feet elevation, with mountains and deep valleys. The climate, which is dry with extremes in temperatures, was probably favorable for the development of the long, lustrous mohair. Today, the leading countries in mohair production are the United States, Turkey, and South Africa.

Angoras were first imported into the United States in 1849. They spread westward after the Civil War, particularly to Texas and California. Texas has been the leading state ever since. In recent years, Texas, with more than 4 million head, has had about 95 per cent of the Angoras in the United States. Other leading states are Arizona, New Mexico, Missouri, Oregon, California, and Utah.

The Angoras, with their long locks of mohair, present a distinctive appearance (Figure 13.11). The locks are of different types, such as ringlet, flat, or web, as described by Gray (1959). The mohair locks grow 6 to 12 inches per year, and the goats are normally clipped twice each year. Mohair becomes coarser with age, and the fine kid mohair is most valuable. Quality, covering, weight of fleece, and freedom from kemp is emphasized in selection. The average mohair clip per goat has in-

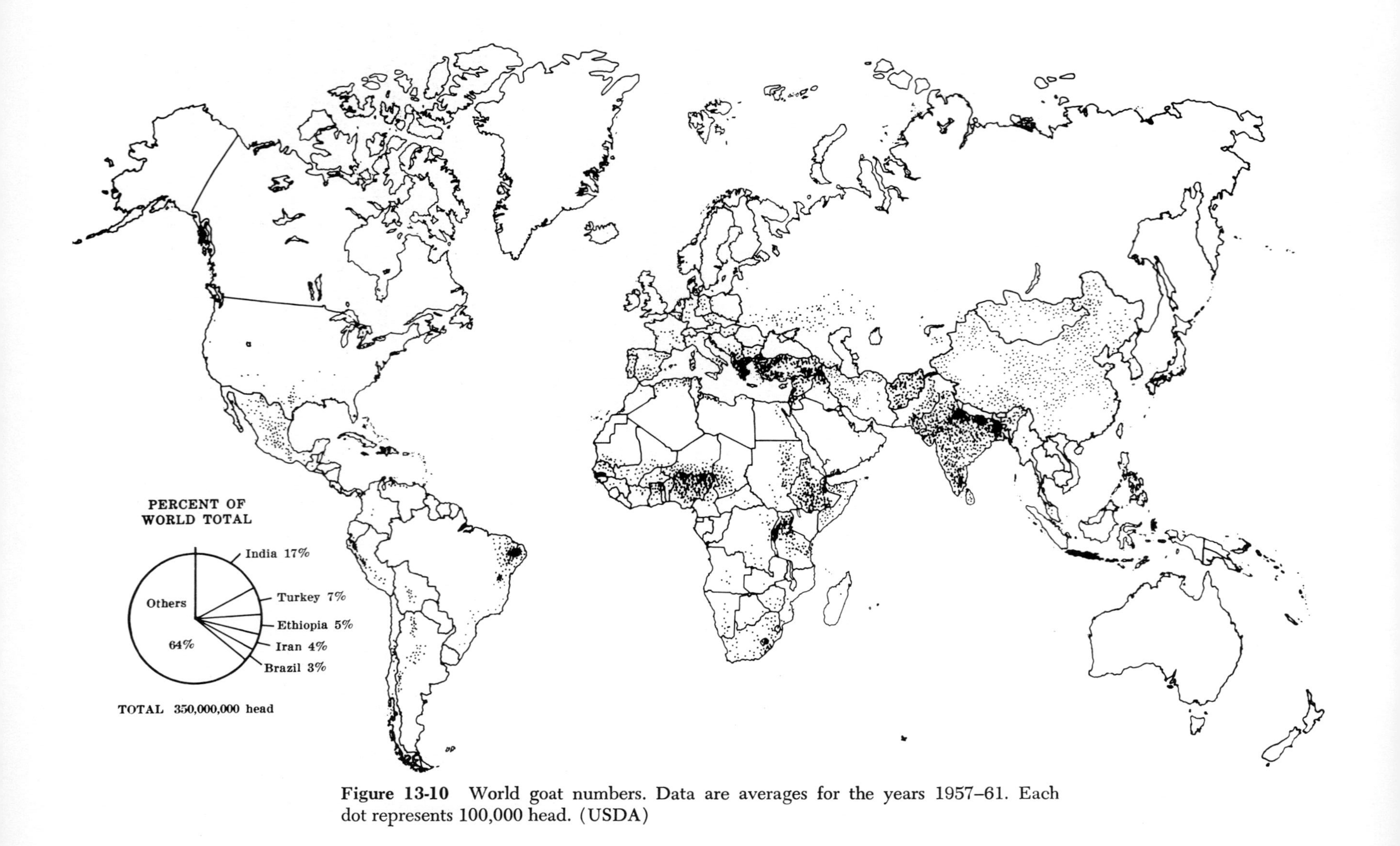

Figure 13-10 World goat numbers. Data are averages for the years 1957–61. Each dot represents 100,000 head. (USDA)

creased from about 5 pounds in the late 1940's to about 6.5 pounds in the early 1960's.

Mohair has a much smoother surface than wool, with a very smooth, thin scale; so it does not have the felting properties of wool. It is used for making upholstering material and is often mixed with other fabrics to make men's summer suits if additional strength is desired.

Both sexes of Angoras have horns. Open faces are associated with higher kid production (Shelton, 1960). Size and weight for age are also emphasized in selection. A yearling buck should weigh at least 80 pounds, mature bucks from 125 to 175 pounds. Yearling does should weigh at least 60 pounds, and mature does from 80 to 90 pounds.

Angoras and sheep are often grazed together, and total production is increased because Angoras can browse brush more effectively than sheep. They do this by standing on their hind legs to reach as high as possible. They can be grazed on brushy areas unsuitable for sheep where grass is too sparse, and they are sometimes used to clear land where heavy brush needs to be eliminated.

Figure 13-11 Angora buck. This is the important Mohair breed in the U.S. (Courtesy Sheep and Goat Raiser.)

Cashmere Goats

Cashmere goats are long-haired and famous for their fine silky wool. They are raised in Tibet and northern India. The fleece is heaviest on the goats that live on high plateaus and mountains. The actual cashmere wool is a fluffy undercoat called *pasha* and lies under the goat's long hair. An average size goat gives only three ounces of this wool, which gives it its high value. The long outer coat is used in making tents, clothes and coarse carpets. They are also used to produce meat and milk.

Cashmere wool is long, straight, soft, and silky. It has a world-wide reputation and is used for making shawls, suits, cardigans, and hats. This goat has large lop-ears, twisted horns, and a rather small body, though size varies with environment. The goats which live in the higher altitudes are deep yellow in color; those raised in the valleys are pure white.

The Camelidae

Most of the other fiber-producing animals of the world belong to the camel family, Camelidae.

Camel

There are only two kinds of camels in the world. The Arabian camel, *Camelus dromedarius,* has one hump on its back and is found in the Middle East, India, and northern Africa. The Bactrian camel, *Camelus bactrianus,* has two humps and nearly all are found in Asia. There are many more Arabian camels than Bactrian camels in the world.

The long hair of the Bactrian camel is woven into cloth and fine blankets. The Arabs make their tents of camel's hair cloth. Other important uses of the camel, of course, include transportation, meat, milk, leather, and bones as a substitute for ivory.

Llama

The llama is used as a pack animal and is the largest member of the camel family found in South America (Figure 13.12). It has no hump and may stand 4 or 5 feet tall.

The llama's hair is coarse, and may be brown, white, black or gray. The Indians of Peru and Bolivia sometimes use the hair to make garments. It has little commercial value, though it is sometimes mixed with alpaca wool. The meat of the llama can be used for food.

Alpaca

The alpaca lives in the mountain regions of Peru, Bolivia, and Chile. It thrives at altitudes of 8,000 to 12,000 feet, and it produces true wool. The Indians of Peru raised alpacas and made the wool into cloth many years before America was discovered.

The alpaca looks much like the llama, but is not as tall. The wool is 8 or more inches long, with a count as high as 70, which is finer than most sheep but not as fine as that of the vicuña. The color may be various shades of brown, black or white. It is one of the best fibers known for making beautiful cloth that is soft and warm. Many thousands of alpacas are kept for their wool. They are sheared once a year.

Vicuña

Whereas the llama and alpaca are domesticated members of the camel family, the vicuña is wild. None have humps on their backs. The vicuña thrives in the Andes Mountains of Ecuador, Peru, and Bolivia in altitudes of 12,000 to 15,000 feet.

The vicuña is small, weighing 75 to 100 pounds, and, because of its alertness and speed, is difficult to hunt. There is now some domestication because of the value of the wool. The wool fibers are the finest of any wool-bearing animal, less than one-half as

Figure 13-12 Llamas and alpacas grazing in Bolivia. These species are closely related and can be crossed successfully. (Courtesy FAO, United Nations.)

thick as the finest sheep's wool. The fleece grows until it hangs below the knees and only the inner fleece is used. It is especially good for high-quality worsteds.

The color of the upper body is reddish-yellow to deep tan or reddish-brown. The belly and lower legs are white. Because of the value and demand for vicuña wool, the Peruvian government controls the slaughter of these animals and the sale of wool.

FURTHER READINGS

Botkin, M. P., and Leon Paules. 1965. "Crossbred ewes compared with ewes of parent breeds for wool and lamb production." *J. Animal Sci.*, 24:1111–1116.

French, M. H. 1970. *Observations on the Goat.* Rome, Italy: FAO.

Scott, G. E. 1970. *The Sheepman's Production Handbook.* Denver, Col.: SID, Inc.

Sidwell, G. M., D. O. Everson, and C. E. Terrill. 1962. "Fertility, prolificacy, and lamb livability of some pure breeds and their crosses." *J. Animal Sci.*, 21:875–879.

Thompson, G. F. 1901. *Information Concerning the Angora Goat.* USDA Bull. 27.

Wentworth, E. N. 1948. *America's Sheep Trails.* Ames: Iowa State College Press.

Fourteen

Poultry

"The cock, astrologer in his own way,
Began to beat his breast and then to crow,
And Lucifer, the messenger of day,
Began to rise and forth her beams to throw
And eastward rose, as you perhaps may know."

Chaucer,
Troilus and Criseyde

According to Hegner, some 14,000 living species of birds are classed into 25 orders. Domestic birds are classified into three orders under *Carinatae* (vertebrates with a keel): *Anseriformes* (ducks and geese); *Galliformes* (chickens, turkeys, guinea fowl, and pheasants); and *Columbiformes* (doves and pigeons).

Of greatest economic importance is the order *Galliformes*, which contains the largest number of domesticated species. Domestic poultry is further classified into families, including: *Phasianidae* (chicken and pheasants, of oriental origin); *Numididae* (turkeys and guinea fowl, of African origin); and *Meleagrididae* (turkeys, of American origin).

Domestic chickens are included under the species *Gallus domesticus*. The genus includes four wild species: *Gallus gallus*, the Red Jungle Fowl of India, Burma, and other southeast Asia countries (Figure 14.1); *Gallus lafayetti*, the Ceylon Jungle Fowl; *Gallus sonneratti*, the Grey Jungle Fowl, found in southwest India (Figure 14.2); and *Gallus varius*, the Javan Jungle Fowl with black plumage, which differs from the other jungle fowls in having a single median wattle and an unserrated comb.

The domestic fowl, *Gallus domesticus*, resembles most closely the Red Jungle Fowl (*Gallus gallus*) in the shape of comb and wattles and in the sound of the voice; the latter readily crosses with domestic stock, producing fertile hybrids. For these reasons Darwin argued that the Red Jungle Fowl was the sole ancestor of all domesticated breeds. Whether other wild species of *Gallus*, either extinct or still in existence today, have contributed to our domestic breeds is open to speculation.

Chickens

The origin, physical characteristics, and economic uses of some of the common breeds of

Figure 14-1 Red Jungle Fowl, *Gallus gallus.* (From Beebe, "A Monograph of the Pheasants, Vol. 2, H. F. and G. Witherby, London, England.)

poultry are given in Table 14.1, and the distribution of domestic poultry species in several countries of the world is given in Table 14.2. The domestication and development of types of chickens to suit man's needs goes back beyond recorded history. The earliest records show that chickens were raised in India at least 1000 B.C. Perhaps cock fighting has had as much to do with domestication as has use for food. Cock fighting as a sport is many centuries old but is still permitted in various parts of the world, including Mexico, Puerto Rico, and the Philippines. Cock fighting was popular in England, especially during the reigns of James I and Charles II, but was made unlawful by an act of parliament in 1849.

Breed Development

Poultry shows played an important part in the development of breeds both in America and many European countries. However, the origin of most breeds of chickens is not well-documented historically. In contrast to most of our other breeds of livestock, breed registry associations for poultry are lacking. In general, we say that body type, size, and contour determine the breed. Within each breed, different varieties are recognized, as distinguished by variations in plumage color, patterns, and comb shape.

The American Poultry Association, organized in 1873, had the primary objective of standardizing the breeds and varieties of domestic poultry shown for exhibition (Figure 14.3). The organization publishes the *American Standard of Perfection,* which was first printed in 1875. It serves as a guide for breed characteristics in judging at shows.

Today official recognition of so-called *standard breeds,* are of little significance in the commercial poultry industry. Thus, we find that the many strains used for the production of layers or broilers, are synthesized from one or more *standard* breeds. Usually, the exact breeding would be guarded by the breeder as a trade secret.

Breeds of Economic Importance

Since World War II the influence of the poultry fancier on poultry production has sharply declined. Instead there has been increased emphasis on meat and egg production with special systems of breeding used for each. The number of breeds used commercially today in most countries of the world is relatively few. Interest in developing new breeds has all but disappeared in America and Europe. However, in Russia efforts are being made to develop better-adapted breeds from crosses between imported and local breeds under government-controlled breeding programs.

Leghorn The Single Comb White Leghorn is by far the most widely distributed egg breed in the developed countries of the world. The Leghorn matures early, and is an excellent layer of chalk-white eggs. Because of its relatively small body size, it is an efficient converter of feed into eggs. It adjusts well to hot climates, and yet produces well in cold climates when reasonable housing is provided. The chicks feather out early and grow rapidly, and the pullets commence laying at six months of age or earlier.

Leghorns have increased in importance as the principal supplier of eggs to the urban centers of the world. More total effort has gone into breeding high-production strains of Leghorns than any other breed. Leghorns were first imported into America in 1835 from Italy, and since then many well-known strains have been developed. The Leghorn is hardy and perhaps less disease-prone than most other breeds. In part, this is probably a reflection of the improvement resulting from the systematic selection for high viability that many individual strains have undergone. As a meat bird, the Leghorn is inferior. In America, most of the hens, after completing their production year, go into the manufacture of chicken soup and other prepared foods.

From a commercial egg-production standpoint, the White Leghorn is not used as a pure strain, per se, but rather as a strain cross, a breed cross, or an inbred hybrid. In Russia, the purebred Leghorns have been largely replaced by the *Russian White,* a breed developed by government breeders from foundation crosses of the Leghorn breed with local varieties.

Figure 14-2 Grey Jungle Fowl, *Gallus sonneratti.* (From Beebe, "A Monograph of the Pheasants, Vol. 2, H. F. and G. Witherby, London, England.)

Rhode Island Red This breed is second in importance to the Leghorn as an egg-laying breed in America and most of Europe. It is widely used as a parent line in brown-egg type crosses. The breed was developed mainly in Rhode Island and Massachusetts in the period 1850–1900. The foundation stock was based on crossing local fowls with Cochins and Malays imported from the Orient.

Table 14.1.

Origin, physical characteristics, and economic uses of some common breeds of poultry.

Kind *Breed*	*Origin*	*Plumage color*	*Leg (shank)*	*Egg color*	*Typical mature body wt. (kg)* ♂	♀	*Economic use*
Chickens							
White Leghorn	Italy	White	Yellow	White	2.5	2.0	Egg production
Russian White	Russia	White	Yellow	White	3.0	2.2	Mainly eggs
Rhode Island Red	U.S.	Red	Yellow	Brown	3.5	2.7	Egg production
Plymouth Rock	U.S.	White, barred	Yellow	Brown	4.5	3.5	Crossbreeding for meat
Light Sussex	England	Columbian[a]	White	Brown	3.5	2.7	Crossbreeding for eggs
Black Australorp	England	Black	Slate	Brown	3.2	2.5	Crossbreeding for eggs
Dark Cornish	England	Black, brown[b]	Yellow	Brown	5.0	3.5	Crossbreeding for meat
Turkeys				Brown			
Large Bronze	England, U.S.	Bronze	Slate	Speckled	16.0	9.0	Roasters, fryers
Large White	U.S.	White	Pinkish white	"	15.5	8.5	Roasters
Medium White	U.S.	White	" "	"	13.0	7.0	Roasters
Small White	U.S.	White	" "	"	9.7	5.3	Small roasters
Black Norfolk	England	Black	Slate	"	9.5	6.5	Small roasters
Moscow White	Russia	White	White	"	12.6	6.6	Roasters
Ducks				Tinted			
Pekin	China	Creamy white	Orange	White	3.5	3.0	Roast duckling
Aylesbury	England	Pearly white	Pinkish orange	"	3.5	3.0	Meat
Rouen	France	Mallard plumage	Orange	"	3.5	3.0	Meat
Khaki-Campbell	England	Brown (khaki)	Orange	"	2.0	2.0	Eggs
Geese							
Toulouse	France	Grey	Reddish	White	8.9	6.9	Meat, feathers, liver
Embden	Germany	White	Orange	"	7.9	5.5	Meat, feathers, liver
Pilgrim	U.S.	Grey, white[c]	Orange	"	6.4	6.0	Young roasters
White Chinese	China	White	Orange	"	4.4	3.5	Young roasters
Kaluga	Russia	Piebald, grey, and white	Orange	"	6.6	5.8	Meat, feathers
Guinea Fowl							
Pearl	Africa	Grey	Grey		1.7	1.6	Specialty meat ("game")

[a] Columbian is a plumage pattern of white with black feathers in the wings and tail. [b] The male is solid black, and the females are dark brown.
[c] Sexes can be distinguished by color: males are white, females are grey.

Table 14.2.
The distribution of domestic poultry species in 20 countries.

		Number grown per year in thousands					
		Chickens					
		Egg type	*Meat type*	*Ducks*	*Geese*	*Turkeys*	*Guinea fowl*
Americas	—Brazil	130,103	141,596	8,343[a]		4,479	very few
	Canada	20,397	134,000				
	United States	279,415	2,000,000	12,500	1,000	120,000	250
W. Europe	—Denmark	8,000	60,000	2,000	1,200	1,200	
	Norway	3,700	3,445	2	6	60	—
	W. Germany	78,000	20,000	1,592	495	783	very few
	United Kingdom	55,200	300,000	6,300	330	15,000	250
	Italy	51,496	319,710	6,000	—	—	—
	Spain	40,000	260,000	475	50	750	few
E. Europe	—Hungary	4,700	89,000	30,000	6,000[b]	4,500	3,000
	E. Germany	25,200[c]	—	10,000	400	400	few
	Poland	117,000	43,000	19,000	8,900	3,000	—
	U.S.S.R.[d]	161,547	52,450	40,124	7,081	1,049	—
Middle East	—Egypt	23,930[c]	—	3	2	—	—
	Israel	7,000	63,000	—	250	3,500	—
Far East	—Japan	170,000	266,000	72	10	37	—
	Philippines	43,000	50,000	—	—	—	few
	Taiwan	3,058	11,763	2,034	70	127	few
Africa	—Kenya	11,377[c]	725	83	—	20	—
Australia		17,000	84,644	968	15	1,331	10

[a] Includes all waterfowl. [b] About ⅓ of these are grown primarily for liver production. [c] Includes all chickens.
[d] Estimates from Penionzhkevich (1962).

A

B

C

D

Figure 14-3 Some breeds of fowl. A, White Leghorns, Mediterranean class. B, Light Brahma, Asiatic class. C, Rhode Island Red, American class. D, Cornish, or Indian Game, English class. (A, B, and D, from Wright's *Book of Poultry,* London, Cassell; C from *Poultry World Annual 1925,* London, Butterworth.

The Rhode Island Red, developed first as a general-purpose breed, has yellow skin, red ear lobes, and deep rich red plumage color, with black primary feathers in the wing. They are quiet in disposition and lay a medium-brown shelled egg. In America this breed, like the Leghorn, has been subjected to intense selection, especially by New England breeders during the first half of this century, concentrating on high egg production with non-broodiness. In Russia, the Rhode Island Red has been extensively used as foundation stock in the development of general-purpose breeds with attempts toward adaptation to different regions of the country. One of the more important Russian breeds using Rhode Island Red blood is the Zagorsk, which, however, has white plumage.

New Hampshire The New Hampshire was officially admitted to the American Standard of Perfection in 1935. It seems to have been developed entirely from the Rhode Island Red with selection especially emphasizing early feathering, early maturity, and rapid growth. The plumage of the New Hampshire is of a lighter shade than the Rhode Island Red. Selections have largely been dictated by the requirements of the American broiler industry. Although the New Hampshire is second in importance to the White Plymouth Rock, undoubtedly it has contributed importantly to the gene pool of many parent flocks of meat-type chickens used today. In some strains the dominant white plumage gene has been introduced to overcome objections to colored plumage in the commercial broiler progeny.

Australorps The Australorp was developed in Australia mainly from the English breed, the Black Orpington, with probably some infusion of blood from Black Langshan strains selected for high egg production. The Australorp is mainly important in the production of crossbreds. In Australia and New Zealand about 90 per cent of the commercial layer-type hens are White Leghorn × Australorp crossbreds. The breed has also been used in these countries in the development of meat-type strains for meat production. This is also an important cross in South Africa.

White Plymouth Rock In America the modern White Rock is literally the mother of today's enormous broiler industry. One reason for this is the consumer preference for white feathers. White-feathered birds dress out into a nicer appearing carcass, because there are no dark pin feathers, often found in colored breeds. To meet the requirements of the broiler industry, various strains have been selected for rapid growth, early feathering, and good feed conversion. Since mature body weight is correlated with early growth, modern broiler strains are large in body size. Mature hens frequently weigh as much as 3.5 kg and cockerels 5 kg. Usually such heavy strains are relatively poor egg producers, so that feed cost of producing hatching eggs is usually high. In an attempt to overcome this, some breeders currently are experimenting with a dwarf strain. The sex-linked recessive dwarf gene, *dw*, has been introduced into such strains. This reduces body size by about ⅓ in the breeding flock, which, in turn, improves feed conversion in the production of hatching eggs. When dwarf hens are mated to normal sires, normal commercial type broiler chicks are produced, since the dwarf gene is recessive to the normal allele.

Cornish The Cornish fowl, also known as the English Game, was originally bred in Cornwall, England, from selections between crosses involving the Malay and the Old English Game. In conformation, both sexes of the Cornish are similar, in being closely feathered, compact, and with a broad, plump breast. They have a small pea comb consisting of three blades fused into one but with separate serrated edges. Three varieties recognized by the *American Standard of Perfection* are the Dark, the White, and the White-Laced.

The Cornish is a major breed used in the broiler industry for the production of commercial broilers. Most of the male lines used today for meat production have a high percentage, perhaps 50% or more, of Cornish blood (Figure 14.4). These commercial strains closely resemble the original Cornish type but most have white plumage.

Other Breeds

The *Barred Plymouth Rock* is perhaps of most importance today as the foundation breed from which the White Rock was derived. The Barred Plymouth Rock was developed as one of the first general-purpose breeds of the American Class from selections among crosses between the Black Java, Black Cochin, and the American Dominique (barred plumage). Body type and plumage color was fixed by inbreeding certain selected families of the Essex strain, which was popular about 1875.

Barred Plymouth Rocks are used to a limited extent today for the production of crossbred brown-egg layers. Since the Barred Rock carries the sex-linked barring gene, certain crosses with it produce chicks with sexes distinguishable at hatching. For example, when a Rhode Island Red male is mated to a Barred Rock female, the male progeny are barred, and the female progeny are black with neck feathers tinged in red.

A

B

C

Figure 14-4 Meat-type chickens. A, A cockerel breeder from a male line. B, A pullet layer from a female line. C, The commercial broiler progeny from the cross of A ♂ × ♀ B. (Courtesy, Goto Hatchery, Inc., Gifu City, Japan.)

In England the Light Sussex is of some importance in the production of crossbred layers, typically with the White Leghorn.

In Russia, general-purpose breeds are relatively more important than strictly egg-type or meat-type chickens. Two such breeds developed at government breeding stations are the Zagorsk Salmon (Figure 14.5) and the Kuchino Jubilee.

Figure 14-5 Zagorsk Salmon; a general purpose breed developed at a government breeding station in Russia. (Courtesy Prof. E. E. Penionzhevich, Worlds' Poultry Science Assoc., U.S.S.R. National Branch, Moscow.)

Commercial Types of Chickens

Practically all commercial type chickens produced today, except in the Soviet Union, can be divided sharply into those bred for egg production and those bred for broiler meat production. In either case, the final product may be a pure strain, a strain cross, a breed cross, or an inbred hybrid.

Pure strains A strain of chickens generally takes the name of the breeder who developed it. A pure strain would not necessarily be any more "pure" genetically than a pure breed; however, if a strain has remained closed to outside blood over a period of years, then, in this sense, such a strain might be called pure. Pure-strain breeding has been an important method for the production of both meat and eggs in the past. However, the last 20 to 25 years has witnessed marked changes in systems of producing commercial types of chickens. Pure strains are now used almost exclusively as parents of commercial crosses, so that it is virtually impossible to buy breeding stock today of any of the well-known pure strains of chickens developed in the United States.

Strain crosses The progeny of the cross of two different strains of the same breed is called a strain cross. Most commercial egg-type chickens produced today are Leghorn strain crosses. At the same time these may still be classed as purebreds. When the strains crossed are inbred or if they differ in origin, the progeny usually show some heterosis in egg production. Consequently, strain crosses are superior to pure lines for egg production. Strain cross Leghorns are popular not only because of the higher production from hybridization but also because they retain the qualities of uniform white-shelled eggs, white plumage color, and other characteristics ordinarily associated with a purebred. Commercial strain crosses may involve two, three, or even more pure strains.

Breed crosses Progeny from crosses of strains representing different breeds are called crossbreds. Crossbreds usually exhibit heterosis for egg production and rate of growth. Some of the more important commercial crossbreds are discussed below.

Sex-linked cross When a Rhode Island Red or a New Hampshire male is mated to a

Barred Plymouth Rock female, the male progeny are barred like their mother while the female progeny are nonbarred like their father. This cross has been used extensively, over the past years, especially in the New England area, where cockerels are sometimes raised for meat production and the pullets are kept as layers. The crossbred pullets are mostly black and carry a variable amount of red or gold in the terminal feathers of the neck hackle. They have been called "golden-necks."

The reciprocal mating of this cross—that is, the Barred Rock male mated to a Rhode Island Red or New Hampshire female—produces all barred crossbred progeny. This was a popular cross for the production of broilers in the early days of the broiler industry.

Leghorn-Red cross Crosses using the Leghorn male on a Rhode Island Red female have been used extensively in the U.S. for the production of medium-weight layers. These have proved to be exceptionally good layers, and when marketed command a better price than straight Leghorns. However, the eggs have a tan color intermediate between the white of the Leghorn parent and the brown of the Rhode Island Red parent (Figure 14.6). Because certain markets discriminate against the tan colored or tinted egg, this has restricted the use of this cross.

An interesting fact is that the performance characteristics of the cross of a Leghorn male on a Rhode Island Red female is quite different from its reciprocal mating. In the former case the crossbreds show a tendency toward early sexual maturity, nonbroodiness, and rather high adult mortality. In contrast, the crossbreds from a Rhode Island Red male mated to a Leghorn female mature more slowly, are more inclined to broodiness, but more important still, have less adult mortality. Whether the difference in adult mortality is due to a sex-chromosome effect or to a maternal effect transmitted through the female Rhode Island Red parent has not yet been definitely established.

Leghorn-Australorp cross The cross of an Australorp male with a Leghorn female (Austra-white) has proved to be a good layer and especially capable of withstanding the stress conditions usually found with cage management. It seems to have been almost completely abandoned in the U.S. probably because the market discriminates against the tinted egg.

The *White-Austra* is a cross of a Leghorn male on an Australorp female. Most of the

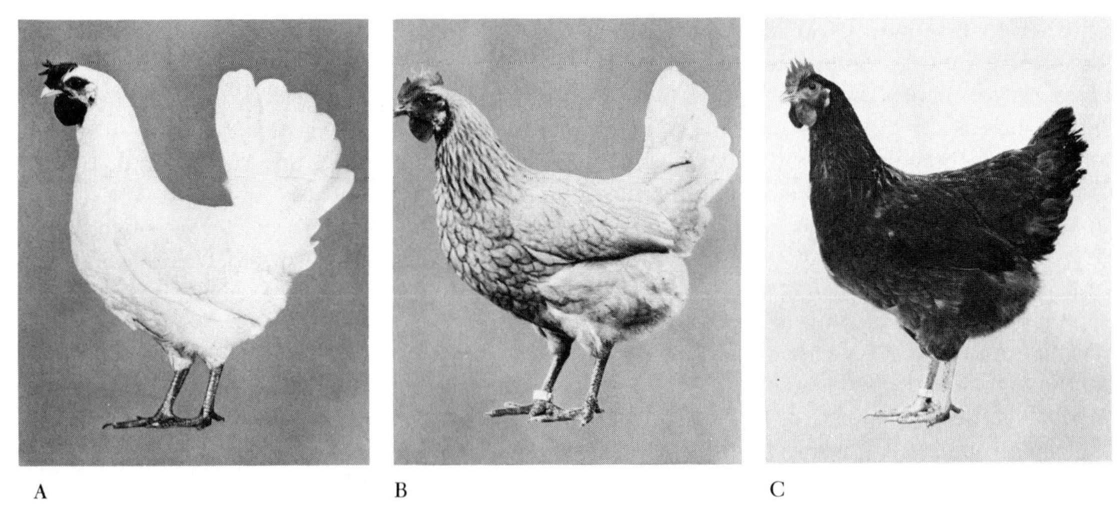

Figure 14-6 Commercial layers. A, A white-egg layer. B, A tinted-egg layer. C, A brown-egg layer. (Courtesy, Goto Hatchery, Inc., Gifu City, Japan.)

commercial egg layers in Australia, New Zealand and South Africa are White-Austras. These have been preferred over the Austra-White because they are less broody and have white rather than dark shanks Now, however, with the increased use of laying cages, the main objection of broodiness is not so important. Broodiness is not a problem with birds in cages since there is no nest and no accumulation of eggs. Tests show that more hatching eggs can be produced from mating of Australorp males to White Leghorn females than the reciprocal mating. Also the hatcheryman can expect a better feed conversion from a White Leghorn than an Australorp breeding flock because of the difference in body size.

Inbred hybrids An inbred hybrid is defined as the progeny produced from a mating of two inbred lines or from a mating of first-generation inbred line crosses. However, the term hybrid is also used in describing crosses between breeds and varieties. Because of the possible confusion that may be associated with the word hybrid, the United States Federal Trade Commission, in its fair trade practice rules for the poultry industry, recommended that the word *hybrid* should be qualified by stating, "in immediate conjunction therewith the type of cross used in the production of the industry product, such as 'inbred line-cross hybrid' or 'inbred hybrid,' 'cross-bred hybrid,' 'strain-cross hybrid,' 'line-cross hybrid,' etc." The FTC also defines an inbred line as "a group of inbred chicks resulting from breeding closely related poultry and in which the individuals in question have an average coefficient of inbreeding of at least 37.5 per cent (equivalent to two generations of brother-sister matings)."

Some of the outstanding commercial chickens in America have been produced by methods similar to the system used in the development of commercial hybrid seed corn. The use of inbreeding as a tool for developing commercial egg-laying chickens attained considerable importance after World War II, particularly in the Midwest. Actual breeding details of commercial inbred hybrids are trade secrets of the companies producing them. Generally, hybrid varieties are sold by a number designation, and a description of the product is given in terms of plumage color, color of the eggs, and other characteristics of the hybrid.

The inbred hybridization method appeared to gain momentum through the 1940's, but some of this was lost after the results of random-sample egg-laying tests, starting about 1952, showed that the best performing entries were frequently strain crosses and not necessarily inbred-hybrids. Thus, whether intensive inbreeding, as used by hybrid seed-corn producers, is justified with chickens is still an issue not wholly settled. Commercial breeders today, whether strain cross producers or inbred-hybrid producers, are more concerned with the performance of their product rather than the system of breeding. If there is any trend among American breeders today, it seems that the inbred-hybridizers are looking more closely at the advantages of strain crossing while the strain crossers, at least some of them, are experimenting with inbreeding methods.

Turkeys

The modern domesticated turkey is thought to be descended from two differing wild subspecies, one found in Mexico and Central America and the other found in the United States. The southern species is small, while the species native to the United States is larger and has a characteristic bronze plumage.

Darwin held to the view that the turkey was domesticated by the original inhabitants of America. He suggested that the wild forms found in Mexico and Central America were first domesticated. However, since the modern strains of Bronze turkeys correspond closely to the wild subspecies in the United States, it is probable that the latter were used largely in the development of American varieties.

Only one breed of turkeys is recognized by the *American Standard of Perfection* and hence it has become a practice to speak only of different varieties. The "standard" varieties are the Bronze, White Holland, Beltsville Small White, Narragansett, Black, Slate, and Bourbon Red.

The most important varieties of turkeys in the U.S. today are nonstandard i.e., not recognized by the *American Standard of Perfection.* These are the Broad-breasted Bronze, the Broad-breasted Large White, and the Medium White. The only commercially important standard variety is the Beltsville Small White. The latter having also been selected for broad-breast conformation, and because essentially all commercial turkeys in the U.S. are now of the broad-breast conformation, the different commercial varieties are now commonly distinguished only by size and color. The term "breed" is now commonly used to identify color and size types, while the term "variety" now usually means different commercial brands offered for sale.

Of some 115 to 120 million turkeys raised for the commercial market each year in the U.S., most have white plumage. In 1971 there were 46 different commercial varieties of turkeys available from different American breeders and hatcheries listed in *Turkey World,* a trade publication. Of these, 40 were white varieties and only six were bronze. Of the 40 white varieties, 23 were designated as *Large Whites,* eight were *Medium Whites,* and nine were *Small Whites.*

White-plumaged turkeys carry a recessive white gene which almost completely prevents the appearance of pigment in the plumage, shanks, and other parts of the body. Eye color, however, being dark, proves this is not an albino condition. White turkeys seem to withstand hot summer sunshine better than colored varieties. Also, white-plumaged birds generally grade better when dressed than colored birds because their pin feathers are less conspicuous. However, some authorities claim that white turkeys are more densely feathered than colored varieties making them more difficult to dry-pick. If this can be considered a disadvantage, it disappears when dressing plants use modern scald-pick methods.

Broad-breasted Bronze (Large Bronze)

This breed has almost completely replaced the original bronze variety; in the U.S. market it is commonly called the *Large Bronze.* England is credited with the origin of the broad-breasted type. In the late 1920's such English turkeys were imported into Canada and from there into the United States about 1935. In five or six years it became the most widely grown type. After World War II the broad-breasted turkey was further selected for rapid growth, high feed conversion, and broad conformation. Very probably because of the increasing emphasis on extreme conformation and size, this type became seriously deficient in fertility, hatchability and egg production. Artificial insemination has now become rather standard practice among both Large Bronze and Large White turkey breeders in an attempt to overcome poor fertility.

In color the modern bronze varieties are similar to the standard-bred bronze but they tend toward buffy white instead of pure-white feather tips. Also they lack the brilliant copper bronze in the plumage of the standard bronze variety. Because the plumage is black, the pin feathers are dark which is a disadvantage when the birds are marketed.

Broad-breasted White (Large White)

This variety was developed mainly from crosses of the Broad-breasted Bronze to the White Holland and also by taking advantage of naturally segregating recessive white genes in bronze-colored flocks. In effect, the Broad-breasted Bronze variety has been virtually transformed into a white variety by successive back crosses to bronze and subsequent selection of the white-plumaged segregates with large size and compact conformation. These are commonly known as *Large Whites* in the American market. Selection for size and conformation has continued, so that for all practical purposes both the Large Bronze and the Large White are not only equal in performance in the production of meat but they are equally poor in reproductive performance.

Most *Large White* turkeys are marketed as heavy roasters at 23–26 weeks but some are marketed at earlier ages.

Beltsville Small White

The Beltsville Small White Turkey was developed by the U.S. Department of Agriculture at the Agriculture Research Center at Beltsville, Maryland. Emphasis has been on relatively small size but with good conformation and breast development. Beltsvilles are essentially similar in color and conformation to the Large White, but they average only about 60 per cent as large. In American markets they are also called *Small Whites*. Under average to good conditions on medium-energy diets, the toms will weigh about 7 kg (15.4 lb.) at 15 weeks and about 10 kg (22 lb.) at maturity. The hens will weigh about 4.3 kg (9.5 lb.) at 15 weeks and about 5.3 kg (11.7 lb.) at maturity. At this age most birds of both sexes will be ready to market as "fryer-roasters," or, as they are sometimes called, turkey "broilers." If Beltsvilles are kept to 21–24 weeks, the age when most large strains are marketed, they develop into high-quality medium to small roasters. The toms then compete with the large-type hens for the medium roaster retail trade, while the hens would be called small roasters.

The Beltsville Small White is much superior to the large breeds in egg production, fertility, and hatchability. They will usually outlay the larger breeds by 20 to 30 eggs per season. Hence, the cost of producing poults would be substantially lower. On the other hand, the feed conversion of the faster growing large type would be less. For example, at 16 weeks of age, Beltsvilles require about 2.9 kg of feed per kg of live body weight under favorable conditions compared with about 2.5 kg of feed per kg of body weight for Large White turkeys at 12 weeks. This difference of about 0.4 kg in favor of the Large White is partly offset by the higher poult costs, about 15 to 20 cents more. However, because the large-type turkeys reach comparable weights in three or four weeks less time, even though at somewhat less finish, they are usually more profitable to grow.

Lately certain American breeders of Beltsville stock have developed a somewhat heavier bird. These are called Medium Whites although they would be classified as a light breed.

Other Breeds of Turkey

In England and in many other European countries, the *Black Norfolk* is recognized as a distinct breed. It is a medium sized turkey. In Russia the *North Caucasion* (Figure 14.7) has been bred since 1932 and subsequently improved by outcrossing to the Broad-breasted Bronze imported from the U.S. Also the *Moscow White* has been developed as a medium to small breed from foundation matings using the Beltsville Small White and other stocks.

Waterfowl

Ducks

The wild mallard duck (*Anas bosches*) is generally regarded as the ancestor of all domestic breeds of ducks, with the exception of the Muscovy, which is a different species originating in South America.

Mallards seem to have been domesticated independently in different countries mainly for meat, not eggs. The Pekin is by far the most popular breed around the world (Figure 14.8). It is early in maturity, very hardy, and subject to few diseases, and develops a good

Figure 14-7 The North Caucasian breed of turkeys, bred in Russia. (Courtesy, Professor E. E. Penionzhevich, Worlds' Poultry Assoc., U.S.S.R. National Branch, Moscow.)

carcass. Being white in color, it dresses out well, with no dark pin feathers.

In England, the Aylesbury duck is the most popular. It is a large meat-type duck with white plumage and is similar to the Pekin, but with a deeper body and somewhat slower growth.

The Indian Runner duck, originating in southeast Asia, is an egg-type breed. It has a distinctive carriage, standing very erect and penguin-like. It may produce over 300 large eggs per bird per year. The Khaki-Campbell is also an egg-laying breed, derived from native ducks of England crossed to the Indian Runner.

Another well-known breed is the *Rouen*, a large meat-type duck from France. In plumage it is identical to the wild mallard. In the U.S.S.R. the Moscow White breed was developed from crosses of the Pekin with the Khaki-Campbell. The object has been to combine both meat and egg production into a single breed.

The major duck-raising industry in the United States is concentrated on Long Island in New York. This area produces the bulk of young roasting ducks mainly for the New York market.

Compared with chickens, ducks are of minor importance in most other countries of the world. In England, West Germany, Italy, and Denmark, ducks represent no more than about 2 or 3 per cent of the total poultry business. However, in Poland meat-type ducks represent about 10 per cent of the poultry produced. In the Far East we find that duck production is important in some countries. For example, about 28 per cent of the poultry produced in Taiwan are ducks used for both meat and eggs. Egg-type duck production is of minor importance in the Philippines, amounting to about 2 per cent of the total poultry raised, whereas in Japan, with a large poultry industry of over 400 million birds grown annually, the ducks grown are less than .01 of 1 per cent of the total.

Geese

There is reason to believe that the goose was the first bird domesticated, even before the chicken. More than 4,000 years ago it was regarded as a sacred bird in Egypt. Domesticated geese are all thought to be descended from the wild grey goose of Europe. In general, the domesticated breeds are much larger than their wild ancestors, and they have almost lost their ability to fly.

Domestic, as well as wild geese, usually mate for life. Usually the hatchability of goose eggs is poor.

The Romans learned to use goose feathers for filling mattresses and cushions. The Romans also regarded goose liver as a delicacy, and large numbers were reared in pens and fed especially for their liver. Even today in Europe some geese are fed mainly for their liver. This is done by "force-feeding": a soft

A B

C

Figure 14-8 Breeds of ducks. A, Pekin. B, Indian Runners. C, Rouen. (From Wright's *Book of Poultry*, London, Cassel.)

mash is crammed into their gullet which produces an enlarged liver weighing as much as two pounds. The liver is made into a delicately flavored paste called *pâté de foie gras*, used for specialty sandwiches or hors d'oeuvres. Geese are good grazers, and the early breeds were developed on the meadows of France, Germany, and Russia. Goose raising is still an important enterprise in several European countries, especially in Poland and in the U.S.S.R.

Geese have never been produced in concentrated areas in the United States as have other classes of poultry. They are raised in rather small numbers on farms widely scattered throughout the United States and Canada. The total number of geese in the United States has consistently declined since about 1930. According to the 1950 census, there were

A

B

Figure 14-9 Pilgrim geese. A, Adults; the gander is white and the geese (females) are grey. B, Day-old goslings: the sexes are distinguishable at hatching; the males have a lighter shade of down than the females. (Courtesy, Animal Rehearch Institute, Central Experimental Farm, Ottawa, Ontario, Canada.)

Figure 14-10 Kaluga geese, a Russian breed. (Courtesy, Prof. E. E. Pevionzhevich, Poultry Breeding Institute, Zagorsk, Moscow Province, U.S.S.R.)

only slightly more than 1 million geese raised in 1949, with considerably fewer being raised today.

Breeds most commonly found in the U.S. are the Toulouse, Embden, and Chinese. Other breeds are the African, Roman, Pilgrim (Figure 14.9), and Sebastopol.

In Russia several breeds were early developed as fighting geese, in much the same way that game breeds of fowl were developed for cock fighting. The most aggressive individuals were selected as breeders. One such Russian breed is the Tula, which means "game" geese, and another ancient breed is the Arzamas. Game geese were very popular in Russia in the seventeenth and eighteenth centuries as used in fighting contests for royal entertainment. The Arzamas is said to have been a breed of unusual size, almost as large as swans, but much more bellicose and aggressive. These were trained for cruel fights, where watchers laid wagers, just as they did in England in the days of cock fights. In 1906 cock fighting and goose fighting were banned from Russia. However, since these early breeds were selected for large size and good muscling, they have provided a foundation for a number of productive meat-type breeds of geese in modern Soviet Russia. A highly regarded meat breed in Russia derived from fighting geese is the Kaluga (Figure 14.10).

Guinea Fowl

The guinea fowl, native to Africa, was brought to Europe by the Portuguese toward the end of the Middle Ages. There are three domestic varieties, the Pearl, the White, and the Lavender. The most common is the Pearl, which appears to be the original variety developed from the wild West African species. Guinea fowl are bred only in very limited numbers in most all countries of the world. They are valued mostly for the delicate and somewhat wild flavor of their meat.

FURTHER READINGS

American Poultry Association. 1958. *The American Standard of Perfection.* Box 337, Great Falls, Mont.: APA.

Brown, E. 1906. *Races of Poultry.* London: Arnold.

Ives, Paul. 1947. *Domestic Geese and Ducks.* New York: Orange Judd.

Jull, M. A. 1927. "The races of domestic fowl." *National Geographic Magazine,* 51:379–452.

Marsden, S. J. 1971. *Turkey Production.* Agric. Handbook No. 393. Washington, D.C.: U.S. Govt. Printing Office.

Penionzhkevich, E. E. 1962. *Poultry Science and Practice, Vol. I: Biology, Breeds, and Breeding.* Moscow. Translated from Russian by Israel Program for Sci. Transl., 1968.

Fifteen

Horses, Mules and Asses

A gigantic beauty of a stallion, fresh and responsive to my caresses,
Head high in the forehead, wide between the ears,
Limbs glossy and supple, tail dusting the ground,
Eyes full of sparkling wickedness, ears finely cut, flexibly moving.

Walt Whitman,
Song of Myself

The evolutionary history of the horse is more completely known than that for any other living species. The earliest known horse, Eohippus, lived in America some 45 million years ago, and was about the size of a fox. The front feet had four toes and the hind feet three toes. Among the more important changes evident in the phylogenetic development of the horse were (1) a great increase in size, (2) a change from a broad-footed animal that walked upon a pad beneath the toes to one that walked on the end of his toes, and finally on the end of a single toe with a broad hoof, and (3) a change from an animal with a short leg of two sections to one with a longer leg, the foot being extended to form a third section of the leg, and with the muscles gathered near the body to give greater capacity for running. Perhaps the most important modification with respect to the horse's running and jumping capability was the development of complex digital ligaments that support the first two phalangeal joints. From a rudimentary system in Eohippus, the ligaments reached a maximum degree of complexity in Merychippus some twenty million years ago. Maximum size and strength of this system, however, was reached in the modern horse (Camp and Smith, 1942). As a result of these skeletal and ligamentous modifications, the limbs of the horse are superbly adapted to running and jumping, with a large degree of automatic springing action involved.

Early ancestral forms of the horse are known to have existed in both the western and eastern hemispheres, but apparently evolution in the eastern hemisphere ended in early extinction. Thus the evolution of the modern horse from primitive forms took place largely in the Americas, and from there the most recent in line of descent migrated to Asia and Europe by way of land connections through the Bering Sea. Later, during the Ice

Age, the horse became extinct in the Americas but survived in the eastern hemisphere (Riggs, 1932).

Source of Primitive Stocks and Their Relation to Modern Breeds

The immediate predecessors of the modern horse have not been determined, but a widely accepted view is that three distinct primitive stocks have contributed most to present-day breeds (Ridgeway, 1905). The Celtic horse, which was found in the Hebrides and part of Ireland, is thought to have been the progenitor of modern pony breeds. The ancient horse of Europe, the Near East, and Britain probably derived from another wild stock that was widely distributed in Europe and Asia. A third type is thought to have originated in the deserts of North Africa and to have given rise to the horses of the Barbary States, Egypt, Arabia, and Persia. There is some historical evidence that the horses of North Africa were introduced into Europe by invaders from North Africa as early as the third century B.C. and were probably imported often thereafter, mainly by way of Spain.

Improvement of Primitive Horses and Their Use in Breed Formation

North African Horses

All the important light horse breeds throughout the world have descended from or have been largely improved by descendants of the North African horse. These horses, the Barbian of North Africa and the Arabian and related strains (the so-called Eastern or Oriental horses), were greatly improved during several centuries by intelligent selection and breeding practices. The most refined of these, the Arabian, is not a large horse, but is characterized by a beautiful head and neck, strong shortly coupled back, good chest capacity, and excellent feet and legs (Figure 15.1). Though not unusually fast, the Arabian has always been noted for great refinement and unusual stamina. The earliest introduction of these horses to Spain, and the subsequent importations during the years of the conquest and occupation by the Moorish invaders, resulted in great improvement of the native stocks, so that the horses of Spain became recognized as being of excellent quality. These were the horses, brought to the Americas by Spanish conquerors and colonists, that provided the seed stock for the western mustang and the horses that eventually spread throughout the eastern part of the United States, as well as for the feral horses of South America.

At a somewhat later time, the Arabian and related horses were used in a more methodical program of improvement of horses native to Europe and Britain and in the formation of new breeds, including the English Thoroughbred. The Thoroughbred horses resulted from the breeding of Barbian, Arabian and Turkish stallions to mares indigenous to the British Isles at that time (1660–1750), as well as to imported mares of these breeds. The Thoroughbred has played a large and important role in the development of all major modern breeds of light horses in every part of the world, the Arabian being the only important exception.

Number of Breeds of Horses

Goodall (1965) has prepared an illustrated survey of types of horses found throughout the world, with over 300 photographs and a listing of over 150 breeds. How many of the breeds listed by Goodall have Registry Associations is not stated, but, generally speaking, registration is very important to insure maintenance of reasonably well-fixed characteristics as well as to provide a record of ancestry. Since all breeds arose from a few primitive stocks, there are many similarities between breeds found in different countries. Eleven selected breeds of horses and ponies found in the United States are listed in Table

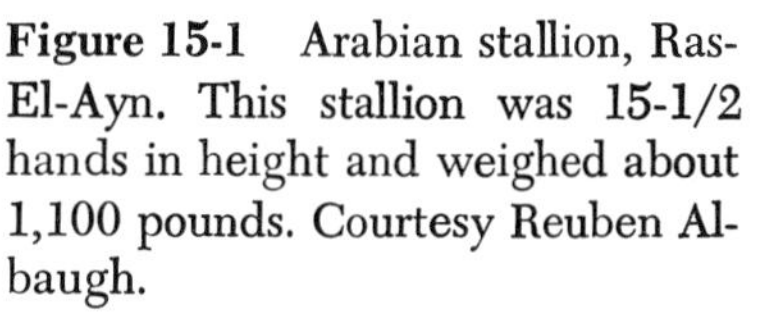

Figure 15-1 Arabian stallion, Ras-El-Ayn. This stallion was 15-1/2 hands in height and weighed about 1,100 pounds. Courtesy Reuben Albaugh.

15.1. Fifty years ago the mainstay of the horse industry in the United States was the draft horse; the draft breeds of greatest importance at that time are listed in Table 15.1, but now they are little more than a novelty in this country.

Stimuli for Breed Formation

Most horse breeds have developed in response to fairly specific stimuli, for example, the Thoroughbred in England in answer to the requirement for speed and stamina in warfare or on the race course. The development of the Morgan breed occurred in America in response to the need for an all-around useful horse for farm work, for riding, or as a roadster at about the time of the American Revolution. As roads improved, the demand for speed in harness—both for transportation and in racing—increased, and the larger, faster American Standardbred was developed, drawing heavily upon the Thoroughbred for size, speed, and stamina. Similarly, the American Saddle Horse breed was formed to meet the need for a good, smooth, and easy traveling horse under saddle. Impetus for the greatest development of the American Quarter Horse as a breed arose from the need for a horse suitable for working cattle. The characteristics desired were at the same time compatible with tremendous speed for short distances. Often the requirements that were largely responsible for the formation of a breed have disappeared or have become relatively unimportant; for example, the need for horses solely for transportation, either on horseback or in horse-drawn vehicles, no longer exists. For this reason the Standardbred owes its continued existence to its use in harness racing and in the horse-show roadster classes. Principal uses of American Saddle Horses are now in the horse shows and as pleasure horses. In more recent years, uses of Quarter Horses have tended to become categorized, so that most are bred either for racing short distances or for stockhorse work, including showing at halter and in western performance classes, rather than for both.

Some of the more recently formed breeds and registries have been established not on the basis of functional requirements, but on the basis of characteristic color or color patterns. Some, such as the Palomino are, in fact, color registries only, and horses registered in any of several other breed registries may also be listed in them; thus they include an extremely heterogeneous genetic group. The Appaloosa registry, which also falls in this general category, lists three acceptable basic color patterns, but these cover a wide range. An extreme range in conformation is also permitted. A factor of importance in selection of foundation stock for formation of breeds

Table 15.1.
Characteristics and uses of some important breeds of horses found in the U.S.

Breed	*Approximate period of breed formation or most important development*	*Colors*	*Height (hands)*	*Weight (lbs)*	*Most important uses*
Arabian	A.D. 100–600	Bay, grey, chestnut, brown	14–2″ to 15–2″	850–1,000	General pleasure
Thoroughbred	1660–1800	Bay, chestnut, brown, black, grey, roan	15–2″ to 17	1,000–1,300	Racing, hunting, polo, general pleasure, stock horse
Standardbred	1800–1875	Bay, chestnut, brown, black, gray, roan	14–2″ to 16–2″	850–1,200	Harness racing, horse-show roadster
American Saddle horse	1840–1890	Chestnut, bay, brown, black, other	15 to 15–3″	1,000–1,150	Gaited saddle and fine harness horse shows, pleasure riding
Tennessee Walking horse	1890–1935	Bay, chestnut, black, brown, roan, gray, sorrel, white	15 to 16	1,000–1,200	Horse-show walking horse, pleasure riding
American Quarter horse	1850 to date	Bay, chestnut, brown, black, roan, gray, dun, palomino	14–2″ to 15–2″	1,000–1,250	Stock horse, rodeo, short racing, pleasure riding
Palomino	1946	Golden with white silvery or ivory mane and tail	15 to 17	100–1,300	Pleasure
Percheron	1800–1885	Black, gray, brown, chestnut, bay	15–2″ to 17	1,600–2,200	Draft
Clydesdale	1720–1880	Bay, brown, black, other	15–2″ to 17	1,700–2,000	Draft
Belgian	1850–1900	Chestnut, roan, bay, brown, gray	15–2″ to 17	1,900–2,400	Draft
Shetland	1870–1871	Black, brown, bay, chestnut, mousy, spotted	9–2″ to 10	300–400	Child's mount, horse shows
Hackney	1760–1885	Bay, chestnut, black, brown	11–2″ to 14–2″	450–850	Heavy harness pony in horse shows

for special purposes has been the natural tendency of certain individual horses or breeds to perform specific natural gaits well, or to learn acquired gaits more readily.

Sources of Foundation Stock

Whenever the need for a horse to perform a job has been a force for breed formation, the foundation stock has been selected from any suitable source available. For this reason, all major breeds of light horses developed in America have a number of other breeds, and usually a considerable number of individuals of unknown breeding, in their foundation stock. Usually the Thoroughbred or breeds tracing strongly to the Thoroughbred have provided the main basis for refinement, intelligence, speed, and stamina.

Formation of Breed Organizations and Registries

Most of the breed organizations and registries were formed late in the developmental stages of the breeds, and selection and admission of so-called foundation stocks took place "after the fact." These organizations have pursued various policies in order to develop the type of horse considered desirable for the breed.

In the early stages of the development of a breed, provisions are frequently made for allowing the registrations of animals whose parents have not been registered. The Quarter Horse Association, for example, maintains an open registry. Prior to 1962, eligibility for registration of outside animals was based on a complicated system which included the evaluation of performance and conformation (Briggs, 1969).

Since 1962 no outside animals may be entered in the registry except those which have as one parent a registered Quarter Horse, as the other a registered Thoroughbred. Such animals must achieve a specified level of performance and pass inspection for suitable conformation prior to permanent registration in the American Quarter Horse Stud Book and Registry.

Examples of other breeds that retain procedures whereby desirable individuals may be admitted to registry are the Standardbred and the American Saddle Horse. In practice relatively few horses are admitted to the Standardbred and American Saddle Horse Registries. In 1969 about 20 per cent of racing Quarter Horses that qualified for register of merit as two-year-olds were sired by registered Thoroughbreds. The rate of infusion of Thoroughbred blood into the Quarter Horse breed as gauged by this measure appears to have been relatively constant for a number of years. The reciprocal cross is also popular and, thus, many racing Quarter Horses are essentially of Thoroughbred breeding. This should result in an increasing rate of infusion of Thoroughbred blood into the stockhorse type of Quarter Horse with the passing of time. There are still, however, many Quarter Horses that have had no outcrosses for at least several generations. As a result there are two distinct types of Quarter Horses, the racing and stockhorse types.

The American Quarter Horse breed must be considered to be still in its formative stage. While breeders of stock-horse types have relied to a great extent upon animals already in the registry for breeding stock for several generations, racing Quarter Horses are commonly not more than a generation or two removed from the Thoroughbred. No doubt if the stud books were closed to Thoroughbreds, dilution of racing Quarter Horse strains with stockhorse strains would result in a slower breed, and recently set track records might stand for a long time. The problem of fixing the breed type appears to be a serious one, because of the divergent ideas of members of the breed organization on just what the uses of the Quarter Horse should be (Figure 15.2).

Thus the degrees of homogeneity of type and genetic constitution vary greatly from breed to breed. Generally it may be said that progress toward homogeneity is greater in breeds that are selected on the basis of one performance characteristic, especially when objective measures of performance are available. The Thoroughbred, by virtue of its long-closed stud book and selection for racing, probably has the most homogeneous genetic base of any breed. The American Standardbred, American Saddle Horse, and Tennessee Walking Horse have been rigidly selected for specific purposes, and reasonable constancy of type has resulted, at least as far as the specialty of the breed is concerned. The conformation of Standardbred or the Tennessee Walking Horse is relatively less important to breeders than is performance; as a result there are wide variations in size and conformation in these breeds. Because it has long been first and foremost a show horse, conformation and quality of performance of varied gaits have been important in selection of breeding stock in the American Saddle Horse breed. Today this breed has no equal in the number and style of gaits it performs, nor does any breed equal it in elegance and refinement of conformation.

Figure 15-2 An excellent example of the working stock horse type of Quarter Horse. Courtesy Reuben Albaugh.

Desirable Characteristics for All Breeds of Horses

Wayne Dinsmore (1935) has listed some characteristics which are desirable in all breeds of horses:

"1. A strong heavily muscled back, which seems short.
"2. A short, wide, strong, heavily muscled loin.
"3. A deep chest, wide through from side to side.
"4. A roomy middle, due to long, well-sprung ribs, and a capacious abdominal region.
"5. Well-set legs, pasterns, and feet; that is, they should be correct in position, viewed from front, side, or rear.
"6. Strong leg joints, deep from front to rear, that are clearly defined with dense bone of good quality.
"7. Straight action and good wind.
"8. Good head, eyes, and temper."

The draft-horse body conformation differs from light horses in that a relatively larger proportion of the animals' total weight rests upon the forelimbs because of a deeper and wider neck and greater depth of body.

The correct and some incorrect feet and leg positions are shown in Figure 15.3.

In addition to these defects in conformation, there are many types of unsoundnesses of the feet and legs which impair the usefulness of individuals of all breeds: corns, contracted feet, ring bones, founder (laminitis), and quartercracks are a few examples of blemishes or unsoundnesses of the feet. Leg blemishes or unsoundnesses include splints, stiffled condition, bone spavin, and curbs.

Dinsmore refers to "good wind." Two conditions which interfere with normal respiration and thus limit animals in their performance are heaves (forced expiration) and roaring, sometimes correctable by surgery. Dinsmore refers to "good temper," which includes a tractable disposition, intelligence, and lack of viciousness. Animation is a desirable characteristic for most breeds, but for certain tasks where trustworthiness is of paramount importance, as in horses used for general farm work or for hauling delivery wagons, docility and calmness under a variety of disturbing situations may be more important than animation.

Traits for Specific Purposes

The primary uses of the horse as a source of power for work or as a means of transportation—either for pleasure, sport, or facilitating performance of the rider's task—require that horses be of good substance and properly conformed so that they function efficiently and without defects in their manner of travel or performance. Because the tasks they perform are so varied, horses must possess special traits that suit them to particular jobs.

Traits of Draft Horses

The draft horse is almost unique because the primary requirement for it is the development of a great deal of power. Some of the most important factors in fulfilling this requirement are relatively heavy weight, heavily muscled back, loin, and quarter, and—as in all horses —sound feet and legs. The chief means of developing the power required is the forward displacement of the center of gravity. This is achieved by the support of much of the weight of the forequarters by the muscles of the back, loin, and rear quarters and by the extension of the rear legs by muscles of the rear quarters. It may be seen that for efficient function the requirement for conformation of head, neck, and forequarters is such that a relatively larger proportion of the animal's total weight must rest upon the forelimbs than in light horses. For the same reason, a slightly greater length of back may be tolerated in draft horses, provided there is ample muscling in that region.

Since great power for relatively long periods is required, stamina is important. Most horsemen look for indications of this almost intangible quality in a deep heart girth and roomy middle. This seems reasonable because such configuration is assumed to provide ample respiratory capacity, space for vital organs, and the ability to handle adequate amounts of nutrients.

Although appearance is secondary to utility, most of today's few purebred draft horses in the United States are used for show and, thus, balance of conformation, refinement, and breed type are emphasized. In many parts of the world smaller draft breeds are found (Figure 15.4). Since strict utility, usually in teamwork, has always been demanded of draft types, a calm, tractable disposition and intelligence are essential.

Traits of Stock Horses

The stock horse of the kind found on western cattle ranches and in rodeo competition and horse shows is required to perform a great variety of tasks: he must be able to start and stop quickly; he must be able to show a great deal of speed for short distances; he must be extremely agile and capable of changing direction rapidly; he must have weight and strength enough to hold a steer on the end of a rope. Since he is usually expected to put in a full day at these activities, he must have great stamina and must necessarily be of calm, even, and tractable disposition, yet alert and ready for action.

Probably most important for rapid maneuvering, starting, stopping, turning, early speed, and so on, are powerfully muscled hindquarters and short heavily muscled back and loins, since muscles of the rear quarter provide propulsive power, and shortness and strength of the back maximize lightness and handiness of the forequarters. Good feet and legs are important. For maximum speed there should be considerable length of limb. However, it appears that most often sheer power makes up for a lack in this respect in horses used solely for stockhorse purposes.

The ideal appearance of stock horses re-

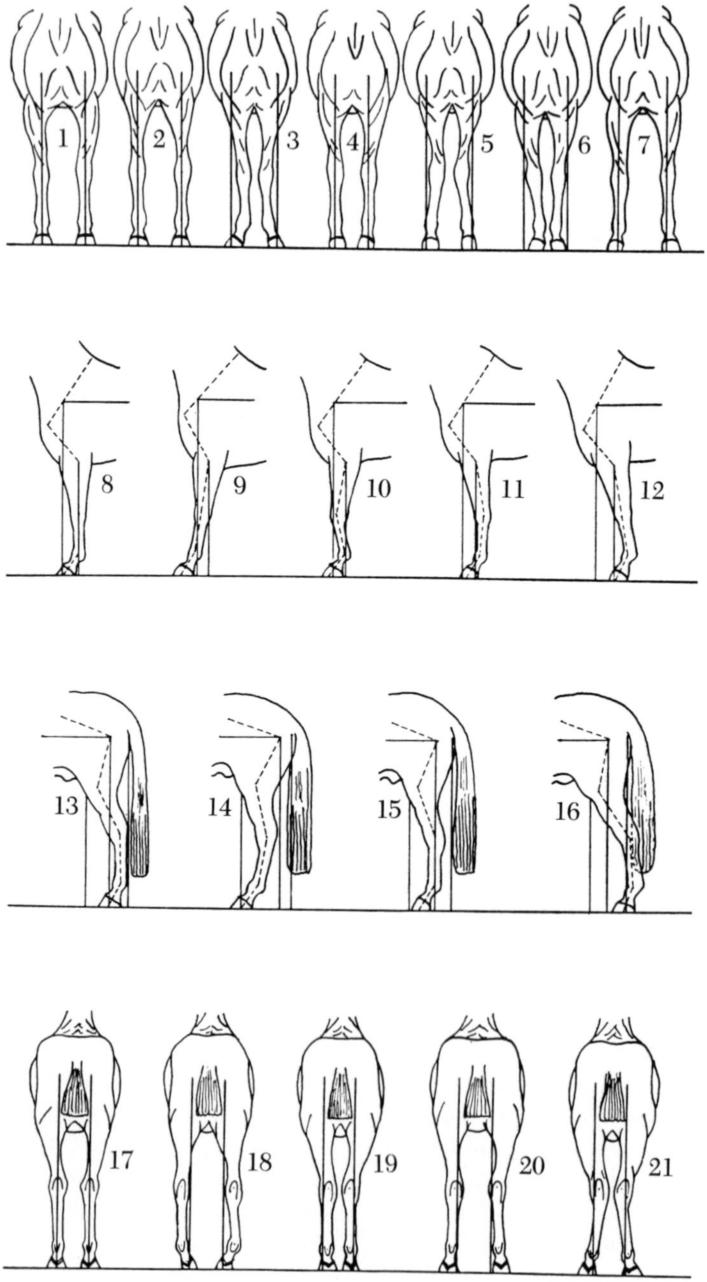

1 2 3 4 5 6 7

8 9 10 11 12

13 14 15 16

17 18 19 20 21

sults when the conformation requisite to desired characteristics is achieved—that is, a relatively short, compact, powerfully muscled, but well-balanced horse, with well-defined withers to hold a saddle, with well-muscled, well-sloped shoulders for maximum support and absorption of concussion resulting from violent maneuvers, and with head carried moderately low to provide visibility and space for action of the rider and, incidentally, to enhance balance and speed.

Obviously the complexity of the tasks performed by stock horses makes a high degree of intelligence and learning ability a necessity. The great majority of stock horses in use on ranches and in sports and shows are of American Quarter Horse breeding, with a few Thoroughbreds, Arabians, Morgans, and Appaloosa constituting the small remainder.

Traits of Race Horses

The demands made by racing are undoubtedly among the most strenuous of any made on the horse, especially on his feet and legs. Speed, great stamina, intelligence, and certainly feet and legs of the soundest kind are prerequisite (Figures 15.5, 15.6). In general appearance there is considerable variation in conformation of horses used for racing; for instance, compare the Thoroughbred distance horse with the racing Quarter Horse or harness racing Standardbred. Nonetheless, certain characteristics are common to each and distinguish them from other types of light horses. As a rule height is somewhat greater in proportion to length in racing horses than in other types, because of the requirement for speed (Figure 15.7). A good slope of shoulder is necessary for maximum length of stride as well as for absorption of concussion at racing gaits. Flat, smooth muscling without bulkiness of the shoulder is also essential for clean, unhampered action of the forelimb, and to reduce burdensome weight. Racing horses tend to have somewhat longer necks than other types, but they are carried moderately low for best balance. Characteristically, muscling should be long, flat, and smooth in horses that are raced for longer distances of a mile or more. This tends to be true for Standardbred horses and for Thoroughbreds racing these

Figure 15-3 Correct conformation and defects in conformation of the limbs of horses. Front view of forelimbs: a perpendicular line drawn downward from the point of shoulder should fall upon the center of the knee, cannon, pastern, and foot. Cut 1, represents the correct conformation. Cuts 2 to 7, inclusive, represent common defects. Cut 2, slightly bowlegged. Cut 3, close at knees and toes out. Cut 4, toes in. Cut 5, knock-kneed. Cut 6, base narrow. Cut 7, base wide. Side view of forelimbs: a perpendicular line drawn downward from the center of the elbow point should fall upon the center of the knee and pastern, and back of the foot, and a perpendicular line drawn downward from the middle of the arm should fall upon the center of the foot. Cut 8, represents the right conformation. Cut 9, leg too far forward. Cut 10, knee sprung. Cut 11, calf-kneed. Cut 12, foot and leg placed too far back. Side view of hind limbs: a perpendicular line drawn downward from the hip point should fall upon the center of the foot and divide the gaskin in the middle, and a perpendicular line drawn from the point of the buttock should just touch the upper rear point of the hock and fall barely behind the rear line of the cannon and fetlock. Correct position of the leg from this view is most important in a horse. Cut 13, represents the correct conformation. Cut 14, leg too far forward and hock crooked. Cut 15, entire leg too far under and weak below hock. Cut 16, entire leg placed too far backward. Rear view of hind limbs: a perpendicular line drawn downward from the point of the buttocks should fall in line with the center of the hock, cannon, pastern, and foot. Cut 17, represents the correct conformation. Cut 18, bowlegged. Cut 19, base narrow. Cut 20, base wide. Cut 21, cow-hocked and toes out, very serious fault. (From Dinsmore, 1935.)

Figure 15-4 Not all draft horses are as large as the typical draft breeds found in the United States. These Fjord horses at work in Denmark are deep and clean-boned, but do not usually exceed 14 hands in height; see Table 15.2. (Courtesy Dr. Henning Staun.)

distances. Muscling often tends to look heavier, rounder, and thicker in Thoroughbreds used for sprinting the shorter distances, and this type of muscular development reaches a maximum in racing Quarter Horses, where extreme speed for distances no greater than a quarter of a mile is most important. Thoroughbreds, particularly sires, that have shown great speed as sprinters have been used extensively in the development of the racing Quarter Horse, and thus the two breeds tend to have many characteristics in common.

Traits of Pleasure Horses

Many pleasure types of breeds are produced mainly for show. The American Saddle Horse (Figure 15.8) and the Tennessee Walking Horse, for instance, are probably used more in horse shows than in pleasure riding.

Figure 15-5 A Standardbred in action. Nevele Pride became the world champion trotter with a mile record of 1:54 4/5 in 1969 and was retired to stud after that season. Note that diagonal front and rear legs move in unison. (Courtesy Horseman and Fair World.)

Figure 15-6 Best Of All, a pacer (right) is shown winning a heat of The Little Brown Jug. The movement in unison of lateral front and hind legs is beautifully demonstrated. Compare action with Fig. 15-9. (Courtesy Horseman and Fair World.)

Because the requirements for pleasure horses vary so markedly from those for horses serving other purposes, it may be helpful to discuss briefly some of the desired traits, such as gaits, conformation, and color. Let us first consider gaits.

The natural tendency noted earlier of certain individuals and/or breeds to perform specific natural gaits well, or to learn acquired gaits readily, is a factor of special significance in the developing of some breeds of pleasure horses. The natural gaits which most breeds can perform are:

(1) The walk, a slow 4-beat (each foot strikes the ground separately) gait;

(2) The trot, a rapid 2-beat gait in which the diagonal front and hind legs move in unison; and

(3) The gallop, a rapid 3-beat gait. The canter is a restrained gallop. The American Saddle Horse performs this gait in a rocking-chair manner.

The following gaits are natural to or acquired by only a few individuals.

(1) The pace, a rapid 2-beat gait in which the laterals move in unison, is natural to only a few individual horses, especially Standardbreds.

(2) The rack, a fast and animated 4-beat gait, is a striking gait, characterized by extreme knee action. Much stress is placed on the performance of this gait in judging 5-gaited horses.

(3) The slow gait. There are actually sev-

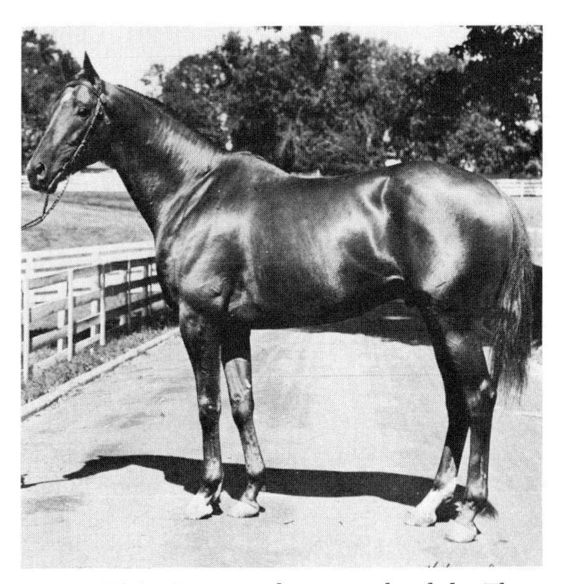

Figure 15-7 Swaps, a fine example of the Thoroughbred, Kentucky Derby winner, with winnings of over $800,000. He has sired many stake winners including Chateugay, winner of the 1963 Kentucky Derby. (Courtesy John W. Galbraith.)

eral slow gaits that may constitute the fifth gait of the 5-gaited horse. They include the running walk, fox trot, and stepping pace. The running walk is a slow 4-beat gait in which the horse appears to be breaking out of the walk into the run. The fox trot is a slow, slightly broken, rhythmic trot. The stepping pace is the most common. There is less swaying than in the true pace, and the cadence of the laterals is slightly broken.

To the extent that conformation is related to performance, it is important in all breeds of pleasure horses, or, for that matter, in all breeds and classes of horses. In some breeds, such as the Tennessee Walking Horse, the main emphasis is on performance. On the other hand, refinement (especially in the head, neck, and topline) is of special importance in the American Saddle Horse.

Owners of parade pleasure horses take special pride in the appearance of their animals (Figure 15.9). Stylish action, animation, and spectacular coat color are prized especially in parade horses. The golden body color and white, silvery, or ivory mane and tail of the Palomino, together with appropriate trappings, presents a striking appearance. According to Ensminger (1956), the perfectly marked representative of the Pinto is half color or colors and half white, with many spots well placed. Colorful spots are also characteristic of the Appaloosa.

Figure 15-8 Plainview's Julian, winner of the $10,000 five gaited stake at the Kentucky State Fair in 1960 shows the extreme refinement and elegance of the American Saddle Horse. The long, sloping shoulders and pasterns, so important to a smooth comfortable ride, are well demonstrated. (Courtesy the American Saddle Horse Association.)

Traits of Ponies

With the exception of the Hackney Pony, whose sole use is in horse shows as a heavy harness pony, the primary uses of ponies in the United States are for children's mounts and in horse shows. Characteristics desirable in a child's pony are about the same as those in general pleasure horses, with distinct emphasis on gentleness, tractability, and reliability. Since many of these ponies are shown in competition, they are also required to have an attractive appearance and to show breed characteristics. There has been a rather large increase of interest in ponies recently, with a resulting increase in numbers registered and, generally, improvement in quality.

Traits of Utility Horses

Throughout the world, and particularly in the less well-developed areas, large numbers of horses are used for general utility purposes, including transportation by riding or by horse-drawn vehicles, farm use, and cartage of various types. No uniform type or breed of horse serves these purposes, and the derivations of these animals are diverse. Some appear to have been derived with little improvement from the primitive horses of Europe that also

Figure 15-9 Mares of the Lippizan breed with their foals. Representatives of this breed are noted for their drill team maneuvers. Note that the mares are branded; only those appropriately branded are considered as members of the breed. (Courtesy Lichtbildstelle, BMFLuF.)

gave rise to the draft breeds. Some, such as the Scandinavian ponies, are derived from primitive pony types with or without improvement by the more highly developed light-horse breeds previously mentioned. Others are presumably derived from feral horses of the respective areas in which they are found, for example, the cart horses of South America.

Figure 15-10 A mare with mule colt. (Courtesy Frank C. Mills.)

Table 15.2.

Horses not commonly found in the United States.[a]

Class	*Name of breed*	*Country where located*	*Color*	*Height (hands)*	*Use*
Heavy draft	Rhineland Heavy Draft	W. Germany	Sorrel, chestnut and red roan	16 to 17	Draft
	Seine Inférieure	France	Bay, brown, roan	16	Draft
	Suffolk Punch	England	All grades of chestnut	About 16	Draft
	Vladimer Heavy Draft	U.S.S.R.	Hard colors	16	Draft
Medium heavy draft	Ardennes	France, Belgium	Many colors allowed	15–3	Draft
	Jutland	Denmark	Usually chestnut	About 15–3	Draft
	Swedish Ardennes	Sweden	Usually brown or black	15 to 16	Draft
	North Swedish	Sweden	Dun, brown, chestnut, black	15	Draft
	Finnish Draft	Finland	No specific color	15–2	Draft
	Trait du Nord	France	Usually bay or chestnut	15 to 15–3	Draft
	Russian Draft	U.S.S.R.	Usually chestnut, bay, roan	14–2	Draft
	Sokólsk	Poland, U.S.S.R.	Usually chestnut	15–16	General draft
	Masuren	Poland	All colors except piebald and skewbald	16	Draft and saddle
Light draft	Avelignese	Italy	Chestnut, light mane	About 14	Draft
	Norman Cob	France	Any color	15 to 16	Draft and saddle
	Friesian	Holland	Black	15	Draft and saddle
	Groningen	Holland	Usually black	15 to 16	Draft
	Fjord	Norway, Denmark	Dun, dorsal stripe, black points	14	Draft
	Frederiksborg	Denmark	Usually chestnut	15–2 to 16	Draft and saddle
	Toric	U.S.S.R.	Chestnut and bay with white markings	15–1	Draft
	Lithuanian	U.S.S.R.	Usually bay	16 to 17	Draft
	Welsh Cob	Great Britain	Bay, black, chestnut, cream, etc.	14 to 15	Draft and saddle
	Trakehner	W. Germany	No odd colors	16	Draft and saddle
Light horses	Hanoverian	W. Germany	No odd colors	16 to 17	Draft and saddle
	Holstein	W. Germany	Usually brown, bay, or black	16	Draft and saddle
	Nonius	Hungary	Black and dark brown	15–3	Harness and saddle
	Sorraia	Spain	Dun, dorsal stripe and stripes on legs	17 to 18	Saddle

	Andalusian	Spain	Grey, [illegible]	[illegible]	[illegible]
	Lusitano	Portugal	All solid colors	14 to 15	Saddle
	Lipizzan	Austria	Usually grey	14 to 16	Parade and school
	Salerno	Italy	All colors	About 16	Saddle
	Noram Trotter	France	All solid colors	15	Racing
	Einsiedler	Switzerland	Most solid colors	15 to 16	Saddle
	Knabstrup	Denmark	Spotted	15–3	Saddle
	Dole Trotter	Norway	No limitations; registry based on speed	15	Racing
	Orlov Trotter	U.S.S.R.	Any solid color	16	Racing
	Barb	Algeria	Bay, brown, chestnut, grey	14 to 15	Saddle
	Llanero	Venezuela	Dun, yellow with dark mane	Below 15–2	Saddle
	Criollo	S. America	Dun, grey, sorrel, etc.	13 to 14–5	Saddle
Ponies	Konik	Poland	Yellow, grey, blue dun	13–1	Dual purpose
	Haflinger	Austria	Chestnut, waxen mane	14	Dual purpose
	Peneia	Greece	Most colors	10–2 to 14–1	Draft and pack
	Iceland Pony	Iceland	Usually grey or dun	12 to 13	Saddle
	Connemara	Ireland	Dun, dorsal stripes and black points	13 to 14	Dual purpose
	Sumba	Indonesia	Usually dun with dorsal stripe	12 to 12–5	Saddle

[a] We have taken some liberties in characterizing the breeds listed, but have not wilfully misrepresented them. No attempt has been made to prepare a complete list.

Figure 15-11 "Barnyard ballet" by mules owned by Reese Brothers of Gallatin, Tenn. (Courtesy of Nashville Tennessean.)

Asses and Mules

The mule is an interspecific cross between the mare and the jack, and therefore is seldom fertile (Figure 15.10). Jacks were introduced into the United States late in the eighteenth century. The reciprocal cross, between the jennet and the stallion, is known as a hinny, but this cross is seldom found. The mule has been a major source of power in the southern states because of its heat tolerance, its ability to adjust its habits of work and eating to the prevailing conditions, and its all-around hybrid vigor (Figure 15.11). The burro is a small donkey used for packing in many parts of the world. It, also, is admired for its ability to withstand hardships and to thrive under adverse conditions, as well as for its capacity for carrying heavy burdens.

Breeds of Horses Not Common in the United States

All the draft breeds found in the United States were imported from Europe. As described above, the Thoroughbred and Arabian breeds were also imported as were the ancestors of the western mustang. Considering the large number of breeds throughout the world, it is surprising that so few have been introduced into the United States. A selected list of breeds not commonly found in the United States is given in Table 15.2. In assembling this list, we have leaned heavily upon information supplied by Goodall (1965). Our main purpose in this listing is to provide the student with some inkling of the wide number of breeds which exist. As the horse becomes replaced in cities and on the farm by automobiles, trucks, and tractors, numbers of draft animals will no doubt decline. This replacement has become almost complete in the United States and is progressing rapidly in many parts of Europe. In the meantime, there has been an upsurge in interest in riding horses in the United States.

With few exceptions, the characteristics of the breeds listed in Table 15.2 do not differ significantly from those found in the United States. There are a number of breeds, such as the Fjord, specifically designed for light draft purposes, which have no counterpart in this country; earlier, however, the Morgan and crosses between the light and heavy draft breeds did provide horses of this type. In summary, one can say that, for the most part, types of horses found in Europe do not differ greatly from those present in the United States. In most of the less-developed countries, there is a tendency for light horses to predominate, and in many instances power is provided by cattle or by buffaloes rather than by horses (see Chapter 9).

FURTHER READINGS

Dinsmore, Wayne. 1935. *Judging Horses and Mules.* Book no. 219. Chicago, Ill.: Horse and Mule Assoc. of America.

Ensminger, M. E. 1969. *Horses and Horsemanship.* Danville, Ill.: Interstate.

Goodall, Daphne M. 1965. *Horses of the World.* New York: Macmillan.

Ridgeway, W. 1905. *The Origin and Influence of the Thoroughbred Horse.* Cambridge, England: Cambridge Univ. Press.

Riggs, E. S. 1936. *The Geological History and Evolution of the Horse.* Leaflet no. 13. Chicago, Ill.: Field Museum of Natural History.

Sixteen

An Examination of Systems of Management of Wild and Domestic Animals Based on the African Ecosystems

A mixed population will produce more pounds per acre—I think this is well-known. The problem is whether one can, in fact, economically harvest that produce and make use of it.

H. P. Ledger

An extensive review has recently been published by Cuthbertson (1970) in which the role of the ruminant in the world food supply was discussed at length; consideration was given not only to cattle and sheep, but also to water buffalo in Asia, yak in northern Asia, game cropping in Africa, and the deer family in the northern hemisphere. However, there are two important aspects that were not dealt with by Cuthbertson, nor have they been addressed fully by anyone else. One is the effect of husbandry on nutritive value of the animal product. Cuthbertson states that husbandry does not alter the nutritive value, which is quite incorrect under most management systems. The other is the basic ecological importance and reasoning for reevaluating our present narrow attitude to livestock for human food. The ecological background was first crystallised by Fraser-Darling (1960), and developed in his 1969 Reith Lecture. Cuthbertson did not discuss the ecology in any depth. In consequence, we intend to confine this chapter to a discussion of these points with emphasis on one situation, the African environment. In this way it is hoped that questions concerning the potential of nondomestic animals for food and the basic ecological principles applicable to future animal production will be exposed.

Feasibility of Utilization of Nondomestic Species

If there is a case for continuing livestock production, then there is a case for incorporating new species into our agriculture, since there are many parts of the world's land mass not suited to our conventional livestock. In assessing the feasibility of this proposal, it is worth considering the wide variety of species available in Africa. Skinner (1970), in reviewing the question of wildlife utilization, has presented negative aspects based on the following two points.

1. Sheep and cattle are so integral to the

tribal laws and customs of the peoples of Africa that it would be difficult to envisage their replacement by game.

2. Doubts will remain concerning the spread of disease.

The impact of education and communication throughout the world, and particularly in Africa today, has changed and is changing peoples' attitudes. It is also true that no advocate of game utilization suggests "replacing" cattle, sheep, and goats. The suggestion is to employ species indigenous to Africa in areas where they are most productive, not as replacements, but as complements to existing methods. Cattle are not indigenous to Africa, but originated in Asia, and more recently stock has been imported from Europe. Furthermore, sheep and cattle play a part in customs only of certain tribes, whereas other tribes, like the Acholi in Uganda and the Wakamba in Kenya, are traditional hunters (Figure 16.1). It was only the introduction of game laws and the creation of the luxury hunting market by European colonialists which curtailed hunting activity by the local Africans. Indeed even today a considerable amount of what is now termed "poaching," but is in fact traditional food gathering, still occurs in Africa (Figures 16.2 and 16.3). In some instances legislation has been passed to permit a limited amount of local hunting for food. On the other hand, quite extensive game cropping has already been practiced in South Africa, southern Rhodesia, and Zambia, backed by carcass inspection, abbatoir, and cold storage facilities. Similarly, commencing in 1961 in Uganda, the National Parks and individual cropping schemes of the Game Department have contributed over 10,000 tons of meat to the economy (Table 16.1). Surprisingly, in Ghana, investigations by FAO in 1966 revealed that wildlife contributed 27,740 tons of meat, whereas only 15,000 tons were obtained from domestic stock and poultry. The ecological implication of such resources to the changing food structure and disease patterns is not generally understood by agricultural planners, and is seldom appreciated by expert nutritionists, who rarely take the contribution of wild fauna into consideration. Such food sources are of a high nutritional standard (Crawford, 1969), and their importance both quantitatively and qualitatively should not be overlooked.

Reservations which concern the attitude of Africans, and indeed of Europeans living

Figure 16-1 Dissection of a young buffalo on the Tonia-Kaiso Flats, Uganda.

Figure 16-2 Poachers photographed in Tanganyika carry wire snares, a bow and poisoned arrows. Poachers customarily butcher the animals themselves and sell the meat. (Courtesy F. Fraser-Darling.)

among Africans, toward "game" meat are usually based on ignorance and lack of firsthand knowledge. Sometimes the impression is given that people eat only cattle. Apart from the misunderstanding inherent in such a statement, it does raise the interesting converse. There are many people, particularly in Asia, who for religious reasons will not eat cattle, but have no objection to other meats! Unfortunately, food supply and malnutrition can present serious, large-scale problems in such areas as in India, where the biomass of cattle is of the same order as the biomass of people. One solution which could have dramatic results would be to limit the number of cattle so that they could be maintained in good health. Development of other species suitable to the environment and, like the buffalo, capable of producing meat and milk, could provide an additional high-quality, easily managed food resource.

Animal Disease Problems

Just as disease control is vital to cattle husbandry, so also will it be essential to any proper management of wildlife or new domestic species. Table 16.2 gives some indication of the parasitic loads of wild species. Small animals like the Oribi carried light infestations, whereas the bovids (buffalo) were

Figure 16-3 Typical technique for cooking meat by those who are forced to poach.

the most severely affected. In our own experience, the species inhabiting high-rainfall areas were found to be more infested with ticks and flukes than those from the dry areas, and it is likely that seasonal variations will also be important.

The following fact demonstrates that the disease problem is not the prerogative of any one species, but rather a question of management. Muscle cysticerci were found in 18 per cent of the wild animals examined by a Royal Veterinary College team; they comment that the techniques employed involved complete carcass dissection and were far more exhaustive than conventional meat inspection methods; they also point out that, under routine conditions of examination, it is likely that all carcasses would have been passed. On the other hand, the 1968 report of the Ministry of Animal Industry, Game, and Fisheries of Uganda indicates that routine inspections lead to the rejection of some 25 per cent of the cattle carcasses.

Rinderpest and foot and mouth disease provide perhaps the best examples of the danger in attempting to introduce foreign elements to an ecosystem without proper control. As far as is known, neither rinderpest nor foot and mouth disease was to be found in Africa south of the Sahara. These diseases were brought to Africa with the importation of cattle. The precise toll of the rinderpest epidemic which swept from northeastern Africa to the southern tip of the Continent at the turn of the century is not known. From the writings of the many explorers and hunters in Africa just prior to this time, it appears that the toll was in the order of several million metric tons of livestock and wildlife (Simons, 1962).

The impact of diseases of wild species upon domestic animals may be over-rated, excepting that many species of game animals act as host to the trypanosome. The trypanosome seems to be nonpathogenic to the indigenous mammals, but is highly pathogenic to cattle of Asiatic and European origin. Surveys carried out by the Royal Veterinary College team on wild species in areas of endemic trypanosomiasis uncovered only light parasitaemias in the wild species, suggesting tolerance of and

Table 16.1.

Game cropping in Uganda and Zambia, 1966 (From data of Ruhweza, 1968, and Steel, 1968.)

	Total edible yield (lbs.)	
Main species cropped	*Uganda*	*Zambia*
Buffalo	288,750	56,360
Eland	55,000	—
Kob	99,375	—
Elephant	490,000	211,273
Hippopotamus	175,000	183,640
Total metric tons	560.47	205.12

Table 16.2.

Some examples of parasite infections in game species in Uganda.[a]

		Parasite:	*Setaria sp.*	*Liver fluke*	*Lungworm*	*Tapeworm cysts*	
		Organ:	*Peritoneal cavity*	*Bile ducts and gall bladder*	*Lung tissue*	*Viscera*	*Muscles*
Location	*Host*	*No. inspected*	*Per cent infected*				
	Buffalo	68	67.0	63.0	0	0	0
Zoka river	Kob	100	30.0	50.0	0	1.5	3.0
	Hartebeest	43	5.0	42.5	12	9.0	12.0
	Oribi	29	72.0	0	0	0	0
	Eland	43	12.0	12.0	0	2	2
Greek river	Giraffe	8	12.0	0	0	0	0
	Topi	16	12.0	0	6	0	0

[a] Data compiled from Bindernagel (1968), Royal Veterinary College East African Research team (1970), and our own observations. Cunningham (1968) reports on levels of 40–50% Trypanosomiasis infection in wild species inhabiting endemic areas. The Royal Veterinary College team examined six each of eland, hartebeest, topi, and zebra and three Grants gazelle in Karamoja, and found no anthrax as evidenced by blood smears. Serum agglutination tests for *Brucella spp.* gave three doubtful positives of 1/20, 1/40, and 1/80, which were not confirmed. Bacterial examination yielded no Salmonella species. Liver flukes were found in four eland and three zebra, and muscle cysterci in four eland and one hartebeest. In general, tick infestations of wild species are lighter than are commonly found in cattle from the same area which have not been dipped. Bindernagel (1968) found 100% tick infestation in 86 buffalo, 12% in 59 kob, 69% in 42 hartebeest, and 50% in 20 oribi.

resistance to the infection. It is fair to say that insufficient work has been done on trypanosomiasis in wild species to define the mechanism of resistance, but it is clear that wild species breed and thrive in the presence of the Tsetse fly (*Glossinia* Sp.) and the trypanosome. However, cattle in the same area succumb to the infection (Reid *et al.*, 1966; Lambrecht, 1966), although there are strains of cattle, such as the Ndama of West Africa, which can tolerate trypanosomiasis.

It seems that East Coast fever (Brocklesby and Vidler, 1966) and African swine fever are again pathogenic to foreign species, and not to the indigenous population.

With the prevailing unlikelihood of these diseases being conquered in the near future, research on this discrepancy between the sensitivity of indigenous and foreign species should be given a high priority.

Animal Disease and Human Ecology

The impact of animal disease on the human ecology of Africa following introduction of cattle is not usually considered. Just as the introduction of the game laws in seventeenth-century England had serious sociological and nutritional implications for the peasants (Trevelyan, 1944), so the overnight destruction of game as well as domestic stock in Africa must have immediately created serious human nutritional problems and, in the long-term, affected attitudes toward food selection. The loss of high-quality animal food resources certainly entrenched in East Africa, and probably elsewhere, the agricultural development of the plantain (*Masa* sp.), the sweet potato (*Ipomeas batatas*), and cassava (*Manihot* sp.). Like the rinderpest, these plants were brought to Africa from outside. They were introduced from Asia and South America when the trade routes were opened within recent historical time. Because of the ease of culture of these crops and their low value for protein, structural lipids, and other nutrients, their engrainment into the local customs undoubtedly was a major factor which contributed to the now prevalent malnutrition and repetitive infections among the peoples dependent on these foods.

In view of the human effects, the introduction of bananas, sweet potato, and cassava can be seen as a practice with unfortunate consequences. Similarly, in view of the biological evidence, it might well be argued that the case was weak for the introduction of cattle to Africa. In our view, both were bad practices in the biological sense, but that is academic, since the practices are there to stay.

Wild-life is earning foreign currency for Kenya at the rate of £28.3 million per annum. This is the most important single revenue of foreign exchange and far in excess of the return from cattle. If rinderpest were to eliminate the Kenya wildlife, the loss to the country would be more serious than the loss of its entire cattle population. Consequently, the view that wildlife use cannot be rationalized because of the disease transmission problem is out of date. The proper attitude is the correct management of both resources in their proper context.

Management and Nutrient Value

We would like to comment on the food-selection practices of free-living animals in relation to nutritional quality of their products. The criterion on which the quality of domestic animal production is based is usually weight gain. Gain in weight can be due to either the deposition of fat or an increase in muscle tissue. (Figure 16.4).

Modern intensive beef-production systems generally produce carcasses with about 30 per cent fat, which can be readily trimmed, and about 50 per cent "lean." However, even the so-called lean contains 7 to 20 per cent of its fresh weight as adipose fat (Crawford *et al.*, 1970). Therefore the animal may at best produce about 45 per cent of the carcass as actual

Figure 16-4 These African buffalo appear fat, but in fact the heavy-looking body is composed of lean meat and very little fat.

muscle cells, which means that if the muscle water is discounted only 9 per cent of the carcass is protein. On this basis of calculation, the carcass is shown to provide more than three times as much adipose fat as protein. The same calculations on free-living domestic or wild species not subjected to intensive feeding give 15 per cent protein from the carcass and 2 to 5 per cent adipose fat (Table 16.3).

Another aspect of the nutritional management of animals and their food value to man arises from the biochemical composition of meat itself. Meat amino-acid composition of free-living and domestic species is identical, but the fats are different (Crawford, 1968, 1973). In Table 16.4 it can be seen that the total fat extractable from muscle tissues of domestic species is 90 to 98 percent saturated and monounsaturated, i.e., fat which is a nonessential energy store. The fat extracted from muscle tissues of wild animals is 30 to 40 per cent polyunsaturated, i.e., of essential nutrient value. This difference can be attributed largely to the triglyceride infiltration generated by the conditions of high-energy nutrition for intensive production in the case of domestic animals. However, examination of the phospholipids also show differences, in that the ethanolamine and choline phosphoglycerides of wild species are richer in the essential polyunsaturated fats than in tissues of domestic species fattened for market. This suggests dietary influences other than simply the plane of nutrition. The nutritional significance of these differences in human diets have not been fully explored.

The dietary role of essential, fat-soluble nutrients in ruminant metabolism and pathology is poorly understood. It is, however, clear that these constituents may be available to a relatively greater extent in free-living systems than in a domestic food structure based on our observations. Polyunsaturated fatty acids deteriorate with storage of food. We have found in fresh dry hay that while as much as 20 per cent of the lipids may be linolenic acid, after six months of storage this

Table 16.3.

Meat analyses of free-living and domestic species expressed as grams per 100 grams of fresh tissue or as calories.

	Ash	*Fat*[a]	*Protein*	*Calories*
Eland[b]	1.1	1.9	23	125
Free-range cattle	1.1	2.0	22	120
Intensive fat stock	0.9	15.0	20	230

[a] The carcass fat content from free-living species will vary with seasons. In general it will be increased in the rainy season when much fresh, young grass is eaten.

[b] The meat of the hartebeest, topi, giraffe, buffalo, and warthog has been found to be similar.

Table 16.4.
Per cent of individual fatty acids expressed as total lipids in muscle and in adipose tissue of free-living and domestic species.

	Muscle tissue				*Adipose tissue*			
Type of fatty acid[a]	*Eland*	*Lean beef*	*Warthog*	*Pork*	*Eland*	*Beef*	*Warthog*	*Pork*
16:0	14.0	26.0	14.0	20.0	29.0	33.0	18.0	21.0
16:1	1.0	3.4	0.7	2.2	2.3	5.1	2.0	3.0
18:0	21.0	34.0	8.0	15.0	38.0	18.0	22.0	24.0
18:1	16.0	42.0	7.0	38.0	32.0	49.0	28.0	53.0
18:2	24.0	6.0	33.0	6.0	4.3	3.5	17.0	8.0
18:3	5.0	0.2	5.0	1.0	3.2	02.6	17.0	1.0
20:3	0.3	1.1	0.8	0.7	—[b]	—	—	—
20:4	9.0	3.2	6.3	3.3	0.9	—	0.7	—
20:5	2.4	0.5	1.9	0.8	—	—	—	—
22:5	4.2	1.1	3.2	0.9	00.8	—	00.9	—
22:6	0.8	0.6	1.2	0.3	—	—	—	—

[a] Number of carbons in fatty acid: number of unsaturated bonds.
[b] Blanks indicate that fatty acid was not detected.

may be as low as 2 per cent. In green barley, linolenic acid fell from 11 per cent to 4.6 per cent over six months. Thus, by the time feedstuffs have been harvested, stored, and compounded into finishing rations, the polyunsaturates may be substantially lower than in the fresh-crop material. We analysed a pelleted cattle finishing ration, and found it had 3 per cent lipid, of which only 1.8 per cent was linolenic acid. This compared to intensively grazed pasture grass with a similar level of lipid, but which contained 55 to 75 per cent linolenic acid. Furthermore, ingredients of rations compounded for intensive meat production are subjected in varying degrees to various processing procedures, such as crushing and rolling grains. This breaking of seeds' capsules will make them more susceptible to hydrogenation in the rumen, a process which in effect increases "saturation" of fatty acids.

Therefore, free-living animals who themselves harvest plant materials as they graze or browse may have a higher level of polyunsaturates presented in the food they consume. Furthermore, because of the coarseness of leaf material, especially characteristic of diets of wild animals, much of it may pass through the rumen undigested escaping hydrogenation and becoming extracted by the lipid-solubilizing secretions of the small intestine. Seeds being small and light float through the rumen and also bypass the hydrogenating influence of the anaerobic microorganisms in the rumen (Wilde and Dawson, 1966).

Application of the high-energy concept of animal feeding to systems of wild specie management could be expected to produce the same emphasis on "excessive fat production" as in modern intensive animal production, which in our view is wrong. It is doubtful from the above data if intensification of production with high-energy feeds have resulted in a net gain in meat yield from the carcass by comparison with the performance of the free-living. Furthermore, processed and compounded diets affect the biochemical composition of meat, which may not be in the best interest of its food value for man. At any rate, the high-energy producing systems applied at present to our fat-stock would be unsuitable for Africa or Asia, where the important need is for nutrient-rich rather than calorie-rich food. If eland reach the same weight in the same time as the Holstein (Crawford, 1968), but have more meat and less adipose fat, it is clear that the free-living food structure is worth examining in greater detail (Table 16.5).

Biological Adaptations

There are two biological reasons for incorporating new species into our domestic livestock system which broadly come under the headings of physiological, i.e., species-specific adaptations to heat stress, and ecological, e.g., plant/animal interactions on water economy.

Physiological studies by Taylor and Lyman (1967) have demonstrated that eland and oryx can tolerate a heat stress of 45°C with relative ease. This degree of tolerance is achieved not by any difference in water requirement but by permitting the body temperature to rise, thereby avoiding the loss of water in thermoregulation. Diurnal variations in body temperature of more than 10°C have been recorded. Efficient oxygen extraction by the lungs also plays a part in water economy. A high oxygen-extraction efficiency implies a smaller loss of water in expired air.

The response of the eland to the semi-arid environment illustrates the manner in which physiological adaptations must be considered in livestock management. During the heat of the day the animal rests in the shade, allowing its body temperature to rise. At night the animal eats the leaves, which by then are replenishing their water supplies lost during the heat of the day. As the air temperature falls, the body temperature falls; this permits increased activity, and increases the energy requirement. This change encourages food consumption, and because the food is rich in water the animal can obtain its water purely from leaf material and its own metabolic water production (Taylor and Lyman, 1967; Taylor, 1968a, 1968b, 1969; Rogerson, 1968).

The Ecological Case for Nondomestic Species

The ecological case for utilization of nondomestic species is simple, but rests on an understanding of ecosystems in which the prevailing conditions involve little rain and much sun. This is quite in contrast to the opposite conditions of much rain and little sun under which the familiar agriculture of Europe evolved. Thus, we have come to accept that European livestock management is today based on grass and cattle, but this is very different from the semi-arid ecosystem. For example, in marked contrast to the shallow turf in Britain, the tap roots of which seldom reach further than one-half meter, perennial grasses in East Africa send down tap roots which may reach to depths of five meters. An appreciation for the significance of such differences and an understanding of the interplay between the variety of vegetation will reveal the potential productivity of the semi-arid regions, which represent a third of the world's land mass. Techniques different from those to which we have become accustomed are needed to develop and exploit the potential.

Maintenance of Water Economy

The viable ecosystem of the semi-arid tropics includes a diverse spectrum of vegetation, the function of which can be defined. An interplay exists between trees, bushes, sedges, and grasses in the maintenance of the water economy. The large trees send their root systems deep into the water tables; they maintain humidity and affect climate conditions for the

Table 16.5.

Biological data relevant to the productivity of some African herbivores.[a]

Species	*Age (years) at sexual maturity*		*Gestation period (days)*	*Birth weight (lbs.)*	*Mature adult weight (lbs.)*	*Edible products as per cent of live weight*	*Carcass fat as per cent of carcass*
	Male	*Female*					
Warthog	1.7	2.0	175	1.0–1.7	208	62	1.8
Oribi	0.5	0.5	220	4.0	37	70	1.9
Uganda Kob	1.0	2.0	240	11–12	209	65	2.6
Jackson's Hartebeest	1.4	2.3	242	20–24	450	63	2.2
Buffalo	3.5	6.5	340	108–118	1,660	59	5.6
Eland	3.0	4.0	255	69–71	1,540	68	4.2

[a] Adult weights are those of young adults, as opposed to "trophy" animals, which could be considerably larger. The long gestation period of buffalo by comparison with cattle might appear to be a disadvantage commercially. However, Grimsdell (personal communication and Ph.D. thesis, University of London) comments that the long gestation means that the calves are dropped when the rains are established; domestic calves may be dropped before the onset of the rains. Most of the data in this table were derived from Bindernagel (1968). Reviews on game production in other parts of Africa have been presented by Joubert (1968), Retief (1970), and Roth and Osterberg (1971).

small plants and animals both by transpiration from leaves and by the provision of shade from their umbrella-type structures.

We have observed a marked diurnal swing in the water content of the leaf from *Balanites aegyptica,* which means that water is being released into the atmosphere during the day and replenished at night. This not only provides a mechanism for humidity control but also supplies water for animals adapted to eating such vegetation, for in such regions the animals rest during the heat of the day and feed in the evening and early morning. Usually the clarity of the African night provides sufficient light from the stars and particularly the moon to permit continual feeding during much of the night. Overcast skies only occur during the rain seasons, when water is available in greater abundance. Schmidt-Nielson (1964) has suggested that the hygroscopic nature of certain plants adapted to the semi-arid environment might be an important water source for animals feeding at night. Taylor (1968b) demonstrated this to be true for *Disperma sp.*, the leaves of which are capable of absorbing water from the atmosphere. At midday temperatures and humidities, the leaves contained 1 to 5 per cent water, but 40 per cent during the night because of water absorption. Hence small plants with relatively shallow roots could be obtaining water from the deep tables indirectly via the transpiration from the larger deep-rooted plants. Plants with succulent leaves also contribute to the water supply; *Sansevaria* are apparently eaten by eland but not by cattle (Spinage, 1962).

Utilization of Soil Mineral and Nutrient Resources

A second and important point regarding the function of the large deep-rooted vegetation is that it is tapping the mineral and nutrient resources of the soil in a three-dimensional manner on a far more extensive basis than can be accomplished by a grass pasture. This extensive cull of the soil nutrients means a greater potential productivity than from grass alone, but also means that quite different techniques will be required for maintaining the productivity through fertilization and for utilization of the forage.

Seasonal Changes in Vegetation and Possible Relation to Behavioral Patterns

Seasonal variation in plant growth buffers the dry season for the animals. The end of the dry season, like the end of the winter in the northern hemisphere, sees little vegetation growth and only some of the trees still retain green leaves. When the rains commence, it is the grass which first provides green foods for the animals. As the season progresses, the grass seeds develop. The trees take longer to produce new growth and, by the time the grass dries off, the trees are bearing leaves which will carry the animals through the dry season to the next wet season (Figure 16.5). The details of the seasonal alteration in feeding patterns have not been fully investigated, but the principle is known.

Because of seasonal changes in vegetation, there are changes in the availability of nutrients, most easily seen in the seasonal availability of lipids. The nutrient contents of the leaf and the seed, for example, are quite different, and this is most easily seen in Table 16.6 by comparing the lipid composition of the *Balanites* leaf with that of the kernel. The seeds provide linoleic acid and the leaf linolenic acid. The seeds are rich in lipids, and many small seeds will float through the rumen, bypassing the hydrogenating influence of the rumen microorganisms. Quite apart from the contrast in the leaf and seed, we also know that the flowers are eaten and in particular we have found the yellow pigment of the acacia liberally distributed in the stomach contents of eland when the plant is in flower. There are likely to be other nutrient differ-

Figure 16-5 A gerenuk browses a thornbush obtaining both food and water. (Photo © 1972, National Geographic Society.)

ences which have not yet been studied. In other words, the animals are eating leaves, flowers, woody vegetation, and seeds, and there are obviously seasonal variations in the type of food and its chemistry.

Physical Adaptations of Animals in Relation to Food Selection

Finally, it should be appreciated that many of the edible trees are thorn-bearing. Species like the eland, kudu, and giraffe, through the combination of a narrow mouth and long tongue, are able to strip the branches of leaves and fruit in a remarkable sweeping motion that avoids the sharp thorns, which are capable of puncturing a Landrover tire (Figure 16.6). The development of the neck, such as in eland, gerenuk, and giraffe, aids these species in their browsing behavior; in the giraffe the long neck provides the animal with a food-harvesting structure which is more than 90 per cent dependent on trees; the eland has a longer neck than most other antelopes and bovids, and can also use its horns to hook down branches which would otherwise be out of reach. In contrast, bovids, such as cattle, which are adapted to grass and non-thorny vegetation, have wide mouths and short tongues which are not suited to this type of food-selection pattern. As Fraser-Darling (1960) and Lamprey (1963) have emphasized, the selective feeding habits of the different species provide a spectrum of food-selection patterns which cover all aspects of the vegetation.

Figure 16-6 A giraffe browsing on *Balanites;* note the size of the thorns, which demand a long and dexterous tongue.

Table 16.6.
Crude composition of some plants eaten by eland. Figures are percentages of dry weight.[a]

Plant	*Part eaten*	*Pro-tein*	*Lipid*	*Fiber*	*Ash*
Acacia nigrescens	Leaves	16.0	4.0	23.0	6.0
Acacia senegal	Leaves	20.0	3.0	27.0	6.0
	Beans	18.0	20.0	5.0	6.1
Balanites aegyptica	Leaves	20.0	11.0	22.0	7.0
	Kernels	36.0	42.0	13.0	4.0
Cordia gharaf	Leaves	11.0	2.0	37.0	10.0
Codrophospermum mopane	Leaves	12.6	5.5	21.0	5.0
Commelina benghalensis	Leaves	19.4	1.5	21.0	26.0
Combretum apiculatum	Leaves	14.0	4.0	20.0	5.0
Diplolophium africans	Leaves	22.0	5.0	14.0	14.0
Euclea divinorum	Leaves	10.0	11.0	14.0	14.0
Fagava chalybea	Leaves	10.0	11.0	14.0	14.0
Gomphrena celosioides	Leaves	29.0	3.0	16.0	
Grewia subspathulata	Leaves	13.4	6.1	36.2	7.0
Terminalia catappa	Leaves	11.0	5.0	23.0	4.2
	Nuts	24.0	60.0	3.0	4.3
Tepherosia emeroides	Leaves	24.0	1.8	31.0	5.7

[a] The authors are grateful to Dr. F. Busson for the analyses. It is of interest that many of the plants eaten by animals are also eaten by man.

Use of Complementary Food-selection Patterns

The food-selection patterns are complementary, so that different species eat different portions of the vegetation. This means that optimum use can be made of the three-dimensional vegetation growth in operation both below and above the ground.

Consequently we have two potential techniques whereby the vegetation could be converted into animal products: (1) *monoculture,* the use of a single species, such as eland, which is adapted to eating both grass and browse foods; and (2) *polyculture,* the use of a wide spectrum of herbivores utilizing different portions of the vegetation. The complementary food-selection patterns could ultimately be developed into a system of crop rotation at a secondary level.

Both the use of a browsing monoculture and the development of a polyculture system would seem preferable to the alternative method of land use suggested by many experts, which is irrigation with bore-hole water. The rapid rate of surface water evaporation makes such a method very expensive in terms of water, inefficient, and unviable if water is the limiting factor. The consequent sinking of the water table, drying of surrounding areas, and climatic changes, with the high risk of dust-bowl formation, is too well-established in principle to recommend irrigation except where water is in abundance on the surface.

Monoculture would appear to be less reliable than polyculture, since monoculture is more likely to lead to stagnation of soil and vegetation (Fraser-Darling, 1960). Monoculture, however, has the advantage of simplicity. The species of choice for this would be the eland (Figure 16.7; Tables 16.7 and 16.8), be-

Figure 16-7 Eland on a ranch at Kilimanjaro. They and the Guernsey cow are all 3.5 years of age. (Courtesy of Russell Kyle.)

cause it has already been domesticated, can be milked (Posselt, 1963; Treus and Kravchenko, 1968), is amenable to management, and has highly palatable meat (anonymous, 1971). In this instance, the suggestion is simply to farm eland and bush, in semi-arid regions in a manner analogous to cattle and grass in temperate, well-watered zones.

Polyculture is ecologically more viable because it involves a broad spectrum of vegetation which theoretically could be managed to maintain or even improve soil condition. It is more difficult because the number of variables increases with the numbers of plant and animal species employed.

Polyculture in Semi-arid Zones

Basically, a polyculture in East Africa could consist of the following:

Giraffe,	top browser;
Eland,	mid-level browser;
Kudu,	thick woodland;
Oryx,	open grassland;
Hartebeest,	low-level browse and grazing;
Grants Gazelle,	grazing and some browsing;
Oribi,	open grazing.

There are obviously many variations on this general theme, which could be used. Furthermore, there are two separate ways in which the polyculture could be developed. Use could be made of the ability of the oryx and oribi to tolerate high mid-day temperatures and exposure to full sun, whereas eland and giraffe seek the shade and the woodland. Or, by rotating species which feed at different levels through successive plots, one could "crop rotate" at a secondary level.

Polyculture in High-rainfall Areas

In the same vein as the polyculture proposal for the semi-arid zones, it is true that the riverine and high-rainfall areas have a different spectrum of species which could similarly lend their complementary food-selection patterns to productive management. For example, the buffalo with its sharp teeth can eat the tufted grasses like *Sporobulus;* the hippopotamus only plucks with its lips, collecting creeping grasses like *Cynadon* and avoids the tufted species (Field, 1968).

Vesey-Fitzgerald (1965) has commented on what he calls successional grazing. The *Vossia* grows so rapidly in the seasonal swamps during the rains that it is impossible for small species to eat it. The buffalo can eat the tall grasses; in so doing the grass is trampled down and small shoots sprout from the nodes that have been pressed into the ground. This means that the smaller animal species can follow after the buffalo.

Table 16.7.
Eland carcass composition.

	Body weight (kg.)	*Edible products as per cent of live weight*	*Carcass as of per cent live weight*	*Brain (kg.)*	*Heart (kg.)*	*Lung (kg.)*	*Liver (kg.)*	*Spleen (kg.)*	*Kidneys (kg.)*	*Intestine (kg.)*
1 year										
Male	224	70	60	—	0.65	2.30	1.20	0.25	0.30	8.00
7–9 years										
Female	510	69	61	0.40	2.00	4.90	2.50	0.45	0.83	17.50
Male	700	68	59	0.43	3.10	7.20	6.30	0.84	1.10	24.50
						In g per 100 g of fresh weight				
Protein				18.00	23.00	18.00	20.00	21.00	24.00	18.00
Fat				61.00	2.00[a]	3.20	7.00	4.00	3.10	4.30

[a] Excludes heart mantle.

Table 16.8.
Eland milk amount and composition (per cent).[a]

Lactation Period (Days)	200–300
s.g.	1.0309–1.0364
Dry matter	22 g/100 ml.
Protein	7–10
Fat	8–11
Lactose	3.60–4.48
Calcium	0.29–0.36
Phosphorus	0.22–0.29
Total Ash	1.086–1.164
Highest milk output (kg/day)	5.56–7.0
Total protein (kg/lactation)	200
Total fat (kg/lactation)	220

[a] Treus and Krevchenko (1968).

In the high-rainfall areas, the species with potential are as follows:

Elephant,	Forest and woodland management;
White rhinocerus,	High-rainfall, high productivity grassland;
Hippopotamus,	Riverine zones, creeping grasses;
Buffalo,	Tufted grass and bush;
Topi,	Plains grazing, slow browse;
Waterbuck,	Riverine grazing, sedges;
Kob,	Grazing, short grass;
Warthog,	Rhizomes, Bulbils, and grazing.

The advantage of exploiting the high-rainfall tropical areas is the remarkable productivity of up to 60 tons of the vegetation per acre. This represents the highest order of productivity the world land mass has to offer. We have as yet no method to harness the yield. Clearly the large mammals could be very effective tools.

Judging by the successful results of management in Queen Elizabeth National Park and similar areas, it is possible that Africa is capable of doubling the present world meat production of about 75 million metric tons by using only half the available semi-arid and high rainfall zones. It is likely that the present biomasses do not represent optimum conditions, for much of the present land is degenerate through improper use (Laws and Parker, 1968). Given a broad ecological approach to livestock management, it appears that Africa could become the world's foremost meat producing area (Table 16.9).

Table 16.9.
Potential of Africa for animal production.[a]

Area	*Pounds of live wt. per sq. mile*	*Edible yield of 64.5 per cent live weight at 25 per cent cull*
Favorable biomass (Queen Elizabeth, Parc Albert)	200,000	32,250 lbs.
Unfavorable degenerate Acacia-commiphora bushland)	30,000	4,837 lbs.
One-fourth of African land mass with favorable biomass		43,794,034 metric tons
One-fourth of African land mass with unfavorable conditions		6,569,105 metric tons

[a] Africa has an area on 11,950,000 square miles. Approximately 40 per cent of the African land mass is semi-arid, and 20 per cent is high-rainfall tropical forest and woodland. Under favorable conditions, on one-fourth of the African land mass, a 25 per cent cull could yield annually a total edible output of more than half the present world meat production of 70,000,000 metric tons. Using the vegetation, animal management and communications could lead to considerable increases in cull rate and productivity.

FURTHER READINGS

Crawford, M. A., ed. 1968. *Comparative Nutrition of Wild Animals.* New York and London: Academic Press. See especially the articles by Field, by Laws and Parker, by Rogerson, by Taylor, and by Treus and Krevchenko.

Crawford, M. A., and S. M. Crawford. 1973. *What We Eat Today.* London: Neville Spearman.

Crawford, M. A. 1969. "Dietary prevention of atherosclerosis." *Lancet,* II, 1419.

Crawford, M. A., M. M. Gale, M. H. Woodford, and N. M. Casperd. 1970. "Comparative studies on fatty acid composition of wild and domestic meats." *Int. J. Biochem.,* I, 295.

Crawford, S. M. 1970. "Wild protein: a vital role for Africa." *Animals,* 12, 540.

Cuthbertson, D. 1970. "Role of the ruminant in world food supply." *World Rev. Nutr. Diet.,* 24, 414.

Fraser-Darling, F. 1960a. "Wildlife husbandry in Africa." *Scientific American,* 203, 123.

Fraser-Darling, F. 1960b. *Wildlife in an African Territory.* Oxford, England: Oxford University Press.

Holman, R. T. 1968. "Essential fatty-acid deficiency." *Prog. Chem. Fats Lipids,* 9, 123.

Joubert, D. M. 1968. "An appraisal of game production in South Africa." *Trop. Sci.,* 10, 200.

Schmidt-Nielson, K. 1964. *Desert Animals: Physiological Problems of Heat and Water.* Oxford, England: Clarendon Press.

Simons, N. 1962. *Wildlife in Kenya: Between the Sunlight and Thunder.* London: Collins.

Skinner, J. D. 1970. "An appraisal of the eland as a farm animal in Africa." *Trop. Anim. Hlth. Prod.,* 2, 151.

Spinage, C. A. 1962. *Animals of East Africa.* London: Collins.

Taylor, C. R. 1968b. "Hygroscopic food: a source of water for desert animals." *Nature,* 219, 181.

Taylor, C. R. 1969. "Metabolism, respiratory changes, and water balance of an antelope, the eland." *Amer. J. Physiol.,* 217, 907.

Taylor, C. R., and C. P. Lyman. 1967. "A comparative study of the environmental physiology of an East African antelope, eland, and Hereford steer. *Physiol. Zool.,* 40, 280.

Wilde, P. F., and R. M. C. Dawson. 1966. "The biohydrogenation of a linolenic acid and oleic acid by rumen microorganisms." *Biochem. J.,* 98, 469.

Chromosome bridges between generations—one pair only shown

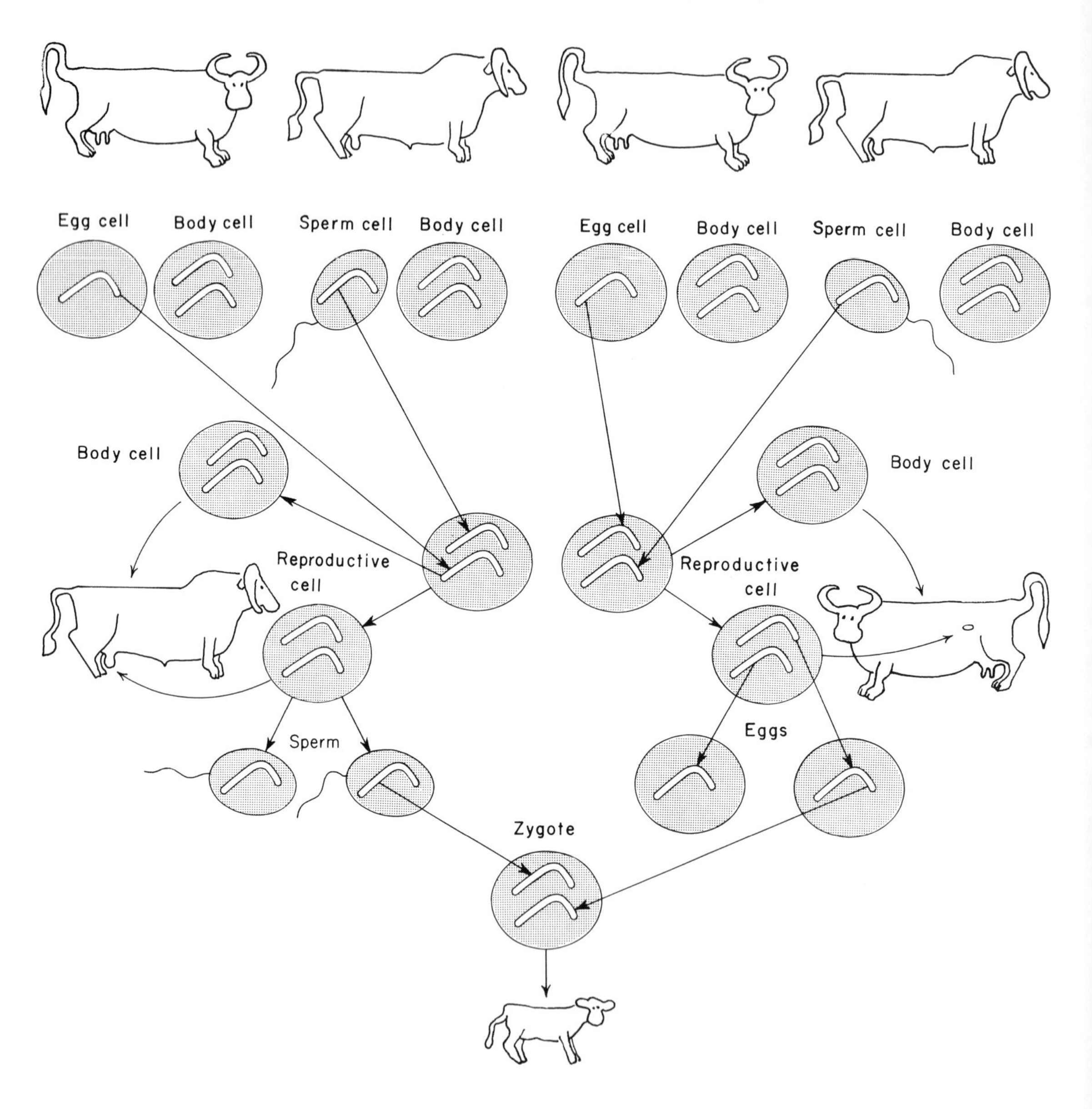

Section Four

Inheritance and Animal Improvement

How one pair of chromosomes acts as bridge of heredity between generations. One of each pair of chromosomes is in each sperm and egg. When the two come together to form the zygote which will grow into a new individual, they re-form each pair of chromosomes. Thus a sample of all the heredity of each pair of parents is transmitted to each offspring.

Seventeen

Heredity and How It Works

"Nothing from nothing ever yet was born."

Lucretius

Our knowledge of heredity, though far from complete, has grown tremendously since the nineteenth century, when spontaneous generation of life from nonliving material was widely accepted as a way for life to begin. Pasteur and Tyndall, late in the last century, finally produced conclusive evidence that life comes only from preexisting life. Even after spontaneous generation of life was disproven, the belief was widespread that the heredity, particularly the heredity of quantitative or "measurement" traits, of an individual organism, such as a calf, was a blend of the heredities of his two parents, rather like the blending of equal parts of two salt solutions differing in concentration. Such a system would obviously result in ultimate elimination of all differences among members of a given interbreeding population, such as a breed or species. Until the work of Mendel, published in the middle of the last century, but practically unnoticed until this century, no good explanation was available for the *fact* that differences do *not* disappear.

Even though genetics, the science concerned with heredity, is relatively young among the biological sciences, sufficient information has accumulated for the writing of many volumes. An interesting and informative chronology of genetics is given by Cook (1937). Many details and refinements have been added, but few basic concepts have changed during the intervening 35 years. To visualize how our knowledge of heredity has evolved to its present state, it is helpful to consider also certain major developments in the pre-Mendelian history of biology (Table 17.1).

Mendel and His Laws

An Austrian monk, Gregor Mendel, performed breeding experiments in his monastery garden using garden peas. His results concerning the mode of inheritance of several different traits were first made public in 1865,

Table 17.1.
Some important high points in pre-Mendelian history of biology.

Worker(s)	*Date(s)*[a]	*Contribution*
de Graaf	1672	Discovered that mammalian ovaries produce eggs equivalent to eggs of birds.
Leeuwenhoek and Hamm	1677	Discovered in mammalian semen *animalcules* (supposedly containing a complete organism) known now as sperm (Preformationism). Leeuwenhoek is credited with development of first microscope.
Redi	1621–1697	Produced circumstantial evidence against spontaneous generation of life.
Linnaeus	1707–1778	Developed beginnings of taxonomy. Promoted notion of "fixity of species."
Spallanzani	1729–1799	Showed feasibility of artificial insemination.
Wolff	1738–1794	Published work which resulted in the replacement of preformationism by the idea of *epigenesis*.
Lamarck	1744–1829	Published first attempt at a comprehensive theory of evolution based on inheritance of acquired characters and "use and disuse" hypothesis.
Brown	1773–1858	Named and described the cell nucleus.
von Baer	1792–1876	Credited with laying first foundation for what we know as embryology.
Schleiden Schwann	1804–1881 1810–1882	Credited with first statement of the cell theory.
Pasteur Tyndall	1822–1895 1820–1893	Produced conclusive evidence against spontaneous generation of life.
Darwin	1809–1882	Published theory of evolution based on modification of species by natural selection which allows "fittest" individuals to reproduce at highest rates.
Weismann	1834–1914	Produced conclusive evidence against inheritance of acquired characteristics.
Strasburger Hertwig	1844–1912 1849–1922	Showed the nucleus to be directly involved in fertilization: Strasburger with plants, Hertwig with sea urchins.
Schneider	1873	Showed that chromosomes divide longitudinally during cell division, permitting the regular systematic pattern of heredity to operate.

[a] Ranges of dates are life spans; single dates are dates of publication.

and printed the following year, but the significance of his work was not appreciated until 1900, when three other Europeans, De Vries, Correns, and Tschermak, independently "rediscovered" Mendelism, as the general process by which heredity operates is often called.

The Law of Segregation

Though our primary interest here is heredity in animals, let us show well-deserved respect for Mendel and his work by using his peas as examples in gaining some insight into the workings of heredity. The important basic principles are the same in animals as in plants, the major difference between them being that most plants can be self-fertilized, whereas, to obtain the same sort of genetic result in animals, we must mate females to males of the same hereditary makeup.

When Mendel crossed peas from a true-breeding tall strain with those from a true-breeding short strain, the resulting offspring were all tall. When these hybrid offspring were self-fertilized, the resulting progeny were of two kinds: ¾ were tall, ¼ were short. The short plants from this generation produced only short offspring when self-fertilized. However, among the ¾ which were tall, only

⅓ of them bred true for tall when self-fertilized. The other tall plants of this generation reproduced just like their parents had, yielding offspring ¾ of which were tall and ¼ short. This experiment can be better described by using letters to represent the hereditary factors involved: T for the factor responsible for tallness, t for the factor for shortness (see Table 17.2).

What actually happens when organisms, such as Mendel's peas, reproduce is that the reproduction is accomplished by the union of special cells produced by the parents. The special cells are called germ cells or *gametes*, each of which contains only one member of each set of hereditary factors. (Later, we shall consider some details of the mechanism of gamete formation.) For plant height, the set we are concerned with contains only two factors, called alleles, T and t. In some allelic sets, there may exist more than two kinds of factors in a population, but an individual organism possesses only two of these, which may be identical, as in the true-breeding tall (TT) peas and in the true-breeding short (tt) peas. Such individuals are said to be homozygous or are homozygotes. The two allelic factors possessed by an individual may be different, as in the non-true-breeding tall (Tt) plants. Such plants are said to be heterozygous or are heterozygotes. The TT, Tt, and tt symbolic representations of the three genetic kinds of plants are referred to as genotypes, whereas the outward appearances, tall or short, are called phenotypes.

Table 17.2.
Mendel's experiment with garden peas. Here *TT* represents tall plants that breed true, *Tt* tall plants that do *not* breed true, and *tt* short plants that breed true.

Generation	*Produced by*	*Contains*
First	(Given)	½ TT, ½ tt
Second	Crossing TT and tt.	All Tt
Third	Self-fertilizing Tt plants of second generation	¼ TT, ½ Tt, ¼ tt

When more than one kind of gamete is produced by reproducing parents, such as Mendel's Tt peas, chance determines, according to the laws of probability, which gametes will unite to produce new individual organisms or zygotes. Probability is verbally defined as the likelihood of a chance event. Numerically, it is usually a fraction between 0 and 1. The values 0 and 1 are the probabilities of impossible and certain events, respectively. When the outcomes of a random process, such as the reproduction of Tt peas, can be classified, and each class enumerated, each outcome class is thought of as an event. The probability of each event is the ratio of the number for that class to the total number for all classes. For the Tt peas, each plant produces two kinds of gametes, those containing a T and those containing a t, in equal numbers. Thus, the probability that the pollen grain involved in a chance or random union with a female gamete (an ovum) will contain a T is ½ (since half of all the pollen grains will contain a T), and likewise the probability that such a pollen grain will contain a t is also ½.

Familiarity with two basic principles of probability is essential if we are to acquire clear insight into the mechanics of the hereditary process. The first of these concerns the simultaneous occurrence of two chance events, and the second concerns the alternative occurrence of two chance events on a given occasion. These principles are as follows.

(1) If an event A has a probability of occurring a and another event B has a probability of occurring b, and both events *can* occur together, the probability of such simultaneous occurrence is the product of the probabilities, that is, ab.
(2) Considering the same two events, A and B, the probability of either A or B occurring on a given occasion is the sum of the probabilities, that is, $a + b$.

Novices in genetics can best visualize the application of these principles to an example, such as the reproduction of Mendel's heterozygous (*Tt*) peas, by making use of a Punnett square, named for R. C. Punnett, a noted early geneticist. It is simply a device to guide us in applying the multiplication and addition principles of probability. The symbolic representations of the paternal (male parent) gametes, and their frequencies or probabilities of occurrence, are written across the top side of the square, and those for the maternal (female parent) gametes are similarly placed along the left side of the square. In the cells of the square are written the symbolic representation of the zygotes and their frequencies or probabilities of occurrence (Figure 17.1). The ¼'s are products, ½ × ½, according to the multiplication principle of probability. Applying the addition principle gives the ½ *Tt* in the distribution of genotypes, since both the gametic union represented in the upper right-hand quarter of the square and the one represented in the lower lefthand quarter of the square yields a *Tt* individual; that is, ¼ + ¼ = ½ gives the frequency or probability of the *Tt* genotype.

Perceptive readers have probably recognized the fact that the Punnett square we have just used is a geometric way of depicting the application of the binomial theorem, first encountered in beginning courses of algebra. The binomial expansion is an algebraic way of automatically applying the two principles of probability to a situation in which the outcomes or events in question can be classified in two categories with fractional probabilities adding to 1.0. In the present application of the expansion, the power is 2, since we are interested in calculating the probabilities of random unions of gametes, *two* at a time, one paternal gamete and one maternal gamete in each union. To obtain the same result as we obtained with the Punnett square, we would simply square a binomial in which the two fractions are each ½:

$$[(1/2)T + (1/2)t]^2 = (1/4)TT + (1/2)Tt + (1/4)tt.$$

		Paternal Gametes 1/2 *T*	1/2 *t*
Maternal Gametes	½ *T*	¼ *TT*	¼ *Tt*
	½ *t*	¼ *Tt*	¼ *tt*

Figure 17-1 A Punnett square for *Tt* peas.

Now let us take advantage of the same example to introduce certain essential genetic concepts and terminology. Note that Mendel's plant-height example is a case of complete phenotypic dominance of one allele over the other; that is, both genotypes, *TT* and *Tt*, exhibit the same phenotype, namely, tall. The *t* factor is said to be recessive to *T*. We shall later introduce the fact that in some cases the action of hereditary factors is something other than a dominance effect. However, since the traits with which Mendel worked in developing the law of segregation involved dominance, dominance was implicit in his statement of the law, which can be paraphrased as follows:

The hybrid generation resulting from crossing two parent types homozygous for contrasting alleles will exhibit the phenotypic effect of one or the other of the two allelic factors, even though possessing one of each of them, but in the next generation, the two phenotypes reappear in definite numerical proportions.

The association of the *T* factor with the *t* factor in the hybrid individuals does not in any way alter the factors. In other words, there is a "clean" separation (segregation). *T* continues to be *T*, and *t* continues to be *t*.

The Law of Independent Assortment

At least two different allelic pairs are required to illustrate this principle. A simple and concise statement of it would be:

The factors in each allelic pair segregate independently of those in other pairs.

When Mendel crossed true-breeding plants having round seeds and yellow cotyledons with true-breeding plants having wrinkled seeds and green cotyledons, the resulting hybrids all had round seeds and yellow cotyledons. When these hybrids were self-pollinated, the resulting progeny were of four kinds, with frequencies closely approximating these fractions:

9/16 had round seeds and yellow cotyledons;
3/16 had round seeds and green cotyledons;
3/16 had wrinkled seeds and yellow cotyledons;
1/16 had wrinkled seeds and green cotyledons.

Mendel was able to recognize in his results the action of two independently segregating pairs of alleles. This 9:3:3:1 phenotypic ratio is commonly called the classical dihybrid Mendelian ratio, just as the 3:1 ratio (seen earlier in the plant-height example) is called the classical monohybrid ratio. Each requires dominance as the mode of phenotypic action of one member of each pair of factors. The major difference between the monohybrid and dihybrid cases is in the number of kinds of gametes produced by the hybrids. The monohybrid produces only two kinds of gametes in equal numbers. This fact is responsible for the denominator, 4, in the expected frequencies of the phenotypes produced by the self-fertilized hybrids. A dihybrid produces four kinds of gametes in equal numbers, thus making the denominator 16 in the expected phenotypic frequencies of offspring of dihybrids. Letting R and r represent the factors responsible for roundness and wrinkledness of seed, respectively, and Y and y represent the factors responsible for yellowness and greenness of cotyledons, respectively, we can use a Punnett square to illustrate how the 9:3:3:1 ratio is produced (Figure 17.2).

	1/4 RY	1/4 Ry	1/4 rY	1/4 ry
¼ RY	(1) 1/16 $RRYY$	(2) 1/16 $RRYy$	(3) 1/16 $RrYY$	(4) 1/16 $RrYy$
¼ Ry	(5) 1/16 $RRYy$	(6) 1/16 $RRyy$	(7) 1/16 $RrYy$	(8) 1/16 $Rryy$
¼ rY	(9) 1/16 $RrYY$	(10) 1/16 $RrYy$	(11) 1/16 $rrYY$	(12) 1/16 $rrYy$
¼ ry	(13) 1/16 $RrYy$	(14) 1/16 $Rryy$	(15) 1/16 $rrYy$	(16) 1/16 $rryy$

Figure 17-2 A Punnett square for dihybrids.

The 1/16 in each cell in Figure 17.2 results from the multiplication principle of probability. The addition principle says that, since the zygotes represented in cells numbered 1, 2, 3, 4, 5, 7, 9, 10, and 13 all have round seeds and yellow cotyledons, the probability or expected frequency of this phenotype would be 9/16, the sum of the frequencies or probabilities of these nine types of gametic union. The expected frequency or probability of the "round, green" phenotype, 3/16, is the sum of the values in cells numbered 6, 8, and 14; the 3/16 "wrinkled, yellow" comes from cells numbered 11, 12, and 15; and the 1/16 "wrinkled, green" is from the one kind of gametic union depicted in cell 16.

It is possible, by extending the application of these same techniques, to calculate corresponding phenotypic ratios or frequencies for examples involving three, four, or more pairs of independently segregating factors, but the manipulation and arithmetic involved become increasingly cumbersome as the number of pairs increases. Little of practical significance is accomplished by the calculation of such ratios. All that matters is that we be aware of the existence of such ratios and have a general idea as to how they might be calculated, if needed, by a series of Punnett diagrams or appropriate application of the binomial theorem. Pertinent to such extension to more than two pairs of independently segregating factors are the basic notions incorporated in Table 17.3.

Algebraically, the probabilities or expected frequencies among the offspring of polyhybrids with n pairs of heterozygous factors (and showing a single phenotype which is the combined phenotypes of the dominant alleles of all pairs) would be given by successive terms in the product of n binomials, in each of which the two fractions are (¾) and (¼). Since the n binomials are numerically identical, this product is equivalent to one binomial raised to the nth power. This would be using the binomial expansion to guide us in the application of the probability principles in calculating probabilities or expected frequencies of all possible phenotypes, of which there would be 2^n with respect to n pairs of factors. In any polyhybrid phenotypic distribution, the probabilities or expected frequencies of all phenotypes have the same denominator, namely, 4^n, the number in the third column of Table 17.3. The successive numerators are 3^n, $3^{(n-1)}$, $3^{(n-2)}$, . . . , 3^0 (where, of course, $3^0 = 1$). The number of phenotypes with each probability or expected frequency is the numerical coefficient of the appropriate term in the expansion of $(p + q)^n$. In the present application, $p = ¾$ and $q = ¼$.

Table 17.3.
Expectations in progeny of hybrid individuals having different numbers of pairs of segregating factors.

Number of pairs of segregating factors	*Number of kinds of gametes produced by hybrids*	*Number of possible gametic combinations*	*Number of phenotypes when dominance is complete*	*Number of genotypes*
1	2	4	2	3
2	4	16	4	9
3	8	64	8	27
4	16	156	16	81
⋮	⋮	⋮	⋮	⋮
n	2^n	$(2^n)^2$ or 4^n	2^n	3^n

Table 17.4.
Heredity of polledness vs. hornedness.

First generation	Polled (PP) bulls × Horned (pp) cows
Second generation	All polled (Pp)
Third generation (from mating Pp cows to Pp bulls)	¼ PP, ½ Pp, ¼ pp (¾ polled, ¼ horned)

Before moving from Mendel's laws to other considerations, let us examine two Mendelian traits in animals which conform to Mendel's laws. In cattle there is a pair of Mendelian factors responsible for polledness (having no horns) and hornedness, with the factor for polledness (P) completely dominant over the factor for hornedness (P). The segregation of these factors is exactly analogous to that of Mendel's factors for tallness and shortness of pea plants. Table 17.4 show that the heredity of polledness-hornedness is exactly like that of Mendel's tallness-shortness.

Another example of simple Mendelian inheritance in cattle is provided by the segregation of a pair of factors responsible for black (B) and red (b) coat colors. Anyone familiar with Angus cattle is aware of the existence of red Angus, of which a few are occasionally born in black Angus herds, and there is an association of breeders of Red Angus. The matings in Table 17.5 could conceivably be made.

Table 17.5.
Heredity of black and red coat color.

First generation	Black (BB) bulls × Red (bb) cows
Second (hybrid) generation	All black (Bb)
Third generation (from mating Bb cows to Bb bulls)	¼ BB, ½ Bb, ¼ bb (¾ black, ¼ red)

Now, if we pretend for a moment that appropriate cattle are available in large enough numbers, we could produce dihybrids for these two pairs of factors, among which the phenotypic frequencies would closely approximate the classical 9:3:3:1 ratio (Table 17.6).

Unfortunately, we know the details of the heredity of relatively few qualitative Mendelian traits in farm animals. Much of what we know has been gleaned from the results of matings made for purposes other than to produce information on the Mendelism of various traits. Description of a large number of specific qualitative traits is beyond the scope of our discussion here, but perhaps this presentation of basic principles will permit readers to study with interest and understanding the numerous descriptions of Mendelian traits in most of the familiar animal species which have been published through the years in the *Journal of Heredity* and in various textbooks, such as *Animal Genetics* by F. B. Hutt.

The Underlying Mechanism of Heredity

Thus far we have been sort of "putting ourselves in Mendel's shoes," though his intellectual "shoes" were unusually large, and few of us could fill them. Mendel had practically no factual knowledge of the basic mechanism responsible for the outwardly observable phenomena he studied; yet he was able to reason from his experimental results and give us the laws of segregation and independent assortment. The fact that his genius was not recognized for more than thirty years after the publication of his pea breeding results is truly regrettable. While conjecture as to what might have happened had his work been widely recognized and accepted in 1865 is fruitless, many present-day geneticists have indulged in it. Had the impressive experiments in genetics which were conducted in the early 1900's been initiated in 1865, we would undoubtedly be much farther along in our pursuit of genetic truth, and we would probably now have answers to many presently still perplexing problems.

Now let us enjoy the advantage we have over Mendel and examine what is known of the basic cellular mechanism responsible for Mendelian heredity. The hereditary "factors" of Mendel are now commonly called genes, and the science of heredity is called genetics, as already mentioned. This name for this branch of biology was suggested by Bateson, a prominent British pioneer in the field, in 1906. Bateson's contributions were truly significant. In 1902 he published the results of his studies of poultry, which supplied the first evidence of the operation of Mendel's laws in animals. He found that genes for pea comb and rose comb were each dominant to a gene allelic to them, one responsible for single comb, yielding 3:1 phenotypic ratios in the offspring of heterozygous pea-comb and rose-comb birds resulting from crosses between true-breeding pea-comb birds and true-breeding single-comb birds and from crosses between true-breeding rose-comb and true-breeding single-comb birds.

Table 17.6.
Heredity of dihybrid cattle.

First generation	Polled black (*PPBB*) bulls × Horned red (*ppbb*) cows
Second generation	All polled black (*PpBb*)
Third generation (from mating *PpBb* cows to *PpBb* bulls)	9/16 polled black, 3/16 polled red, 3/16 horned black, 1/16 horned red

Division of Body Cells

Genes are the units of Mendelian heredity. They normally occur in pairs in each body (nongerm) cell of individual organisms, such as our farm animals. They are located on or in (or are integral parts of) distinct, roughly rod-shaped bodies called chromosomes, which are carried in the center or nucleus of each cell. The chromosomes occur in sets of two, the two of each set being visually identically (an exception to this will be discussed later). When body cells divide, which is the process by which tissue growth and renewal is accomplished, each chromosome is normally duplicated exactly along its entire length. One of these duplicates normally goes into each of the two cells resulting from the division of one body cell.

The process of division of body cells is called mitosis. This regular process keeps the number of chromosomes constant in body cells throughout the organism. The essential features of mitosis are illustrated in Figure 17.3, in terms of only two pairs of chromosomes.

The number of chromosomes in each body cell of all members of a species is normally the same, and is a constant characteristic of the species. All our farm animals are normally *diploid;* that is, their chromosomes are in sets of two. The diploid chromosome numbers of some familiar mammalian species are: cattle, 60; swine, 38; sheep, 54; horse, 64; mouse, 40; rat, 42; man, 46.

Gamete Formation

The cell-division process involved in the formation of gametes, gametogenesis, provides the mechanism responsible for Mendelian heredity. Some geneticists use the term meiosis in referring to the entire series of divisions resulting in the transformation of primordial germ cells into functional male or female gametes. Others retain this term for the one critical division which is distinctly different from a mitotic division. The occurrence of this critical division and the essential role it plays in heredity are far more important than what we call it. In this division, which occurs at one stage during gametogenesis, the chromosomes of each pair align themselves together on the division plane with homologous parts of the members of each pair precisely adjacent to one another. Then, when the nucleus divides, one member of each chromosome pair goes into one of the two "daughter" nuclei. Thus, gametes are haploid cells, containing only half the normal diploid number of chromosomes, one member of each chromosome pair. The two members of each pair of genes are located in the precisely corresponding positions (loci) on the two members of a pair of chromosomes, and genes are arranged linearly along the length of the chromosomes, one gene of a given allelic set occupying a specific site (locus) on a chromosome. Figure 17.4 depicts essential distinguishing features of gametogenesis. The next-to-last event shown in the figure, fertilization or union of a male gamete and a female gamete, restores the diploid chromosome number which was halved by the reduction division of the diploid primary gametocytes to produce the haploid secondary gametocytes.

The fact that only one potentially functional female gamete (ovum or egg) is produced from each primary oocyte is an interesting and impressive arrangement which nature has made to allow the female parent to supply the early nutritional requirements of

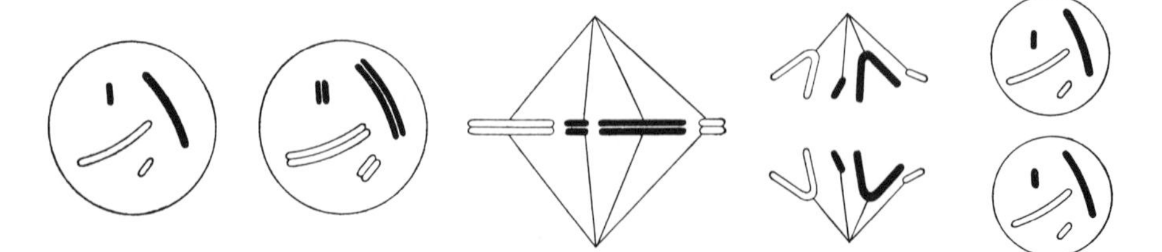

Figure 17-3 Mitosis. Each chromosome reproduces itself exactly, and the duplicates are separated, one going into each daughter cell. (After Sharp, *Fundamentals of Cytology.* © 1943 McGraw Hill Book Co.)

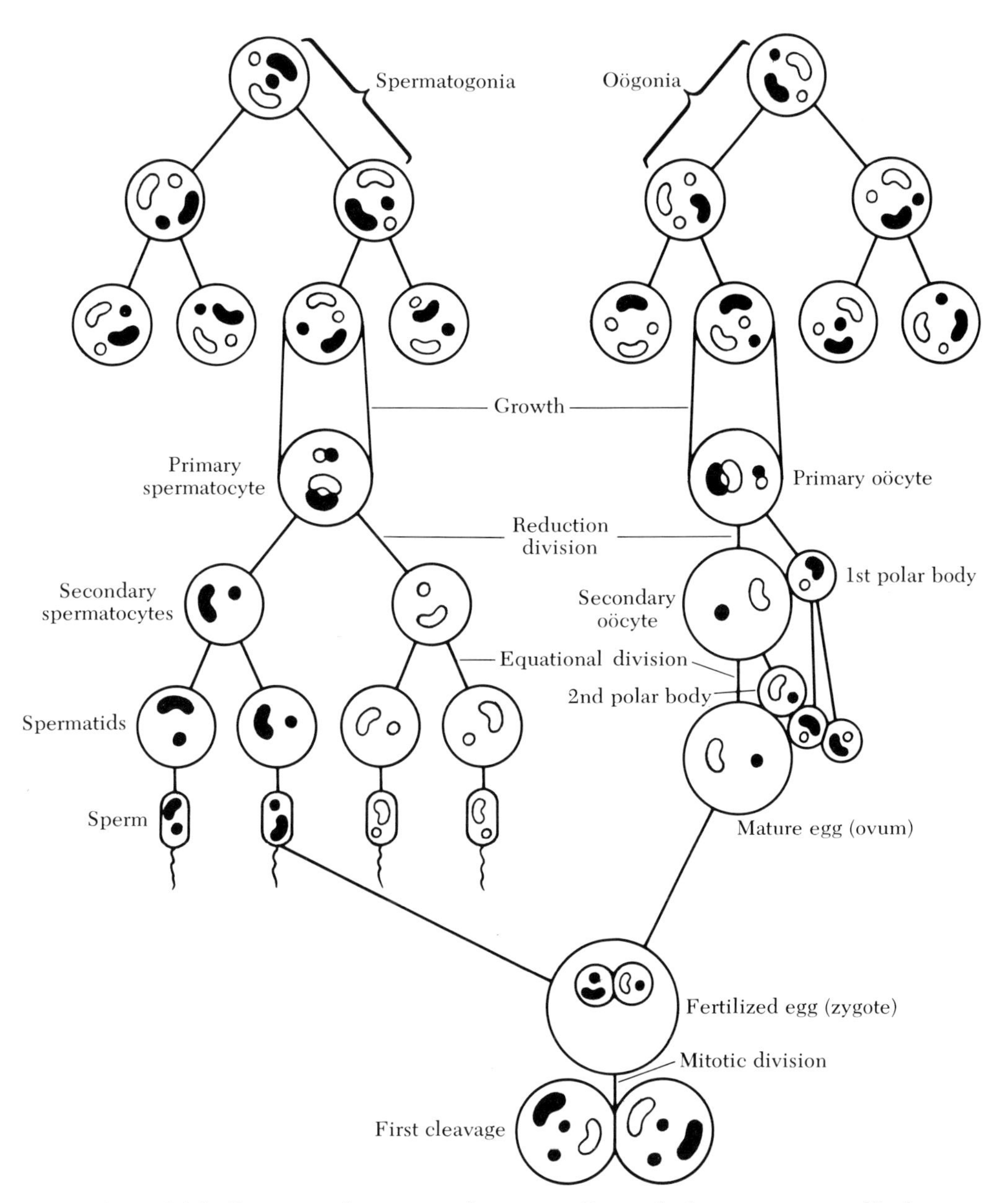

Figure 17-4 Diagram of a gamete formation. Paternal chromosomes are black, and maternal chromosomes are white. Chromosomes in the zygote received from the mature egg are thereafter maternal, and those received from the sperm are thereafter paternal, regardless of what they were in the mature gametes.

newly produced zygotes as a result of her having such relatively large cytoplasms (nonnuclear parts of cells) in her eggs. The polar bodies have very little cytoplasm and are normally nonfunctional, enjoying only a brief existence. The sperm cells are similar to the polar bodies in this respect, possessing very little cytoplasm, but four potentially functional sperm normally are produced from each primary spermatocyte. A sperm cell has need for only enough nutrient material to sustain it until fertilization.

Although only one sperm and one egg are involved in a single fertilization, considerable surpluses of both kinds of gametes are produced, especially of sperm, which, in bulls, for example, number in the billions in a single ejaculate. Apparently, a large number of sperm must accompany the one sperm which ultimately reaches the nucleus of the egg, for each sperm seems to carry a minute amount of egg-penetration-assisting substance or substances without which no penetration of the egg cell wall could be accomplished. In a single insemination many sperm may penetrate the egg-cell wall and get into the outer fringe of the egg cell, but only one sperm normally succeeds in reaching the nucleus of the egg and fusing with it.

Exceptions to the Law of Independent Assortment

Mendel's statement of the law of independent assortment was based on his observations of the simultaneous hereditary transmission of different pairs of alleles which are now known to be located on different pairs of chromosomes. If the sites (loci) occupied by the members of two pairs of genes are on the same pair of chromosomes, segregations of the two pairs of genes are likely not to be independent of one another. Two gene sites located on the same chromosome are said to be linked, for they are, in fact, physically linked to one another. Thus, in the production of a dihybrid, *AaBb*, if *A* and *a* genes are located on the same chromosome as *B* and *b* genes, it would appear that the genes from these two pairs which come into the hybrid together should stay together in the gametes produced by the hybrid. This should cause dihybrids with respect to two linked loci to produce only two kinds of gametes instead of four. However, all four kinds *are* produced but not with equal frequencies when the two loci are linked. The two genes at the two loci which come into the hybrid together do *tend* to stay together, but a phenomenon called "crossing-over" permits the production of the other two kinds of gametic gene combination to occur. Crossing over involves a physical exchange of corresponding parts of homologous chromosomes. The frequency of reduction divisions in which crossing over between two linked loci occurs is directly proportional to the distance between the loci. This fact has made it possible for geneticists working with various animal and plant species to draw chromosome maps showing the relative locations of many genes on the chromosomes. Unfortunately, they have not been drawn for domestic animals of practical importance in agricultural production.

Little is known with precision concerning linkage in farm-animal species. Nevertheless, it is an interesting and important basic genetic phenomenon, and awareness of its existence and consequences helps us understand, or be less mystified by, certain observations concerning the behavior of frequencies of phenotypes with respect to pairs of loci in successive generations of a freely interbreeding population. The subject of linkage is given a great deal of attention in courses devoted exclusively to genetics.

Sex Determination and Sex Linkage

Once again it becomes necessary to point out an exception to an earlier statement, namely, the one to the effect that the members of each pair of chromosomes are visually identical. In animals there is one important pair of chromosomes for which this is not true. The members of this exceptional pair of chromosomes are called sex chromosomes. Female mammals each have a pair of sex chromosomes which *are* visually identical, but mammalian males have only one sex chromosome that is visually identical to the two sex chromosomes possessed by each female. The other sex chromosome of each male is quite different in shape. It has become customary to refer to each of the sex chromosomes of a female mammal as an *X* chromosome and to the differently shaped sex chromosome of a male animal as a *Y* chromosome. Thus, female mammals are said to be *XX*, males to be *XY*, with respect to sex-chromosome endowment. Considering the sex chromosomes carried by gametes, female mammals produce only one kind of egg, those containing an *X* chromosome, but male mammals produce two kinds of sperm, those containing an *X* chromosome and those containing a *Y* chromosome. In mammals, then, females are said to be homogametic, males to be heterogametic. The sex of the zygote produced by an individual egg is thus determined by the kind of sperm which unites with it at fertilization. If the sperm contains an *X* chromosome, the individual produced will be a female. If the sperm contains a *Y* chromosome, the individual produced will be a male. Thus, the expected ratio of males to females is, at least at fertilization, 1:1. In birds, the situation is exactly the reverse of that in mammals, in that female birds are heterogametic, but the basic mechanism of sex determination is the same. If a chicken egg contains an *X* chromosome, the individual produced from it by fertilization will be a male. If the egg does not contain an *X* chromosome, the individual produced will be a female.

Genes carried on the *X* chromosome are said to be sex-linked. In female mammals and male birds, the heredity of such genes is analogous to the heredity of genes on the non-sex chromosomes which are called autosomes. However, in the heterogametic sex an obvious difference prevails with respect to sex-linked genes as compared to autosomal genes. A single sex-linked recessive gene possessed by a member of the heterogametic sex will show its phenotypic effect; that is, if a member of the heterogametic sex receives a recessive gene from its homogametic parent, the phenotype will be that of the recessive gene. However, for a recessive sex-linked gene to show its phenotypic effect in the homogametic sex, the individual must have received the gene from both parents; that is, the individual must be homozygous with respect to this recessive gene, just as is the case with autosomal recessive genes. It appears that any active genes which may be carried on *Y* chromosomes are genes which contribute something to the determination of maleness in mammals.

Few sex-linked genes in farm animals are known with certainty, but surely they do exist and possibly in large numbers, though they may individually have only small influences on phenotype.

Undoubtedly, there are some individual genes on the autosomes as well as on the sex chromosomes which are involved in causing variation in degrees of maleness and of femaleness among members of the same sex. The degree of expression of the so-called secondary sex characteristics probably results in part from the action of such genes. Every breeder knows of variation in fertility and reproductive performance of his animals. The causes of such variation are quite complex and must result partly from differences in environment (the complex situation in which each individual develops and lives) and partly from differences in individual heredity.

The Biochemical Basis of Heredity

A living organism is actually a complex chemical factory, in the cells of which myriads of chemical (called *bio*chemical since they are in a living organism) reactions are constantly occurring. None of us involved in the present effort to gain some understanding of the way in which heredity operates are sufficiently well versed in basic organic chemistry to understand fully the biochemistry of heredity, but we can acquire a broad general notion of what takes place. A pioneer in this area, J. D.

Watson, has given a quite understandable account of biochemical genetics in his book *Molecular Biology of the Gene.*

It is now known that the basic material of heredity is a substance called deoxyribonucleic acid, commonly referred to as DNA. This substance has the remarkable ability to reproduce itself, molecule by molecule, by joining together appropriate ingredient parts, chemical building blocks, so to speak, which it finds in solution in the cell contents. DNA is the material of which genes are made. Many kinds of DNA molecules with different reaction capabilities can be produced from a relatively small number of basic building-block materials by varying the arrangement of these in the molecular structure of the DNA molecule.

It appears that Mendel's factors, later called genes, are actually complex enzyme systems which are transmitted as units but involve the action of many biochemical genes, a distinguishing one of which differs between alleles to make the final products of two such allelic enzyme systems different and make possible the phenomenon of Mendelian segregation as we observe it in reproducing organisms.

Kinds of Gene Action

In addition to complete dominance, which was the mode of gene action characteristic of the genes Mendel studied, there are genes which act in a simply additive manner, genes which exhibit some degree of dominance but not complete dominance over their alleles, and genes which interact with genes which are not allelic to them to produce different phenotypic effects depending on which genes of the other allelic series are present.

To visualize these various kinds of gene action or phenotypic effect of genes, let us pretend that we could substitute an *A* gene for an *a* gene in the genotype of an individual. If we were to make such a gene substitution in an *aa* individual, that individual would be changed to a heterozygote. If the action of gene *A* is additive, the phenotype of the individual would move exactly halfway up the scale of phenotypic measurement, the extreme positions of which are occupied by *aa* and *AA* individuals. If we were then to substitute *A* for the *a* in a heterozygote, the genotype would be changed to *AA*, and the phenotype would be moved the rest of the way up the phenotypic scale if the action of *A* is additive.

If the action of *A* is a dominance effect, substituting *A* for *a* in *aa* individuals would cause the phenotype to move more than halfway up the phenotypic scale, all the way to the *AA* position, if the dominance of *A* over *a* is complete. If dominance is partial, changing *a* to *A* in *Aa* individuals would move their phenotype from where it was for a heterozygote (somewhere between the midpoint of the scale and the top or *AA* position) to the *AA* position on the scale.

If nonallelic gene interaction occurs, the only general statement we can make is that changing *a* to *A* in *bb* individuals produces a different effect on phenotype than does changing *a* to *A* in *Bb* or *BB* individuals.

What we have attempted to verbalize above is briefly stated diagrammatically in Figure 17.5.

Some nonallelic gene interaction is qualitative in nature and amounts to dominance between nonallelic genes. That is, such interaction causes the phenotypic effects of genes in one allelic series to be masked by genes in other allelic series. Such qualitative interaction is given in general the name "epistasis," and results in various modifications of classical phenotypic ratios in the offspring of polyhybrids.

Basic Notions in Population Genetics

Animal breeders are interested in characteristics of whole groups or populations, and are concerned with individuals and particular

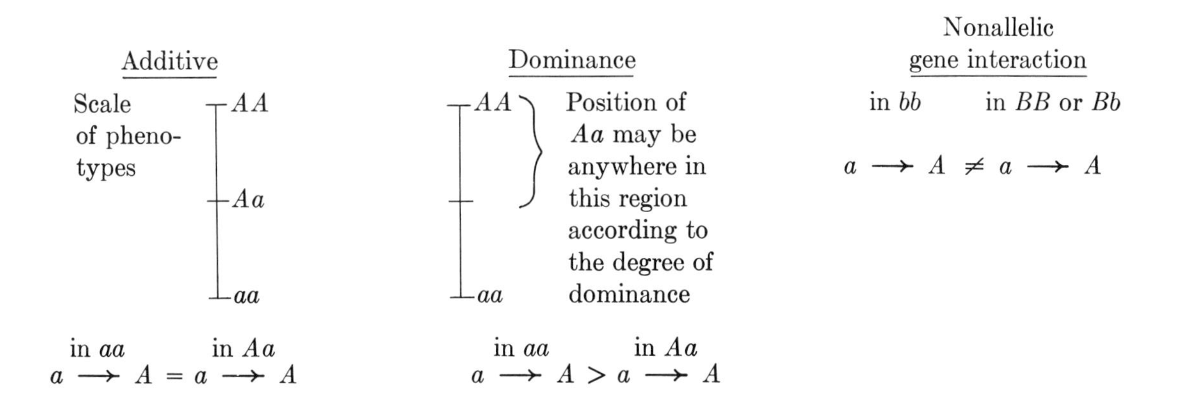

Figure 17-5 Three kinds of genetic interaction.

matings only because they make a population what it is. A concept basic to consideration of the genetics of a population is that of gene frequency. This is a value characteristic of a particular gene in a population, and is simply the fraction that the number of genes of that kind makes up of the total number of genes in that allelic series which exist in the population. We shall use in this discussion the letter q to represent a gene frequency and attach a subscript to it to identify the gene to which it pertains. Thus,

$$q_A = \frac{\text{number of } A \text{ genes in the population}}{\text{number of } A \text{ genes} + \text{number of } a \text{ (non-}A\text{) genes}}$$

and

$$1 - q_A = \frac{\text{number of } a \text{ (non-}A\text{) genes in the population}}{\text{(same denominator as above)}}.$$

A gene frequency may have any value between 0 and 1, or may have either of these two values in populations in which all individuals are homozygous, as can and does happen. The hybrids we use in studying Mendelian genetics are special cases: groups in which q is 0.5.

To illustrate a gene frequency other than 0.5, let us consider the genes responsible for coat color in Shorthorn cattle, in which *RR* individuals are red, *Rr* individuals are roan, and *rr* individuals are white. Suppose that a herd consists of 600 red animals, 200 roan animals, and 200 white animals. In this herd q_R would be 0.7:

$$q_R = \frac{1{,}200\ R \text{ genes of } RR \text{ cattle} + 200\ R \text{ genes of } Rr \text{ cattle}}{1{,}200 + 200 + 200\ r \text{ genes of } Rr \text{ cattle} + 400\ r \text{ genes of } rr \text{ cattle}} = \frac{1{,}400}{2{,}000} = 0.7.$$

The frequency of r in this herd would be $1 - q_R$ or 0.3. Thus we have a binomial expression in which the two fractions are 0.7 and 0.3. If these frequencies are the same in males and females, and mating is random with respect to coat color, it is obvious that the distribution of zygotes or genotypes in a calf crop thus produced would be a binomial distribution obtained by squaring the binomial expression $q_R + (1 - q_R)$, which is called the gametic array:

$$\begin{array}{lll} & RR & Rr & rr \\ [q_R + (1 - q_R)]^2 = & q_R^2 & + 2q_R(1 - q_R) + (1 - q_R)^2, \\ & RR & Rr & rr \\ (0.7R + 0.3r)^2 = & 0.49 & + 0.42 & + 0.09. \end{array}$$

This is using the binomial expansion to apply the probability principles, just as we did in explaining the 1:2:1 genotypic ratio of offspring of monohybrids, but here our two fractions are 0.7 and 0.3 instead of both being 0.5 as they were in the monohybrid case. In both instances we have squared a gametic array to obtain a zygotic array. If we were to use a Punnett square to illustrate this, dividing the top side and the left side in proportion to the two gametic or gene frequencies, 0.7 and 0.3, the area of the cell representing red calves would make up 49 per cent of the area of the entire square; roan calves would be represented by two cells, each of which would contain 21 per cent of the area of the entire square; and the lower righthand cell, representing white calves, would contain 9 per cent of the area of the entire square.

Gene frequencies are basic quantities in population genetics and are ultimately responsible for the values of population averages with respect to various phenotypes and for the amount of variation or differences in phenotypes of individuals in the population.

Permanent genetic changes in populations are produced by changes in gene frequencies. Factors involved in gene-frequency changes are as follows.

1. Selection: individuals of some genotypes being permitted to leave more offspring than individuals of other genotypes.

2. Migration: introduction of groups of individuals in which the gene frequencies are different from those existing in the population before the introduction.

3. Mutation: changes in the molecular structure of genes, such as changing *a* to *A* or vice versa by such alterations.

4. Chance: chance or sampling differences in the gene frequencies of individuals which are parents of the next generation.

Selection is the one of these over which a breeder can exercise most control, though he can make some changes in gene frequency through migration, which is exemplified in animal breeding by importation of breeding stock. Mutation and chance are not likely to produce desirable changes in gene frequencies, but they can be real causes of gene-frequency changes in actual breeding situations.

In addition to efforts to change gene frequencies in desired directions, the structure of populations can be changed by employing various nonrandom systems of mating. This general topic of mating systems is given more attention in another chapter, but enumeration of them here is appropriate. There are two general classes of nonrandom mating systems. In the first, mating is of like or similar individuals on the basis of genotype (pedigree), which is inbreeding, or on the basis of phenotype, which is positive assortive mating. In the second, mating is of unlike individuals on the basis of genotype (pedigree), which is outbreeding, and may range from mating very distantly related members of the same breed to crossbreeding, or on the basis of phenotype, which is corrective mating, using males which complement phenotypically the females which are bred to them.

The genetic consequence of inbreeding is an increase in the frequency of homozygotes, that is, an increase in homozygosis and a consequent decrease in heterozygosis. Breeders can take advantage of this to make groups within a breed more different from one another and thus to permit distinguishing between them genetically. Outbreeding in the extreme is employed to take advantage of the generally familiar phenomenon called hybrid vigor (phenotypic superiority of highly heterozygous individuals) which may result from a high degree of heterozygosity.

Heredity of Quantitative or Measurement Traits

The mechanism of Mendelian heredity is learned by considering traits which are qualitative; that is, the phenotypes are distinct classes, the members of each of which have some attribute characteristic of the class. The variation exhibited by quantitative or measurement traits, such as weaning weight or average daily gain, is continuous. There would theoretically be no distinct classes if we could measure each individual precisely. Every value within the observable range encompassing an infinite number of values is theoretically possible. Quantitative traits occupy most of the attention of animal breeders because the greatest economic improvement can be attained through modifying herds with respect to such traits. Production of a marketable product, such as milk, meat, wool, or eggs, is obviously one of these quantitative traits, the recorded individual values of which for a herd or flock are phenotypes and constitute the array or distribution of a continuous variable.

In considering quantitative traits, our first realization is that all we can know about an individual animal with respect to a quantitative trait is his phenotype. We cannot know his genotype (combination of genes which influence the trait in question). Yet we are certain that he has a genotype for a particular quantitative trait. The problem is that hereditary differences in a quantitative trait are produced by the segregation of a large number of sets of alleles, individual genes in which each may have only a very small effect on phenotype. In addition to the complication of genes at a large number of segregating loci influencing a quantitative trait, we know that much of the variation of quantitative traits is produced by differences in the environments of individual animals. There are many random influences in these individual environments which make them different even for a pair of genetically identical twin calves nursing the same cow.

The genes responsible for the hereditary differences in quantitative traits are transmitted from generation to generation by exactly the same mechanism as are genes responsible for qualitative differences. They undergo segregation at each locus and assortment between loci. In other words, they "Mendelize."

As we have already observed, the distribution of genotypes for a segregating pair of alleles is a binomial distribution. As mentioned in the discussion of polyhybrid ratios, we can arrive at a composite distribution of phenotypes for pairs of genes involved in a given case by multiplying together the binomials for all loci, thus obtaining a distribution of probabilities or expected frequencies of the various phenotypic combinations. We shall use the same reasoning here for quantitative traits, traits affected by many pairs of segregating genes.

The distribution of genotypes for any number of segregating pairs of genes would be given by the product of their separate distributions. That is, the complete zygotic or genotypic array or distribution for N pairs of segregating genes would be given by

$$[q_A + (1 - q_A)]^2 \cdot [q_B + (1 - q_B)]^2 \cdots [q_N + (1 - q_N)]^2.$$

Now let us make some simplifying assumptions, the possible falsehood of which in the real world will not nullify our present effort to understand the heredity of quantitative traits. First, let us assume that $q_A = q_B = \cdots = q_N = 0.5$. This is not a bad assumption, since the average gene frequency may well be 0.5 in the population, though surely all gene frequencies are not equal. Second, let us assume that each gene designated by a capital letter has the same additive effect on phenotype as does each other capital-letter gene. That is, genotypes *AAbbcc . . . nn* and *AaBbcc . . . nn* would produce the same phenotype, namely, the phenotype of individuals whose genotype contains two capital-letter or "plus" genes. Third, we shall assume, but only temporarily, that environmental influences have no effect on phenotype.

Now since we have assumed that all N gene frequencies are equal, the N binomials are numerically identical, and the product above is actually

$$[q + (1 - q)]^{2N}.$$

Each term in the expansion gives the probability or expected frequency of a particular phenotype. For example, the first term, q^{2N}, is the probability of the phenotype of individuals homozygous for all N "plus" genes; the second term, $2Nq^{(2N-1)}(1 - q)$, is the probability of the phenotype of individuals whose genotypes contain any combination of $2N - 1$ plus genes, and so on to the last term, $(1 - q)^{2N}$, which is the probability of the phenotype of individuals whose genotypes contain no "plus" genes. We can draw a vertical bar graph to represent such binomial distributions as this, letting the height of each bar be proportional to the probability or expected frequency of a particular phenotype; that is, the heights of the bars are proportional to the values of the terms in the expanded binomial. Such graphs are called histograms, and a series of them for various values of N is shown in Figure 17.6.

The superimposed curves in Figure 17.6 represent the familiar bell-shaped distribution, which is called the normal curve of "error" or simply the normal distribution. The mathematical function which this curve graphically depicts describes the way in which random events congregate about a central, most-probable event. (The term "error" as used here denotes a random deviation from the central value, and does not mean "mistake" as it does in common everyday usage.) The graphs of Figure 17.6 demonstrate the mathematical fact that the binomial distribution approximates or approaches the normal distribution as the power of the binomial expansion increases.

This mathematical approximation of the normal distribution of phenotypes for a quantitative trait is further improved in an actual situation by the random effects of environment, which we assumed not to exist for the purpose of drawing the histograms. It is known that such random effects do help determine the phenotypes individuals actually attain. These random environmental influences would shift an individual's actual phenotype to the right or to the left on the phenotypic scale from the phenotype his genotype would produce if there were no environmental influences. The shift would be to the right if the net effect of random environmental influences were favorable or positive, and it would be to the left if the net environmental effect were unfavorable or negative. Such shifting would smooth the tiny corners of the vertical bars not contained by the superimposed normal curve.

The assumption of additive gene action was invoked to make our histograms symmetric. The fact that this assumption is not actually true does not mean that the binomial distribution would not approach the normal if gene action were not additive. If the equally untrue and extreme assumptions were made that each capital-letter gene is completely dominant to its allele, and that all capital-letter genes produce the same effect on phenotype, and then histograms of the binomial expansion $(\frac{3}{4} + \frac{1}{4})^N$ were drawn, the resulting skewness (lack of symmetry) of the distribution, while quite pronounced when N is small, would tend to disappear with increasing N, and the approach to the normal distribution would actually be quite close with larger values of N.

When frequency distributions of large actual samples of phenotypes are graphed, these distributions closely resemble the true normal distribution. The fact that the underlying genetic distribution is a binomial with a large

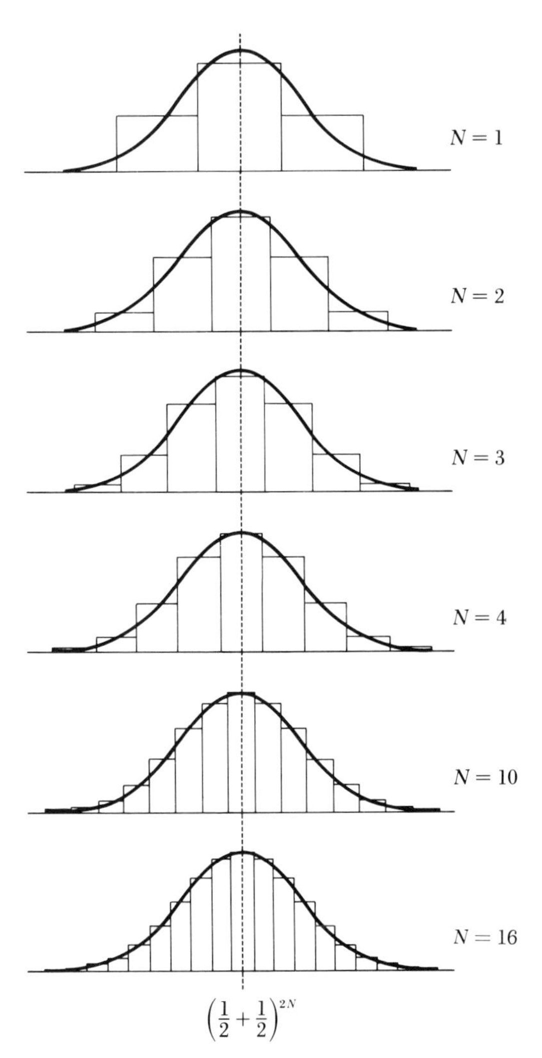

Figure 17-6 Histograms of binomial distributions for N pairs of genes with equal and additive effects and equal frequencies (all 1/2 or 0.5) with superimposed normal curves to illustrate the approach of the binomial distribution to the normal distribution as N increases.

exponent and the random effects of environment make this a logical expectation.

There are three important values which characterize a normal distribution. These are the mean (average), the variance, and the standard deviation. The mean is a familiar expression to everyone. It is a single number which summarizes all the information concerning magnitude or size which all the individual values in the distribution have in them. The variance is a similar summary value which condenses or summarizes all the information concerning differences between values in the distribution. The variance is expressed in squared units of the measured trait or variable. The standard deviation is the square root of the variance and thus is expressed in the original units of measurement used in recording individual phenotypic values. The general shape of a given normal distribution is well-described by using the mean and the standard deviation. The mean is in the very center of the scale of measurement. Whether the distribution is relatively wide or narrow is conveyed by the fact that about 68 per cent of the individual values are within one standard deviation of the mean on either side, about 95 per cent are within two standard deviations of the mean, and more than 99 per cent, or practically all of the values in the normal distribution, are within three standard deviations of the mean on either side (Figure 17.7).

In a population, the total observed phenotypic variance of a quantitative trait ($\sigma_p{}^2$) has two main parts, variance due to hereditary differences ($\sigma_H{}^2$) and variance due to environmental differences ($\sigma_E{}^2$). The hereditary variance contains two main parts, additively genetic variance ($\sigma_G{}^2$) and variance due to nonadditive effects of genes ($\sigma^2{}_{D+I}$), that is, variance due to phenotypic effects produced by dominance and by nonallelic gene interaction. The fraction which $\sigma_G{}^2$ makes up of the total observed phenotypic variance is called heritability. Methods have been developed for calculating estimates of this fraction from samples of observed values of phenotype of a quantitative trait. The variance, $\sigma_G{}^2$, is the variance among phenotypes calculated as though each conceivable substitution of a gene for its allele throughout a population produced in an additive manner an effect on phenotype equal to the average effect of all such gene substitutions.

In terms of selection of breeding stock based on phenotype, heritability is the fraction of the average phenotypic superiority of

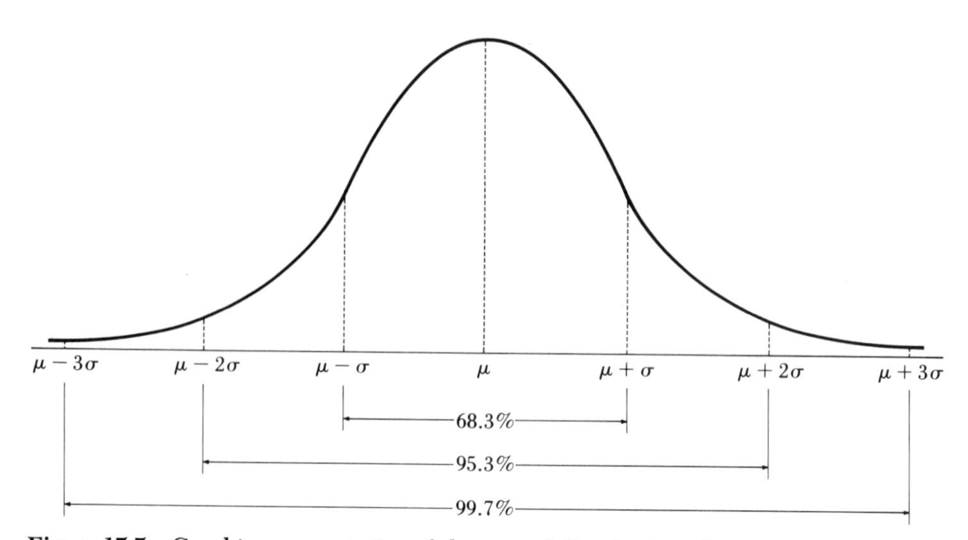

Figure 17-7 Graphic representation of the normal distribution of a quantitative variable in a population using μ to represent the mean and σ to represent the standard deviation.

selected breeding animals over the population average which is transmitted to their offspring, assuming no change in average environment between the two generations. A commonly used term for this average phenotypic superiority of animals selected for breeding is "reach." Defining as progress the difference between the average phenotype of offspring and the average phenotype of the population from which their parents were selected, the concept of heritability is commonly conveyed by the statement,

$$\text{Progress} = \text{heritability} \times \text{reach}.$$

Non-Mendelian Heredity

Various cytoplasmic entities are, of course, transmitted from parent to offspring, but this transmission does not follow a regular pattern in the way that transmission of genes does. Little specific knowledge is available on this subject in farm animals. However, certain maternal influences which are not environmental in nature, either prenatal or postnatal, may result from cytoplasmic transmission. The large cytoplasm of an egg (as compared to that of a sperm) allows the possibility for a female parent to have considerable influence on her offspring by way of the cytoplasm. Because of its irregularity, cytoplasmic transmission does not offer much possibility for useful practical applications by breeders of farm animals. Awareness of the possibility of maternal effects, however, should help breeders avoid being confused by them and possibly avoid making serious errors in selection which could result from such confusion.

Summary

Heredity is accomplished by units called genes, which occur in pairs in the body cells of individual animals, but singly in germ cells. Genes are carried by chromosomes, which also occur in pairs in the body cells of individual animals, but singly in germ cells. The regular pattern of chromosomal behavior during cell division provides a predictable mechanism for the operation of the laws of heredity. Both qualitative and quantitative characteristics of animals are determined by the action of genes, which are transmitted from generation to generation according to regular and consistent laws. Variation among animals is partly hereditary and partly environmental in origin. Methods are now available to permit breeders to partition observed variation into parts attributable to these two main causes of difference and utilize this information in conducting effective breeding programs.

Although certain essential ideas concerning heredity and how it works have been covered in this chapter, much else, of equally great importance, has had to be left out of such a brief introductory presentation. However, this introduction should provide a firm foundation on which the reader can continue building a more nearly complete and usable understanding of what is currently known concerning heredity.

FURTHER READINGS

Beadle, G. W., and E. L. Tatum. 1941. "Genetic control of biochemical reactions in *Neurospora*." *Proc. Nat. Acad. Sci.*, 27:499–506.

Cook, R. 1937. "A chronology of genetics." *Yearbook of Agriculture*, pp. 1457–1477. Washington, D.C.: USDA.

Hutt, F. B. 1964. *Animal Genetics*. New York: Ronald

Watson, J. D. 1970. *Molecular Biology of the Gene*. 2d ed. New York: Benjamin.

Eighteen

Selection and Mating Systems

"Understanding the origin, nature, and potentialities of germ plasm is the most important activity of man on this earth, infinitely more important than the probing of space with rockets."

Henry A. Wallace

Powerful means for developing more efficient farm animals lie in the hands of the breeder. Just as the designer and engineer may blueprint and develop more efficient tools and machines, the breeder may outline programs for developing improved animal tools, better equipped to produce food and fiber of higher quality and at less cost. Genetic "engineering" is a field of development barely tapped in the improvement of most farm animals.

Much has changed in the total field of genetics. New dimensions have been created through the knowledge of the chemical and physical bases of inheritance. DNA and RNA are now in the vocabulary of biology courses even in the junior high school. Obviously, we must look to these developments as they relate to our attempts to produce better livestock.

There has also been a revival of interest in irradiation in speeding up the results from selection. Buzzati-Traverso (1950) early reported that there appeared to be much greater selection progress in irradiated fruit flies than in non-irradiated ones. The results, however, are still inconclusive. If irradiation of spermatozoa could be applied to induce genetic mutations that could be selectively used to improve our livestock, it would be a very exciting possibility.

Among the pioneer scientists who have developed theories and experimental testing of procedures that are the most useful in applying the results of Mendelian genetics to animal improvement are Sewall Wright of the University of Wisconsin, R. A. Fisher of Cambridge University, and J. L. Lush of Iowa State University. They have developed and explained many of the concepts concerning selection and mating systems that apply to the genetics of herds and populations. George Harrison Shull and Edward M. East caused a genetic revolution in the breeding of plants, which has affected animal breeding as well, by their discoveries of the power of inbred lines when hybridized to increase corn yields by 20 per cent or more (Kiesselbach, 1951).

Population genetics is the term used for this special area of genetics so important to animal improvement. Although its biological soundness is rooted in Mendelian genetics, it differs from the classical studies of the effects of single genes in F_1 and F_2 generations. Population genetics is concerned with genetic changes in the total herd, the effect on the average of all animals, and the variations that occur. The population approach was found to be necessary because early experiments indicated that ratios of phenotypes such as 3:1 and 9:3:3:1 did not occur for most of the economically important traits. Rather, a continuous distribution of phenotypes over a wide range was observed, which was best explained by assuming many pairs of genes to be involved. For example, variations in milk production indicate involvement of 7 to 200 pairs of genes. Given estimates of 1,000,000 genes in man (and we can assume farm animals are about as complex), it becomes obvious that the breeder is dealing with large populations of genes as well as with large populations of animals. Appraising individual actions of genes in these animals seems an impossible task. The most useful avenue, to date, has been to study the gross effects of the actions of many genes under effects of selection and mating systems.

Genetic Effects of Selection

Selection is a simple process: it is managing the herd so that the better animals produce more offspring. The genetic results of effective selection are also very simple: decreasing the frequency of the less-desired animals decreases the frequency of the less-desired genes in the herd, and increasing the frequency of more-desired animals in the herd increases the frequency of more-desired genes. Whereas it is difficult to demonstrate these principles using traits affected by many genes, they can be clearly demonstrated with simply inherited characters, such as color, in the following example.

In the Cauca Valley of Colombia, South America, a new breed of dual-purpose cattle, the Lucerna, is being formed with the goal of producing a solid-colored red breed that will have good milk and beef production. Parent stock included the native Spanish-derived breeds or Criollo, which, while well-adapted to the local environment, are poor producers. Genetically superior European dairy breeds, such as the Holstein, on the other hand, do not perform well because they are poorly adapted to the environment. When these breeds were crossed, however, the offspring produced so outstandingly better than either parent breed that they have become the bases for new breeds being formed.

In the first crosses between black Holstein, *BB*, and the brownish-red Criollo, *bb*, the progeny are black, showing the dominance of the black color. They, of course, would be heterozygous, *Bb*, for red color. The frequency of the black gene is reduced from about 100 per cent in the Holstein to 50 per cent in the first cross, since half of the genes in the calf crop would be for black, the other half for red. In the case of the Lucerna, a later cross was made to red Milking Shorthorn cattle, which further reduced the frequency of the black gene. By further selection the frequency of the black gene was reduced to about 10 per cent, and the breeder, Dr. Carlos Durán, began using nothing but red bulls (Figure 18.1). In another ten years, or less, there likely would be no black cattle remaining in the herd, the frequency of the black gene having been reduced to zero and that of the red gene increased to 100 per cent.

Differential expression in traits most important to cattle breeders is controlled by differences in environments and genotypes due to many pairs of genes. We measure the improvement resulting from a given selection pressure for a trait as its "heritability," which,

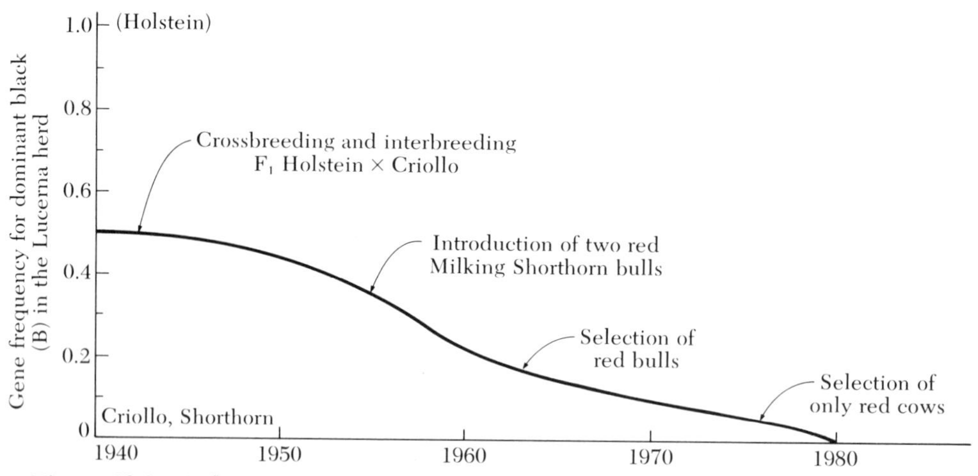

Figure 18-1 A diagrammatic example of the effects of crossing and selection on the elimination of the black gene in establishing the Lucerna, a tropical red breed of dual-purpose cattle.

experience has taught us, differs for different traits.

Heritability Estimates

Further complexities in applying Mendelism to the improvement of economic traits result from the variation caused in these traits by environment as well as by inheritance. Even identical twins, animals that have the same genotypes, differ. These differences illustrate conclusively the effect of environmental variation on the expression of traits and contribute to the popularity of debate about which is more important, heredity or environment? "Which is more important" is generally interpreted to mean which contributes more to the individual's own deviation from the general average.

Population geneticists and animal breeders (Lush, 1948) have found mathematical and statistical tools that have helped them appraise how much difference between animals is attributable to genes and how much to environment. The relative importance of heredity in explaining these differences, they have found, varies with the trait. In traits such as variations in the amount of white on Holsteins, 90 to 95 per cent of the differences have been found to be due to heritable differences between animals. In traits such as litter size of pigs, heritability is only 10 to 15 per cent.

Heritability indicates the theoretical percentage of response expected from exertion of a given amount of selection pressure on a trait in his herd. For example, if a breeder's unselected sheep flock averages 4 kg of wool and rams averaging 7 kg are selected from the flock to mate with selected ewes averaging 5 kg, the average selection differential between that selected group and the general average is $[(7+5)/2] - 4 = 2$ kg. If heritability were 100 per cent, the progeny of the selected group should average 2 kg above the mean of the progeny of an unselected flock. However, the average of the offspring is usually nearer the mean of the herd than is that of their parents. That is, heritability is less than 100 per cent and the progeny of selected parents do not, under the same conditions, on an average produce as well as the selected parents. In observations of many experimental flocks, about 40 per cent of the selection differential or "reach" in wool production has been reflected in the progeny of the selected group. The heritability then is 40 per cent. One method of computing heritability is by noting the regression of each offspring's record on the average of its parents. The selection differential must be multiplied by the heritability to estimate the progeny average. Since $.4 \times 2 = .8$, .8 kg improvement in wool yield in the next generation is expected, rather than 2 kg.

Much of this regression toward the mean is attributed to the selection of animals that have had, by chance, a better environment than the average within the herd. The environmental contribution to their superior phenotype is not inherited.

Since, for a given amount of selection, more is gained by selecting for traits with high heritabilities than for those with low, the breeder might simply conclude that these are the traits to emphasize. This is true only if they have approximately the same economic worth, for the breeder is an applied economist as well as an applied geneticist, and must combine these considerations in a comprehensive selection program. In Hereford cattle, the shade of red is highly heritable, about 70 per cent, but variations in the trait have yet to be shown to be important in the production of beef. On the other hand, weaning weight is much less heritable but is one of the more economically important traits in beef cattle. Therefore a breeder might well decide to emphasize selection for a trait with only moderate or even low heritability if its economic importance is high.

Table 18.1.

Estimated ranking of importance of single traits to the breeder based on economic worth and heritability.[a]

Traits	*Relative economic worth* (r^2)[b]	*Heritability* (g^2)	*Index of importance* $(r^2 \times g^2)$
Weaning weight	0.64	0.30	0.19
Size of dam	0.10	0.70	0.07
Daily gain	0.14	0.45	0.06
Days to finish	0.21	0.25	0.05
Percent of calf crop	0.64	0.07	0.04[c]
Feed per pound of gain	0.04	0.39	0.01
Carcass cut-out value	0.08[c]	0.25–0.50	0.02–0.04[c]
Slaughter grade	0.21	0.00	0.00

[a] From Lindholm and Stonaker, 1957.

[b] Squared correlation coefficient between each trait and net income.

[c] Estimated.

Heritability values thus are not abstract or theoretically derived. They are based on actual observations from selection experiments or resemblances of relatives in herds. The ranking in Table 18.1 was obtained in a single experiment and may not be applicable in other situations. The example shows, however, that one trait may deserve five to ten times more emphasis than another. Although single-trait selection is not advocated as the most efficient procedure for genetic improvement, emphasis on a trait or traits must take the factors of economic worth and heritability into consideration when they are to be incorporated into a total score or selection index for an animal.

Selecting for one trait may often affect other traits, for genes may influence more than one characteristic. For example, selecting for rate of gain in swine will cause a genetic improvement in efficiency of feed conversion, but will also slightly increase fatness. A clearer illustration of this resulted from using mice as a model laboratory animal to simulate rate of gain selection in large animals (Figure 18.2). It was found that lines selected for gain showed a marked increase over unselected animals. Unexpectedly, perhaps, the fat content of the "carcasses" also increased in the gain-selected mice and remained approximately constant in the unselected controls.

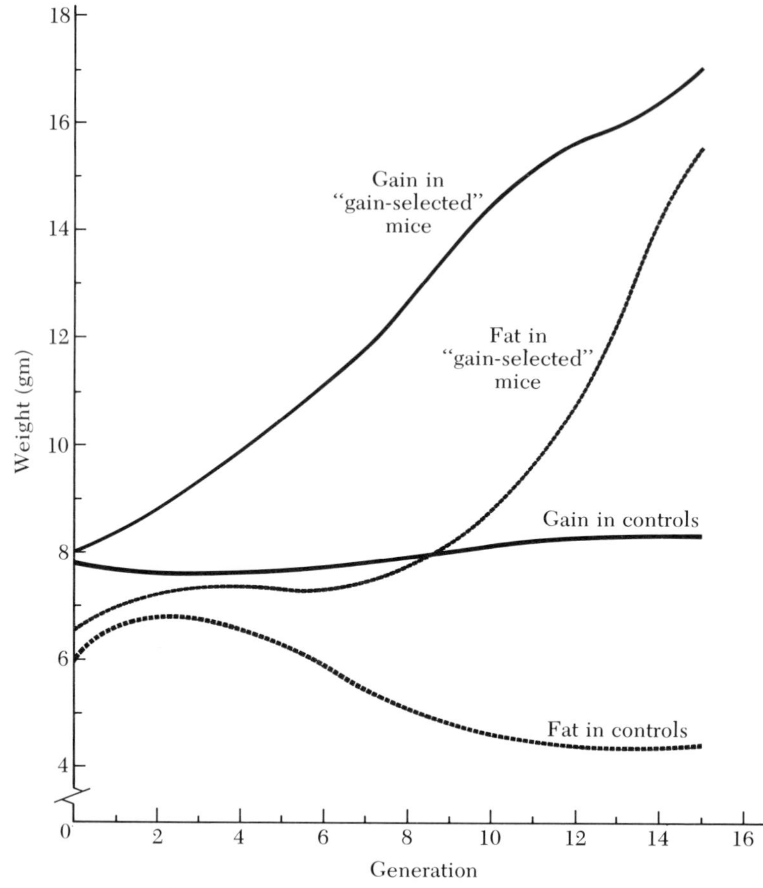

Figure 18-2 Gain-selected mice changed as well in fat content (ether extract) of their carcasses. An example of an unexpected correlated change resulting from gain selection. (Adapted from Biodini *et al.*, 1968.)

Selection Systems

Selection on Basis of Progeny Tests

Unrevealed traits, such as carcass characteristics, or sex-limited traits, such as milk production, cannot be accurately predicted from the animal's appearance. Evaluation of a breeding animal for these traits must be delayed until the progeny have been tested. On the other hand, several important economic traits in sheep, such as growth rate, fleece character, and face covering can be observed directly. Thus the selection of breeding rams on the basis of phenotypes will result in a shorter generation interval and thereby hasten improvement. This point is discussed in Chapter 17.

In dairy cattle, progeny testing of sires based on the production of their daughters is essential. Rapid advances in procedures have been made in recent years. The long-followed sire index—based on the comparison of daughters' records with their dams' records—has practically been discredited because of the confounding effects and important influences that changes in the herd environment have had upon the daughters' records.

C. R. Henderson of Cornell University and Alan Robertson of the University of Edinburgh, Scotland, independently devised a new successful system for dairy-bull selection, which provides the most accurate measure of total genotypic value ever devised for a production characteristic of any animal.

The estimated daughter superiority (EDS) of a given sire is calculated as follows:

$$\begin{aligned} \text{EDS} = b\ &(\text{daughters' records} \\ &\quad - \text{contempory herd mates}) \\ &- 1/8\ (\text{dams' records} \\ &\quad - \text{contemporary herd mates}) \\ &+ 1/10\ (\text{herd mate average} \\ &\quad - \text{breed average}). \end{aligned}$$

Here b is a weighting factor that gives increased emphasis to differences between progeny groups as the number of daughters (d) increases, and equals $\frac{d}{d + 12}$; 1/8 is based upon 50 per cent relationship between daughters and dams and 25 per cent heritability; and 1/10 is the estimate of the level of genetic differences between herds.

Progeny testing of sires based on sufficient numbers of offspring from selected mates and raised concurrently under similar environmental conditions can theoretically reveal precisely the additive genotype of the parent. No other system can do this. The number of progeny required to reveal this additive genotype varies with the heritability of the trait and relative sampling errors. Generally speaking, however, greater genetic gains can be expected by subordinating the accuracy of the sire's progeny evaluation to opportunities to select among more sires. If 100 cows are available for the progeny testing of bulls, it is usually better to mate a few cows to each of a relatively large number of bulls than many cows to few bulls.

These relationships are illustrated in Table 18.2. In the example, 100 cows are available for progeny testing and 2 progeny-tested bulls are ultimately needed. The 100 cows could be mated to only 2 bulls. This would give a very accurate appraisal of the 2 bulls, but since 2 bulls are needed both must be used, and no selection on progeny test is possible. Thus, a zero selection differential is obtained, as shown in column 4. The greater selection differentials are obtained by progeny testing more bulls, even though each individual progeny test is a less accurate appraisal of that particular bull's genotype.

Column 5 gives an example of the expected correlations between a sire's genotype and the average of varying numbers of offspring. This shows that, for a trait of a given heritability, the reliability of a progeny test increases with

Table 18.2.

An example of genetic progress as a result of using different numbers of sires in progeny tests with 100 females.

Number of bulls to progeny test	*Number of progeny per male*	*Per cent of males to be retained*	*Selection differential on sires*[a] *(A)*	*Correlation between sire's genotype and daughters' av. performance*[b] *(B)*	*Relative expected progress* $(A \times B)$
2	50	100	0	1	0
4	25	50	0.8	0.82	0.66
8	12	25	1.3	0.73	0.95
16	6	12	1.6	0.60	0.96
32	3	6	2.2	0.47	1.05
100	1	2	2.4	0.30	0.73

[a] Lush, 1945. [b] Lush, 1931.

increased numbers of progeny. The over-all relative rate of progress from progeny testing is shown in column 6 as the product of the selection differential and the correlation between sire's genotype and offspring average. As can be seen, a peak in rate of genetic improvement was achieved by using the test herd of cows to produce about three progeny per sire.

Selection on Basis of Pedigree

Pedigrees are useful aids in selection but are limited in predictive value because of the sampling nature of inheritance, the influence of environmental factors on traits, and, to a degree, the reliability of the pedigree itself, for an appreciable number of errors in pedigrees are found when blood types of calves are compared with those of their parents.

Pedigrees are most useful for traits that are sex-limited, low in heritability, or greatly influenced by inbreeding and hybridizing. In recent years they have been widely used for lowering the incidence of such undesirable recessives as dwarfism in cattle. The probability of an animal being a heterozygote can be indicated from pedigree information as shown in Table 18.3. This table indicates that a breeder could better his chances of avoiding dwarfism, which has been not uncommon, if the trait had not been found "close up" in the pedigree. Pedigrees thought to be clear of dwarf production have been worth a considerable premium.

Stressing ancestors many generations removed usually approaches faddism; that is, in promoting the sale of an animal, a breeder may emphasize the animal's relationship to some famous ancestor even though, because of the halving of relationships each generation, the relationship is too remote to be on any significance. Exceptions occur in linebreeding, where repeated lines of relationship even to rather remote ancestors are highly regarded in breeding systems devised for that purpose; examples of such exceptions are given below, where linebreeding is discussed in more detail.

Table 18.3.
The probability of an animal being a dwarf carrier (*Nn*) based on its relationship to a recessive dwarf (*nn*).

Relationship to a recessive dwarf	*Probability of being a carrier (Nn) (per cent)*
Parent	100
Full brother or sister	67
Half brother or sister	over 50
Son or daughter of a half sib[a]	30–40
Average, normal-appearing animal in major beef breeds	15–25

Lush and Hazel, n.d.
[a] Above 25 per cent because of chance of obtaining a recessive gene from the other parent.

Mating Systems

There are traits which give little response to selection. This might seem to be very discouraging to the breeder, but the fact that a trait has low heritability does not mean that it offers little chance for genetic manipulation or control. In fact, the greatest commercialization of genetics in farm animals, thus far, has been with low heritability traits: egg production in chickens and litter size in swine. This is because fertility characteristics generally have a low heritability but show a considerable amount of heterosis, or hybrid vigor, when breeds or inbred lines are crossed. Heterosis is that extra performance obtained in the cross above the average of parents raised under a comparable environment.

The study of mating systems has to do with an aspect of genetics and animal improvement that is quite apart from the selection process. There are many ways in which breeding animals may be paired for matings. They may be paired by relationship in order to produce progeny that are more inbred than the average, for example brother to sister, sire to daughter, son to dam.

Royal families for a number of reasons have practiced close marriages. Pride was probably a factor, but political and economic pressures also favored keeping it "all in the family." Brother-sister marriages were common in ancient Egypt and the Polynesian islands. With these there were no "mother-in-law problems"! Figure 18.3 is a reputed pedigree of that famous Egyptian queen, Cleopatra.

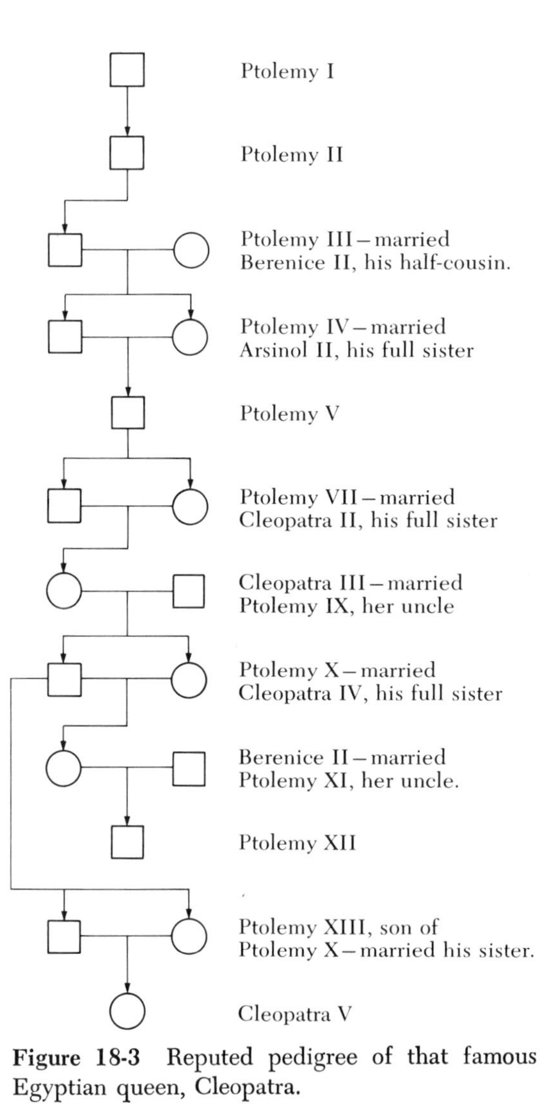

Figure 18-3 Reputed pedigree of that famous Egyptian queen, Cleopatra.

Breeding animals may be paired to be different genetically and to produce outbreds, as by mating a Shorthorn bull to a Brahma cow. They may be paired in order to be dissimilar in appearance, such as a racing-type Quarter Horse to a "bulldog" type: this tends to produce intermediate types of offspring, and is sometimes called corrective mating. Phenotypically similar animals may be paired; such pairs cause more extreme types in the total population than are found otherwise. If tall men were to select tall wives, and short men short wives, we would have an example of assortive mating, or the mating of likes. Assortive mating would tend to create greater extremes among the offspring within the population than if mating were random for that trait. Different mating systems are used to accomplish different ends. In most programs, selection is used in conjunction with a mating system.

Inbreeding and outbreeding (hybridizing) are the best-known mating systems. They are the opposite of one another in pairing procedure and in their genetic effects: inbreeding makes animals more homozygous, outbreeding makes them more heterozygous. A strong incentive to investigate inbreeding vs. hybridizing effects in farm animals has resulted from the independent findings of Shull and East (Kiesselbach, 1951). Shull (1909) reported the phenomenon of hybridizing from using two inbred strains of corn for crossing. Later, in noting results of crosses of eight inbred strains, he found some crosses that produced markedly more than outbred open-pollinated corn. This hybrid vigor in corn has led to widespread experimentation in hybridization of most farm animals. Successful commercial application has been achieved with chickens and swine.

Why should the hybrid be able to produce more than such rigidly selected outbred populations? This question has not been completely answered, but there are two theories that are important in explaining heterosis.

In the study of lower forms of life, such as the molds and bacteria, clues are developing which indicate that the life processes may be enhanced by the biochemical action of different alleles, each contributing something to an increased efficiency of the organism's development. An example in man is the greater resistance to malaria of individuals heterozygous for the sickle-cell gene, which causes hemoglobin cells to have a distinctive sickle shape. In malaria-ridden sections of Africa, these heterozygotes have a higher survival rate than individuals without the sickle-cell gene. Individuals that are homozygous for the sickle-cell gene, however, have a high rate of death from anemia. Thus the heterozygote is more fit than either homozygote. Geneticists call this overdominance, or the extra performance of the hybrid due to heterozygosity at a given locus.

Another type of hybrid vigor results because favorable genes often have a degree of dominance and unfavorable genes are likely to be recessive. Hybrids thus have more dominant genes at the many loci involved than do their more inbred parents.

Traits that require considerable heterosis for their fullest expression show an increase in the amount of vigor with a decrease in the relationship between animals in the cross. Crosses between breeds give greater hybrid vigor than crosses between families within a breed. Crosses between inbred strains from different breeds give even greater heterosis. But if crosses are made between extremely unrelated animals—as between different species or different genera—there may be high rates of embryonic loss and other incompatibilities that limit the ability to cross. For example, in crossing bison bulls on domestic cows, hydramnios, the development during pregnancy of excessive amniotic fluid, frequently occurs. Also, F_1 males from the cross are often sterile.

Various types of matings listed are shown, in order of probable increasing heterozygosity produced in the progeny.

Inbred The progeny resulting from the mating of closely related animals. Linebreds are inbred, but with a high relationship to a particularly admired animal.

Outcross The mating of relatively unrelated animals within the same breed or variety.

Topcross The mating of a male of a specified family to females of another family of the same breed.

Topincross The progeny resulting from the mating of inbred sires with noninbred dams of the same breed.

Incross The progeny resulting from the crossing of individuals of inbred lines within same breed.

Crossbred The progeny resulting from the mating of different breeds.

Topincrossbred The progeny resulting from the mating of inbred sires with noninbred dams of different breeds.

Incrossbred The progeny resulting from the crossing of individuals from inbred lines of different breeds.

The maintenance of a crossing system requires breeders of parent seed stock to maintain isolated stocks from which crosses can be made. This is how the purebred breeder and the registry society fulfill needs of the commercial producer. They serve to maintain a closed population that is not permitted to cross with other populations. Although the degree of homozygosity obtained is not high —probably not more than 8 to 12 per cent in many breeds—it is high enough to maintain a source of material that makes crossbreeding commercially feasible in some parts of the livestock industry.

Many breeders obtain further control over the inheritance of their herds by linebreeding to the most admired animals within their herds. The idea here is not to inbreed pur-

Table 18.4.
Inbreeding of chickens.

Mating	*Relationship between mates*	*Inbreeding of progeny*
Uncle-niece	.125	.062
Half brother-sister	.250	.125
Full brother-sister	.500	.250
Full brother-sister, 2 generations	.600	.375
Full brother-sister, 3 generations	.727	.50

posely, but to maintain a close relationship to the best animals. Homozygosity increases more than it would in the usual purebreeding methods, but the prepotency of the breeding stock from the selected group of animals should be enhanced in the process. In the history of many herds, linebred families have become famous for their breeding performance and individual excellence as well.

Robert Kleberg purposely linebred the King Ranch Quarter Horses to Old Sorrel, a son of a Thoroughbred mare and Old Hickory. The pedigree of a Quarter Horse colt bred by King Ranch illustrates the results of this long-continued linebreeding program (Figure 18.4). A high degree of relationship to Old Sorrel (.45) has been maintained, and yet the inbreeding of the colt ($F = .09$) has not been greatly increased.

The development of inbred families within the breed serves to increase or maintain the genetic uniformity within the strain and to increase the breeding predictability of that strain within itself and in crosses with other strains (Figure 18.5), but experiments indicate that the conformation and producing ability within inbred strains is hurt by inbreeding. In chickens, and to a lesser degree in swine, there has been considerable development of inbred lines to be used in crossing to produce useful hybrids. Brother-sister, sire-daughter, and son-dam matings are the closest that can be made in livestock. These cause a decrease in heterozygosity of 25 per cent per generation in contrast to the 50 per cent reduction that can be made by selfing, or self-fertilization, in corn. Relationships from matings of relatives are shown in Table 18.4.

To compare the systems of mating and selection, we rely today largely on experimental evidence from laboratory animals. It has

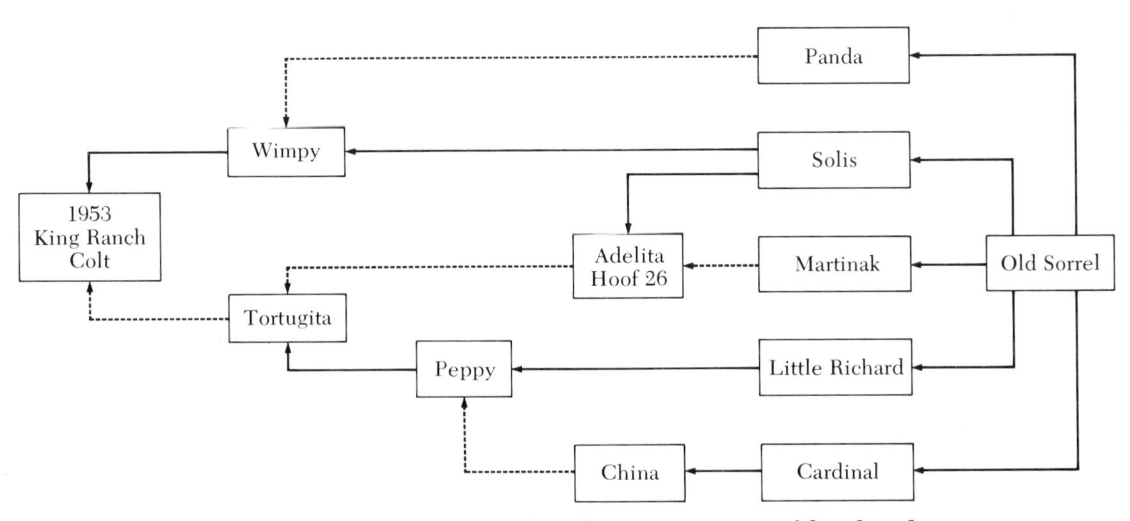

Figure 18-4 A pedigree of a Quarter Horse. This illustrates a system of line breeding with a minimum of inbreeding.

been found that "elite" hybrids in Drosophila have consistently outproduced strains carefully selected without inbreeding for egg production, a trait low in heritability. Elite hybrids are superior individuals obtained by crossing specific inbred lines; the best inbred lines to use to produce elite hybrids are determined by trial and error. For traits higher in heritability, such as egg size, selection has been highly effective. Where both high and low heritability traits with both high and low heterosis are involved, a combination of selection and crossing seems indicated. In Table 18.5, the single crosses are crosses within two inbred lines. Reciprocal crosses are between two large noninbred populations that have been selected specifically to combine well with each other. Recurrent crosses are crosses between two strains in which one strain is selected to cross well with a specific inbred tester strain.

Figure 18-5 Inbred herd bulls from two distinct inbred families are the products of many years of inbreeding coupled with selection for heavy weaning weights and fast feedlot gains. (Top) Linecross sons of this bull were the fastest-gaining cattle in a feed test involving 21 progeny groups. (Bottom) This bull, used through artificial insemination, increased weaning weights in a commercial herd by 40 lb. and sired high-gaining progeny in several feeding tests. For commercial beef production, linecross daughters of one strain should be bred to bulls in the other strain. This results in hybrid vigor within a breed. (Courtesy Colorado State University.)

Commercialization of Genetics in Livestock Breeding

The extent and rapidity with which commercial genetic exploitation has been made of the application of genetics and physiology of reproduction to livestock industry have been spectacular. The hybrid-corn breeders soon learned that what worked for corn might be applied commercially in poultry. Thus practically all the genetic stock for both egg production and meat production are now centralized in the hands of a relatively few companies with their integration of scientific, production, and sales staffs. Selection and strain crossing with much use of inbred lines have been the tools of these companies, much as they had used in the development of corn hybrids. Somewhat later to enter this field but rapidly expanding have been the artificial breeding companies. These began as cooperatives among dairymen who wanted to eliminate the danger and cost of maintaining a bull. Extension, teaching, and research departments associated with universities and governments often maintained a service unit for artificial insemination. This has changed as well, with the major part of the artificial insemination in dairy cattle, which accounted for about 8,000,000 dairy animals in the U.S.A. in 1965 and 600,000 beef cattle, being handled by a few companies. Capability for utilizing, more economically, the scientific processing, distribution, and sales put this as well into the

Table 18.5.
Relative performance in various selected traits after 16 generations of selection under different methods.[a]

Methods of selection	*Number*	*Mean daily fecundity*	*Mean egg size*	*Performance index*
Closed population	582	90.8	38.93	2,309.2
Reciprocal cross	573	97.7	38.42	2,314.5
Recurrent cross	544	102.1	39.16	2,366.9
Single cross I	544	102.1	39.16	2,346.0
Single cross II	164	101.3	38.38	2,324.1

[a] Bell, Moore, and Warren, 1955.

organization of large, well-capitalized companies. Swine breeding too, has become a part of the business operation of some hybrid-corn companies and they are developing in the field of beef cattle as well.

Technological changes bring about new types of jobs and demands for new skills. The president of the major company in artificial insemination earned his Ph.D. in animal breeding and population genetics. Geneticists are employed by every major organization involved in these activities. This also means more employment for specialists in sales, business management, public relations, journalism, statistics, computer science, biochemistry, biophysics, nutrition, physiology, and all the other components that are essential to the smooth operation and success of new businesses emanating from applications of science to people's food needs.

FURTHER READINGS

H. T. Fredeen. 1958. "The genetic improvement of swine." *Animal Breeding Abstracts,* 24:314–326.

I. M. Lerner and H. P. Donald. 1966. *Modern Developments in Animal Breeding.* New York: Academic Press.

J. L. Lush. 1948. *Genetics of Populations.* Ames: Iowa State Univ. Mimeo.

W. S. Spector. 1956. *Handbook of Biological Data.* Philadelphia: Saunders. See pp. 111–113.

Nineteen

Significance of Breeds

Happy is he who lives to understand,
Not human nature only, but explores
All natures,—to the end that he may find
The law that governs each; and where begins
The union, the partitions where, that makes
Kind and degree, among all visible Beings;
The constitutions, powers, and faculties,
Which they inherit,—cannot step beyond,
And cannot fall beneath; that do assign
To every class its station and its office,
Through all the mighty commonwealth of things
Up from the creeping plant to Sovereign Man.

William Wordsworth,
The Excursion, Book IV (composed 1795–1814, published 1814)

In early childhood, we learn to distinguish between cats and dogs and cattle and sheep and pigs. A little later we notice that there are differences between dogs, that cattle vary in color, that some have horns and others don't. Later on, some of us learn that there are differences between groups of cattle or groups of sheep in economically important traits such as milk production or kind and yield of fleece. We also recognize differences in a multitude of traits between individuals within such categories as breed. It is unfortunately true that what we come to know so clearly as a basis for distinguishing our breeds of livestock does not provide us with a brief, scientifically precise definition of "breed."

Meaning of Breed

Breeds are not "pure" in the sense of having a high degree of homozygosity for different genes in different breeds. In fact, for most breeds the loss in heterozygosity since the breeds were founded seems to be between 1 and 11 per cent, or a loss of about .005 per generation. Animals of each breed are, however, sufficiently homogeneous in certain traits to allow us, as a rule, to determine the breed to which each individual belongs.

The distinctions we notice may be primarily in superficial traits or in those which are of fundamental importance to the producer or consumer. Some of the differences we recognize between breeds may result from one breed having genes which another one does not have, or from differences in the frequency with which the same genes occur in the different breeds (i.e., when the gene frequency is not zero or one). In the former case there will be no overlap in the traits between the breeds. In the latter, the traits will overlap, the difference being between the averages of the breeds, as can be seen, for example, in Figure 19.1. This graph illustrates the clear-cut difference between two breeds in fat per cent in milk (midpoint 3.6 per cent for Holsteins, 5.2 per cent for Jerseys) even though the distributions overlap (some Holsteins and Jerseys have the same test). About 95 per cent of the Holsteins tested less than 4.4 per cent, whereas only about 3.6 per cent of the Jerseys did.

That genetic differences exist between breeds in traits other than those commonly known to distinguish one breed from another (coat color and pattern, presence and absence of horns, fat per cent, milk production, wool quality and quantity) is demonstrated by the hybrid vigor which occurs when certain breeds are crossed and by differences between them in the frequencies of blood group (red-cell antigens) alleles. For example, Maijala (1969) and his coworkers have shown that in West-Finnish and Ayrshire bulls about 30 per cent of the B blood-group alleles were shared by the two breeds, about 20 per cent of the alleles were found in the Ayrshire only, and about 50 per cent in the West-Finnish only. In addition, for those genes which the two breeds share, the frequencies are usually widely different. Similar results have been reported for other breeds.

There is little evidence that the identified blood markers show much, if any, relationship with economically important traits or that they have demonstrable adaptive value. They do, however, point to a genetic divergence both between and within breeds, probably as a result of random sampling of the genes in the formation of the germ cells. Presumably, similar random sampling effects have pro-

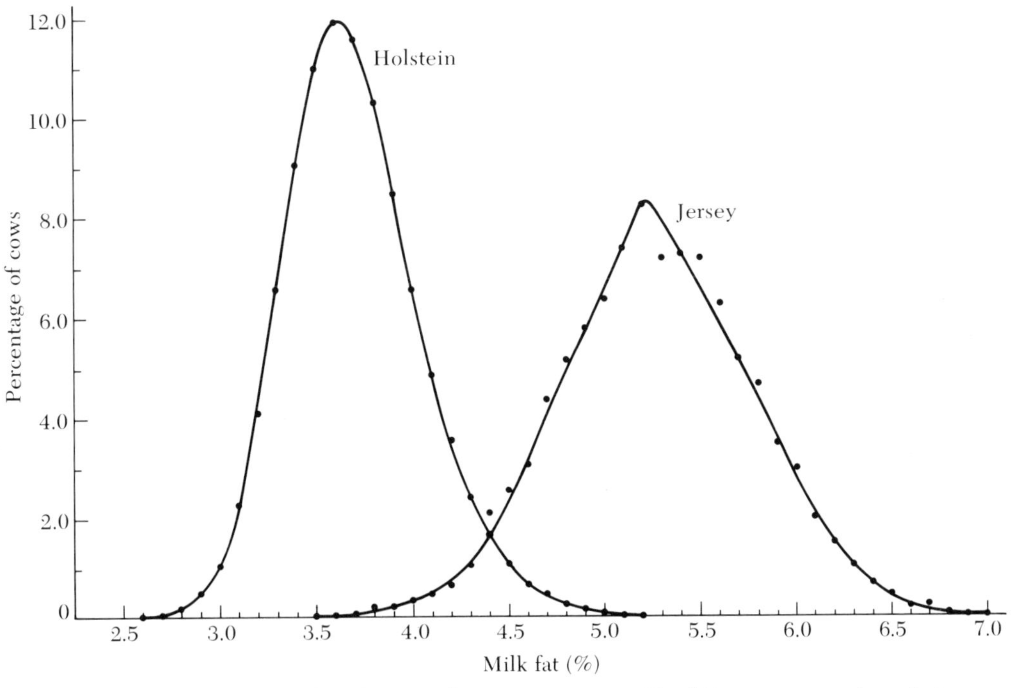

Figure 19-1 Frequency distribution of Wisconsin Dairy Herd Improvement Association official milk fat percentages for Holsteins and Jerseys, 1960–1969 inclusive. (Based on summary by G. E. Shook, Diary Science Department, University of Wisconsin.)

duced genetic differences between breeds for other traits in addition to those differences which selection has produced as a result of different emphasis on the traits within the different breeds.

It would appear that, for most economically important traits, the breed distributions overlap, that the differences are in gene frequencies other than zero or one, that a large part of the observed variation within breeds is environmental, and that most of the differences between the averages of breeds is genetic. Obviously, this brief summary statement on the nature of breed differences does not contain the qualifications needed for it to apply to all genetic and environmental circumstances in which comparisons might be made. It is intended to depict what is usually found in studies of the economically important traits of different breeds under somewhat comparable conditions.

How Did Breed Differences Originate?

How did these breeds come about, and what keeps them as distinct breeding groups? There are several factors which seem to have been responsible for producing the distinctly different breeding groups which have come to be known as breeds. First, there was geographic separation, which allowed different groups to develop either as a matter of chance or as a result of differential natural or artificial (man's) selection. Second, there was the development of breeding groups which were kept separate by man, even though geographically they were fairly close together. These groups which were kept separate could have drifted apart gentically by chance or been pushed apart by artificial selection.

The extent to which differences in gene frequencies between the older breeds have been caused by selection and by chance is difficult to say. Undoubtedly, superficial breed characteristics, such as color, set of the ears, presence or absence of a white belt, and amount of white spotting, have been maintained by man as identifying marks of the breed. Also, some traits of economic importance, such as milkfat content, length of staple, and fineness of fleece, have been emphasized differently within different breeds, and some of the newer breeds have been developed by crossing of breeds, followed by selection to blend desirable traits. For many other traits the roles played by natural and artificial selection and by chance in producing differences between breeds are not clear.

Breed Record Associations

It is clear, however, that once certain differences were found to exist between breeding groups, it was desired to conserve some of these groups as separate entities. The mechanism for maintaining the "purity" of certain stock was the pedigree registry or herdbook. The prototype of the modern Record Association was one started by Coates for Shorthorn cattle in 1822.

This desire to protect breeding groups from introduction of outside inheritance, or sometimes from introduction of any but specified stock, and to promote the breed has resulted in the formation of purebred record associations throughout the world. In the U.S. alone there are approximately 100 registry associations: 17 for beef cattle; eight for dairy cattle; three for dual-purpose cattle; 26 for sheep; three for goats; 20 for swine; and 26 for horses, ponies, and jacks. The requirement that both parents be registered in a particular association before an individual can be registered is the device used to protect the "purity" of the breed.

The effect that the breed societies have had on the genetic improvement of livestock is hard to measure. The encouragement and programs provided for members to improve their stock in those traits which have to do with more efficient production of the quality products needed by the consumer have undoubtedly had a beneficial effect. The emphasis they have given to traits which are not known to bear any relationship to efficiency of production of the products needed by the ultimate consumer is indeed unfortunate because of its limiting effect on the improvement otherwise possible in the more important traits.

As a general rule, more rapid progress can be made by selecting for the economically important and heritable traits themselves, rather than for those traits which are correlated or are thought to be correlated with them. This dictum does not mean that one should fail to cull those animals which have handicapping abnormalities. The important principle is that the larger the number of traits selected for, the less effective is the selection for each one. The number of really important traits and the low heritability of most of these means that nature has already placed us under a severe handicap in making genetic improvement; don't let us be a party to imposing any more handicaps.

There is a clear call for breed associations and for all those associated with the livestock industry to ask these questions: What are the traits which lead to efficient production of the desired end-products? What are the heritabilities of these traits? Then, with whatever answers are available to these questions, we should proceed to use the information to maximize genetic improvement for these important traits.

Role of Breeds in Genetic Improvement

What are the ways in which purebred animals might be used to make over-all genetic improvement? There is, of course, the way just mentioned: to improve the breed by selecting judiciously for those traits which lead to efficient production of what the consumer needs. There is also the opportunity to use the breeds in a systematic crossbreeding program in an attempt to combine the best of each breed in the progeny and to capitalize on whatever hybrid vigor (heterosis) there is when two or more breeds are crossed.

Breeds may also be used to form a crossbred foundation out of which selection may be able to mold new breeding groups superior to the parental stocks, or which might play the role of a stock to which other less improved groups could be graded up. It may well be that the so-called "inferior" stock is not inferior in all respects, and that part of its inheritance should be preserved by these schemes. Adaptability of indigenous stock to local conditions, but lack of adaptability of exotic stock which shows superiority in other traits, would be an example of the need for conservation of some of the inheritance of the indigenous breed while trying to introduce the other desirable features of the exotic breed.

There are many examples of combining breeds successfully by the plans given above, and yet there is need for clear documentation of the results likely to be achieved when using specific breeds in each of these combinations under a variety of management and other environmental conditions.

Mating Systems

Effects in General

The general picture of the effects of mating systems (including crossbreeding) and of selection in livestock is clear in broad outline. The average effects of different mating systems on swine provide an illustration of what appears to be generally true in livestock. The extent of these effects varies from one group to another, between and even within species of livestock, for a particular trait as well as between traits, but the general trends are similar for each species.

Figure 19.2 illustrates the points made by Dickerson (1952): that traits which have been subject to long-time undirectional natural or artificial selection (fertility and viability traits) would be expected to be low in heritability; that any genetic variance remain-

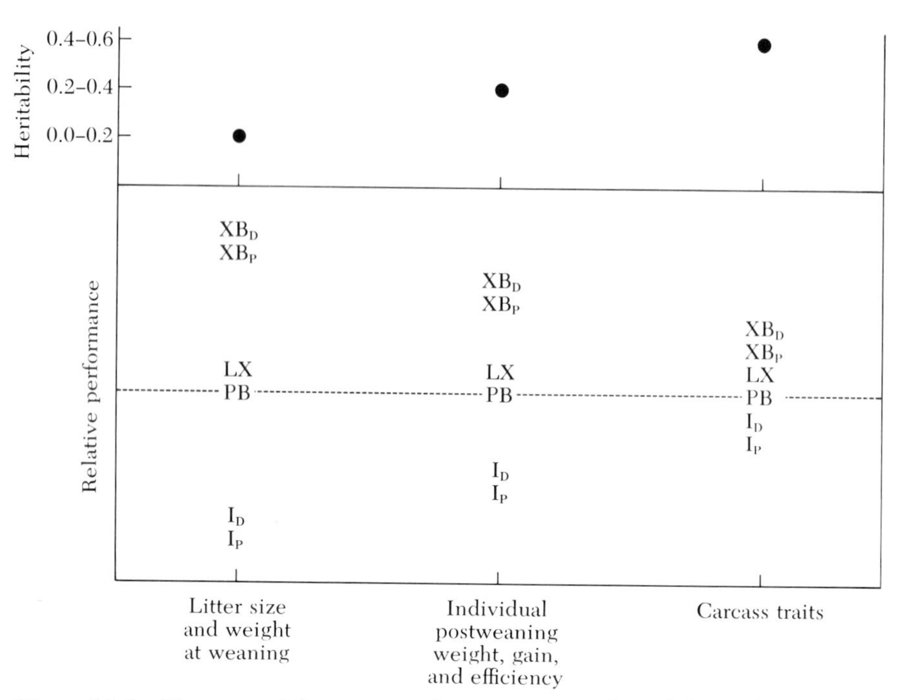

Figure 19-2 Top part of figure gives the approximate heritabilities (*o*), for the traits listed below the abscissa; the bottom part gives the effects of inbreeding of pig (I_p) and of dam (I_D), of linecrossing (*LX*) and crossbreeding of pig (XB_p) and of dam (XB_D), relative to non-inbred purebreds (*PB*). (Chapman, 1969.)

ing in these traits, including dominance and overdominance, would likely be nonadditive and hence lead to depression with inbreeding (I_D, I_P); that lost vigor would likely be restored or enhanced with crossing of inbred lines within breeds (*LX*); and that with wider crosses, involving different breeds, inbred or non-inbred, even more hybrid vigor would result (XB_P, XB_D). At the other end of the scale are those traits (carcass) for which selection does not appear to have occurred over many generations. These traits would be expected to have more of their genetic variation in additive form, and hence be more responsive to selection and show less effect from inbreeding and crossing.

Value of Breeds in Crossing

The value of breeds when used to produce crossbreds is indicated by the average results found from crossbreeding experiments with different species of livestock. These averages do not depict what will happen to the particular traits in each class of livestock every time matings of this kind are made. Just as with all other biological phenomena, there is variation around the average, and in many cases this variation is large enough to encompass "no difference" between crossbred and purebred averages, and even includes reversal of the average differences in some cases.

Swine In pigs, the kinds of results that have emerged from crossbreeding experiments are shown in Table 19.1 The figures summarize the results from a number of experiments on crossbreeding in the U.S. and abroad. The ranges of values in the table for each trait encompass most of the results found for those traits, but can only be looked on as guides for any particular cross. As a rule, the wider the cross, the greater the crossbreeding effects. Even though there is a great deal of variation in the crossbreeding results for each trait, the over-all performance of crossbreds, in terms of amount of pork produced per litter, would be expected to show considerable superiority over each of the purebred parents in the vast majority of cases.

Beef Cattle In summarizing the results of crossbreeding in beef cattle (Table 19.2), Warwick (1968) points to the following advantages of crossbreds over purebreds.

The crossbred beef cattle of British breeds do not always excel the purebreds in individual traits. If, however, over-all merit is considered in terms of calf crop, weaning weight, post-weaning performance, and carcass value, it was found that the crosses usually exceeded the best purebred. One recent estimate of the value of systematic crossbreeding of British breeds is that it can increase production per cow exposed for breeding by about 20 to 25 per cent, with about half this advantage coming from heterosis in the calf and half from a combination of heterosis in reproduction and maternal ability of crossbred cows (Cundiff, 1970).

Crosses between breeds of diverse origin, such as between Brahman or Africander and British beef breeds for use in hot climates, show even larger heterotic effects and also allow the combining of the traits for adaptability to the environment with those for productivity. The calves from these crosses under tropical conditions have shown gains of 10 to 15 per cent in weaning weight over the British stock, and the crossbred cows have shown similar improvements in the weaning weight of their calves as well as a larger calf crop when bred back to bulls of British breeds.

Dairy Cattle The dairy cattle crossbreeding experiments have involved Ayrshire, Brown Swiss, Guernsey, Holstein, Jersey, Milking Shorthorn, Red Dane, and Red Polled. The general picture (Legates, 1966) is that the crossbred cows have exceeded the average of the purebreds in fat yield by between 8 and 18 per cent. Milk yield increases of the F_1 over the parental stock tend to be in about the same range of values, with one experiment reporting considerably less advantage for milk yield. The observed difference in favor of the crossbreds is greatest in the first lactation, suggesting that the hybrid vigor is through early maturity. In Illinois experiments (Touchberry, 1970), yield differences between purebreds and crossbreds in the same generation tended to decrease as the animals aged. In crosses involving Holstein, the crossbreds have not usually exceeded the purebred Holstein in either production or body weight, even though heterosis in terms of a difference between F_1 and the parental average was found.

Sheep Crossbreeding in sheep has been used successfully for capitalizing on hybrid

Table 19.1.
Advantage of crossbred performance in swine as a percentage of purebred parental performance.

Dam	Offspring	Number		Pig weight weaned or at 154 days	Litter weight weaned or at 154 days	Daily gain or feed efficiency
		Farrowed	*Weaned*			
Purebred	Crossbred	0 to 5	5 to 15	5 to 10	10 to 25	0 to 5
Crossbred	Crossbred	5 to 15	10 to 30	5 to 15	15 to 40	0 to 5

vigor and for combining traits peculiar to certain breeds with those distinctive of others (Sidwell and Miller, 1971; Sidwell, Wilson, and Hourihan, 1971). Selection based on a crossbred foundation has been used to combine desirable traits from two or more breeds into what has developed into another breed.

Results on crossbreeding give the same kind of information you would have if you had some loaded dice, but did not know at first which number was favored. If much empirical data on these dice gave the frequency of a 6-spot to be 60 per cent, it would certainly be worth putting your money on the 6-spot, even though 40 per cent of the time you would expect other numbers to show. When matings are made to produce crossbreds, the outcome is loaded in favor of what the evidence from crossbreeding experiments shows on the average. It does not mean that the outcome will be like that every time.

One should decide whether or not to embark on a crossbreeding program not only in the light of the average results and the variation in the results from crossbreeding experiments, but also after considering: the superiority of the available stock produced by a non-crossbreeding approach; the possibility of securing suitable replacements for continuance of the crossbreeding plan; and the extent to which it is feasible to carry out a profitable crossbreeding program, not forgetting the problems of overlapping generations in a rotational scheme and any additional costs that might be involved.

Table 19.2.
Advantage of crossbred performance as a percentage of purebred performance in crosses of Angus, Hereford, and Shorthorn beef cattle.[a]

Breeding	*Per cent of advantage*
Purebred cow, crossbred calf	
Calving rate	1.3
Calf survival rate	3.0
Calf crop weaned	4.1
Weaning weight	4.9
Post-weaning gain	2 to 4
Slaughter or yearling weight	2 to 4
Feed efficiency, weaning to slaughter	0.7
Conformation score	0
Age at first heat, heifers	slight
Crossbred cow and calf	
Calf crop raised	4.3
Calf weaning weight	5.3

[a] Adapted from Warwick (1968).

Future Role of Breeds

What, then, has been and still can be the significance of breeds in the genetic improvement of livestock? They have been and still are significant as sources of improved stock which supply a variety of products to meet the many different needs of the consumer, and at the same time to fit into the many ecological niches which nature's diversity has provided. In addition, they are a storehouse on which we can draw: for grading up poorer stock; for combining in crossbreds desirable traits, none of which are the legacy of one breed (maybe as a basis for another breed); and for combining the genes for those traits which show hybrid vigor when breeds are crossed.

The breeders responsible for the breeding programs in each of the breeds of livestock should be aware of the tremendous responsibility they have for looking ahead and planning their programs so that the important roles mentioned above can be fulfilled by the livestock of the future. It is not enough to rest on the laurels which come from past performance. Intrabreed improvement in efficient production of what the consumer needs is a "must" for breeds if continuous progress is to be made, not only for breeding within the breeds, but also for their use in crossing programs.

It is also a "must" if breeds are to have a reasonable probability of survival. There is no reason why stock which does not carry a breed label cannot provide the same opportunity for improvement and also be improved by the same means that the purebred breeders can use. If stock of any origin can be demonstrated to be genetically superior for economically important traits, it is highly likely that the demand for it will reflect its merit. It is not just any purebred or nonpurebred that provides the basis for livestock improvement. It is the best of these on which our future depends. The proof of the pudding is in the eating, not in the label it carries.

For those readers who are entering the field of animal science, it should be stated that there are many challenging questions about the future of animal agriculture which still need answering, not only in this country but around the world. The future of breeds involves deciding on the most effective way of improving them as producers of desirable products when bred straight and when used in crosses. The theory on how to do this is very extensive, but the experimental work on how it actually works in practice is less extensive, and the implementation of what is known is far from complete.

Need for Conserving Breeds

Do we actually need all the breeds that exist at the present time, or could we accomplish as much in the future with fewer breeds? We cannot, of course, be sure of the answer to this question until the future arrives. We cannot be certain because we do not know what the demands of the future for products are going to be; we cannot be sure of the kinds of environments (kinds of feed, method of feeding, housing, management, etc.) to which our animals will have to adapt; we do not know enough about the genetic makeup of each of our breeds or their value in specific crosses under a variety of environments. There is obviously the chance that what appears useless to us now may be of value under changed circumstances; so, theoretically, to be on the safe side we should try to conserve all breeds. Practically, of course, this would be an impossible task. As an alternative, it would appear wise to evaluate as many of them as we can when bred straight and when bred in crosses under a variety of conditions, and then try to ensure that those which hold the most hope for efficient production of what is now (or possibly likely to be) important to us be preserved.

This really becomes a worldwide problem because of the potential value of stock in this country for improvement in other countries; furthermore, we should not be complacent about the possibilities that foreign stock offers us. It may well be that there is inheritance in foreign stock that could prove most valuable under certain conditions in this country as a breed and in combinations with other breeds.

Evaluating Breed Resources

To lay the background for the "most effective use of world breed resources would seem to require the following," according to Dickerson (1969).

"1. Careful definition of biological objectives in performance, considering economic effects of both consumer preferences and variable production costs under probable future conditions.

"2. Determination of management systems required for more efficient livestock production.

"3. Identification of the more promising breeds, indigenous and exotic, based on existing information and using criteria consistent with 1 and 2 above.

"4. Importation of adequate samples of the more promising exotic breeds for experimental evaluation.

"5. Experiments designed to show which breeds

and what methods of utilizing the better ones will permit maximum efficiency of production. . . .

"Choice of the most efficient breed for a specific type of production requires reliable estimates of relative performance for the more promising pure breeds, three-breed crosses from crossbred dams, and F_1, F_2, and F_3 generations of two-breed crosses. These would identify superior breeds and the most effective methods of utilizing them, except for gains from further selection within existing or newly developed breeds and from heterosis in male reproductive performance.

"The awesome task of testing n breeds, $n(n-1)(n-2)/2$ three-breed crosses, and $n(n-1)/2$ combinations in F_1, F_2, and F_3 generations can be made more manageable by:

"1. Limiting the direct experimental comparisons of pure breeds for the chosen objectives and management conditions to those found to be superior on the basis of existing performance data in their present areas of distribution.

"2. Restricting the number of breeds (n) included in F_1 crossbred comparisons to the six or eight which were most promising in purebred performance (except usually include all breeds chosen under 1 which were available initially only as male parents).

"3. Limiting breeds to be tested as male parents of three-breed crosses to the two or three (n') found to be superior for individual performance (i.e., growth, carcass, milk, wool) in the pure breed and F_1 comparisons. This would reduce numbers of three-breed crosses by n'/n to $[n'(n-1)(n-2)]/2$ instead of $n(n-1)(n-2)/2$.

"4. Testing the potential of new breed development in advanced generations of *inter-se* matings (e.g., F_2 and F_3) for only a few of the better breed combinations, chosen on the basis of purebred and F_1 cross performance."

As stated earlier, one use made of breeds in crossing is to put together in one group the best of each parental breed. One approach to improvement within the parent breeds might be to select specifically for traits in one breed that are different from those selected for in the other(s) for the express purpose of combining their traits in the final product. This is a subject which has received considerable theoretical treatment. The conclusions reached are that if selection is for specialized sire and dam lines and they are then combined, the "rate of improvement . . . is never less than in a single line and can be considerably greater" (Smith, 1964). There is need for experimental information on this, just as there is on so much of what has been indicated to be true theoretically. Simplifying assumptions have to be made to allow one to work through problems of this kind on paper. Nature's complexity may involve some factors which invalidate these conclusions. Experimental tests should be made to expose these.

The hope would be that testing of breed resources in the several ways suggested would eventually be carried out worldwide, so that all countries, especially the developing ones, could benefit from this information and thereby put themselves in a position to provide more of the high-protein food so many of them so desperately need.

Breeding Structure for Maximizing Improvement

After the merit of the breeds has been evaluated in straightbreds and in crosses under a variety of conditions, how should the inheritance of the superior one be conserved? Should the breeds be merged into a small number of large breeding units with selection for desirable inheritance within each? Should a large number of breeds and smaller groups within breeds be maintained with selection being practiced within each? The answer to this would appear to be clear, based on the results of studies on the ideal breeding structure for genetic change in geologic time (evolution) and for genetic improvement of animals and plants (under man's guidance).

Selection within a large number of partially isolated groups (breeds) and within breeds, which are tested in various combinations from time to time, seems to offer most hope for genetic improvement.

Specific systems of mating and selection procedures which can be used within and between these groups in the different species of livestock have been discussed in the preceding chapter. Extensive further treatment of these topics can be found in Lush (1945, 1948), Lerner and Donald (1966), and Johansson and Rendel (1968).

With this background, the aspiring animal scientist should be able to see immediate problems urgently needing solution if animal agriculture is going to do its part in helping to feed and clothe an inadequately fed and clothed world. The challenge is there; our very survival demands that we meet it.

FURTHER READINGS

Johansson, I., and J. Rendel. 1968. *Genetics and Animal Breeding.* W. H. Freeman and Co.

Lerner, I. M., and H. P. Donald. 1966. *Modern Developments in Animal Breeding.* New York: Academic Press.

Lush, J. L. 1945. *Animal Breeding Plans.* Ames: Iowa State Univ. Press.

Lush, J. L. 1948. *The Genetics of Populations.* Ames: Iowa State Univ. Mimeo.

Twenty

Breeding for Milk Production

"If the dairyman wants milk, let him pursue the milking tribe; let him have both bull and cows of the best and greatest milking-family he can find."

George Culley,
Observations on Live Stock (1786)

The average annual milk production of the U.S. dairy cow has increased from 4,508 pounds in 1930 to 9,388 pounds in 1970 (Table 20.1). During the same period, production of cows in Dairy Herd Improvement Association (DHIA) herds increased from 7,642 pounds to 12,750 pounds each. Obviously part of this increase has been due to increased genetic merit for milk yield.

Table 20.1.
Production (in lbs.) of all U.S. dairy cows and DHIA cows.[a]

Year	*U.S. cows*	*DHIA cows*
1930	4,508	7,642
1940	4,622	8,133
1950	5,314	9,172
1960	7,026	10,561
1970	9,388	12,750

[a] King *et al.* (1971).

The reason for the existence of the dairy cow is the production of milk; this trait justifiably has been the subject of more genetics research than any other trait of the dairy cow, and perhaps even more than any other single economically important trait of farm animals. Genetic studies in dairy cattle, however, have also been directed toward reproduction, physical traits and appearance, disease resistance, and other characteristics.

Measurement of Production and Factors Affecting It

Recording techniques

The standard measure of milk yield is the 305-day record, although records of different length have been used in the past and are still used occasionally for research purposes. Rarely, however, is milk weighed daily; more often the weight for one day, A.M. and P.M. milking, is recorded once monthly, with samples of these two milkings combined for chemical analysis for milk fat, and perhaps also for solids-not-fat (SNF) or protein. Other tests for quality may also be made. The total yield of milk for the month is then calculated, and accumulated with weights from other months for the estimate of 305-day yield. If 10 months were represented in the 305-day record, e.g., 20 actual weights, an excellent estimate of the true record is obtained, with a considerable saving in expense. The efficacy of other schemes, such as weighing every six weeks, or weighing the milk from the A.M. and P.M. milkings only on alternate months, has also been studied, and these approaches are being used to a limited degree in the U.S. and other countries.

The Dairy Herd Improvement Association Testing Program

Performance of an individual animal contributes to genetic progress only if recorded and analyzed. A major deterrent to improvement in many dairy cow populations is the low number of cows actually on an organized milk-recording program. Although more than 58,000 U.S. dairy herds with about 3,200,000 cows were on DHIA test in early 1971, this represented less than 26 per cent of the total population; percentages were greater in some countries, considerably less or essentially zero in others.

The first U.S. "cow testing" association in history was organized at Fremont, Michigan, in 1905 and was patterned after milk-control societies started in Denmark in 1895. The evolution of testing programs, including those sponsored by the purebreed associations, to the present time is a fascinating story in itself as chronicled by Becker (1971).

The program has lent itself well to the electronic computer. From a beginning in Utah in 1950, virtually the entire program now utilizes computers on state-wide or regional bases. The dairyman benefits first from the short-term management information and recommendations provided him, for example, in feeding and culling. In addition, annual summaries, running averages, and multiyear statistics are provided the dairyman on his own herd and other herds identified only by their general characteristics to make them valuable for comparison.

Long-term and indirect benefits accrue also from the use of DHIA data for research in management, reproduction, genetics, and nutrition. The use of DHIA records in the USDA sire-proving program may represent the most extensive use of the principles of quantitative genetics for the improvement of livestock in the world today.

The dairyman receives computer printouts, usually monthly, describing the productive status of every cow in the herd, with recommendations such as for feeding, breeding, suggested dry dates and due dates. Lists of cows suggested for special action (e.g., culling) are also provided. The program varies slightly from area to area of the U.S., depending on the wishes of the cooperating dairymen and the available computer capabilities. Furthermore, improvements are made continually with additions to the programs being made when possible. Detailed descriptions of programs are available from the Extension Dairyman in each state.

Environmental effects

Innumerable nongenetic factors affect production; the cow has an incredibly high workload, and is very sensitive to factors such as level of nutrition, management techniques, age, and disease. Many of these can be measured and, to the degree that they are systematic, performance can be adjusted to take them into account for genetic evaluation. Many others, however, to date defy quantification.

During a given lactation, yield usually increases for a time postpartum, say four to seven weeks, and then gradually declines. Considerable variation exists in the shape of the curves for individual cows, and for the same cow in different lactations. Age of the cow, season of freshening, and pregnancy have important effects, for example.

Lactation yield can be described to some degree by knowledge of the point of maximum production and by the slope of the curve, the latter indicating persistency of lactation. Among cows in the same lactation, 74 per cent of the variation in yields was attributed to these two factors by Lennon and Mixner (1958).

Knowledge of performance during a portion of the lactation can be used to predict the total lactation. This has the advantage of making an evaluation of a cow or her sire available earlier, and permits use of records cut off before completion. A number of sets of projection factors are available for milk and fat yields as well as yields of SNF and protein. Different specifications apply to different breeds and ages of cows. A portion of one set is shown in Table 20.2. A young Holstein which had produced 5,000 pounds of milk by the 95th day of lactation could be expected to produce about 13,400 (5,000 × 2.68) by 305 days, if she followed the pattern of the average cow.

In modern times selection for milk yield has most commonly been based on 305-day production. In evaluating these milk records, consideration must first be given to the systematic factors which affect such performance; e.g. little would be gained by comparing a young cow in a poor environment with an old cow in an excellent environment, since both factors influence yields to an important degree.

With 305-day records, age and contemporaneity are systematic environmental effects of major importance, the latter term referring to location as well as time. The herd in which the cow lives, and the year and season in which she freshens, are now considered in genetic studies; e.g., as much as 50 per cent of the variation in age-adjusted milk records may be attributable to herd-year-season effects. Such estimates necessarily pertain only to tested herds; so the range of environments and their effects would be considerably greater if all dairy herds were included.

Milk yield changes with age in curvilinear fashion. As the young cow matures, her milk yield increases to a maximum at about five to eight years and declines gradually thereafter. Slight but real differences exist among breeds and among areas. To the degree that age effects are systematic, production can be adjusted, by using factors such as those shown in Table 20.3, where the adjustment is to the point of maximum production, e.g. maturity. Such records are denoted M.E. (mature equivalent). Jerseys freshening at 24 months and giving 9,000 pounds of milk containing 350 pounds of milk protein would be expected to give about 12,000 pounds of milk and 473 pounds of protein at maturity, other things being equal. In a population of contemporary herdmates, 15 to 30 per cent of the variation in yields of milk and its major constituents are due to variations in age. An alternative to the use of adjustment factors would be to compare directly cows of similar ages.

Advancing pregnancy results in decreasing yield during the lactation, and variation among cows in the number of days open (or pregnant) leads to some nongenetic variation in yields, which should be considered in evaluation of performance. Likewise, an unusually short dry period preceding parturition results in decreased yield following that parturition, on the average. Usually, these two environmental effects, though real, do not account for major portions of environmental variability.

Table 20.2.
DHIA 305-day projection factors (Holsteins less than 30 months of age).[a]

Days in milk	*Milk yield*	*Fat yield*	*Days in milk*	*Milk yield*	*Fat yield*
35	7.13	6.87	185	1.47	1.49
65	3.85	3.81	215	1.30	1.31
95	2.68	2.68	245	1.17	1.18
125	2.08	2.10	275	1.07	1.08
155	1.72	1.74	305	1.00	1.00

[a] McDaniel *et al.* (1965).

Table 20.3.
Age conversion factors for Jersey milk and protein yields.[a]

Age		*Factor*	
Years	*Months*	*Milk yield*	*Protein yield*
1	10	1.37	1.39
2	0	1.33	1.35
2	6	1.24	1.25
3	0	1.17	1.18
3	6	1.12	1.13
4	0	1.08	1.09
4	6	1.05	1.06
5	0	1.03	1.03
5	6	1.02	1.02
6	0	1.01	1.01
6	6	1.00	1.00
to			
8	0	1.00	1.00
9	0	1.02	1.02
10	0	1.03	1.03
11	0	1.05	1.05
12	0	1.05	1.06

[a] Wilcox *et al.* (1971).

In given situations, specific nutritional and management practices also are major sources of environmental variation. In general, however, population studies would rely primarily on statistical control of effects due to age and herd-year-season plus perhaps days open and length of dry period. Hopefully, the remaining environmental variance is random in nature and does not lead to biases in genetic selection decisions.

Repeatability and Heritability of Production Traits

Repeatability of lactation milk yield and composition has been studied by many researchers. Knowledge of repeatability assists the dairyman in making decisions on which animals to cull, since he knows that performance, either superior or inferior, tends to repeat itself, but the degree of repeatability is far from perfect. Although outstanding individuals tend to remain that way, there is still considerable change in rank of the cows in a herd from year to year. Obviously, several records of an individual characterize the individual better than a single record.

Recent research has shown that the repeatability of 305-day M.E. milk yield is about 0.5; this estimate represents the average correlation among records of the same cow. Adjacent records, that is, first and second, or second and third, generally are slightly more repeatable, with the estimate decreasing as the time between records increases. This was well-demonstrated by Butcher and Freeman (1968), as shown in Table 20.4. Nearly 130,000 records from two sources were involved in this study.

Considerable variation exists in heritability estimates of the many traits of dairy cattle.

Table 20.4.
Repeatability of milk and fat yields between various pairs of lactations.[a]

	Analysis I		*Analysis II*	
Lactation pair	*Milk yield*	*Fat yield*	*Milk yield*	*Fat yield*
1 and 2	0.56	0.54	0.49	0.46
1 and 3	0.50	0.41	0.46	0.40
1 and 4	0.43	0.33	0.39	0.32
1 and 5	0.38	0.27	0.35	0.26
2 and 3	0.63	0.56	0.54	0.51
2 and 4	0.55	0.46	0.49	0.47
2 and 5	0.50	0.39	0.43	0.40
3 and 4	0.64	0.56	0.55	0.52
3 and 5	0.57	0.48	0.48	0.45
4 and 5	0.62	0.54	0.57	0.56

[a] Butcher and Freeman (1968).

Some variation would be expected from the fact that estimates have been made in different populations. Sampling errors would account for considerable variability, since many early studies have been based on small samples of data. Lastly, some estimates are doubtless biased, because of the inability of the researcher to account for many environmental factors which could inflate or deflate estimates of genetic variability. Dairy-cattle data frequently suffer severely from extreme disproportion in the various subcells (classification categories), and such disproportion is still troublesome today mathematically. For example, in comparing several sires, all may have an arbitrary minimum number of daughters and records. However, they may have widely disproportionally varying numbers of daughters, records per daughter, number of herds, or number of daughters per herd. Such disproportionate data are still troublesome to analyze even with the increased availability of electronic computers and sophisticated statistical techniques. Knowledge of the heritability of traits other than milk yield and composition is necessary for efficient selection programs. The values shown in Table 20.5 represent generally accepted levels of heritability for the traits listed.

An interesting controversy, still unresolved, has existed for some time between U.S. and European researchers. Several of the latter have shown estimates of heritability of milk yield to be higher in good environments than in poor; that is, in herds where average production is high, estimates are high. Most U.S. research, however, has suggested that the heritability of milk yield is essentially unaffected by herd production level.

Most heritability estimates for the various measures of reproductive performance have been zero or close to zero, although suggestions have been made in years past that estimates might be higher in poorer environments (the opposite has also been suggested, however). To date, only a limited amount of research suggesting appreciable heritability in most reproductive traits has been presented. This does not mean that variations in reproductive performance dependent upon inheritance do not exist, but, rather, it indicates that environmental factors, such as disease, influence reproductive performance to an extent which masks the role played by inheritance.

Table 20.5.
Heritability of various traits of dairy cattle.[a]

Trait	*Heritability*	*Trait*	*Heritability*
Milk yield	0.2–0.3	Mature weight	0.4–0.6
Milk-fat yield	0.2–0.3	Wither height	0.4–0.6
Protein yield	0.2–0.3	Heat tolerance[d]	0 –0.2
Total-solids yield	0.2–0.3	Conception rate	0 –0.1
Milk-fat percentage	0.5–0.6	Reproductive efficiency	0 –0.1
Protein percentage	0.5–0.6	Calving interval	0 –0.2
Persistency	0.3–0.5	Life span	0.1–0.3
Peak milk yield	0.2–0.4	Feed efficiency	0.3–0.4
Milking rate[b]	0.3–0.6	Mastitis resistance	0.2–0.3
Gestation length	0.3–0.5	Over-all type score	0.1–0.3
Birth weight[c]	0.3–0.5	Dairy character score	0.1–0.3

[a] Compiled from various research sources; most published estimates fall within the range indicated.
[b] Peak or average flow. [c] Nonmaternal. [d] Response to standard test.

Relationships of Yield and Composition of Milk

Present pricing systems for milk in the U.S. are based essentially on its weight and fat composition, although in some areas other constituents, such as SNF or protein, may be included in the pricing formula. Selection programs necessarily must take into account the relationships of composition and yield to result in a product that will meet legal and consumer acceptance standards.

Perhaps the most convenient way to consider these interrelationships is to look at the genetic correlations shown in Table 20.6. There is some variability among estimates by different researchers, but, in general, yields of milk and its constituents are positively and highly correlated genetically. Very similar estimates have been obtained involving the yields shown in the table and yields of SNF, total solids, and the lactose-mineral fraction.

Note the negative correlations between milk yield and the percentage of protein. This shows the difficulty in selecting for both increased yield and percentage composition. Furthermore, the milk constituents percentages are generally highly correlated with each other (including SNF and total solids), pointing to possible difficulties in increasing protein content while decreasing fat content, for example.

Table 20.6.

Genetic correlations of milk composition and yield.[a]

	Fat yield	*Protein yield*	*Fat percentage*	*Protein percentage*
Milk yield	0.70	0.82	−.30	−.30
Fat yield		0.81	0.46	0.17
Protein yield			0.13	0.28
Fat percentage				0.55

[a] Based on Holstein lactation records; from Wilcox *et al.* (1971).

Relationships of Milk Production with Other Traits

Reproductive Performance

Reproductive performance is a trait of major economic importance in dairy cattle because the cow must first reproduce in order to lactate. The probability of a conception to any particular service can be thought of as the product of the male's ability to fertilize and the female's ability to conceive, assuming no interaction between the two. There are differences in the ability of the male to fertilize and these are at least to some degree heritable. However, the ability to fertilize is essentially 100 per cent with bulls of high conception rates in artificial service, if the females ovulate on schedule and other conditions are normal.

The forces of natural selection doubtless have had long-term effects on reproductive performance. Simply enough, animals which were sterile or of low reproductive efficiency left no offspring or relatively few. Whether or not appreciable additive genetic variance exists in most reproductive traits is somewhat in question today, but most heritability estimates are low and close to zero, for example, conception rate and calving interval as shown in Table 20.5.

It can be shown mathematically that zero additive genetic variance would also mean a zero covariance and genetic correlation. In other words, if heritability of reproductive traits is zero or near zero, a genetic correlation between milk production and reproductive efficiency of any consequence would not be expected. Phenotypically, there appears to be only a slight relationship, if any at all, between milk yield and reproductive efficiency. The phenotypic correlation between milk yield during early lactation and days open, though positive, is certainly below 0.10 and perhaps even below 0.05.

Dairymen have observed that some of their highest-producing cows are afflicted with cystic ovaries; research has suggested that these cows are not cystic because they are high producers but perhaps the reverse. Cause and effect are frequently difficult to separate, but inability to conceive in a reasonable time has a depressing effect on lifetime production. Production for the particular lactation in which the cow is open, however, would be expected to be slightly higher because of the absence of demands of a growing fetus and a more satisfactory hormonal balance for lactation.

Body Conformation

Desirable body type has appealed to dairymen for centuries, and major changes in appearance have occurred over the years concurrently with increases in milk yield. This fact does not by itself mean that the two traits are appreciably genetically correlated. Further, regardless of what has gone before, the breeder must concern himself with the present facts of life in his selection program.

That milk yield and type and conformation are heritable, albeit not highly so, is known. Estimates of the genetic correlations between most measures of conformation and yield have been low though positive. There have been a number of changes in the pure breed-type classification programs and policies in recent years, and the evaluation of such changes can perhaps not yet be made. A number of scoring systems also have been developed by others, such as universities and AI studs. It would seem that the final value of desirable type must be in its relationship with production and longevity of the cow.

Genetic correlations between milk yield and the over-all type score for an animal have generally fallen between 0.15 and 0.35. Most of the type subcategories (general appearance, body capacity, mammary system, feet and legs and rump) have similar genetic correlations, or perhaps are slightly lower. Dairy character has been reported to have a slightly higher relationship.

Longevity

Longevity may be defined as the ability to avoid death or culling. The former results from disease or accident, and the latter is based on factors such as reproductive failure, low milk yield, or mastitis. Few cows die simply from old age. Of the various possible reasons for loss, some have a genetic basis, but few, including longevity, are highly heritable. High-producing cows remain in the herd longer than low producers; so part of the additive genetic variance observed by researchers for longevity is due to variation in yields. Natural selection will also work in favor of the ability to resist death or culling.

The interrelationships of longevity, body conformation, and milk yield have been studied. First-lactation milk yield appears highly correlated genetically with longevity, perhaps 0.5 to 0.7. At least two factors are involved in this high correlation. First, for a cow to produce at a high level, she must have a good constitution which will favor longevity. Second, poor producers are culled at an early age by the dairyman and thus have no opportunity to demonstrate longevity. The genetic relationship between body conformation and longevity has not been delineated well, but there does appear to be a positive phenotypic relationship, though not as high as the one between longevity and first lactation milk yield.

Body Size and Weight

Body size is a trait which has received considerable attention in the show ring and in type-classification programs. Most classifiers and judges have traditionally given preference to large cows, other points being equal. On the average, large cows give more milk than small cows of the same breed, but they also require more feed for maintenance. Apparently the additional expense of feed for larger cows about equals the additional income. To obtain the same total yield of milk from a herd of smaller cows will require more overhead (labor, housing, veterinary expense). In the case of two cows of equal production but different size, the smaller cow will be more profitable. There is a problem in measuring body size. Body weight is certainly imperfect as a single measure, and so is a single skeletal measure, such as wither height. Heart girth may have some of the features of both, but is highly correlated with body weight.

Most measures of size show appreciable heritability and variability; so larger cows could doubtless be developed if desired. With this additional size would be expected additional milk, since the two traits appear to be genetically correlated, positive but low. Putting as many of the variables together as possible, and realizing that progress in the future is based on the cattle alive today, one can make an intelligent judgment on attempts to increase body size. It would be very difficult to summarize the situation more accurately or succinctly than was done by McDaniel and Legates (1965):

> "We can conclude from a study of the relation between the body weight and production traits of Holstein cows subjected to good feeding and management conditions that:
>
> a. Larger cows give more milk, but it is questionable if the increase offsets the additional maintenance costs.
>
> b. The genetic variance in body weight is largely independent of the genetic variance in milk yield and fat percentage.
>
> c. Selecting for milk yield will produce little if any change in body weight; conversely, selecting for body weight would be expected to produce little change in milk yield.
>
> d. An appreciable opportunity exists to increase milk yields without materially increasing body size.
>
> e. Little attention should be given to differences in weight of cows within a breed, except to check the stamina and productivity of cows markedly above or below the breed average in weight.
>
> f. Undue emphasis on larger cows in the short run could lead to the development of less profitable animals."

Other Traits

Many other traits are economically important and are heritable to some degree. Many qualitative ones, ranging from lethals on one extreme to esthetic traits on the other, can be important to an appreciable number of dairymen under certain circumstances. The advent of artificial insemination, particularly now with frozen semen, has caused concern because of the possibility of widespread dissemination of undesirable genes as well as desirable ones. The genetics of the situation, and its practical and theoretical aspects, have been studied, however, and with application of present knowledge potential dangers can apparently be avoided.

Quantitative traits other than those mentioned are also important. Mastitis resistance, for example, is of extreme economic value and has measurable heritability. Selection for high milk yield, however, will automatically result in some selection pressure against those animals susceptible to mastitis. Furthermore, a certain amount of direct pressure for resistance will be exerted from culling of mastitic animals, although clinical cases may remain in a dairy herd for years under proper treatment.

Inbreeding and Crossbreeding

Inbreeding Depression

Use of inbreeding as a breeding system shows little promise for most dairymen. A few registered cattle breeders have practiced linebreeding in the past with commercial success but probably with little genetic success. Long generation intervals, small population sizes (as compared with plants), large investments per animal, inbreeding depression, and other disadvantages rightfully dissuade nearly all dairymen from inbreeding, although inbreeding research projects have shed considerable light on the genetics of the dairy cow.

The indictment of inbreeding, or even line-breeding, as a breeding system for dairy cattle, seems severe enough to suggest that its use should be limited to certain circumstances, such as for research and perhaps for registered cattlemen who could profit from the vagaries of the marketplace.

Heterosis

Attempts to utilize the phenomenum of heterosis would logically involve crossbreeding. Though little research has been done on the crossing of dairy cattle lines within breeds, that available does not show it to be promising. It is hard to demonstrate that some lines cross well with certain others, though a certain number of breeders still so state. Progeny resulting from crossing Holsteins with other breeds common in the U.S. have not been superior in milk production to those of straight Holsteins.

Of course the problems of which breeding program to use depends on which is most profitable. The questions seem to be whether or not a system of crossbreeding is preferable to one of straightbreeding. However, if a dairyman is doing well with Holsteins, it does not seem reasonable to expect him to do better if he started crossbreeding. It seems true that when all the small benefits of crossbreeding are put together, the crossbreds look better than they might at first glance. Yet, whether or not they are more profitable than straightbreds remains to be seen.

Selection for Milk Yield and Composition

Finding Superior Parents

The novice breeder soon discovers that finding superior parents is harder than it appears. He first makes the profound discovery that direct measures cannot be made on the male, that pedigree selection has been shown to have very disappointing results, and that the obvious direct measure of the breeding value of a male (measuring his offspring) takes years to make. The female frequently has only a few lactation records on which to make an estimate of her breeding value. With the heritability of milk yield fairly low, the accuracy of selection for yield is also low; e.g. if $h^2 = 0.25$, $h = 0.50$.

Again, perhaps the most widespread search for superior animal phenotypes in the world may be that arising for the U.S. DHIA program. A DHIA cow-performance index list is published periodically to identify the top 2 per cent of the registered progeny of AI and natural service sires. Information going into the index includes the performance of the cow herself, her lactation records being weighted according to their number. In addition, information from the cow's paternal half-sibs is also incorporated. Rather sophisticated adjustments for the major environmental effects are also made, to avoid bias in the estimates.

DHIA records also provide the data bank from which the USDA can evaluate the breeding values of AI and natural-service sires periodically. The measure used is called the Predicted Difference. The Predicted Difference is the expected average deviation in production of a sire's progeny from their herdmates in a herd producing at the average level for the breed. A typical Sire Summary List might include over 11,000 individual sires, more than half of which would be Holsteins. A number of modifications and refinements have been made in the program in recent years. Major efforts are made to avoid environmental biases, and an estimate of the reliability (repeatability) of each proof also is given. These estimates of male and female breeding values are for milk and fat yields only. The techniques and concepts can be used for other traits, and have been to some degree. These efforts by necessity have not been nearly as extensive, however. The problems encountered by an individual dairyman who would hope to evaluate his own herd himself, without the aid of an organized testing program, quickly become apparent, not only in the difficulty the dairyman would have, but also in the marginal accuracy of single herd evaluations.

Response and Correlated Response

As shown in Chapter 18, knowledge of certain phenotypic and genetic parameters permits the estimation of possible responses to selection, both in the trait on which selection pressure is exerted and in other traits. Though traits other than those included in Table 20.7 are economically important, the major milk and composition traits are included.

The pattern for multitrait selection is set here. If selection pressure is placed on traits other than milk yield, some sacrifice in yield generally must be made. The decision left to the breeder is what is the most profitable combination of traits to select for, and the weight is to be applied to each. Such estimates can be made by the use of index selection. Unfortunately, we cannot pursue this point further here, but the requisite mathematical techniques have been developed and are being expanded and refined.

Potential and Observed Genetic Progress in Milk Yield

If milk yield is given maximum emphasis in the selection program, with little or no pressure being exerted on other traits, and efficient use is made of an AI sire-proving program, the maximum attainable genetic change in milk yield is slightly over 2.0 per cent per year. Most of this, over 90 per cent, would come directly or indirectly from sire selection. In a Holstein population averaging 15,000 pounds per cow, this would amount to 300 pounds per year. Though this estimate has appeared disappointingly small to some breeders, it is more than the actual total (genetic plus environmental) change which has been attained in the U.S.

Table 20.7.
Direct and correlated response for single trait selection.[a]

Trait	*Milk yield (pounds)*	*Fat yield (pounds)*	*Fat percentage*	*Protein yield (pounds)*	*Protein percentage*
Holsteins					
Milk yield	607	23.3	−.036	13.7	−.018
Fat yield	443	34.7	.058	14.1	.010
Fat percentage	−287	24.1	.190	3.4	.051
Protein yield	428	23.2	.014	14.3	.014
Protein percentage	−231	7.2	.084	5.9	.075
Jerseys					
Milk yield	460	13.7	−.110	14.4	−.050
Fat yield	271	18.6	.040	11.7	−.004
Fat percentage	−434	8.1	.330	−8.9	.106
Protein yield	379	15.6	−.059	14.6	−.013
Protein percentage	−378	−1.6	.202	−3.6	.137

[a] Wilcox *et al.* (1971). Change per generation estimated after an intensity of selection of 1.0; estimates of direct response on diagonal; correlated response, off diagonal. To amplify, direct selection for fat percentage in Holsteins which would result in an increase of 0.190% would be expected to have these correlated responses: milk yield, −287 pounds; fat yield, +24.1 pounds; protein yield, +3.4 pounds; and protein percentage, +0.051%.

Though the theoretical upper limits of genetic change have been known for some time, it is only in recent years that reliable estimates have been made. A major problem in the actual measurement of such progress is that environmental and genetic changes are confounded in such a manner over time as to make their separation extremely difficult. Results obtained have been compatible with present theory, however. In the main, genetic improvement in milk yield has been shown to average around 0.7 per cent per year, or about one-third of the maximum theorized. The challenge to breeders and researchers alike is now apparent.

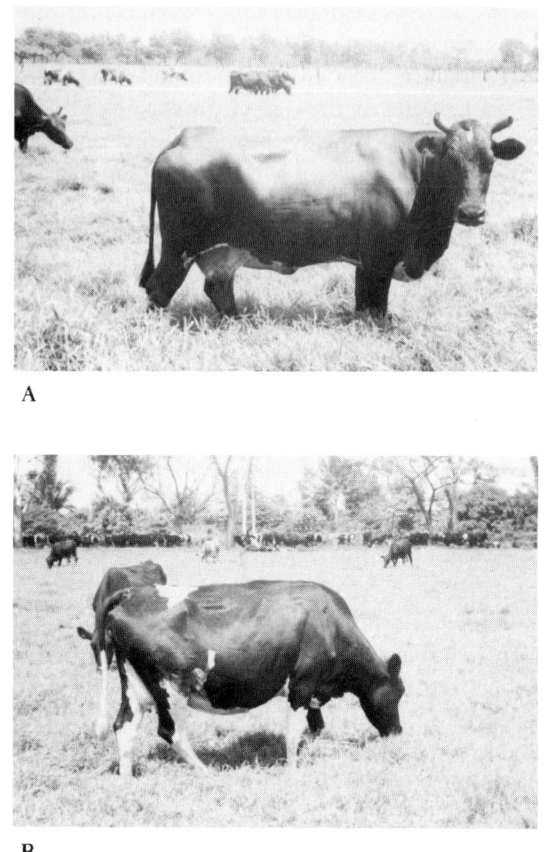

Figure 20-1 An AI upgrading program in action in El Salvador, 1971: A, a Holstein-Native crossbred; B, the second cross to Holstein.

Genetics of Production in Developing Countries*

The dairy industry is beginning to develop at a dramatic pace in many new areas of the world. As over-all conditions for dairying improve, and as dairymen become specialized, questions on breeding and selection programs arise. Further, economic and environmental conditions are so diverse that general recommendations, such as might be made in highly developed dairy areas, are more difficult to formulate.

In many areas, for example, a large portion of the national milk supply comes from animals of native breeding, oftentimes of nondairy origin. These animals may be dual-purpose or even tri-purpose. In times of high wholesale milk prices and other favorable circumstances, they might be milked once or twice daily, perhaps producing only two to three pounds daily. As conditions deteriorate, milk produced is made into cheese for sale or local consumption, or calves are permitted to nurse, or the cow dried off completely. Under these conditions, the dairyman is probably justified in placing only a minimum emphasis on selection for milk production. A cow of superior merit might well be unable to survive under these conditions. The native cattle not only stay alive but reproduce well and provide an important source of income.

As the dairy industry develops, however, there frequently occurs a trend toward the specialized dairy breeds. In this instance, a dairyman who owns a mixed group of dairy cattle, possibly with some native and even beef blood, might select a purebreed and grade up to this breed by artificial insemination with frozen semen (Figure 20.1). After several generations of up-grading, his herd will essentially be a straightbred herd of the breed of his choice. Dairymen presently pro-

* For further discussion of this topic, see Chapter 18.

viding a low level of environment, that is, low feeding levels and below-average management, would have a chance to improve their practices gradually and in concert with the up-grading in the genetic merit of their herd. The time is already at hand in many developing areas where only the specialized dairy cow can make appreciable profit.

FURTHER READINGS

Becker, R. B. 1971. *Dairy Cattle Breeds: Origin and Development.* Gainesville: Univ. of Florida Press.

McDowell, R. E., C. V. Richardson, B. E. Mackey, and B. T. McDaniel. 1970. "Interbreed matings in dairy cattle, V: Reproductive performance." *J. Dairy Sci.,* 53, 757.

Roman, J., C. J. Wilcox, and F. C. Martin. 1970. "Milk production of tested Holsteins in Ecuador." *J. Dairy Sci.,* 53, 673.

Verley, F. A., and R. W. Touchberry. 1961. "Effects of crossbreeding on reproductive performance of dairy cattle." *J. Dairy Sci.,* 44, 2058.

Wilcox, C. J., J. A. Curl, J. Roman, A. H. Spurlock, and R. B. Becker. 1966. "Life span and livability of crossbred dairy cattle." *J. Dairy Sci.,* 49, 991.

Young, C. W., W. J. Tyler, A. E. Freeman, H. H. Voelker, L. D. McGilliard, and T. M. Ludwich. 1969. *Inbreeding Investigations with Dairy Cattle in the North Central Region of the United States.* N. C. Tech. Bull. 266 (Minn. Agr. Exp. Sta.).

Twenty-One

Breeding for Meat Production

"Variation—differences between individuals—is the raw material on which the breeder works. It is not necessary that the animals vary widely enough that the breeder can at the very start find some perfect ones to select, but merely that some of them will be closer to his ideal than others are."

J. L. Lush

Improvement through breeding is possible because animals that differ in growth and meat qualities are expected to differ in genetic makeup. Success in breeding for improved production depends on the relative importance of heredity in causing differences in expression of growth, on identifying animals that carry desired genes, and on mating selected animals to produce optimum genetic combinations in offspring. The premises for improvement through breeding are simple, but execution is more difficult.

Breeding Goals in Meat Production

Commercial vs Seedstock Production

Market animals that return a maximum profit over time are the goal of commercial meat production. This goal is concerned with rate and efficiency of production as well as quality of product, since long-term consumer demand is determined by price and desirability. From the standpoint of animal expression, individual productivity and quality are paramount.

Specialized seedstock herds, usually purebreds, provide sires used in commercial herds. Commercial herds normally produce the dams required for market-animal production and to this extent perform part of the seedstock function. The genetic content of commercial herds is determined by the succession of sires used; so seedstock producers become the primary source of genetic improvement. Superior breeding value is the goal of seedstock production in contrast to emphasis on individual performance by commercial producers. Breeding stock should be judged by the productivity of their offspring rather than their individual merit.

Future crossbreeding programs calling for specific combinations of sire and dam breeds may require specialized seedstock herds to provide dams as well as sires for use in commercial herds. Economics will dictate the ultimate pattern.

Efficient Meat Production

Efficient meat-animal production depends on costs and returns associated with (1) reproduction, (2) production from females, and (3) growth of the progeny (Dickerson, 1970). Increased rate of reproduction is important because it lowers costs of maintaining breeding females per animal marketed. Costs would be reduced by a factor of $1/n$, n being the number of animals weaned, except that there may be additional feed costs during gestation and suckling associated with carrying a larger number of progeny. Also, some increased costs to attain market size and condition may be expected from a greater number born because of smaller size at weaning. The relative advantage of changing the number of animals weaned per female each year from 1 to 2 in cattle is about 8.5 times greater than changing number weaned per year from 16 to 17 in pigs. It is obvious that increasing reproductive capability has more potential influence on efficiency in the less prolific cattle or sheep than in swine.

Increasing inherited milk or wool production of females lowers fixed and operating costs per unit of production. However, if increasing inherent production leads to increased body size, the extra feed for maintenance must be taken into account. Emphasis should be placed on increased rate of production, independent of body size. Increased milk production of females affects meat production by heavier weaning weights of offspring or reduced days to market.

Efficiency of growth is determined by the number of days and total feed required to reach market weight and value of edible meat produced. Optimum market weight for breeds or breed crosses will vary according to the relative growth rate of lean and fat. In general, the optimum will be just past the point where heavier market weight markedly reduces carcass value. Efficiency of feed conversion declines at heavier weights, but there are advantages in prorating unit costs of raising and of slaughtering over more weight marketed. As fixed costs of production and slaughter increase, we can expect an increase in optimum market weight for total industry efficiency. Changes in marketing meat, such as the sale of trimmed wholesale or retail cuts by packing plants, will lead to larger optimum market weights than are currently popular.

Efficient gains involve differences in feed required for maintenance and energy associated with creating a unit of muscle, fat, or bone. Energetic considerations of the biochemistry of metabolism suggest that within a species, the efficiency of feed conversion above maintenance for a given tissue is not likely to change as inherent production rate increases. This leads to the conclusion that some opportunity for increasing efficiency may come from decrease in maintenance requirements but primarily from differences in the type of tissue, i.e., relative amount of fat or muscle produced per unit of feed consumed.

Selection for Efficient Meat Production

Traits to Select

Meat production can be improved through breeding by selection and the mating system. Selection is the primary tool for breeders to make continued improvement in the genetic makeup of herds and flocks. Change through selection is slow, but tends to be permanent. Breeding for efficient meat production requires careful attention to selecting only for those traits of economic value and using selection opportunities wisely. The relative importance and opportunity for improvement varies among the economic traits of beef cattle, sheep, and swine, yet a similarity exists for traits to be considered. Most economic traits can be classified under (1) reproductive performance, (2) preweaning growth, (3) postweaning growth, (4) efficiency of gain, (5) carcass merit, (6) conformation, (7) longevity, and (8) disease resistance or defects in function.

An array of economic traits of cattle, sheep, and swine, and their heritabilities, is shown in Table 21.1. These heritability values indicate the relative importance of transmissible genetic variation to the total phenotypic variation. Values in the table are averages of estimates obtained in many different experimental herds.

Table 21.1.
Heritability estimates of economic traits.[a]

	Cattle	*Sheep*	*Swine*
Reproductive performance			
Number born		15	10
Number weaned		15	15
Services per conception	10		
Gestation length	20	45	20
Date of birth or interval	10	35	
Growth traits			
Birth weight	40	30	5
Gain, birth to weaning	30		
Weaning weight of individual	30	30	10
Weaning weight of litter			15
Postweaning gain, full fed	45	30	35
Postweaning gain, limited fed	30		
Mature weight	60	40	
Efficiency of gain	45		40
Conformation	40	40	30
Backfat probe			40
Carcass traits			
Carcass grade	40	15	
Marbling	40		30
Loin-eye area	50	55	55
Fat thickness	45	25	45
Per cent edible portion or lean cuts	40	50	45
Per cent fat trim or fat cuts	40		60
Tenderness, shear value	60		
Cancer eye susceptability	30		

[a] All figures are percentages. Data is from: Altman and Dittmer (1962); Craft (1958); Cundiff and Gregory (1968); Omtvedt (1968); Scott (1970).

Reproductive performance High reproductive performance is probably the most important factor affecting efficient meat production. Productive differences begin at the birth of a live animal, and reproductive capacity has important effects on average costs per animal. High reproductive rate means that a smaller fraction of the young females are needed for replacements, allowing more intense selection among replacement females and permitting a larger fraction to be marketed. Heritability of most measures of reproductive performance are low in all species. A generally accepted hypothesis is that, over the years, nature has automatically selected for good reproductive performance. More fertile animals left more offspring than less fertile ones. Sterility was self-eliminating. As a result, within any interbreeding group (species, breed, or strain), nature tended to exhaust the additive genetic variation. However, different gene combinations leading to acceptable reproduction were fixed in different breeds or species. When these different genetic groups are crossed, new genetic combinations occur which lead to improved reproduction. Some attention to selection for differences in reproductive performance seems warranted because of the economic importance of the trait. Primary improvement of reproduction through breeding is likely to be achieved by crossing breeds or strains within a breed. High production and high reproduction have a tendency to be antagonistic. Both have an optimal nutrient requirement and an orderly sequence of events must be maintained. Under conditions of stress or limitations in nutrients, genetic regulation may establish priorities for available metabolites, resulting in lowered production of milk or growth of muscle, fat deposition, or an impairment of reproduction. Efficient production and reproduction may require a different genetic constitution for optimum production under limited feeding conditions as compared to abundant feeding conditions.

In beef cattle single births are customary, with occasional twinning and very rarely triplets or quadruplets. There seems to be little chance to improve the twinning rate in cattle by selection under normal circumstances. Hormone therapy offers promise of a means to increase the number of calves born, but it brings a new set of problems such as freemartins, problems of rebreeding, and rearing of smaller calves. Solution of these problems will be the challenge for future breeders.

Reproductive rate of sheep is somewhat better than that for cattle. In most breeds single births predominate, but twins are frequent, with triplets produced occasionally. Breeds differ in the average frequency of twinning, and a few breeds such as the Finnish Landrace and the Romanov regularly have litters of three to five lambs.

Although reproductive rate in swine is much higher than in cattle or sheep, attention to reproductive performance is still an important consideration. Number of pigs born is commonly 8 to 12 per litter, but about two pigs are lost before weaning. When optimum nutrition, management, and disease control has been provided to prevent loss between farrowing and weaning, increased numbers of pigs weaned can only come from improved inherent ability. Selection aimed at specific phases, such as ovulation rate or embryo survival, might be more successful than selection on number born or weaned.

Preweaning growth Birth weight has little economic merit other than the fact that a heavier animal at birth requires less gain to reach market weight. Birth weight is the first indicator of growth rate and is useful as a starting point for measuring subsequent growth. Species differ in the relative fraction of market weight represented by birth weight. At birth, cattle weigh about 8 per cent of market weight, sheep about 10 per cent, and swine 1 to 2 per cent. The small percentage for swine indicates the immature status at birth, which accounts for the greater problem of death loss from birth to weaning than for calves or lambs. In cattle and in sheep, birth weight is positively related to growth potential of the animal, although the relationship is not high. Sire averages for birth weight serve as an early indicator of breeding value for growth. Even though birth weight is associated with growth potential and is of medium heritability, selection should not be directed toward increased birth weight because of possible increases in lambing or calving difficulty. The antagonism between growth potential and need to keep birth difficulty at a low level means a compromise must be sought. Emphasis on rapid growth will lead to an indirect increase in birth weight, but it will be less than if selection were directed at increasing birth weight directly. Selection for increasing the size of pelvic opening and for conformation which favors less difficulty at birth can help to prevent excessive losses at birth as selection for rapid growth continues.

Growth from birth to weaning depends to a large extent on the maternal environment (milk and maternal instincts) provided by the dam as well as growth potential of the individual. Breeding for increased gain from birth to weaning involves selection for increased mothering ability and growth potential. Suckling gain of pigs is of less relative importance to total market weight than for cattle or sheep. Suckling gain of pigs accounts for about 15 to 20 per cent of total gain to market weight, in cattle about 35 to 40 per cent, and in sheep 60 to 90 per cent, depending on the management system. Heritability of gain to weaning is low in swine (15 per cent), and emphasis should be placed on total litter weight rather than on individual weaning weights in evaluating the mothering ability of the sow. The heritability of weaning gain of cattle or sheep is in the medium range (30 per cent) and should respond to selection. Gain of the calf or lamb serves as an indicator of differences in mothering ability of dams and as a criterion for both genetic merit of mothering ability and growth potential of the animal itself. Although selection for increased gain during the suckling period seems important in any breeding program, the relative emphasis to place on milking ability or growth potential is not easily answered. Beef cattle, sheep, and swine have been selected primarily for meat- or wool-producing capabilities and are relatively inefficient converters of feed into milk, which must then be converted by the offspring into growth of muscle, fat, or bone. Increased attention is being given to early weaning and using the feed to raise the offspring while restricting the feed supply to the dams. An alternative to selecting for increased milking ability in commercial breeding situations is the use of high-milking breeds of cattle or sheep as a part of a crossbreeding program. Milk production can be increased immediately in this manner, in contrast to the slow change associated with selection. In situations where feed supplies are very limited, a high level of milk production may be detrimental, and a compromise may be necessary between milk production, reproduction, and growth in maximizing total returns from the enterprise.

Postweaning growth Meat animals are grown for considerable periods after weaning on pasture or in the feedlot before marketing. Rapid-gaining animals tend to make efficient gains and have a higher percentage of lean at comparable ages or weights than slower-gaining animals. Rapid growth provides the alternatives of marketing at younger ages with fewer days on feed or at heavier weights, depending on the degree of finish desired. Tenderness of meat is associated with younger ages, and rapid gains aid in marketing meat when it is more tender. The heritability of postweaning growth rate is quite high in cattle, sheep, and swine (30 to 45 per cent) and should respond readily to selection. Differences in genetic potential for postweaning gain can best be measured by providing a nutritional level and a time period long enough to permit differential ability to be expressed. Errors of weighing are a problem with ruminants because of fluctuation in fill, which for cattle can amount to differences of 30 to 60 pounds. Weighing several times and averaging the weights and shrinking animals are often used to reduce these fluctuations. As a general rule, using as wide a range in weight as permitted by the limits of weaning and marketing is to be preferred to short feeding intervals, since fluctuations in temporary environmental conditions during the growing period have an opportunity to cancel each other, and weighing errors represent a smaller fraction of the gain compared, providing a more accurate evaluation of genetic potential. Male or market females are usually fed *ad libidum* on a full-feeding program. In swine, replacement gilts are often fed until they reach about 150 to 175 pounds and then are limited fed to prevent excessive fatness. Replacement heifers are not usually fed on a high-concentrate ration but on a growing ration, one that will enable them to reach puberty and be bred at 14 to 15 months of age without becoming fat. Gains made through summer pasture season can be considered as part of the postweaning gain evaluation. Full feeding of heifers is not recommended because of its cost, but, more importantly, there is evidence that a rate of growth that promotes fattening may impair future reproduction and milking ability.

Selection for rapid postweaning growth is associated with increase in mature size. Larger mature size raises the maintenance costs of females in the breeding herd. For this reason selection procedures are needed which stress rapid growth to market weight but a plateauing of growth and early maturity after the desired market weight is attained. Methods for accomplishing this are being investigated in research herds.

Efficiency of gain Variation among animals in converting feed into edible meat has an important effect on the net income from meat production in all species. Efficient growth in meat-animal species can be described as the amount (or value) of edible product produced for a given feed intake (or cost). This measure combines differences among animals in the conversion of feed into meat with differences in feed used for maintenance or the production of nonedible products. Factors affecting comparisons of efficiency between animals are differences in weight, differences in composition of gain, and variation in environmental conditions. Animals measured at different weights but having similar ability to convert feed into meat will show different apparent efficiencies because feed required to maintain body functions varies with a power of body weight. As animals increase in age and weight, the relative growth of lean, fat, and bone changes. Muscle and bone tend to grow at a uniform or slightly decreasing rate from weaning until market weight, while rate of fat deposition tends to increase, particularly under full feeding. Since the caloric value of fat is about 1.6 greater than protein, animals which have a higher proportion of gain as protein would be expected to show more efficient gains. The importance of genetic differences in maintenance requirements among animals of the same size and composition are not clearly established. There is evidence that large differences exist among animals in their relative growth of different kinds of tissue, i.e., fat, lean, or bone. The relative variation of fat seems to be greater than muscle or bone. A study of genetic variation in edible product, fat trim, and bone from cattle carcasses adjusted to the same weight, indicated the relative variation in fat trim was four times larger than that for edible product (Cundiff *et al.*, 1969).

Various methods of measuring efficiency include feeding over a constant interval of (1) weight, (2) age, or (3) time, or (4) to a constant degree of finish. All methods have problems of interpretation. Dickerson (1970) suggests measuring growth to the same market live weight, one that is optimum for the particular breed or breed cross. In this situation, edible lean growth per unit of feed consumed takes advantage of fewer days of maintenance feed and less feed for fat deposition, as well as any advantage in metabolic efficiency of maintenance or tissue synthesis. In practice it is difficult to accurately determine differences in feed consumption and composition of gains for individual animals. It is also very expensive. Fortunately, rate of gain seems to be related genetically with efficiency of gain. Selection for rate of gain should cause an associated increase in efficiency of gain. A study of cattle indicated that selection for rate of gain would lead to 80 per cent as much genetic improvement as direct selection for efficiency (Koch *et al.*, 1963). Thus, rate of gain, adjusted for live measures of fatness, seems a satisfactory method of selecting for improved efficiency within a breed or strain.

Carcass merit Carcass merit refers to relative amounts of muscle, fat, or bone, and items related to the eating qualities of meat. Increasingly consumers demand meat that is lean with a small amount of outside fat, but that is juicy, tender, and flavorful. Consistent with these consumer demands, breeding programs should stress selection for maximum amounts of muscle, particularly in the preferred cuts of the rib, loin, and round. Because growth of one body part tends to be correlated with growth of other body parts, there is less opportunity to change proportions of preferred to less-preferred cuts than there is to change amount of muscle relative to amount of fat.

Quality in meat is evaluated by differences in (1) marbling, (2) texture, (3) color of lean, (4) firmness of lean and fat, (5) color and density of bone as a measure of maturity, and (6) the amount of force required to shear a core of meat of a given diameter. Many research studies indicate low correlations between the quality items mentioned and taste-panel evaluation of different cooked-meat samples. Some have interpreted these low correlations as meaning the indicators of quality have no value. Such a view seems extreme, since the average preferences of taste-panel members do follow the direction suggested by the quality indicators. The results may also be interpreted as showing that under present systems of feeding and marketing, our domestic meat animals produce meat that is widely acceptable or that differences in preference vary a great deal. Measures of meat quality closely related to palatability are needed and research is continuing in this area.

Heritability of most carcass traits is quite high (40 to 60 per cent), suggesting differences in average genetic merit even among animals fed and managed similarly. The problem of selection for improved carcass merit is the difficulty of evaluating these differences in the live animal. The use of backfat probe in pigs, along with selection for growth rate, offers an effective means of improving the percentage of lean cuts from pork carcasses (see Figure 12.6, p. 166). In cattle the use of ultrasonic measures of tissue differences offers a similar but less accurate appraisal of differences in fat thickness or rib-eye area. There are no accurate indicators of differences in quality items of carcass merit that can be assessed in the live animal. Tenderness of meat is not a problem in pork or in lamb. Emphasis on growth rate and marketing at young ages provides an acceptable solution to producing beef that has high market acceptability.

Progeny tests or records on sibs afford the breeder a basis of selecting for improved carcass merit. Progress from selection based on progeny or sib evaluations will be slow, and the relative emphasis to be placed on such programs must be weighed against the use of less accurate indicators of carcass merit, such as conformation or estimated fat, and the improvement associated with selection for growth rate.

Conformation Conformation concerns differences in shape of various body parts and is usually evaluated by visual appraisal. There has been much said and written on the virtues and wrongs of using conformation for selecting breeding animals. There is little debate that differences in shape occur or that these differences can be evaluated with a reasonable degree of precision. The tendency has been to establish a form or pattern that describes the "ideal" animal for most profitable production. However, the concept that animals which gain rapidly and efficiently and cut out a high per cent of quality meat also conform to one pattern is just not borne out by the evidence. Good performance comes in many sizes and shapes. Conformation serves as an indicator of development that has already taken place, but is a poor predictor of future performance.

Conformation is of value in determining differences in structural soundness, particularly of the feet and legs, but there is no quantitative measure of the degree of impairment due to structural defects, such as buck knees, post legs, or sickle hocks. Conformation items related to carcass merit include thickness of muscling (weight of muscle in relation to length of long bones) and estimates of thickness of outside fat. Where objective measures of outside fat (such as the backfat probe or ultrasonics) are available, there is little need to estimate it visually. Ultrasonic evaluation is not generally available to breeders; so visual estimation may still be the best alternative for the live animal. Long (1970) has described a cattle scoring system based on (1) predisposition to waste, (2) muscling, (3) size of frame, (4) soundness of structure, and (5) breed and sex character. Separation of the components of conformation for scoring has much to recommend it over the use of a single score.

Thickness of muscling among cattle of similar finish has some value in predicting differences in shape and percentage of steak and roast meat in cattle (Martin *et al.*, 1966), but is not likely to be of much value in predicting total lean. Based on preliminary work with cattle, Gregory (1969) has suggested that muscle score may be inversely related to differences in mature size. Muscle scores used in conjunction with rapid early growth may aid in selecting animals that have rapid growth, but do not grow to large mature size.

Longevity A long productive life is a useful attribute for breeding females. Longevity is more important in beef cattle and sheep, where the average age of breeding animals is four to five years and animals can remain productive for 10 to 15 years. There is little opportunity or need to pay much attention to longevity in swine, since boars and sows are removed from the herd after a few seasons. Longevity is affected by fertility and sterility factors, unsound feet and legs, udder unsoundness, loss of teeth, and other infirmities.

When animals remain in the herd a long time, fewer replacements are needed. This increases the number available for marketing at young ages when they will command higher prices, amortizes the cost of raising females to breeding age over more years, and permits more intense selection among those animals saved as replacements. A disadvantage of longevity is the increase in generation interval, which tends to reduce annual rate of progress from selection. This is of greater importance in seedstock herds than in commercial herds. There is little opportunity to select directly for longevity, since only a few animals achieve old age. Some selection for longevity will occur automatically as a result of the increased likelihood of selection of replacements from animals leaving more descendents in the herd.

Disease resistance and genetic defects Examples of genetic differences in disease resistance are common in plants, poultry, mice, or rats. Little is known about genetic variation in disease resistance of meat animals, but there is no reason to doubt its existence. An investigation reported by Blackwell *et al.* (1956) concerning susceptibility to cancer eye in Hereford cattle led to the conclusion that heredity did play an important role in pigmentation and other factors related to the incidence of cancer eye. However, breeding for disease resistance is difficult and slow, relying strongly on family selection procedures. The livestock industry has relied on preventive hygiene rather than genetic resistance to control the incidence of most diseases of meat animals.

Defects that have a genetic basis have an impact on productive efficiency similar to a loss in number of animals raised. Many hereditary defects are conditioned by one primary gene difference, frequently recessive. Lists and references detailing these defects can be found in genetics or animal breeding textbooks (Johansson and Rendel, 1968; Lasley, 1972; Rice *et al.*, 1967; and Srb, Owen, and Edgar, 1965). When defects occur, reactions of breeders are frequently guided more by emotion than reason. If the defect occurs in a commercial herd, replacement of the sire or sires with unrelated animals will often be sufficient. In seedstock herds, there is reason to remove females carrying the gene as well as the sire. Removing all sires and dams having defective calves, and their half-sibs, will greatly reduce the gene frequency in the herd. Known carriers or their offspring may be useful as a group on which to progeny-test suspected sires, but such groups are maintained only in very large herds.

Individual, Ancestor, Progeny, and Sib Records as a Basis for Selection

Individual performance Individual performance expressed as litter size, growth rate, or conformation, is the most commonly used basis for selection. Heritability (actually the square root of heritability) is a measure of the accuracy of individual records in indicating differences in genetic value. Individual performance is usually the main basis of selection for traits that have heritability of 25 per cent or more and can be measured on the individual. Heritability of performance traits was given in Table 21.1.

Since reproductive performance has low heritability, progress from selecting on individual performance is expected to be slow, but limiting the breeding season and saving only pregnant females and sires which have good conception records will aid selection for regular reproduction. Growth traits, including birth weight, preweaning gain, postweaning gain, and efficiency of gain, have heritability values that are high enough to make individual selection the major basis for breeding improvement within breeds for all species. A long-term selection experiment by the USDA (Hetzer and Harvey, 1967) provides evidence that individual selection based on backfat probe can make marked changes in fatness of swine (see Figure 12.6, p. 166). With the exception of fat thickness, measures of carcass merit cannot be accurately estimated in the live animal and other selection bases must be used.

Ancestor performance Selection of animals on the basis of records of ancestors is often referred to as pedigree selection. In this case a performance pedigree is required instead of the usual history of ancestral names. Performance of ancestors is a guide to the genes they were expected to transmit to the animals in question (see Chapter 18). Pedigree selection is useful for characters expressed late in life such as maternal traits, longevity, or cancer eye.

Progeny records The average records of progeny may be used as a basis of selection for sires or dams. Progeny represent a random sample of half the breeding value of a parent. Repeating the sample a sufficient number of times and averaging provides an accurate estimate of the parent's breeding value: the larger the number of progeny, the more accurate the estimate becomes. The essential features of an accurate progeny test are explained in Chapter 18. In meat animals the progeny test should be used as a supplement to individual performance on traits expressed by the individual serving to correct errors made in estimating breeding value from individual performance. Progeny and sib tests are the primary bases for selection of carcass traits.

Sib performance Sibs (brothers and sisters) are useful as a basis for selection because the average performance of sibs is a progeny test of one or both of the parents. Sib averages provide an estimate of the breeding value the parent(s) could have transmitted to the individual being evaluated. The two types of sibs are half-sibs, having a common sire or dam, and full sibs, where both parents are common. In swine, full-sib litter mates are the most frequently used sib groups. In cattle and sheep, performance measures are usually made on contemporary paternal half-sib groups. Sib records are particularly useful for providing selection information on carcass traits without increasing the generation length.

Breeding Programs

Seedstock Breeders

Numerous options are available to the seedstock breeder for developing a breeding program, including specific goals, traits, methods of selection, and the system of mating to be followed. Accurate records of the performance traits are basic to all improvement programs. Typical values and the phenotypic standard deviations for a number of economic traits of cattle, sheep, and swine are presented in Table 21.2. These values are guidelines, since many factors, including heredity, feeding, management, and climate, affect the values observed in specific situations. Variation in expression is universal for biological traits even within the same feeding and management. The standard deviation is a measure of this variation. An approximation to the upper and lower limits of normal expression can be determined by subtracting or adding three times the standard deviation to the typical or mean value. Guidelines for record of performance programs are available through agricultural extension, stockgrowers or breed associations.

Records require an accurate system of iden-

Table 21.2.
Typical values and standard deviations of economic traits of cattle, sheep and swine.

	Cattle		*Sheep*		*Swine*	
Trait	*Typical value*	*Standard deviation*	*Typical value*	*Standard deviation*	*Typical value*	*Standard deviation*
Number born	0.97		1.3		9.5	
Gestation length, days	285	6.0	147	2.0	114	2.0
Birth weight, lbs.	76	8.0	10	1.0	3	0.3
Daily gain, birth to weaning	1.80	0.20	0.60	0.06	0.70	0.07
Daily gain, postweaning, full fed	2.25	0.23	0.45	0.05	1.50	0.15
Gain per 100 lb. TDN	16	2.0	20	2.0	36	3.0
Backfat probe, inches					1.2	0.16
Per cent edible meat or lean cuts	66	2.5	66	3.2	53	2.0
Per cent fat trim or fat cuts	19	2.5	21	5.0	33	2.5

tifying each individual, its dam, and its sire. Animals should be given an equal opportunity to express their performance, and standardizing records by adjusting for conditions known to affect performance will improve comparisons. For example, correction for variation in age of animals at weaning or for differences in age of dam. Feeding and management conditions should be similar to those under which their progeny or descendants are expected to develop. Records should be obtained on all normal animals in the herd, because the herd average is an important basis for comparisons.

Direct comparison of adjusted records is appropriate for contemporary animals treated similarly within a herd. Comparisons among animals born in different seasons or years, raised under different conditions, or raised in different herds are best made by expressing records in terms of deviation from or a ratio to the contemporary herd average. Deviations or ratios express performance relative to the group average, but leave unanswered the causes for differences between averages. Central testing stations, where animals from many herds are assembled to evaluate performance under uniform conditions, offer a partial solution to evaluating animals from different herds. However, pretest levels of feeding and management have some effect on subsequent gains; therefore central tests do not remove all environmental differences between groups, but they are minimized. If pretest and test gains are considered together, they provide a reasonable measure of an animal's growth characteristics. Records have no practical value unless they result in selection decisions that differ from those that would have been made without them. Each record and trait must have a place in planned selection schemes. Multiple trait selection, using an index of a total genetic merit which considers the relative economic merit, the heritability of the trait, and the genetic correlation between traits, will maximize rate of improvement (Hazel and Lush, 1942).

As the average merit of the herd or flock increases, attention should be given to using a high percentage of sires raised within the herd. A breeder knows his own livestock best, and has the first opportunity to select the truly outstanding progeny. Genetic improvement implies the new generation is superior to the old and a rapid turnover in sires and dams is justified in seedstock herds.

Commmercial Producers

Using records of performance Maximum productivity, taking into account quality and cost of production, is the goal of commercial breeding programs. Ultimately the genetic merit of the herd is determined by the succession of sires used. As a consequence, the genetic constitution of commercial herds are determined by the choice of breeds, sires within these breeds, and whether or not crossbreeding is used.

Continued improvement in commercial herds, as in seedstock herds, depends on record of performance and selection. A simple but effective record of performance program consists of determining which breeders are using records effectively to improve their herds and purchasing sires only from such breeders, using their records to select sires. Over a long period of time, choice of breeders is likely as important as sire selection. To select replacement females in herds where individual identification is not practical, producers can weigh all heifers at replacement age and select the heavy, structurally sound heifers. This simple procedure puts selection pressure on one or both of two important traits, early breeding and growth rate. Records over a seven-year period at the Fort Robinson Beef Cattle Research Station indicated that selection of heifers for actual unadjusted weight after one grazing season following weaning would lead to 90 per cent as much selection differential as using adjusted 550-day weights (unpublished data). All degrees of complexity, from the simple program described to a complete record program as carried out by the seedstock breeders, can be undertaken. After females have been raised to breeding age, primary attention should be given to maintaining only those females that are pregnant at the end of the breeding season, with some culling of the very poorest producers.

Crossbreeding The choice of breed or breeds for crossbreeding is an opportunity producers have to influence genetic merit of their herds. Breeds differ in performance, being strong in some traits, weak in others, and near average for most. Net productivity of breeds varies far less than individual traits. However, using these differences to match breeding objectives is important. The decision to crossbreed or not depends largely on whether crossbred animals excel the best breed available. Reasons for crossbreeding are (1) to take immediate advantage of average differences among breeds for various traits rather than wait for slow change associated with selection, and (2) heterosis as measured by the performance of crossbreds in compari-

Table 21.3.
Heterosis of economic traits of cattle, sheep, and swine expressed as a percent of the parental breed average.[a]

	First cross			*Three-breed cross*		
	Cattle	*Sheep*	*Swine*	*Cattle*	*Sheep*	*Swine*
Number or per cent born	102	102	101	105	102	111
Number or per cent weaned	104	102	107	105	117	125
Weaning weight per animal	105	110	108	105	117	110
Weaning weight per female bred	109	113	—	114	138	—
Growth to market weight	104	—	114	104	—	113
Weight marketed per litter	—	—	122	—	—	141

[a] Data is from: Cundiff (1969); Hazel (1963); Sidwell *et al.* (1962, 1964); Warwick (1968).

Figure 21-1 Examples of how purebred boars are used on crossbred sows in three-breed rotation. (University of Nebraska Extension Service.)

son to the average of the parent breeds. Table 21.3 shows the performance of crossbreds expressed as a percentage of the parental average for some important economic traits of cattle, sheep, and swine. In swine, marked economic advantages have been demonstrated, and it is estimated that 80 to 90 per cent of the market hogs in the corn belt of the United States are crossbreds. Evidence was slower to accumulate in beef cattle, but several extensive experiments indicate significant advantage in net productivity justifying the increased use of crossbreeding. In sheep the practice of crossbreeding has been used for many years in market-lamb production by crossing mutton breeds on the "hill" sheep in the British Isles or the use of Down breeds on western wool-type sheep in the United States. Crossbreeding among the mutton breeds has received less attention, but the suggestions outlined by the Sheep Industry Development, Inc. (Scott, 1970) call for increased attention to systematic crossbreeding in the future.

Crossbreeding of swine usually follows a two-, three-, or four-breed rotational scheme. A two-breed rotation uses boars of two breeds in alternate generations. Crossbred gilts are saved and mated to boars of the alternate breed to form the next generation. A three-breed rotation system is the most commonly followed program. Three-breed rotations result in slightly more hybrid vigor than the two-breed rotation but require the location of top performing boars of a third breed. Crossbred gilts and sows are mated successively to boars that are least represented by percentage of blood in the females (Figure 21.1).

Two-breed or three-breed rotations may be used in beef cattle also. The length of generation in cattle complicates programs by causing an overlapping of generations. The system requires at least as many breeding pastures as the number of breeds in the rotation. If artificial insemination is used, there is no added problem of extra pastures, but semen of the various breeds must be available. Rotational crossbreeding takes advantage of 67 per cent of the maximum individual and maternal heterosis in two-breed systems, and 87 per cent of maximum in three-breed crosses. Rotational crossing leads to the average merit of the breeds involved for the traits not exhibiting heterosis.

Specific three-breed crosses in which crossbred females are mated to sires of a third breed offer a means of using maximum individual and maternal heterosis and the opportunity to take advantage of breed characteristics which are more favorably used through dams or through sires. This matching of maternal and sire breeds has been termed complimentarity (Cartwright, 1970). Traits desired in the dam breeds include small mature size, early maturity, adequate milking ability, freedom from calving difficulty, high fertility, adaptability, and carcass quality. Sire breeds emphasize rapid growth rate, feed efficiency, and maximum lean that is palatable and tender. The use of F_1 cows or crosses of dam breeds requires a special program of crossing the parental dam breeds and selling all heifers to commercial producers who will mate them to bulls of the third breed. All offspring from specific three-way crosses are marketed. Future advances in sex control could have an important effect on the feasibility of extensive use of this system.

Western ewes of the Rambouillet, Merino, Columbia, Corriedale, or Targhee crossed with mutton breeds have long been the source of F_1 crossbred ewes sold into the farm areas for further crossbreeding to terminal sire breeds producing market lambs. The availability of western ewes cannot be taken for granted in the future. Breed combinations for market-lamb production will likely stress polyseasonal breeding as exemplified by the Rambouillet, Merino, and the Dorset, and litters as typified by the Finnish Landrace or Romanov. Traits desired in sire or in dam breeds of sheep are similar to those discussed for cattle.

FURTHER READINGS

Cartwright, T. C. 1970. "Selection criteria for beef cattle for the future." *J. Animal Sci.*, 30:706–711.

Craft, W. A. 1958. "Fifty years of progress in swine breeding." *J. Animal Sci.*, 17:960–980.

Cundiff, L. V., and K. E. Gregory. 1968. "Improvement of beef cattle through breeding methods." *No. Central Regional Publ. 120* (Neb. Res. Bul. 196).

Hazel, L. N., and J. L. Lush. 1942. "The efficiency of three methods of selection." *J. Hered.*, 33:393–399.

Johansson, I., and J. Rendel. 1968. *Genetics and Animal Breeding.* San Francisco, Ca.: W. H. Freeman and Co.

Koch, R. M., *et al.* 1963. "Efficiency of feed use in beef cattle." *J. Animal Sci.*, 22:486–494.

Lasley, John F. 1972. *Genetics of Livestock Improvement.* 2d ed. Englewood Cliffs, N.J.: Prentice-Hall.

Lush, Jay L. 1945. *Animal Breeding Plans.* 3d ed. Ames: Iowa State College Press.

Petty, R. R., and T. C. Cartwright. 1966. "A summary of genetic and environmental statistics for growth and conformation traits of young beef cattle." *Dept. Anim. Sci. Tech. Rep. No. 5.* Texas A & M University.

Rice, V. A., F. N. Andrews, E. J. Warwick, and J. E. Legates. 1967. *Breeding and Improvement of Farm Animals.* New York: McGraw-Hill.

Sidwell, G. M., *et al.* 1962. "Fertility, prolificacy, and lamb livability of some pure breeds and their crosses." *J. Animal Sci.*, 21:875–879.

Warwick, E. J. 1958. "Fifty years of progress in breeding beef cattle." *J. Animal Sci.*, 17:922–943.
Warwick, E. J. 1968. "Crossbreeding and linecrossing beef cattle experimental results." *World Review of Animal Prod.*, 4 (no. 19–20):37.

Twenty-Two

Improvement of Animal Fibers Through Breeding

"An old black ram is tupping your white ewe."

Shakespeare,
Measure for measure

Man has for centuries used animal skins and fibers to provide comfort. Originally, sheep pelts were used for clothing and shelter. The prehistoric time is unknown when man changed from animal pelts to wool pelts and then to spun wool garments. An ancient folk tale of the first spinning of yarn tells of a nomad shepherd lad who twisted together stray lengths of wool fiber to bind a bundle of firewood (Leggett, 1947). Since the discovery of spinning, wool and other animal fibers have been used in innumerable fabrics for a multitude of purposes, such as apparel for comfort or style, blankets, mattresses, rugs, and carpets. The versatility of wool for providing comfort regardless of temperature is related to its ability to take up and give off moisture. Wool fabrics entrap air to provide warmth at low weight. Wool is resistant to burning. Its natural resilience in both wet and dry states leads to wrinkle resistance. The many chemically active constituents of the fibers give wool the qualities needed to react to finishing treatments and to be dyed many colors. Wool is durable and resistant to soiling. Its unique felting ability extends its many uses. Wool is being used to advantage in blends with less expensive fibers where it imparts its valuable physical qualities to the resulting fabrics.

In general, the valuable characteristics of wool are also found in other animal fibers, although each has special uses. The long mohair fibers from the Angora goat bring a softness and luster to fabrics which permits unusual decorative effects. Cashmere from the fine, soft undercoat of the cashmere goat is renowned as the softest animal fiber. It is used in high-quality knitwear and coats. Common goat hair is often used in inexpensive felts and carpets.

Animals of the ungulate family, *Camelidae,* are important in the production of specialty hair fibers. One of the best known is camel hair, which is used in overcoats, knitwear, rugs and industrial fabrics (von Bergen, 1963). Fleeces from llama (Figure 22.1), alpaca, vicuna, guanaco, and related animals from South America have a variety of textile uses. Hand spinning, knitting and weaving for

Figure 22-1 A llama. (Courtesy, National Geographic Society.)

clothing, blankets, rugs, ponchos, carpets, bags, ropes, etc., are among their many uses in the traditional handcraft and textile industry (Villarroel, 1966).

Wool and specialty animal fibers in the early 1970's are facing decreased demand and increasing competition from man-made fibers. World production of wool is still rising but tending to level off. It has exceeded consumption for several years. Both wool and mohair production are declining in the United States. Mohair fluctuates widely in price because demand varies with fashion, but both wool and mohair reached lower price levels in the early 1970's than for many years. There is a trend to give more attention to other uses, especially meat production, for all fiber-producing animals. Animal fibers will probably continue to be fully used because they still have favorable properties in comparison to other fibers, they are widely used in blends with synthetic fibers, and their low price levels will be more competitive with other fibers. United States and some other government-support programs for wool and mohair also provide a stabilizing factor. The species and breeds of animals producing animal fibers are discussed in Chapter 13.

Improvement of Wool Through Breeding

Sheep are undoubtedly much more efficient producers of wool now than when they were first domesticated some ten to twelve thousand years ago. In fact, much of the progress has been made in the last few hundred years. In 1810, the average wool clip in the U.S. was reported as 2 lbs. per head (Connor, 1918). USDA statistics show that this had increased to 7 lbs. near the beginning of this century, and now approaches 9 lbs. In Australia, fleece weights have increased from about 4 lbs. for Macarthur's sheep in 1821 to almost 11 lbs. in recent years (Belschner, 1965). Gains in fleece weight have been accompanied by improvement in staple length, density, uniformity of fiber diameter, and clean yield. Much opportunity for further improvement remains if systematic and effective selection practices are applied.

Improvement of wool cannot be separated from needs for gains in other traits which affect total production, such as reproductive rate and related attributes. Furthermore, since efficient sheep production often depends on more than one product, improvement plans must take into account all of the traits which are related to net income (Scott, 1970).

Selection is the most important way in which the sheep breeder can improve his products and efficiency of production. It has been repeatedly demonstrated that effective selection methods will lead to permanent gains, not only in quantity but also in quality of lamb and wool produced. The amount of selection which can be practiced for any one trait is limited. Therefore, it is important that traits for which the greatest progress can be made and that are most valuable are emphasized in selection.

Estimates of heritability, which give the proportion of gain made in selection of parents which is passed on to the offspring, are useful in determining the relative progress that can be made in selection to improve various traits. Thus, emphasis can be given to the traits with which the most progress can be made. In sheep, estimates of heritability have usually been obtained from relationships among relatives. Estimates are available for a large number of traits, but many are based on relatively small numbers under varying conditions, and, therefore, show considerable variability. Nevertheless, rough groupings can be made according to their relative heritability (Terrill, 1958; Terrill, 1962; Turner and Young, 1969). These are presented in Table 22.1. In general, average heritability estimates over 40 per cent have been classified as high, those from 20 to 40 per cent as moderate and those under 20 per cent as low. Most wool traits are moderately or highly heritable.

Traits to Consider in Selection

Traits important to income must be emphasized if selection is to be effective in producing more profitable sheep. Fleece weight is the most important wool trait. Clean fleece weight is preferable, but grease fleece weight is usually closely related to clean fleece weight. Neale's device (Neale *et al.*, 1958) for estimating clean fleece weight by measuring the volume of the fleece under constant pressure provides a quick, easy method for identifying high-producing sheep. Improvements in staple length may be desirable if the staple

Table 22.1.
Relative estimates of heritability of traits in sheep.

High	*Moderate to high*	*Moderate*	*Low to moderate*	*Low*
Face covering	Grease fleece weight	Fur traits Birth weight	Weaning weight	Multiple births
Staple length	Clean fleece weight	Index of over-all merit	Milk production	
Birth coat	Clean wool yield	Average daily gain		Lambs weaned
Crimps per inch	Fiber diameter			Conformation
Skin folds	Fiber density Body weight	Carcass traits		Fatness

Figure 22-2 Measuring staple length on mature Rambouillet ram. (Courtesy, USDA.)

is too short for the grade. Increased staple length also leads to increased clean fleece weight (Figure 22.2). Gains in fleece density and uniformity of length and fineness may also be of economic advantage. Fleece quality will no doubt deserve more emphasis by producers when prices paid for wool are more commensurate with quality.

Traits which are related to over-all production should receive greatest emphasis in selection. Number of lambs produced per ewe per year is of greatest importance because of its high relationship to net income, especially where lamb meat accounts for the majority of income. Ewes with open faces produce more lambs and more pounds of lamb per ewe (with only slight loss in fleece weight) than ewes with covered faces (Terrill, 1949). Larger ewes will produce more and heavier lambs but probably at proportionately higher feed costs. Increased twinning will generally pay. Ewes having twins (Figure 22.3) can be expected to wean an average of about 40 lb. of lamb per ewe-year more than ewes of the same age having singles (Sidwell, 1956). Breed differences in twinning can be quite marked. Some breeds have unusually high lambing rates; an example is the Finnsheep, with an average of about 2.5 lambs per birth. Performance recording of sheep was used first in Finland as early as 1918 (Owen, 1971). This may help explain in part the present high

Figure 22-3 Morlam ewe with two sets of twins born 8 months apart at Beltsville. (Courtesy, USDA.)

prolificacy of the Finnsheep. Twinning has low heritability, but small improvements may be worthwhile. Selection for twinning has been quite effective. To select for twin production, one should favor young ewes that have had twins and rams born as twins from young mothers. Twinning is less frequent at first lambing than in older animals; thus mothers having twins at an early age are more likely to transmit this trait.

Relationships among wool traits are generally favorable, in that improvement in one trait leads to improvement in another. There are some exceptions. Selection for finer fiber may reduce fleece weight. In crossbred types, selection for longer stapler may increase fiber diameter. The Drysdale breed of New Zealand (Figure 22.4) with the *N*-type gene discovered by Professor Dry produces a high-quality carpet wool. The *N*-type gene causes a hairy birth coat and a coarser fleece in the adult animal. Skin folds are negatively related to clean fleece weight, staple length, and lamb production. Body weight tends to be positively related phenotypically to wool production, but not genetically (Turner and Young, 1969). High lamb and wool production are positively related genetically, but high lamb production may tend to depress wool production.

There are many somewhat simply inherited traits which must be considered in sheep improvement. These have been reviewed by Rae (1956) and Terrill (1958). Color other than white is generally undesirable. Lethals and sublethals, which usually result in early death, are generally inherited as simple recessives, and include muscle contracture, earlessness, cleft palate, paralyses, rigid fetlocks, dwarfism, nervous incoordination, congenital photosensitivity, and blindness.

Cryptorchidism, or failure of one or both testicles to descend, appears to be due to recessive genes, and is sometimes linked with the polled trait. Entropion, turned-in eyelids, is inherited but not as a simple recessive. Wattles, which appear in some sheep and goats, appear to be inherited as a dominant. Jaw inequalities, particularly a short lower jaw, appear to be due to several pairs of genes.

Selection against recessive defects is most effective if relatives of the affected animal which may carry the recessive gene are culled. This should generally include the parents, sibs, and other relatives with at least 50 per cent probability of carrying the recessive gene.

Any selection program should include the culling of unsound and unthrifty as well as low-producing animals. Since unthrifty animals would not leave many offspring, their elimination may not greatly increase genetic progress. However, the remaining animals will give higher average production and there will be less risk of transmitting defects.

Figure 22-4 Drysdale Ram. (Courtesy, Bob Barton, New Zealand.)

Inheritance of horns in the Rambouillet and Dorset breeds is of interest, although horns, especially scurs, may also appear in polled breeds. This is true of breeds with Rambouillet blood, such as the Columbia and Targhee. Horns are apparently recessive to the polled trait in both Rambouillets and Dorsets, although Rambouillet ewes homozygous for the horn gene have horn knobs and those carrying the polled gene have depressions in the skull in place of the knobs. The inheritance of scurs has not been well-defined, but scur or horn growth appears to be greater in rams heterozygous for horns than in heterozygous polled rams.

Studies of blood antigens and other blood traits have laid a foundation for identifying parentage and for checking on changes in homozygosity for traits that are not evident to practical breeders. Only slight, if any, relationships have been shown between blood types and productivity of sheep.

Chromosome number seems established as 54 for the diploid number for sheep (Borland, 1964) and 60 for goats (Makino, 1951).

Improvement of Angora Goats

The same practices which are used to improve sheep through selection will also apply to Angora goats (Figure 22.5). Gray (1959) has outlined many important considerations in selecting for mohair and kid production. Most of the income from Angora goats comes from the sale of the mohair; so weight of fleece is the most important trait to emphasize in selection. Individual fleece weights, at least those for males, should be recorded by breeders. Length of staple contributes to fleece weight and thus is important. Goats with short staple, with fleeces that are light or kempy, or with colored fibers should be discriminated against. The body and belly should be well and densely covered. The web lock of intermediate type of fleece holds up well in production and is often preferred by commercial producers. Angora goats have more of a tendency to shed than do sheep, and animals with this tendency should be culled.

Heritability estimates (Shelton and Bassett, 1970) are high for staple length, body weight, neck cover, and clean fleece weight, moderate for face cover, moderate to low for clean fleece weight, and low for fiber diameter and belly cover.

Size, vigor, and fertility are important to efficient production and should be emphasized in selection for both sexes. Open faces should be stressed in selection because they are related to the production of both kids and mohair. Does with open faces are larger, produce heavier fleeces, and drop a much higher proportion of kids than does with covered faces (Shelton, 1969). Meat conformation should receive some attention in selection, because many Angora goats, especially wethers, are sold for meat.

Improvement of Other Animal Fibers

The principles for improvement of fibers from any animal are much the same as for sheep and Angora goats. Obviously considerable selection in the past has resulted in the development of special fibers, such as those produced by the Angora rabbit. Other animals like the camel, cashmere goat, and common goat have probably developed their present fiber coats largely through natural selection. Breeding and selection techniques could undoubtedly be used to improve the yield of downy undercoat of the cashmere goat and to eliminate the troublesome coarse outer-coat fibers. Similar improvements could likely be made in other specialty fibers. However, the by-product nature of many of these fibers, fluctuating demands of fashion, and the increasing use of man-made fibers do not support the expense of improving the fiber production of some of the animals involved. Some selection for improvement may be justified for other traits, such as in the reproductive rate. Other im-

Figure 22-5 Angora buck. (Courtesy, *Sheep and Goat Raiser.*)

provements may involve very little cost, such as the selection of sires with white fibers or those with absence of fiber defects.

Breeding for Improved Fiber Production

Breeding practices for improving fiber production as well as for over-all merit have been developed for application to sheep, but they apply equally well to other fiber-producing animals.

The identification of individual animals and the accurate measurement of important traits is fundamental. Objective and direct measures of the desired traits should be used if available. Measures taken early in life before any selection is practiced are most effective, unless the trait may be measured less accurately or is lowly repeatable at the early age. Repeated observations may increase accuracy of measurement and repeatability, and may also be helpful in selecting for lowly heritable traits.

Adjustments for environmental effects such as age, type of birth, and age of dam will usually increase effectiveness of selection, since these may tend to confuse nongenetic with hereditary effects. Single lambs may weigh 8 to 10 lbs. more at weaning than twins. Lambs from mature dams excel those from young dams by 6 to 8 lbs. When lambs are gaining over 0.5 lb. per day at weaning, a few days difference in age can make an appreciable difference in weight. If single lambs are favored in selection programs because they are bigger, it is probable that no progress will be made in improving weaning weight because twins, even though smaller, might still be better genetically than single lambs.

Age of the dam has important effects on body weight, fleece weight, staple length, fineness, and pounds of lamb raised per ewe. Thus, when comparing dams of different ages, one should consider that body weights, fleece weights, and lamb production increase until the animal is three to five years of age or older. Generally, staple length decreases and fleeces become coarser with age. Effects of age on phenotype are fully as important for sires as dams. Lack of adjustment for age will favor selection of three- to five-year-old rams over those one and two years of age. Younger rams of the next generation will generally be genetically superior if improvement is being made.

If these nongenetic effects are not taken into account, the selected sheep may owe their advantages to favorable environmental factors and thus no genetic improvement will be made. Years may make quite a difference. There are small genetic differences from year to year in sheep because only some of the parents are changed each year, but there may be large environmental yearly changes. Clean fleece weight can vary as much as 1.5 to 2 lbs. in different years, simply because feed or other environmental conditions are better in one year than another.

One way to minimize these environmental effects is to select within groups of the same age or within singles or twins by saving the same proportion of each. Statistical adjustments are more precise and can easily be applied to an index.

Generations should be turned rapidly for greatest gain, as genetic progress is made only from one generation to the next. Length of generation is defined as the average age of the parents when the offspring are born, and is about four years in sheep. Ram generation length can be reduced to 2 years by using only the best yearling rams each year for breeding or to one year by using only the best ram lambs each year. Reducing generation length by one-half or one-third would double or triple the genetic gain per year from selection of sires. If improvement is being made and a reasonably large number of offspring are produced, the best son should be better than his sire. The more quickly the change is made, the more rapidly one may take advantage of this gain. Thus, in many circumstances, more rapid progress can be made by selection solely on the basis of phenotype than by the slower progeny-test procedure. The breeder often tends to make repeated use of an outstanding sire when a quick change to his best son might give greater improvement.

Progeny Testing

The sheep or goat breeder must often choose between the frequent selection of outstanding young males on their own merit and the slower procedure of selecting sires on the performance of their progeny. Progeny testing is most effective in selecting for traits that are low in heritability such as conformation or fattening ability. It is usually unnecessary in selecting for highly heritable traits, such as face covering or staple length, where the offspring simply confirm what has already been observed in the parents.

Progeny testing is usually desirable when selecting among sires whose own records were made in different herds or farms, since adjustment of individual records for environmental herd differences is difficult. Selection for crossing ability between lines or breeds necessitates the use of progeny tests. Similarly, progeny testing is important where a trait can be observed in only one sex, such as lamb production, or where it can best be measured in a group of offspring, such as twinning or lamb mortality. Progeny testing may be most important in selecting for carcass traits, which cannot be adequately measured in the live animal. Artificial insemination is safer with progeny-tested sires, since the greater accuracy in selection of individual sires is more justified where sires are to be used so widely. The use of inferior rams to sire large numbers of offspring, as from artificial insemination, can be very costly.

Records of other relatives, such as parents or sibs, may sometimes be used advantageously. This is particularly true for the dam's production, multiple births, or time of lambing. Half or full sibs may be used for carcass traits or fertility. Caution must be used in paying much, if any, attention to relatives when the trait can be directly and accurately measured in the individual under selection. Progress is generally slower if selection must be based on the records of relatives than if the trait can be measured directly in the animals to be selected.

Maximizing Selection Differentials

The aspect of selection which can be most influenced by the breeder is the selection differential (Figure 22.6). This represents the difference in a trait, or a group of traits, between the herd average and the average of the selected animals which produce the next generation. Any progress made is always the heritability fraction of the selection differential. Selection is aimed at obtaining as large a selection differential as possible for important traits. A measure of the effectiveness of a selection index is the size of the selection differentials for the respective traits resulting from its use, as compared with those of alternative methods of selection. In comparing selection differentials in terms of pounds of wool or inches of staple, it may be desirable to convert the selection differential to some standard unit of measure, such as a percentage of the mean or a multiple of the standard deviation.

Selection of superior animals usually involves the use of an index of some kind, because one must combine the values of various important traits simultaneously, or balance the strong points against the weak points in order to rank the animals from best to poorest. This is always true of fiber production because more than one quality trait must generally be considered, as well as reproduction and usually meat production traits. A calculated index which combines measures will, generally, be more accurate and effective than any mental combining of values by a breeder ranking animals.

The use of an index, as shown by Hazel and Lush (1942), is a more effective way of selecting for several traits at the same time than the use of independent culling levels or tandem selection for one trait at a time. An index permits a constant and objective degree of emphasis on each trait considered in selection. Without an index, ideals or emphasis on different traits are more likely to shift from year to year. Also, one is likely to overemphasize more obvious traits, such as body type or color, and to underemphasize traits of greater economic importance, such as growth rate or multiple births. Lush (1945) has presented an objective basis for determining emphasis on each of several traits in an index by giving weight to these characters in proportion to their heritability times economic importance. Environmental adjustments and attention to records of relatives can be conveniently included in the index.

The amount of selection that can be practiced is limited by the size of the flock. For example, at least three rams usually need to be used within a closed flock to avoid inbreeding. As each ram can be mated to 50 to 75 ewes, the total flock size must be in the range of 150 to 225 ewes. A smaller flock with a minimum of three rams will cover the selection differential for rams. Selection takes advantage of genetic variation, and the larger

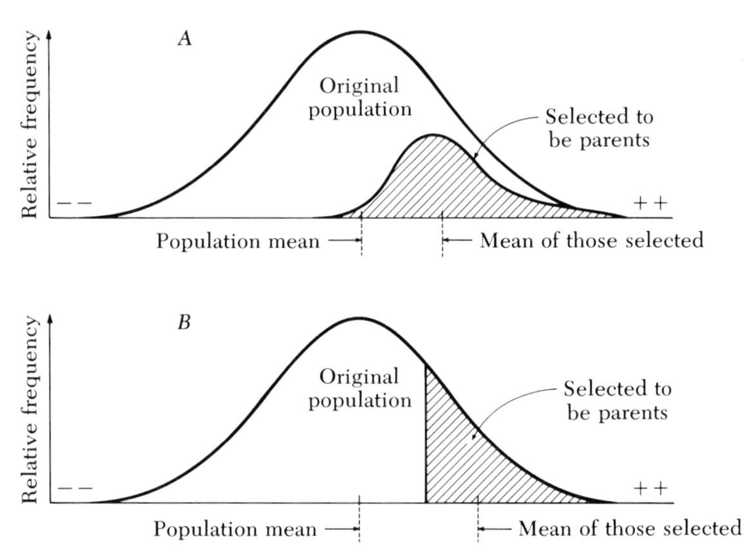

Figure 22-6 Two ways in which the merits of those chosen to be parents by rather intense selection might be distributed with respect to the merit of the original population from which they were taken. The better individuals are to the right, the poorer to the left. A indicates the usual kind of selection, in which at least a few mistakes are made and in which some attention must be paid to characteristics other than the one for which merit is indicated here. B is the most extreme form of selection conceivable. No mistakes are made, and selection is entirely for the characteristic for which degrees of merit are indicated along the horizontal scale. (Reprinted by permission from Animal Breeding Plans, 3rd Edition, by Dr. Jay L. Lush, © 1945 by The Iowa State University Press, Ames, Iowa.)

the breeding group, the larger will be the total range of variation. Furthermore, in large flocks the best rams can be mated to the best ewes to produce rams for the entire flock. However, this super flock should be large enough that considerable selection can be practiced among the males produced.

Higher reproductive rates permit greater selection differentials, because the number of replacement animals needed remains constant. Feeding and management practices which increase lambing rate are aids to greater gains from selection.

It pays to select for as few traits as possible. Opportunity for selection is limited. Attention to unimportant traits, such as color on the legs or shape of the head, can reduce attention to valuable traits, such as pounds of fiber or meat produced. The decrease in progress for any one trait, as the number of traits selected for increases, is equal to one divided by the square root of the number of traits involved (Lush, 1945). If a breeder selects for four traits he makes only one-half as much progress in each one as he would if he selected for one trait only. It is not profitable to select for fancy points or other things for which the lamb and wool producer is not paid.

Aids to Selection

It is also important to select under the feed and environmental conditions under which sheep are going to produce. Unfortunately, conditions under which purebred sires are generally raised are usually better than the range or farm conditions under which their offspring will be raised. Thus they may be selected for attributes which are not desirable in the ordinary environments in which most market lambs and wool are produced.

Maximum selection pressure should be applied to both males and females. However, selection of males is of greater importance, because the number of progeny from a single male may be so much greater. About 80 to 90 per cent of the gains that can be made in a flock closed to outside breeding comes from the selection of rams. One ram can be mated to 50 to 75 ewes in a limited breeding period, thereby requiring the use of only 3 to 4 per cent of the rams born as sires for the next generation. More than half the ewes produced must be retained to become mothers of the next generation. Thus, much larger selection differentials can be obtained for rams than for ewes, and hereditary or permanent gains from selection will come largely from the choice of rams to become sires.

Selection differentials for sires can be increased further by the use of artificial insemination. A record of 17,000 ewes artificially inseminated by semen from one ram during a 115-day period has been reported from the Soviet Union. Artificial insemination of sheep and goats is advantageous chiefly because it permits more intense selection of outstanding sires (Terrill, 1968). Justification is difficult on any other basis. Any advantage of the breeding value of a sire can be extended to a much greater number of offspring by artificial insemination—but so can any of his disadvantages. Rams that have not been thoroughly tested may spread undesirable traits far more widely under this method, especially recessive defects, which may not be revealed under ordinary use.

Performance Recording

Performance testing offers a means of increasing attention to selection. Ram tests, such as those conducted in Texas and Iowa, permit the breeder to compare rams under standard conditions and especially to compare his own rams with those of other breeders. Such tests of the breeder's own rams alone may be done by on-the-farm tests, as is done in Colorado, Ohio, Wisconsin, and other states. Recommendations for uniform sheep-performance records have been presented in the National Extension Sheep Committee Report of 1968 (USDA, 1968c). This program gives primary emphasis to number of lambs born and weaned per ewe exposed to service and to high growth rate of lambs. Wool production and other records are provided for as needed.

Performance recording in sheep over the world has been reviewed by Owen (1971). He lists the benefits of performance recording in selecting rams, in selling breeding stock, in culling ewes, and in improving management. Recording schemes for 16 countries show considerable similarity. Costs of recording programs are paid at least in part from public funds in most cases, and in some cases largely or entirely from government of industry funds. This is reasonable, since the benefits from improvement resulting from performance recording will generally come to the industry in the future and often not to the individual breeder. These benefits will often provide products at lower cost to the public.

Rate of Progress from Selection

Progress from selection for improvement of sheep and goats may not be as rapid as it is for improvement of animals that reproduce at a faster rate, such as poultry or swine. However, improvements can be made, especially if factors which influence the rate of progress are optimum. Selection ideals must remain constant for a period of years for selection to be effective. The exhibition of sheep at fairs and shows has not only stimulated interest in sheep breeding, but also served to set breed ideals and standards. Unfortunately, the show ring has sometimes led to emphasis on traits that are not economically important, such as head shape, scurs, or complete face covering (rather than lack of face covering) or on traits of low heritability, such as conformation or fatness. Greater emphasis on production traits that cannot be assessed in the show ring is needed.

Better measures are needed for many traits in order for them to receive deserved attention in selection. The rapidity and economy of taking the measure is important. Attention is being given to development of better measures for surface area, body form, carcass traits, and wool quality.

Rate of progress expected from selection on an annual basis can be estimated by multiplying heritability times the selection differential and then dividing by the generation length. Because the last two generally differ for sires and dams, it is convenient to add the selection differentials and then multiply by heritability. The result must be divided by the sum of the generation lengths for sires and dams to obtain the expected progress per year. For example, if selection differentials for fleece weight were 2 pounds for rams and 0.4 pound for ewes, with generation lengths of 2 and 4 years, respectively, and if heritability is estimated at 40 per cent, then $[(2.0 + 0.4) \times 4] \div 6 = 0.16$ pound of gain per year. These estimates cannot be very precise if several related traits are considered in selection, because exact and strictly applicable estimates of heritabilities and phenotypic and genetic correlations among traits are generally not available. Even generation length may be difficult to calculate when it differs among traits. However, such estimates may still be very useful as guides to the best selection procedure to follow in a given situation.

Actual progress from selection is usually determined by trends of values for each trait over a period of years. This method is often unsatisfactory, because genetic trends are often confounded with environmental trends. The use of unselected control groups or random repeat matings are thought to be essential for selection experiments to obtain estimates of annual environmental changes. Even such control groups may be subject to bias by natural selection and by genetic drift.

Crossbreeding

The gain in production or hybrid vigor from crossing breeds is an important advantage in commercial production of meat and wool. Sheep producers have long recognized that crossbreeding results in progeny that are more fertile and have higher growth rates than the average of the purebred parents. The crossing of a third breed on two-breed-cross ewes or a fourth breed on three-breed-cross ewes has given added gains (Sidwell *et al.*, 1964). Added advantages of the two-, three-, and four-breed crosses were 13, 23, and 13 per cent, respectively, with a total gain in the final cross of more than 64 per cent over the original purebred parents. Furthermore, improved meat-type offspring of wool-type ewes, such as the Rambouillet or Merino, are obtained by crossing with meat breeds, thereby increasing the efficiency of production of both lamb and wool. Heterosis from breed crosses is also shown in wool traits.

Interbreeding of crossbred animals has not resulted in a marked increase in variability, which is often cited as a disadvantage of crossbreeding (Rae, 1956). Offspring of breed crosses may be as uniform as, or even more so than, purebreds. The coefficient of variability of fiber diameter within fleeces decreased with crossbreeding in a majority of crosses (Sidwell *et al.*, 1971).

Use of crossbreeding in commercial production does not reduce the demand for high-quality purebred sires. In fact, improvement in purebreds is passed on to their crossbred offspring (Galal *et al.*, 1969). Further improvement in productivity of crossbreds must come from selection of the purebred parents.

Crossbreeding, particularly to produce the wool-type highly fertile mother of the market lamb, has not been as fully exploited in the United States as has the use of the meat-breed sire on the wool-type ewe. Experimentation is needed to compare breed sequence in crossbreeding to bring about maximum efficiency in production of high-quality lambs and wool.

FURTHER READINGS

Belschner, H. G. 1965. *Sheep Management and Diseases.* 8th, rev. ed. Sydney, Australia: Angus and Robertson.

Connor, J. J. 1918. *A Brief History of the Sheep Industry in the United States.* Annual Report of the American Historical Association.

Gray, James A. 1959. "Selecting angora goats for increased mohair and kid production." *Texas Agr. Expt. Sta. MP-385.*

Leggett, William F. 1947. *The Story of Wool.* Brooklyn, N.Y.: Chemical Publishing.

USDA. 1968c. *Recommendations for Uniform Sheep Selection Programs.* Washington, D.C.: USDA.

Terrill, C. E. 1958. "Fifty years of progress in sheep breeding." *J. Animal Sci.,* 17, 944.

Turner, H. N., and S. S. Y. Young. 1969. *Quantitative Genetics in Sheep Breeding.* Ithaca, N.Y.: Cornell University Press.

Villarroel, L. J. 1966. "The production and industry of alpaca, llama and vicuna in Peru." *Proceedings of International Symposium on Technical Economic Problems of the Production of Sheep, Goats, Auchenids, and Fur Animals,* Milan, Italy.

Twenty-Three

Breeding for Eggs and Poultry Meats

"It is the possession of certain characters by the parent that determine the development of like characters in the offspring."

Aristotle

The Industry

Perhaps more has been accomplished from the scientific application of genetics to poultry than from its application to other classes of livestock. To develop high-performing genetic stocks, breeders have made applications of Mendelism, inbreeding, hybridization, the theory of quantitative inheritance, and even blood-typing techniques. Although the U.S. is the largest supplier of breeding stock for worldwide commercial use, a growing supply of scientifically bred strains is being developed in other countries. The typical commercial poultry breeding establishment, be it located in the United States, Canada, Great Britain, Japan, or Australia, is directed by a person with a knowledge of methods of recordkeeping, high-speed computers, and the application of statistics for the evaluation of genetic stocks and progress from selection.

The commercial breeding of any economically important species, including poultry, is directed toward three broad objectives: (1) increased product output per animal; (2) increased efficiency of production; and (3) improved quality of an existing product. In poultry the improvement of fertility, hatchability, growth rate, body conformation, egg yield, meat yield, feed conversion, egg quality, meat quality, and viability are all facets of these three objectives. Interest in breeding for type, breed characteristics, and freedom from defects is secondary, because breeders realize that undue emphasis on characteristics of questionable economic value slows down the possible rate of genetic improvement in the more important economic traits.

The poultry industry includes three main segments, which deal separately with the production of eggs, broilers, and turkeys. Breeders of egg-type chickens are concerned with the development of more efficient layers in terms of feed and chick costs. To this end they direct their improvement toward higher egg production, earlier maturity of layers, smaller body size, greater disease resistance, and higher-quality market eggs of acceptable size.

In general, the problem of breeding for meat production is more complex than breeding for egg production, because the meat breeder must be concerned not only with growth rate, feed economy, viability, and meat quality of the commercial product, but also with reproductive performance: egg production, hatchability, fertility, and feed economy of the parent strains.

The Foundation Breeder and the Hatcheryman

The breeding sector of the poultry industry includes the foundation breeder, concerned with genetic improvement, and the hatcheryman, who multiplies the stock supplied by the foundation breeder. The producer, or farmer, buys his chicks or poults from the hatcheryman, and then grows them out as commercial egg layers, broilers, or market turkeys.

The foundation breeder and the hatcheryman are typically associated in a franchise arrangement. The breeder provides the hatcheryman with breeding stock for his hatchery supply flocks. Such supply flocks are called parent flocks, because they produce the commercial chicks the hatcheryman sells. The males going into the parent-flock matings are usually of different breeding than the females, so that the commercial chicks produced are a cross. These may be strain-crosses, crossbreds, or inbred hybrids.

The breeder labels the chicks he produces for sale with his copyrighted trade name to prevent infringement by competitors. Since the pure line stock is never sold, the foundation breeder maintains complete control of his breeding stock.

Exclusive franchises are less widely used today in the poultry breeding and hatching business than a few years back. Many hatcherymen sell stock from more than one breeder. In Iowa, 64 of the 72 hatchery members of the Iowa Hatchery Association in 1971 operated under a franchise arrangement. Most, however, sold more than one brand of chicks.

Figure 23-1 The H & N chick, the commercial product of the breeding scheme known as "reciprocal recurrent selection." The parents of each of two lines are selected on the basis of cross-line performance of their progeny. (Courtesy, Heisdorf and Nelson, Inc., Redmond, Washington.)

Commercial Breeding Methods

Essentially all poultry breeders today make use of the principle of hybridization. Two methods are used. The first method is the crossing of distinct but mildly inbred strains. If the cross is between strains of the same breed, it is called a strain cross. If it is between strains of different breeds, it is called a breed cross. The second method is the development of inbred lines by intense inbreeding and then crossing them to produce incross hybrids. Both methods have proved to be highly successful for the production of commercial egg-layer stock.

Pure strains* used for commercial crosses are improved either by selection of superior in-

* By convention, a "pure" strain means any strain, which when crossed with another, produces a strain cross or a breed cross. In this sense the word "pure" has no exact genetic meaning.

dividuals based on pure-line performance of each strain or on the selection of individuals of one strain which best crosses or "nicks" with a tester sample of the other strain. The latter, based on the principle of "breeding for cross-line performance," is called Reciprocal Recurrent Selection (Figure 23.1). In brief, *A* males are selected as breeders for the *A* line on the basis of how well they cross with line-*B* females. Likewise, *B* males are selected as breeders for the *B* line on the basis of how well they cross with *A*-line females. The method is designed to take advantage of gene interactions or "nicking" effects which may occur between the two lines *A* and *B*. Table 23.1 gives a comparison of pure-line and strain-cross performance which also demonstrates heterosis. Although the particular breeding techniques practiced by different commercial breeders vary, and often are held as trade secrets, one or the other of these two options of pure-line improvement or a combination of them are followed by all strain-cross breeders.

The Inbred Hybrid

During the years 1940–1945 a different approach to poultry improvement began to take shape. Because the procedure of first developing inbred lines and then crossing them to form hybrids was borrowed from the hybrid-corn breeders, it is natural that the two larger commercial seedcorn companies in the U.S.* were the first in the field with chicken hybrids. The steps in the procedure include, first, the formation of inbred lines by intensive inbreeding, and second, the screening of the inbred lines by comparing the performance of test cross combinations. Finally, after enough consistent field-test comparisons with an established commercial variety have been obtained, the breeder decides whether the new experimental variety is good enough to replace the established commercial variety.

It may seem rather strange that no one has yet proved scientifically that the inbred-hybrid approach is superior to the strain-cross approach using mildly inbred strains. All we can be sure of is that both methods are capable of producing high-performing layer-type chickens. So far, the inbred-hybrid scheme has not yet proved to be commercially feasible for commercial broiler and turkey production.

Table 23.1.

Average egg production of two pure lines and their cross, in percentages of days when eggs were laid. The pure strain breeders have been selected on the basis of their strain-cross half-sibs' performance. Note heterosis in the cross and also the evident improvement from selection over the years. (From *Proc. National Poultry Breeders' Roundtable*, 1969, Kansas City. Data from Heisdorf and Nelson, Inc.).

Year	*Av. of pure lines* (*P*)	*Av. of strain cross* (*X*)	*Heterosis* (*X* − *P*)
1954	57	62	5
1963	66	71	5
1968	74	81	7

Chromosomes, Genes, and Gene Pools

The rediscovery of Mendelism in 1900 has had a tremendous impact on plant and animal breeding. Chickens were instrumental in proving the generality of the principles of Mendelian genetics. For about 15 years (1905–1920), students of heredity were polarized into two camps: those who believed that Mendelism explained all heredity; and those who believed that Mendelism was an exception to the general rule of the law of ancestral heredity, which was explained statistically by the regression of progeny on parents. It remained for a school teacher, R. A. Fisher of England, in 1918, to formulate a unified theory of heredity which fitted the inheritance of

* Pioneer Seed Corn Company, Des Moines, Iowa, and the De Kalb Agricultural Association of De Kalb, Illinois.

both simple as well as complex quantitative traits. The theory was framed on biometrical principles with an underlying Mendelian basis of inheritance.

Genes were originally defined as single particles which segregate in the progeny according to Mendel's laws. Today genes are also defined as the smallest segment of a chromosome, bounded on each end by a point of crossing over. On a biochemical basis, genes produce specific enzymes for protein synthesis. On a molecular level, the gene is composed of one to three nucleic acids. In population genetics, the gene is thought of as a statistical entity whose complete expression is determined by the environment of the organism. Part of the environment of an individual is represented by other members of the population. Thus, in dealing with populations, the genes and their interactions with both environment and other genes can be sorted and classified only by statistical methods.

Genes are carried on chromosomes. The exact number of chromosomes in the fowl is not known. Counts reported by recent studies from Minnesota (Shoffner, 1968) indicate 39 pairs in the chicken and 41 in turkeys. In all avian species there are two distinct size groups. In the chicken and turkey a set of five or six pairs of large chromosomes with consistent features are recognized (Figure 23.2). In addition, each species has a set of "microchromosomes," which are small and without distinguishable features, but seem to have all the basic characteristics of chromosomes. Whether they carry genetic information is not known.

In mammals, including man, the male is the heterogametic sex (*XY*) and the female is the homogametic sex (*XX*); in avian species the opposite is true. The sex chromosomes in chickens are designated Z and *W*, and correspond to *X* and *Y*, respectively, in mammals; the cock is homogametic (*ZZ*), and the hen is heterogametic (*ZW*).

The Z chromosome seems to be especially important in chickens. It is the fifth largest and comprises nearly 10 per cent of the total

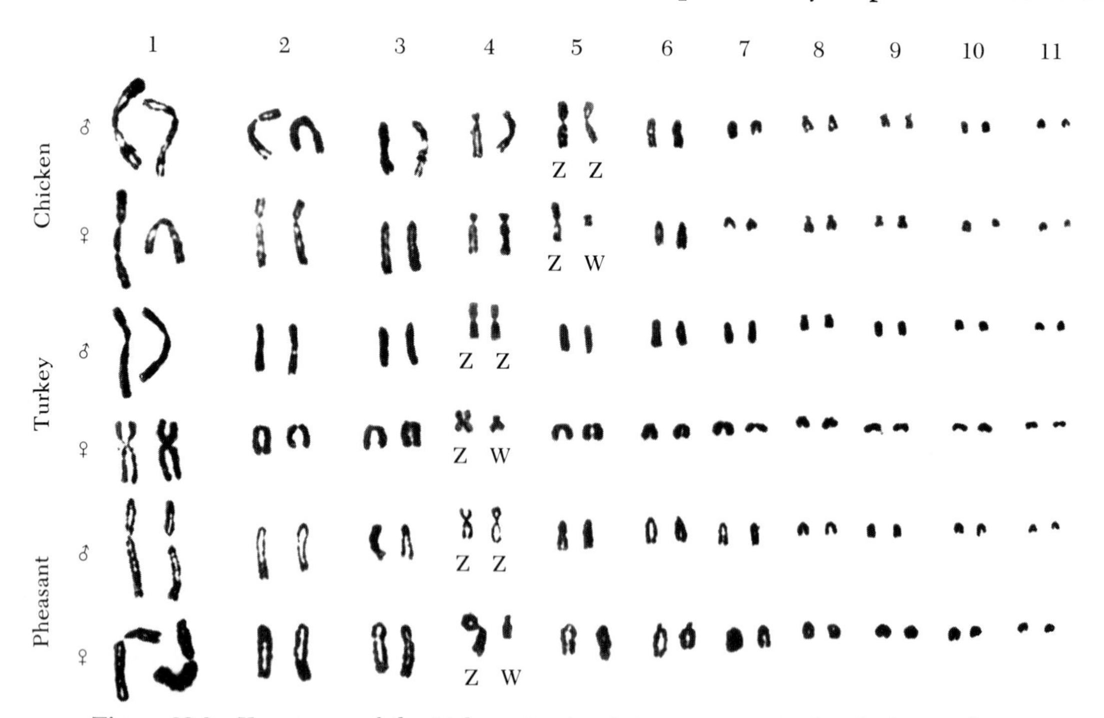

Figure 23-2 Karyotypes of the 11 largest pairs of chromosomes in the chicken, turkey, and pheasant. (Courtesy, R. N. Shoffner, Univ. of Minn.)

DNA material. Of the six known linkage groups, the sex-linked group has the most known segregating loci. The *W* chromosome, corresponding to the *Y* chromosome in man, seems not to carry much, if any, genetic information.

Traits Showing Mendelism Segregation

In the fowl, there are many traits including comb shape, plumage color, and skin color which show typical Mendelian segregation. In general, these are easily fixed so that commercial breeders are not usually concerned with them. On the other hand, certain Mendelizing traits, including blood groups, dwarfism and rate of chick feathering, may be of real or potential economic importance to the breeder.

Blood Groups

For the past several years commercial poultry breeders, as well as cattle breeders, have shown an intense interest in red-cell blood groups. About twelve different red-cell blood groups are recognized in chickens. The usefulness of red-cell groups and other biochemical variants of the blood to animal and poultry breeders depends on how much such variants aid in better characterizing an animal's total breeding value measured in terms of economic productivity. Although biochemical genetics is not likely to replace conventional methods of selection, it could augment these methods. Blood-group information might permit culling the least desirable genotypes at an early age, and thereby reduce breeding costs. Blood-group information is useful in detecting pedigree errors and to verify the genetic make-up of widely distributed flocks of some far-flung commercial poultry-breeding enterprise.

The most important blood-group system known in chickens is the *B* system. Some *B*-locus alleles are "bad," others are "good." For example, recent studies at Iowa State University show that the homozygous B^1B^1 genotype is consistently associated with lower egg production and much higher adult mortality (Table 23.2). The adult mortality of the B^1B^1 pullets was nine to 13 times greater than that of the other blood-group genotypes. Studies made to determine the differential causes of mortality have not yet clearly shown why the B^1B^1 genotype has poorer livability than the others. It is possible that general vigor genes or specific genes for resistance to certain diseases, such as Marek's disease, may be linked with *B*-locus blood-group genes. Some commercial breeders are now selecting chickens with specific blood-group genotypes at the *B*-locus in an attempt to develop strains with higher egg production and lower adult mortality. However, additional research is yet required to answer certain basic questions before broad use of blood groups in commercial chicken production can be recommended.

Table 23.2.

Influence of the *B*-locus blood type on egg production and laying-house mortality. The B^1 gene is a "bad" gene, the B^2 a "good" gene. B^1 is recessive to B^2. These are average results of six years of data from Iowa State University.

Blood type	*Egg production (per cent)*	*Laying house mortality (per cent)*	*Number of birds tested*
B^1B^1	42.2	50.0	352
B^2B^2	57.2	3.8	170
B^1B^2	57.7	5.5	356

Sex-linked Dwarfism

Dwarfism, which causes an animal (or plant) to be subnormal in size, is well-known in a wide variety of species, including chickens. A number of different mutant genes are known in each species. In chickens, certain bantam breeds carry a sex-linked dwarf gene. The same (or similar) recessive gene has been found in the Leghorn and the New Hampshire. Hens that carry the sex-linked dwarf gene (*dw*) are about ⅓ smaller than normal body size, but their egg size is only about 8 per cent smaller. Otherwise, the dwarf chickens seem not to be seriously handicapped compared to their normal sisters. Recently, the *dw* gene has been incorporated into laying flocks on an experimental basis. The idea has been to lower body-maintenance requirements and thereby increase the efficiency of feed use. However, the better feed efficiency has proved to be a simple consequence of reduced body size; so if a breeder wished to reduce body size by conventional methods of selection for smaller body size, an equivalent improvement in feed efficiency could be expected.

Another possible application of the *dw* gene is in broiler production, where a persistent problem has been the high cost of producing hatching eggs. Broiler breeder flocks tend to be rather poor layers. The hens are necessarily large-bodied to insure that their broiler progeny have high growth rate. As a consequence, feed costs of hatching eggs are high. However, when the dwarf gene is introduced into the female parent line, body size is reduced about 30 per cent (Figure 23.3). Since such hens require less feed, the cost of hatching eggs and broiler chicks is correspondingly reduced. When dwarf female breeders are then mated to a normal nondwarf male line, the progeny are normal because the dwarf gene is recessive. Broiler progeny from dwarf mothers show a slight reduction (2 or 3 per cent) in growth to market age, because they are hatched from smaller eggs. This method of breeding broilers is being evaluated by a number of commercial breeders and experiment stations in the U.S. and other countries, notably France.

Figure 23-3 A dwarf and a normal broiler breeder hen. (Courtesy, Indian River Farms, Lancaster, Pa.)

Early Feathering

Early feathering is especially important in broiler production, because birds ready to be processed should be well-feathered and free of pinfeathers. The sex chromosome contains a major genetic locus which controls early feathering of young chicks. Gene *K* produces late or slow feathering, and its allele *k* determines fast or early feathering. In general, the Mediterranean breeds are predominantly early-feathering while the Asiatic and English breeds were originally slow-feathering. Most of our American breeds of forty years ago were also slow-feathering. During the 1930's and 1940's, the White and Barred varieties of Plymouth Rocks and the Rhode Island Reds were selected and fixed for the early feathering.

On the other hand, in Leghorns, where the early-feathering gene is fixed, some breeders are now introducing the slow-feathering gene, *K*, into one of their strains. The purpose is to make it possible to produce baby chicks which can be autosexed at hatching. The unwanted slow-feathering males, having no economic value, are then discarded.

The method of sexing based on the (*K*,*k*) locus requires an examination of the primary wing feathers (Figure 23.4). The problem is to first incorporate the *K* gene into a female parent line; this is done by starting with an outcross to a stock carrying *K*, and then successively backcrossing to the female line until it becomes essentially a reconstitution of the original line but maintains segregation at the (*K,k*) locus.

The fraction of commercial chicks sold today using the (*K,k*) genes for sex determination is not yet large. Most hatcheries still use the Japanese method of vent sexing. This increases the cost of baby production because it can be performed only by a trained sexor. The development of parent strains capable of producing autosexing baby chicks would not wholly eliminate the cost of sexing, but should reduce the cost by 50 to 60 per cent.

Selection for Economically Important Quantitative Traits

Quantitative inheritance had its origin with Francis Galton of England, about 1876. Galton was a cousin of Charles Darwin, and like him was interested in the nature of inheritance. To study the inheritance of human stature, a quantitative trait, he developed the much-used statistical method of regression.

Some quantitative traits are called "metric" traits because they can be measured with a scale, say, in grams or centimeters. Others are enumerated or counted, and are usually expressed in percentages, e.g., egg production. Those classifiable in only two categories are called all-or-none traits: for example, an egg either will or won't hatch, and a chicken either lives or dies at a given age.

The application of the principles of quantitative genetics to selection in poultry goes back to the early 1940's, and following World War II several poultry-breeding operations developed on a large scale. Perhaps the one person who has had the most significant influence on all of modern animal and poultry breeding is J. L. Lush of Iowa State University through his book, *Animal Breeding Plans*, first published in 1938. This book draws heavily on the ideas of Sewall Wright and R. A. Fisher, who are the principal architects of the science of quantitative genetics. Perhaps the most important single concept dealt with in Lush's book, and one which has had a great impact on the commercial field of breeding, is the fundamental equation for genetic gain by selection: that genetic gain depends not only on the level of superiority of the selected parents, but also on how highly heritable the selected trait is. Because some traits, such as body size and growth, are highly heritable, they are readily improved by selection. On the other hand, reproductive traits, such as hatchability and egg production, not being highly heritable, are much more sensitive to environmental influences and are therefore more resistant to change by selection.

Body size, growth rate, and conformation: these traits are highly heritable, and probably have a good deal of underlying genetic mechanism in common. Pleiotropic genes probably control both rate of growth of body parts and ultimate body size.

Body Size

Body size is of special importance to broiler and turkey breeders, because it places a ceiling on growth. Broilers and turkeys hatched from large strains grow the fastest and are the most efficient in feed utilization. On the other hand, intermediate body size is preferred in egg-laying strains of chickens. Even though small birds have the lowest maintenance requirements, they do not produce market eggs of satisfactory size.

Because differences in adult body size between individuals of the same strain usually prove to be highly heritable, the breeder may change body size readily by mass selection. This is demonstrated in a two-way selection experiment for high and low body weight in Leghorns shown in Figure 23.5. Over the 11-generation interval, body weight was increased 1.0 kg. in the high body-weight line and decreased .7 kg. in the low body-weight line. These results also show that the direction of selection may influence gains from selection, since selection downward approaches a limit (zero), whereas selection upward has no theoretical limit.

Body size is commonly measured by simply weighing birds. Usually no distinction is made between size and weight, but the latter is determined both by bone size and by degree of fleshing (muscle and fat). Because feeding, management, and disease have a lesser effect on the bone size than on the fleshing, the former is more highly heritable. Shank length, therefore, should be a better measure of body size (bone frame) than body weight.

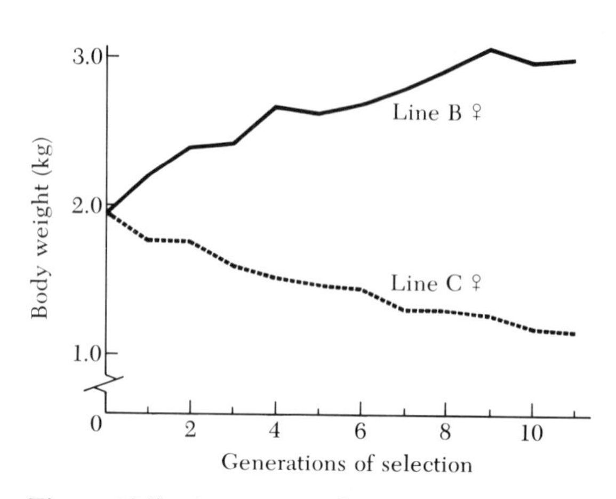

Figure 23-5 A two-way selection experiment for high and low body weight at 32 weeks of age in Leghorns. After 11 generations of selections, these lines differed by almost 2 kg. Clearly selection is effective both up and down. (Data from Iowa State University.)

Growth Rate

Growth is of special importance to breeders of broilers and turkeys, because it determines the amount of time, labor, and feed consumption required for the production of meat. For all animals, rate of growth is characterized by a period of acceleration followed by a period of deceleration. In broilers the periods of acceleration and deceleration are roughly divided at 12 weeks of age, and in turkeys at about 16 weeks of age. The most economic growth takes place during the accelerated-growth period. For this reason, broilers marketed at 12 weeks or less, or turkey fryers at 16 weeks of age, can be produced on less feed per pound of gain than, say, more mature roasting chickens.

Figure 23-4 Determining sex of baby chicks using the sex-linked feathering locus. When early feathering males (*kk*) are mated to slow feathering females (*K*–), the cockerel chicks, being heterozygous *Kk*, show little primary feather development (left). The pullets, being *k*–, show well-developed primary feathers (right). (Courtesy, John Tierce, Iowa State University.)

Growth rate is about 30 per cent heritable in both broilers and turkeys, and readily responds to selection. A two-way selection experiment for 12-week body weight in broiler-type chickens carried out in Oklahoma demonstrates this fact (Figure 23.6). Two groups of chickens were selected from a base population. Individuals in the high line were selected with body weights above the population mean, and those in the low line below the population mean. Selection in the high line increased body weight of the males from 1.2 to about 2.7 kg. Upward response to selection was greater in the high line than was downward response in the low line, as might be expected, because the latter approaches a limit of zero.

Conformation

Conformation is determined by body proportions and is a function of both bone structure and fleshing. In chickens the sex-linked dwarf gene alters conformation by a shortening of the shank bone. In normal chickens variations in conformation are mainly due to differences in fleshing, and are therefore largely a reflection of feeding, management, crowding, and disease rather than of genetic effects.

Conformation is of special importance in turkeys. The broad-breasted turkey, common on the market today but unknown 35 years ago, is a singularly important accomplishment of modern turkey breeders. However, the great emphasis on broad breasts has produced some undesirable side effects; in particular, the poor reproductive performance common to most large strains of turkeys today is probably a consequence of the extreme emphasis on broad breast conformation.

Body conformation in broiler chickens, owing to the practice of marketing broiler meat in the cut-up and packaged form, is of importance only because it relates to amount of fleshing. In egg layers, conformation per se is of little direct economic importance, because of the low salvage value of the hen carcass when marketed at the end of the laying year, to be used mainly for making canned chicken soup.

Egg Production, Egg Size, and Egg Quality

The fowl, through hundreds of years of domestication, has developed into the high egg-laying machine that it is today. An important factor of domestication affecting egg production has been the removing of eggs regularly from the nest. This acts as a physiological stimulus inducing the hen to continue laying. Even the sparrow has been reported to have the capacity to lay 50 eggs in a season if the eggs are taken from the nest. Needless to say, proper care and feeding are also necessary to obtain the high production records that we find today.

It is interesting to note that the chicken is probably not our most prolific egg-producing domestic species. Ducks of the *Khaki Campbell* breed have equaled or exceeded the best records obtainable in chickens, with flock averages of 320 eggs per bird per year having been reported. Unfortunately, there is the problem of consumer preference. If we are accustomed to eating chicken eggs for breakfast, we resist duck eggs as a substitute. Also, chickens convert feed into eggs as efficiently as ducks.

Egg production may be measured in terms of either the flock or the individual hen. In

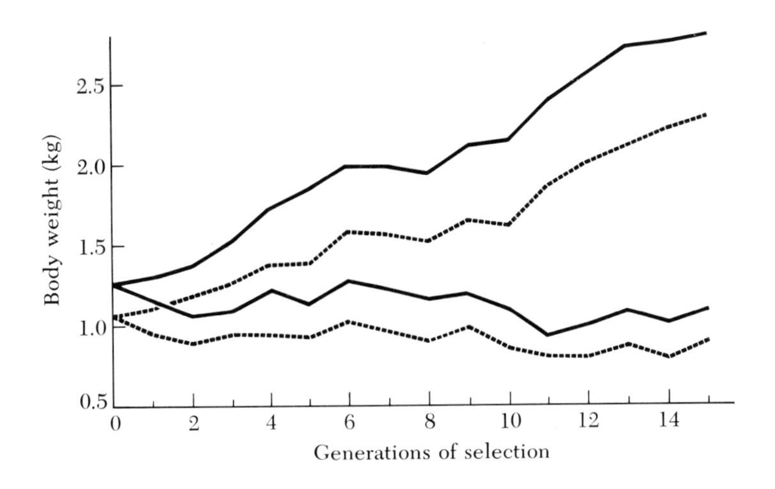

Figure 23-6 Selection for high and low 12-week body weight in two broiler strains. (Data from Maloney *et al.*, 1967.)

order to obtain individual hen records, trapnesting is necessary, or the hens must be kept in individual cages.

Hens begin laying at approximately 20 to 22 weeks of age, and continue for about a year. Usually, a flock will reach peak egg production about six to eight weeks after coming into lay; production will then gradually decline to the end of the laying year. When flock management or feeding is faulty, or when there is a disease outbreak, the production curve will be correspondingly depressed.

Egg yield is the product of two factors: intensity or rate of laying, and the time interval in the laying period. The interval between the first egg that a hen lays and the last one she lays before she goes into a molt is known as the "biological year." Factors that tend to increase rate of production or the length in the biological year result in greater total egg production.

The components of the annual egg production record are: (1) the age at which a hen becomes sexually mature and starts laying; (2) the length of time for which a hen continues to lay (persistency); (3) the rate (intensity) of egg production; (4) the number of pause periods during which the hen stops laying; and (5) the number of broody periods during the biological year. Of these, the first, third, and fifth are usually given special attention by commercial breeders.

Age at sexual maturity A pullet is considered sexually mature when she lays her first egg. The earlier the pullet commences laying, the longer will be her biological year and the more eggs she should produce. On the average, Leghorns become sexually mature at between 170 and 185 days. Heavier breeds, such as the Rhode Island Red, usually reach sexual maturity two or more weeks later. Strains of the same breeds may differ significantly in age at sexual maturity. Differences between reciprocal crosses of early- and late-maturing strains demonstrate the existence of sex-linked genes for maturity. For this reason, whenever a cross is planned between two strains differing in maturity, it is wise to mate the males of the earlier-maturing strain to females of the late-maturing strain. The resulting female progeny then receive the sex chromosome from the sire with the early sexual-maturity genes.

Intensity or rate of production This is measured as the number of eggs laid during a standard time interval or as the percentage of eggs produced during a variable time interval. Rate of egg production is highly correlated with the profit potential of an egg-laying strain, but since this trait is highly subject to environmental influences, it has low heritability, usually no higher than 10 per cent (Figure 23.7). Because this trait is not highly heritable, most breeders use family records in their attempt to make selection more effective.

Broodiness In general, the white-egg (Mediterranean) breeds of fowl, such as the Leghorn, are characteristically non-broody, but the brown-egg breeds, including the

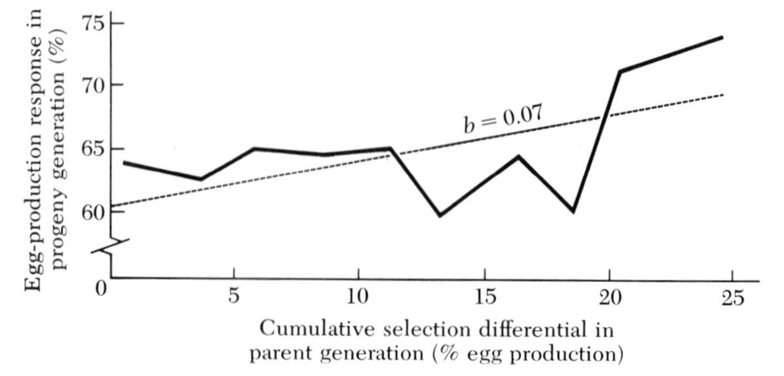

Figure 23-7 Progress in selecting for rate of egg production in Leghorn Line *A* is slow because the heritability of this trait is low. In this experiment, no gains from selection were evident until the eighth generation. The realized heritability of this trait was only 7 per cent. This corresponds to the regression of the progeny on parents, $b = 0.07$. (Data from Iowa State University, Ames.)

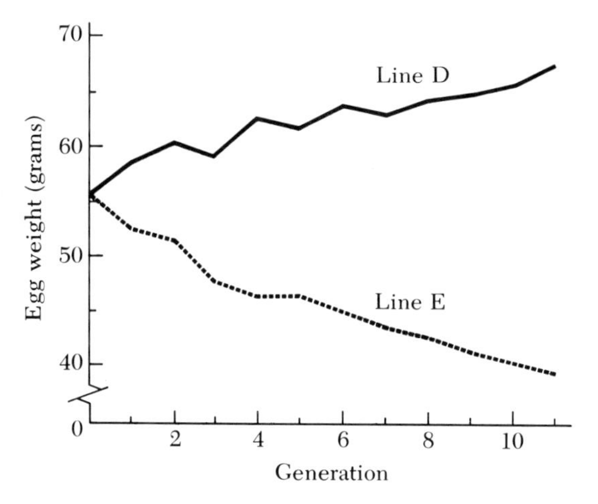

Figure 23-8 A two-way selection experiment for egg weight at 32 weeks of age in Leghorns. Selection is highly effective in either direction. In 11 generations, the averages of lines D and E differed by 29 grams. (Data from Iowa State University, Ames.)

Rhode Island Red, White Plymouth Rock, Sussex, and Australorp, show variable amounts of broodiness. Since the hen takes time out from laying when she goes broody, this tends to reduce the annual egg yield. Broody hens are also a nuisance because of their excessive occupation of the nests in the laying house.

Broodiness seems to be determined by complementary interaction of genes. The prominent strains of Rhode Island Reds developed in Massachusetts between 1915–1950 were consistently selected for freedom from broodiness. Yet today, when these are crossed with Leghorns, the progeny are usually more broody than the parents, indicating gene interactions. Broodiness also has a sex-linked basis. Experiments have demonstrated that the crossbred progeny of Rhode Island Red males mated with Leghorn females have more broodiness than the reciprocal cross. Likewise, the relatively high broodiness of the Austra-White (Australorp male × Leghorn female) compared to the reciprocal cross is a manifestation of sex linkage.

To eliminate broodiness from a cross, breeders would probably need to select pure-line birds from records of their cross-line progeny.

Egg size The first egg laid by a pullet weighs about 75 per cent of the maximum reached at maturity. The speed at which a pullet reaches mature egg size is influenced by the season when hatched and by the age at maturity. Birds hatched in the spring reach maximum body size and egg size faster than chickens hatched in the fall. The increasing lengths of days in the spring months is most favorable to rapid growth and consequently to rate of egg-size increase. Egg size, being rather highly heritable, can be changed readily by selection (Figure 23.8).

Egg quality In addition to egg size, breeders are concerned with several egg-quality traits. These are classified for convenience into exterior and interior quality groups. The former includes the shell characteristics of color, texture, strength (resistance to breakage), and shape. The interior quality group includes freedom from blood and meat spots, amount of thick white, and proportion of white to yolk.

In Leghorns and other white-egg varieties, hens which lay off-white or tinted eggs are selected against. In brown-egg varieties, some breeders strive for a more uniform color. As a rule, brown eggs tend to fade as the laying season advances. Breeders discriminate against hens which lay rough or poorly textured eggs or which otherwise show thin or weak shells. Some breeders have developed special devices to measure shell strength.

Interiorly, the egg should be free of blood spots or meat-like inclusions. In addition, many breeders break sample eggs to measure the amount of thick albumen. A "fresh" egg usually has considerable thick white, which decreases progressively as the egg becomes stale. In fact, the amount of thick white is used as an index of freshness. Eggs of different hens vary in the initial amount of thick albumen, which can be increased by selection. How far the breeder should go in increasing the amount of thick white is questionable, because the highest-producing birds tend to have somewhat thinner albumen.

Economically, the yolk is the most valuable part of the egg. The egg-breaking industry prefers eggs which have a high proportion of yolk to white. In general, the yolk makes up about 30 per cent of the total weight, the albumen about 60 per cent, and the shell 10 per cent. Small eggs will average about 2 per cent more yolk and 2 per cent less albumen than large eggs.

Egg production and feed utilization Of special economic importance in world animal agriculture is the efficiency with which chickens, as well as other species of livestock, convert feed into an animal product. The principal dry-matter nutrients of a feed are energy and protein. Nutrients consumed by a hen are used for body maintenance and for the production of eggs. The balance of the nutrients consumed are excreted in the feces and urine. The fraction of feed nutrients converted into eggs is defined as efficiency. Species vary widely in their capacity to convert feed into animal products.

Efficiency can be measured in terms of specific nutrients, such as the conversion of the protein in a feed to egg protein or the conversion of total energy in a feed to total energy in eggs. One hundred grams of eggs contain about 37.5 grams of nutrient energy (mainly protein and fat). One hundred grams of a typical layer feed (10 per cent moisture and 3 per cent fat) contain about 94 gms. of energy. For a chicken that averages an output of 35 gm. of egg mass on 125 gm. of feed per day, the efficiency of protein conversion is about 23.5 per cent and of energy conversion is about 11.2 per cent. Table 23.3 compares the efficiency of egg production of modern high-producing chickens today with layers of more than 30 years ago.

Table 23.3.

Efficiency of egg production in a 1938 flock of New Hampshire hens compared with a 1964 flock of straincross Leghorns. The great improvement in the 1964 flock is the result of modern methods of breeding, feeding, and management.

	1938	*1964*
Breed	New Hampshire[a]	White Leghorn[b]
No. birds	88	545
Test period	12 weeks	8 weeks
Body weight (kg)	2.4	1.88
Egg mass/day (gms)	24	43
Per cent of egg production	42	76
Feed/day (gms)	119	126
Per cent of energy efficiency	10.3	13.6
Per cent of protein efficiency	15.5	28.6

[a] From Brody (1945), p. 884.

[b] Data from Heisdorf and Nelson (Germany) presented at British Poultry Breeders' Roundtable (Cheltenham, 1964) by M. Von Krosigk and F. Pirchner.

Genetic Resistance to Disease

Although great progress has been made in the control of disease of baby chicks and growing birds, mortality is still a major problem of the commercial poultry producer.

The mortality picture for a particular laying flock naturally will vary a great deal depending on the health conditions of the flock and the care the flock receives. Under reasonably good conditions, a producer can figure from .5 to 1.5 per cent mortality per month. In the experiment shown in Figure 23.9, mortality for approximately 11 months of the laying year averaged 17.3 per cent with a mean flock size of 700 pullets. By years, total mortality ranged from 12.7 to 26.4 per cent, the monthly average being about 1.5 per cent. The specific causes of mortality varied from year to year, although no major epidemic was observed during the 11-year period. The most frequent cause of mortality was due to the various forms of the leukosis complex.

Leukosis is a cancer-like disease which can affect most body tissues of a chicken; it produces tumors or carcinomas. The disease is initiated by two kinds of viruses. One is a Herpes virus, which typically leads to a cancer of the nerves known as Marek's Disease. Recently, a vaccine prepared from a strain of Herpes virus found in turkeys has been developed by the USDA, and gives about 85 per cent effective protection. The other virus, a Myxo virus, produces lymphoid leukosis in chickens.

Significant progress is being made today toward a greater understanding of the nature of leukosis viruses as they are able to specifically infect certain genetically susceptible chicken-cell types found only in certain flocks. It is now clearly established that genetically different chicken-cell types in tissue cultures are susceptible or resistant to different races of virus, and that resistance of different cell types is controlled by genes showing Mendelian segregation. As further basic knowledge of disease resistance increases, better control of leukosis should be possible, and perhaps, at the same time, we may come a step closer to controlling cancer in man.

Correlated Responses from Selection

In general, if selections are made for trait X, the response in the progeny on trait X is called a direct effect from selection and on trait Y is called a correlated effect from selection. The magnitude of a direct effect is a function of the degree of heritability of the trait selected; that of a correlated effect is dependent not only on the heritability of the trait selected, but also on the genetic correlation between the two traits. The genetic basis of a correlated response is controlled either by pleiotropic genes which influence both traits X and Y or by genes, separately influencing each trait, that are linked on the same chromosome.

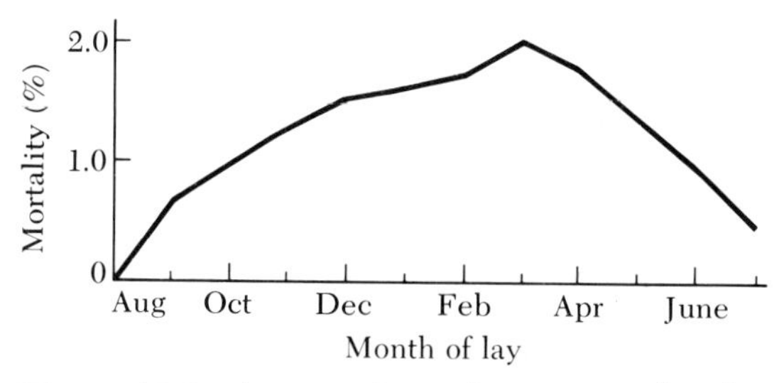

Figure 23-9 Average laying-house mortality by months, based on data collected for 10 years on over 5,000 hens of a Leghorn Line A, selected for high egg production. Mortality was highest during the sixth and seventh months of the laying period. (Data from Iowa State University, Ames.)

As an example, if a breeder selects for lower body size, this will reduce egg size automatically. Conversely, if a breeder selects for larger egg size, this will increase body size. A recent study at Iowa State University has shown that genetically increasing body size by 10 per cent increases egg size by 2 per cent, whereas a genetic increase in egg size of 10 per cent increases body size by about 6 per cent. When traits are tightly correlated, correcting an undesirable correlated response from selection can only be made gradually. When traits are loosely correlated, corrections can be made more readily.

The influence of selection for body weight, egg weight, and rate of egg production on feed efficiency (Figure 23.10) provides practical examples of correlated responses. Small-bodied birds are the most efficient because they have lower feed requirements for maintenance. Selecting for large egg size reduces feed efficiency less than selecting for high body weight. Selecting for high rate of egg production improves feed efficiency because this increases egg mass output.

Another economically important correlated effect from selection for body weight and egg weight is the effect on reproductive fitness. Usually chickens with intermediate body weights and egg weights have the highest hatchability and egg production. Selection for extremely large eggs lowers hatchability more than selection for small eggs. Recent work in Iowa shows that a 10-gm. increase in egg size above the optimum of 50 gm. lowers hatch-

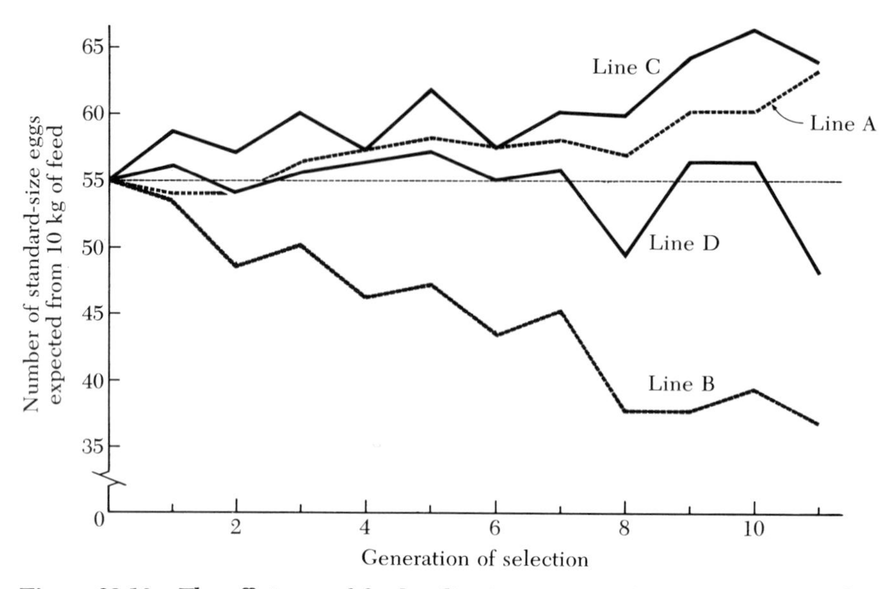

Figure 23-10 The efficiency of feed utilization as a correlated response to selection for high egg production (line A), high body weight (line B), low body weight (line C), and high egg weight (line D). High egg production and low body weight favors high efficiency, but large egg size seems to have little consistent effect. (A standard-size egg weighs 1 oz or 57 gm.) (Data from Iowa State University, Ames.)

ability about 10 per cent, but a 10-gm. decrease lowers hatchability only about 4 per cent. Selecting for large body size seems to depress rate of egg production more than selecting for large egg size (Figure 23.11).

In turkeys, the evident decline in reproductive fitness due to selection is a major breeding problem. Flock managers usually prefer to keep the largest individuals with the broadest breast for their breeding flocks. This conflicts with natural selection which favors the intermediate sized individuals with better fertility, hatchability, and egg production. To overcome poor fertility, artificial insemination is widely used for turkey flocks; yet poor fertility often remains a serious problem.

Selection and Inbreeding

Certain side-effects in populations undergoing selection are a consequence of inbreeding. Because closed strains or flocks are of finite size, they will undergo some inbreeding even though we may try to avoid inbred matings each generation. The more intensely we select, the smaller is the fraction of breeders selected. This reduction in number of breeders increases the rate of inbreeding. Likewise, when we select between families, we limit the number of families selected and hence inbreeding is increased. If the breeding population is small, this restricts the genetic variation and the opportunity to make improvement by selection. The breeder, therefore, is forced to use moderate rather than intense selection in order to maintain a reasonably effective breeding population size, especially if the populations or lines are small in the first place.

As an example, a Leghorn line was selected for high egg production at Iowa State University and maintained each generation by 16 sires and 144 dams. A pedigree study revealed that, on the average, only 12 of the 16 sires left progeny in the next generation's breeding flock. Of the 144 hens used as breeders each year, an average of only 55 left progeny in

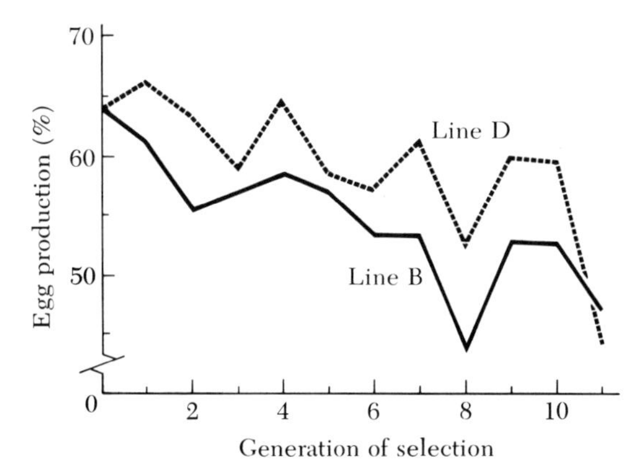

Figure 23-11 The correlated response in egg production to selection for high body weight (line B) and high egg weight (line D) in Leghorns. On the average, egg production declined about 1.6 per cent for each generation of selection for high body weight, about 0.8 per cent per each generation of selection for high egg weight. (Data from Iowa State University, Ames.)

the subsequent generations. As a consequence, the amount of inbreeding increased by 1.6 per cent per generation; after 10 years this amounted to 16 per cent, which is equivalent to more than the expected inbreeding in a half-sib mating. Inbreeding took place even though close matings such as brothers with sisters were avoided. In another line (Leghorn line D) which was selected only for large egg size, inbreeding increased 2 per cent per generation, or for the 10-year period, 20 per cent. The line was maintained each year with 8 sires and 60 dams but an average of only 38 of the 60 dams left progeny each year which were selected as breeders. Thus, inbreeding is an unavoidable consequence of small population size; furthermore, the greater the intensity of selection, the more the population accumulates inbreeding.

Inbreeding causes deterioration in reproductive traits. In the Iowa study on eight Leghorn populations selected for single traits, for each 10 per cent increase in inbreeding, egg production declined an average of 3.5 per cent. The different lines ranged in inbreeding from 18 to 26 per cent at the end of the experiment, and at this time egg production had declined by 6 to 9 per cent.

Reproductive traits are especially sensitive to inbreeding. Thus, when breeding populations are small, some decline in reproduction can be expected because of the unavoidable inbreeding. Fortunately, when two unrelated populations are crossed, the inbreeding is wiped out and as a consequence, the original reproductive performance will be regained. This is really the basis for hybrid vigor. Strains tend to be homozygous for different combinations of genetic loci which lower reproductive fitness. Upon crossing, these loci become heterozygous, which is interpreted as hybrid vigor or heterosis, or, conversely, as a recovery of inbreeding depression.

FURTHER READINGS

Hutt, F. B. 1949. *Genetics of the Fowl.* New York: McGraw-Hill.

Johansson, I. and J. Rendel. 1968. *Genetics and Animal Breeding.* San Francisco: W. H. Freeman and Co.

Lerner, I. and H. P. Donald. 1966. *Modern Developments in Animal Breeding.* New York: Academic Press.

Lush, J. L. 1945. *Animal Breeding Plans.* Ames: Iowa State Univ. Press.

Twenty-Four

Genetic Improvement of Work and Pleasure Animals

"The story has been told of how little eohippus became the ancestor of Dobbin. . . . The proportions of eohippus and its whole running mechanism were similar to those of a . . . dog of today."

George Gaylord Simpson
Horses

In considering the genetics of animals kept for work or pleasure, that branch of the subject known as behavioral genetics looms increasingly large, because behavior is intimately related to the ease of handling and the training of our domestic animals. Studies in the field of behavioral genetics cover many species of laboratory and domestic animals. To mention a few, there are drosophila (the fruit fly), chickens, rats, rabbits, cats, cattle, dogs, and horses. Although this chapter will focus on only the last two of these, it will also deal briefly with the inheritance of "bravery" in Spanish fighting bulls (a well-documented story of the response to man's selection in a direction opposite to his usual goal, the development of gentleness in his animals), with inheritance of racing ability in horses, with the genetics of coat color in dogs, which has been relatively well worked out (a similar treatment of coat color in horses has been made by Rollins, 1966), and with the genetics of size and body shape (conformation). On this last topic, the requisite methods and results for dogs and horses are quite similar to those described in Chapters 20 to 22 for cattle, sheep, and swine. However, here special attention is directed to the importance of the shape of horses' feet and legs.

Modes of Inheritance and Response to Selection

The modes of inheritance of traits of interest in work and pleasure animals fall into two broad classes, referred to by geneticists as simple Mendelian (classical genetics) and polygenic (quantitative genetics), respectively. In the former, the variation being observed is discrete; i.e. specimens can be classified into well-defined groups associated with specified genotypes. In the latter, the variation is continuous; consequently the observed specimens cannot be classified into distinct groups, but can only be ranked from "best to poorest" or "most to least." This continuity of variation is (in part) due to the action of many pairs of genes, each with small effect; hence the term polygenic. In contrast, the discrete variation in the simple Mendelian case is due to the segregation of alleles of one or at most a few genes with large effect.

Traits exhibiting polygenic inheritance respond to selection, but less spectacularly than those with simple Mendelian inheritance. Furthermore, the methods of predicting and evaluating response to selection differ in the two instances.*

Most of the beautiful and striking coat-color patterns found in our domestic animals are inherited in the simple Mendelian fashion. Traits of interest in our domestic animals that fit the polygenic pattern of inheritance are behavior patterns, temperament, ability to learn and perform specialized routines, body size and shape (conformation), and health. However, many forms of dwarfism and specific organic and structural diseases and abnormalities are inherited in a simple (usually recessive) Mendelian fashion.

Genetic Improvement of Dogs

Burns and Fraser (1966) state, in their excellent book, *Genetics of the Dog*, "Undoubtedly the very special position which the dog, amongst all domestic animals, holds in relation to mankind developed because of the behavioral and mental attributes of the canine species. If many dogs are now valued more for their appearance than their behavior, this is a relatively recent and perhaps a passing phenomenon; the firm basis of the dog's place in civilization, his privileged position as 'man's best friend,' is dependent on his extraordinary adaptability rather than on any physical characteristic, except his companionable range of body size."

* Modes of inheritance and responses to selection are developed more fully in Chapters 17 and 18. Chapter 19 also affords background for this chapter.

The Genetics of Performance, Size, and Shape

Significant scientific studies of the genetics of dog behavior have been reported by Humphrey and Warner (1934) and Scott and Fuller (1965). In both the behavioral studies were associated, directly or indirectly, with successful programs of selecting and training dogs to perform complex specified tasks, e.g., guide dogs for the blind.

A prerequisite for successfully training dogs is to develop or exploit behavioral patterns in the dog that enable it to socialize to humans. A brief description of the comprehensive 13-year experiment conducted by Scott and Fuller (1965) will set the stage for a discussion of some of their significant findings in this area.

The five breeds chosen by Scott and Fuller (1965) for intensive study, Basenji, Beagle, Cocker Spaniel, Shetland Sheep Dog, and Wirehaired Fox Terrier, are representative of the major groups of dogs as recognized by dog breeders (omitting only the toy and nonsporting breeds). They also encompassed an adequate amount of genetic variability (Figure 24.1).

The experiment showed that a puppy's development could be divided into periods based on major changes in social relationships (Figure 24.2). Immediately after birth, the puppy establishes the nursing relationship with its mother, marking the beginning of the neonatal period. The transitional period consists of rapid changes in this relationship, partly as a result of the puppy's growing awareness of his surroundings (eyes and ears opening). At its end the puppy, now able to walk and thus able to approach or avoid another individual, is capable of forming a new type of social relationship which will persist into adult life. This third period, the so-called period of socialization, is critical for conditioning the dog for successful training

Figure 24-1 The five pure breeds. Left to right, Wire-haired Fox Terrier, American Cocker Spaniel, African Basenji, Shetland Sheep Dog, and Beagle. (From Scott and Fuller, *Genetics and the Social Behavior of the Dog*. University of Chicago Press. 1965.)

since, to quote the experimenters, "it determines what species and individuals will become the chief adult relatives of the puppy. A puppy taken from its litter early in development and raised by hand will form its paramount relationships with people, becoming an 'almost human' dog and paying little attention to its own kind. Removed a little later in the period, it forms strong relationships with both dogs and human beings. Still later it has already formed strong relationships with dogs and its ties with human beings tend to be relatively weak." This period lasts from about three to ten weeks of age.

It is important to a trainer to know precisely when the socialization period begins for a dog, because this is a critical period in its successful training. The experimenters developed a test, called the startle test, consisting of observing the age at which a puppy is first startled by a sound. This event accurately indicates the beginning of the socialization period.

The experimenters developed an objective measure of socialization in puppies, called the handling test. This test involves all the things which people ordinarily do with puppies. In recording data they marked on a check list whatever the puppy did as an immediate response to each type of stimulation. The data were analyzed under the following headings: escape and avoidance behavior (Figure 24.3), aggressive behavior, and social investigation and attraction (such activity may be divided into initial attraction and subsequent investigatory behavior).

In conclusion, the experimenters interpreted the patterns of breed differences for the above test results to show that genetic factors have important effects on the process of socialization.

A distinction should be made between learning ability (intelligence?), per se, and willingness to perform assigned tasks. Humphrey and Warner (1934) found a positive correlation of .44 between these two attributes (far from a perfect association). In more graphic terms, they state, "certain dogs learned very readily, but were instructed with great difficulty. They were able to learn how to avoid doing the work demanded by the instructor without, at the same time, incurring a serious correction."

Successful dog programs capable of exploiting genetic differences are built around selection indexes based on several measurable behavioral patterns reflecting emotional and motivational qualities. Humphrey and Warner (1934) report on the training of German Shepherds for such work as herding, trailing, guiding the blind, serving with police (frontier and penitentiary), Red Cross (sanitation), and the military (liaison), and being companions. Kelly (1949) reports on the training of Border Collies for sheepherding, and Pfaffenberger on that of German Shepherds (and some retriever breeds) for guiding the blind.

In the Kelly and Pfaffenberger programs, linebreeding proved very effective. Pfaffenberger, a trainer, indicates the success of his program as follows: "Fourteen years ago [1946] we found that only 9 per cent of the dogs who were started in training for guide dogs at our school [in San Rafael, California] could be trained to become responsible guides. In 1958 and in 1959 all the dogs who had been bred and developed by the new

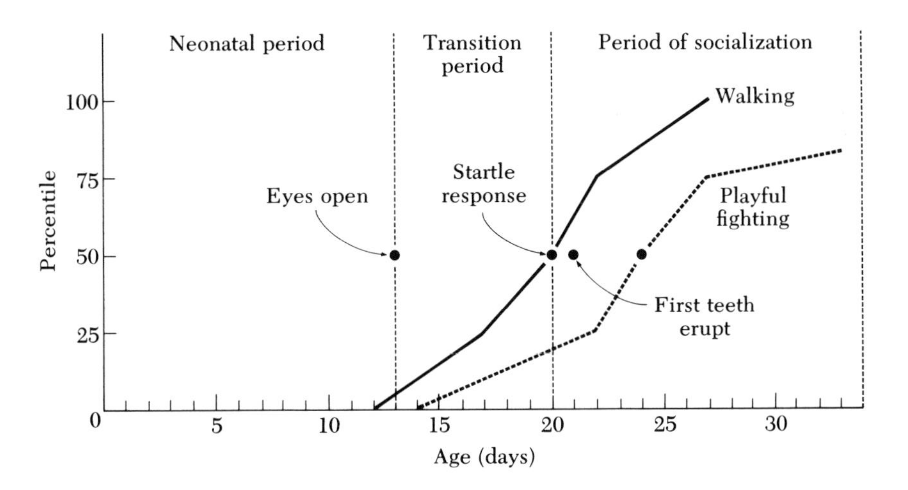

Figure 24-2 Development of walking and playful fighting in relation to the opening of the eyes, startle response to sound, and eruption of first teeth. The graphs represent cumulative figures of first occurrences in animals observed in 10-minute daily periods. The zero point is the day before the first animal was observed walking or showing playful fighting. (From Scott and Fuller, *Genetics and the Social Behavior of the Dog*. University of Chicago Press. 1965.)

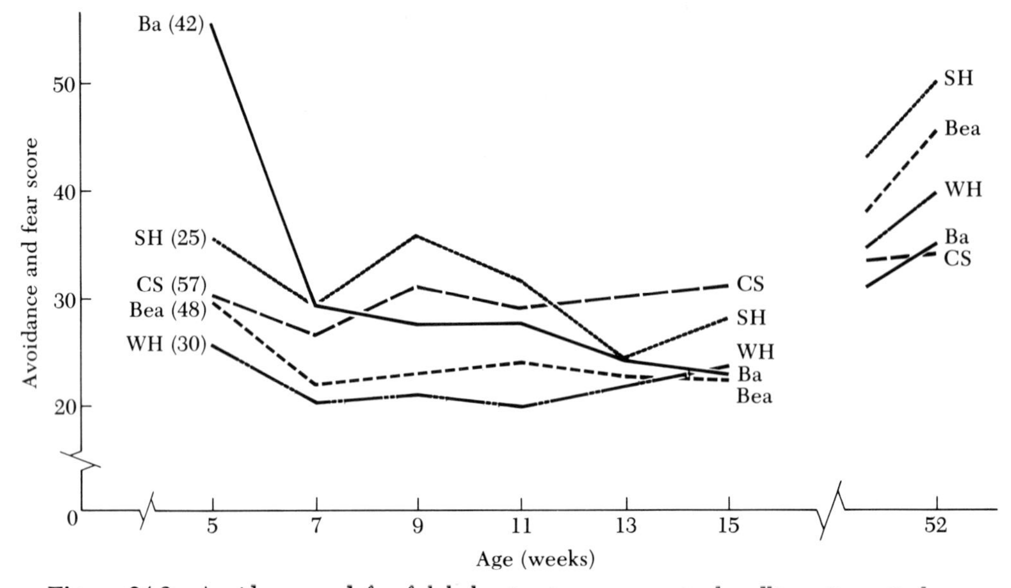

Figure 24-3 Avoidance and fearful behavior in response to handling. Basenjis have much higher scores at the outset, and scores of all breeds fall during the next 2 weeks. At one year of age dogs respond to this test as they would to catching, and the scores of all breeds are markedly higher except for cockers. Ba, Basenji; Bea, Beagle; CS, Cocker Spaniel; SH, Shetland sheep dog; WH, wirehaired terrier. (From Scott and Fuller, *Genetics and the Social Behavior of the Dog.* University of Chicago Press. 1965.)

knowledge [acquired in collaboration with J. P. Scott, Geneticist, at the Roscoe B. Jackson Memorial Laboratory, Bar Harbor, Maine (Scott and Fuller, 1965)] were as good or better than the best dogs graduated in 1946. Even with our stricter requirements, 90 per cent became guide dogs."

The genetics of dog size and shape (conformation) is similar to that of other mammalian species studied, e.g., rabbits, swine, sheep, and cattle. With such a variety of dog breed types for size and shape, crossbreeding affords a ready means of synthesizing an almost limitless array of combinations of size and conformation.

Too rigid a selection with respect to size or conformation (or coat color) seriously limits effective selection for other important traits. Scott and Fuller (1965) have suggested that, "Breed standards should include regulations relating to health, behavior, vigor, and fertility as well as body form. These can perhaps best be accomplished by introducing tests of performance and emotional reactions as well as appearance. Obedience trials and field trials are a valuable step in this direction."

The Inheritance of Coat Color

Little (1957) has given the genotypes for typical coat colors of 86 breeds of dogs. A sample of his work is presented in Table 24.1.

In Cocker Spaniels, a bicolored coat, i.e., black with tan points, arises from the presence of the allele a^t. The extent of the areas seems to depend on modifying genes. The usual particolored or tricolored cocker is genetically piebald (s^ps^p). Roughly between 30 and 80 per cent of the coat is pigmented. This is the normal range of expression of the s^ps^p type.

The red in dachshunds is usually due to the action of the a^y allele. When the red animal carries a^t as a recessive, the color may become darker and a band of dark pigment may cover the mid dorsal line and sides. The basic color varieties are, therefore: a^ya^yBB, clear red, black nose; $a^y a^t BB$, dark red-sable, black nose; $a^t a^t BB$, black with tan points. The allele for black (B) can also be replaced by that for brown (liver), in which case the three types are: $a^y a^y bb$, clear light red or yellow, brown nose; $a^y a^t bb$, clear red, brown nose; $a^t a^t bb$, liver with tan points.

The characteristic spotted pattern of Dalmatians is due to the combined action of two genes, s^w and T. At birth the coat is extreme white piebald ($s^w s^w$), and later the pigmented spots appear due to the action of T. The genotype tof a Dalmatian with black spots is $A^s A^s BBCCDDEEs^w s^w TT$.

That some breeds have a more restricted variety of coat-color patterns than others reflects only the restrictions set up by the breed associations. For example, only a black dog with tan points is eligible for registration in the Toy Manchester Terrier breed.

Genetic Improvement in Horses and Ponies

The Genetics of Performance and Shape

Let us next consider behavior, temperament, and performance in light riding horses and

Table 24.1.
Alleles involved in the inheritance of typical coat colors in dogs.

			Alleles carried by		
Series name	*Allele*[a]	*Effect*	*Cocker Spaniel*	*Dachshund*	*Dalmatian*
Agouti	A^s	Self, uniform dark color	x		x
	a^y	Restriction of dark color	x	x	
	a^w	Agouti			
	a^t	Tan points (bicolored)	x	x	x
Black	B	Black	x	x	x
	b	Liver (brown)	x	x	x
Albino	C	Full color	x	x	x
	c^{ch}	Chinchilla	x	x	
	c^e	Extreme dilution	?		
	c^a	Albinism			
Dilute	D	Intense color	x	x	x
	d	Dilute color	x		
Extension	E^m	Black mask (super extension)			
	E	Normal extension of dark pigment	x	x	x
	e^{br}	Brindle		x	
	e	Yellow (restriction of black)	x	?	x
Spotting	S	Self, complete pigmentation	x	x	
	s^i	Irish spotting[b]	x		
	s^p	Piebald spotting[c]	x		
	s^w	Extreme white piebald	x		x
Ticking	T	Pigmented spots in white areas[d]	x		x
	t	Normal	x	x	

[a] The first five series, A, B, C, D, and E, are widespread in mammalian species (Searle, 1968). In each series, the alleles are listed in descending order of dominance except as otherwise indicated.

[b] Pigmentation is typically absent from part of the chest and ventrally between and behind the front legs.

[c] S is incompletely dominant over piebald. Ss^p dogs usually have slight spotting of the Irish type.

[d] The effects are not visible at birth, but appear gradually later on.

ponies. In selecting a riding horse consideration of temperament and behavior is of utmost importance.

Although no comprehensive studies in the genetics of behavior in the horse have as yet been reported in the U.S.A., the potential for genetic improvement in this area exists.

Experience with dogs indicates the effectiveness of collaboration between geneticists (and other biologists) and trainers in devising selection and training plans to improve behavior, temperament, and learning ability as integral parts of a program to produce a better performing animal. Computerization is a useful component of such a program, since it facilitates the ready and continuous analysis of the test information collected, thus forming a basis for the elaboration of the most efficient selection procedures.

To carry out such a program, a large enough organization is needed. It might be a forward-looking breed association or a cooperative of breeders and owners similar to the numerous beef-cattle improvement associations that have been established during the past decade in the U.S.A. to select breeding animals on the basis of performance tests.

An example of the useful genetic information such a program could generate is afforded by Varo (1965) who gives heritability estimates, for 32 traits, based on the performance records of about 6,000 mares tested by the Finnish Stud Book Association during 1952–1963. A few of these heritability estimates for test scores for the following traits are: temperament, .23; gaits, .41; conformation, .23; soundness of feet and legs, .25.

In the type of program envisioned, the geneticists and other biologists should work closely with the trainers in developing tests of behavior patterns that correlate highly with total performance. Here we can think of total performance as a "supertrait" being selected for in order to achieve the goal of a satisfactorily performing riding horse with a good disposition and a pleasing appearance.

Hildebrand's (1959, 1965) study of gaits is an example of a biological study that might well serve as a basis for setting up a practical scoring test to aid in selecting horses with improved gaits. His initial interest in this area was to analyze and contrast "the running motions of these champions"; the horse and the cheetah. The cheetah is conceded to be the fastest of all animals for a short distance (maximum estimated speed, 70 m.p.h.) but lacks the endurance of the horse (maximum estimated speed, 40 m.p.h.).

Obviously speed is the product of stride rate times length. Figures 24.4 and 24.5 depict the relation of footfalls to stride for the galloping horse and cheetah. The estimated stride lengths of the horse and cheetah were equal (23 feet), despite the great disparity in body size (at the withers the horse was twice as high).

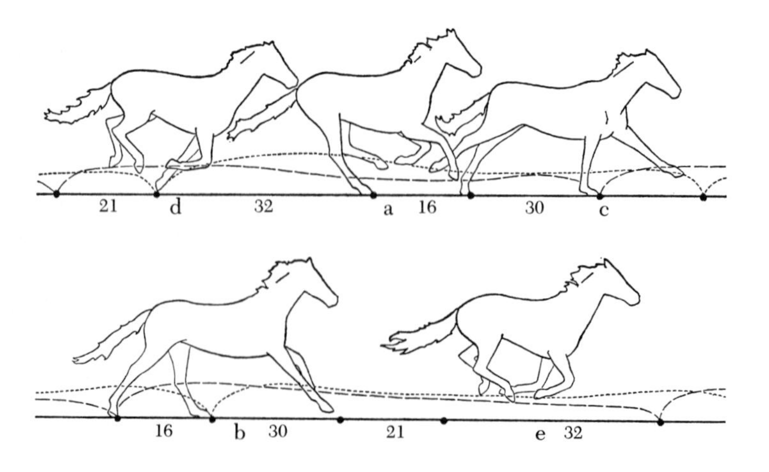

Figure 24-4 Five positions of a galloping horse shown in correct spatial relationship. Trajectories followed by the front feet are indicated above, those by the hind feet, below, long dashes for right feet and short dashes for left feet. Positions of footfalls are shown by spots on the ground line. Figures below ground line give for each interval its percentage of total stride distance. (From Hildebrand, 1959, p. 481.)

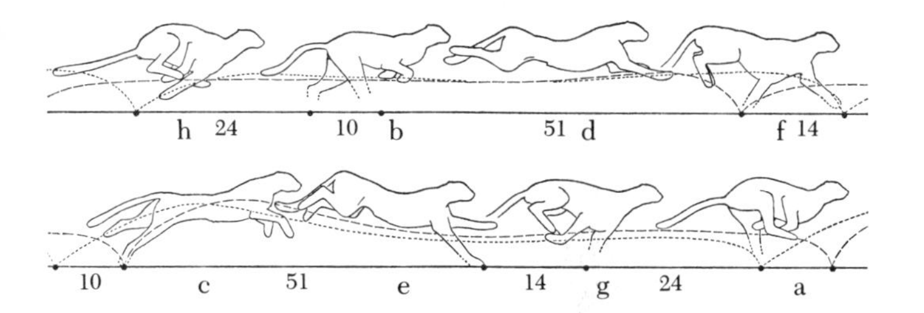

Figure 24-5 Eight positions of a galloping cheetah, shown in correct spatial relationship. Symbols and figures as for Figure 25-4. (From Hildebrand, 1959.)

The greater relative stride length of the cheetah was attributed to: (1) two principal suspension periods per stride instead of one; (2) greater proportion of suspension in total stride; (3) greater swing of limbs, to allow them to strike and leave the ground at more acute angles; and (4) flexion and extension of the spine, synchronized with action of the limbs, to produce progression by a measuring-worm motion of the body.

The faster rate (about 1.5 times that of the horse) of the cheetah's stride was attributed to: (1) its smaller muscles having faster inherent rates of contraction; (2) its limbs being moved simultaneously by independent groups of muscles; (3) its feet moving farther after starting their downstroke before striking the ground, thus developing greater backward acceleration; (4) the forelimbs having a negligible support role and probably actively drawing the body forward; (5) the limbs being flexed more during their recovery strokes; and (6) the shoulders and pelvis moving forward slower than other parts of the body at the times that their respective limbs are propelling the body.

This study led Hildebrand to a comprehensive analysis of the gaits of tetrapods in general and horses in particular. In the introduction to his 1965 paper he states: "The objectives of my research are to devise precise methods for describing and contrasting quadrupedal gaits, to survey the gaits of vertebrates, and to interpret the gaits used by particular species in relation to speed, body conformation, body size, maneuverability, and ancestry. This article presents results for the best-documented species, the horse. This master cursor has received particular attention because it is readily available, its locomotion is more controllable than that of other animals, and at the hand of man it has learned to be versatile in the selection of gaits and also to use gaits (termed artificial) that are unnatural to the species and unique to itself."

From a study of motion pictures (taken at a known speed), frame by frame, a quantitative method of analyzing and classifying gaits, called the gait formula, was devised. It is based on the footfall formula (Figure 24.6) and the duration of each support pattern. Two horses with the same footfall formula could be moving quite differently because of relative differences in the duration of their support patterns.

In the case of symmetric gaits, the walk, the trot, and the pace, the gait formula has only two variables (Hildebrand, 1963): "One is expressed as the fraction of the duration of a complete cycle of motion (or stride) that each foot is on the ground. This is expressed as a percentage figure. A very slowly moving horse may have each foot on the ground 85 per cent of the time. A fast-trotting horse has each foot on the ground about 22 per cent of the time. This variable is not a direct measure of speed, but it relates to speed.

"The other variable can be expressed as the

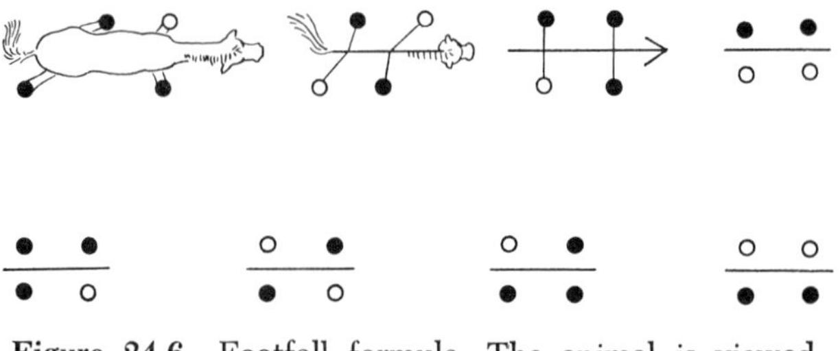

Figure 24-6 Footfall formula. The animal is viewed from above as it moves from left to right. The feet are represented by circles which are black when the respective foot is weighed. The cycle repeats after the eight support patterns shown. (From Hildebrand, 1966, p. 9.)

percentage of the stride that the footfall of a forefoot lags behind the footfall of the hindfoot on the same side of the body. If there is no lag, the feet on the same side swing together and the animal is said to be pacing. If the lag is about 25 per cent, the gait is a four-beat gait similar to a walk or single-foot. If the lag is 50 per cent, diagonal opposite limbs swing together and the animal is trotting.

"It follows that every symmetrical gait can be expressed by two percentage figures; no diagram is needed. A 35–50 gait is a moderate trot, a 55–22 gait is a fast walk, a 28–5 gait is a running pace, and so on." *

Two potential uses of the gait formula in selection and training programs come to mind. Namely, it should aid one in evaluating the consistency of a horse's gait and in obtaining some insight into how its performance deviates from a desired (ideal) gait.

For asymmetric gaits, the gallop and the canter, more than two variables are needed in the gait formula. Hildebrand plans to publish the details on these gaits at a later date.

The appropriate genetic methods for the study of racing ability are those used in the study of other quantitative traits, such as fleece weight in sheep, egg number in chickens, or milk yield in dairy cattle.

Eight heritability estimates of racing ability reported in four studies average .34, and range from .19 to .68. These results imply that the racing performance of a stallion or mare (due allowance having been made for nongenetic factors) is a useful criterion in selection. Estes (1967) points out that studies of this nature have, among other things, refuted a once-popular belief among horsemen that a mare's successful racing career indicated her *poor* genetic value as a breeding animal. One of the studies (Gilmore, 1947) in addition produced evidence that "nicking" (a mating of a particular stallion to a particular mare that results in especially outstanding progeny) may also be of importance.

The late J. A. Estes, former director of research for the American Thoroughbred Association and editor of *The Bloodhorse*, developed a useful selection index for racing ability called the Average-Earnings Index (Estes, 1958, esp. p. 86). As the name implies, it makes use of the correlation between racing ability and earning power. It is applicable to both stallions and mares (also to geldings in progeny tests of stallions). In a computerized selection program, its usefulness should be enhanced, since larger amounts of data on earnings could be analyzed.

In a performance-oriented improvement program for livestock, body shape (conformation) should receive the attention that its correlation with over-all performance justifies (e.g., a draft-horse-type conformation would undoubtedly hinder the performance of a horse being considered for jumping). Addi-

* From a talk given by Hildebrand as part of the Horse Day Program at the University of California at Davis, April 27, 1963.

tionally, the strenuousness of the horse's locomotion, along with the weight it carries (its own and its rider's), warrants the paying of careful attention to anatomical malformations in feet and legs (e.g., sickle hocks). In horses, these traits as a rule seem to have a polygenic type of inheritance. A survey of the literature revealed no well documented cases of simple Mendelian inheritance.

The Inheritance of "Bravery" in Spanish Fighting Bulls

The livestockman consciously or unconsciously selects against wildness in his stock. In Spain during 1962 an opportunity to study the effects and efficacy of selection in the opposite direction presented itself to the author. A Spanish genticist, Dr. M. Odriozola, who is studying the inheritance of bravery in fighting bulls, discussed his work and arranged an itinerary that permitted the visiting of several of the outstanding breeding herds of fighting bulls in Spain. Herds or breeds producing fighting bulls are called "brava."

Typically the "brava" bull is characterized by a muscular, well-balanced body with a trim middle, fine skin, silky hair and switch, fine bone, and a small head. It has a dressing percentage in the 60's. Alert in temperament, it also possesses a quick, graceful action. Up to 50 yds. the "brava" is faster than a Thoroughbred horse.

The management of the "brava" herds is

Figure 24-7 A two-year-old bull being tested for "bravery" (Courtesy of Don Eduardo Miura, Seville, Spain.)

similar to that used in Western range herds in the U.S.A., except that a conscious effort is made to bring the animal into contact with humans as little as possible.

Bulls and cows for the breeding herd are selected mainly on the basis of a test of bravery, at two years of age: *"Tienta en campo abierto"* for bulls and *"Tienta en plasa"* for heifers. In the former test a bull, in an open field, is maneuvered into confronting a horseman armed with a pike (Figure 24.7). His fitness as a fighting bull is judged by the manner in which he charges the man and horse. In the latter test a heifer is confronted with a cape in an enclosed corral (Figure 24.8) simulating the conditions of the actual bullfight. The bull is not tested under such realistic conditions because experience has shown that in even so short a time he would learn the subterfuge of the cape and thus become too dangerous for a man to fight. Hemingway (1932), in Chapter 11 of his book on bullfighting, vividly describes these tests and also relates several anecdotes about the ferociousness of Spanish fighting bulls.

Figure 24-8 A two-year-old heifer being tested for "bravery" (Courtesy of Don Eduardo Miura, Seville, Spain.)

Linebreeding is practiced widely, and two of the largest and most important herds, those of Miura and Ramirez, have been closed to outside blood since the 1840's and 1880, respectively. Each breeder maintains his own herd book, there being no breed societies.

On the basis of daughter-dam regression (500 pairs) and son-dam regression (400 pairs), Dr. Odriozola (personal communication) has estimated the heritability of "bravery" to be 30 per cent. The trait is measured on a scale from 1 to 15, based on the evaluation of performance in the *tienta* or *corrida* (the bullfight itself). On this scale the average for unselected heifers is 4 plus, for cows in the breeding herd 7.4, and for bulls used in the *corrida* 8, showing that there is substantial selection for replacement heifers. This is consistent with Miura's estimate of 10 per cent replacement per year in his cow herd. Neither the highest-testing nor the lowest-testing bulls are used in the *corrida*. The former are saved for breeding and the latter are sent to the butcher.

FURTHER READINGS

Burns, Marca, and Margaret N. Fraser. 1966. *Genetics of the Dog*. 2d ed. Philadelphia: Lippincott.

Gilmore, Robert O. 1947. "Statistics on sires and dams." *The Thoroughbred of California,* 6: no. 8; 24–25, 66; and no. 9; 22–23, 58.

Hildebrand, Milton. 1959. "Motions of the running cheetah and horse." *J. Mammalogy,* 40:481–495.

Hildebrand, Milton. 1965. Symmetrical gaits of horses. *Science* 150:701.

Kelly, R. B. 1949. *Sheep Dogs.* 3d ed. London and Sydney: Angus and Robertson.

Little, Clarence C. 1957. *The Inheritance of Coat Color in Dogs*. Ithaca, N.Y.: Cornell Univ. Press.

Pfaffenberger, Clarence. 1963. *The New Knowledge of Dog Behavior*. New York: Howell.

Scott, John Paul, and John L. Fuller. 1965. *Genetics and the Social Behavior of the Dog*. Chicago, Ill.: Univ. of Chicago Press.

Section Five

Animal Functions and Their Physiological Control

Portable equipment for measuring gaseous and other excretory products. (USDA photo.)

Twenty-Five

Regulation of Body Functions: Hormonal and Nervous Control

"The human mind is often so awkward and ill-regulated in the career of invention that it is at first diffident, and then despises itself. For it at first appears incredible that any such discovery should be made, and when it has been made, it appears incredible that it should so long have escaped men's research. All of which affords good reason for the hope that a vast mass of inventions yet remains."

Francis Bacon
Novum Organum

The major function of the nervous and ductless gland systems is to initiate the necessary adjustments that enable an animal to respond to changes in its external and internal environments. Changes in the external environment, such as in the availability of feed, temperature, light intensity, altitude (reduced oxygen-carrying capacity of the blood at high altitudes), or impending physical danger, are changes to which an animal must be able to adjust for its own well-being. Similarly, the animal must be able to adjust to changes in the internal environment, such as in body chemistry, body temperature, and physiological state (pregnancy or lactation), if it is to survive. When defects occur in either the nervous or ductless gland system, then the animal is not able to respond to changes in its environment, and the animal may die. Hence, both the nervous and ductless gland systems are essential for life. An understanding of nervous and hormonal control of body functions, therefore, is a prerequisite to an understanding of those physiological activities such as body motility, egg-laying, growth, and lactation which make domestic animals of economic importance to man.

How does the nervous system respond to changes in the internal or external environment? This question can best be answered by two illustrations of how the nervous system responds to such changes. If one steps on a sharp object, receptors in the foot respond by sending a signal to the brain. The brain responds to this signal by sending out signals of its own, which cause some muscles in the leg to contract and others to relax, thus removing the foot from the sharp object. In a similar manner, the nervous system can respond to changes in the internal environment, such as the increase in carbon dioxide in the blood which results after a 100-yard dash or any other short burst of strenuous activity. Specialized receptors in the neck or brain, which are called chemoreceptors, are stimulated by the increase in carbon dioxide in the blood. These receptors send signals to those centers of the brain concerned with regulation of heart and respiratory rates. These centers respond by sending signals to the heart and to the muscles involved in respiration, causing an increase in respiratory and heart rate, thus increasing the rate of removal of carbon dioxide from the blood. The above sequence of events, which leads to a decrease in the carbon-dioxide content of the blood, can be briefly summarized as follows:

Increased CO_2 content of blood —stimulates→ Chemoreceptors in carotid body in neck

—sensory nerve fibers→ Inspiratory center activity increased —motor nerve fibers to inspiratory muscles→ Respiratory rate increased

This schematic diagram is a simplification of the events which actually take place following an increase in carbon-dioxide content of the blood, since changes in circulation and heart rate would probably also take place.

From the two illustrations given above, we can see that the first function of the nervous system is to detect changes in the internal or external environment by means of specialized receptors. The next function of the nervous system is to send information regarding such changes to the brain, where the information is analyzed. After the brain centers have determined what is the proper response to the changes, they send this information to the proper tissues or cells which will bring about the desired results.

The ductless gland system, more properly called the endocrine system, responds to changes in the environment in a somewhat different manner from the nervous system. This can best be illustrated by an example of how the endocrine system responds to a specific change in the environment, such as body temperature. A specialized area in the hypothalamus, known as the temperature-regulating center, is able to detect changes in body temperature. If the body temperature decreases, this center causes the release of a substance called thyrotropin-releasing factor (a hormone). The thyrotropin-releasing factor in turn causes the pituitary gland to secrete another hormone, called thyrotropin. Thyrotropin causes another endocrine gland located below the vocal cords in the neck, the thyroid, to secrete its hormones, thyroxine and triiodothyronine. One of the major functions of these two hormones is to increase metabolic rate. This increase in metabolic rate results in an increase in heat production and hence an increase in body temperature. This increase in body temperature is then detected by the temperature-regulating center in the hypothalamus, and further secretion of the thyrotropin-releasing factor is inhibited. This somewhat complicated scheme is illustrated in Figure 25.1.

The Nervous System

Now, let us consider the nervous system in greater detail to see how it functions to bring about body adjustments in response to changes in either the internal or external environment. First we will examine the basic units of the nervous system and find out how they function. Next we will consider the broader aspects of the nervous system and explore how certain parts of the nervous system control particular body functions.

Basic Units of the Nervous System

Receptors For an animal to survive, it must be able to respond to the world around it and must also be able to regulate its own internal environment. In all animals this ability to respond and regulate its internal environ-

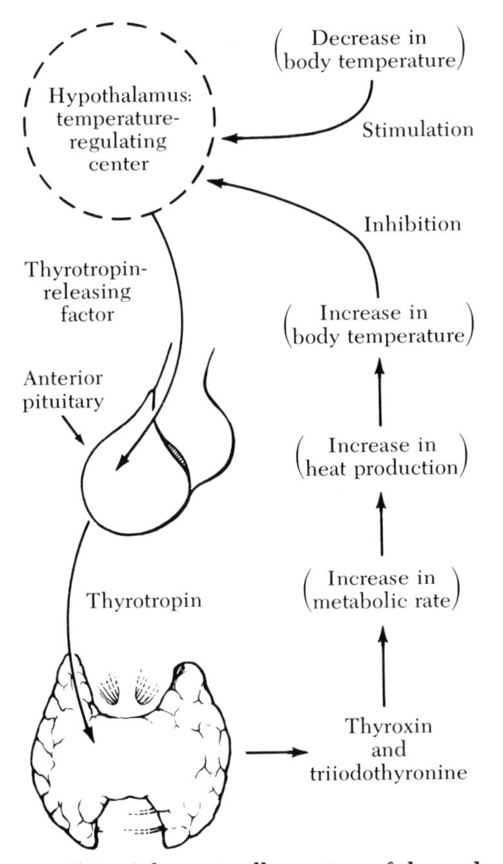

Figure 25-1 Schematic illustration of how the ductless gland system (endocrine system) responds to a decrease in body temperature.

ment depends on specialized receptor cells that are sensitive to a wide variety of physical, chemical, and mechanical stimuli. The primary function of the receptor is to detect changes in either the internal or external environment and generate a signal, known as a nerve impulse, which is passed through the neurons.

Since there are many diverse types of stimuli to which the animal must be able to respond, it is not surprising that there are also many different types of receptors. Some of the receptors are very simple, like free nerve endings which detect mechanical distortion, whereas other receptors are very complex, such as the eyes and ears (Table 25.1). Figure 25.2 shows some of the different types of receptor cells found in vertebrates. Just as there are specialized receptors which detect various external stimuli, there are also specialized receptors which respond to internal stimuli. The most prominent type of internal receptors are those which detect changes in the oxygen or carbon-dioxide content of the blood. These are called chemoreceptors, chemical receptors. There are also internal receptors which respond to mechanical distortion or stretch in muscles, and others which detect pressure changes, such as changes in blood pressure.

Neurons The function of the receptor is to detect changes in the environment and to determine if the change (stimulus) is of sufficient magnitude to generate a signal. If the animal is going to be able to respond to the stimulus, the signal must be transferred to the proper location so that the animal may make the necessary response. The neuron's function is to carry the signal generated by the receptor

Table 25.1.

List of some of the types of receptors found in mammals and the type of stimulus to which they respond.

Type of receptor	*Stimulus for receptor*
Exteroceptors	External stimuli
Pacinian corpuscle	Pressure
Meissner corpuscle	Touch
Krause corpuscle	Cold
Ruffini endings	Heat
Free nerve endings	Pain
Proprioceptors	Stimuli from muscles, tendons, and other internal tissue
Muscle spindles	Stretch
Golgi tendon organs	Stretch
Interoceptors	Stimuli from the respiratory and gastrointestinal tract
Specialized receptors	
Taste buds	Taste
Eye	Light
Ear	Sound
Semicircular canals	Movement

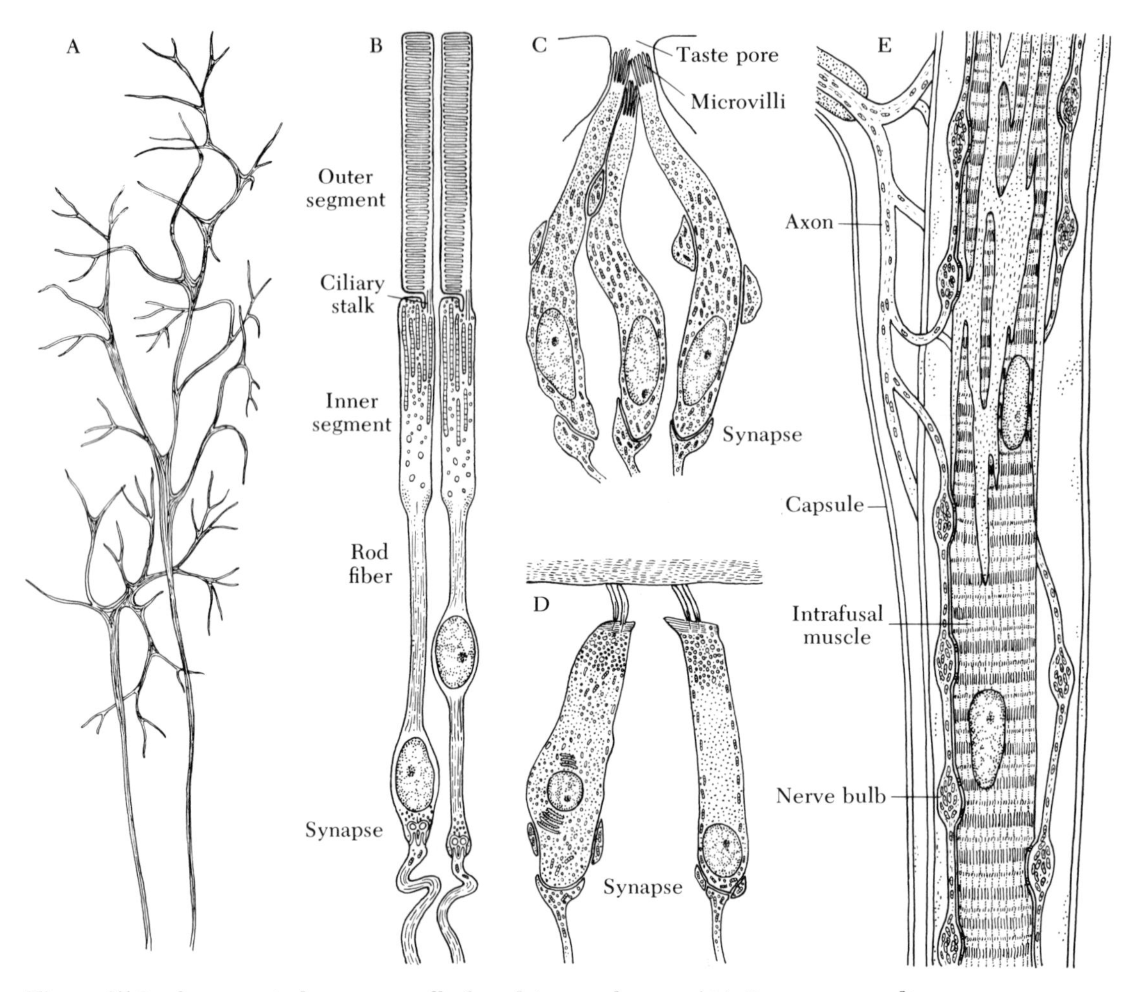

Figure 25-2 Some typical receptor cells found in vertebrates. (A) Free nerve endings found on the head of the pit viper which are sensitive to heat and help the viper locate its prey. (B) Rods, which are light-sensitive cells in the retina of the eye. (C) Taste buds, which are chemoreceptors embedded in the tongue. (D) Corti cells of the inner ear, which contain hair-like projections that are sensitive to sound vibration. (E) Muscle spindle found in muscles which are sensitive to stretching of muscle fibers. (From Miller, Ratliff, and Hartline, "How Cells Receive Stimuli." Copyright © 1961 by Scientific American, Inc.)

to the cell or tissue (effector) where the necessary response is made. Thus we can think of the neurons as somewhat similar to telephone lines, which conduct the message from the mouthpiece of one telephone to the receiver of another telephone. If a neuron is cut, then the signal is blocked, just as when a telephone line is cut the message fails to get through. But in reality, the nervous system is much more complex than the telephone, for the signal generated by the receptor may be modified in various ways before it reaches its intended destination. The signal may be either amplified or inhibited depending upon the circumstances. We will consider this in greater detail when we discuss the function of the higher nervous system.

There are three different types of neurons,

the sensory or afferent neuron, the motor or efferent neuron, and the association or interneuron. The sensory or afferent neurons conduct the signal or impulse from the receptors to the spinal cord and brain. The sensory neuron may connect directly with a motor neuron or may connect with an interneuron. The interneuron or association neuron makes connections between sensory and motor neurons within the spinal cord and brain. There may be any number of interneurons between the sensory and motor neurons. The motor neuron conducts the signal from the spinal cord or brain to the effector organ.

The neuron contains a cell body and typically two types of cell projections, dendrites and axons. Neurons differ in the location of the cell body, number of dendrites they possess, and the length of the axon. The cell body of the sensory neuron is located outside the spinal cord, while the cell body of the motor neuron is located within the spinal cord or brain. A typical myelinated motor neuron is illustrated in Figure 25.3.

Neurons do not make direct connections with one another; instead, there is a very small gap between them, of approximately 200 Ångstroms (Figure 25.4). This gap is called a synapse, and for the signal to be propagated it must be transmitted across this gap. This is known as synaptic transmission. The propagation of the signal across the synapse differs from the propagation of the signal along the neuron. Transmission of the signal across the synapse involves the release of a chemical substance (acetylcholine in some cases) and the diffusion of this substance across the synapse where it combines with a specialized receptor, which generates the signal in the other neuron. Because transmission of the signal is slower across the synapse than along the neuron itself, the more synapses a signal must cross, the longer it takes for the signal to reach its ultimate destination. Thus, when a quick response is required there are very few synapses between the sensory or motor neurons, but in other cases such as memory there may be hundreds of synapses between the sensory and motor neurons.

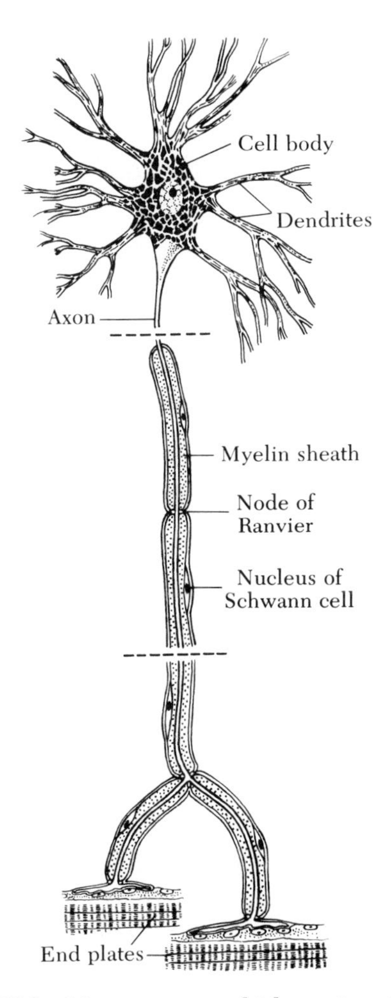

Figure 25-3 Motor neuron which carries signal to muscle fibers. The cell body fans out into a number of "twigs," the dendrites, which connect with other neurons. The nerve impulse or signals travels from the cell body through the axon to the motor end plate. (From Katz, "How Cells Communicate." Copyright © 1961 by Scientific American, Inc.)

Effectors The effector organ is not really a part of the nervous system, but for our discussion we will consider it a basic part of the nervous system, since it is what responds to the initial stimuli. The effector organ may be a muscle, such as the muscles involved in removing your hand from a hot object after you

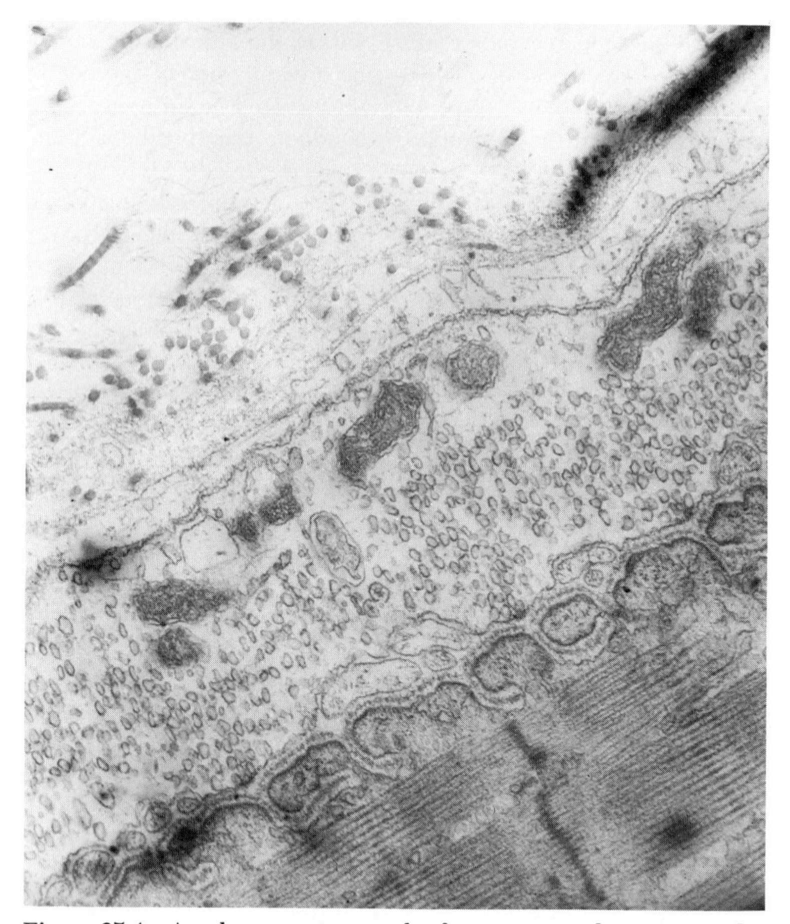

Figure 25-4 An electron micrograph of a nerve-muscle synapse. The motor neuron terminal runs diagonally from the lower left to upper right of the picture. The muscle fiber is the dark stricted area at the lower right, with a folded membrane. (Micrograph by R. Birks, H. E. Huxley, and B. Katz.)

have touched it, or it may be some type of a gland, such as the salivary glands. The heart and respiratory muscles may also be considered effector organs, since they respond to various impulses by either increasing or decreasing heart or respiratory rate.

The Control of Voluntary Muscle

So far we have considered the basic units of the nervous system, the receptor, neuron, and effector. Let us now consider how these units function together in regulating muscles involved in such diverse action as standing, sitting, walking, writing, lifting, or throwing a ball. The simplest type of regulation of voluntary muscle involves very few neurons, usually sensory, association, and motor neurons. This is known as the somatic reflex arc or spinal reflex arc, and since it involves very few synapses, the response is very rapid.

We have previously considered one example of the somatic reflex arc or spinal reflex arc, the withdrawal of the foot from a sharp object, or painful stimulus. Let us now consider how the basic units of the nervous system function together to cause withdrawal of the foot. The basic components of this spinal reflex are illustrated in Figure 25.5. First, pressure receptors in the sole of the foot are stimulated when the foot steps on the sharp object. If the stimulus is of sufficient strength or duration, then a signal or impulse will be generated in the sensory-nerve fiber. This signal travels the entire length of the sensory nerve into the spinal cord where it may synapse with an interneuron as shown in Figure 25.5 or may cross the spinal cord to the ventral horn where it synapses directly to a motor neuron. This signal is then transmitted across the synapse as described previously, to the interneuron or motor neuron. The interneuron may pass across the spinal cord to the ventral horn, where it synapses with the motor neuron as shown in Figure 25.5, or it may pass up the spinal cord to synapse with higher brain centers; this latter possibility will be considered later. The signal is then transmitted through the motor neuron to the skeletal muscle and causes it to contract, thus removing the foot from the sharp object.

In reality the picture is much more complicated, but the basic components of the reflex are the same for all reflex actions of the body. Instead, many receptors would be activated and the sensory neurons would synapse with several interneurons, some of which might pass across the spinal cord to synapse with several motor neurons. Other interneurons might pass up the spinal cord to synapse with motor neurons in other segments of the spinal cord. Others might synapse with higher brain centers, such as the medulla, cerebellum, or cerebral cortex as illustrated in Figure 25.6. These higher brain centers would then send modifying signals down the spinal cord which could either excite (stimulate) or inhibit the motor neurons. Thus, the higher brain centers act to modify the response. They may increase the excitability of certain motor neurons, making

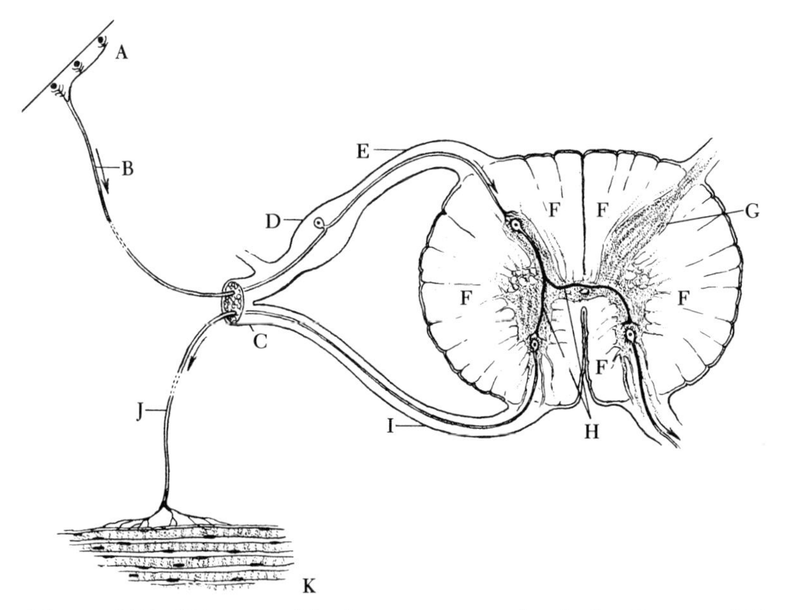

Figure 25-5 Diagram of basic components of spinal reflex arc. (A) Pressure receptor in the foot, (B) sensory neuron, (C) spinal nerve, (D) sensory nerve cell body, (E) dorasl root of spinal nerve (F) white matter of spinal cord, (G) Gray matter of spinal cord, (H) Interneuron, (I) ventral root of spinal nerve, (J) motor neuron, (K) muscle of leg, (L) spinal cord. (From Cole, *Introduction to Livestock Production*. Copyright © 1966 by W. H. Freeman and Company.)

them easier to stimulate, or they may decrease the excitability of other motor neurons, making them more difficult to stimulate. In our example of reflex withdrawal of the foot from a sharp object, some muscles are stimulated to contract, while other muscles are prevented from contracting due to inhibitory signals from higher brain centers. The modifying signals from higher brain centers are responsible for the smooth and coordinated withdrawal of the foot. If the spinal cord were cut above the entry point of the sensory neuron, the reflex would not be abolished, but the withdrawal of the foot would no longer be smooth and coordinated.

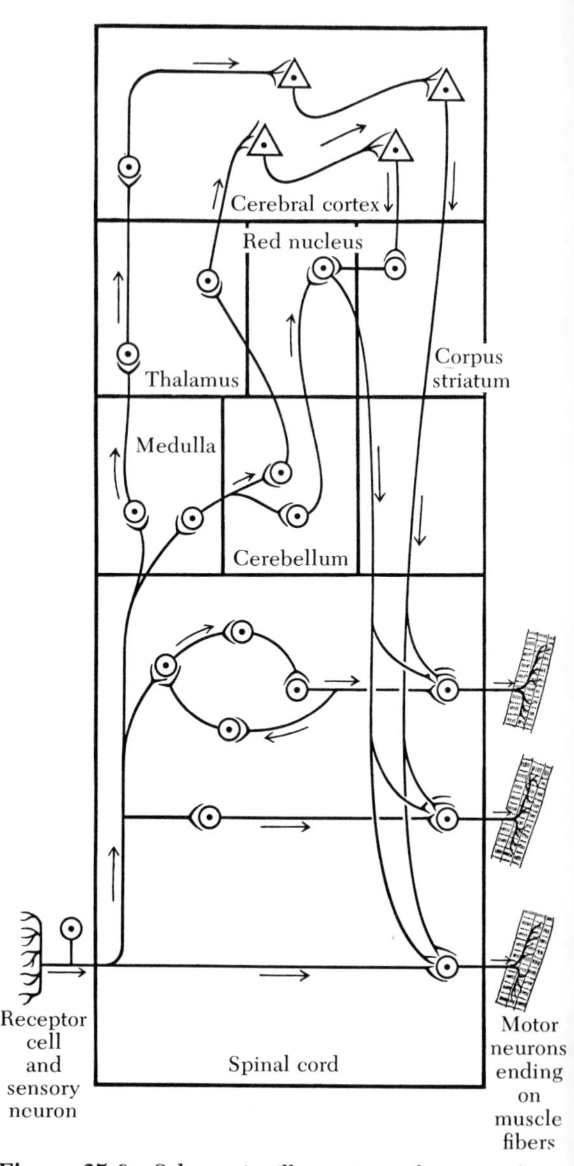

Figure 25-6 Schematic illustration of some of the possible routes a nerve impulse may take from the receptor cell to the effector organ. The impulse may pass directly across the spinal cord to connect with a motor neuron; or it may travel up the spinal cord, and connect with a motor neuron, or may travel on up to various centers in the brain, where the impulse may be either amplified or inhibited. These centers in the brain in turn send signals down the spinal cord which connect with the motor neurons (From Katz, "How Cell Communicate." Copyright © 1961 by Scientific American, Inc.

The Control of Involuntary Muscles and Glands

The regulation of involuntary muscles (smooth muscle) and glands of the body are controlled by the autonomic nervous system, as opposed to the somatic nervous system, which regulates voluntary or skeletal muscles. The autonomic nervous system differs structurally from the somatic nervous system in that it has two motor neurons instead of one. Whereas the somatic nervous system is mainly concerned with regulation of the body responses to external stimuli, the autonomic nervous system is more concerned with regulation of the internal environment of the body. The autonomic nervous system regulates such diverse tissues as the heart, salivary glands, eye movement, and the digestive, excretory, and respiratory systems (Figure 25.7).

The autonomic nervous system consists of two divisions, the sympathetic and parasympathetic. In many cases these two divisions of the autonomic nervous system innervate the same tissue, but have opposite physiological actions. For an example, stimulation of the sympathetic nerve fibers to the heart causes dilation of the heart vessels and an increase in heart rate, whereas stimulation of the parasympathetic nerve fibers to the heart causes constriction of the heart vessels and a decrease in heart rate. The sympathetic and parasympathetic divisions of the autonomic nervous system also are quite anatomically different from each other. These anatomical differences are summarized in Table 25.2.

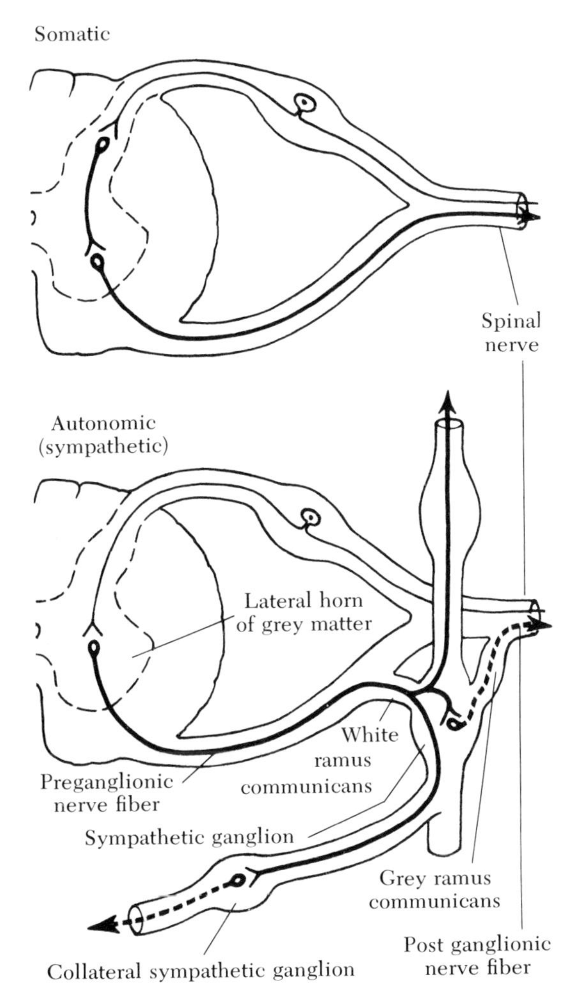

Figure 25-7 Diagram illustrating a somatic reflex arc (left) and an autonomic reflex arc (right). Note that only a single motor neuron is involved in the pathway from the CNS to a voluntary muscle, whereas there are two neurons in the efferent pathway to involuntary muscles and glands. The preganglionic neuron of adrenergic nerve fibers (sympathetic) terminate around a postganglionic neuron in a ganglion just outside the spinal cord in most instances, whereas preganglionic cholinergic fibers (parasympathetic) usually terminate in a peripheral ganglion in close apposition to some organ.

Table 25.2.

Comparison of anatomical differences between the sympathetic and parasympathetic nervous systems.

Sympathetic nervous system	*Parasympathetic nervous system*
Cell bodies of first motor neuron located in thoracic and lumbar region of spinal cord	Cell bodies of first motor neuron located in brain and sacral region of spinal cord
Cell bodies of second motor neuron located in single ganglionic chain	Cell bodies of second motor neuron located in ganglia scattered throughout body
Synapse between first and second motor neuron occurs some distance from target organ	Synapse between first and second motor neuron occurs close to target organ
Second motor neuron releases norepinephrine	Second motor neuron releases acetylcholine

The parasympathetic system is mainly concerned with regulation of body function under resting conditions, and regulates such functions as secretion of salivary and digestive juices, normal movement of the digestive tract, and diameter of peripheral blood vessels. The sympathetic system, on the other hand, is concerned with regulation of body functions under emergency conditions. Stimulation of the sympathetic system produces a response commonly referred to as the "fight or flight" response. Heart rate is increased, blood is diverted from nonactive areas of the body to active areas, stores of body energy are mobilized, the eyes accommodate for near vision, motility and secretion of the digestive tract are inhibited, and hair becomes erect when the sympathetic system is stimulated.

Let us now consider two examples of how the autonomic nervous system functions. First we will consider autonomic nervous system regulation of rumen motility, and next the control of salivary secretion. The rumen receives both parasympathetic and sympathetic innervation. Parasympathetic innervation is by way of the dorsal and ventral trunks of the vagus nerve, whereas sympathetic innervation is by way of the splanchnic nerve. Stimulation of the vagus causes contraction of the rumen and rumen motility, whereas stimulation of the splanchnic inhibits all rumen motility. If the vagus nerve is cut, all rumen motility is abolished. The rumen becomes distended and the animal will bloat and die.

Like the rumen, the salivary glands receive both parasympathetic and sympathetic innervation. However, unlike the autonomic innervation of most smooth muscles and glands, the parasympathetic and sympathetic do not have opposite effects on salivary secretion. Stimulation of the parasympathetic nerve to the salivary glands results in the secretion of a large volume of dilute saliva and vasodilation. The latter, in turn, results in increased blood flow through the glands. Stimulation of the sympathetic innervation causes the secretion of small quantities of a thick saliva containing mucoproteins and vasoconstriction, which causes decreased blood flow through the glands.

The Central Nervous System

The central nervous system, or the higher nervous system as it is sometimes called, consists of the spinal cord and the brain. We have already considered the spinal cord in some detail in our discussions of control of voluntary muscles. It should be remembered that afferent neurons from the periphery pass up the spinal cord to the brain and efferent neurons of both the autonomic and the somatic nervous system pass down the spinal cord to innervate different glands and muscles. Thus the spinal cord may be thought of as a sort of freeway for the travel of nerve impulses.

The brain itself is a vastly complex tissue, consisting of billions of nerve cells, which are interconnected by interneurons. It is these many interconnections between different nerve cells which makes the brain a difficult tissue to study. But certain areas of the brain have been found to have specific functions and we will now consider these.

The Hindbrain

The hindbrain consists of two distinct structures, the medulla and the cerebellum. The medulla is that portion of the brain which is connected with the spinal cord (Figure 25.8). The medulla contains centers which regulate breathing, the heart, blood flow, the secretion of several glands, coughing, sneezing, vomiting, and the adjustment of posture. Since the medulla contains many centers which regulate vital life functions, the importance of this area of the brain should be apparent.

The lesser brain, as the cerebellum is sometimes known, is located just anterior to the medulla. The size of the cerebellum is related to body size, being largest in elephants and whales. There appears to be good reason for this relationship, since one of the major functions of the cerebellum is coordination of body movements. The cerebellum receives signals from receptors located in muscles, as well as from the motor center of the cerebral cortex. The information received by the cerebellum is somehow analyzed, and the cerebellum sends back modifying signals to the motor center of cerebral cortex and motor neurons. This results in a smooth and coordinated response by the muscles (Figure 25.6). In addition to its function in coordinating skeletal muscle activity, the cerebellum also serves a similar function in maintaining the equilibrium of the body, and in coordinating the sensations of touch, hearing, and sight. An example of this last type of coordination is when an animal hears a sound and turns its eyes to look in the direction of the sound.

The Forebrain

In mammals, the forebrain consists of the thalamus, the hypothalamus, and cerebrum. The cerebrum, with its prominent olfactory bulbs, composes the anterior portion of the forebrain, while the thalamus and hypothal-

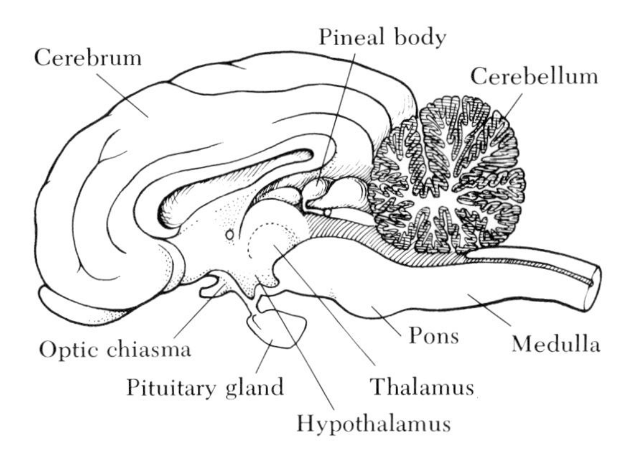

Figure 25-8 Diagram of the major structures of the sheep brain and location of the origin of the cranial nerves: A, lateral view; B, dorsal view; C, ventral view.

amus make up the posterior portion of the forebrain.

The thalamus serves as the major relay center for sensory impulses to the cerebrum. It sorts the various sensory impulses which it receives from all over the body and directs them to the appropriate areas of the cerebrum. The thalamus also contains part of a neural formation known as the reticular system. One of the functions of the reticular system is the arousal response, or the ability to awaken an animal. Thus the reticular system acts as a sort of an alarm clock to awaken an animal; it does this by making the animal aware of external stimuli, such as early morning light shining through the window.

The hypothalamus is located directly below the thalamus, and is an important control center for visceral functions of the body. Centers which control hunger, thirst, body temperature, water balance, blood pressure, reproductive behavior, rage, and pain are also located in the hypothalamus. The hypothalamus also controls the synthesis or secretion of hormones from the anterior pituitary gland, a subject we shall consider in more detail when we discuss control of hormonal secretion.

In birds and mammals, the cerebrum constitutes the major part of the brain; in fact, it represents about three-quarters of the nervous tissue of the body. Only a small area of the cerebrum is devoted to purely motor or sensory functions, but definite areas of the cerebrum can be shown to have control of different motor functions of the body, such as movement of the fingers, toes, or lips. Similarly, there is a definite area of the cerebrum which regulates the various sensory functions of the body. The major portion of the cerebrum is concerned with such diverse functions as speech, learning, memory, thought, and purposeful actions, which cannot be easily explained in terms of reflex areas and reflex centers.

The Endocrine System

So far we have considered in some detail how the nervous system allows an animal to respond to changes in its environment, but the endocrine system also plays an important role in regulating an animal's response to changes in its environment. The response of the endocrine system, however, is slower and more diffuse compared to the nervous system. We will now consider in some detail the ductless gland or endocrine system and how it responds to changes in the environment.

The locations of the various endocrine organs in the cow are illustrated in Figure 25.9. The hypothalamus and pituitary glands, which exert considerable control over the other endocrine organs, are located within the skull, in close association with the brain. As we shall see later, this close association with the brain is an important factor in regulating the secretion of the various endocrine glands. It is also another indication of the close relationship between the nervous and endocrine systems.

We will not be able to consider all the various physiological actions of the endocrine glands. Some of the endocrine glands such as the gonads (the ovaries and testes) and the pituitary will be discussed in more detail in other chapters in this section. Therefore, we will not discuss each endocrine gland separately, but consider the overall picture of the endocrine system.

Hormones

Hormones are secretory products of the endocrine system. They are carried by the general circulation to some distant target tissue or organ where they exert their effects. But not all hormones have to travel to some distant site to exert their effect; for example, testosterone, produced by the interstitial cells of the testes, has a local action on the seminiferous tubules of the testes. Some hormones excite their target tissue or organ; other hormones have an inhibitory effect on their target organs. Still other hormones may have either excitatory or inhibitory effects, depending on their concentration.

The hormones produced by the various endocrine glands of the body, the type of tissue or organ they affect, their major physiological action, and their chemical nature are listed in Table 25.3. As can be seen from this table, the anterior pituitary produces six different hormones. No single cell in the anterior pituitary has the ability to produce all six hormones. Instead, each hormone appears to be produced by a different type of cell. The adrenal gland also produces several hormones which have very different physiological actions. The adrenal medulla, the inner portion of the adrenal gland, produces epinephrine and norepinephrine, whereas the adrenal cortex, the outer portion of the adrenal gland, produces the glucocorticoids and mineralocorticoids. The thyroid and the pancreas, on the other hand, produce only two hormones: thyroxine and triiodothyronine from the thyroid; insulin and glucagen from the Islets of Langerhans in the pancreas. The parathyroid glands produce only a single hormone, the parathyroid hormone. Although there is some debate about the source of calcitonin, recent evidence indicates that calcitonin is found in the ultimobranchial cells embedded in the thyroid.

Although the various hormones secreted by the different endocrine glands have different physiological effects, it is believed that most hormones have a common mechanism of action. Scientists do not agree on the nature of this underlying mechanism of action. Some believe that hormones act by increasing the synthesis of a specific enzyme; others believe that hormones act to increase the permeability of a given tissue to some substrate or ion.

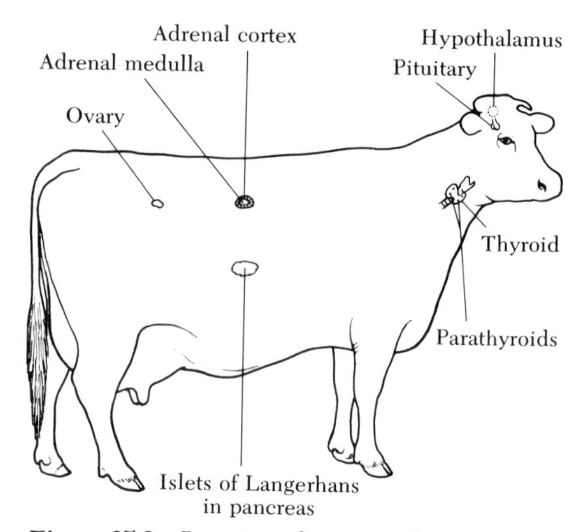

Figure 25-9 Location of major endocrine glands in the cow.

Means of Controlling Hormone Secretion

Although certain hormones are probably secreted continuously in small amounts, the secretion of hormones is controlled by three means: (1) the nervous system; (2) other hormones; and (3) changes in the chemical composition of the blood supplying the endocrine organ.

Nervous control One example of nervous control of hormonal secretion is the secretion of epinephrine by the adrenal medulla. Stimulation of the sympathetic nerve to the adrenal medulla will cause the secretion of epinephrine. It should be remembered from our discussion of the autonomic nervous system that the sympathetic nervous system is stimulated in response to emergency situations, and the adrenal medulla secretes epinephrine to aid the animal in meeting such a situation.

In certain instances the nervous and endocrine system work together to elicit a certain response. An example of this type of joint action between the nervous and endocrine systems would be the regulation of ovulation in the rabbit. The act of mating or any type of mechanical stimulation of the cervix will excite sensory neurons, which in turn will bring about the release of gonadotropins from the anterior pituitary. The gonadotropins then will bring about ovulation. Other examples of the endocrine and nervous system working together to elicit a certain response, such as milk let-down, as well as the nervous control of hormonal secretion, will be discussed in other chapters in this section.

Hormonal control In the introduction we considered one example of hormonal control by another hormone, the control of thyroxine secretion by thyrotropin from the anterior pituitary gland. The anterior pituitary also secretes three other hormones: adrenocorticotropin (ACTH), which controls the secretion of glucocorticoids by the adrenal cortex; and luteinizing hormone (LH) and follicle-stimulating hormone (FSH), which control the secretion of the gonadal hormones by the testes and ovaries. Hormonal control of the secretions of the gonads will be considered in more detail in the chapters on reproduction. Therefore, we shall consider the control of glucocorticoid secretion by ACTH.

Low levels of glucocorticoids in the blood act to stimulate the anterior pituitary to secrete ACTH, which in turn causes the adrenal cortex to secrete more glucocorticoids, thereby raising the level of glucocorticoids in the blood. This type of hormonal regulation by the anterior pituitary is referred to as the negative-feedback mechanism. As was the case with the hormonal regulation of thyroxine, the hypothalamus is also involved in the control of glucocorticoid secretion by the adrenal cortex. Low levels of glucocorticoids act on the hypothalamus to cause the secretion of a substance called the corticotropin releasing factor, which in turn causes the release of ACTH from the anterior pituitary.

There is another means by which the secretion of glucocorticoids can be regulated, that is, by the secretion of epinephrine from the adrenal medulla. Any stressful stimulus which causes the release of large amounts of epinephrine from the adrenal medulla will also cause the secretion of ACTH from the anterior pituitary and so increase the secretion of glucocorticoids. Thus, both epinephrine and ACTH act to regulate the secretion of glucocorticoids, but epinephrine exerts its control of glucocorticoid secretion through ACTH.

Control by changes in blood chemistry The secretion of several hormones is controlled by changes in the chemical composition of the blood, such as insulin by the level of blood sugar, aldosterone by the sodium ion concentration of the blood, and calcitonin and parathyroid hormone by blood calcium levels. As an example of this type of control, let us consider the regulation of parathyroid hormone. When blood calcium levels fall below a certain value, the parathyroid gland is stimulated to secrete parathyroid hormone. Parathyroid hormone acts to mobilize calcium stores from the skeleton and thereby increases blood-calcium levels. This means of regulating parathyroid hormone secretion is of practical interest, since one of the causes of milk fever in dairy cattle is believed to be the inability of low blood-calcium levels to stimulate the secretion of parathyroid hormone. It has been postulated that this is due to the fact that parathyroid is unresponsive because the calcium needs are low during late lactation and the dry period.

Interaction Between Various Hormones

Most endocrine responses usually involve the action of more than one hormone. Hormonal regulation of reproduction, for example, involves hormones from both the anterior pituitary and gonads. Hormonal regulation of lactation involves hormones from the anterior and posterior pituitary, the adrenal cortex, and the gonads. Similarly, the hormonal regulation of various metabolic processes of the body in-

Table 25.3.

List of hormones produced by the various endocrine glands of the body, their source, type of tissue they influence, their chemical nature, and major physiological actions.

Endocrine organ[a]	*Hormone*	*Source of hormone*	*Chemical nature of hormone*	*Tissue influenced*	*Major physiological action*
Anterior pituitary	Growth hormone	Acidophilic cells	Protein	All tissues	Stimulates general body growth
	Adrenocorticotropin (ACTH)	Unknown	Polypeptide	Adrenal cortex	Maintenance of function of adrenal cortex
	Thyrotropin (TSH)	Small basophilic cells	Protein	Thyroid	Stimulate secretion of thyroxine
	Prolactin (LTH)	Acidophilic cells	Protein	Ovaries and mammary gland	Corpus luteum maintenance and progesterone secretion in some species, milk secretion
	Luteinizing hormone (LH)	Large basophilic cells	Protein	Ovaries and testes	Androgen secretion by testes, maturation of follicles and ovulation
	Follicle stimulating hormone (FSH)	Large basophilic cells	Protein	Ovaries and testes	Oogenesis and spermatogenesis
Intermediate lobe of pituitary	Melanocyte stimulating hormone (MSH)	Basophilic cells	Polypeptide	Melanophore cells	Controls cutaneous pigmentation
Posterior pituitary	Oxytocin	Supraoptic and paraventricular nuclei of hypothalamus	Polypeptide	Mammary gland and uterus	Milk release and contraction of uterus
	Vasopressin or antidiuretic hormone	Supraoptic and paraventricular nuclei of hypothalamus	Polypeptide	Peripheral blood vessels and kidney tubules	Constriction of blood vessels and water reabsorption from kidney tubules
Thyroid	Thyroxine and triiodothyronine	Thyroid follicles	Amino acid	All tissue	Increased rate of cellular metabolism
	Calcitonin	Ultimobranchial cells	Polypeptide	Bone, kidney	Decreases blood calcium levels
Parathyroid	Parathyroid hormone	Chief cells	Polypeptide	Bone, kidney and intestine	Increases blood calcium level

Iodine

		zona reticularis			increase blood sugar levels, anti-flammatory action
	Mineralocorticoids	Zona glomerulosa	Steroid	Kidney, all tissues indirectly	Salt and water balance
Adrenal medulla	Epinephrine	Chromaffin cells	Amino acid	Heart, skeletal muscle and liver	Increases strength of heart contraction and heart rate, increases blood sugar
	Norepinephrine	Chromaffin cells	Amino acid	Peripheral blood vessels, smooth muscle	Constriction of peripheral vessels and contraction of smooth muscles and glands
Pancreas	Glucagon	Alpha cells of Islet of Langerhans	Polypeptide	Liver	Raises blood sugar levels
	Insulin	Beta cells of Islet of Langerhans	Polypeptide	Muscle and adipose tissue	Lowers blood sugar levels
Kidney	Renin and angiotensin	Juxtaglomerular cells	Polypeptide	Zona glomerulosa	Increases blood pressure, aldosterone secretion
Ovary	Estrogens	Follicles	Steroid	Mammary gland, genital tract	Stimulates development and maintenance of female secondary sexual characteristics
	Progesterone	Corpus luteum	Steroid	Mammary gland, genital tract	Required for implantation of fetus
	Relaxin	Unknown	Polypeptide	Cartilage and ligaments of pelvic girdle	Relaxation of cartilage and ligaments to facilitate parturition
Testes	Androgens	Interstitial cells	Steroids	Accessory sex organs, testes	Development of accessory sex organ and secondary sex characteristics of male
Gastrointestinal tract	Secretin	Pyloric mucosa of stomach	Polypeptide	Pancreas	Stimulates alkaline secretion for digestion
	Pancreozymin	Mucosa of duodenum	Polypeptide	Pancreas	Stimulates enzyme secretion for digestion
	Cholecystokinin	Mucosa of duodenum	Polypeptide	Gall bladder	Evacuates bile into intestine
	Enterogastrone	Mucosa of duodenum	Polypeptide	Stomach	Inhibits motility and acid secretion of stomach
	Gastrin	Mucosa of duodenum	Polypeptide	Stomach	Stimulates secretion of HCl

[a] The hypothalamus may also be considered an endocrine organ, since it is the source of the various releasing factors.

volve the interaction of two or more hormones (Figure 25.10). We do not have the space here to consider all these various hormonal interactions, and some of these will be discussed in later chapters.

Use of Natural and Synthetic Hormones in Animal Science

Many of the hormones and other biological and synthetic compounds which possess hormonal activity have been found to have practical application. These applications have not been limited to medical uses only; in fact, a number have practical application to the livestock industry. We will not be able to discuss all these applications here; some will be discussed in greater detail in other chapters.

Many of the hormones and compounds which possess hormonal activity have found practical application due to their ability to either affect the general metabolism of the animal or to promote or alter the growth rate. An example of some of these compounds are iodinated casein, the goitrogens, and stilbestrol. Other hormones and hormonal-like substances have found practical application due to their

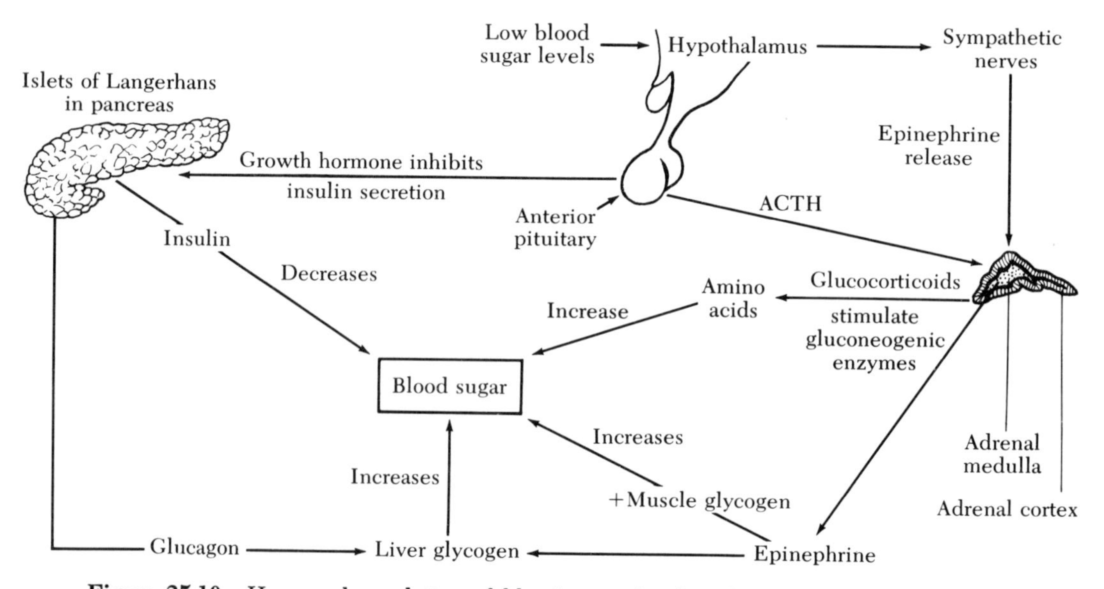

Figure 25-10 Hormonal regulation of blood-sugar levels: The Islets of Langerhans in the pancreas secrete insulin (*A*), which decreases blood sugar; low blood-sugar levels (*B*) stimulate the hypothalamus, which in turn stimulates the adrenal medulla via the sympathetic nervous system (*C*) to secrete epinephrine (*D, H*); epinephrine (*D*) stimulates the anterior pituitary to secrete ACTH (*E*), which stimulates the adrenal cortex to secrete glucocorticoids (*F*), which cause amino acids to catabolize to glucose, thereby increasing blood sugar (*G*); epinephrine (*H*) also acts upon glycogen in muscle and the liver to convert it to glucose, thereby further increasing blood sugar (*I, J*). The Islets of Langerhans in the pancreas also secrete glucagon (*K*), whose immediate effects are opposite those of insulin: just as epinephrine, glucagon acts on the glycogen in the liver to convert it to glucose, thereby increasing blood sugar (*J*). Growth hormone (*L*), which is secreted by the anterior pituitary, inhibits the secretion of insulin. Note that only one hormone, insulin, is required to lower blood sugar, but that the sympathetic nervous system and as many as five hormones—glucagon, epinephrine, ACTH, the glucocorticoids collectively, and growth hormone—may be involved in raising blood sugar.

ability to regulate reproductive processes. These hormones and compounds have been used for estrus synchronization, to improve fertility in both males and females, and to promote multiple births in sheep and beef cattle. An example of some of the hormones and compounds used for this purpose are melengestrol acetate and 6-chlor-Δ^6-dehydro-17-acetoxy progesterone, which are synthetic progestins, pregnant mare serum gonadotropin, and human chorionic gonadotropin.

Finally, let us consider two applications of substances possessing hormonal activity to areas outside the livestock industry, specifically, to birth control. The "pill" usually consists of some synthetic progestin and a small quantity of an estrogen. It is still not understood exactly how the pill works, but it is believed that the progestin in the pill prevents ovulation, the release of the egg, by preventing the secretion of enough LH to induce ovulation. Another area where hormones and synthetic hormones have found practical application is in controlling rodent and other wildlife populations. Progesterone and progestins have been used to control reproduction in these animals by preventing either ovulation or implantation.

FURTHER READINGS

Davidson, E. H. 1965. "Hormones and Genes." *Scientific American,* 212 (no. 6): 36. Available as Offprint No. 1013 from W. H. Freeman and Company.

Grayson, J. 1960. *Nerve, Brain, and Man.* New York: Taplinger.

Turner, C. D., and J. T. Bagnara. 1971. *General Endocrinology.* Philadelphia, Pa.: Saunders.

Walsh, E. G. 1964. *Physiology of the Nervous System.* London: Longmans, Green.

Twenty-Six

Reproduction in the Female

"And this shows that the seed of the mother has power in the embryo equally with that of the father."

Leonardo da Vinci
Quaderni d'Anatomie, III, 8V.

Reproduction in the female is a phenomenon involving three cycles: the life cycle; the annual breeding cycles; and the estrous or ovarian cycle. The sequence of estrous cycles is broken by pregnancy, which, generally speaking, occurs annually.

Cyclic Reproductive Phenomena

The Life Cycle

The eggs (ova), which when fertilized by the spermatozoa of the male begin the next generation, are produced in the ovaries of the female. These eggs are present at birth and, although many thousands are potentially available, relatively few—only some 20 to 30 in most cows—are shed during the entire reproductive life. Each egg is surrounded by a layer of flattened cells, and in later life some of these structures (primary follicles) develop into Graafian follicles. These follicles, each of which contains only one egg, are filled with follicular fluid and may be from one to several millimeters in diameter.

The development of the follicles to a point where they are mature, and their rupture to shed their eggs, is dependent upon hormones produced by the anterior pituitary gland. The functioning of this gland is regulated by a special region at the base of the brain called the hypothalamus. Two hormones, follicle-stimulating hormone (FSH) and luteinizing hormone (LH), are presumably involved in all species in controlling ovarian activity, whereas a third hormone, prolactin, appears to be involved in the maintenance of *corpora lutea* in certain species, e.g., the rat. The anterior pituitary produces a series of hormones, all of which are proteins, and during the period of rapid growth in early life it is fully occupied producing the hormones responsible for growth, growth hormone or somatotropin, and the thyroid-stimulating hormone, thyrotropin. When growth slows down, at about two-thirds of adult weight, increasing amounts of the gonadotropic hormones, FSH and LH, are produced. This results in the development of follicles in the ovaries. These follicles grow and regress in waves until one or more rupture. This dramatic point in the life of the female animal is called puberty, and thereafter she will exhibit a regular pattern of reproductive behavior.

Puberty in pigs is a function of age, and is little affected by nutrition; it is necessary to restrict feed intake by 50 per cent in order to delay the age of mating. In sheep and cattle, on the other hand, the age of puberty is markedly affected by nutrition. In sheep puberty is also affected by the time of the year when lambs are born, since the pituitary of the sheep responds to short hours of daylight. Lambs born very early in the spring, and well-grown when the days begin to shorten in the following summer-autumn period, will respond to the short-daylight/long-dark regime and will ovulate, and so commence their breeding life at six or seven months of age. Lambs born later will not be sufficiently mature to respond in the autumn following their birth, and must await the following autumn, when they are 15 or 16 months old, before attaining puberty.

The young ewe or heifer does not exhibit mating behavior, called estrus or "heat," at the first ovulation. However, a series of ovulations at regular intervals is initiated, and at the second and subsequent ovulations the animals mate. The series of events occurring in the ovaries and accessory reproductive organs between two successive ovulations and the associated behavioral changes are referred to as an estrous cycle. The length of this cycle of events may vary from 4 days in the rat to 21 days in the cow. If a ewe is not bred, her series of estrous cycles will extend approximately from September 1 to February 1 in the northern hemisphere, which period constitutes the breeding season. In the cow and sow, on the other hand, cycles may be continuous throughout the year.

During the first breeding season of the life cycle, breeding efficiency is relatively low. In multiparous species, fewer eggs are shed at each period of estrus than later in life, and the percentages of young born and of females producing young are lower. Breeding efficiency, as measured by size of litter or frequency of twinning, and percentage of animals producing live young, increases to a maximum in the first half of the expected life span and then falls gradually. Toward the end of life, as the animal approaches senility and death, reproductive efficiency drops markedly. These characteristics of the life cycle are reviewed by Hammond and Marshall (1952) and Hammond (1971).

The Annual Breeding Cycle

Natural selection of the progenitors of our modern breeds of livestock was toward birth of their young at a time of year when their chance of survival was greatest, as is mirrored today in the different breeding seasons of the many breeds of sheep. Those which originated in particularly harsh environments, at extremes of latitude or of altitude, have short breeding seasons which restrict the birth of their young to the spring months. Those, such as the Merino, which originated in more benign environments have a much less restricted period of breeding and hence of lambing. Domestication, associated with artificial conditions of housing and feeding, has greatly modified this evolutionary pattern, but even in those classes of livestock which breed the whole year round, breeding efficiency is greatest in those months corresponding to the production of young in the spring. An exception is the domestic pig, which shows no obvious seasonality.

This annual rhythm is under the control of the external environment. Environmental factors which show an annual rhythm are daylight/dark hours, mean daily temperature, nutritive conditions, and rainfall patterns. Of these, the daylight/dark ratio is of prime importance. Ewes placed in a blacked-out room in springtime with their daylengths regulated to autumn timing will come into estrous and breed after a latent period, despite high temperatures, so that young are produced in the autumn. The goat likewise has been shown to respond to short-daylight/long-dark hours. The mare, on the other hand, responds to a long-daylight/short-dark regime and breeds in the spring and summer months. The pig and most breeds of cattle do not appear to be significantly light/dark sensitive.

The mechanism whereby the daylight/dark ratio operates is obscure, and the obscurity is increased by the fact that natural selection has caused animals, such as the sheep and goat, which have short gestation periods (five months) to respond to short daylengths, whereas those, such as the mare, which have long gestations (11 months) respond to long days. All are geared to produce young in the spring. Regardless of whether animals respond to long or short days, the effect of photoperiod in regulating the annual breeding cycles is mediated via the anterior pituitary gland and its production of the gonadotropic hormones (Figure 26.1).*

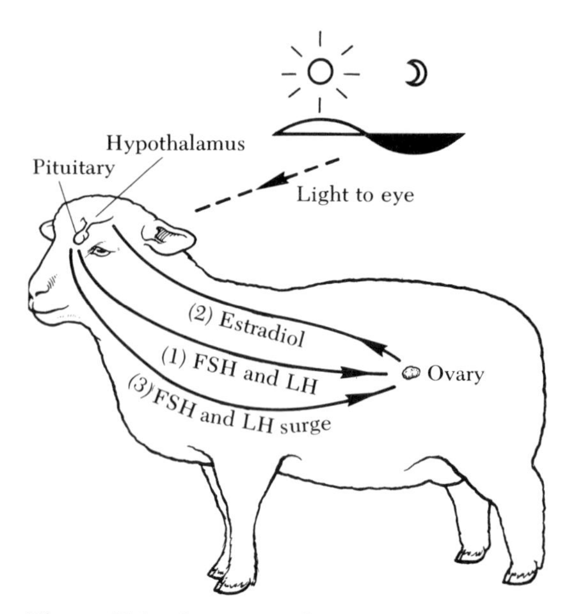

Figure 26-1 Sequence of events at puberty and at commencement of each annual breeding season. *A*, stimulus of light/dark ratio acts on receptors in head region; *B*, stimulus is transferred to hypothalmic region at base of brain; *C*, chemical message from hypothalmic region to anterior pituitary stimulates production of FSH and LH, which are released into bloodstream; *D*, FSH and LH in bloodstream cause a follicle to develop in one of the ovaries; *E*, the ripening follicle produces estradiol, which is released into bloodstream; *F*, estradiol in bloodstream stimulates anterior pituitary to release more FSH and LH into bloodstream; *G*, surge of FSH and LH causes final ripening of follicle, its rupture to form the corpus luteum, and hence the commencement of regular estrous cycles.

The Estrous Cycle

The first estrous cycle in the life of a young animal commences with the development of one (cow, mare, ewe, goat-doe) or several (sow) follicles near the surface of the ovary as a result of the action of the pituitary gonadotropins, FSH and LH. The classical concept of the role of the two gonadotropins was that there was a sequential release of FSH and LH from the pituitary. FSH was believed to be responsible for the development of follicles and the consequent production of estrogen. The estrogen exerted a positive feedback mechanism on the pituitary, resulting in release of LH and consequent ovulation. There is a good deal of evidence in the literature in support of this concept. For example, purified FSH will induce the growth of follicles in the normally inactive ovaries of sows which are suckling young.

However, some elegant, recently developed radioimmunoassay techniques for FSH, LH, and prolactin in the blood, applied to studies in women and rats, have shown no evidence for sequential release of FSH followed by LH. Both appear to be released simultaneously and, in the rat at least, prolactin is released at the same time. It is clear that one cannot consider the action of FSH or LH independently. Both are probably involved in follicular development and ovulation. The early development of follicles preceding estrus and the ini-

* For details of the duration of the breeding cycle in all species, see Asdell (1964); for a discussion of the mechanism of its control, see Marshall (1956), Amoroso and Marshall (1960), and Clegg and Ganong (1969).

tiation of secretion of estrogen is probably due to the base level or tonic secretion of FSH and LH. Increasing levels of estrogen may trigger the surge in secretion of FSH and LH which concludes the maturation process. Finally, high levels of estrogen may reduce the release of the gonadotropin to tonic levels.

The surge of gonadotropin, particularly LH, initiates the production of progesterone by the thecal cells of the developing follicles. This is believed to activate the ovulatory enzyme, which acts on the follicle wall, increasing its distensibility. The follicle swells and wall tension increases with consequent thinning of the wall on the surface of the ovary (Figure 26.2). This is the ovulation point (Figure 26.3), from which the egg with the follicular fluid of the follicle will be shed. Ultimately the follicle bursts and its contents slowly ooze out and are trapped in the distended end of the oviduct, or Fallopian tube, called the fimbriated funnel.

This maturation process of the follicle is accompanied by maturation of the egg or ovum. In the resting state its nucleus contains the normal number of chromosomes characteristic of the species, and, the egg cell being large, so too is the nucleus. As ovulation becomes imminent, the nucleus undergoes a reduction or heterotypic division, as a result of which the number of chromosomes in each of the resultant pronuclei is halved; the nucleus is reduced from the diploid to the haploid state. One of these haploid pronuclei, with some associated cytoplasm, is extruded from the egg as a small cell-like body called the first polar body. This process, the first maturation division, is usually completed just before ovulation, and is followed immediately by commencement of the second maturation division, which will not be completed unless the egg is fertilized and will be a homotypic division, in which the number of chromosomes is not further reduced but in which the total nuclear material is halved.

As mentioned earlier, the first ovulation in the life of the ruminant (ewe, goat-doe, heifer) is not accompanied by mating behavior, and the egg which enters the oviduct perishes. However, the cavity of the follicle from which the egg was shed undergoes a dramatic change. With the release of pressure following rupture, the follicle walls collapse and there is some hemorrhage into the cavity. The granulosa and thecal cells lining the cavity of the follicle, begin to multiply rapidly and to grow

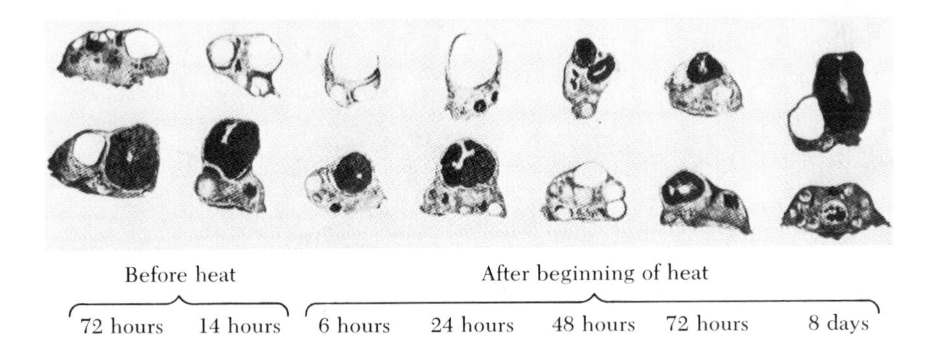

Figure 26-2 Ovaries of cows at different stages of the estrous cycle The two ovaries (right and left) of an animal are shown one above the other. In the top row the ripening follicle is seen which ruptures between 24 and 48 hours after the beginning (or 14 hours after the end) of heat and forms the new *corpus luteum* which is at first dark from a blood clot (48 hours) and later becomes rather paler and increases greatly in size (8 days). In the bottom row, which shows the ovary with the old *corpus luteum* of the previous heat period, the stages of its degeneration are seen. (Hammond 1927).

in to fill the cavity, taking with them associated blood vessels. These are secretory cells, and within four days the ruptured follicle is transformed into a highly active endocrine gland producing the hormone of pregnancy called progesterone, which, like estradiol, is a steroid (Figure 26.4). This is produced in huge quantities, several milligrams per day from a gland weighing only a gram or so. The regressing gland is yellow to orange in color, and accordingly is called the yellow body or *corpus luteum.*

Unless pregnancy intervenes, the *corpus luteum* has a highly predictable life (Table 26.1), and this determines the length of the estrous cycle. In only a small proportion of animals does this vary by more than a day or two, and this has led to an intensive study of the manner whereby the life of the *corpus luteum* is maintained or terminated. There is now evidence that the secretion of progesterone is maintained in many species by low levels of pituitary LH, but in the rat, at least, prolactin plays a role. Further, the nonpregnant uterus may actively secrete some substance which terminates the life of the *corpus luteum* after the appropriate time.

It can be seen that the estrous cycle of the farm animals is characterized by two distinct phases, one in which the animal is under the influence of estrogen and exhibits mating behavior, estrus, and one in which progesterone is the dominant ovarian hormone and she will not mate, diestrus. Between these phases there are two brief, ill-defined transition periods, during which the animal is coming into and going out of estrus, proestrus and metestrus (Figure 26.5).

Progesterone is important for three main reasons. First, it conditions the female reproductive tract to respond to the estrogen produced before the next ovulation, so preparing it for the reception of the spermatozoa introduced by the male (or by artificial insemination) and of the egg. Second, it exerts a "feedback" mechanism on the anterior pituitary, preventing a sudden release of the gonadotropic hormones, so resulting in a build-up in

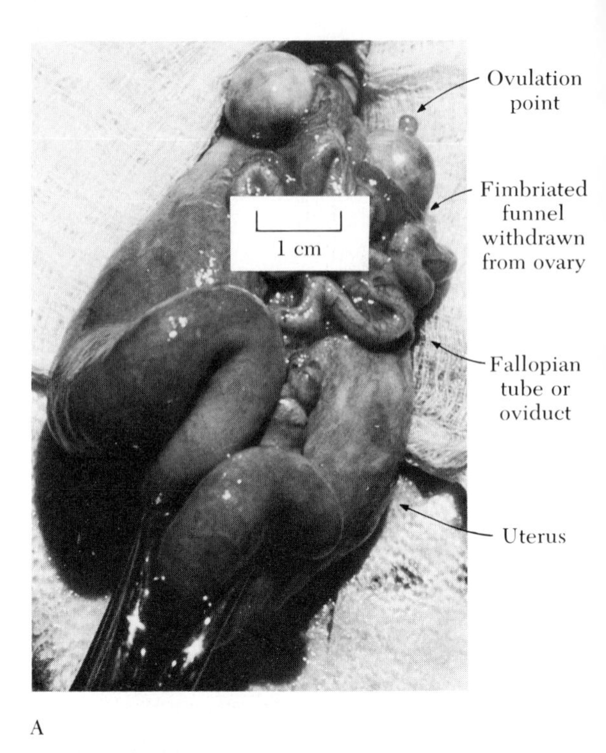

A

B

Figure 26-3 Ovary of ewe at time of ovulation showing (a) ovulation point, (b) follicular fluid oozing from ruptured point.

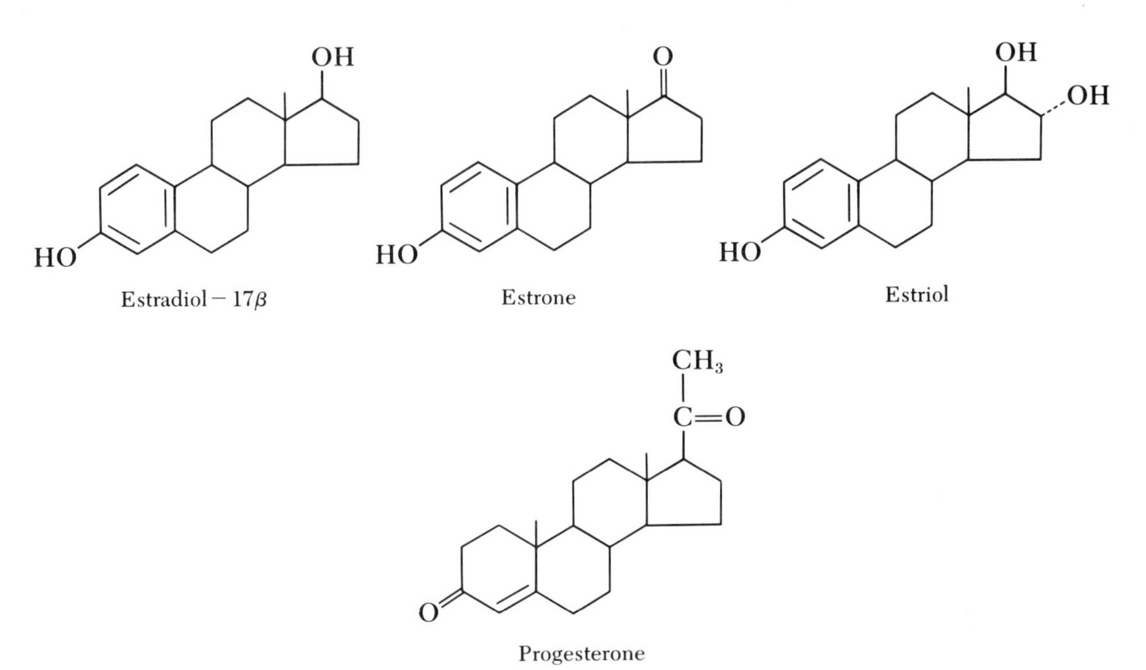

Figure 26-4 Structural formulae for the most common steroid hormones produced by the ovary. Estradiol-17β, estrone, and estriol are estrogens: estradiol-17β is the active form; estrone is relatively inactive and is a probable precursor; and estriol has very low activity.

their level for subsequent release on the decay of the *corpus luteum.* Finally, in the ewe and cow (and probably in the goat-doe, but not in the sow), the progesterone is necessary to condition the animal to exhibit estrus under the influence of the estrogen produced by the next developing follicle. The sequence of events is shown in Figure 26.5 and the effects of the hormones in Table 26.2.*

Pregnancy

Mating

At the second and subsequent ovulations following puberty in the ruminants (cow, ewe, goat-doe) and possibly the first ovulation in others, the behavior of the female undergoes a dramatic change (Figure 26.6). She indulges in courting behavior and will mate. The intensity of estrus, as measured by its duration and the number of times mating occurs, varies with age, season of the year, and nutritional status. Generally speaking, mating intensity, and hence efficiency, is low in young, poorly nourished females, particularly at the beginning and end of the breeding season in ewes and mares and in winter in heifers. For a full discussion, see Hafez (1969a).

Transport and Survival of Spermatozoa

The place where spermatozoa are deposited and stored in the female tract varies with the species. In the ewe, cow, and goat-doe, the highly concentrated semen is deposited in the vagina in the region of the cervix. As shown in Table 26.2, one effect of the estrogen produced by the developing follicle is to stimu-

* For the characteristics of the estrous cycle in all species, see Asdell (1964) and Eckstein and Zuckerman (1956); for discussions of the endocrinological and other characteristics of the cycle, see Austin (1969), Catchpole (1969), Clegg and Ganong (1969), and Nalbandov (1964).

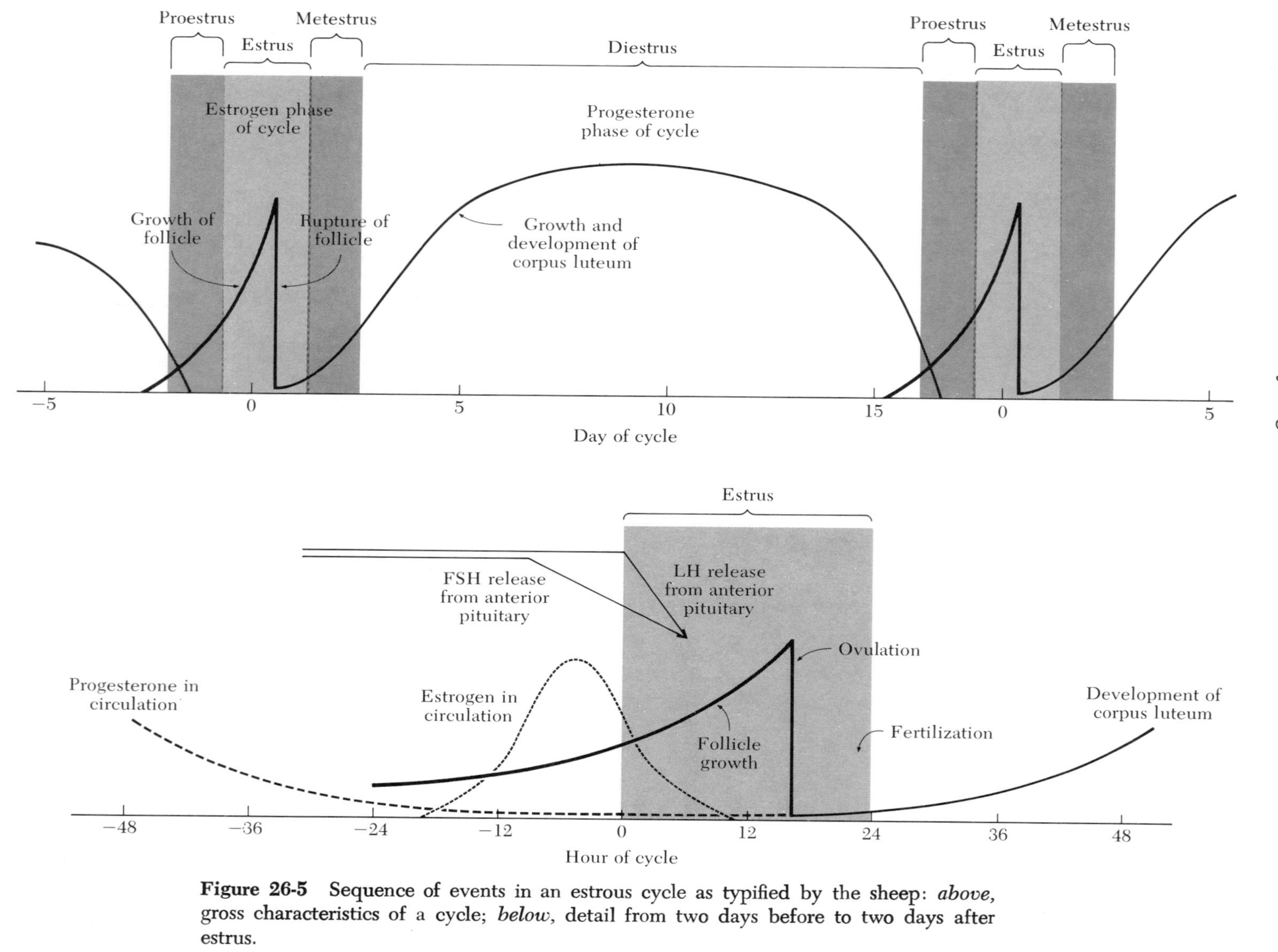

Figure 26-5 Sequence of events in an estrous cycle as typified by the sheep: *above*, gross characteristics of a cycle; *below*, detail from two days before to two days after estrus.

Table 26.1.
Characteristics of the estrous cycle and duration of gestation.

Species	Length of cycle		Duration of estrus		Life of corpus luteum (days)	Time of ovulation relative to estrus	Number of eggs shed	Duration of gestation (days)
	Mean (days)	Range (days)	Mean	Range				
Mare	21	18–24	5 days	2–8 days	16	1–2 days *before* end	1	330–345
Cow	21	18–23	18 hr	13–27 hr	18	10–15 hr *after* end	1	277–290
Ewe	17	15–19	36 hr	24–48 hr	14	Toward end	1–3, depending on breed and season	144–152
Goat-doe	20	18–22	48 hr	30–60 hr	16	Toward end	2–3	147–155
Sow	21	19–23	2–3 days		18	Second day	10–20	112–116

Table 26.2.
Sex hormones in the female.

Origin	*Name*	*Nature*	*Function*
Anterior pituitary	Follicle stimulating hormone (FSH)	Protein	With LH, stimulates follicles in the ovary to develop and to produce estrogen.
	Luteinizing hormone (LH)	Protein	With FSH, stimulates developing follicles to mature, rupture, and form *corpus luteum*. May be involved in maintaining secretion of progesterone by the *corpus luteum*.
	Lactogenic hormone (Prolactin)	Protein	Stimulates secretion of milk by mammary gland. May be involved in secretion of progesterone.
Ovary, Placenta[a]	Estradiol-17β	Steroid	Stimulates (a) growth of the vaginal wall and oedema (water retention) in the reproductive tract, (b) production of thin watery mucus by the glands of the uterus and cervix, (c) courting behavior and mating, (d) uterine motility.
	Estrone, Estriol	Steroids	Precursors and/or degradation products of estradiol-17β.
	Progesterone	Steroid	(a) With estrogen, stimulates development of the glands of the uterus. (b) Causes thickening of the mucus in the cervix. (c) Decreases uterine motility and maintains pregnancy. (d) In ruminants, conditions the animal to respond to estrogen.
	Relaxin	Protein	Relaxing of pelvic ligaments.

[a] In nonpregnant female, source is the ovary; in pregnant females, the placenta becomes increasingly important as pregnancy advances.

late the production of mucus in the cervix. The molecular structure of the mucus is arranged in lines parallel to the axis of the cervix, which facilitates entry of the spermatozoa, which form a reservoir close to the wall, where they are protected by the mucus. Over the next 24 hours the numbers of spermatozoa in the oviducts, where fertilization of the eggs occurs, gradually increase as a result of transport by their own motility and by peristaltic movements of the tract. Many spermatozoa perish during transport from the cervix to the oviduct, and this is an important cause of reproductive failure in certain conditions, such as infertility in sheep fed on pastures rich in plant estrogens. High chance of fertilization is associated with the presence of some 3,000 or more spermatozoa in the tubes at the time of fertilization.

In the sow and mare, the large volume of highly diluted semen is ejaculated through the cervix directly into the uterus and is almost immediately evenly distributed throughout the whole tract. It is not known if there is any specific region of the tract which acts as a reservoir of spermatozoa, but, by analogy with some laboratory animals, the region of the junction of the uterus and the oviduct may fulfill this role.

The duration of estrus and the time when the eggs are shed varies with the species (Table 26.1). Mating normally occurs early in estrus and may be repeated continuously, giving ample opportunity for the build-up of an adequate population of spermatozoa in the oviduct to ensure a high chance of fertilization when the eggs are shed late in estrus or, as in the cow, after estrus has ceased. Although the

A

B

Figure 26-6 Mating reflex in the sow.
A. Sow in heat stands rigid when pressure is applied to the region of the loin.
B. The immobile sow permits great pressure to be applied to the back region. (Photos courtesy of J. P. Signoret.)

duration of fertilizable life of both eggs and spermatozoa is limited (Table 26.3), spermatozoa require a period in the female reproductive tract before they are capable of penetrating the shell membrane of the egg (the *zona pellucida*) and the egg membrane (the vitelline membrane). This phenomenon is known as capacitation.

Although there are many excellent treatises on capacitation and the process of fertilization (Austin, 1969), information is limited on the mode of transport of spermatozoa through the female tract, the numbers necessary for fertilization, storage sites in the tract, and the effect of the female environment on the survival and fertilizing capacity of spermatozoa.

Fertilization and Differentiation

Fertilization of the egg occurs in the upper third of the Fallopian tube, usually within six or eight hours of the egg being shed. The capacitated sperm attach to the *zona pellucida,* and one or more penetrate and traverse the zona and enter the space between the egg membrane and the zona, the perivitelline space. The head of one, and only one, spermatozoon then attaches to the vitelline membrane, and by a process of dissolution its nuclear contents enter the egg cell mass to form the male pronucleus. This activates completion of the second maturation division, following which one half of the nuclear material of the ovum is extruded as the second polar body.

Table 26.3.
Fertilizable life of eggs and spermatozoa.

	Fertilizable life in the female tract	
Species	*Egg*	*Sperm*
Mare	20 hours	2 to 4 days
Cow	6 hours	24 hours
Ewe	24 hours	36 hours
Sow	15 hours	30 hours

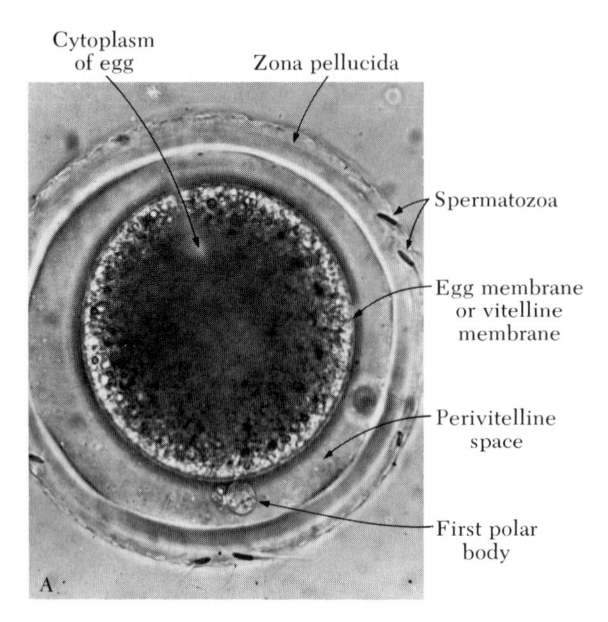
Cytoplasm
of egg
Zona pellucida
Spermatozoa
Egg membrane
or vitelline
membrane
Perivitelline
space
First polar
body
A

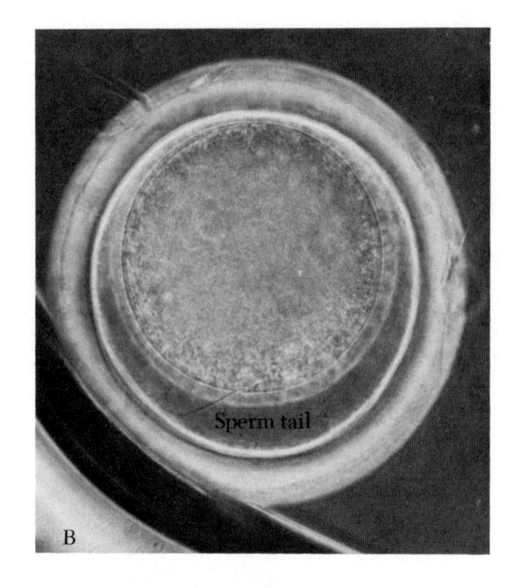
Sperm tail
B

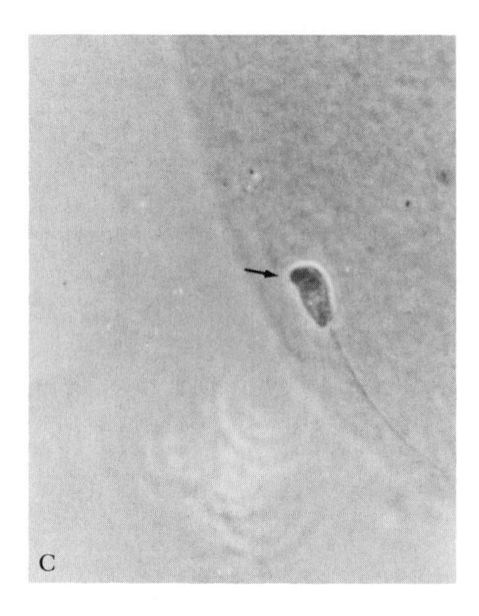
C

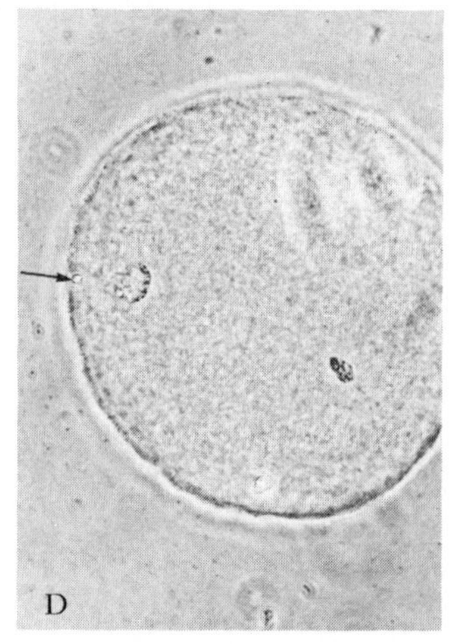
D

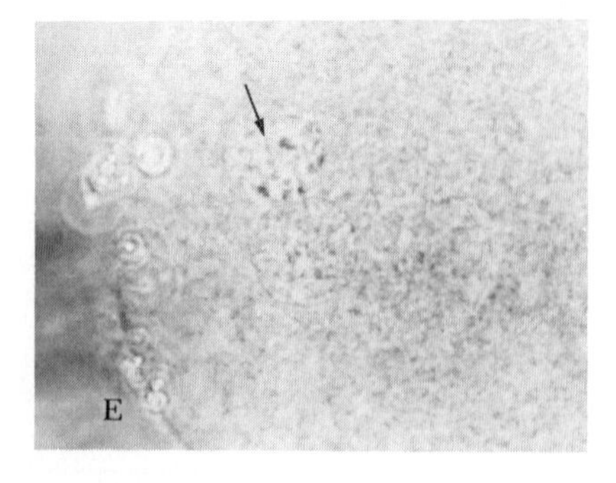
E

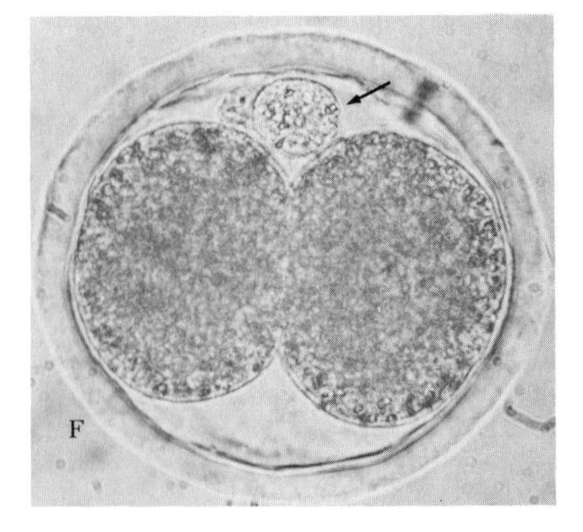
F

The haploid male and female pronuclei then fuse to form the diploid nucleus of the fertilized egg.

The stages of fertilization are shown in Figure 26.7. First cleavage occurs some 20 hours after fertilization, and the second cleavage, to provide a four-cell egg, some 24 hours later. The egg enters the uterus some 24 hours later again, at about the eight-cell stage, that is, about three days after being shed. The egg is still enclosed in its *zona pellucida,* so that most, if not all, its energy requirements for early division are derived from its own resources. Cell divisions continue within the *zona pellucida* for a further three days, by which time the egg mass has developed into a hollow sphere or blastocyst and shows differentiation into two types of cells, those which will become the fetus and those which will form the outer membrane which attaches to the uterus, the trophoblast. The zona is then lost.

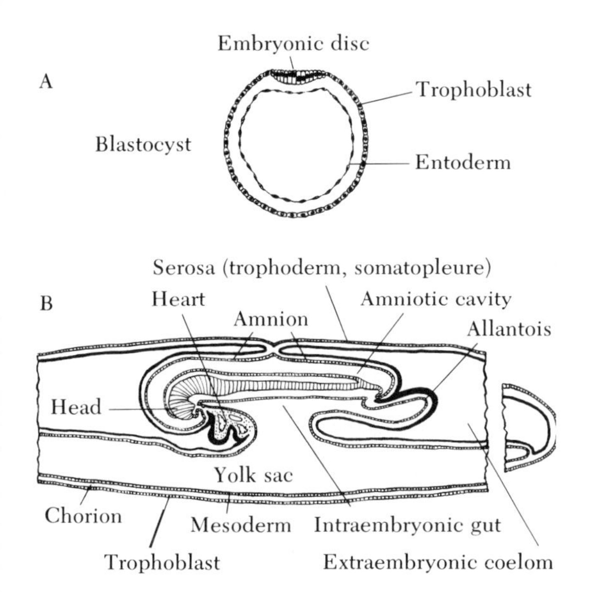

Figure 26-8 Fetal membranes of pig embryos, shown in saggital section.
A. Blastocyst, at eight days (× 80). (From Arey, *Developmental Anatomy,* 7th ed. W. B. Saunders Co., 1965.)
B. Diagram, at eighteen-somite stage (× 8); most of the long chorionic (or serosal) sac has been omitted. (From Patten, *Human Embryology,* 3d ed. McGraw-Hill Book Co., 1968.)

The blastocyst rapidly elongates to form a hollow trophoblastic tube with its clearly discernible germ area or embryonic disc, which within a few days is a recognisable embryo (Figure 26.8). Mesodermal cells, which in the embryo give rise to the vascular, muscular, and skeletal systems, grow out from the embryonic disc inside the trophoblast to form the chorion. In the ewe, cow, and doe, if more than one elongating and developing chorion is present, these fuse together to form a common chorion, within which the embryos and their later developing membranes, the amnion and allantois, are enveloped. By 21 days, the chorion of even a single embryo occupies most of the length of both tubes of the uterus.

In the pig, by contrast, the neighboring fetal membranes do not fuse. The blastocysts are spaced more or less evenly throughout both uterine horns and each increases in size to

Figure 26-7 Living sheep eggs showing stages of fertilization and first cleavage.
A. Attachment of spermatozoa to the *zona pellucida* within one hour of surgical introduction of spermatozoa into the Fallopian tubes.
B. An egg showing a sperm tail inside the *zona pellucida* ½ hour later. A spermatozoon has penetrated the zona and has entered the vitellus but the head has not yet expanded.
C. The expanded sperm head in the vitellus one hour later, still with the tail attached.
D. One hour later the sperm tail is detached, and the female pronucleus is apparent.
E. After another hour (4½ hours after introduction of spermatozoa) the male and female pronuclei are formed prior to fusion.
F. Fifteen hours later, first cleavage has occurred to provide a two-celled fertilized egg. Note the clearly defined second polar body and the remnants of the first. (Photos courtesy of I. D. Killeen and N. W. Moore.)

form an individual chorion with its specialized embryonic area.*

Development of the Fetus and Placenta

Development now proceeds along two coordinated lines. One is the differentiation of the embryonic disc area to form the embryo with its various structures and its enveloping membrane, the amnion. The other is the elaboration of the embryonic placenta, which ultimately has three major components. The first is the outer cell layer, or trophoblast, which attacks and attaches to the maternal wall of the uterus, which reacts to this attack by thickening and by increasing its blood supply, so forming the maternal placenta. The second is the mesodermal layer which grows out from the primitive embryonic disc area inside the trophoblast to form the chorion. The third component develops from the embryo itself. About the 18th day in the ewe, sow, and goat-doe, and a day or two later in the cow and mare, a sac grows out from the gut area of the young embryo and rapidly increases in size until it invests almost all the internal wall of the chorion. This membrane is called the allantois. It carries blood vessels which transport nutrients and gases, and provides the main source of exchange between mother and fetus in later pregnancy. The fused allantois and chorion is called the allantochorion or fetal placenta.

By the fourth to sixth weeks, differentiation of the embryo into its major components is completed, and it has become a fetus. At this stage, growth of the fetal and maternal placentas is rapid, but attachment, which is a gradual process, is not completed until about a third of the way through pregnancy. The nature of the attachment varies. In the sow and mare it is general, over a large area of the uterus, and is said to be "diffuse"; the classical classification of Grosser is "epitheliochorial," meaning that the surface epithelium of the chorion (the trophoblast) is attached to the intact epithelium of the uterus. In the ruminants (ewe, cow, goat-doe, and wild species), attachment takes place in specialized areas in the uterus called caruncles. These areas of attachment develop into complex structures called cotyledons, and attachment is said to be "cotyledonary." The surface epithelium is eroded by the trophoblast, and the placenta is classified as "syndesmochorial," meaning that the surface cells of the chorion are in contact with the connective tissue of the uterus (Figure 26.9).

The early stages of development of the embryo and of its membranes are critical, and loss of embryos within the first three weeks of pregnancy is high. This loss is exaggerated when the number of eggs shed is increased above normal by injection of gonadotropic hormones (Figure 26.10). After three weeks, pregnancy is well-established, and later loss of the conceptus is relatively rare in pigs, sheep, cattle, and goats. The mare differs, in that there is a critical period, as discussed later, at about the sixth week. In the pig there is a second period of loss later in pregnancy. It is probable that losses in early pregnancy are due to defects in the environment of the maternal uterus, whereas later losses are due to genetic defects of the embryos. The classic treatise on this subject is Amoroso (1952).

Parturition

As pregnancy advances, the fetus and its membranes make rapid growth. Toward the end of pregnancy, growth of the fetus outstrips that of the placenta which actually begins to degenerate and decrease in size. This imposes a stress on the fetus, and its movements increase. At the same time there is a shift in the balance of circulating hormones, particularly in estrogen and progesterone. This combination of factors, maternal and fetal, results finally in a violent rejection of the fetal placenta and its contents at a time which is remarkably uniform for each species. (For further details, see Catchpole, 1969.)

* Detailed accounts of fertilization and early differentiation, together with source references, are to be found in Parkes (1960, 1952) and in Cole and Cupps (1959, 1969). For recent exciting advances in the field, see Raspe (1969).

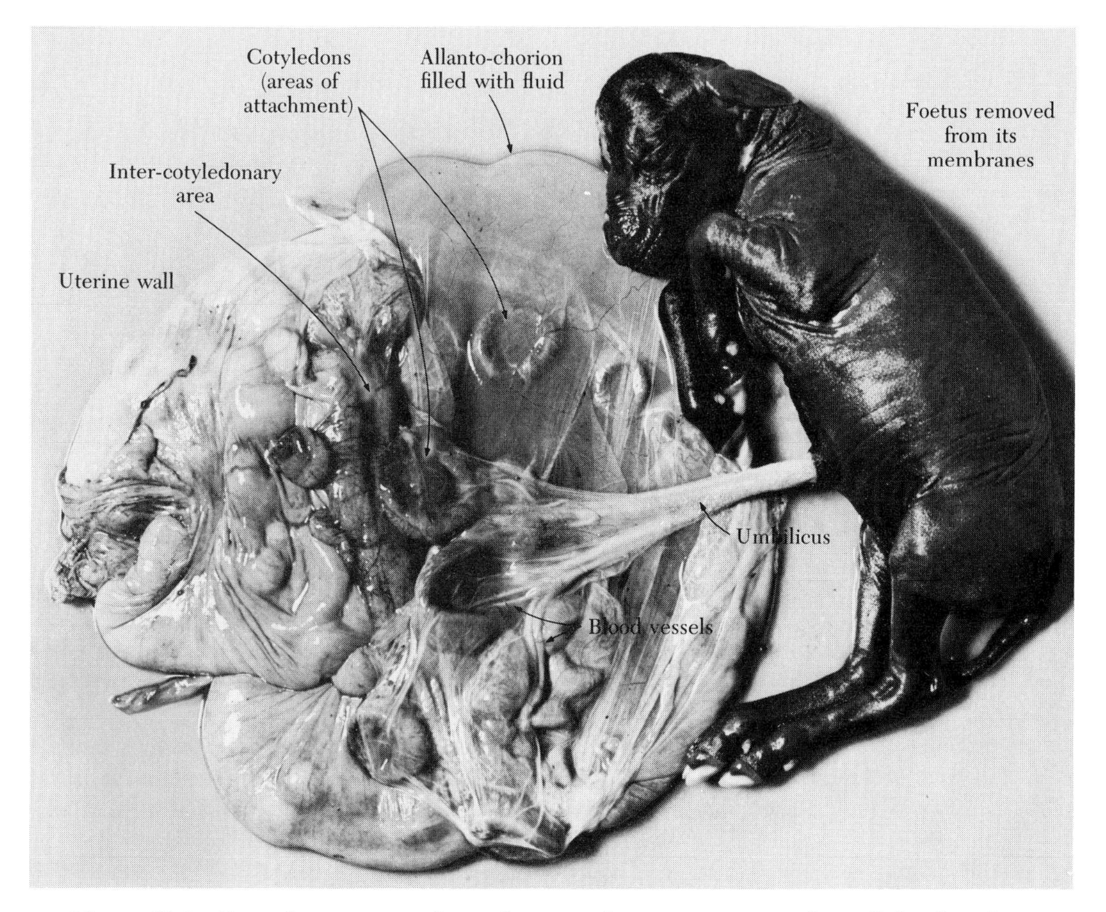

Figure 26-9 Reproductive tract of cow four months pregnant. Note the well-developed areas of attachment of the fetal and maternal placentae and the highly developed vascular system.

Survival

The last few weeks of pregnancy, particularly in mares and in the ruminants, are important for the chances of survival of the newborn young. The nutritional demands of the fetus are very great, and, unless these are met by adequate nutrition of the dam, its growth is retarded and its reserves of energy needed to meet the stresses of living in the outside environment are reduced. Further, undernutrition in late pregnancy retards the physiological development of the fetus and its ability to mobilize its limited reserves. Finally, development of the udder is retarded and potential milk production is reduced (Figure 26.11), so that even if the poorly developed young survive the first critical few hours after birth, subsequent growth (particularly if they are twins) will be less than that of young born to well-nourished dams. (Details of the experiments on which these conclusions are drawn are in Hammond, 1971.)

The Hormones of Pregnancy

The presence of an embryo and its membranes in the uterus prevents the regression of the *corpus luteum* which formed in the follicle from which the egg was shed. The proges-

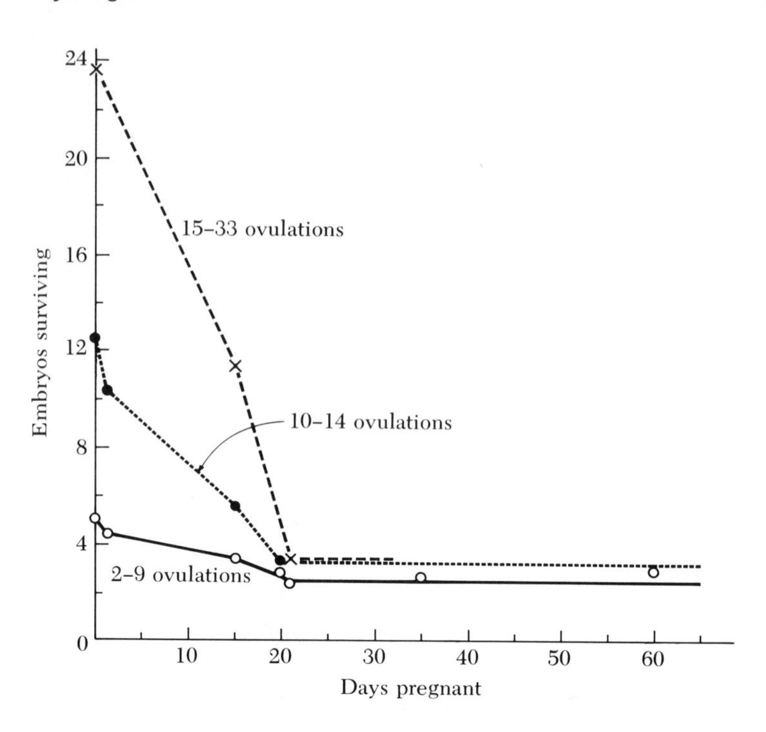

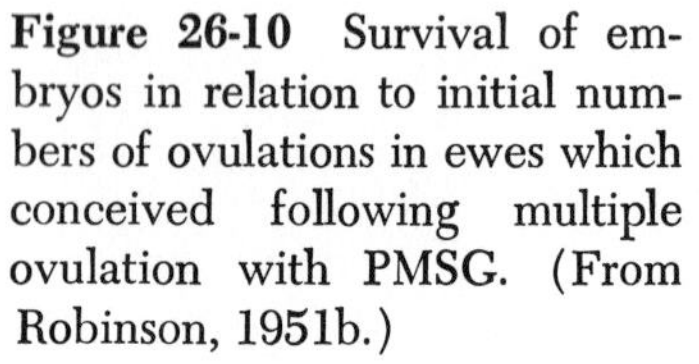
Figure 26-10 Survival of embryos in relation to initial numbers of ovulations in ewes which conceived following multiple ovulation with PMSG. (From Robinson, 1951b.)

terone secreted by the corpus is essential for the maintenance of pregnancy, and if it is removed in early pregnancy, the conceptus is resorbed. The *corpora lutea* persist throughout pregnancy in some species, and in the sow their removal at any stage of gestation results in abortion. By contrast, in the ewe, cow, and goat-doe, the corpus is not necessary in later pregnancy. The mare is unusual, in that accessory *corpora lutea* are formed during the second month of pregnancy as the result of the production of large quantities of a unique

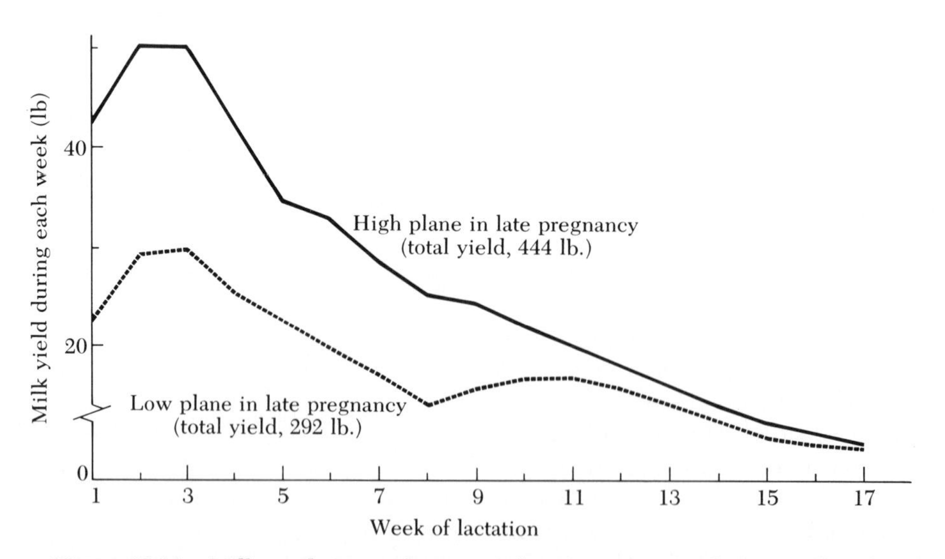

Figure 26-11 Milk production of ewes producing twins in relation to the plane of nutrition in late pregnancy. (Data from Wallace, 1948b.)

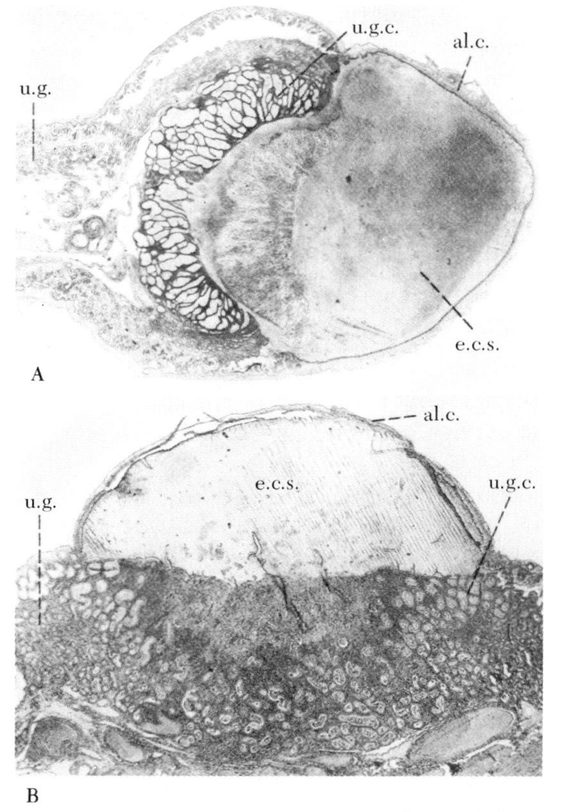

Figure 26-12 (A) Section through pendulous endometrial cup from mare at 105th day of pregnancy. Note the enormous enlargement of the uterine glands in cup area (ugc). alc, allantochorion; ecs, endometrial cup secretion; ug, uterine glands outside the cup area (× 6).
(B) Section through endometrial cup from mare at 63rd day of pregnancy. Note the absence of uterine epithelium, and autolysis of tissues in centre of cup (× 12). (From Cole and Goss, *Essays in Biology*. University of California Press, 1943. Reprinted by permission of The Regents of the University of California.)

gonadotropin called "pregnant mare serum gonadotropin" (PMSG), produced by special structures in the maternal placenta (Figure 26.12). The blood serum of the pregnant mare is very high in this substance between the 50th and 140th day of pregnancy, and its effect on the ovaries of young rats can be used as a test for pregnancy. The accessory *corpora lutea* in the mare appear to be important for the maintenance of pregnancy during a critical stage in the development of the placenta, and, as mentioned earlier, abortions about the 60th day are common.

The ability of the ruminants to maintain pregnancy in the absence of the *corpus luteum* is due to the production of progesterone by the placenta, which also produces estrogen in ever-increasing quantities as pregnancy advances. It is characteristic that estrogen production, mainly of placental origin, increases to very high levels toward the end of pregnancy, and one view is that the shift in the balance between the levels of progesterone and estrogen is involved in the onset of labor. Hence the placenta is not only an organ of respiration (transfer of gases) and alimentation (transfer of nutrients), but is also an important endocrine gland involved in the maintenance and ultimate termination of pregnancy.*

Fertility and Infertility

Definitions

It is necessary to define several terms in common usage: sterility, complete inability to produce offspring; fecundity, ability to form and produce fertile germ cells; fertility, ability to conceive and to produce live young; prolificacy, relative number of live young produced in a given time, such as a lifetime.

Sterility in farm animals is relatively rare, although temporary infertility due to disease or some environmental stress is quite common. Sterility is generally due to some gross anatomical or physiological defect, as in the freemartin in cattle. Heifers born twin to a bull are generally sterile because of the mixing of their blood caused by the fusion of the allantochorion. As a result the female fetus is masculinized by hormones produced by the male calf.

* For a recent review of hormones in pregnancy, see Catchpole (1969); Barcroft (1952) should be consulted with reference to the placenta as an organ of transfer.

Fecundity of most farm animals is high during the breeding season. Failure to produce eggs is rare except in the nonbreeding or anestrous season.

Fertility of healthy adult farm animals is generally high. About 80–85 per cent of sows, 75 per cent of British breed ewes, and 65–70 per cent of cows will conceive to a single service. Mares are less fertile; a 50 per cent conception rate in studs is common.

Prolificacy varies between species and breeds, with only pigs among the farm animals being truly prolific. With modern methods of early weaning and artificial rearing of piglets, two litters of 8 to 12 piglets per year can be produced during a lifetime.

Relative Infertility

Failure to produce live young is due to failure at one of several critical stages of the reproductive process.

Failure to mate to a fertile male This is a common cause of failure under range conditions when relatively few males are available. These males tend to form harems and, if a male is sterile, cows or ewes in this particular harem will not conceive.

Failure of fertilization Generally speaking, failure of fertilization is not a serious cause of infertility. Eggs shed by animals mated to fertile males are highly fertilizable; 90 to 100 per cent of eggs of sows, pigs, ewes, and goat-does are generally fertilized. In mares, the proportion may be lower, because of the long interval which may intervene between mating and shedding of the egg. However, under some circumstances, such as at the beginning and end of the breeding season, and in animals which have been grazing pastures rich in plant estrogens, the percentage of eggs fertilized can be low, not because of faults in the eggs, but because of failure of spermatozoa to survive in the female reproductive tract.

Early embryonic death Loss of embryos during the first critical three weeks of pregnancy is high, and generally is estimated as about 25 per cent of all fertilized eggs. In cows and mares, and in those sheep and goats which shed only one egg, this represents loss of the pregnancy. In pigs, and in those sheep and goats which shed more than one egg, pregnancy may persist, but the size of the litter is reduced.

Fetal death Losses in late pregnancy in cows, ewes, and goat-does are relatively rare, except in disease states, such as contagious abortion. In sows, fetal death due to genetic defects is a common cause of reduction in litter size.

FURTHER READINGS

Amoroso, E. C. 1952. *In* A. S. Parkes, ed. *Marshall's Physiology of Reproduction,* II. London: Longmans, 1952. Pp. 127–311.

Asdell, S. A. 1964. *Patterns of Mammalian Reproduction.* 2d ed. Ithaca, N.Y.: Cornell Univ. Press.

Hafez, E. S. E., ed. 1969. *The Behaviour of Domestic Animals.* 2d ed. London: Bailliere, Tindall, and Cassell.

Hammond, J., Jr., J. T. Robinson, and I. L. Mason. 1971. *Hammond's Farm Animals.* London: Arnolds.

Raspe, G., ed. 1969. *Advances in the Biosciences.* New York: Pergamon.

Robinson, T. J. 1954. *Endocrinology,* 55:403.

Twenty-Seven

Male Reproduction and Artificial Insemination

"And thus it comes that individuals, in procreating their like for the sake of their species, endure forever."

William Harvey, 1628.

Physiology of Reproduction in the Male

Although the male organs of reproduction in domestic animals differ in size and occasionally in shape, essentially all are similar to those of the bull (Figure 27.1).

The primary organs of reproduction in the male are the two testes, which are located outside the body cavity in an external sac, the scrotum, in most mammals. Attached to each testis is an enlarged convoluted tube, the epididymis, which leads upward into a tube called the *vas deferens.* Each *vas deferens* passes from the scrotum into the pelvic cavity, where it joins the *vas deferens* leading from the other testis. At the point of juncture, the tube again enlarges and, referred to as the ampulla, empties into the urethra. The point at which the ampulla is attached to the urethra is close to that where the bladder empties, and from this point the urethra carries the products of both through the penis to the exterior.

Attached to the excurrent ducts—the *vas deferens,* ampulla, and urethra—are three types of glands, generally called the accessory sex glands. These are the paired seminal vesicles, the prostate, and the Cowper's or bulbourethral glands.

Functions of the Reproductive Organs in the Male

The testis The testes, like the ovaries, have a dual function; they produce spermatozoa (sperm), and they secrete testosterone, the male sex hormone. The production of sperm takes place in tubular structures the seminiferous tubules, that comprise most of the testicular mass (Figure 27.2). These tubules each contain rapidly dividing cells which lead to the production of sperm. The testes of a healthy rabbit may produce 200 million sperm a day, and those of a healthy boar may produce more than twenty billion a day.

Lying between the seminiferous tubules are clumps of interstitial tissue containing specialized cells, the Leydig cells, which produce tes-

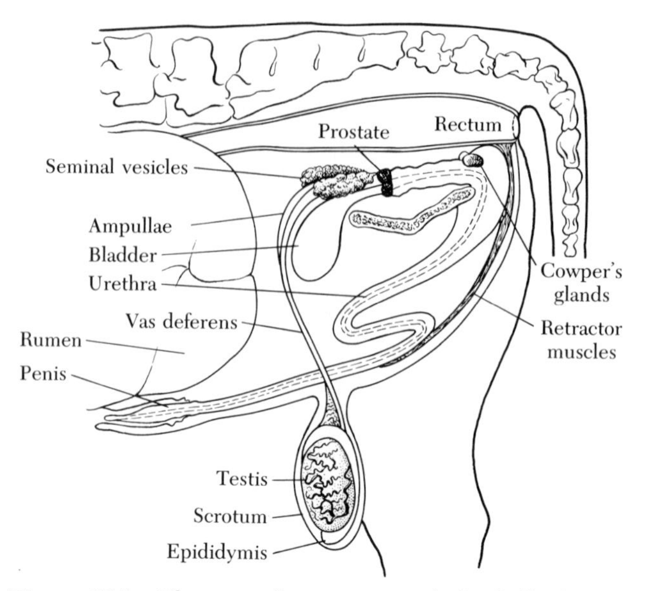

Figure 27-1 The reproductive tract of the bull. (From Salisbury and VanDemark, 1961.)

tosterone (Figure 27.2). Testosterone is essential for production of sperm by the testis and maintains the accessory sex organs. In addition, testosterone promotes the development of the secondary sex characteristics that make a male obviously different from the female of the same species. These include body build, distribution of body hair, growth of horns (in deer, for example), differences in color of hair or feathers, changes in the voice and behavior.

The testes of most mammals do not function properly at normal body temperatures. Testes retained in the body cavity of animals, a condition called cryptorchidism, are unable to produce sperm, and do not produce normal amounts of testosterone. If only one testis is retained in the body cavity, the other may produce sperm and the animal may be fertile, but if both testes are cryptorchid, the male is sterile.

The scrotum Normally, males are not affected by the adverse effects of body temperature on the testes, because they are kept in the scrotum, outside the body cavity. The scrotum primarily maintains the testes at a temperature 2° to 6°C (4° to 11°F) below that found in the body, by virtue of muscles which allow the testes to remain relatively further from the body cavity in hot conditions and closer to the body's warmth in cold environments, thereby allowing better regulation of testis temperature.

The epididymis When sperm cells are formed in the seminiferous tubules of the testis, they are not yet capable of fertilization. After they are formed, they all progress to a common duct, the epididymis, where they accumulate and mature prior to ejaculation. During their progress through the long (more than 100 feet in the bull) epididymis, they undergo little-understood changes which bring them to a stage of maturation sufficient for fertilizing ova.

The accessory sex glands As the sperm cells pass up the *vas deferens* from the epididymis to the urethra, they are mixed with fluids from the seminal vesicles and prostate gland; the total mixture is the semen. The seminal vesicles provide sugar (fructose), citric acid, amino acids, and buffering proteins to the seminal plasma, and the prostate

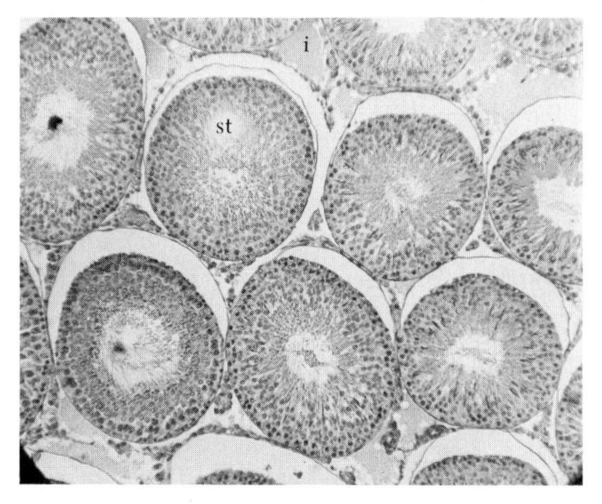

Figure 27-2 Cross section of testes of the mature rat, showing seminiferous tubules (st) and the interstitial area (i), containing hormone-producing cells.

adds necessary minerals. The semen is preceded at ejaculation by a lubricating fluid from the bulbourethral gland. In some animals, semen coagulating factors (rat) or a gelatinous material (boar) may be added to the semen by accessory organs.

Artificial Insemination (AI)

The most important development in reproductive physiology has been the successful development of artificial insemination and its wide use in animal breeding. The male reproductive cells (sperm) are deposited into the female reproductive tract by mechanical means. The procedure of artificial insemination, along with techniques for preserving and transporting semen, makes possible the improvement in farm animals through use of superior sires.

History and Growth of Artificial Insemination

In 1677, Johan Hamm, a student, and Anton van Leeuwenhoek, his teacher, discovered sperm under a crude microscope. The minute, moving sperm cells were described as miniature men or "animalcules."

One hundred years later, Lazarro Spallanzini, an Italian physiologist, began experiments leading to the successful artificial insemination of several reptiles and, subsequently, reported a litter of three puppies born after artificial insemination of a dog.

The first serious study of artificial insemination was undertaken in Russia by Elias Ivanoff. Ivanoff extended the work in horses and dogs, and was the first to use artificial insemination successfully in cattle and sheep (Perry, 1968). So widespread was the application of artificial insemination by Ivanoff and his workers that the Russians inseminated over 100,000 mares, 1,200,000 cows, and 15,000,000 ewes in 1938, the year in which the first artificial-breeding association for cattle was organized in the United States. Since 1938, however, the increase in use of AI in cattle has been rapid in the United States.

The most recent data available for several nations (Nishikawa, 1964) and the United States (USDA, 1970a) show that the USSR continues to lead in numbers of cattle inseminated artificially, but, in per cent of total cows, Denmark, Israel, and Japan use the practice most widely (Table 27.1). On the other hand, the utilization of the practice is only minor in Africa, South America, and the Arab nations, and only about 5 per cent of the cattle in the world are bred artificially.

Artificial insemination of sheep, goats, and swine is extremely limited in the United States (3,230 swine and 235 goats in 1969), but this is not true for the rest of the world. About 50,000,000 sheep, 1,000,000 swine, over 125,000 mares, and 50,000 goats are bred artificially each year in the world, as reported by Nishikawa (1964); artificial breeding of these animals in the United States is predominantly experimental. Although artificial insemination of beef cattle is rarely reported separately from that of dairy cattle, it is well known that the practice is not as widespread in beef cattle. For example, 45 per cent of dairy cattle in the

United States and 2.3 per cent of the beef cattle were bred artificially in 1969, but as early as 1962, 97 per cent of the dairy cattle and 89 per cent of beef animals were bred artificially in Japan.

Collection of Semen

The artificial vagina The foundation for the rapid expansion of artificial insemination described above was provided in 1914, when the Italian, Guiseppe Amantea, invented the artificial vagina. The use of the artificial vagina to collect ejaculated semen directly from males increased tremendously the amount of semen that could be collected from a single ejaculation and decreased the possibility of disease transfer from cow to cow. Although there are a variety of modifications of the artificial vagina, and different sizes and shapes for different species, all are designed to provide proper temperature, pressure, and lubrication for the penis in order to evoke normal ejaculation (Figure 27.3).

Maximum amounts of highest-quality semen are obtained by allowing a bull a period of sexual preparation and permitting several "false mounts" (mounting without mating and ejaculation) prior to semen collection. Some bulls and most boars can be trained to mount a dummy female; false mounts even on a dummy are beneficial to semen volume and quality.

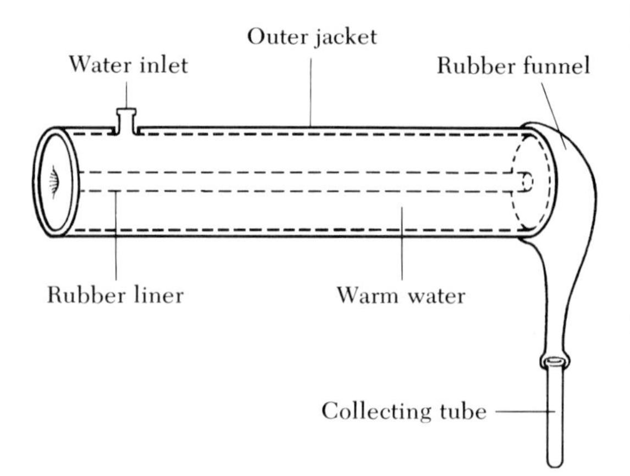

Figure 27-3 Artificial vagina used for collecting semen from bulls.

Table 27.1.
Use of artificial insemination in cattle in several countries.

Country	*No. of cows artificially inseminated yearly*	*Per cent of total cows*
USSR	18,700,000	65
USA	8,200,000	44
France	5,850,000	54
Germany	4,300,000	45
Czechoslovakia	2,400,000	85
Poland	2,350,000	39
Great Britain	2,100,000	67
Denmark	1,600,000	99
Japan	1,400,000	93
Netherlands	1,250,000	63
Italy	1,000,000	23
Israel	75,000	95

Electroejaculation Some males which have been injured or which suffer from arthritic or other diseases are unable to serve the artificial vagina, but are of such superior genetic merit that collection of their semen is worthwhile. In these animals, in animals which refuse to serve the artificial vagina (generally preferring natural mating), and in those species which do not serve an artificial vagina (e.g., chinchillas, hamsters), the preferred method for collecting semen is electroejaculation.

A rhythmic application of electrical charges to male animals, usually via a rectal probe, results in excitation and ejaculation. Using this technique, good semen samples can be collected from bulls, rams, cats, several laboratory rodents, and monkeys. Less optimal results have been obtained in boars, rats, and mice, but the technique has been successfully used in these species.

Other methods of collection Boars and dogs often give less than satisfactory semen samples using the artificial vagina. In these cases, semen may often be collected by manually deviating the penis from a receptive female or from a dummy and grasping the base (dog) or end (boar) of the penis with the gloved hand. Semen is collected into a funnel and collection flask.

Semen Evaluation

When the semen is collected, it must be protected from the direct sunlight and from inclement weather. The container should be stoppered to prevent bacterial contamination, and the semen should be rapidly evaluated.

It is beyond the scope of this chapter to discuss in detail the methods used to determine semen quality. The ultimate test of semen quality is its ability to fertilize the ovum. However, more immediate measures must be used before the semen is used. Volume and sperm concentration are important (Table 27.2). Microscopic examination will provide information on concentration, motility, and normal appearance of sperm (Figures 27.4 and 27.5). Oxygen and fructose utilization are also useful tests of normalcy.

Extension and Preservation of Semen

One of the main advantages of artificial insemination is that semen can be diluted and that each ejaculate from a superior sire can be used to inseminate many cows. As pointed out by Foote (1969), an average bull produces enough sperm cells in a week to inseminate 4,000 or more cows with maximum fertility, when semen is properly extended and inseminated, since only 5 million motile sperm need be used per insemination, whereas 30 billion cells can be collected weekly.

The average sire in American breeding organizations provided semen for breeding 250 cows in 1939, 3,500 cows in 1969. One Holstein Friesian sire, Shaws DCT Una Creamelle, produced semen for 207,699 artificial insemination services in less than eight years, over 26,000 services per year.

Use of a good semen extender not only increases the volume of semen that can be used for insemination, but also provides the chemical agents necessary to sustain sperm life under the storage conditions employed.

Semen extenders Salisbury and VanDemark (1961) summarized the characteristics of a satisfactory semen extender as one which maintains the proper pressure and buffering capacity for sperm cells, provides nutrients and minerals for sperm, protects the sperm from cold shock, and is free of bacterial or other products which may be harmful to the sperm, to the female tract, or to the developing offspring. When chemicals to meet these specific requirements are known, completely synthetic extenders will be developed. At pres-

Table 27.2.
Characteristics of semen from mature males of several species.

Animal	*Average volume (ml per ejaculate)*	*Sperm concentration (millions per ml)*	*Sperm collectable (billions per week)*	*Maximum use of semen per male (inseminations per week)*
Bull	8	1,800	30	4,000
Ram	1.2	3,000	25	350
Boar	200	250	110	40
Stallion	100	150	30	15
Dog	2	300	2	15
Rooster	0.7	3,500	10	150
Monkey	1	500	3	10

ent, however, all widely used extenders contain one or more biological products such as egg yolk or milk, which help to protect sperm against cold shock and allow semen to be cooled to refrigerator temperatures with excellent preservation of cell life. Buffering chemicals are added to extenders to prevent undesirable changes in *p*H of the semen during utilization of nutrients and production of byproducts by sperm. Buffers such as sodium citrate and sodium bicarbonate are most common. Although the egg yolk or milk used in most extenders contains nutrients and minerals that can be utilized by sperm, most extenders contain small amounts of sugar such as glucose or fructose for additional nutrition. To reduce growth of microorganisms and to prevent the spread of bull-borne diseases that may be carried in the semen, antibiotics such as penicillin and streptomycin are normally added to the extender.

During the last several years, Hafs *et al.* (1969), at Michigan State University have demonstrated marked improvement in the fertility of bull semen when enzyme preparations were added. Included are the enzymes amylase and β-glucuronidase.

Other chemicals may be added to semen extenders when they meet specific purposes. For example, vegetable dyes are often added to differentiate between samples from different breeds of animals or from different experimental groups.

Preservation of semen at refrigerator temperatures As mentioned above, the use of egg yolk or milk in semen extenders protects the sperm cells from cold shock during cooling. When semen is extended in this manner and stored at about 4–5°C, sperm remain viable for four or five days. Packing such semen in thermos bottles or ice chests allows for shipping to any point that can be rapidly reached, even by air. Using liquid semen in essentially this form, millions of cows were inseminated around the world until the widespread use of frozen semen occurred. Even though frozen semen has completely replaced

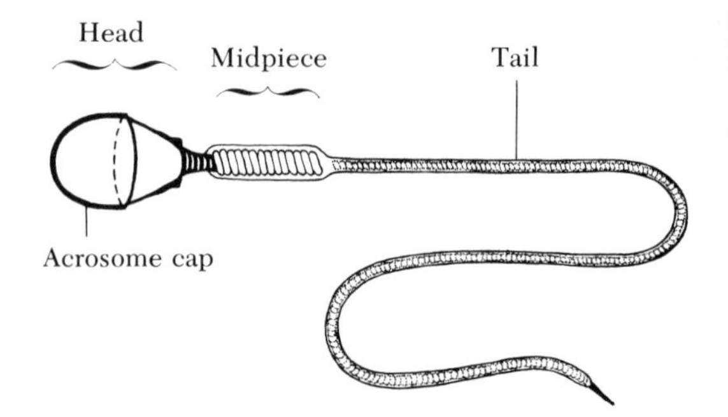

Figure 27-4 Morphology of a normal bull sperm.

liquid storage for commercial United States bull studs, liquid semen is still widely used throughout parts of the world (almost exclusively in the USSR; Turton, 1969).

Frozen semen Although earlier observations suggested that sperm cells could survive freezing, the real idea of freezing semen for later use was not accepted until the use of glycerol in extenders was accidentally discovered by Polge and Parkes in 1949. As related by Polge (1968), contents of a flask thought to contain fructose protected both chicken and bull semen from being damaged by freezing. However, a subsequent flask of fructose did not, so the material in the first flask was chemically analyzed. This analysis showed that the flask contained not fructose, but glycerol and protein, a mixture routinely used for preparing tissues for microscopic examination. Thus, the protective action of glycerol was discovered. This led to the extensive use of frozen semen in artificial insemination of cattle in this country and around the world. Since glycerol does not protect sperm from all kinds of animals against freezing, and is even harmful to some (e.g., boar semen), the search continues for other chemical agents which have this beneficial property.

After the slow addition of glycerol and extender to the semen and gradual cooling to just above freezing temperatures, the semen is packaged in single insemination units and frozen. Containers presently in use include ampules and straws.

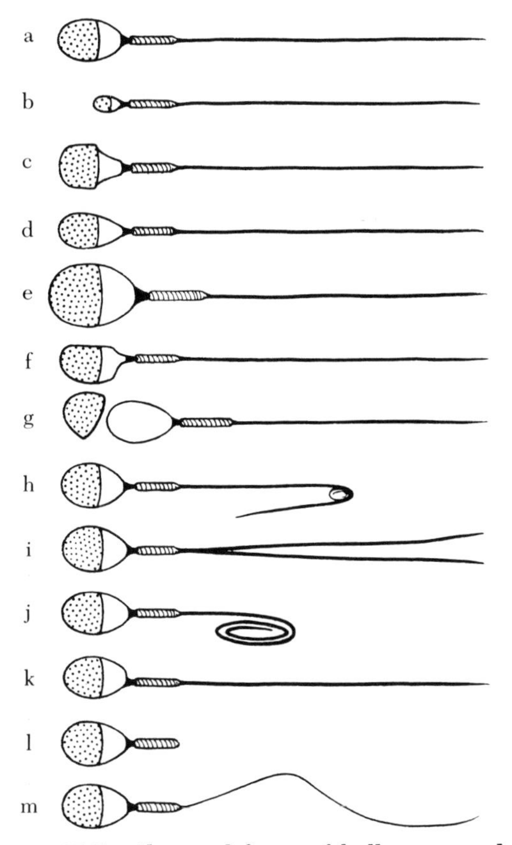

Figure 27-5 Abnormal forms of bull sperm and the relative frequency of their appearance in a normal ejaculate: A, normal sperm (91%); B, microhead (0.2%); C, pyriform head (0.4%); D, tapered head (2.2%); E, giant head (0.2%); F, asymmetric head (0.5%); G, detached or abnormal acrosome (0.5%); H, protoplasmic droplet in tail (3.8%); I, double tail (0.3%); J. coiled tail (0.6%); K, abaxial attachment (0.1%); L, head and tail separated (0.6%); M, filiform tails (0.1%).

As pointed out by Graham (1966), the ampule is the familiar package in the American artificial-insemination industry today, and American equipment, procedures, technology, and skills are oriented toward use of the ampule. The individual containers can be clearly labeled and can be stored in alcohol, dry ice, or liquid nitrogen. Because they are sealed glass, added gases may be entrapped if desirable. On the other hand, it is difficult to achieve a consistent freezing rate in ampules, so fewer sperm cells live through freezing, compared to pellets and straws. Because of the shape of the ampule and the insemination pipette, it is also difficult to transfer all of the sperm from the ampule to the cow.

In 1950, Cassou adapted an earlier Danish method of semen storage by freezing semen in small plastic (polyvinyl chloride) straws, 13.5 cm (about 5.5 in.) long. The straws are filled with semen, plugged with a polyvinyl sealing powder (which forms an airtight seal upon contact with water), and rapidly frozen in liquid nitrogen vapor. The straws are available in several colors, as is the polyvinyl sealing powder, allowing many combinations of color to be used for cataloging semen. In addition, information such as the name and identification number of the bull, ejaculate number and date, and other data can be imprinted right on the straw.

Since "French straws" are inserted directly into an insemination syringe, transfer losses, such as occur with ampules, are overcome. Furthermore, the rapid freezing and thawing rates with straws result in a greater proportion of healthy sperm following freezing; these two factors may result in a net saving of 25–30 per cent of the sperm frozen.

After semen is frozen, no matter what the package used, it must be kept at temperatures equal to or below that of dry ice (−79°C) to markedly slow deterioration of fertilizing capacity. According to extensive studies conducted by Salisbury (1968) at the University of Illinois, sperm survival at −79°C, −88°C, or even −196°C (the boiling point of liquid nitrogen) is extended tremendously, but not indefinitely. Viability of sperm cells begins to decrease after about eight months in storage and reaches a very low point within 30 months.

Storage at ambient temperature About the time that frozen semen was becoming widely used in the United States, VanDemark and his colleagues (Salisbury and VanDemark, 1961) discovered that gassing of extended liquid semen with carbon dioxide would inhibit metabolic activity of sperm and allow their storage for several days at room temperature. Since this discovery was overshadowed by the advent of frozen semen, extension of this technique has been slower than one might hope. However, it has led to the development of several ambient temperature extenders that will maintain semen for several days, including one developed by the Illinois workers using carbon-dioxide gassing of an egg yolk and citrate type extender. This method should be useful in developing countries.

Artificial Insemination in Cattle

Although the earliest successes in artificial insemination were achieved using dogs and horses, the procedure has gained its most widespread use and its most striking successes with dairy cattle. In general, semen is deposited into the uterus and front part of the cervix of the healthy cow (see Chapter 26) using either a hollow plastic rod or a stainless steel holder for polyvinyl straws. Insemination should take place 6–10 hours after the end of the heat period, making certain that conditions are as sanitary as possible. Insemination records should be carefully and accurately kept.

Regulations governing artificial insemination Under regulations set by the Purebred Dairy Cattle Association (PDCA) of the United States, the responsibility for honesty and accuracy rests primarily on the organization supplying semen. The few regulations which have been adopted by the PDCA, along with health standards described by The American Veterinary Medical Association and the National Association of Animal Breeders, serve as guidelines for all of the breed associations governing dairy-cattle registrations in the United States.

Regulations adopted by the PDCA require that a bull used in artificial insemination outside his own herd must be bloodtyped if offspring are to be registered. Semen collected from the bull must be kept in carefully marked containers, showing the name and number of the bull. When a bull dies, or ownership is changed, inventories of frozen semen on hand must be reported. Semen-producing businesses must obtain a permit from the PDCA, keep accurate inventories of all semen, and record transfers of semen in more than 10-ampule lots, if offspring from their bulls are to be registered.

Unlike dairy-cattle associations, which all adhere to PDCA regulations, beef-cattle associations vary in their regulations governing registration of calves following artificial insemination; all are more rigid than the PDCA. Beef-cattle regulations generally prohibit the registration of calves unless the artificially inseminated cow and at least 25 per cent of the bull are owned by the same person.

Canadian associations have recently relaxed beef-cattle regulations for superior Angus sires (Canadian Aberdeen Angus Association) and for all Charolais bulls (Canadian Charolais Association).

Artificial breeding organizations Although a relatively large number of individual organizations own bulls and produce frozen semen, a relatively limited number of such organizations (associations, centers, or studs) handle the bulk of the business. For example, 33 bull studs bred 92 per cent of the cows artificially inseminated in the United States in 1969, and virtually all the cows inseminated artificially in England and Wales are bred with semen from bulls at 28 centers.

Status of Artificial Insemination in Other Species

Most of our knowledge of artificial insemination has resulted from its use in dairy cattle, but is directly applicable to beef cattle and water buffalo. Much recent progress has been made in its use in other animals; since 1968, for example, reports have emerged of the successful recovery of viable cells after freezing semen from dogs, boars, and fish.

Sheep and goats In the United States, artificial insemination of sheep and goats is limited almost entirely to experimental work. In other parts of the world, however, the practice is extensive. As high as 65 per cent of the 56 million ewes in Russia are bred artificially each year, and 3 to 4 million ewes are artificially inseminated yearly in Bulgaria and Rumania. The highest rate of artificial insemination of sheep is in Albania, where the procedure is used on 75 per cent of all ewes. Although it is still limited to large individual farms in Australia and in South America, rapidly increasing numbers of ewes are being bred artificially (Nishikawa, 1964).

Although rams and billies ejaculate less semen than bulls (Table 27.2), the sperm-cell concentration is greater, so the number of sperm available per ejaculate is nearly as high.

Successful pregnancies were reported as early as 1951 following insemination of ewes with frozen ram semen, but fertility rates have generally been extremely low. Recent studies in Australia have shown that careful experimental use of the pellet-freezing method can markedly increase fertility in sheep, though not to the level achieved with natural mating or unfrozen semen.

Unlike the sheep, goat semen can be successfully frozen with high fertility rates in inseminated nannies.

Swine During recent years, the practice of artificial insemination of swine has grown rapidly, reaching the stage where it is a significant industry in the Netherlands, Finland, Japan, Taiwan, and Russia, and a developing industry in France, England, and the United States. Techniques for storing semen and for inseminating sows with small volumes of semen are lacking (25–30 ml of extended semen containing 2 billion sperm cells appear minimal at present). Rates of conception and litter size are not as great as with natural service, partially because heat detection in sows and gilts is difficult.

The problem of semen storage may soon be overcome, however. Recent news releases from the laboratory of Graham and Rajamannan (anonymous, 1971) have reported successful freezing and thawing of boar semen with resulting pregnancies and live pigs. This confirms and expands reports from more than a decade earlier (Hess *et al.*, 1957; Hoffman, 1959). Development of this finding could help the industry materially, but the method must first stand the careful scrutiny of intensive research.

Horses Even though horses were one of the first species in which artificial insemination was used, the growth of horse artificial breeding has not been particularly great. In 1939, a quarter-million mares were mated artificially in Russia alone, but in the 1961–1962 period, only half that number were artificially inseminated per year in the entire world, including Russia.

During recent years, successful means of freezing stallion semen have been worked out, so that this procedure is available to horse breeders. Earliest use of frozen semen in the field suggests that it will be of good fertility.

Poultry Artificial insemination is a widely used practice in experimental work with fowl and, in some cases, in the poultry industry.

In turkeys, and particularly in the meat breeds raised in the United States, artificial insemination results in a marked increase in fertility compared to natural service. Insemination of turkey hens with as little as 0.01 ml of semen (a tom produces about 0.3 ml per ejaculate) may result in fertility as high as 85 per cent; fertility from natural service rarely exceeds 60 per cent. For this reason, over 90 per cent of the turkey hens in this country are inseminated artificially.

Storage of turkey semen has been extremely disappointing thus far. The best extenders and procedures available maintain sperm viability only four hours or less, even though chicken semen can be successfully frozen or kept viable in liquid form for several days.

The use of artificial insemination in chickens is relatively limited in the United States, but is much more widespread in other countries (notably Japan). At present, fertility and hatching rates are not quite as high as with natural breeding.

Artificial insemination has been used in Japanese quail, pheasants, ducks, and geese, with procedures and results differing little from those in chickens.

Honeybees Artificial insemination of bees is relatively easy, since only one queen bee need be inseminated from a colony of 50,000 or more bees. In this fashion, rapid progress has been made in producing superior strains of hybrid bees.

Farm and house pets Generally speaking, artificial insemination is rarely used in dogs or cats, but is possible in both.

In dogs, artificial insemination is used when disease or physical deficiencies, psychological problems (as in trained watchdogs), or geographical distance between the male and female make natural mating impractical. However, little success was made in storing and preserving semen from dogs until Saeger (1969) announced a litter of puppies whelped after insemination with frozen semen.

Experimental work has shown that cats can be inseminated artificially with very high fertility and good litter size but storage of semen from cats for extended periods of time has not been accomplished.

Miscellaneous and laboratory animals The use of artificial insemination in fur-bearing animals, laboratory animals, and other pet-store species has met with variable success, depending on the species and the scientific interest. In rabbits, artificial insemination is relatively easy, and many studies have been made with rabbit sperm cells. Semen has been successfully frozen with good fertility in recovered cells. Chinchillas can also be bred artificially (and with frozen semen). In gerbils, hamsters, rats, and mice, artificial insemination can be used, but appears to be only of scientific interest.

Artificial Insemination in Humans

It is beyond the scope of this chapter to consider the many nonbiological factors (legal, moral, emotional, and cultural) which have a significant bearing on the use of artificial insemination in humans. From the biological point of view, artificial insemination of women, using liquid or frozen semen, is workable. Timing of insemination to coincide with ovulation is possible, and reasonable fertility is common.

In practice, artificial insemination of women falls into three potential categories. In the first, the husband supplies the semen and artificial insemination is used to deposit the semen in a more favorable environment, or to deposit the sperm cells from several ejaculates at once, if concentration is normally too low.

In the second type of procedure, semen is provided by an anonymous donor (or semen from several anonymous donors is mixed), and the woman may become pregnant even though her husband is infertile.

Third, Muller (1961) has suggested that "banks" of frozen sperm cells be stored so that people may "select" a genetic base superior to their own for their children. Muller based his idea, in part, on the mistaken notion that sperm do not deteriorate when deep frozen.

FURTHER READINGS

Foote, R. H. 1969. "Physiological aspects of artificial insemination." *In* H. H. Cole and P. T. Cupps, eds., *Reproduction in Domestic Animals*. New York: Academic Press. Pp. 313–345.

Mann, T. 1964. *Biochemistry of Semen and of the Male Reproductive Tract*. New York: Wiley.

Nishikawa, Y. 1964. "History and development of artificial insemination in the world." *Proc. Fifth Intern. Congr. Reprod. Artificial Insemination, Trento, Italy,* pp. 163–256.

Perry, E. J., ed. 1968. *The Artificial Insemination of Farm Animals.* 4th rev. ed. New Brunswick, N.J.: Rutgers Univ. Press.

Salisbury, G. W., and N. L. VanDemark. 1961. *Physiology of Reproduction and Artificial Insemination of Cattle.* San Francisco, Ca.: W. H. Freeman and Co.

wenty-Eight

Egg Laying

"There are some hens that lay so many eggs that they produce two a day." *

Aristotle

[Success or failure of a species is measured by its ability to preserve itself and undergo some numerical expansion.] The success of birds as a class is largely due to the fact that they have evolved physiological mechanisms that cause them to lay eggs at a time of season when such factors as weather and food supply are optimal and when maximal survival of the young can be expected. [The numbers of eggs laid and incubated are commensurate with the physical capabilities of the hen for brooding the eggs and caring for the chicks;] the species that are most successful are those that tax these capabilities to the utmost and allow for the highest reproductive rate. There are mechanisms that cause the hen to *want* to brood eggs and care for chicks; finally, provisions are made for the return of the ability of hens to lay eggs after a cycle of mothering chores has been completed. Obviously such a complex situation requires a system of signaling mechanisms that tell a hen what time of the year it is, when it has laid enough eggs to start brooding them, and when the chicks no longer need parental protection so that she can again start the cycle of laying eggs. Such a signaling system cannot be simple, and neuroendocrine feedback mechanisms have evolved in both mammals and birds. The nervous system maintains communications among parts of the animal body, and the endocrine system, using hormones as messengers and as local organ representatives, sees to it that this or that part is stimulated to the right degree at the right time. The neuroendocrine control systems have the tasks of synchronizing events within the body and relating the internal events to the external environment in which the population lives. Unfortunately, it will not be possible to treat in detail the many elegant methods that have evolved to permit optimal functioning of animal organisms, but a brief synopsis should encourage the interested student to go deeper into the fascinating problems of reproductive physiology.

The eyes of birds serve as receptors of light

* Rarely, if ever does this occur with normal-shelled eggs, but if eggs are gathered only once daily, this conclusion could be reached erroneously.

intensity. Birds can thus tell the difference between seasons: increasing intensity and duration of light foretell spring, and decreasing light signals the approach of fall and winter. In either event, the excitations resulting are transmitted via the optic nerve to the hypothalamus, one of the major control centers of the neuroendocrine system (Figure 28.1). From the hypothalamus, chemical substances are released and eventually reach the anterior lobe of the pituitary gland. If the amounts of light are increasing, the anterior lobe responds by increased secretion of the hormones responsible for gonadal growth.

Many interrelated events occur in the reproductive cycle of hens, and there are many different messages received by and transmitted from the hypothalamus. For instance, in the hen's wild ancestor, which laid about twelve eggs in a nest, the pressure of the eggs against the breast initiates a message to the effect that about as many eggs have been laid as a hen can hatch—that is, cover with her body. These messages, when properly decoded and re-

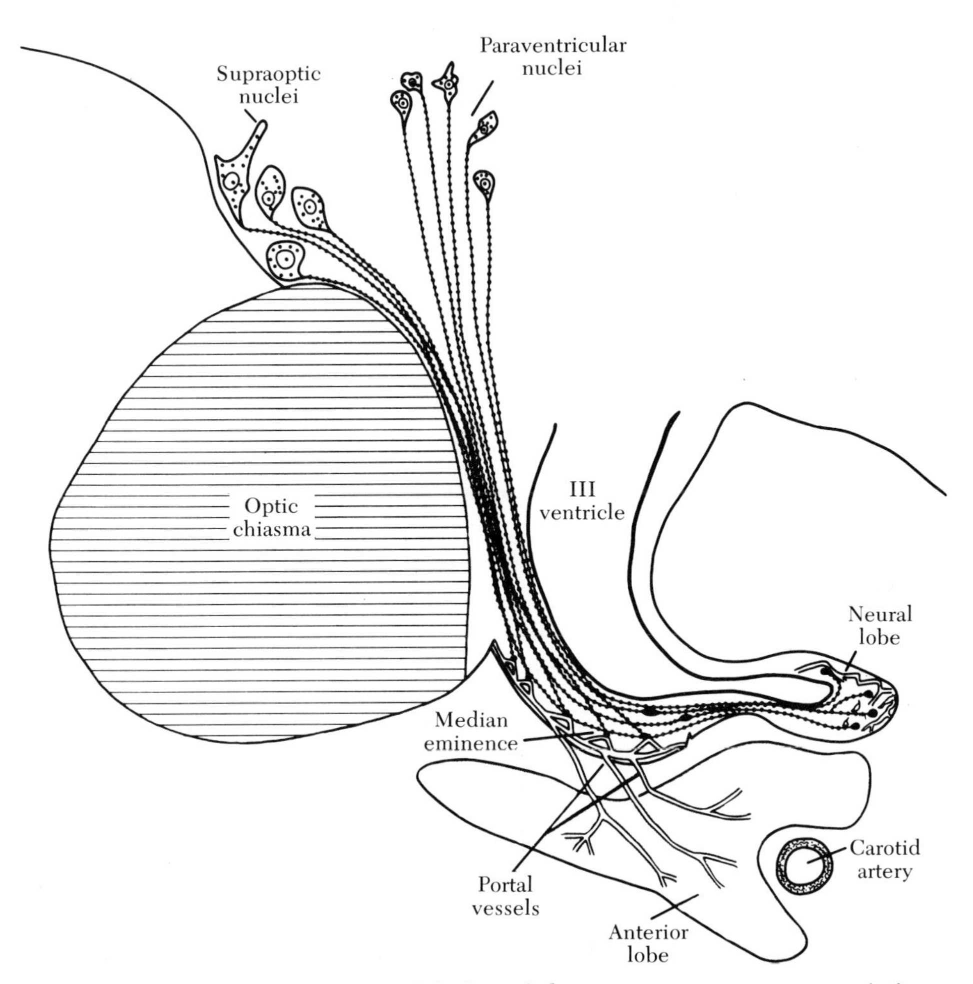

Figure 28-1 Schematic diagram of the hypothalamic neurosecretory system, pituitary gland, and hypophyseal portal vessels. Note that the headed neurosecretory fibers arising in the hypothalamus innervate the median eminence and neural lobe of the pituitary, but that the only connection between the hypothalamus and the anterior lobe is vascular.

layed, cause the pituitary gland to switch from secretion of gonad-stimulating hormones to the secretion of the hormone prolactin, which is responsible for the manifestation of the maternal instinct. The flow of prolactin is at first maintained by the continued tactile messages caused by the pressure of eggs against the breast, and later by the presence of the hatched chicks. As the chicks become more independent, the chirping and crowding around the mother for shelter diminishes, and the signal for continued secretion of prolactin weakens. The hen's pituitary gland may now go back to secreting gonad-stimulating hormones and the laying of another set of eggs, *provided* messages from the outside still tell her that enough time remains to raise a second brood; that is, provided there has been no marked decrease in amount of light. If the brood becomes self-sufficient at a time when the amount of light that impinges on the eye is decreasing, the pituitary gland is not sufficiently stimulated to cause secretion of adequate amounts of gonadotropic hormone; the ovary remains regressed until the following spring, when the whole cycle begins again.

The only connection between the hypothalamus and the anterior lobe of the pituitary gland is humoral, and the connection between the hypothalamus and the posterior lobe of the pituitary gland is neural (Figures 28.1 and 28.2). Thus, the nerves connecting the shell gland (or uterus) to the hypothalamus may carry the message that the shell gland contains a hard-shelled egg that is ready to be expelled. This information is transmitted directly via the connecting nervous system from the hypothalamus to the posterior lobe, which instantly responds by releasing appropriate amounts of the hormone oxytocin. This hormone causes contraction of the muscles in the shell gland, which causes expulsion of the egg.

In a very general and sketchy way, this short discussion should introduce the kinds of problems we will examine in the hope of understanding the physiology of egg production. We must, of course, keep in mind that the domesticated hen is a development from its ancestor, the Indian Jungle Fowl, which was modified extensively by man before it achieved its present commercial usefulness. In the concluding section of this chapter we shall return to the problem of how this hen—whose ancestors laid at most 50 eggs a year—became an organism capable of producing as many as 366 eggs in 365 days.

The Ovary and the Laying Cycle

Components of the Egg

The prime aim in life of mammals and birds alike is procreation, but the reproductive problems faced by birds are completely different from the problems faced by mammals. In the mammal, one or more young not only begin their existence in the uterus, but must continue to develop and be nourished in it. Thus the mammalian egg can be small, since the developing fetus can obtain all of its nourishment from the maternal organism, to which it becomes attached by means of a placenta. The avian embryo develops completely outside of the maternal organism, and for this reason the bird egg must be sufficiently large to contain all the nutrients that will be needed by the growing embryo throughout its development. Avian eggs also must be enclosed by a shell rigid enough to withstand the vicissitudes of the incubation period. The nutrients contained in the egg are the lipoproteins of the yolk, the protein of the egg white (albumen), and the calcium carbonate of the egg shell. All these substances come from the maternal organism, and birds have mechanisms by means of which the different building materials are made available to the proper organ at the proper time (Table 28.1).

The chemical components of the yolk, the albumen, and the shell are mobilized either directly from the gut or from maternal body reserves such as the skeleton, the liver, or the fat depots. These components must be transformed into chemical compounds that can readily pass through cell membranes and that

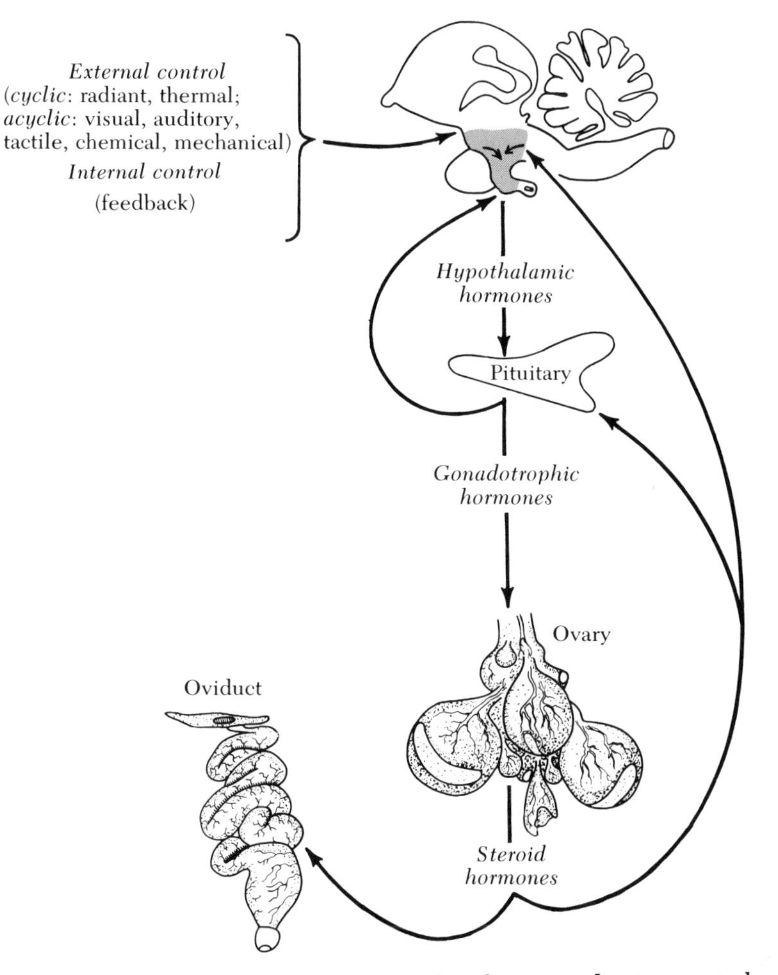

Figure 28-2 Diagram of the principle of neuroendocrine control, indicating types of environmental stimuli that may impinge on the hypothalamic control center and possible sites of action of feedback hormones.

can be transported through the blood stream. After the precursors of yolk and albumen appear in the blood stream, how are the lipoprotein globules suspended in the blood stream consistently directed to the growing follicles where they will be transformed into yolk and deposited into the ovum, and why are the precursors of egg albumen selectively removed from the blood stream in the oviduct and nowhere else? Similarly, how is the calcium carbonate filtered out by the shell gland and deposited around the finished egg? These questions remain, for the time being, without definite answers. Endocrinologists call the ovary, the oviduct, and all other hormone-supported structures, "genetically conditioned end organs," but this term is meaningless since it explains neither the physiology nor the biochemistry of these processes. At best, it simply calls attention to the fact that the ovary has the built-in ability to distinguish between yolk and calcium carbonate, to accept and to use the yolk and to reject the calcium carbonate. In the same mysterious way the ovary can differentiate among the multitude of hormones that are carried by the blood stream. It re-

Table 28.1.
Time of egg formation in the chicken.

Part of egg formed	*Site of formation*	*Time of formation*
Yolk	Ovary	7–9 days
All others	Oviduct	24–28 hours
Thick mucin	Infundibulum	15–30 minutes
Albumen	Magnum	2–3 hours
Shell membranes	Isthmus	1.5 hours
Watery solution	Shell gland	3–5 hours
Shell	Shell gland	19–20 hours
Bloom (mucus)	Vagina	1–10 minutes

sponds only to certain hormones, and remains completely unaffected by all the others that bathe it.

Processes of Egg Formation

Near the time of ovulation, the ovary of a mammal contains one or two ripe follicles if the female is monotocous (producing a single offspring at one time), or several follicles of ovulatory size if the female is polytocous. In contrast, the ovary of the laying hen consists of a series of follicles (called the follicular size hierarchy), ranging from microscopic size to the size of the follicle that is destined to ovulate next (Figure 28.3). This folicle weighs about 15 g, the next member of the size hierarchy about 14 g, and the third about 10 g. Below this size there may be several follicles of the same weights, the different weight classes ranging down to folicles weighing less than 1 mg, and eventually down to a multitude of follicles of microscopic size.

The interesting and significant fact about this size hierarchy is that only one follicle reaches ovulatory size each day. Only after ovulation of the largest follicle do the smaller follicles move up one notch in size and reestablish the hierarchy as it existed just before ovulation. The interval between the ovulation of two successive eggs is usually 24 to 28 hours. As a rule hens lay one egg a day on several successive days before there is a day on which no egg is laid. The uninterrupted series of successive eggs laid is called a clutch. The tendency of some hens to lay short clutches (1 to 3 eggs) and others to lay longer clutches (6 and up, to 100 or more) is a heritable characteristic, and each hen tends to repeat her typical clutch length throughout most of her productive life. The hens with the longest clutches will of course be able to produce the largest number of eggs each laying year, because they have the fewest number of nonproductive days; the hens with 1- or 2-egg clutches cannot produce more than 180 or 240 eggs a year, even if there is no time off for molt or winter pause. In order for the hen to be able to produce 300 or more eggs annually, the clutch length must be 5 or more eggs (Figure 28.4).

The laying cycle of the hen is, to some degree at least, related to light. Under normal day-night conditions, ovulations always occur early in the morning. The yolk then begins its trip down the oviduct, where it acquires the egg white, the soft shell membranes, and even-

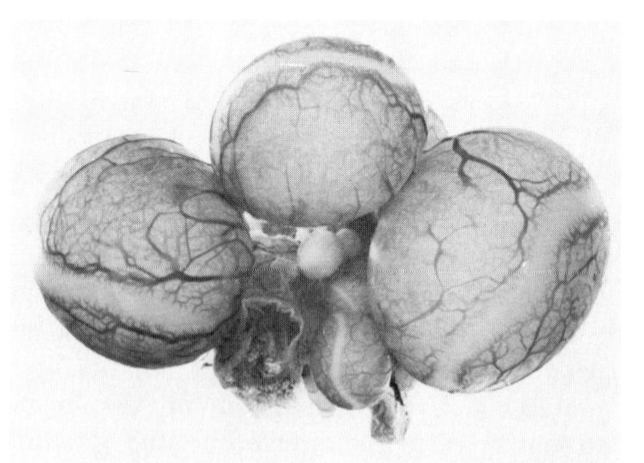

Figure 28-3 Ovary of laying hen showing hierarchy of rapidly growing follicles. Note the extensive network of blood vessels in the walls of the large follicles and the nonvascular area, known as the stigma, along which rupture occurs to discharge the ovum. The empty follicle at the center of the ovary released its ovum just before the ovary was removed for photography.

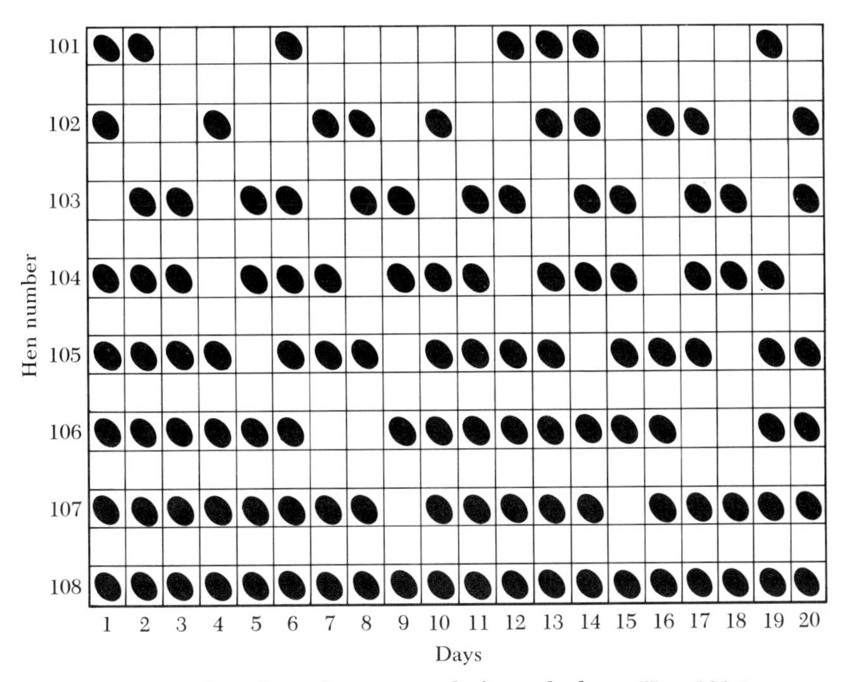

Figure 28-4 Selected egg laying records from chickens. Hen 101 is an unusually poor producer. The record of hen 102, showing clutches of 1 or 2 eggs frequently separated by intervals of 2 days, is indicative of an abnormal condition found in broiler-type hens, in which ovulation rate is high, but the oviduct does not always pick up the egg. Records of hens 103 and 104 are representative of hens commonly used in research on ovulation, primarily because, in repeated short clutches of constant length, the beginning and end of a clutch can be accurately predicted. Hens 107 and 108 show long clutches characteristic of good producers.

tually the calciferous shell. This whole process from ovulation to oviposition (laying the egg) takes an average of 26 hours. Because in the majority of hens ovulations are held in abeyance until the previous egg has been laid, each subsequent ovulation and oviposition occurs a little later in the day than did the previous one, so that eventually the next scheduled ovulation would have to occur late in the afternoon. For unknown reasons, however, this ovulation does not take place and there is a break in the clutch of 24 to 36 hours. Hens with a very short interval between oviposition and ovulation lay longer clutches, while hens in which the interval between oviposition and ovulation is long—2 or more hours—tend to have short clutches and a lower total annual egg production.

These facts raise two very important and interesting problems. One is the problem of the mechanisms involved in establishing and maintaining the follicular size hierarchy for prolonged periods of time. The hen must be able to distribute the hypophyseal hormones circulating in the blood stream in such a way that some follicles get a larger quantity of hormones, so that they grow faster and attain larger size, and other follicles receive a smaller quantity. The net result of this rationing system is that there is established and maintained a follicular size hierarchy in which the position of the individual follicle is determined by the amount of hormone stimulating it. The second problem concerns the possible mechanisms involved in the timing of the intervals between ovulations.

The following information is necessary before we can try to provide answers to these problems.

Endocrine Control of Follicular Growth and of Ovulation

Hormones Involved in Follicle Growth

The endocrine control of follicular growth and of ovulation is by no means simple and is not yet completely understood. Much of what will be said in this section can be documented by data, but such proofs are beyond the scope of this discussion and the interested student should consult recent symposia and textbooks if he wishes to separate scientific facts from educated guesses.

The neuroendocrine mechanisms governing reproduction in birds are very similar to those of mammals. In birds, the anterior and posterior lobes of the pituitary gland are two anatomically distinct structures that are separated by a septum. Each of these lobes is connected by a separate stalk to the hypothalamus. Through the stalk of the posterior or neural lobe run nerve fibers connecting this lobe to the hypothalamic nuclei. The anterior lobe is not innervated and its only connection to the hypothalamic nuclei is through the vascular system, in which the blood flows only in one direction, from the hypothalamus to the anterior lobe of the pituitary gland. The posterior lobe serves as a storage reservoir for the posterior pituitary hormones (vasotocin and oxytocin), which are formed in the area of the hypothalamus and pass into the neural lobe along the nerve fibers connecting it to the hypothalamus.

The hormones of the anterior lobe are probably formed directly in the anterior lobe and released from it into the peripheral circulation. Although the anterior lobe secretes five or six different hormones—the somatotropic, adrenotropic, thyrotropic, gonadotropic, and lactogenic hormones—we will be concerned here mainly with the gonadotropic complex. Frequently this is subdivided into the follicle-stimulating hormone, FSH, associated predominantly with follicular growth, and the luteinizing hormone, LH, mainly responsible for causing ovulation (Figure 28.5). For the sake of simplicity we shall use the term "gonadotropic complex," GTC, throughout this discussion.

Proofs of Hormonal Control of Follicular Growth

The gonadotropic complex from the pituitary gland is responsible for the growth and maturation of the ovarian follicle. That the ovary and its follicles depend upon the gonadotropic hormone is demonstrated by the fact that removal of the pituitary gland (hypophysectomy) leads to a rapid and complete degeneration of the ovarian follicles. This process of follicular degeneration is called atresia and is illustrated in Figure 28.6. Conversely, one can prevent follicular atresia in hypophysectomized hens by the injection of gonadotropic hormones (Figure 28.7). Experiments of this type allow us to conclude that the rates at which hens lay eggs depend on the amount of the gonadotropic complex secreted by the pituitary gland.

The vascular networks covering the larger follicles (Figure 28.8) are much more extensive and intricate than the networks supplying the smaller follicles. Ovulation of the largest follicle of the hierarchy results in an abrupt shutting down of the very extensive vascular network supplying that follicle. Thus, following ovulation, the amount of blood flowing through the vascular networks of smaller follicles of the hierarchy increases and more hormone becomes available for the lesser members of the hierarchy. How some follicles gain more than their mates is not quite clear, but it is reasonable to think that this is a simple matter of chance and may depend on the initial proximity of some of the follicles to a larger blood vessel, allowing them to grow faster than some of their neighbors.

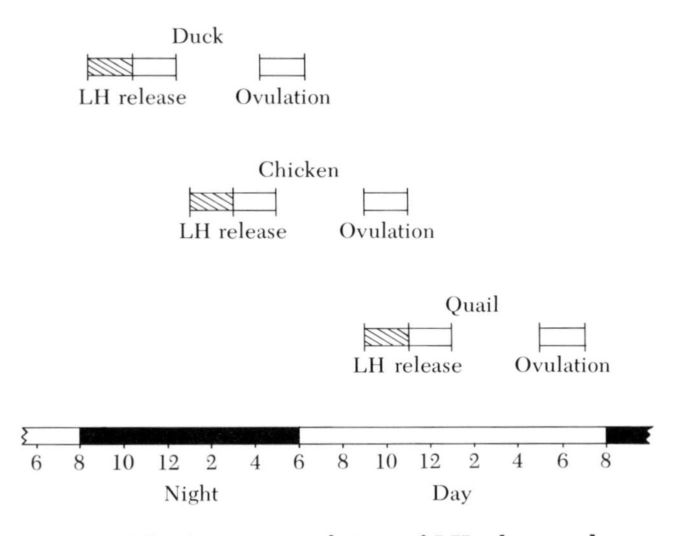

Figure 28-5 Comparison of times of LH release and ovulation in three domesticated birds exposed to a constant 14-hour light-day. At the base of the figure, black bars indicate period of darkness; open bars the period of light. Open devices labeled "ovulation" show time of ovulation for the first egg of a clutch. Diagonally hatched portion of device labeled "LH release" indicates period in which LH release for the first ovulation is initiated; clear portion shows period in which release is completed.

Hormonal Control of Ovulation

Next we can examine the question of how ovulation is triggered and how the intervals between ovulation are kept at about 24 to 28 hours. We already know that the normal laying hen ovulates only one follicle each day, and that without hormonal support the follicles degenerate. We can study the mechanisms involved in follicular maturation and in ovulation either by injecting gonadotropic hormones, GTH, into normally laying hens, or by hypophysectomizing hens and then replacing the missing hormones by injection in an attempt to initiate the normal sequences of events involved in follicular maturation and in ovulation. If GTH is injected subcutaneously for several days into normally laying hens, the normal follicular size hierarchy is abolished; four or even ten or twelve follicles can be caused to reach ovulatory size simultaneously. (This fact is an excellent argument in favor of the assumption that the total amount of hormone normally available for the whole ovary as well as for the individual follicle is limited. When the ovary and the individual follicles are flooded with exogenous hormone—introduced artificially from the outside—one can increase the number of follicles reaching ovulatory size in accordance with the amount of hormone injected.) If hens pretreated with GTH are given an intravenous injection of this hormone, there will be ovulation of as many follicles as reached ovulatory size under the pretreatment. In laying hens with a normal follicular hierarchy, an intravenous injection of GTH usually hastens the ovulation of the largest follicle, which normally would have ruptured a few hours later; but it is never possible to cause the ovulation of the second- or third-largest follicles of the hierarchy. This demonstrates that follicles that have been nurtured to ovulatory size by proper endogenous or exogenous GTH can be made to ovulate if the organism is flooded by GTH at the proper time. This can be done experimentally by the intravenous injection of GTH. This suggests that in the normal laying hen the anterior pituitary releases increased quantities of GTH at certain times during the maturation of the largest follicle, and that this causes ovulation of this follicle.

It has been established that the interval between release and endogenous GTH and ovulation (or the injection of exogenous GTH and ovulation) is between 8 and 12 hours. During this interval the hormone causes certain changes in the wall of the largest follicle, permitting it to break open and to release the ovum. What these changes are is not known, but a possible answer may be provided by the data presented in Table 28.2. These data

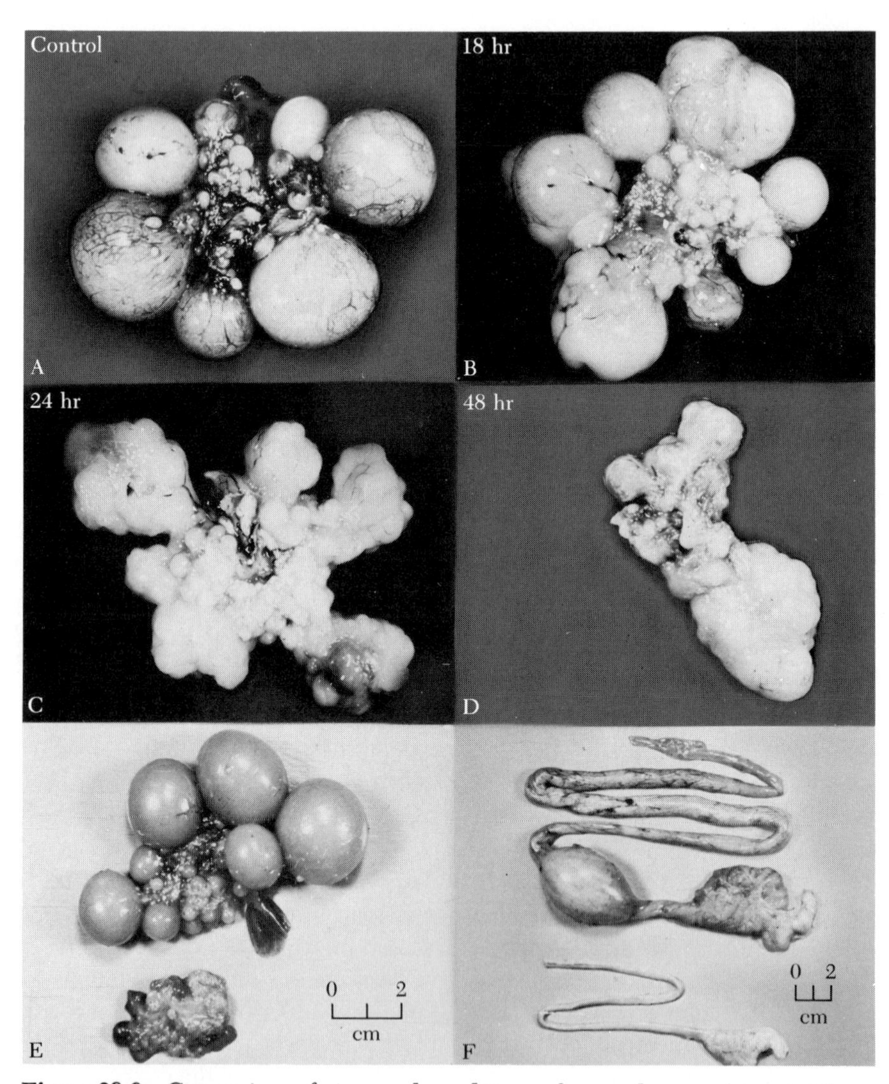

Figure 28-6 Comparison of sizes and conditions of normal ovary (*A* and *E*) under the effects of hypophysectomy. Note the rapidity with which the ovary degenerates 18, 24, and 48 hours after hypophysectomy (*B*, *C*, and *D*, respectively). In *E* and *F* the ovaries and oviducts of normal and hypophysectomized chickens are shown six days after hypophysectomy. (From Opel and Nalbandov, *Proc. Soc. Exptl. Biol. and Med.* 107:233, 1961.)

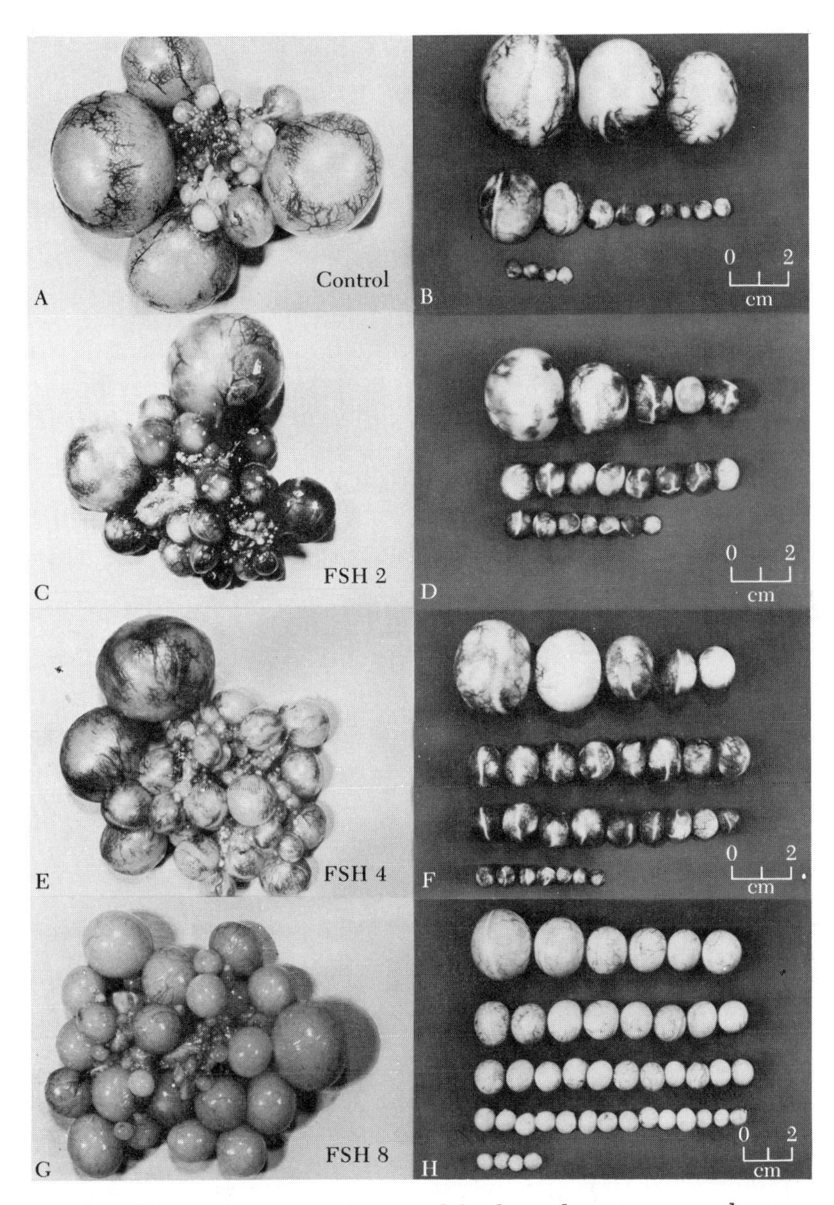

Figure 28-7 Follicular atresia caused by hypophysectomy can be prevented. (A) Shows the ovary of a normal hen. In (B) the size gradation of follicles (hierarchy) is illustrated. Attempts to substitute for the hen's own pituitary by injection of mammalian GTH (C–H) are only partly successful; atresia is prevented, total number of follicles is increased, but the hierarchical distribution of follicular sizes is different (compare A with C, E, and G, and B with D, F, and H). (From Opel and Nalbandov, 1961a.)

were obtained from laying hens, which were hypophysectomized and given a single injection of GTH at different intervals after the operation. It was found that when the GTH was injected fewer than three hours after hypophysectomy, the hens ovulated only one ovum; but when the GTH injection occurred three or more hours after the operation, an increasing number of hens ovulated two or more ova. In other words, the longer the ovary remains without support of GTH from its own pituitary gland, which was removed, the easier it is to cause multiple ovulations by exogenous GTH. By comparing (Table 28.2) the weights of the yolks, whose ovulation was induced by hormone injection (I_1 and I_2), with the weights of ova released prior to hypophysectomy (C_1 and C_2), it is found that, in the absence of the hen's own pituitary gland, not only the largest follicles, but also immature members of the follicular hierarchy, can be caused to ovulate. This is in distinct contrast to normal intact hens, in which injection of GTH can only hasten the ovulation of the largest follicle but can never cause the rupture of the second or third members of the follicular hierarchy. (On the basis of this information, try to formulate a theory concerning the mechanism of ovulation before reading the explanation given below.)

These data suggest that the changes preceding follicular rupture occur more rapidly in the absence of the pituitary gland; that is, in the absence of follicular stimulation by GTH. Thus, ovulation may be normally an

Figure 28-8 Resin cast of a follicle of near ovulatory size, showing the intricate venous system. The arterial system is shown in black. (From Nalbandov, 1964.)

Table 28.2.

Effect of increased interval between hypophysectomy and LH injection on ovulability of follicles.[a]

Interval (hours)	*No. of hens*	*No. of single ovulations*	*No. of double ovulations*	*Average weight of ova (gm)*			
				C_1	C_2	I_1	I_2
0	8	8	0	17.8	17.8	16.9	
2	8	8	0	18.4	17.7	16.3	
3	8	6	2	17.0	16.3	14.9	7.7
4	8	6	2	18.4	17.4	15.0	11.1
6	8	1	7	17.6	17.0	15.4	13.7
12	8	2	5	17.5	17.2	14.9	10.2

[a] From Opel and Nalbandov, 1961b.

aging process, which is initiated by the absence of GTH of hypophyseal origin. This would mean that as long as a follicle is supported by some GTH it is incapable of ovulating. Therefore, in the normal hen, so long as the blood stream carries adequate amounts of GTH, the smaller follicles are stimulated to grow actively and are incapable of ovulating. The largest follicle approaching ovulatory size has the most extensive vascular system and receives the largest amount of hormone. As this follicle reaches its maximum size, the pressure of the accumulating yolk inside the follicular membranes partially squeezes shut some of the blood vessels supplying the follicle walls. This leads to diminishing blood flow, hence to diminishing hormone stimulation, hence to a breaking down of parts of the follicle wall, especially the stigma. All of these events finally culminate in ovulation. The smaller follicles of the hierarchy do not ovulate because their vascular systems carry enough GTH not only to maintain but to stimulate the follicle wall to further growth. So long as they receive such stimuli, the follicle wall cannot break down (or "age" or become ischemic) and ovulation cannot occur.

There remains the question of the mechanism involved in the timing of intervals between ovulation. We can present a plausible theory for which there is good evidence, but much additional work remains to be done before we can be certain of all the factors involved in the timing mechanism.

In the great majority of normal ovulations, the largest follicle ruptures within a few minutes, or at most a few hours, *after* the previous egg has been laid. We have learned earlier that the interval between release of ovulatory doses of hypophyseal GTH and ovulation is about 8 to 12 hours. The time sequences in Tables 28.1 and 28.2 show that this GTH release occurs during the time when the egg is in the process of acquiring the hard shell in the uterus.

There is experimental evidence to show that the oviducal nervous system participates in signaling instructions to the pituitary gland, and that instructions *not* to release amounts of GTH adequate for ovulation may originate in the oviduct. We already know that such signaling systems are frequently involved in phenomena controlled by hormones and that they are part of the neuroendocrine feedback mechanism. In the hen it seems to operate, in brief, as follows. If a yolk is present in the albumen-secreting portion of the oviduct, a nerve-conducted signal goes from the oviduct to the hypothalamus, and from there to the pituitary gland, informing the latter that there is already one ovum in the oviduct and that no new ovulations should be permitted (that is, no ovulatory doses of GTH should be released) until the oviduct is ready for the next egg. Soon after the egg leaves the oviduct and enters the uterus or shell gland, these signals stop and the pituitary gland now releases the amounts of GTH needed to accomplish ovulation some 8 to 12 hours later; that is, after the hard-shelled egg is laid (Figures 28.9 and 28.10). As soon as the next ovum enters the oviduct, the ovulation-blocking signal again becomes effective and holds further ovulations in abeyance.

Function of the Oviduct

The oviduct consists of five anatomically distinct portions, four of which are histologically different. The vital statistics of these areas, as well as their function, are summarized in Table 28.1. It is amazing that the magnum is able to contribute 32.9 g of a proteinaceous substance to the egg in only three hours. Even though the albumen is mostly water, the task of filtering this amount of albumen from the blood stream into the glands of the magnum, and from there into its lumen through which the yolk is passing, seems truly phenomenal. It is interesting that the magnum is incapable of distinguishing between a yolk and any other foreign body. Thus, it will deposit albumen around a cork or a wadded-up piece of paper, or even around a cockroach, as occurred in one experiment when one somehow found its way into the magnum. The size of the laid egg is largely determined by the size of the ovum passing down the oviduct. Relatively less albumen will be deposited around a small yolk (as in the case of pullet eggs), and the resulting finished egg will weigh considerably less than the egg laid by a mature hen, which tends to ovulate larger yolks. There is a slight tendency for successive eggs of a given clutch sequence to become smaller, because the yolks ovulated late in the clutch sequence are likely to be smaller than those ovulated earlier. This relationship holds statistically true for populations of birds, but individual hens may lay progressively larger eggs in a clutch.

The physiological function of the oviduct is controlled by the ovarian steroid hormones. The ovary normally secretes estrogen, androgen, and perhaps progesterone. Estrogen alone is able to cause the morphological enlargement of the oviduct, which is tremendously sensitive to estrogen. Even small doses of exogenous estrogenic hormone are capable of enlarging the oviduct of immature chicks several hundred per cent. This hormone alone, however, is incapable of stimulating the development of the oviducal glandular apparatus. To develop the glands, and to cause them to secrete albumen, the shell membranes, or the calcareous shell, requires the interaction of at least two of the ovarian hormones. One of these is certainly estrogen, but it remains unclear whether the second cooperating hormone is androgen or progesterone. The ovary is known to secrete the male sex hormone, androgen, which is responsible, among other things, for comb size and its bright red color in laying hens. Because both androgen and estrogen are normally secreted by avian ovaries, it appears plausible that the interaction of these two hormones is normally responsible for the size of the oviduct, its glandular development, and its ability to secrete the various components of the completed egg. Under experimental conditions, it is possible to substitute progesterone for androgen and to obtain oviduct growth and albumen secretion even in sexually very immature female chicks, from either the estrogen-androgen or the estrogen-progesterone combination.

Figure 28-9 Chicken oviduct. Egg is entering magnum where it will assume its characteristic shape.

Having acquired the layers of albumen in the magnum, the egg is enclosed in the soft-shell membranes in the isthmus, and the calcareous shell is deposited around it in the shell gland or uterus. It is possible to cause the premature expulsion of the partially calcified egg from the uterus by the injection of the posterior pituitary hormone, oxytocin, which causes contractions of the musculature of the shell gland. It seems probable that this hormone is normally responsible for the expulsion of the finished egg and that uterine nerves are involved in signaling the posterior lobe of the pituitary that the egg is fully formed and can be laid. It remains unknown whether it is the thickness of the hard shell or the time spent by the egg in the uterus that is responsible for the initiation of this neural signal. There is a striking similarity between the process of oviposition in birds and parturition in mammals: essentially the same hormones and the same mechanisms (uterine musculature) are involved in the expulsion of the fetus. In neither case is it known exactly how the signal for release of oxytocin is timed.

Concluding Remarks

The ancestor of the domestic hen laid from 25 to 50 eggs a season, hatched them into chicks, and took care of them as long as they needed protection. The enormous plasticity of the species, which could be converted by domestication from such primitive beginnings into an efficient egg-producing machine capable of producing up to 366 eggs a year, is astonishing. The physiologic principles underlying this transformation have been discussed earlier and need only be summarized. (It is suggested that the student attempt to make such a summary before reading what is to follow.)

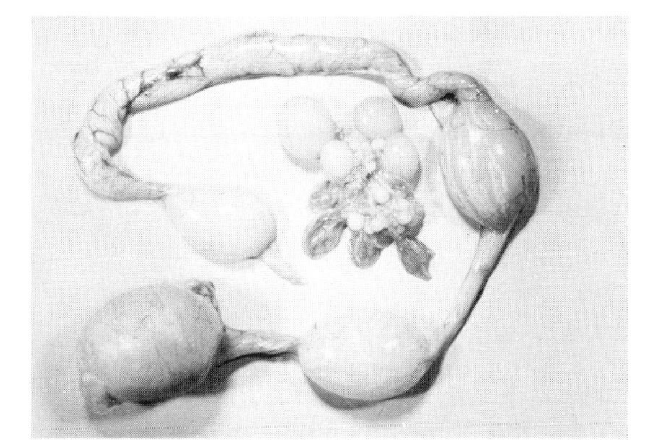

Figure 28-10 Oviduct of chicken containing 4 ova, each forming a separate egg. Multiple ovulations were obtained following injection of mammalian LH into a hypophysectomized hen. In many hens given this treatment, two or three ova, discharged from the ovary within a period of one hour, entered the oviduct separately and formed individual eggs. At oviposition, each egg contained a normal amount of albumen and both shell membranes, but only the first to reach the shell gland was fully calcified. (From Opel and Nalbandov, 1961b.)

When the chicken was taken into the household of early man, it was noticed that by removing the eggs from the nest of the hen, the onset of broodiness could be postponed and more prolonged periods of egg laying could be obtained. For a long time, this was probably the only way in which an increase in productivity was obtained, but it is unlikely that the productivity of the individual hen could have been pushed much beyond 100 or, at most, 150 eggs per year. It is only within the present century, and especially since the advent of incubators, that the maternal instinct has become an unnecessary and even undesirable attribute of hens, because hens do not lay during the time of incubating eggs and caring for chicks. It was noted that those hens that were better mothers, and that spent more time on being broody, consequently laid fewer eggs. By giving the hens with a poorly developed mothering instinct a greater opportunity for reproduction, it became possible to increase productivity of the population by gradually eliminating the broody instinct. We know now that by eliminating the broody individuals, the early breeder was actually eliminating those hens and strains of hens that had genes capable of causing the pituitary gland to secrete copious quanties of prolactin.

The ability of hens to lay the largest number of eggs is due to the presence or absence of at least four genetically controlled characteristics. High-producing strains must be free from broodiness, they must have a short winter molt, they must be persistent (that is, able to lay throughout most of the year), and finally, they must have the ability to lay long clutches. The intensity of the expression of these characteristics is determined by the rates of production of certain hormones, which are more or less prominently concerned in determining the degree to which these characteristics are manifested. The early breeders, who were interested in improving domestic chickens, unknowingly selected individuals whose pituitary glands secreted less prolactin, more gonadotropic hormone, and reduced amounts of thyroxine, known to control rate and intensity of molt. Like most other quantitative characteristics, such as milk production or growth rate, high egg production requires the complex interaction of a great number of genes, which determine not only the rates at which glands secrete their hormones but also the sensitivity with which end organs respond to the hormonal or neural stimuli acting on them.

FURTHER READINGS

Farner, D. S., F. E. Wilson and A. Oksche. 1967. "Neuroendocrine mechanisms in birds." In L. Martini and W. F. Ganong, eds. *Neuroendocrinology,* vol. II, chap. 30. New York: Academic Press.

Fraps, R. M. 1955. "Egg production and fertility in poultry." In A. McLaren, ed., *Progress in the Physiology of Farm Animals,* vol. II, chap. 15. London: Butterworth.

Gilbert, A. B. 1967. "Formation of the egg in the domestic chicken." In A. McLaren, ed., *Advances in Reproductive Physiology,* vol. II, chap. 3. New York: Academic Press.

Nalbandov, A. V. 1964. *Reproductive Physiology.* 2d ed. San Francisco: W. H. Freeman and Company.

Opel, H., and A. V. Nalbandov. 1961. "Follicular growth and ovulation in hypophysectomized hens." *Endocrinology* 69:1016–1028.

van Tienhoven, A. 1968. *Reproductive Physiology of Vertebrates.* Philadelphia: Saunders.

Twenty-Nine

Lactation

"When suckling their young, swine—like all other animals—get attenuated."

Aristotle
History of Animals, Book 8.

The development of knowledge of lactation occurred in three relatively discrete stages. Early studies, largely completed prior to 1950, were directed at study of mammary development, identification and characterization of gross and microscopic anatomical structures in mammary glands, and study of general physiological and nutritional aspects of lactation. The studies of mammary anatomy led to identification of the alveolus (Figures 29.1 and 29.2) as the basic structural element for milk synthesis and secretion; descriptions of the mammary glands of a wide range of mammalia, ranging from those of the duckbill (platypus), which are very elementary in structure, to those of the human (Figure 29.3) and the cow (Figure 29.4), which are highly developed; elucidation of stages in mammary development; and evaluations of relationships between structure and function, including, prominently, clarification of processes involved in milk letdown.

Although some notable progress toward identification of hormonal requirements for mammary development and function and of metabolic pathways involved in milk synthesis was made prior to 1950, the primary period during which information regarding these aspects of lactation was developed was between 1950 and the early 1960's. These developments were facilitated by advances in endocrinology and neuroendocrinology, the availability of "pure" preparations of hormones, and the development of techniques for the culture of mammary tissue in the laboratory. The hormonal requirements for lactation are summarized in Figure 29.5, and include the ovarian hormones estrogen and progesterone, the pituitary hormones oxytocin, prolactin, growth hormone, and adrenal glucocorticoids. Identification of the metabolic pathways of mammary energy metabolism and milk synthesis was accomplished through the use of radioisotope tracer techniques, which indicated the patterns and extents of conversion of compounds from blood to milk components; biochemical techniques which indicated feasible pathways for conversion of blood constituents to milk components; and measurements of

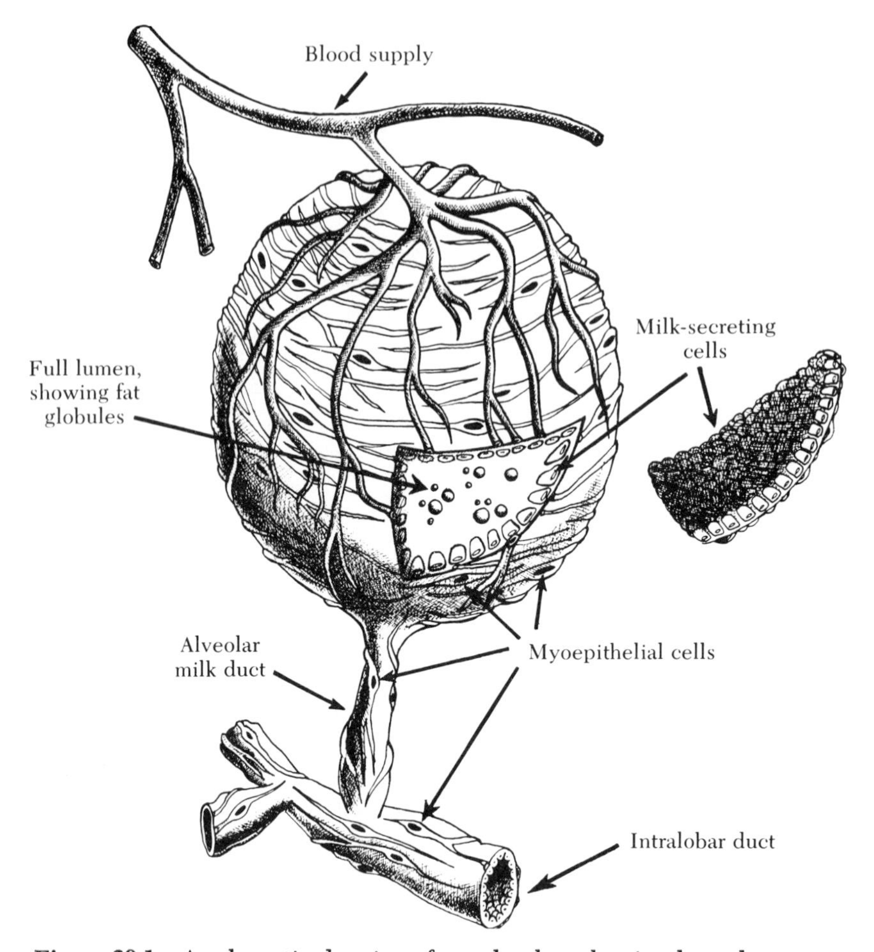

Figure 29-1 A schematic drawing of an alveolus, showing how the mammary secretory cells are arranged to form a saclike structure, how the lumen, into which milk is secreted by the secretory cells, is connected to an intralobar duct, the close relationship between blood capillaries and the alveolus, and the myoepithelial cells which contract to expel milk from the alveolus. (From Turner, 1969).

mammary uptakes of blood constituents, which indicated the amounts of each nutrient available to mammary glands for milk synthesis. Considerable species differences exist in milk composition (Table 29.1), in amounts of different blood constituents used for milk synthesis, and in pathways of synthesis of milk components.

The third period in the development of knowledge of lactation started in the early 1960's and extends into the present. Research during this period has been heavily oriented toward identification and evaluation of mechanisms of regulation of mammary development and function. Reasonable progress toward determinations of the specific actions of insulin, prolactin, glucocorticoids, and the ovarian hormones, and identification of systems and sites of regulation of mammary metabolism during lactation, has been made during this period. Information regarding regulatory processes will hopefully provide the basis for determination of physiological limits of milk yield and facilitate attempts to increase milk production through selection and management of better cows.

In the subsequent sections, an attempt will be made to summarize our knowledge of the development and structure of mammary glands, of the hormonal requirements for mammary growth and function, of the unique physiological and metabolic functions associated with milk synthesis, and of practical aspects of milk letdown and milking management.

Table 29.1.
Compositions of milks of several species.[a]

	Lipid	*Pro tein*	*Lac- tose*	*Ash*	*Dry matter*
Holstein cow	3.4	3.2	4.6	0.7	11.8
Goat	3.5	3.1	4.6	0.8	11.7
Human	3.5	1.3	7.5	0.2	12.0
Dog	8.3	7.5	3.7	1.2	23.0
Pig	9.6	6.1	4.6	0.9	21.2
Rat	15.0	12.0	2.8	1.5	31.0
Whale	36.6	10.6	2.1	1.2	50.0

[a] All figures are percentages.
See Chapter 10 for data on composition of milk in some other species.

Morphogenesis, Development, and Structure of the Mammary Gland

Mammary Development and Growth Through Puberty

Early in fetal development, two light-colored "milk" lines, extending from the thoracic to the abdominal-inguinal areas, become apparent in all mammalian species. Soon thereafter, epidermal thickenings called mammary buds form at points along the milk lines. The mammary buds serve as focal points for differentiation of the mammary glands, and correspond in placement to the points at which mammary glands characteristically develop in each species. For example, in humans and elephants, two mammary buds form in the thoracic area; in the cow, four mammary buds form in the inguinal area; and in rats and pigs, 10 to 20 mammary buds are formed along the milk lines in the thoracic, abdominal, and inguinal areas. Formed in later stages of fetal development are teats or nipples and primary epithelial sprouts which will ultimately develop and form the gland duct and secretory systems. Also formed during this period are the connective tissues which will support the fully developed mammary gland, the connective stroma which will surround and support the secretory elements of the fully developed mammary gland, and adipose tissue which surrounds the glandular elements and makes up the bulk of the mammary glands at birth and through puberty. There are species differences in the extent to which development of the gland has progressed by birth, but, in general, duct development at birth is rudimentary, whereas teats and nipples are well-developed, as are the connective elements, the vascular and lymphatic systems, and adipose tissue. Hormones are not necessary for prenatal mammary development, even though mammary elements are responsive to hormones during this period. The high levels of testosterone secreted by male fetuses inhibit teat or nipple development, and cause disruptions in the formation of other glandular elements. Castra-

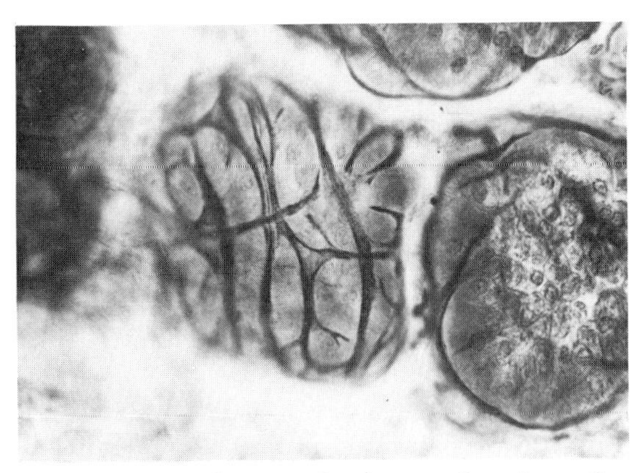

Figure 29-2 Photograph of part of surface of a small contracted alveolus (goat), showing a myoepithelial cell with nucleus and branch process. (From Richardson, 1949.)

tion of male fetuses prevents this inhibition of mammary development, and results in the development of glands indistinguishable from those of females. High doses of estrogens cause abnormal development. In response to hormones formed by mothers during pregnancy, human infants sometimes undergo abnormal mammary development and, at birth, secrete a milklike substance referred to as "witches' milk."

Mammary Development Through Puberty

Mammary development after birth and prior to the onset of puberty is largely characterized by generalized growth at a rate comparable to the growth rate of the whole animal (isometric growth) and maturation of glandular elements not clearly defined or developed at birth. In studies of prepubertal mammary development in calves, an intensive effort was made to correlate udder size as determined by palpation with subsequent lactational performance in the hope that this technique could be utilized to identify at birth calves with high production potential. Unfortunately, although udder size in calves is correlated with subsequent lactational performance, the technique cannot be utilized because individual differences (variance) are very large and because palpation does not clearly distinguish mammary secretory elements from associated adipose tissue.

At the onset of puberty, mammary growth becomes allometric, i.e., faster than general body growth, due to the influence of estrogen. The growth which occurs during this stage of development is primarily characterized by extensions of larger ducts and rapid proliferation of smaller ducts and ductules (Figure 29.4). Growth during this period is cyclic in nature, with the mammary ducts proliferating rapidly during stages of the estrous or menstrual cycle, when estrogen levels are high, and regressing slowly during other phases of the cycle. In each cycle, more ducts are formed during the proliferative phase than are lost during the regression phase, with the net result that a highly branched matrix of ducts leading from the nipple or gland cistern is developed. This matrix represents the first stage of the development of true, lobuloalveolar secretory structures. The proliferation of ducts and ductules is accompanied by development of connective stroma which will surround the fully developed lobuloalveolar structures (Figure 29.4), growth of adipose tissue, and further development of the myoepithelium (smooth muscle fibers) and the vascular and lymphatic systems. During postpubertal development, species differences become evident. In cows, goats, and rodents, considerable growth of adipose tissue occurs and development of the connective stroma lags behind duct growth. In humans, growth of adipose tissue is more limited, and growth of the connective stroma precedes duct growth. In general, species (including humans, cows, and rats) which have short estrous cycles with very short luteal phases exhibit primarily duct growth, whereas species

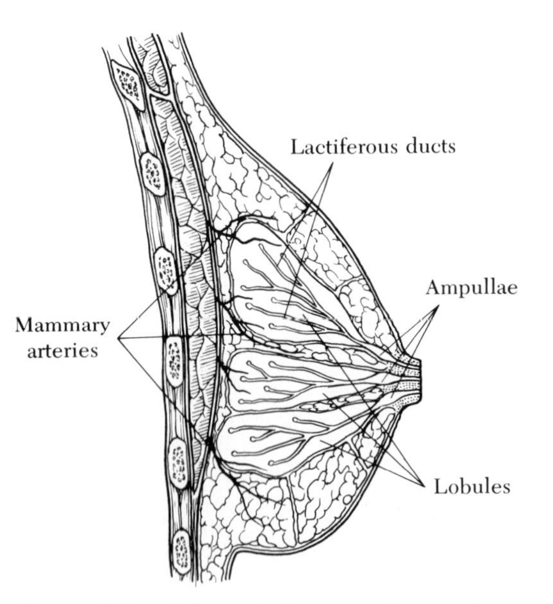

Figure 29-3 Schematic drawing of human breast, showing ducts leading to nipple from ampulla and the intermingling of lobuloalveolar elements with adipose tissue.

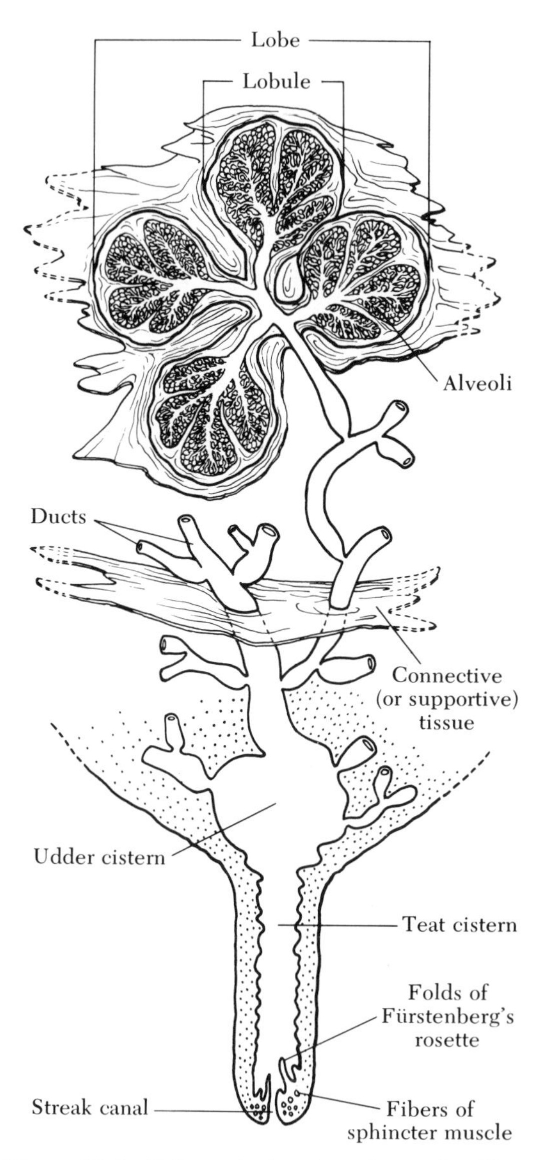

Figure 29-4 Schematic drawing of a cow's udder, showing the major structures of the teat, the gland cistern, the connective stroma, and lobuloalveolar structure (from Turner, 1969).

such as the dog with long luteal phases exhibit duct growth plus significant lobuloalveolar development. This species difference plus other evidence indicate that estrogens stimulate duct growth, and that progesterone, secreted by the corpus luteum during the luteal phase, stimulates lobuloalveolar development.

Anatomy of the Mammary Glands

The alveolus, a spherical or sacklike structure that consists of a single layer of mammary epithelial cells surrounding a cavity or lumen which narrows at one end and leads to a small, capillary duct, is the basic secretory structure of the mammary gland (Figure 29.1). The epithelial cells of the alveolus, often referred to as mammary secretory cells, are the site of milk synthesis. Milk synthesized in the epithelial cells is secreted into the lumen of the alveolus, where the milk is stored until milking or suckling is initiated. When milking is initiated, the myoepithelial (smooth muscle) cells (Figures 29.1 and 29.2) surrounding the alveolus receive a signal (oxytocin) to contract and force milk from the alveolar lumen into the milk ducts. The epithelial cells of the alveolae are very active and require large amounts of nutrients, which must be obtained from blood. In order to provide for this need, a network of very small blood vessels (capillaries) surround each alveolus and are in close proximity to the secretory cells (Figure 29.1). Alveolae are arranged in groups (Figure 29.4) called lobules and lobes. These resemble bunches of grapes, where the alveolae are analogous to grapes, and the intralobar and interlobar ducts (Figure 29.4), through which milk is removed from the alveolae, are analogous to the grape stems. In fully developed mammary glands, the network of ducts required for milk removal is very complex. Each alveolus, lobule, and lobe is surrounded and supported by connective stroma (Figure 29.4). As indicated in the previous section, alveolae and lobules, and their associated connective, myoepithelial, and vascular elements, don't begin to develop until late puberty. In many species, alveolar and lobular (lobuloalveolar) growth and development is not completed until late pregnancy or early lactation.

In the cow, ducts leading from the lobes connect and form larger ducts which lead to the udder cistern (Figure 29.4), where milk collects after being expelled from the alveolae

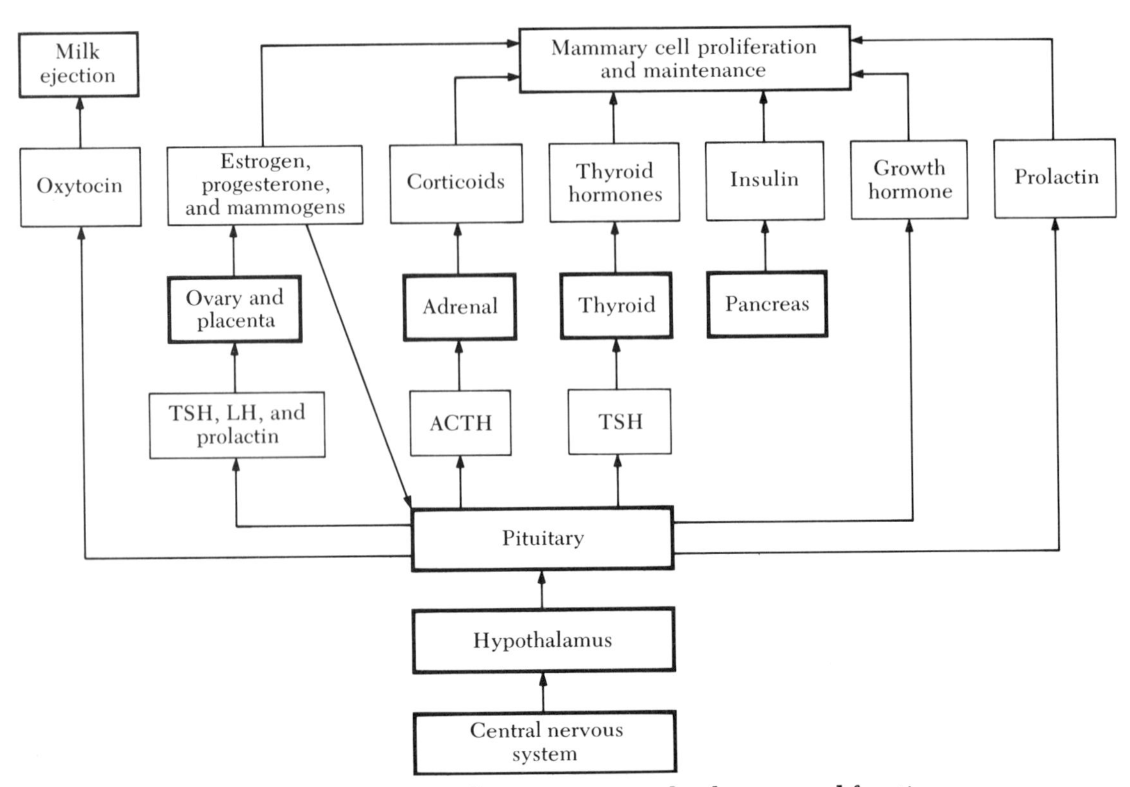

Figure 29-5 Summary of known factors affecting mammary development and function: FSH, follicle-stimulating hormone; LH, luteinizing hormone; ACTH, adrenocorticotropic hormone; TSH, thyroid-stimulating hormone.

and smaller ducts by contraction of myoepithelial cells (milk letdown). A common alternative anatomical arrangement for milk removal is depicted in Figure 29.3. In the human and other species (including the rat), large ducts from a number of lobes lead directly to the nipple. In these species, the only provisions for collection of milk after letdown and prior to removal are slight enlargements called ampullae in the large ducts (Figure 29.3).

As is evident from previous discussion, the mammary glands are very complex structures made up of a number of different cell types, including epithelial cells, muscle cells, connective tissue, and adipose cells. In many species lobuloalveolar structures are surrounded by fatty tissue (Figure 29.3). The fact that the mammary gland is actually a mixture of numerous different cell types causes considerable difficulty in studies of mammary secretory-cell development and metabolism.*

Mammary Development During Pregnancy and Early Lactation

Mammary development during pregnancy and early lactation is characterized by the growth of small ducts and extensive proliferation of lobuloalveolar structures. This growth and proliferation can be considered to be the final stage of mammary development and is essential for lactation. At the end of each lactation, when a mother weans her young or a cow is "dried up," the lobuloalveolar structures, small

* See Turner (1952) and Kon and Cowie (1961) for detailed descriptions of mammary morphogenesis and anatomy, of mechanisms of milk secretion by epithelial cells, of mammary suspension systems, and of nervous connection of the mammary gland.

ducts, and associated structures regress (involution). Therefore, growth and proliferation of duct and lobuloalveolar structures occurs during every pregnancy, and is essential whether the animal has had a previous pregnancy or not.

The most prominent hormonal changes which occur during pregnancy are dependent upon the marked increases in estrogen and progesterone in the blood (see Chapter 26). As indicated in a previous section, these two hormones have also been implicated in the regulation of mammary development during puberty. Numerous studies (summarized by Meites in Kon and Cowie, 1961) have been undertaken with the objective of developing techniques for the artificial initiation of lactation. These studies were based upon the presumptions that estrogen and progesterone, acting directly on the mammary glands and indirectly by affecting secretions of other hormones, cause normal mammary development to occur during pregnancy, and that the abrupt decrease in the levels of these hormones which occurs at parturition causes the initiation of lactation. The most promising experimental method for artificial (without pregnancy) initiation of lactation in the virgin and mature cows resulted from the implantation of pellets containing 100 mg of diethylstilbestrol and 3 g of progesterone for three months, implantation of an additional pellet containing 1.5 g of diethylstilbestrol for one month, removal of the pellets, and the initiation of milking. Milk yields during lactations thus induced are usually below normal and are quite variable. The variability appears to be due, in part, to animal differences in response to the hormones and in the basic physiological (hormonal) background on which the hormone treatments are imposed. Sud, Tucker, and Meites (1968) observed good mammary growth and lobuloalveolar development in ovariectomized heifers injected three times weekly for twenty weeks with 200 mg of progesterone and 800 μg of estradiol-17β. Variability was low in their experiments, presumably because animal differences in ovarian activity and responses to treatment were obviated by ovariectomy. Until researchers are able to obtain better and more consistent milk yields in normal animals, methods for the artificial initiation of lactation will not be practical for extending the productive lives of hard-to-breed and sterile cows.

Treatments with estrogen and progesterone have been utilized in species other than cows to induce mammary growth and lobuloalveolar growth. Barnawell (1965) induced abundant lobuloalveolar development in rats, guinea pigs, and rabbits with daily injections of 0.5, 5.0, and 15 μg, respectively, of estradiol benzoate and 2, 0.5, and 0.1 mg, respectively, of progesterone. The species difference in the ratio of estradiol to progesterone achieving best results should be noted. This ratio is considered more important than the absolute amounts of hormone administered.

Hormonal Requirements for Mammary Development and the Initiation and Maintenance of Lactation

Considerable research emphasis has been placed on determination of the specific hormone requirements for mammary growth, development, and function and of specific hormone actions (see Kon and Cowie, 1960, and Reynolds and Folley, 1969, for summaries of various aspects of this work). Three basic approaches have been utilized in these studies of hormone requirements and actions. Techniques involving incubation in the laboratory (*in vitro*) of small bits of mammary tissue (explants) removed from virgin and pregnant animals have been utilized extensively. These techniques offer the advantage that hormones can easily be added to incubation media in any combination and responses evaluated. Disadvantages of this approach are that the amounts of tissue used are very small and limit the types and number of measurements that can be used to evaluate responses, and that it can only be used to study development. *In vitro* methods for the maintenance of normal secretory activity in explants from lactating animals have not been developed. Difficulties encountered in attempts to culture lactating explants may be attributable to the rapid accumulation of milk and the lack of means for removing milk. A second approach has been to evaluate changes in mammary glands in normal animals during pregnancy and lactation and to attempt to relate these to changes in hormone patterns. A variation of this approach is to inject hormones into normal animals as described above for the artificial initiation of lactation with estrogen and progesterone. The limits or disadvantages of this approach are evident. Specific cause-and-effect relationships cannot be established. A third approach, which has been utilized extensively, has been to remove surgically one or all endocrine glands and to administer hormones in various combinations.

Removal of an endocrine gland accomplishes two objectives. First, if the hormone(s) secreted by that gland is essential for mammary development or function, development or function is impaired unless the hormone is injected (replacement therapy). In this way, cause-and-effect relationships can be established. Second, indirect effects of hormones can be identified. For example, if estrogen acts in intact animals by affecting the secretion of pituitary hormones, which in turn regulate mammary development (which they do, as we shall see shortly), removal of the pituitary breaks the sequence and establishes the fact that a hormone (estrogen) is acting indirectly. Because this approach provides a means for establishing specific cause-and-effect relationships, it is the preferred approach for many purposes. The primary factors limiting the use of this approach are the difficulty of some of the surgical techniques, the extreme care which must be taken in caring for the animals, and the cost of some of the hormones which must be supplied. The latter is a very important disadvantage encountered in studies with large animals.

Until relatively recently, the primary techniques utilized to evaluate mammary responses to hormones were histological and, for *in vivo* studies, functional (milk yield). Much of the available data relate to the ability of hormones to promote duct growth and lobulo-alveolar development and to maintain milk yields. More recent studies have employed biochemical methods to assess specific actions of hormones on secretory cell proliferation, development, and function.

In the section on mammary development during pregnancy and early lactation, it was indicated that development comparable to that which normally occurs during pregnancy can be achieved in non-pregnant, intact animals by administration of estrogen and progesterone. It must be emphasized at this point that estrogen and progesterone cannot stimulate mammary growth and development in animals which are adrenalectomized (adrenals removed) or hypophysectomized (pituitary removed). This observation indicates that estrogen and progesterone do not act alone in regulating mammary development during pregnancy, but rather in combination with pituitary and adrenal hormone(s). Many scientific papers attest to the research efforts which have been directed at determining the roles of the pituitary and adrenal hormones in regulating mammary growth and development during pregnancy. Several significant species differences exist and are ignored in the generalizations presented in the following paragraphs.

The most important pituitary hormone regulating mammary secretory cell proliferation and development is prolactin (lactogenic hormone, mammotropic hormone). The primary actions of estrogen in intact, pregnant animals appear to be: (1) stimulation of prolactin secretion; and (2) enhancement of prolactin action in the mammary glands. Hence, estrogen acts both indirectly through prolactin and synergistically (working with) prolactin in promoting mammary development. The most important adrenal hormone regulating mammary development and function is hydrocortisone (cortisol). Reasonably good duct, ductule, and lobuloalveolar growth can be induced in hypophysectomized, adrenalectomized, and ovariectomized virgin animals with injections of prolactin, hydrocortisone, and estrogen. Administration of additional hormones, including growth hormone, thyroid-stimulating hormone, and adrenalcorticotropic hormone, improves development, but in general these hormones do not fulfill central roles as do prolactin and hydrocortisone. Growth hormone appears to be essential in some species and can mimic prolactin in others. *In vitro* data obtained with mammary explants from virgin and pregnant rats largely confirm the conclusion that prolactin and cortisol are essential for the formation and development of lobuloalveolar structures and secretory cells. Recent *in vitro* studies conducted by Turkington and Topper (1966) have shown that a glucocorticoid must be present to insure the formation of competent (differentiated) secretory cells, and that prolactin is essential for the development of newly formed secretory cells. *In vivo* studies with hypophysectomized rats have demonstrated that prolactin and hydrocortisone are required for the synthesis of the enzymes required for milk synthesis. A prominent feature of *in vitro* studies is a high and absolute requirement for insulin for the formation, maintenance, and development of mammary secretory cells. This requirement has not been investigated thoroughly *in vivo,* and it is not yet certain whether insulin plays a unique role in regulating mammary development or a permissive or nonspecific role analogous to its function in other tissues.

Mammary lobuloalveolar growth is essentially complete late in pregnancy, and at parturition milk synthesis is initiated. Several questions remain regarding the exact nature of the regulatory processes which, first, prevent significant milk synthesis from occurring during pregnancy and, second, provide the signal(s) for initiation of milk synthesis at parturition. The most attractive, simple suggestion is that the high levels of progesterone secreted during pregnancy inhibit milk synthesis, and that the decrease in progesterone secretion which occurs at parturition acts as a "trigger" in initiating milk synthesis. Consistent with this suggestion, recent data have shown that milk synthesis starts in pregnant animals after ovariectomy and hysterectomy (removal of primary progesterone secreting tissues), and that this initiation of milk synthesis can be prevented by administration of progesterone. Alternately, the onset of lactation is considered a very complex process involving interplay of a number of regulatory mechanisms which influence cellular, tissue, and animal metabolism, blood flow rates, and patterns of secretion of several hormones, including oxytocin, prolactin, estrogen, progesterone, thyroxin, insulin, and adrenal glucocorticoids (Figure 29.5). Additional research is required to improve our understanding of the initiation of lactation (Reynolds and Folley, 1969).

Studies with several lactating species indicate clearly that the minimal hormone requirements for the maintenance of lactation in animals which have been hypophysectomized, ovariectomized, and adrenalectomized are prolactin, cortisol, and oxytocin. Prolactin and cortisol are essential for the maintenance (survival) and function of mammary secretory cells. Oxytocin, a pituitary hormone, causes the contraction of mammary myoepithelial cells and is, hence, essential for milk removal. Cessation of milking, weaning, or failure to provide oxytocin to insure milk removal in hypophysectomized animals results in very rapid loss of secretory cells and lobuloalveolar structures (involution). Estrogen, thyroxin, and growth hormone stimulate milk production in triply-operated animals, but, as was the case for mammary development, are not generally considered to be absolutely essential for the maintenance of lactation. Cessation of insulin therapy in lactating diabetic animals results in an immediate depression of milk synthesis and, within 48 hours, involution.

Recent biochemical studies directed at determination of specific hormone functions in the regulation of mammary secretory-cell development and function have been useful in establishing relationships between hormone requirements and the initiation and maintenance of lactation. The initiation of lactation is accompanied in several species by a threefold to fivefold increase in the activities of all enzymes. This general increase in enzyme activities reflects the maturation of mammary secretory cells. At the same time, the activities of enzymes in the pathways of synthesis of milk components increase up to 100 times. This increase is considered essential for lactation. Two hormones, prolactin and cortisol, are required for the formation of these key enzymes of milk synthesis. The primary action of prolactin appears to be stimulation of RNA synthesis, and hence development of the capacity for synthesis of tissue enzymes and milk protein. Cortisol acts in a more specific fashion in regulating the synthesis of enzymes in the pathway of lactose formation, including lactose synthetase, enzymes associated with milk-fat synthesis, including glucose-6-P-dehydrogenase, citrate cleavage enzyme, malic enzyme, fatty acid synthetase, and, possibly, enzymes or proteins required for casein synthesis. In adrenalectomized and/or hypophysectomized animals receiving prolactin, the levels of enzymes whose synthesis is regulated, in part, by cortisol are between 20 and 50 per cent of normal, and milk yield is only 40 to 60 per cent of normal. Administration of cortisol restores enzyme levels and milk yields to normal. In hypophysectomized animals not receiving prolactin, cortisol administration does not restore enzyme levels to normal indicating, that the RNA synthesis regulated by prolactin is essential for cortisol action. An attractive explanation of these observations (supported, in part, by additional data) is that prolactin regulates ribosomal RNA synthesis, and that cortisol regulates the synthesis of messenger RNA's for key enzymes (possibly 14 to 16) involved in milk synthesis. Without the ribosomal RNA formed due to prolactin action, the messenger RNA's formed due to cortisol action cannot be translated to form required key enzymes.

Milk Biosynthesis

The major components of milk are lipid, protein, lactose, and ash (Table 29.1). Milk lipids are secreted as small droplets called globules by mammary secretory cells. The globules are composed, primarily, of triglycerides surrounded by a thin membrane made up of protein and phospholipids. In nonruminant species, the primary precursors of milk triglycerides are blood triglycerides, which provide fatty acids for milk triglyceride synthesis, and glucose, which is converted to both triglyceride glycerol and fatty acids by the mammary glands. In ruminants, the primary precursors of milk triglycerides are blood triglycerides, acetate, ketone bodies, and glucose. Acetate and ketone bodies are converted to fatty acids by the mammary gland, and glucose is converted to triglyceride glycerol. Up to 50 per cent of the fatty acids for milk-triglyceride synthesis are derived from blood triglycerides in ruminants. Lactose, often referred to as milk sugar because it is only found in milk, is formed from blood glucose, and is made up of glucose and galactose. The primary milk protein is casein, which is formed from blood amino acids. Casein is only found in milk. Some blood proteins, most notably the immune globulins, are found in milk. Colostrum, the milk formed during the first few days of lactation, contains large quantities of globulins, and is a primary vehicle for transfer of immunity from mother to offspring in mammals.

Both the amounts and proportions of lipid, protein, and lactose found in milk vary considerably from species to species (Table 29.1). The large difference in lactose content between cows' milk and human milk is the reason why cows' milk is often supplemented with dextrose and maltose for the feeding of human infants.

Lactating mammary glands have exceptionally high metabolic activities. For example, the mammary glands of a dairy cow producing 30 kg of milk daily synthesizes 1.02 kg of lipid, 0.96 kg of protein, and 1.38 kg of lactose daily. Similarly, the mammary glands of a 250-g lactating rat synthesize up to 7.5 g of lipid, 6.0 g of protein, and 1.4 g of lactose. It is interesting to compare the relative amounts of human food produced per day by a lactating cow producing 30 kg of milk per day and by a steer gaining three pounds of weight daily. The cow produces 3.4 kg of dry edible product per day, the steer between 0.3 and 0.4 kg of dry edible product daily, almost a tenfold difference. In order to provide the raw materials (precursors) and energy required for milk synthesis, lactating animals often consume three to four times the amount of food normally eaten while not lactating. The food intakes of cows and rats increase from about 7 kg to 22 kg and from about 13 g to 45 g per day, respectively, during lactation. These increases in food intake show that three to four times more nutrients must be digested, absorbed, processed, and transported within the animal in order to provide for milk synthesis. To provide for the increased flow of nutrients into and through lactating animals, the gastrointestinal tract, liver, and heart increase in both size and activity. Cardiac output and heart work increase markedly at the initiation of lactation in order to provide for increased absorption of nutrients from the gastrointestinal tract and transport of nutrients to the mammary gland. Approximately 400 kg of blood must flow through the mammary gland to supply the nutrients required for the synthesis of 1 kg of milk. This means that 12,000 kg (13 tons) of blood must be pumped to the mammary glands of a cow producing 30 kg of milk per day—a significant amount of work.

The exceptionally high metabolic activities of mammary glands during lactation is expected, and is often noted and referred to, but the increased metabolic activities of other tissues and of the whole animal during lactation is not generally considered. The above discussion may emphasize that lactation is not a simple mammary-gland function, but rather a complex, whole-animal function.

Milk Letdown and Milking Management

Milk secreted by the alveolar cells of the mammary gland cannot be removed until the myoepithelial cells surrounding the alveoli contract and force milk from the alveoli and small ducts to the large ducts and gland and teat cisterns or ampullae. This process, wherein contractions of the myoepithelial cells force milk from the gland, is called milk "letdown." The agent responsible for causing contraction of the myoepithelial cells and milk letdown is oxytocin, a hormone secreted by the pituitary gland and transported to the mammary gland via the blood. The signals which trigger oxytocin release are quite complex, and range from the cry of a baby or the clanging of milk pails to stimulation of the nipples or teats. The former stimuli are conditioned reflexes, whereby nerve impulses arising from higher brain centers are transmitted via the hypothalamus to the pituitary, causing oxytocin release. The primary, normal stimulus leading to milk letdown is suckling or massage of the nipple or teat. Sensory stimuli resulting from suckling and massage are transmitted from the gland to the pituitary via the hypothalamus, and cause oxytocin to be secreted. This neurohormonal mechanism for controlling myoepithelial cell activity and milk letdown is taken into account in recommendations concerning premilking practices for dairy cattle and goats.

The fact that oxytocin causes milk letdown also means that vasoconstriction (as by adrenalin) in the udder prevents milk letdown by preventing oxytocin entry to the gland. Hence this fact explains what has long been known by herders: animals that have been chased, frightened, or hurt, with a resultant secretion of adrenalin, do not let down their milk. Numerous experiment-station bulletins, and an excellent book by C. W. Turner (1962), present detailed recommendations and information on premilking practices providing for milk letdown and proper sanitation. Most of these publications also present information on the principles of milking machine operation and milking management.

FURTHER READINGS

Kon, S. K., and A. T. Cowie. 1961. *Milk: The Mammary Gland and Its Secretion.* 2 vols. New York and London: Acadamic Press.

Reynolds, M., and S. J. Folley. 1969. *Lactogenesis: The Initiation of Milk Secretion at Parturition.* Philadelphia: Univ. of Pennsylvania Press.

Turner, C. W. 1952. *The Mammary Gland, I: The Anatomy of the Udder of Cattle and Domestic Animals.* Columbia, Mo.: Lucas Bros.

Turner, C. W. 1969. *Harvesting Your Milk Crop.* Chicago, Ill.: Babson Bros.

Thirty

Growth and Body Composition

"Growing bodies have the most innate heat; they therefore require the most food, for otherwise their bodies are wasted."

Hippocrates

Scientists interested in the raising of domestic animals for meat production must learn more about the control mechanisms for muscle, bone, and fat deposition in cattle, swine, and sheep. This knowledge will allow us to more efficiently produce animals with a high proportion of muscle and an optimal amount of fat tissue at market weight.

When we discuss animal growth and composition, we must be concerned about hunger in this and other countries. When cereals become scarce, it may be necessary to restrict their use in livestock production. We must be aware, on the other hand, that animals provide human food by utilizing feeds unsuitable for man. The future of the livestock industry will be heavily dependent on the progress made in improving the utilization of non-human foods and in increasing the efficiency of food conversion by domestic animals for production of our meat and protein supply.

A thorough knowledge of animal growth is necessary to improve the efficiency of food conversion in livestock production. Feed efficiency is also highly associated with the body composition of the animal and the particular stage of growth for bone, muscle, and fat deposition. Therefore, this chapter will emphasize the important relationships which occur between growth potential, body composition, and efficiency of food conversion in cattle, swine, and sheep.

Growth can be defined at many structural and biological levels. It can be a correlated increase in the mass of the body during definite intervals of time in a way characteristic of the species. Growth is also an increase in new biochemical units brought about by cell division and by enlargement through incorporation of materials from the environment. A more specific definition refers to true growth as an increase in weight of tissue, such as bone, muscle, and internal organs. This definition doesn't include excess fat deposition as part of the growth process. Fat deposition would therefore be part of the development process which results in alterations in the form of animals from changes in the rate of increase of individual components of the body.

Prenatal Growth

Prenatal growth involves a series of changes starting with a single-celled zygote and resulting in an animal suitable for the external environment after birth. Recent studies show that the same factors which influence prenatal development in domestic animals may have an influence on growth rate, muscle, bone, and fat deposition, and environmental adaptation during postnatal growth. The following discussion deals with tissue and structural changes during prenatal growth and the genetic and environmental factors affecting these changes.

Stages of Prenatal Growth

The mammalian egg contains reserve nutrients in the cytoplasm (yolk) and is much larger than the somatic cells of the body. The union of the sperm with the egg produces the zygote, which in the very early stages of prenatal growth divides and redivides several times, but doesn't increase the volume of the cytoplasm. This process of cellular division without an increase in cell volume is called cleavage. The cleaving cells are called blastomeres and the process decreases the size of cells in early development (Figure 30.1).

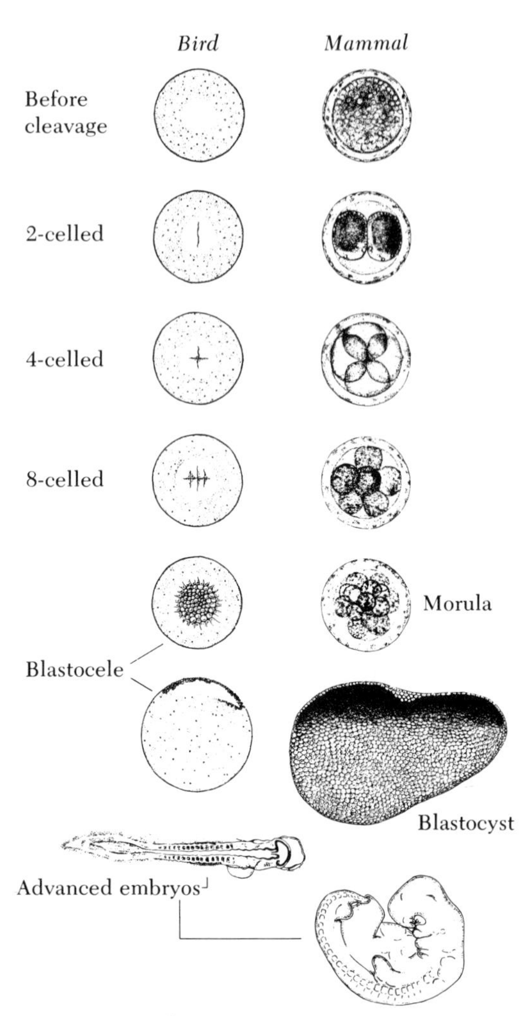

Figure 30-1 The cleavage process which occurs during prenatal development. Note the decrease in size of the cells during the cleavage process. (From Hafez and Dyer, 1969.)

During the early divisions there is little difference among the accumulating cells. Before long, however, the cells, even though not generally dissimilar in appearance, begin to be arranged in three distinct layers, which have been referred to as the three fundamental germ layers. They are named according to their location in the developing embryo. The outer layer of cells is called the ectoderm, the middle layer the mesoderm, and the inner layer the endoderm. The cells of each of the three germ layers divide, differentiate, and group themselves into specialized tissues, which in turn are organized into organs. These steps are divided into three stages, and are characteristic for the species of the meat-producing animal.

Blastocyst stage In the first phase, blastocyst phase, fertilization of the egg occurs, and it develops into the blastocyst (Figure 30.1). In this stage, genetic differences in size between different breeds occur, and nutrition is derived from uterine secretions. Both between species and within a species, some unknown substance or substances in the uterine secretions limit the number of blastocysts that can develop. A ewe, for example, normally produces a maximum of about three lambs, but 30 eggs may be fertilized in the ewe after artificial stimulation with gonadotropins; all but two or three of the fertilized eggs perish in the blastocyst stage. This blastocyst period lasts for about 10 days in sheep and swine, and 11 days in cattle. The two or three remaining blastocysts develop into normal lambs.

Embryonic stage The embryonic stage of development is the second phase of prenatal growth. Cell differentiation into organs and tissues occurs by cell multiplication (Figure 30.2). These cells have a very high priority for nutrients from the blood stream and therefore, the size of the embryo is not greatly in-

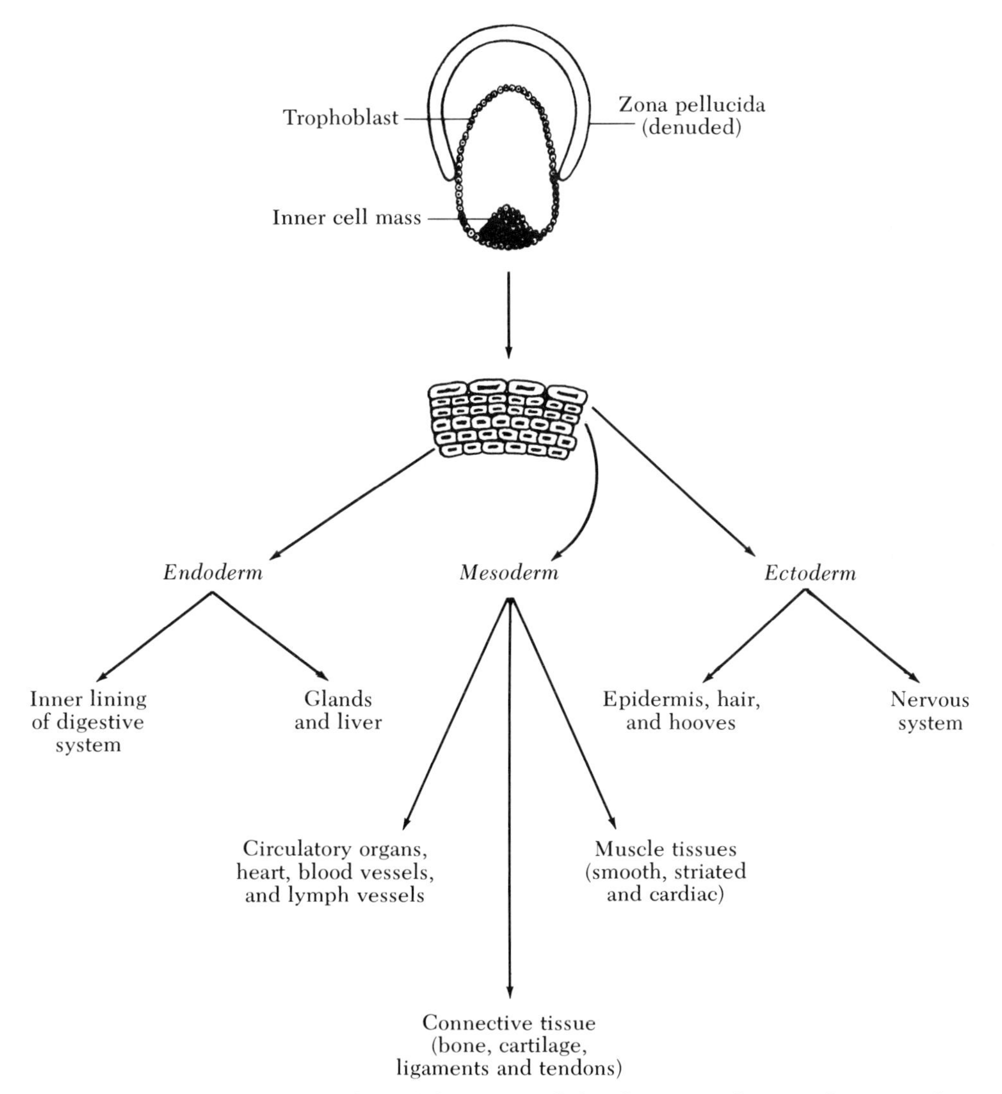

Figure 30-2 Each tissue is derived from one of the three specific germ layers in the prenatal development. This figure shows the source and type of tissue differentiation. (From Hafez and Dyer, 1969.)

fluenced by the level of nutrition of the mother. Very small weight increases occur at this stage. The embryonic phase lasts from 10 to 34 days in sheep, 10 to 36 days in swine, and 11 to 45 days in cattle.

Fetal stage After the embryonic stage, the placenta continues to grow and the extent of this growth determines the amount of nutrition the young animal will receive during the third phase. Major growth in weight and composition changes for organs and tissues occur in this phase. At this stage, nutrition for the fetus is obtained by diffusion from the maternal blood stream.

The diffusion takes place through placental tissue layers which join the fetus with the maternal blood source. The number of cotyledons in the placenta can regulate the growth rate during fetal development and therefore birth weight. An example of this is the difference in birth weights between lamb singles and twins. The average number of cotyledons in the placenta is 83 for single lambs and 57 for twins.

The size of the placenta usually varies with the size of the uterus. Therefore, the size of the mother can influence the size of the animal at birth. This has been shown in the classical experiment by reciprocal crosses between large and small breeds of horses (Walton and Hammond, 1938) in which the weight of the placenta and foal of a large mother is three times that of a small mother. This also has been shown to be true in sheep and cattle. The genetics of the sire influences the upper limit of size at birth in the large mother, but the size of the placenta limits its genetic potential in the small mother.

Recent studies at Iowa State University indicate that growth potential in fetal development is highly associated with growth rate from birth to market weight. Reciprocal crosses between Angus, Hereford, Holstein, and Brown Swiss cattle show that Holstein and Brown Swiss have more growth potential in both prenatal and postnatal development than the Angus or Herefords used in the study (Table 30.1).

Bone Growth and Development

All meat-producing animals generally follow a common pattern in bone development. Some species differences do exist, but these are small when patterns are compared on an equivalent physiological age basis. Bone is an active type of tissue where new cells are replacing old

Table. 30.1.
Relationship between birth weights and weights at 12 months postnatal growth.[a]

	Breed of dam							
	Angus		*Hereford*		*Holstein*		*Brown Swiss*	
Breed of sire	*Birth wt.*	*12 mo. wt.*	*Birth wt.*	*12 mo. wt.*	*Birth wt.*	*12 mo. wt.*	*Birth wt.*	*12 mo. wt.*
Angus	63	805	70	867	86	978	80	979
Hereford	71	881	72	826	89	1,017	85	974
Holstein	75	921	80	922	88	1,001	90	1018
Brown Swiss	72	914	80	964	101	1,060	90	957

[a] Data courtesy, R. L. Willham, Iowa State University, 1971.

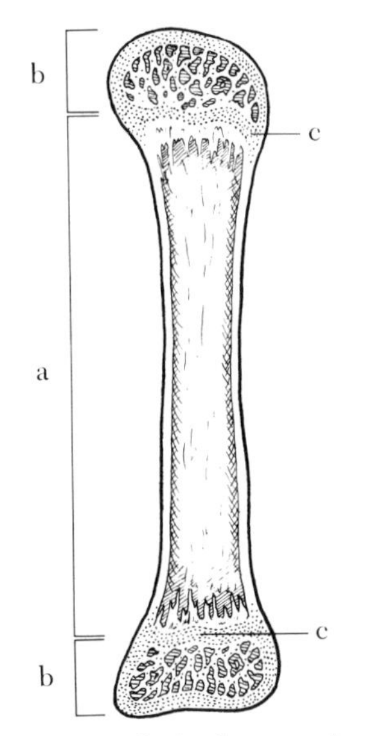

Figure 30-3 Bone shaft showing location of the ossification centers, the diaphysis (a), the epiphyseal centers (b), and the epiphyseal plate (c). (From Hafez and Dyer, 1969.)

cells throughout the life cycle. Only a general discussion of bone development and its importance to meat production will be attempted in this chapter.

Prenatal Development

The skeleton develops from the mesoderm (Figure 30.2) by cell differentation into osteoblasts, which are responsible for bone formation, osteoclasts, responsible for bone reabsorption, and osteocytes, responsible for maintenance of living bone tissue. Biosynthesis from these cells results in skeleton development into two major portions. The axial section includes the vertebrae, ribs, sternum, and skull, and the appendicular portion includes the shoulder, pelvic girdles, and limb bones.

Postnatal Development

Secondary ossification centers At or near birth, the secondary ossification centers (epiphyseal) develop in each end of the long bone (Figure 30.3). The increase in the number of cartilage cells from this phase of bone growth results in nearly equal growth in all directions of the bone. The diaphysis or primary ossification center continues its growth for bone length after the epiphyseal centers appear. The cartilage plate which separates the diaphysis from the epiphysis is called the epiphyseal plate. New cartilage in this area is formed as long as bone growth continues. When the cartilage in the epiphyseal plate is absorbed and no new cartilage is formed, bone growth ceases and the diaphysis is bound to the epiphysis by a bony union. The zone resulting from this union is visible in the mature animal and is called the epiphyseal line. Maturity estimates of carcasses use these bone changes for final judgments (see Chapter 45).

Fat Development and Deposition

Embryonic Development

Primitive fat cell The primitive structures forming the network of adipose tissue start developing in the embryonic phase of fetal development. The differentiation of white adipose cells start from a stem cell which has the same appearance of a fibroblast-type cell. Further differentiation results in a round-type cell and small fat globules start to enter the cell.

After the first lipid appears in the cell, the cell loses the capacity to divide. This is an example of a growth-controlling mechanism for fat deposition and shows genetic control over the formation and maturation of adipose tissue.

Postnatal Growth

At birth, the animal's body contains only a small percentage of lipid material (2 per cent for the pig). As an animal grows, fat deposition occurs by enlargement of the individual fat cells and by the addition of new fat storing cells. The enlargement of the individual fat cells accounts for most of the fat development and expansion as an animal grows.

Fat is deposited in prenatal and early postnatal growth in specific locations and then it develops further from these locations. Liebelt and Eastlick (1952) observed that the chick embryo starts to develop fat in sixteen separate places and as further growth takes place, adipose tissue grows and spreads until a mature weight is reached.

Subcutaneous fat Subcutaneous fat is located under the skin and accounts for the largest quantity of fat in the body. Subcutaneous fat is deposited in layers and connective tissue separates these fat layers.

As an animal grows, the three fat layers are clearly differentiated. Their relative growth is of interest for live animal and carcass evaluation. The outer layer is laid down relatively early in life, and increases little as an animal grows to a more mature weight (Table 30.2) compared to the middle layer. The layer directly over the *longissimus* muscle develops at a later stage of the fat-development process. In a fat-type pig it may be evident at 120 pounds, in a meat-type pig very small at 230 pounds. This reflects the difference in genetic potential for fat deposition at various weights in the growth process of the pig. In the selection for the meat-type pig, we have changed or reduced the middle and inner layers of fat considerably, but we have not extensively changed the thickness of the outer layer (Figure 30.4). Figure 30.5 shows subcutaneous fat-deposition patterns for Angus steers as they grow from 850 to 1,000 pounds live weight. It should be noted that the greatest fat deposition at 850 pounds is in the rump, hindflank, lower loin, brisket, and center of the shoulder. As the animal develops to 1,000 pounds, the fat deposition increases, and more than 0.8 inches of fat extends from the top of the rump to the lower hindflank, and joins with the center portion of the loin and chuck areas.

Kidney fat Fat surrounds the kidney even in young calves and this increases intensively as mature weights are reached. A knowledge of the fat-deposition patterns is essential for evaluation of total carcass fat or retail-cut per

Figure 30-4 An example of subcutaneous fat layers over the loin eye in "meat type" (left) and "fat type" (right) pigs.

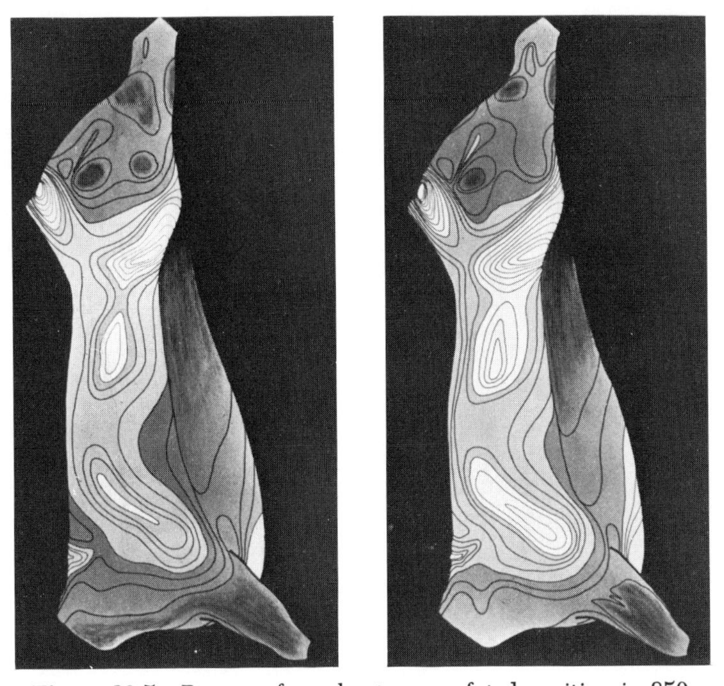

Figure 30-5 Patterns for subcutaneous fat deposition in 850-pound (*left*) and 1,000-pound (*right*) Angus steers. In the lightest areas, the fat is more than 0.8 inch thick; in the darkest areas, it is less than 0.4 inch thick. (Courtesy University of Wisconsin, Meat and Animal Science Department.)

cent in live animal or carcass evaluation. Anatomical locations for fat tissue in the hindleg of cattle (Figure 30.6) show the importance of fat in altering body shape and size.

Intermuscular fat Intermuscular fat is located between muscles and often surrounds moving muscle surfaces and fills the spaces between bone and muscle points of attachment. From birth to a mature weight, intermuscular fat has a much greater relative increase than subcutaneous fat, even though subcutaneous fat accounts for much more of the total per cent of fat (Table 30.3). Large deposits of intermuscular fat often causes problems for the packer and retailer in the merchandising of cuts, particularly from the shoulder in lamb and pork carcasses and from chucks from beef carcasses (Figure 30.7).

Intramuscular fat Intramuscular fat is located between and within muscle fibers and its greatest relative deposition is in the later stages of the growth process. Intramuscular fat is called marbling in the trade industry and

Table 30.2.

Relative changes in layers of subcutaneous fat in the loin and shoulder region of the pig.[a]

Fat location	*Birth*	*4 weeks*	*16 weeks*	*24 weeks*
Outer layer, mm	1.7	4.5	6.2	9.5
Middle layer, mm	1.7	7.0	11.2	25.5
Total, mm	3.4	11.5	17.4	35.0
Total, inches	0.13	0.46	0.7	1.4

[a] Adapted from McMeekan (1940).

Table 30.3.
Weight in grams of subcutaneous and intermuscular fat in the pig.[a]

	4 weeks	*8 weeks*	*16 weeks*	*20 weeks*	*24 weeks*	*28 weeks*
Subcutaneous	792	1,660	5,167	9,824	15,690	23,845
Relative changes	100	209	652	1,240	1,981	3,010
Intermuscular	192	314	1,960	3,362	5,164	10,668
Relative changes	100	163	1,021	1,751	2,707	5,556

[a] From McMeekan, 1940.

is used by the USDA to evaluate palatability for beef carcass grades (see Chapter 45).

Table 30.4 shows the rate of intramuscular fat or marbling deposition in cattle from the time they are placed in the feedlot at approximately 467 pounds and fed until they reached 975 pounds. It should be noted that some intramuscular fat is present in muscle in the early phases of growth. Some intramuscular fat is necessary for normal cell metabolism and growth, and is therefore present in muscle even at birth.

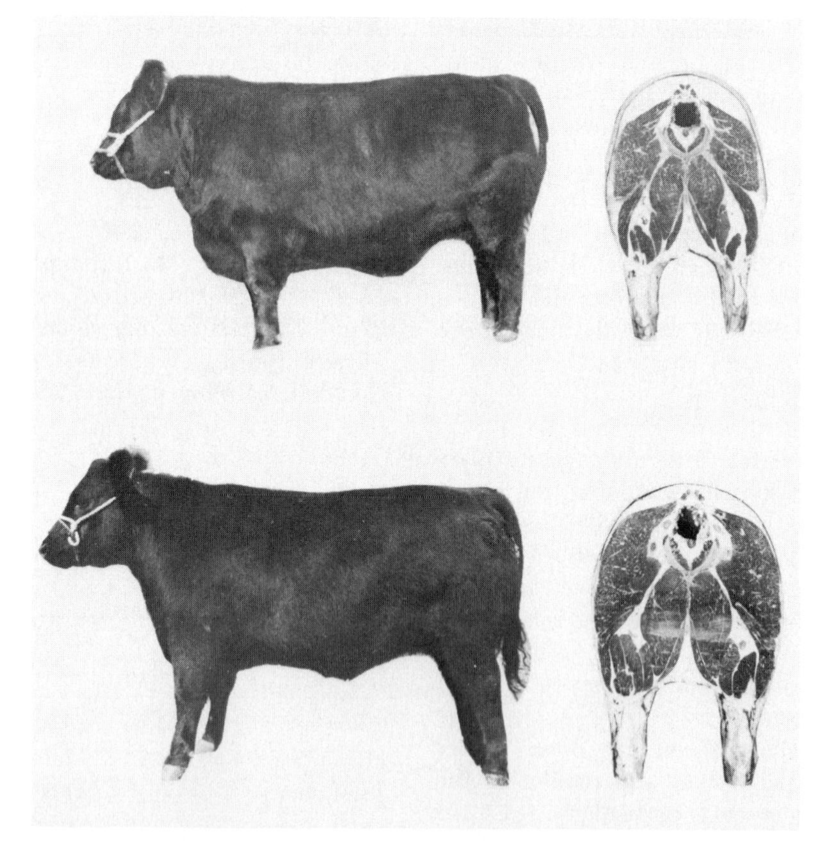

Figure 30-6 The role that fat may have in altering body shape in cattle. *Above,* "fat type"; *below,* "meat type." Note cross-section through the hindleg for each sample. (Courtesy E. A. Kline, Iowa State University.)

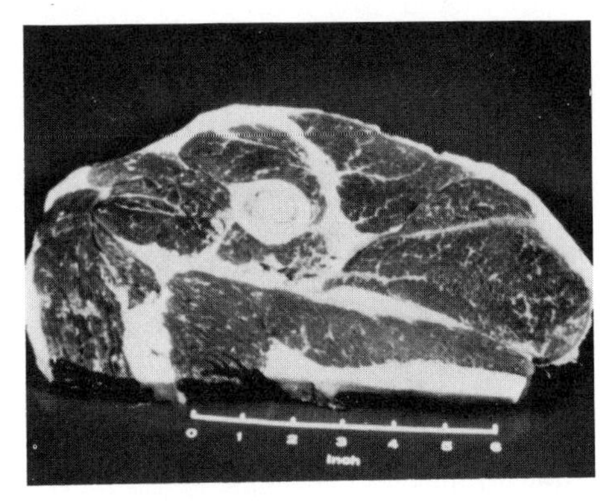

A

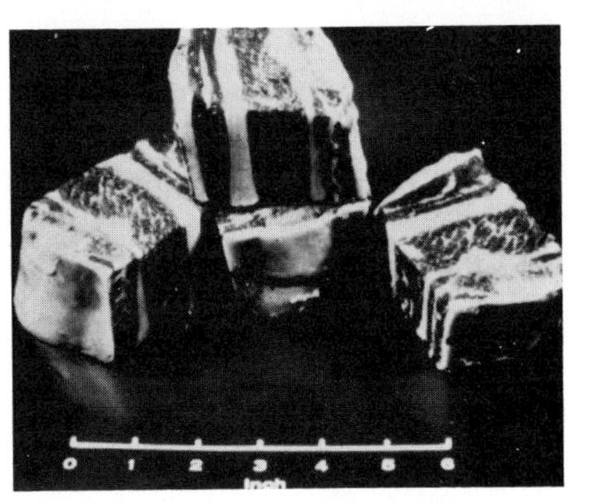

B

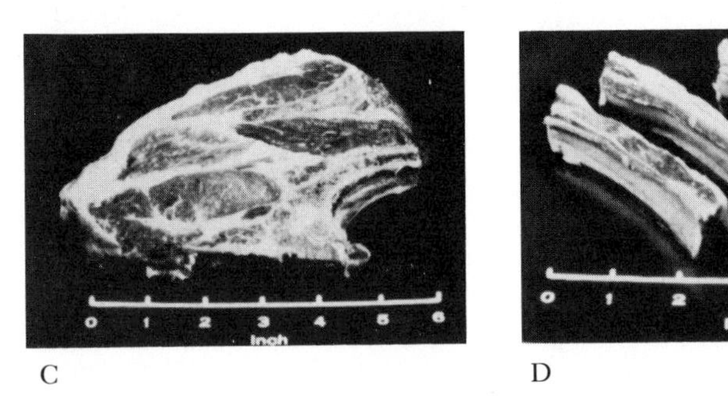

C D

Figure 30-7 Examples of intermuscular or "seam fat" in retail meat cuts. *A,* beef chuck; *B,* beef plate; *C,* pork shoulder; *D,* lamb breast.

Relationship of weight and animal sex on fat deposition Each species has a characteristic pattern for fat deposited as it develops to a mature weight. Young animals grow rapidly and in this physiological stage of growth more muscle and bone are deposited than fat, but the total fat deposition in mature meat-producing animals is similar whether fat is deposited throughout the growth phase where muscle and bone deposition occur or if greater quantities are deposited near maturity after the major portion of muscle and bone is deposited.

The sex of the animal interacts with animal weight to influence rate of fat deposition. In cattle and sheep after comparable periods of time on feed, males have less fat than castrates, and castrates have less fat than females, but in swine gilts have more muscle and less fat than barrows at market weights.

Hormones play a large role in explaining

Table 30.4.
Marbling score for beef cattle *longissimus* muscle as influenced by weight.[a]

Weight, lbs.	*Days on feed*[b]	*Marbling score*[c]
468	0	1.36
518	30	1.91
580	60	1.86
619	90	1.97
712	120	2.66
772	150	3.30
874	180	3.45
906	210	3.75
943	240	6.03
975	270	6.13

[a] Adapted from Zinn *et al.* (1970).

[b] Animals were approximately eight months of age when placed on experiment.

[c] Degrees of marbling equal (2) practically devoid, (3) traces, (4) slight, (5) small, (6) modest.

differences in fat deposition. Testosterone, secreted by the testes, stimulates muscle development and has an inhibitory influence on fat deposition. This natural hormone also stimulates more efficient weight gains and therefore the production of bulls and ram lambs for supplying trim, muscular carcasses will probably have an important role in future production of meat animals.

Testosterone has an inhibitory influence on all types of fat deposition including intramuscular fat. Bulls will not usually possess enough intramuscular fat to grade in the USDA Choice grade when they reach 1,000 to 1,200 pounds. According to some reports, eating qualities of this meat are as acceptable as meat from Choice steers if slaughtered at a young age, usually before 18 months. Bulls are more commonly fed in Europe for beef production than in the United States at the present.

Muscle Growth and Development

Muscle tissue is a highly specialized and differentiated tissue. Contraction is its most important characteristic, and therefore muscle cells are elongate structures and not round.

Prenatal Development of Muscle

The striated musculature of animals is of mesodermal origin. The stages in the development of muscle cells are described in embryology texts.

It has long been recognized that the number of fibers in a muscle increases considerably in the later stages of embryonic growth and during fetal development. In domestic animals, total muscle-cell numbers are genetically established by birth; so increase in muscle size results from development of muscle fibers and not from increase in muscle cells as an animal grows through postnatal life.

Postnatal Muscle Characteristics

Skeletal muscle fiber arrangement Muscle has a very intricate connective-tissue network that is involved in a vital part of muscle function. The major muscle function is to do work by shortening, which requires a large and constant supply of energy. These muscles must therefore be richly supplied with blood vessels to remove the end products and supply required nutrients for muscle contraction. The connective tissue wrappings (Figure 30.8) provide the needed route for the blood vessels and through the endomysium each muscle

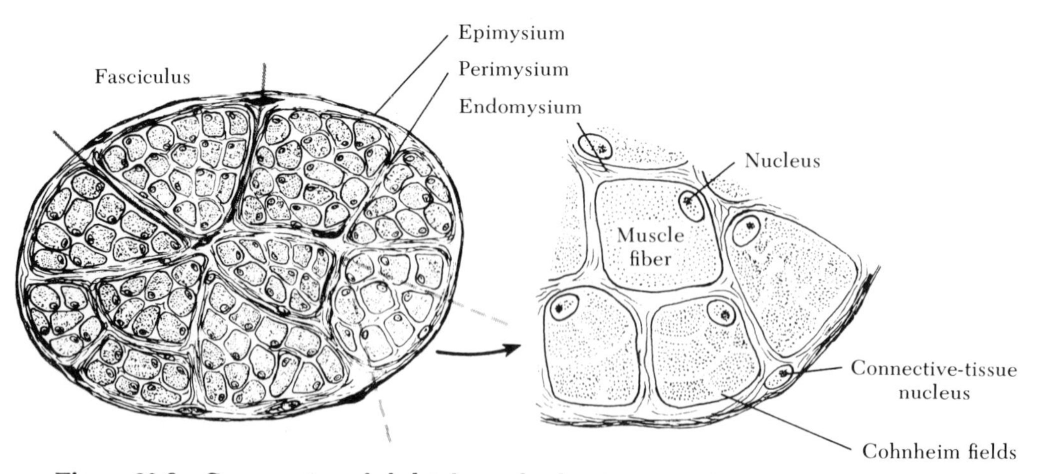

Figure 30-8 Cross section of skeletal muscle showing extensive connective tissue network. (From Patt and Patt, 1969.)

fiber can maintain a close capillary connection. The formation of end products, such as lactic acid, influence meat quality characteristics, especially meat color. Therefore, the number of blood vessels and their function can influence meat quality characteristics. A considerable amount of research is being conducted in this area with stress-prone pigs, which possess an abundance of muscling.

The proportion of connective to muscle tissue and chemical structure of the connective tissue is associated with meat tenderness. The degree of connective tissue is associated with muscle function, and therefore muscle functions may indirectly influence meat tenderness.

Structural changes in chemical characteristics of connective tissue during postnatal growth can also greatly alter meat tenderness. When cattle are compared at 12 and 36 months, the older cattle have less tender muscle and stronger chemical bonds holding the connective tissue together.

Postnatal muscle fiber growth Muscle development after birth is not a simple function of age alone. Other factors, such as plane of nutrition, location of muscle, action of the muscle, animal weight, breed, sex, heredity, and environmental factors also influence muscle fiber growth. A complex interrelationship involving muscle fiber size, fat deposition, muscle moisture depletion, and protein deposition is associated with postnatal muscle development. In the pig at 80 days, more than 50 per cent of *longissimus* muscle-fiber growth has been achieved. As the pig grows and develops from birth, the moisture content of muscle decreases, and at 80 days postnatal growth the decrease in moisture content has nearly reached its plateau, which is between 74 to 76 per cent. At 80 days postnatal development, intramuscular fat gradually starts to increase. Following the initial period of high protein deposition up to 80 days, a transition period occurs, during which protein levels remain rather stable, the fiber growth rate slackens and fat deposition rate begins to increase. At 120 days postnatal life, the fattening period starts to increase rapidly, and intramuscular fat (marbling) content increases as per cent of water and protein decrease. At this time, muscle fiber growth is nearly 75 per cent complete and at 150 days, 95 per cent of the maximum fiber diameter is attained.

Body Composition

Selection for efficient animal production involves an understanding of the magnitude and sequence of tissue deposition at different developmental stages. Therefore, the total body composition of swine, sheep, and cattle should be known in order to adequately balance rations for least-cost production through the growth cycle. When a high degree of muscle is deposited, greater protein requirements are necessary in the ration than when the body has completed its growth cycle for muscle deposition. This body composition section will relate the growth stages for bone, muscle, and fat tissue, which in turn will influence efficiency of nutrient utilization and production costs.

Growth stages The first growth stage, shortly after birth, reflects the early development of the head, neck, and legs. These parts are much larger in comparison to the short, shallow body and poorly developed hindquarters. These characteristics result because, at this stage, weight increases are almost entirely made up of bone and muscle tissue. The second growth phase associated with conformation changes is a proportional increase in body length. Therefore, length increases are more related with early stages of development rather than later stages.

The third general phase of postnatal growth involves a deepening and thickening of the body. This phase is associated with fat deposition and as these dimensions are increased, the degree of fat tissue is also increasing in a proportional degree. This phase starts at approximately 3 to 4 months for lambs, 8 to 12 months for steers, and 3 to 4.5 months for swine. These ranges can vary, depending on the level of nutrition and the animal's growth potential for muscle and bone deposition. The fourth phase involves a more concentrated development of the loin and hindquarters and additional development in depth and thickness of the body. These growth characteristics usually develop at 4 to 6 months in sheep, 12 to 15 months in cattle, and 4.5 to 6.0 months in swine. A knowledge of growth patterns with the characteristic shape and form of the live animal can be helpful in evaluating live animals for composition differences.

Lamb carcasses have more bone and less fat at market weights than swine carcasses (Tables 30.5 and 30.6). Under intense genetic selection pressure, swine have been developed with 58 per cent lean, 15 per cent bone, 7 per cent skin, and 20 per cent fat (compare with figures in Table 30.6). The nature of postnatal growth of cattle varies widely between breeds. Holsteins, for example, have a heavier carcass, more muscle, and more bone than Herefords at 24 months of age. No differences in total fat content were observed, but Holsteins had greater quantities of internal fat, such as kidney and pelvic fat, whereas the Hereford had more subcutaneous fat.

Nutritional influences Animal scientists have conducted extensive research relating nutritional influences on body composition traits of domestic animals. Priorities exist for available nutrients as an animal grows from birth to a mature weight. Vital organs, such as the heart, liver, and lungs, have the first priority. Skeletal or bone tissue has the next priority, followed by muscle tissue. Nutrients left over after growth and development of these tissues are used for fat deposition.

Table 30.5.

Total carcass muscle, fat, and bone percentage for lambs at 70, 100, and 120 pounds.

Weight group	*Per cent muscle*	*Per cent fat*	*Per cent bone*
70	60.7	18.2	21.1
100	54.3	24.7	20.9
120	50.4	32.8	16.8

Table 30.6.

Relationship between body weight and body composition of swine (these values are for swine which have not been selected for lean carcasses).

Live weight, lbs.	*Per cent muscle*	*Per cent fat*	*Per cent bone*	*Per cent skin*
2.2	48.6	0.0	31.8	19.3
20.0	57.1	11.6	19.8	11.2
100.0	57.9	15.5	16.3	9.4
140.0	51.3	27.4	14.0	7.3
200.0	46.4	35.2	11.5	6.7
300.0	42.3	41.7	10.0	5.9

Body composition can be altered in the young pig (birth to 50 pounds) by changing the energy-protein ratios. However, when these animals reach 220 pounds, high protein or low energy levels in the ration during the growth phases has only a small influence on body composition. The animal compensates and alters its tissue-growth patterns if protein-energy ratios are not balanced for maximum growth. The high-priority tissues for nutrient utilization develop first, since they have preference for the limited nutrients. If protein is limited and energy is in excess, more fat tissue will be deposited than with a proper energy-protein ratio. Once a pig reaches a stage in his growth curve where the major portion of his muscle and bone is developed and it is approaching its mature weight for these tissues, the total body composition is only slightly altered by protein-energy ratios fed during growth up to these weights. At mature weights, no differences in essential tissue can be detected.

It is questionable at this time if body-composition traits can be altered in cattle if they are compared on an equal-weight basis. Recent theories state that when cattle reach a certain weight, they will have a composition which is genetically coded for that weight, and a high or low energy ration can only alter the rate at which the animal reaches that weight. It is well-established that energy levels early in the growth phase of cattle have no influence on the composition of cattle when they reach their mature weight. This is due to the compensatory growth of all tissues and organs when a mature weight is reached for muscle and bone tissue. Many steers are marketed at weights close to their mature weight for muscle and bone deposition.

It becomes apparent from this discussion that all domestic meat-producing animals should be fed the least-cost ration which will lend itself to the most efficient production of meat-producing animals. It is not economically feasible to try to alter body composition traits on a commercial basis by levels of nutrition when our meat-producing animals are marketed at weights where they have completed a major portion of their growth for muscle and bone deposition.

Hormonal Influence Upon Growth

Hormones play a major role in determining the ultimate size of birds and mammals (Turner and Bagnara, 1971). As concerns their role in growth, they may fall into the following categories.

1. Some hormones are permissive for growth. Permissive hormones for growth would include hormones concerned in controlling metabolic processes, stuch as thyroid hormone, insulin, and the glucocorticoids and mineralocorticoids secreted by the adrenal cortex. To illustrate, if the thyroid gland in any young mammal is removed, growth is markedly retarded and the injection of adequate amounts of thyroid hormone (thyroxin and/or triiodothyronine) will result in a restoration of growth. On the other hand, the injection of thyroid hormone to a normal individual will not increase the rate of growth above normal. Thus a certain amount of thyroid hormone is essential to permit normal growth to occur; i.e., it has a permissive role in growth.

2. Some hormones are concerned with the growth of specific organs. Thyrotropic hormone, secreted by the anterior pituitary, presumably has as its sole function the growth and secretory activity of the thyroid gland; ACTH stimulates growth and secretory activity of the adrenal cortex; and gonadotropic hormones (follicle stimulating and luteinizing hormones) are concerned with the growth and secretory activity of the gonads. Adequate dosages of gonadotropins into immature female rats, as an example, will increase the size of the ovaries 10 to 20 times over a 96-hour period.

3. Other hormones are concerned with growth of the body as a whole. The pituitary growth hormone stimulates general body growth. If the pituitary gland is removed in mammals, growth essentially comes to a halt. Growth hormone will, in part, restore growth. In an intact immature rat or dog, the chronic injection of growth hormone will stimulate growth beyond that considered to be normal; in rats the body weight may be almost twice that of the untreated mature individual following treatment with bovine or ovine growth hormone. On the other hand, humans respond to human growth hormone but not to ovine or bovine hormones. This illustrates an important point, namely, that protein hormones from different species vary slightly in their chemical structure and in biological activity. This fact is of some importance in the treatment of diabetes in humans with insulin. If antibodies against insulin from one species are built up, use of insulin from another species may obviate the inhibiting effect of the antibodies.

FURTHER READINGS

Boyd, J. D. 1960. "Development of striated muscle." *In* G. H. Bourne, ed. *Structure and Function of Muscle*. New York, Academic Press.

Hafez, E. S. E. 1969. "Prenatal growth." *In* E. S. E. Hafez and I. A. Dyer, eds. *Animal Growth and Nutrition.* Philadelphia, Lea and Febiger.

Hammond, J. 1961. "Growth in size and body proportions in farm animals." *In* M. Zarrow, ed. *Growth in Living Systems.* New York, Basic Books.

McMeekan, C. P. 1940. "Growth and development in the pig, with special reference to carcass quality characters." *J. Agr. Sci.,* 30:276.

Rouse, G. H., D. G. Topel, R. L. Vetter, R. E. Rust, and T. W. Wickersham. 1970. "Carcass composition of lambs at different stages of development." *J. Animal Sci.,* 31:846.

Turner, C. D., and J. T. Bagnara. 1971. *General Endocrinology.* Philadelphia, Pa.: Saunders.

Weiss, G. M., D. G. Topel, R. C. Ewan, R. E. Rust, and L. L. Christian. 1971. "Growth comparison of a muscular and fat strain of swine, I: Relationship between muscle quality and quantity, plasma lactate, and 17-hydroxycorticosteroids." *J. Animal Sci.,* 32:1119.

Thirty-One
Digestion

"In health the muscular activity of the bowel seems often to be soothing to the brain."

W. C. Alvarez
An Introduction to Gastroenterology.

Most food substances, because of their structural complexities, cannot be used by animals immediately after eating. The various categories of chemical compounds of which foods are composed must first be degraded into smaller units which can be absorbed into the cells of which the animal is composed. As a biological process, digestion includes all activities associated with this degradation.

The study of the digestive process is complex because it involves a consideration of more than one scientific discipline, namely, anatomy, physiology, and biochemistry; in some cases, particularly for herbivores, an understanding of bacteriology and protozoology is essential. In the ensuing pages, digestion will be considered from the standpoints of, first, biochemistry, then anatomy, and finally physiology.

Biochemistry of Digestion

Chemically speaking, digestion is a hydrolytic process; in other words, the large molecules are split into small ones by adding a molecule of water to the bond between two adjacent small molecules. For example (Figure 31.1), when two amino acids are bound together to form a dipeptide, a molecule of water is eliminated in the formation of the peptide bond. When this same dipeptide is digested (hydrolyzed), the molecule of water is reinserted between the two amino acids. The same sequence of events can be shown for a disaccharide and a triglyceride (Figure 31.2).

Enzymes

The hydrolytic reactions involved in digestion are accelerated by catalysis. Catalysis is the process whereby a chemical reaction is accelerated by the addition of a substance, a catalyst, which itself is not changed by the reaction. Biological catalysts are called enzymes, all of which are proteins and are synthesized

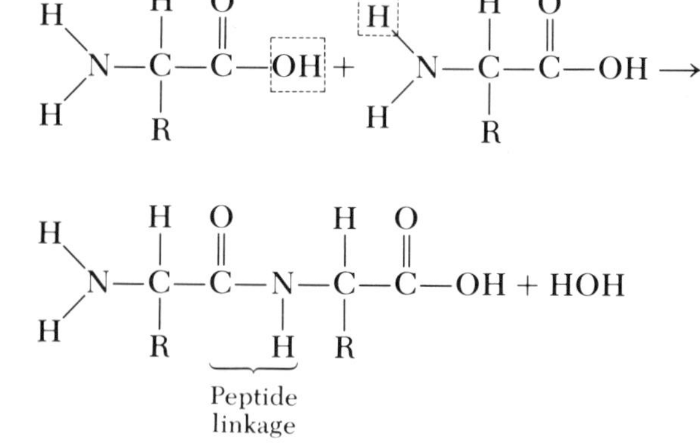

Figure 31-1 The formation of a dipeptide from the condensation of two amino acids. Note the elimination of a molecule of water during the formation of a peptide bond.

and secreted as they are required by the organism. More will be said about the release of digestive enzymes during the discussion of the physiology of digestion.

In the chemical reactions associated with digestion, it is essential to remember that they are hydrolytic in nature. In all cases, the reactants are the substrate (the food substance) and water. The products are the smaller particles, e.g., amino acids from protein digestion. The catalyzed digestion of a protein might appear as follows:

Enzyme + protein + H—OH ⟶

enzyme
/ \\
protein H—OH

⟶ Enzyme + amino acids

Thus, it is shown that the enzyme and the reactants form a complex and highly transitory intermediate structure. Presumably, the active sites to which the reactants are attracted on the enzyme molecule put a stress on the large substrate molecule such that bonds holding adjacent molecules together are easily broken; water is then added to form two molecules from the original one. It is not quite this simple, but it is the end-result nonetheless.

By convention, enzymes are named according to the specific reactions they catalyze, and all end with a terminal -ase suffix. Thus, carbohydrases, proteinases (proteases), and lipases catalyze the degradation of carbohydrates, proteins, and lipids (fats and oils), respectively.

Biochemical Steps in Digestion

The degradation of each food substance does not occur in one step; it is much more efficient, energetically speaking, to carry out the process in small states.

Carbohydrate digestion The enzymes that catalyze the degradation of carbohydrates are collectively called carbohydrases. The complex polyglucose molecule of starch is split into the disaccharide maltose with the assist-

Figure 31-2 (a) The hydrolysis of a molecule of maltose, a disaccharide, into two molecules of glucose, a monosaccharide. A molecule of water was used in the reaction. (b) The hydrolysis of a general triglyceride into a molecule of glycerol and three fatty acid molecules. The R's refer to carbon chain fragments of varying length. Three molecules of water are used in the reaction.

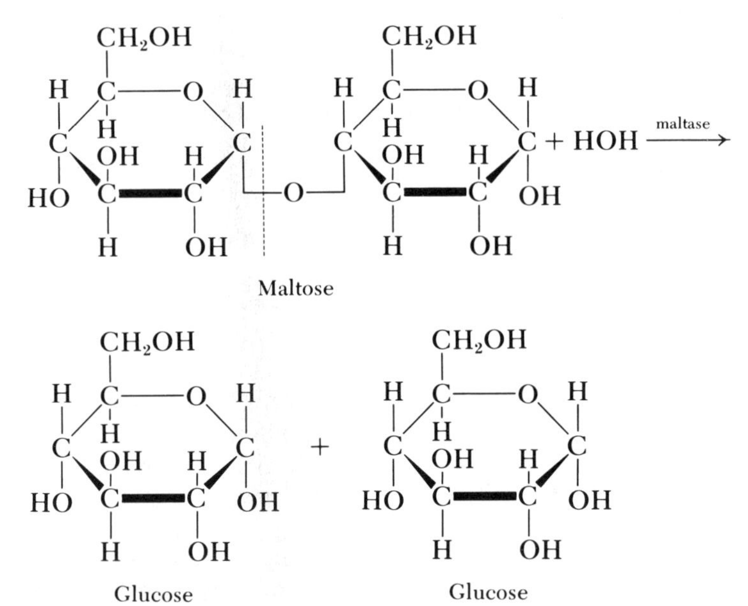
maltase
Maltose
Glucose
Glucose

A

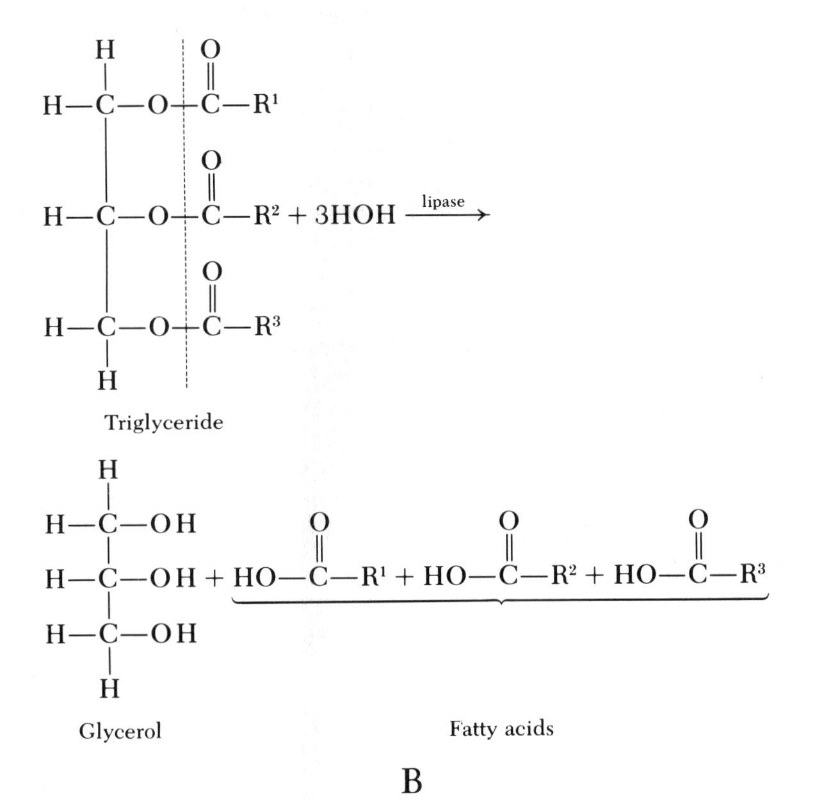
lipase
Triglyceride
Glycerol
Fatty acids
B

ance of an enzyme called amylase. There are several enzymes involved in the amylase step, and as a result several products, but adding complexity at this point will not enhance understanding. Maltose cannot be absorbed from the alimentary canal and must be split into its two constituent glucose molecules. The enzyme which facilitates this reaction is maltase. The over-all starch reaction might be viewed as follows:

—Glucose—glucose—glucose—glucose

$$+ \text{H—OH} \xrightarrow{\text{amylase}} \text{polyglucose (—2 glucose)}$$

$$+ \text{glucose—glucose (maltose)}$$

$$\text{glucose—glucose} + \text{H—OH} \xrightarrow{\text{maltase}} 2 \text{ glucose}$$

Other important dietary disaccharides, such as table sugar (sucrose) and milk sugar (lactose), are split into their respective constituents of glucose and fructose, and glucose and galactose, by the enzymes sucrase and lactase, respectively.

Cellulose digestion in herbivores is essential to their very existence, and yet they secrete no cellulase into the alimentary canal. It has been shown that the hydrolysis of cellulose is dependent upon cellulase, which originates from the bacteria and protozoa inhabiting the alimentary canal. The exact sequence of events of cellulose digestion is not well-understood at this time; however, among the digestive products, free glucose and cellobiose, a disaccharide composed of two molecules of glucose bound together by a beta linkage, have been identified. The hydrolysis of cellobiose to glucose is catalyzed by the enzyme cellobiase, also of microbial origin.

Protein digestion The group of enzymes concerned with protein digestion is called proteases. The first step in the breakdown of a complex polypeptide is to reduce the chain length (Figure 31.3). This step is enhanced by the endopeptidases, such as pepsin, trypsin, and chymotrypsin. The endopeptidases are so named because they act on peptide bonds within the protein molecule, thus, the prefix endo-.

The shorter polypeptide chains resulting from endopeptidase activity are now attacked by the exopeptidase enzymes. There are two of them, carboxypeptidase and aminopeptidase (Figure 31.3). Proteins all have a free amino end of the molecule and a free carboxyl end; the shorter polypeptide chains also have free amino and carboxyl ends. Aminopeptidase and carboxypeptidase begin lopping off amino acids from the constantly shortening polypeptide chain, each from its own end of the molecule. Eventually, a dipeptide is all that remains of the long polypeptide chain. Dipeptidases catalyze the final breakdown of the dipeptides into their constituent amino acids (Figure 31.3).

Fat digestion Of the main dietary consituents, the triglycerides are the least complex to hydrolyze. The lipases, enzymes which catalyze fat digestion, are relatively nonspecific in their action. The products resulting from their activity include glycerol, fatty acids, diglycerides, and monoglycerides (Figure 31.2b). All of these, along with triglycerides themselves, are capable of being absorbed.

The Structure of the Alimentary Canal

In its simplest form, the alimentary canal, or digestive tract, in mammals can be conceived as a simple tube which courses internally from one end of the organism to the other. At prescribed intervals the tube becomes specialized into regions called the mouth, esophagus, stomach, small intestine, large intestine, rectum, and anus. Along the way are two signifi-

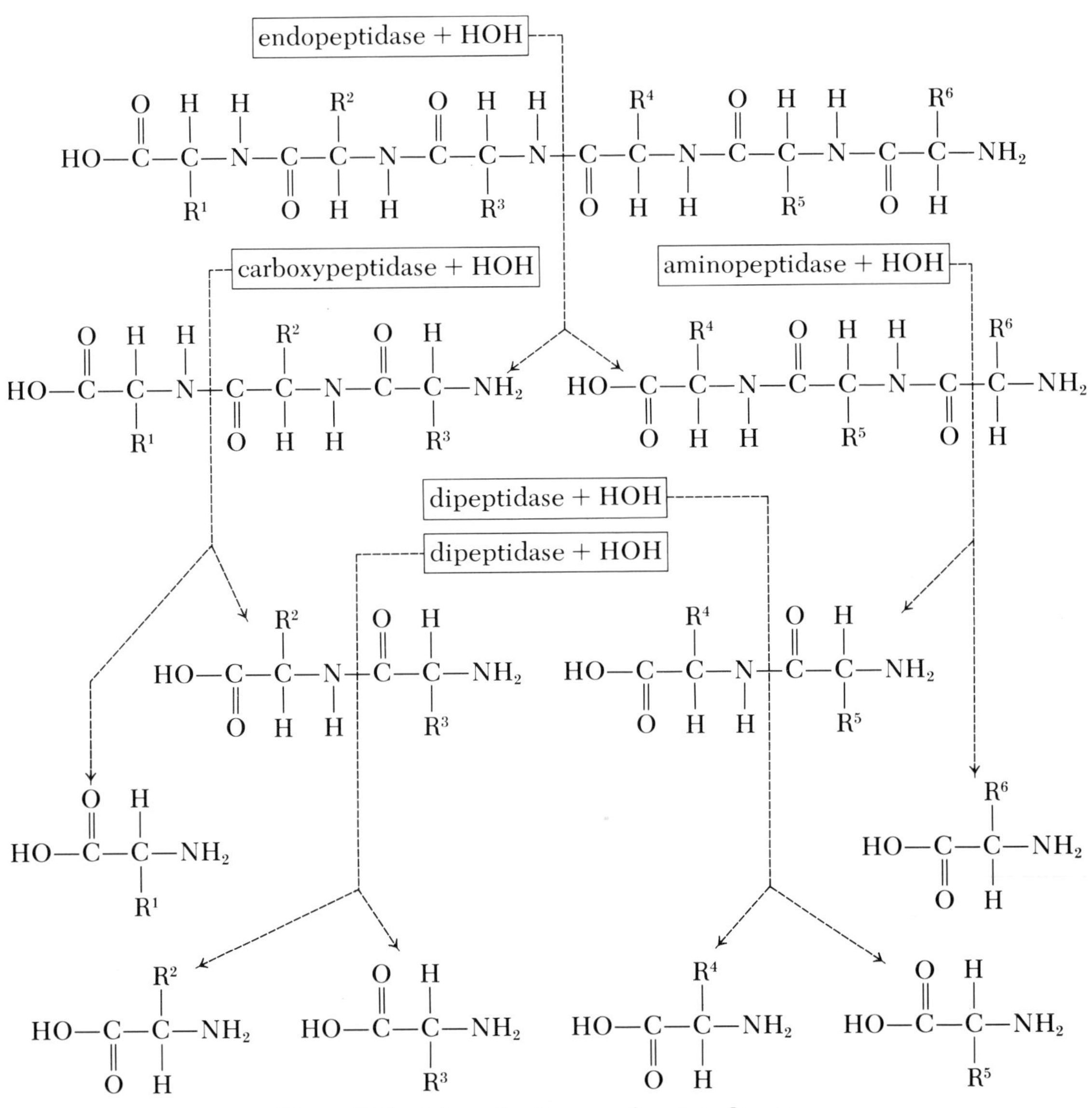

Figure 31-3 Schematic diagram of protein digestion.

cant evaginations in the tract which provide secretory products essential for digestion: the liver and the pancreas.

Beginning at the esophagus, with few exceptions, the alimentary canal has a remarkably similar microscopic structure throughout. The following structures can be easily identified (Figure 31.4): mucosa, submucosa, a layer each of circular and longitudinal musculature, and serosa. The mucosa provides special secretory products, both endocrine (into the blood) and exocrine (into the digestive tract), which are necessary for normal digestion; important absorptive cells are also present at significant points. The submucosa contains veins, arteries, and nerve fibers, which provide the logistical support or supply service for the mucosal activity. The two muscle layers provide the forces required for mixing the ingested food with the secretory products of

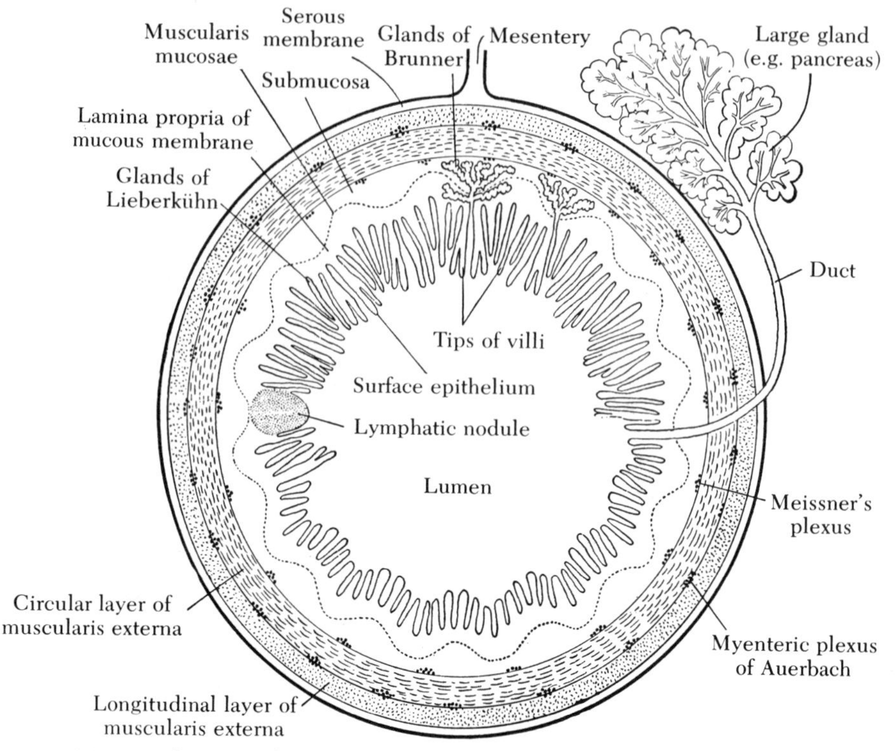

Figure 31-4 A diagram of a generalized cross section of the mammalian alimentary canal. The esophagus, stomach, and large intestine have mucosa which are somewhat folded, but villi, as plentiful as indicated in this diagram, are primarily a feature of the small intestine. The villi vastly increase the surface area available for absorption, an important function of the small intestine. (From Maximow and Bloom, *A Textbook of Histology*. Saunders, 1949.)

the tract and the propulsion of the ingesta from one end of the tract to the other. The serosa enables the organs to slip back and forth without irritation.

The regions between the submucosa and the circular muscle layer, and between the circular and longitudinal muscle layer, are richly endowed with the nerve networks essential for coordinated mechanical and secretory activity. The integrity of this nervous tissue is prerequisite if the intrinsic activity of the alimentary canal is to be maintained.

Structural Variations in the Alimentary Canal

The kind of diet, to a great extent, dictates the structural variations in the digestive tract. Thus, it is possible to make the general statement that carnivores and omnivores have a simple digestive tract and herbivores have a complex digestive tract.

Digestive Tract of Carnivores

Carnivores consume meat as part or all of their diet. Their digestive tracts are similar in structure, in that they all are made up of a mouth, esophagus, one-compartment stomach, small intestine, large intestine, rectum, and anus (Figure 31.5).

The stomach is usually divided into several regions, which include the cardia, fundus, corpus, and pyloric antrum. Separating the stomach from the esophagus is the cardiac sphincter, and from the small intestine the pyloric sphincter (Figure 31.6). These sphincters act as valves when they are strongly contracted and can effectively prevent the back-and-forth movement of ingesta.

The small intestine is divided into three segments, which, starting at the pyloric sphincter, are the duodenum, jejunum, and ileum. In order to increase the area for the absorption of nutrients, small finger-like projections of mucosa, called villi, can be seen with low-power magnification (Figure 31.7). The villi are the most profuse in the duodenum, decreasing in number as the small intestine approaches the large intestine.

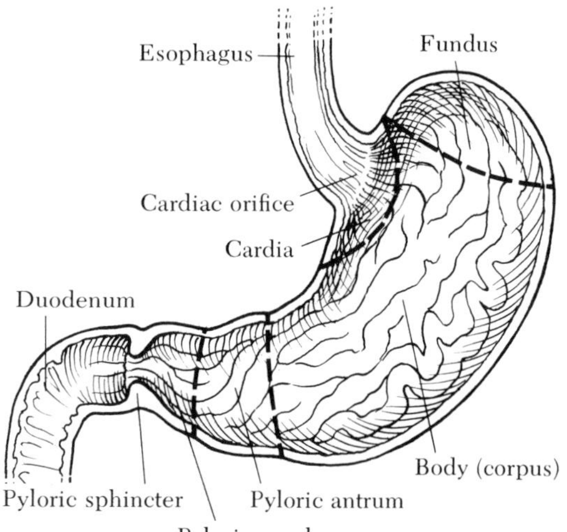

Figure 31-6 Diagram showing the anatomical regions of the typical carnivorous stomach.

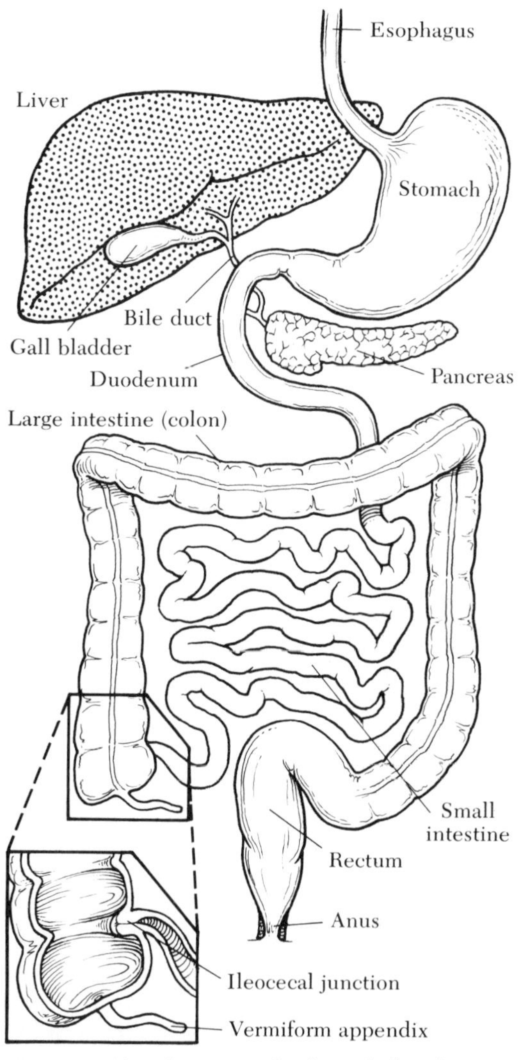

Figure 31-5 The general plan of the digestive tract of carnivores. Note the structure of the ileocecal junction.

The terminal portion of the ileum actually protrudes somewhat into the large intestine and is separated from it by the ileo-cecal valve (Figure 31.5). The valve prevents the regurgitation of the now almost completely digested food mass from the large intestine into the ileum. The large intestine, or colon (Figure 31.5) empties into the rectum. The anus, located at the terminus of the rectum, is controlled by two sphincters, one under involuntary control, the internal anal sphincter, and one under voluntary control, the external anal sphincter. In those animals which have been trained, the external sphincter permits the retention of solid wastes until a convenient time for voiding.

Digestive Tract of Herbivores

Animals such as the cow and horse have modifications of their alimentary tracts which enable them to digest the large amounts of roughage consumed. It is possible to consider the cow, a ruminant animal, as a pregastric digester and the horse a postgastric digester. This is not to imply that all digestive activity occurs in such regions exclusively; however, these are the areas where most of the roughage is digested.

Ruminant alimentary canal The ruminant stomach has several compartments (Figure 31.8). The anatomical names of these compartments and their synonyms are as follows: rumen, first stomach paunch, forestomach; reticulum, second stomach, honeycomb; omasum, third stomach, manyplies; abomasum, fourth stomach, true stomach, secretory stomach.

It is into the reticulo-rumen that the ingested roughage goes first, and it is here that it is reduced in size by the digestive processes to be discussed later.

The reticulum and omasum are in communication by means of the reticulo-omasal orifice; a muscular sphincter surrounds this orifice. Connecting the cardiac orifice with the reticulo-omasal orifice is a groove (the esophageal groove), the musculature of which is capable of forming a tunnel or tube when stimulated properly. The tube becomes an extension of the esophagus and enables the liquid diet of the young ruminant to enter directly into the abomasum, bypassing the reticulo-rumen.

The omasum is a spherical organ which lies immediately to the right of the reticulum and slightly above it. The lumen or interior of the omasum is filled with 75 to 85 leaves or laminae, which are attached to the omasal wall on all borders except the ventral orifice. This large orifice opens into a groove called the sulcus omasi which passes directly to the abomasum. Thus, it is possible for a food particle to pass through the reticulo-omasal orifice and flow directly down the sulcus omasi to the abomasum. An alternate pathway involves the passage of the particle up into the omasum from the sulcus omasi, an eventual return to the sulcus, and then into the abomasum.

In the mature rumen, small tongue-like projections called papillae can be readily identified with the naked eye. The papillae give the rumen a soft fur-like appearance. The mucosal lining of the reticulum is formed into open-topped hexagonal chambers and looks like a piece of honeycomb. The many leaves or laminae of the omasum feel and look like sandpaper on both sides. The abomasum is analogous to the simple stomach of carnivores; it leads via the pylorus to the duodenum (Figure 31.8). The small and large intestines in ruminants are fairly similar to those of all other animals.

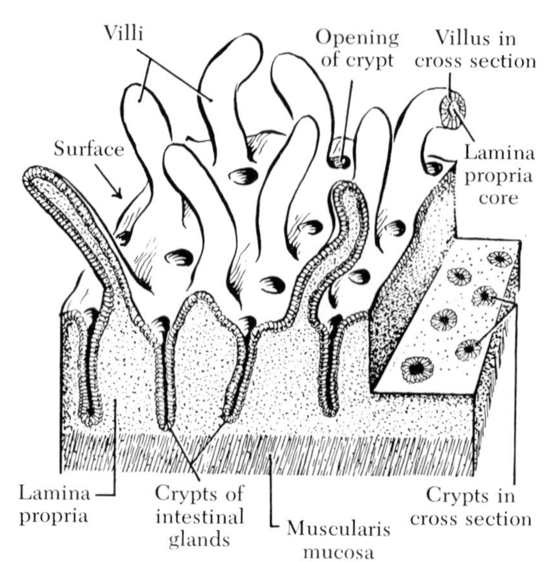

Figure 31-7 Diagram showing the surface of the small intestine. Note the finger-like villi which extend into the lumen to increase the absorptive area and the openings to the intestinal glands at the base of the villi. Compare this diagram with Figure 31-4. (From Ham, *Histology*, 6th ed. J. B. Lippincott. 1969.)

Equine alimentary canal The stomach of the horse is similar to the stomach of those animals possessing a simple stomach as far as general gross conformation is concerned. However, there is a large area, referred to as the saccus cecum, which is lined with stratified squamous epithelium. In this regard, the epithelium is similar to that of the esophagus.

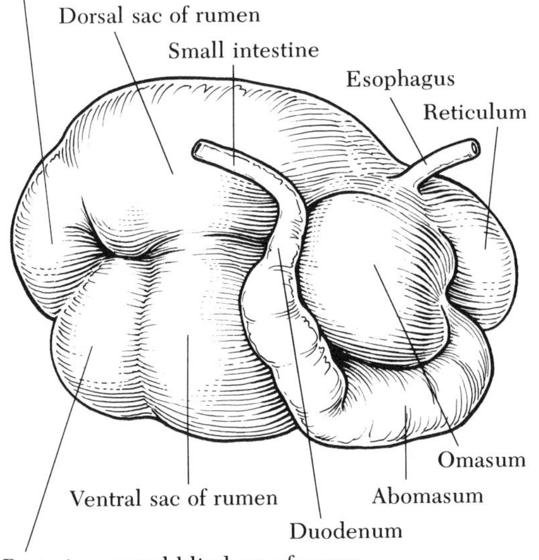

Figure 31-8 Bovine stomach; view from the right side. Note the position of the various compartments. (Adapted from Sisson and Grossman, 1938.)

By studying the remaining areas of the stomach, it is possible to identify cardiac, fundic, and pyloric regions.

The remainder of the alimentary canal is similar to that of ruminants with the exception that the cecum is much larger (19 per cent of the intestinal tract) and more important in the digestive process (Figure 31.9). The cecum is to the horse and other postgastric digesters, such as the rabbit, what the rumen is to ruminants.

Digestive Tract of Omnivores

The stomach of the pig, a typical omnivore, appears externally similar to the stomach of the horse and dog. It is like that of the horse in that a saccus cecum, lined with stratified squamous epithelium, can be identified. Unlike other stomachs so far discussed, the stomach of the pig has a very large cardiac area, which contributes about one-third of the stomach surface area. The usual segments of the small intestine can be identified. At the junction of the small and large intestines, there is a small cecum present which, in size, is about 8 per cent of the total intestinal capacity.

Avian Digestive Tract

In domesticated birds, the alimentary canal is modified to include the following structures: a crop, gizzard, and cloaca (Figure 31.10). The mouth of birds contains no teeth; therefore, if coarse food is ingested, it must be triturated in the gizzard. The crop is an esophageal diverticulum, and serves as a storage area for recently ingested food; it is absent in insect-eating birds. The acid-secreting section of the digestive tract of birds is called the proventriculus, and is completely analogous to the stomach of such animals as the dog. Usually, the stomach opens into the small intestine, but in birds the proventriculus leads into the gizzard which then opens into the duodenum.

The small intestine in birds, as in mammals, varies in length depending on eating habits, being longer in herbivores than in carnivores. Unlike that of mammals, the small intestine in birds is difficult to differentiate into a jejunum and ileum, although the duodenum is clearly defined.

At the juncture of the small and large intestines are located the ceca, which, in the domesticated birds, are paired. The openings to the ceca are surrounded by prominent bands of muscle forming efficient ileocecal valves. The terminal segment of the large intestine opens into the cloaca, the chamber into which the urinary and genital canals also open. The external orifice of the cloaca is referred to as the vent.

Activities of the Alimentary Canal

The activities of the alimentary canal and accessory structures include motility, secretion,

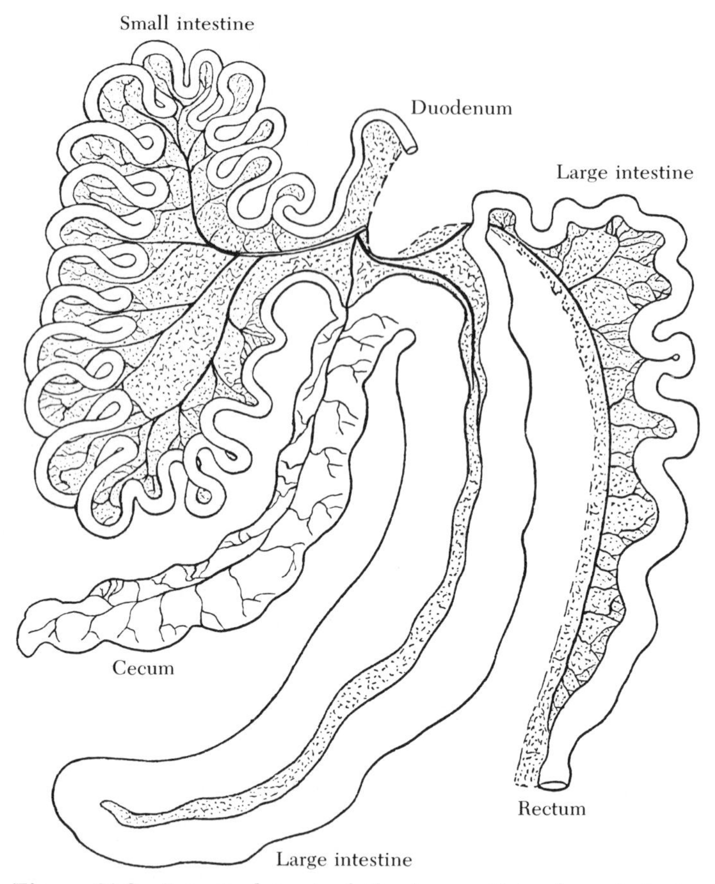

Figure 31-9 Intestinal tract of the horse. Note the relatively large size of the cecum and the capacious uncoiled large intestine. The stipled area is the mesentery containing the blood supply. (From Mitchell, 1905; courtesy of the Zoological Society of London.)

and absorption. Motility and secretion are dependent upon stimuli of neural and/or hormonal origin, whereas absorption is a function of concentration gradients and transporting mechanisms.

Gastrointestinal Motility

The movement of the contents of the alimentary canal is a function of the contractions of the circular and longitudinal musculature of its wall. Waves of muscular contraction, which include peristalsis, force ingesta from the cephalic end to the caudal end. Such waves result from a close coordination between the circular and longitudinal muscles. When the circular muscles contract alone, as in segmentation, ingesta are not progressively moved, merely mixed. Contractions of the longitudinal musculature alone, particularly in the small intestine, produce movements referred to as pendular; these also are primarily concerned with ingesta mixing.

For the most part, the movements of the alimentary canal are under the control of the autonomic nervous system. As you may recall, the autonomic nervous system can be divided into the parasympathetic and sympathetic systems. In general terms, a stimulation of the parasympathetic nerves to the gastrointestinal tract (the vagi and nerve fibers from the sacral region of the spinal cord) results in increased muscular activity. A decrease in activity is associated with a cessation of parasympathetic stimulation or an increase in the level of sympathetic nervous system activity. Thus, as far as the alimentary canal is concerned, the parasympathetic and sympathetic nervous systems are antagonistic to each other.

In most birds and mammals, denervation of the gastrointestinal tract, i.e., severing the parasympathetic and sympathetic nerves to this area, temporarily reduces the activity of its circular and longitudinal musculature. After a variable length of time, coordinated muscular activity resumes as the result of nerve networks (plexes) present in the wall itself, the submucosal and myenteric plexes. These two plexes are referred to collectively as the intrinsic nerves and the parasympathetics and sympathetics as the extrinsic nerves.

Gastric Motility Contractions of the gastric musculature of animals with simple stomachs vary from weak to strong. The weak contractions are only ripples as they pass over the stomach wall, and produce only a modest mixing of the ingesta and gastric secretions adjacent to the stomach wall itself. As these contractions intensify, a more thorough mixing takes place, and, at the same time, semiliquid portions of the ingesta are forced into the duodenum via the pyloric canal. The stomach gradually empties during the feeding interval. In animals with a simple stomach, the amplitude or strength of the gastric contractions increases as the interval between feeding increases. Intense, frequent contractions of the empty stomach have become associated with "hunger pangs."

The proventriculus is the true or secretory

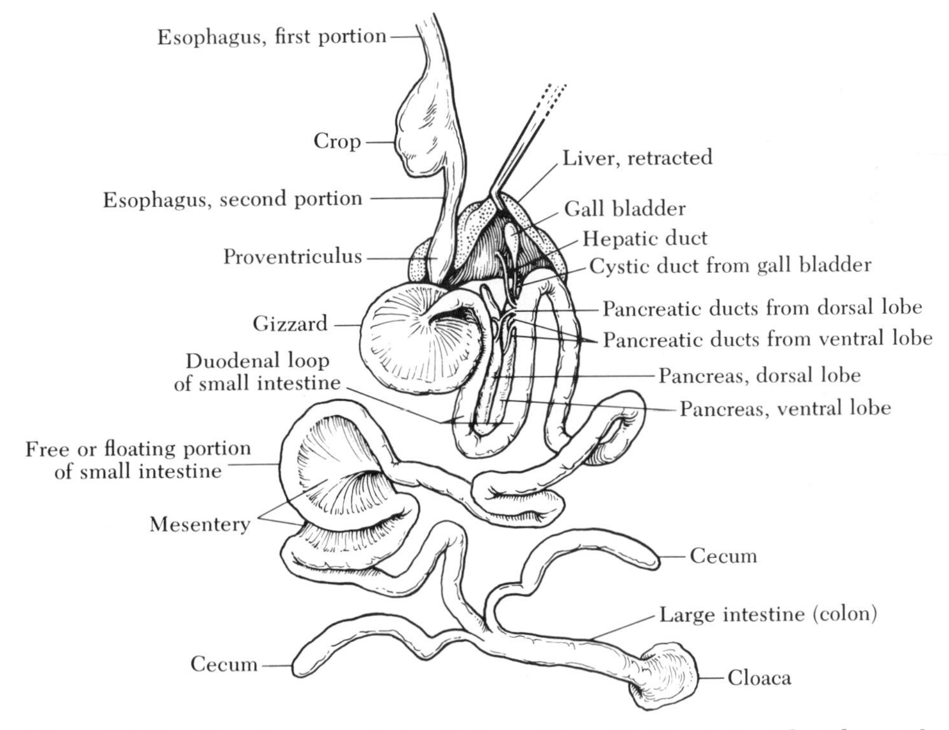

Figure 31-10 Digestive tract of the chicken. (From Sturkie, *Avian Physiology*, 2d ed. © 1965 by Cornell University. Used by permission of Cornell University Press.)

stomach of the chicken. Presumably, the motility of the proventriculus is similar to the gastric motility of simple-stomached animals. Extrinsic innervation does not appear to be essential. Gizzard motility is greatly influenced by the nature of the diet; hard, coarse food causes a more rapid rate of contraction and a greater strength of contraction than soft food.

Contractions which result in the emptying of the crop in chickens begin in the esophagus; they are peristaltic in character. If this region of the alimentary canal is denervated, all motility of the esophagus and crop ceases. The volume of ingesta in the gizzard has a direct influence on crop motility; if the gizzard is distended with food, crop motility is inhibited.

The contractions of the stomach musculature of animals possessing compartmentalized stomachs, such as ruminants, are complex. For the most part, the food of these animals cannot be digested by the animals themselves because they lack the essential enzymes. Instead, there is a convenient symbiotic relationship established between the food-gathering animal and the bacteria and protozoa which inhabit part of the alimentary tract.

The fermentation area of the alimentary canal in such animals as cattle, sheep, and goats is the reticulo-rumen. Two types of contractions of the reticulo-rumen can be identified, mixing and eructation (belching). It is possible also to identify a special contraction of the reticulum alone which is associated with the rumination reflex, about which more will be said later.

Since the reticulo-rumen is a large hollow organ near the skin surface, it is a relatively simple matter to determine its motility by surgically inserting a hollow tube, called a canula, through the wall and connecting it to a pressure-sensitive device for recording pressure changes. The changes in pressure are the result of contractions of the musculature of the reticulo-rumen wall.

When the musculature of the reticulo-rumen contracts in a sequence moving posteriorly, ingesta are forced backward and down into the ventral sac. Such a contraction can be referred to as the "mixing contraction." When the reticulo-rumen musculature contracts in a sequence which begins with the ventral and posterior sacs and then includes the dorsal sac, ingesta and gas are moved in a forward direction. Usually, the cardiac orifice of the esophagus opens whenever the gas pressure caused by the forward-moving contraction is at its maximum. At this time, reticulo-rumen gases pass up the esophagus and are expelled into the atmosphere. This forward-moving contraction is called the "eructation (belching) contraction."

Periodically, the reticulo-omasal orifice opens and ingesta are forced toward the omasum and abomasum by contractions of the reticulum. Thus, in this fermentation area appropriate contractions of the musculature of its wall cause a thorough mixing of the ingesta and microorganisms, expel the gases resulting from the fermentation process, and gradually force the digesta to more distal regions of the alimentary canal.

Ruminants are so-named because they ruminate, they "chew a cud" or rechew coarse ingesta which have been returned to the mouth. The entire physiological event is called rumination, a complex reflex involving four different acts: regurgitation, reensalivation, rechewing, and reswallowing. The reflex is initiated by the scratching action of coarse food on sensitive receptors in the reticulo-rumen wall. During the regurgitation phase of the reflex, the opening between the esophagus and the rumen, the cardiac orifice, becomes flooded with ingesta as the result of a special contraction of the reticulum. At the same time, the animal makes a strong inspiratory effort with the glottis held tightly closed; when this occurs, a negative pressure is created in the thorax. At precisely the right moment, the cardiac orifice opens and a bolus of ingesta is sucked into the esophagus. The cardiac orifice rapidly closes and the bolus is carried toward the mouth by an antiperistaltic wave of contraction of the esophageal musculature. In the mouth, excess fluid is expressed from the bolus and chewing begins, during which ingesta particles are reduced in size and become mixed with saliva. When it is chewed long enough, the bolus is swallowed, and after a brief time the entire process is repeated.

The length of time a cow or sheep ruminates is diet-dependent. On soft diets exclusively, such as young pasture plants and grain, the rumination reflex may not be elicited at all. As the proportion of hay in the diet increases, the time spent ruminating also increases. At the time of the regurgitation phase of the rumination reflex, the rumen musculature is relaxed. This is not true at the time of eructation which normally occurs at the peak of rumen contraction. It is worthy of noting that both of these events involve the passage of substances up the esophagus, in one case a liquid, in the other a mixture of gases. In one instance the rumen is relaxed (regurgitation), in the other (eructation) it is contracted. To make matters more complex, the glottis is closed during regurgitation and open during eructation.

In ruminants a severing of the vagus nerves, which carry the parasympathetic impulses to the stomach, results in a permanent dysfunction of stomach activity. All progressive and meaningful contractions of the stomach cease, ingesta putrefy, and death ensues.

Intestinal movements The mechanical activity of the intestines includes peristalsis, segmentation, and pendular movements. The muscles involved in these events have been mentioned before. The only muscular activity of the intestine which results in the progressive movement of ingesta is peristalsis (Figure 31.11a).

Segmentation contractions (Figure 31.11b) serve to thoroughly mix the ingesta with the enzymes of the intestinal tract and to expose all surfaces of the ingesta to the intestinal wall for absorption.

The full impact of the pendular movements of the small intestine on intestinal events is not well-understood (Figure 31.11c). Some mixing of the ingesta can be visualized but more important may be a periodic stretching of the mucosa so as to expose it to fresh particles of ingesta for more effective digestion and absorption.

Cecal movements The cecum, like the reticulo-rumen, is an area of intense fermentative activity. Its nutritional importance to the animal depends on whether there is another fermentative area preceding it. The motility of the cecum is not well-understood. It is known, however, there is a periodic filling and emptying of this organ which implies a well-coordinated sequence of muscular activity.

Alimentary Canal Secretions

The secretory compounds of digestive importance consist of enzymes, mucus, hydrochloric acid, bile, certain ions, and hormones. All these compounds, with the exception of the hormones, are secreted directly into the alimentary canal or into ducts leading to it; such secretions are called exocrine. The hormones which are of mucosal origin are released directly into the bloodstream and are the result of endocrine secretory activity. The exocrine secretions themselves are controlled by neural and hormonal stimuli.

The liver secretes bile, a liquid containing compounds very important in fat digestion. Bile secretion by the liver is a constant process, but the release of bile into the duodenum is periodic. Between the periods of bile release, it is concentrated by and stored in the gall bladder. Not all animals possess a distinct gall bladder but in those which do not, the bile duct itself may become somewhat dilated and stores bile. The duct leading from the gall bladder, when present, empties into the upper section of the duodenum. The most potent stimulus for bile release is cholecystokinin, a hormone of duodenal origin.

The enzymes secreted by the pancreas include trypsin, chymotrypsin, carboxypeptidase, amylase, lipase, and nucleases. These enzymes, or their precursors, reach the duodenum primarily via ducts which enter the common bile duct; some animals have subsidiary pancreatic ducts which enter the duodenum directly. The most potent stimulus for exocrine pancreatic secretions is the hormone pancreozymin, which has recently been shown to be identical with cholecystokinin.

The source of the other important digestive secretions can be enumerated by a consideration of the various specific regions of the alimentary canal. Most carnivores and omnivores secrete saliva which contains an amylase, a potent starch-splitting enzyme. Salivary amylase is also referred to as ptyalin. Once the ingesta are in the stomach, i.e., the secretory stomach (proventriculus of birds, abomasum of ruminants, etc.), it becomes exposed to the gastric secretions, hydrochloric acid and pepsin. In most animals, these compounds originate from different gastric cells, pepsin (as its precursor pepsinogen) from the chief cells and hydrochloric acid from the parietal (also called oxyntic) cells. Pepsin is a potent endopeptidase, and catalyzes the hydrolysis of proteins into polypeptides of shorter chain lengths.

The secretory products of the cells lining the small intestine, especially the duodenum, are collectively called succus entericus. The succus entericus contains important digestive enzymes without which complete digestive hydrolysis could not occur. More will be related about these enzymes later, but they include disaccharases, dipeptidases, aminopeptidase, lipase, and nuclease, plus related enzymes. Following the action of these intestinal enzymes, compounds of nutritional importance are ready for absorption. Below or more, caudad to the duodenum, the intestinal secretions have little catalytic activity.

The only other major secretory product of the alimentary canal which has not been considered to this point is mucus. Mucus is important because it hydrates dry food materials, lubricates the gastrointestinal tract, and, finally and importantly, protects the delicate mucosa from abrasion and erosion. Because mucus is chemically a mucoprotein, it serves also as a buffer of modest importance. Mucus is secreted by the salivary glands of the mouth (parotid, submaxillary, and sublingual) and by glands of the esophagus, secretory stomach, and intestines.

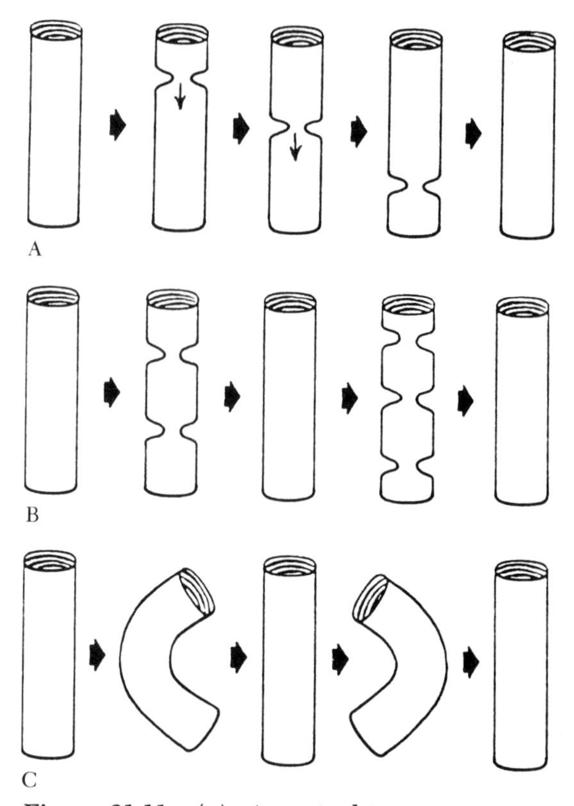

Figure 31-11 (a) A peristaltic wave passing over a section of intestine. It is usually unidirectional and moves ingesta from anterior to posterior. (b) Segmentation movement of the intestinal wall. Such movements are highly localized, nonprogressive and serve to mix ingesta. (c) Pendular movements of the intestinal wall. These movements involve one side of the wall and then the other. They may be involved with mixing of the ingesta and enhancing absorption. (From McCauley, 1971).

The Digestive Sequence

Digestion in Carnivores and Omnivores

The digestive process in most carnivores and omnivores begins in the mouth. When food enters the mouth, there is a copious secretion of saliva elicited which moistens the food and facilitates its passage down the esophagus. In those species possessing a salivary amylase, also called ptyalin, starch digestion begins in the mouth; it continues in the stomach until inhibited by gastric acidity or the ptyalin is deactivated by gastric proteolytic enzymes.

Chewing stimulates sensory fibers in the mouth which carry impulses to the vagal motor centers responsible for preparing the stomach for the arrival of the ingesta. The vagi stimulate (a) the chief cells to secrete pepsin, (b) the parietal cells to secrete hydrochloric acid, (c) the mucous-secreting cells, and (d) the release of gastrin from the pyloric antrum. Gastrin maintains the secretion of hydrochloric acid and pepsin initiated by the vagal fibers. Once in the stomach, the ingesta have a direct stimulating action on the mucosa, and the secretory activity is sustained; in other words, food is acting as a secretogogue (Figure 31.12).

In the stomach, the ingesta are exposed to the strongly acid gastric secretions. The enzyme pepsin, an endopeptidase in an acid media, catalyzes the hydrolysis of protein into polypeptide chains. Gradually, the gastric contents are transformed into a semiliquid mass (chyme), and the stomach begins to empty as the result of increased gastric musculature activity.

The presence of chyme in the duodenum results in the release of the hormones secretin and pancreozymin (cholecystokinin) from the duodenal cells (Figures 31.13 and 31.14). These hormones eventually reach their target organs, pancreas and gallbladder, via the bloodstream. Secretin promotes a copious release of an alkaline juice from the pancreatic duct cells, and pancreozymin causes the acinar (exocrine cells) of the pancreas to secrete the precursors of the endopeptidases, trypsin and chymotrypsin, carboxypeptidase, amylase, lipase, and nucleases into ducts which carry the products to the duodenum. Pancreozymin (cholecystokinin) also stimulates the gallblad-

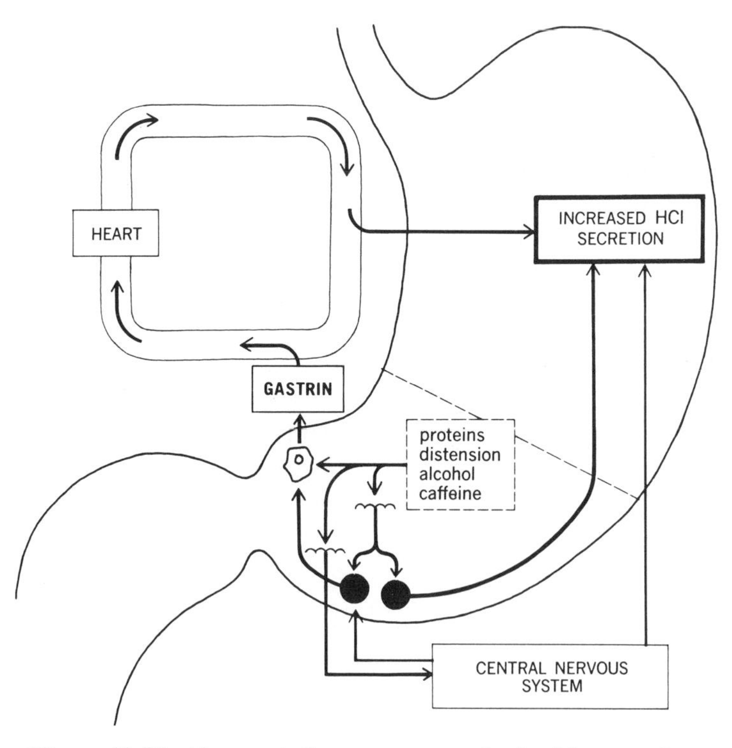

Figure 31-12 Factors influencing gastric hydrochloric acid secretion. (From Vander, Sherman, and Luciano, *Human Physiology*, 1970, by permission of McGraw-Hill Book Co.)

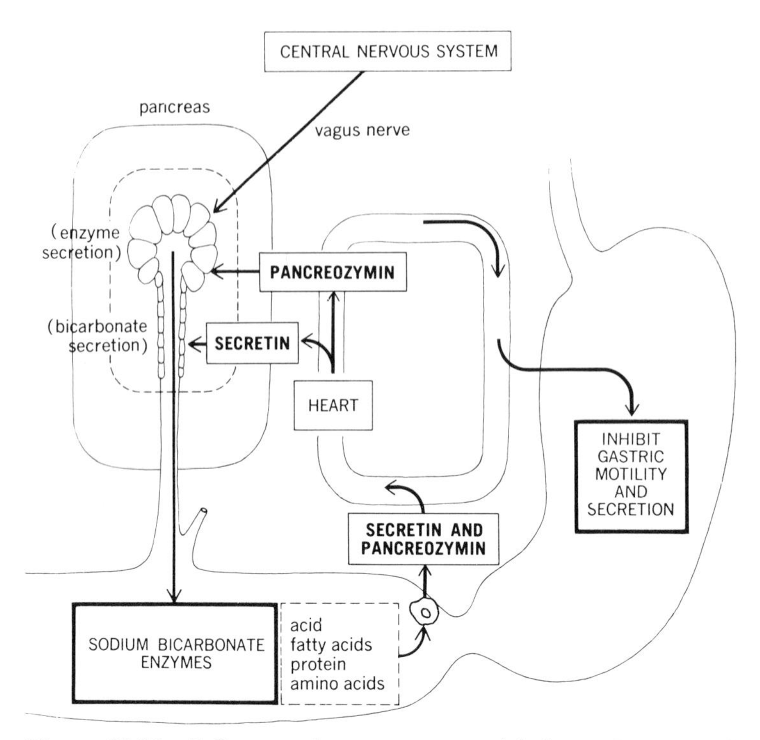

Figure 31-13 Influence of pancreozymin (cholecystokinin) and secretin on pancreatic secretion and the motility and secretion in the stomach. (From Vander, Sherman, and Luciano, *Human Physiology,* 1970, by permission of McGraw-Hill Book Co.)

der, when present, to contract and force bile into the duodenum. In addition to these secretions, the chyme is exposed to the digestive enzymes of the succus entericus. According to one point of view, the hormone enterokrinin, released by the duodenal cells in response to chyme in the duodenum, stimulates the secretion of the succus entericus.

In the small intestine, trypsin and chymotrypsin continue the proteolytic activity begun by pepsin. The polypeptide chains are gradually hydrolyzed to amino acids and dipeptides by the pancreatic carboxypeptidase and the aminopeptidase of the succus entericus. Protein digestion has been completed, and the amino acids are ready for absorption.

The hydrolysis of starch to the disaccharide maltose occurs mainly in the small intestine, despite the presence of a salivary amylase in some carnivores and omnivores. Maltase, of the succus entericus, hydrolyzes maltose to two glucose units. Also, in the succus entericus are the enzymes lactase and sucrase which catalyze the hydrolysis of lactose into glucose and galactose, and sucrose into glucose and fructose, respectively. The monosaccharides are now ready for absorption.

Pancreatic lipase catalyzes the hydrolysis of dietary lipids into glycerol, fatty acids, and mono- and diglycerides. Bile emulsifies the lipids, and in so doing divides the lipid particles into smaller and smaller units. This process vastly increases the surface area available for hydrolysis. Because of bile, lipids can be absorbed as such, or as monoglycerides and diglycerides. Otherwise, lipids are absorbed as their constituent parts, glycerol and fatty acids.

In carnivores and in omnivores, the major portion of digestion ends in the small intestine. The major cite of cellulose digestion in omnivores is the cecum. In the total nutritional economy of these animals, however, cecal digestion contributes only a small part.

Absorption in Carnivores and Omnivores

The end products of digestion in these animals are primarily amino acids, monosaccharides, glycerol, fatty acids, and mono-, di-, and triglycerides. The primary site of absorption of these compounds is the small intestine.

In broad terms, absorption occurs as the result of diffusion or active transport. Diffusion involves the movement of molecules from a region of high concentration of those molecules to a region of low concentration; only physical forces are involved, i.e., the migration of molecules from one place to another. If molecules are moved by active transport from one region to another, the expenditure of energy is required.

Digestion in Herbivores

The digestive sequence in herbivores is one of great complexity and involves a fermentative vat at the beginning or end of the alimentary canal, in some cases both ends. It is in these fermentative areas that the large amounts of rough, coarse dietaries are transformed into products useful to the animal.

Without becoming too involved, one can view these fermentative areas as vats containing many billions of bacteria and protozoa. The host animal, in which the vat is located,

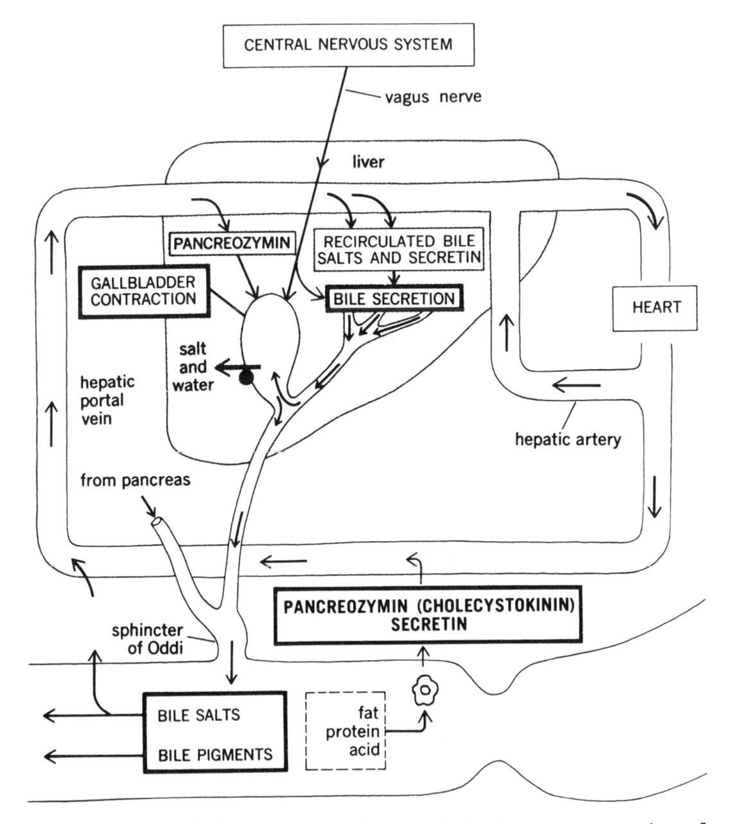

Figure 31-14 Influence of cholecystokinin (pancreozymin) and secretin on bile secretion and gallbladder contraction. (From Vander, Sherman, and Luciano, *Human Physiology,* 1970, by permission of McGraw-Hill Book Co.)

ingests the food, provides a constant temperature, maintains the proper osmotic relationships, supplies fluids when necessary by drinking, mixes the ingesta and microorganisms as the result of muscular activity, and gets rid of the waste products periodically; all in all, a highly satisfactory environment for microorganisms to grow and multiply. In return for the services provided by the host, the bacteria and protozoa digest the complex cellulose, take whatever nitrogen and other substances, such as minerals, that are necessary from the surrounding breis, and synthesize their own protoplasm. As the microorganisms grow and multiply in this fermentation vat, they excrete waste products into the surrounding fluid. The waste products include such compounds as the gases carbon dioxide and methane and short-chain organic acids (also called volatile fatty acids). Small quantities of carbon dioxide may assist in helping to maintain the proper acid-base relationship, but most, along with methane, is voided.

The short-chain organic acids are of great nutritional significance to herbivores. In animals such as cattle, sheep, goats, and camels, these acids are the major source of energy.

"Pregastric" fermentation The pregastric digesters include the true ruminants and pseudoruminants. In such animals, all food is first subjected to fermentation action in the reticulo-rumen, all the gases are voided via the esophagus, and the small molecules, such as the short-chain organic acids and urea, are absorbed through the rumen wall into the bloodstream via diffusion or active transport. Periodically, small quantities of rumen fluid, which now is composed of microorganisms, soluble compounds, indigestible substances, and water, enter the abomasum or true stomach.

In the abomasum, the rumen fluid is acidified and exposed to the pepsin-catalyzed hydrolysis of its protein. The protein content of the rumen fluid comes from two sources: (1) the food protein not degraded by the rumen microorganisms; and (2) the protein of the bacterial protoplasm itself.

From time to time, the abomasum empties, and the chyme becomes exposed to the duodenal enzymes, similar in all respects to those mentioned above for carnivores and omnivores.

In the neonatal ruminant, the reticulo-rumen is nonfunctional because it has not yet been inoculated with microorganisms. Furthermore, the diet of the young consists of milk, which is easily digested by the enzymes present in the stomach and small intestine. As discussed on page 442, functioning of the esophageal groove permits liquids to enter directly into the abomasum of young ruminants.

Rumination and eructation These two physiological processes are essential to the pregastric digestive process. Rumination insures that long, coarse, tough dietary materials, such as hay, are finely divided enough to be attacked by the reticulo-rumen microflora. Eructation, or belching, is absolutely essential to the fermentation process.

The essentiality of the eructation reflex to the digestive process in ruminants and pseudoruminants cannot be understated. Very large volumes of gas are produced by the fermentative activity taking place in the rumen; these gases cannot be allowed to accumulate. If eructation is impeded for as little as 15 minutes during feeding, the animal becomes "bloated." Young succulent legumes produce an intense frothing of the rumen ingesta which is capable of preventing eructation.

"Postgastric" (cecal) fermentation In cecal digesters, the food is exposed to all the gastric and small intestinal secretions before it reaches the cecum. Some of the soluble carbohydrates (starches and sugars) must undergo fermentation in the stomach, because lactic acid, which has been identified in the ingesta as it enters the duodenum, can here only come from fermentative activity. Because ingesta pass through the alimentary canal of the horse rapidly, stomach fermentation is of little importance when compared with ruminants. Small quantities of the lactic acid resulting from stomach fermentation may be absorbed from the small intestine, but most passes to the colon and cecum where it becomes a substrate for bacteria. Volatile fatty acids are the by-products of lactic-acid fermentation.

As is the case for carnivores, the digestion and absorption of most of the soluble protein and carbohydrate fractions of the ingesta occurs in the small intestine. Any compounds, along with cellulose, which are not digested in the stomach and small intestine become substrates for the fermentative processes occurring in the cecum. Although not completely identified, the by-products of cecal digestion are similar to those found in the rumen.

Cecal digesters appear not to benefit greatly from the protoplasm synthesized during the process of cecal fermentation; however, recent studies indicate the contrary, especially with respect to amino acids.

It would appear that the horse is more efficient than the ruminant in the utilization of soluble carbohydrates and proteins in the diet. In cattle, the soluble carbohydrates are rapidly and indiscriminately fermented in the rumen, and as a result there is a great loss of energy in the form of carbon dioxide and methane. In horses, only a slight amount of soluble carbohydrate appears as lactic acid, and most of it is hydrolyzed with the assistance of carbohydrases in the small intestine; in this case, there are no gases evolved.

Avian Digestion

Salivary amylase has been identified in birds, but as in mammals it seems to be of little importance in the digestion of starch.

The proventriculus is analogous to the stomach of carnivorous animals, in that it is here that pepsinogen and hydrochloric acid are secreted. When food enters the proventriculus, the secretion of pepsin and hydrochloric acid is elicited. This compares favorably with the gastric phase of secretion in mammals.

The extent of protein digestion in the proventriculus of birds is not well-established, and reports vary from very little protein digestion in grain-eating birds to extensive digestion of protein in carnivorous birds.

The extent of proteolysis in the gizzard has been reported to be meager despite the presence of pepsin and a satisfactory acid-base relationship. Therefore it would appear that the primary function of the gizzard in digestion is one of the reduction of the size of food particles by a grinding action.

The digestive activity in the small intestine in birds proceeds, as in mammals, under the influence of enzymes of pancreatic and succus entericus origin.

The absorption of monosaccharides and amino acids in birds is not well-understood, but presumably active transport is the primary mechanism. Fat absorption does not take place until it reaches the lower half of the small intestine and is enhanced by the presence of bile.

The large intestine of birds, like the large intestine of carnivores, is concerned primarily with the absorption of water. Birds are able to digest about 18 per cent of the fiber which is ingested, and this occurs in the ceca as the result of fermentation.

FURTHER READINGS

Alexander, F. 1963. "Digestion in the horse." *In* D. P. Cuthbertson, ed. *Progress in Nutrition and Allied Sciences.* Edinburgh: Oliver and Boyd. Pp. 259–268.

Church, D. C. 1969. *Digestive Physiology and Nutrition of Ruminants.* 2 vols. Corvallis, Ore.: O.S.U. Book Stores.

Dougherty, R. W., ed. 1965. *Physiology of Digestion in the Ruminant.* Washington, D.C.: Butterworths.

Glass, G. B. J. 1968. *Introduction to Gastrointestinal Physiology*. Englewood Cliffs, N.J.: Prentice-Hall.

Phillipson, A. T., ed. 1970. *Physiology of Digestion and Metabolism in the Ruminant*. Newcastle upon Tyne: Oriel Press.

Sturkie, P. D. 1965. *Avian Physiology*. Ithaca, N.Y.: Comstock.

Swenson, M. J., ed. 1970. *Duke's Physiology of Domestic Animals*. Ithaca, N.Y.: Comstock.

Thirty-Two

The Environment Versus Man and His Animals

"The world is too small to be used as it has in the past. Man has the power to manipulate that which surrounds him."

N. D. Bayley

With the expanding human population requiring more food, man is more than ever looking on his domestic livestock as "factories" to convert the potential nutrients of feeds into forms useful and appealing to him. Just as for any other factory, the ratio of output of useful production to the inputs of materials, services, and capital investment should be as high as possible. Many of the components of the surroundings imposed by both nature and man influence the input-output ratio of the animals.

Man's increased interest in deriving a better understanding of how elements of the environment affect animals has led to animal scientists' joining forces with other disciplines to develop the field of Biometeorology. It comprises the study of the direct and indirect interrelationships between the geophysical and geochemical environment of the atmosphere, plants, and animals. This chapter deals with the manner in which various elements of the local natural environments directly influence the animal as a "factory"; some of the ways these elements contribute to problems of animal production through their interactions; and the physiological mechanisms the animal has at its disposal to meet the challenges imposed by changes in the surroundings; along with examples of measures which man may employ to aid animals in maintaining their status as efficient factories. Although food is an important environmental factor, it will not be considered in this chapter, except per some generalizations and a mention of the relationship between food and heat regulation.

Environmental Variables

To obtain a near optimum ratio of input to output for cattle, sheep, goats, buffaloes, swine, and poultry, we would like an environment having an air temperature of 13 to 18°C, a relative humidity in the range of 55 to 65 per cent, a wind velocity from 5 to 8 kilometers per hour, and a medium level of sunshine.

Because the optimum environment seldom exists, the successful livestock producer devotes time and efforts to improving his animals' environment in order to obtain satisfactory performance from them in weight gain, milk yield, wool, eggs, work, or other output.

Direct Influence of Climatic Variables

The physical environment is generally thought of as the prevailing climatic conditions, but the latter are only a portion of the animal's environment. (Climate is defined as average weather conditions, whereas weather comprises the day-to-day, changing meteorological conditions of an area. Microclimate is used to describe conditions immediately surrounding the animal, while macroclimate refers to those further removed.) Still, the elements of climate—temperature, sunshine, wind, light, precipitation, and humidity—are important because they influence most of the other elements of the environment in some fashion. The conditions created by the climatic elements not only have direct effects on the animal, but also dominate the growth and quality of feedstuffs, the incidence of diseases and parasites, the efficiency of labor, and numerous other conditions.

The over-riding climatic element which affects the physiological functions of an animal is temperature. Local climates are characterized by variation in temperatures during a given day from about 6°C near the equator up to 16 to 20°C in desert regions. Changes in the average or mean daily temperature from one season to another may be 40°C or more. For most domestic livestock, a mean daily temperature in the range of 10 to 20°C would not be likely to place them under thermal stress; that is, there are no discernible changes in the animal's usual body processes within this range, which is generally referred to as the "comfort zone." Of course, the newborn is much more sensitive to its thermal environment; therefore, the comfort zone of the very young animal is restricted to ±1 or 2°C.

As the temperature conditions fall outside the comfort zone, other elements of the climate assume greater significance in the comfort of animals. The water-vapor content of the air, expressed as humidity, becomes increasingly important as an interacting factor, when air temperature rises or falls, and so does solar radiation and wind velocity. At low temperatures solar radiation will tend to raise the temperature of the animal's surroundings, which will aid it in maintaining thermal balance, but at high temperatures, solar radiation will impose an excess heat load on the animal. The velocity of the wind may have the reverse effect, since a high wind velocity increases the rate of movement of air away from the animal's body, accelerating the rate at which heat from the body will be dissipated into the atmosphere; however when the ambient air temperature exceeds the temperature of the animal's skin surface, the heated air may impose a significant amount of stress on the animal.

In the natural environment, precipitation (rain, sleet, or snow) can be important factors for an animal. If an animal is exposed to rain on a hot day, comfort will be improved, since additional heat will be removed from the body by evaporation. But at low temperatures, rain chills the animal and causes considerable discomfort because of a rapid rise in heat loss from the body as the precipitation evaporates. Generally, snow does not create serious problems for an animal, since it does not often melt on striking the hair or wool.

Altitude influences the prevailing air temperature pattern. Air temperature tends to diminish at the rate of about 0.65°C per 100-meter increase in elevation; however, local topography frequently upsets this relationship. Above 2,000 meters elevation man begins to feel the stresses imposed by insufficient oxygen (Hock, 1970), but the threshold level for low oxygen in sheep, goats, and cattle is higher, 2,500 meters or higher. In addition to low oxygen pressure, the dry atmosphere of high altitude is a further hazard for the newborn, affecting the body's water balance, temperature regulation, respiration, and vulnerability to diseases. Of most significance to livestock is the influence of altitude on the growth rate and nutritive value of plants. In spite of these disadvantages, reasonably productive livestock enterprises, using domestic breeds, are carried on at 2,000 to 3,000 m elevation. Other species such as the yak, llama, alpaca, and vicuña, for example, appear to flourish at altitudes over 3,000 m (Phillips, 1949).

The period of light for a given day is referred to as the photoperiod. This varies with latitude and season. Near the equator there is little change with season, but it is ±2 hours at 30° latitude, and at 60° latitude the photoperiod may reach 19 hours in June. The photoperiod is very critical for plants and has direct bearing on animal performance. Length of photoperiod has a pronounced influence on the breeding season for sheep and buffaloes, and on the onset of sexual activity in many species of fowl. It also appears to dominate the growth and shedding of hair, wool, and feathers of numerous species.

Indirect Influence of Climatic Variables

Figure 32.1 represents the environmental elements that have a major direct influence on an animal, and the important interactions among them that constitute the indirect effects on the animal. When we consider the animal's total environment, the amount and quality of feed available is the single most important factor. The amount of food consumed determines directly the productivity of the animal, and also contributes to its resistance to other conditions, such as disease. The feed supply is markedly affected by temperature, humidity, rainfall, solar radiation, and light, and to an extent by wind, altitude, and soil fertility. It is readily evident from Figure 32.1 that the animal is subjected to numerous factors which have a direct effect on its well-being as well as to those resulting from interactions among the elements. Except in extreme conditions, as in desert or arctic regions, the indirect influences of climate are generally the most important to animal performance. If an animal is subjected to a temperature 8 to 10°C higher than its comfort zone, it will experience some degree of discomfort, causing it to bring into action certain physiological processes to help counter the impact and to change its general behavior. In spite of some discomfort, the animal will still function in a fairly efficient manner. If the high ambient temperature persists, as in tropical areas or during the summer months in the southern U.S., the indirect effects on the animal will be very detrimental. The high temperature causes the crops to mature rapidly with a corresponding decline in nutritive value. Following poor feeding, the animal comes under nutritional stress which leads to greater susceptibility to disease and parasites. The distribution of rainfall and the level of humidity frequently add to the problems of providing adequate feed and maintenance of animal health.

Maintaining Homeostasis

Homeostasis is the relatively constant state resulting from a variety of physiological adaptive actions and counteractions in a fluctuating environment. When animals that are homeotherms leave the intrauterine environment, they face a continuous struggle to maintain a state of homeostasis; the apparent goal of functional evolution has been the maintenance of a constant internal environment and steady body temperature in a variable external environment. To maintain the body temperature within narrow limits calls for very sensitive and quickly acting mechanisms which will balance any change in heat production by an equivalent change in heat loss, or balance an alteration in heat loss through one channel by an equivalent change in the opposite direction through another.

Examination of the manner in which the animal's body temperature, or at least that of its central core, is kept constant, reveals that the process is not a simple one. The body temperature of most mammals is kept at a relatively high level (usually 38°C or above) which is considerably above the environmental temperatures to which the body is exposed most of the time. Control of body temperature is obtained not so much by varying the rate of heat production as by varying the rate of heat loss. Maintenance of a high body temperature is not without disadvantages, however. The various metabolic processes will go on at a relatively high rate, resulting in far more food energy being required for maintenance in homeotherms than in their "cold-blooded" fellows (poikilotherms).

The centers for regulating heat balance are located in the hypothalamus. The plural term is used because functionally the centers are concerned with both heat loss and heat production. The centers get information on the temperature of the central core of the body

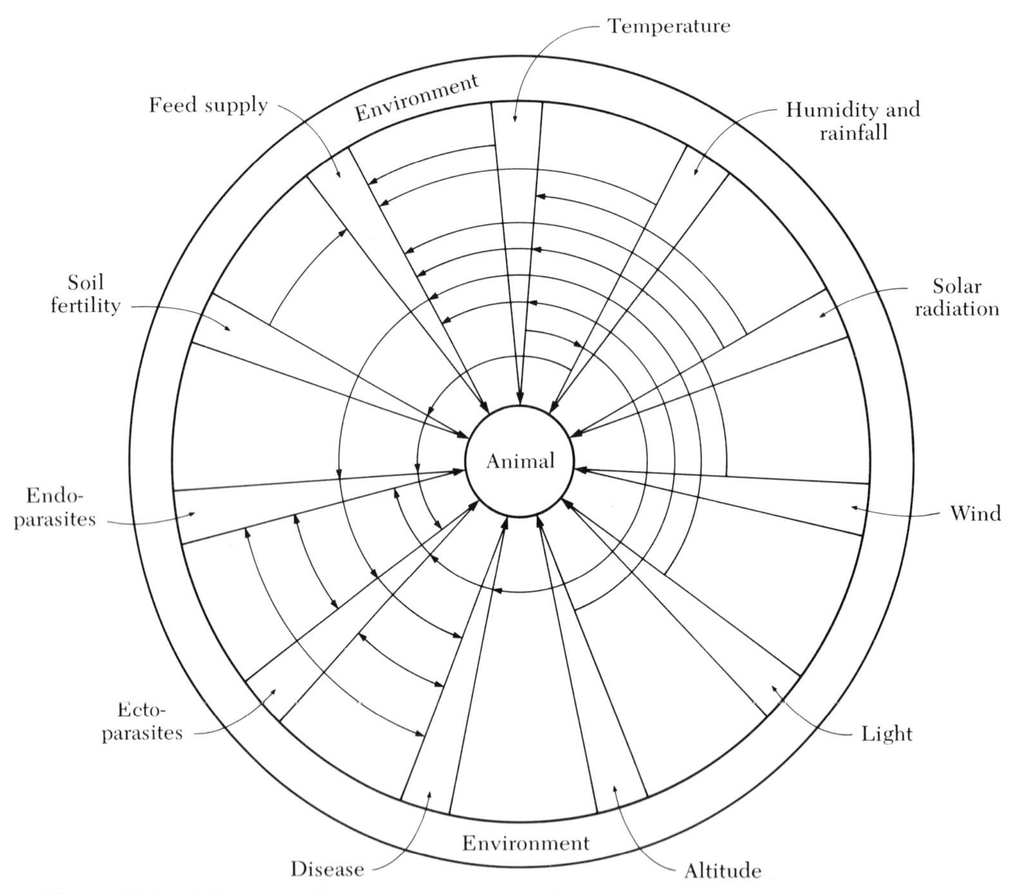

Figure 32-1 Schematic illustration of an animal's environment. The arrows extending from the outer circle to the inter-circle depict those elements which have a direct influence on the comfort and productivity of animals. The concentric arrows represent interactions (indirect actions) between the various environmental factors; such as temperature and rainfall influence the growth of plants used as feed supplies and these same elements often play a major role in the prevalence of diseases and parasites. The indirect affects of climate often constitute the greatest inhibitors to livestock production.

tissues from the temperature of the blood flowing through them and from nerve impulses coming from the skin. Sensations of heat and cold arising in the cerebral cortex are derived mainly this way. In addition to the nervous pathways of control, a chemical pathway may assist, through, for instance, the posterior pituitary gland, one function of which is to maintain the water balance of the body. As such the pituitary is often involved in a secondary fashion in heat regulation, either in water balance or through the thyroid gland, which has a great deal to do with determining the metabolic rate.

The fundamental statement of heat balance in the animal body may be expressed as $S = (M - W) - (R + C + E)$,

where S = rate of change in heat content of the body,
M = rate of energy liberation in metabolism,
W = rate of energy usage for productive processes,
R = rate of heat loss by radiation,
C = rate of heat loss by conduction-convection,
E = rate of heat loss by evaporation-convection.

The object of heat regulation is to keep S as close as possible to zero, or $(M - W) = (R + C + E)$. When the animal is in an environment outside the 13 to 18°C range, it must respond by: (a) shifting the rate of metabolic heat production, upward in low temperature or downward as temperature rises; (b) increasing or decreasing the rate of heat loss; or (c) a combination of the two if S continues near zero.

Changes in Heat Production

Basically, animals have three means of changing heat production. These constitute changes in the basal, behavior, and performance components, some of which are desirable and others which are not. The basal component may be changed by lowering the basal metabolic rate. This has advantages to the animal in a hot environment, but such actions usually result in lowered performance. Heat production may also be changed by modifications of behavior, as a dog on a hot day is more interested in lying in the shade than in chasing a rabbit.

Under hot conditions, lowered rate of performance accounts for the greatest change in heat production. Although this is undesirable from the standpoint of productivity, we can oftentimes do things to help the animal minimize its changes in the performance components, such as reducing the distance the animal must travel for food and water. A cow would need to evaporate 1,000 ml of water per hour to cope with the energy expended to walk over about one hectare of land to secure its food, or about 25 per cent more energy than if she were standing in a stall.

In the cold the animal maintains a relatively constant central body-core temperature by reducing over-all heat loss to a minimum and making up the deficit by increased heat production. The increased heat to maintain thermal balance must eventually come from dietary sources or radiant heat from the environment. Initially the extra heat may come from either muscular activity, including shivering and the heat increment of feeding, or the expenditure of energy directly from body reserves (non-shivering thermogenesis). Unless the animal can seek protection to prevent excess heat loss, the increased heat production to maintain body temperature will bring about lowered performance corresponding to that for hot conditions.

Changes in Heat Loss

Heat is removed from the animal's body through the physical processes of conduction, evaporation, convection, and evaporation. Each of these essentially has a climatic element counterpart: temperature, humidity, air movement, and radiant energy, respectively.

Conduction Conduction serves two roles in heat loss, in the movement of heat from the central core to the external surfaces and in the flow of heat from the skin surface to the surroundings. When the ambient temperature is around 15°C and the body core temperature 39°C, heat passes to the cooler portions of the body and on to the atmosphere rather rapidly. In mild climates the animal loses about 70 per cent of its total heat production in this fashion without undue effort. Some additional heat is dispelled by conduction through warming of inspired air and heating of ingested food or water. The exchange through the latter is quite variable, depending on the temperature of the water or food. If a cow consumes 15 kg of 21°C water, about 15 kcal of heat will be utilized to warm the water.

Evaporation Heat loss by evaporation occurs from the skin and respiratory tract. It is the most efficient means of removing heat, since one gram of water vaporized at 20°C will release 0.6 kcal of heat. The water evaporated from the skin becomes available by simple transudation through the skin, by secretion of sweat by the sweat glands, or by external application of water. Continued loss of heat by evaporation is dependent not only on ambient temperature but also on the rate of air flow about the body and the level of humidity of the air. Low humidity and reasonably high rate of air movement is the reason sweating is seldom perceptible in hot, dry regions; whereas the high humidity and low air movement in the warm, humid regions causes water to accumulate on the skin.

If the air the animal inspires has a low humidity—characteristic of hot, dry climates or extremely cold climates—the amount of heat lost by evaporation from the respiratory tract may be high. Conversely, if the air has a high humidity the rate is much slower. In cold climates the rapid loss of heat through respiration becomes a significant factor in efficiency of animal performance. The significance of respiration is covered in more detail later.

Convection Convection represents the transfer of heat energy by circulatory motion that occurs at a nonuniform temperature. It is an intricate process in both internal and external heat exchange, internally by the circulating blood and externally by the rate of air flow about the body. The thin layer of air adjacent to the skin surface comes into equilibrium with the skin temperature. In still, warm air, convection plays a minor role, but in a cold wind the animal is subject to loss of tremendous amounts of body heat. In fact, a cow exposed to −10°C temperature and 40 km/h wind will lose heat more rapidly than can be generated from intake of feed (Williams, 1967). This illustrates the value of shelters discussed later.

Radiation An animal consistently exchanges heat with its surroundings by radiation. The amount depends on the nature of the surroundings and the amount of body surface exposed. An animal in the natural environment is exposed to: (a) direct solar radiation from the sun, part of which is reflected, with the remainder absorbed as heat; (b) solar radiation reflected from the clouds; and (c) radiation reflected from the ground, other surroundings, and the horizon. Approximately 50 per cent of the animal's heat load while standing in the sun is from (a) and (b) and the remainder from (c), depending on whether conditions are humid or dry. The portion of the radiant energy which is absorbed by the animal's body is changed into heat, thus adding to discomfort under hot conditions, but aiding the animal in maintaining comfort in a cold environment.

The radiant heat load is greatest in the afternoon, because the ground radiation and air temperature are higher. Even though a pig standing in the sun at noon on a summer day in California stands to gain as much as 663 kcal/m^{2}h (Bond, 1967), the net gain is considerably less because of the amount of surface exposed, reflection, and re-radiation. The exchange by radiation is unknown at this time, because of the complexity of determining the net gain; so about all we can say is that from 6 to 12 months of the year, in the lower elevations of North-South 40° latitudes, the intensity of direct and reflected radiation will be such that for 5 to 10 hours per day the animal may receive significant amounts of heat.

External Characteristics in Relation to Heat Balance

Although would-be predators have had a great deal to do with the characteristic coat color and additional traits, like speed of flight, that contributed to survival of a species, environmental conditions have played an important role too. Within the comfort zone the external characteristics of the animal's body are not so important to its maintaining heat balance, but under extreme conditions of hot or cold, nature has exerted quite a bit of pressure toward development of external characters that aid the animal to survive. The camel, for example, has extremely long legs that place its body away from high ground radiation, and spongy pads on its feet which have a low blood supply, thus low sensitivity to heat. These features enable it to move about the hot desert with less discomfort than other species. However, we must not hasten to conclude these traits are solely responsible for the camel's survival in the desert (*cf.* Schmidt-Nielsen, 1964).

Skin, pelage, and appendages For long infrared (thermal) radiation coming from the surface of a building or a hot ground surface, the color of the skin or pelage makes little difference, since the skin surface behaves as a "black body"; that is, all the incident radiation is absorbed. Yet color is important in reflectivity of short infrared radiation (the visible portion of the sun's rays). In a hot environment, a white skin is undesirable because of sunburn problems; therefore, pressures from nature have favored a pigmented skin. A white or light-colored coat has better reflectivity than dark; hence, in a hot climate a dark pigmented skin with light hair is generally most desirable. Of probably more significance than color is the length of the hair and its oiliness, dry hair being more absorptive than oily hair. Pelage color makes little difference to an animal's heat regulation in a cold climate.

Thickness of the skin increases with age. It also varies with different parts of the body. Poor nutrition causes the skin to thicken, as does cold weather. In a hot, dry environment a thick skin helps to reduce the effect of thermal radiation and to prevent dehydration. These attributes are desirable for a cold climate, but a thin skin is desirable when it is warm and humid. There is evidence that breeds of cattle have characteristic skin thickness associated with origin. The South Devon, from northern Britain, has a thick skin, 8.15 mm, and so does the Africander from the hot, dry region of southern Africa, 8.73 mm; whereas the Zebu cattle of India, originating in a warm, humid climate, has a thinner skin, 5.46 mm.

The extent of body surface in relation to mass can be important in heat exchange. If air temperature is below body temperature, a large body surface affords advantages to the animal, but if air temperature is near, above, or markedly below body temperature, a large surface has disadvantages. A large surface area would be helpful under hot, humid conditions in promoting surface evaporation. Many of the cattle indigenous to hot climates have a different configuration from those of the cooler climates (see Figure 32.2). In the past we have assumed that the appendages characteristic of breeds like the Red Sindhi were important in their adaptation to hot climates, but research has not borne this out (McDowell, 1972).

Fat storage Fat stored adjacent to the skin impairs heat loss in a hot climate, but serves as an insulator in cold climates. The polar bear and other species of the arctic generally have a thick layer of fat beneath the skin, whereas sheep native to the hot, dry areas of the tropics have very lean bodies (Figure 32.3) and have their fat reserves in their tail or rump. The camel also has a special place for storage of most of its reserve fat. The hump of cattle native to the tropics, unlike that of the camel, does not serve as a fat

Figure 32-2 Illustration of contrast in phenotypic characteristics of cattle originating in different climates. The Holstein (top) originated in a cool climate and the Red Sindhi (bottom), a breed native to the Indian subcontinent, is a warm-climate animal. The Red Sindhi has a distinct hump, heavy horns, long, wide head, large ears, wrinkled skin on the neck and a pendulous neck fold in contrast to the Holstein, but he is smaller in size and has a shorter body. (Top photo courtesy Eastern Art. Insem. Coop. Inc. Bottom photo, courtesy ARS, USDA.)

Figure 32-3 A Sangsari sheep native to southeast Asia, showing long, thin legs, hair instead of wool covering on the legs and head, and more hair than wool on the body. These are characteristics of several breeds of sheep found in hot, dry regions. (Courtesy F. Olvey, U. S. AID, Iran.)

reserve; the amount of fat in the hump coincides with the general fatness of the animal.

In brief, external body characteristics can be helpful to the animal in a hot or cold environment, but it is not clear at this stage how much emphasis animal breeders should give to these traits in their efforts toward selecting animals which will function most efficiently in a given environment.

Response to Cold

Animals exposed to cold tend to decrease heat loss by reducing the exposed surface area, decreasing respiration rate, avoiding rapid air flows and wetting of the surface, and increasing insulation. Long hair, thick skin, and changes in the composition of certain fat deposits are some of the permanent changes. Those of a more temporary nature are lowered respiration and blood flow to the periphery, and changing the insulation of the hair coat through action of the pilomotor nerve fibers. (Erection of the hair allows for greater trapping of an insulated air mass near the body to reduce heat loss.) Blood flow may be reduced to tissues that are allowed to drop well below body temperature, i.e., the legs and face, and to a lesser degree the surface of the body trunk. For other areas, like the scrotum or teats, blood flow is increased to prevent a serious decline in temperature.

According to Scholander *et al.* (1950), a resting, heavily insulated arctic mammal or bird is able to change its heat dissipation by a factor of 11, and can experience temperatures to −40°C without increasing metabolic rate. Few studies have been made on domestic livestock, but it is likely that cattle and sheep, by growing longer coats or changing the properties of the coats, can lower their heat dissipation by a factor of three to five.

A large ratio of surface area to body mass, dormant methods for increasing heat production, and limited energy reserves make the newborn animal especially susceptible to cold. In general, body temperature declines following birth, depending on ambient temperature. During this stage, the young animal is like a poikilotherm because it does not respond by either shivering or non-shivering thermogenesis. We might interpret this as an adaptive mechanism; however, survival rate is highest among newborn pigs and lambs, which show the least decline in body temperature. Lambs and calves often attain a reasonably stable body temperature in a few hours, but pigs are not in control of body temperature until the sixth to ninth day (Williams, 1967). The amount of food the newborn consumes the first few days has a direct effect on its toleration to cold. Nutritional diseases are more frequent in the cold. All these factors indicate that man can enhance the viability of his young animals, especially under cold conditions, by assisting them with protection and insured feeding.

Calorimetric studies on swine indicate the critical temperature is around 21°C, or about that for man, but air temperature must drop to 0°C for metabolic rate to double. Pigs have an insulting layer of fat and relatively bare skin, so homeothermy is maintained by the fat and reduced blood flow to the skin; also they huddle together or burrow under bedding to gain insulation.

If given adequate feed, dairy cows will tolerate temperatures down to 5°C before milk yield is seriously affected; their body maintenance requirements, however, will be 15 to 20 per cent higher than at 15°C. The beef animals' greatest problems in the winter are precipitation, high humidity, and wind. Singly or in combination, they interfere with the hair coat, which is adequate except for the most extreme conditions. Cattle have a relatively low surface area to body mass, a good protective hair coat and considerable waste heat from rumen fermentation and digestion to use in maintaining heat balance; therefore, they are not nearly as affected by cold as nonruminants.

Sheep are particularly well-suited to withstand cold, provided they have adequate feed and the wool is not dampened or disturbed by strong wind. They can become very susceptible to cold if shorn when inclement weather occurs. The critical temperature for sheep changes from around 33° to −0.3°C as fleece lengthens from 0.0 to 12 centimeters. Marked changes in maintenance needs of livestock in cold climates are of economic significance to livestock producers; therefore, they tend to make changes in their systems of management to alleviate the added needs of feed.

Response to Heat

Rising environmental temperature will cause a lactating cow to become uncomfortable because of her elevated body temperature and high respiration rate; her feed intake will decline, and her milk yield will drop 25 to 50 per cent, depending on the level of air temperature and the duration of the stress period. These changes represent the dramatic consequences which we want to prevent; otherwise, the efficiency of our factory is lowered markedly.

Animals have certain physiological processes which they can employ to minimize the impact of the hot environment. Shifts in the level of function of these, coupled with some help by man, will assist the animal a great deal.

If we sequentially categorize the compensations or attempted compensations by animals, especially livestock, to increased ambient temperature, they take place in about the following order: (1) increased skin blood flow, through vasodilation of the blood vessels near the surface; (2) initiation of sweating; (3) increased but shallow respiration (panting); (4) changes in hormone activity; (5) changes in behavior patterns; (6) increased water intake; (7) increased body temperature; (8) changes in the use of body water; and (9) changes in state of hydration. If these measures fail to renew the animal's equilibrium of heat balance or shift it to a new plateau, then progressive stages of failure of the heat regulation mechanisms will be in evidence, as diarrhea, a general weakness, staggers, convulsions, and death.

Increased blood flow commences with a 1 to 2°C rise in ambient temperature, and helps to increase rate of heat loss by conduction and radiation. If the greater flow is not sufficient, then the heat-regulating centers activate the sweat glands to promote heat loss by evaporation. The glands are stimulated to discharge by nervous impulses reaching them through the sympathetic nervous system. In general, the sweat glands will increase secretion until the resulting film of water on the skin surface is sufficient to restore heat loss to its normal value. Of the domestic animals, the horse is the most highly adapted for sweating, with the donkey, cow, and sheep following (Findley and Robertshaw, 1965). Man is superior to all species in this respect; he can secrete 2,000 cc/hr for short periods and 600–800 cc/hr for several hours. Cattle can sustain a total evaporation rate of 200–300 cc/m²hr, of which 75 per cent is from sweat and 25 per cent from transudation. Species of domestic livestock native to tropical areas usually are able to secrete sweat at a higher rate than those from cooler climates, but as with man there is wide variation among individuals. The rate of evaporation of sweat is highly dependent on the type and extent of body covering (hair or wool), hair density (number of follicles per unit area), whether the hair is erect or lying flat, and the oiliness of the hair. Species without sweat glands, e.g., the rabbit, obtain some evaporation through transudation or evaporating water applied to the hair by licking themselves.

In domestic livestock, increased respiration rate serves as an important means of increasing heat loss under hot conditions, and it is the first visible sign of heat stress observed. At 18 to 20°C a cow will breathe about 20 times per minute and expire 60 liters of air. This dissipates about 1.7 kcal/minute. At 35°C the same cow may have a rate of 115 per minute and a volume of 300 liters per minute, which raises heat loss to 5.6 kcal/minute. Sheep may change their rate from 60 to over 250 breath per minute with corresponding changes in temperature. Increased respiration rate is an efficient means of increasing heat loss for short periods, but high rates for long periods may cause the lungs to become overventilated and the CO_2 content of the alveolar air reduced to where the activity of the respiratory and cardiovascular centers are depressed. High respiration interferes with feeding and rumination. The increased muscular activity associated with high respiration rate utilizes energy and generates heat; so in many respects high respiration rate decreases efficiency of performance.

At present we have little understanding of how high temperatures cause changes in the functioning of the endocrine glands, but we do know that if the animal is exposed to hot conditions for several hours or more, the activity of the thyroid gland declines, which in turn acts either directly or indirectly to reduce appetite, and that continued exposure to cold increases thyroid activity.

Changes in behavior patterns, e.g., the usual posture, activity, and food intake, either to reduce heat production, to promote heat loss, or to avoid adding heat, are among the most significant responses animals make to fluctuations in the environment. Schmidt-Nielsen (1964) describes how many animals manage to survive under extreme conditions of temperature and of little or no water by adopting certain patterns of behavior. The kangeroo rat (*Dipodomys*), found in the deserts of the southwestern U.S., survives in an environment where the day air temperature may be 45°C and there is little or no water by digging burrows 1 to 2 meters in depth. There he remains during the day and comes out to feed only during favorable hours. In this way the rat avoids adding heat and conserves body water. The jack rabbit (*Lepus alleni*), native to the same region, does not adopt such extreme steps of avoidance, but it seeks a shaded depression in the ground during the day to avoid the high heat of both the sun and radiation from the hot ground.

Even though we have little factual evidence on all the things our domestic animals do to avoid heat and cold stress, it is clear that alterations in posture, activity, eating habits and other behavior do affect efficiency of performance. Hafez (1962) describes some seasonal changes in patterns of behavior associated with the way animals combat changes in their environment.

If changes in the processes described, including behavior, do not maintain or restore heat balance, heat builds up in the body. A rise of 0.5°C or less may not prove serious for most of our livestock species, but if the rise in body temperature exceeds 1°C, the consequence is lowered feed intake and production.

Economic Consequences of Changes in Ambient Temperature

There is a mass of reports of experiments which verify the hazards of temperature conditions above and below the "comfort zone" of mammals. Elevated temperature usually creates the greater problems in efficiency of animal performance.

Growth Rate

The optimum temperature for swine depends largely on size (Figure 32.4). When subjected to temperatures outside the comfort range for a given weight, the amount of feed required per unit of gain rises markedly. For best growth, baby pigs need a temperature considerably higher (27 to 29°C) than for the larger pigs.

Gains in lambs are significantly lower at 27°C than at 7 to 10°C, although the feed required per unit of gain is less. Figure 32.5 shows that temperatures in the range of 8 to 27°C do not seriously influence the growth of broilers, but at 38°C growth rate is almost nil. Cattle are not as sensitive to temperature effects as swine and poultry. If feeding is adequate, growth rate is not appreciably influenced until temperature goes below 2°C or rises above 29°C.

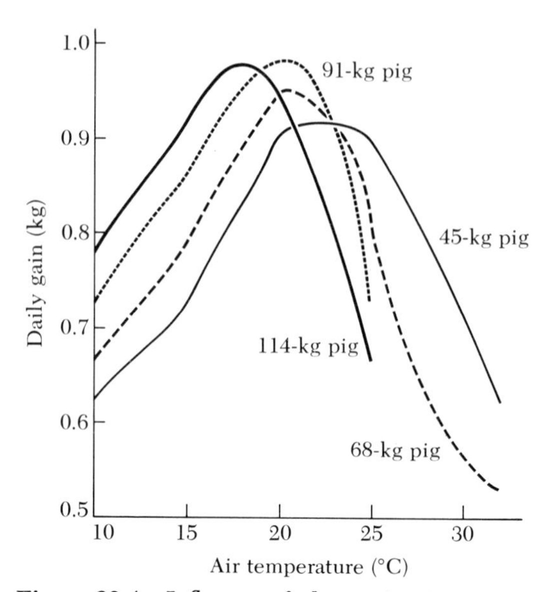

Figure 32-4 Influence of change in air temperature on the daily gains of pigs of varying sizes. (From Morrison *et al.*, 1966.)

Milk Yield

Level of milk yield is very sensitive to elevated temperature conditions, with measureable declines above 24°C or below 6°C. The temperature at which significant depression will occur is dependent on level of production, size of animal, and breed. Very low producers, like the Brahman, show little decline even at 35°C, where yields of Holsteins are down more than 50 per cent (Figure 32.6).

Efficiency of Feed Utilization

Temperature levels outside the comfort range on either the low or the high side increase body maintenance requirements for animals; therefore, efficiency is decreased in the utilization of energy for productive purposes. For example, at 10°C, 68-kg pigs will require 1,090 kg of feed per 100 kg of gain, but at 32°C over 2,000 kg of feed is required for similar gains. Lowered efficiency results because the animals eat less, thereby requiring more time to achieve a given level of weight, and of the energy consumed an appreciable amount, perhaps up to 30 per cent, is siphoned off by the labors exerted to combat thermal stress.

Reproduction

Usually the normal reproductive processes of both the male and the female are impaired by elevated environmental temperature. The breeding efficiency of mature dairy bulls may range from 70 to 73 per cent at 21°C, but will be decreased to 40 to 45 per cent when the bulls are subjected to 29°C for 10 or more hours per day (Bianca, 1965). Elevated temperature appears to influence the estrous cycle, the duration of estrus, and embryonic mortality; for the latter, the first three to five days following fertilization are the most critical. Under heat stress, the frequency of "silent" heat is usually quite high, and the duration of estrus is reduced about 50 per cent in cattle and water buffaloes.

Modification of the Environment

Since the natural environment seldom permits the level of performance desired, the livestock producer has an interest in the creation of an environment which: (a) minimizes all stresses on the animal that may reduce or limit productivity; (b) maximizes the efficiency of labor; and (c) minimizes cost per unit of animal product. This requires careful consideration not only of facilities but also of the development of general management systems and the protection of animals against diseases and parasites. Although some modification of the environment is almost always necessary, the extent required varies markedly from one location to another; more is needed in the higher latitudes or the deep tropics, much less in some of the subtropics. In the northern U.S. or southern Canada, cattle production for either meat or milk would be virtually impossible without storage of feed for the winter months. In the tropics animals can at least survive during the entire year on the available forages; however, the feed supplies may be only of sufficient quality to maintain good levels of animal performance for periods approximately equal to that of the colder regions. Efficient livestock enterprises can be maintained almost anywhere if sufficient environmental modification is made. The establishment of efficient poultry and dairy industries in hot, dry areas of Israel and poultry and swine units in humid tropical areas represent examples.

Virtually all environmental variables can be modified in some fashion with costs and the improvement in productive efficiency being the determining factors. In warm climates, solar radiation can be lessened through the use of rather simple shelters, but if temperature and humidity are to be varied, much more complex equipment is required. The effect of wind can be frequently modified with simple structures, and photoperiod can be adjusted where electricity is available by use of lighting. Air pollutants (dust) are difficult to control in extremely dry climates, but dust

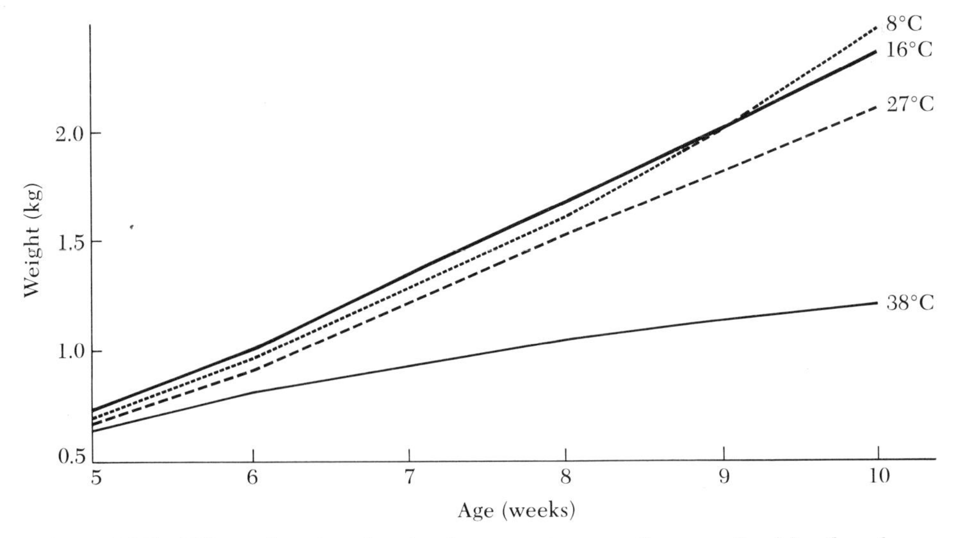

Figure 32-5 Effect of various levels of temperature on the growth of broilers from 5 to 10 weeks of age. (Data from Winn and Godfrey, 1966.)

filters can be employed when mechanical ventilation is practical, such as in poultry houses. Air contaminates, like ammonia that builds up in closed housing, can be reduced by frequent removal of waste or collection of waste under oil. Additional variables, such as nutritional status, can be changed by frequency of feeding or the time animals are permitted to graze.

Some sheltering for livestock is justified irrespective of location to: (a) protect animals and the handlers from the extremes of climate; (b) improve the efficiency of handling animals for feeding, milking, breeding, and reducing health problems; (c) ensure sanitation of animal and product; and (d) give esthetic value in appearance to customers. The type of shelter that best fits the needs is determined by climatic conditions, labor requirements, costs, and available materials. In hot, dry regions, shelters have proven effective in bettering the performance of beef cattle in feedlots, dairy herds, and swine. In the hot dry climate of the Imperial Valley of California, for instance, steers with shade consumed more feed, gained 0.5 kg more per day, and required 25 per cent less feed per unit of gain than animals without shade (Bond, 1967). Contrarily, summer shelters have not rendered much benefit to livestock in the more humid regions, because the rate of air movement is generally lower than for dry climates, and is further reduced when animals are close together under a shade. The decreased rate of cooling through evaporation, because of close proximity, tends to cancel the effect of lowered solar radiation.

Many livestockmen are confining their animals more and more to small spaces and bringing in the feed, which affords the advantages of increased specialization and more efficient use of labor. This usually constitutes a major change in the animal's environment that creates a need for different management practices. It increases the importance of attention to animal health to avoid spread of diseases, but at the same time provides an opportunity for utilizing protective measures.

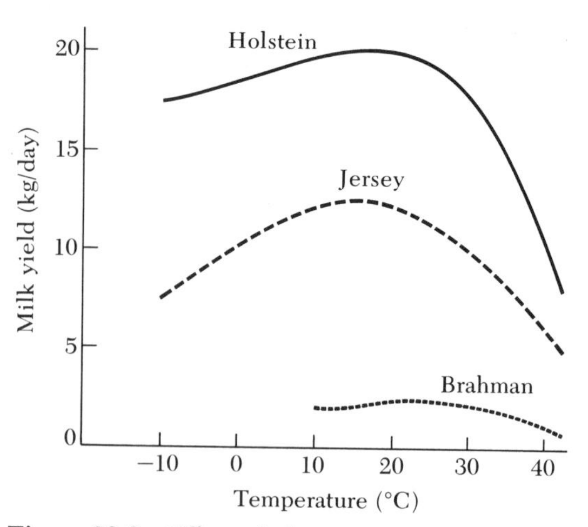

Figure 32-6 Effect of changes in temperature on the milk yield of Holstein, Jersey and Brahman cows. (Data from experiments conducted at the University of Missouri.)

Performance of confined animals can be affected favorably or adversely by the amount of space allocated, accessibility to feed and water, presence of other animals, the amount of shelter provided, and the "peck order" or social relationships. Cattle limited to 3 m^2 of slatted floor space required 20 per cent more feed per unit of gain than others allocated 60 per cent more space. In some studies, type of flooring has affected weight gains and feed efficiency. If the pen has a dirt floor, mud becomes a serious problem. If waste is allowed to accumulate in the penned area, animal performance is lowered.

These few examples of the manipulations of the animal's environment demonstrate the importance of "managing the environment" for livestock production. The alternatives for modifications are large, but investments in facilities, machinery, and labor are required to bring them about. What are the options? If animals are left to their own resources in natural environments, they will place survival and comfort in the highest order of priority. In severe climates, the animal eats enough to get by, and becomes essentially an "escapist," like the rabbit mentioned earlier, resulting in poor performance by man's standards. A second option is to change the animal (see Chapter 18).

The third option is for man to learn how to manipulate the elements of the environment. Research results indicate this is the way to bring about the most rapid changes and attain maximum output of animal products at the lowest costs in all latitudes of the world.

FURTHER READINGS

Bianca, W. 1965. "Reviews in the progress of dairy science, sect. A, Physiology: Cattle in a hot environment." *J. Dairy Res.*, 32:291.

Bond, T. E. 1967. "Microclimate and livestock performance in hot climates." In Shaw, 1967.

Hafez, E. S. E. 1968. *Adaptation of Domestic Animals.* Philadelphia, Pa.: Lea and Febiger.

McDowell, R. E. 1972. *Improvement of Livestock in Warm Climates.* San Francisco, Ca.: W. H. Freeman and Co.

NRC. 1971a. *Animal Environmental Research and Facilities Guide.* Washington, D.C.: Nat'l. Acad. Sci.

Schmidt-Nielsen, K. 1964. *Desert Animals: Physiological Problems of Heat and Water.* New York: Oxford Univ. Press.

Williams, C. M. 1967. Livestock production in cold climates. *In* R. H. Shaw, ed. *Ground Level Climatology.* American Association for the Advancement of Science Publ. 86. Washington, D.C., The Association.

Thirty-Three

Instinctive Behavior and Learning

"Every animal likes its own ways."

William James

The purpose of this chapter is to acquaint the student of animal agriculture with some principles of animal behavior. It is very likely that some students, those who have grown up around farm animals and dealt with them in practical ways, have already accumulated substantial workaday knowledge about animal behavior. The content of this chapter may not add much detail to this fund of information, but at the least it is intended to provide a theoretical framework and set of general ideas that the student may use to organize what he already knows and to develop a keener appreciation of the scientific and applied value of behavior study.

Most of the basic principles which we will consider were not initially developed out of studies using farm animals. As we will see, our current understanding of animal behavior results from the relatively recent synthesis of two schools of thought, one interested in the evolution of behavior in natural environments, and the other interested in general laws of behavior change, particularly learning, which might exist in humans as well as in nonhuman animals. Zoologists have typically studied totally nondomesticated species, while psychologists have relied on species domesticated in the laboratory, with the laboratory rat as the prime example. Fortunately, however, there has been enough work done with common domesticated animals that we will be able to use them to illustrate some of the principles.

Early History of the Study of Animal Behavior

Observations and discussions of animal behavior may be found in the writings of the ancient Greeks, Aristotle in particular. Of course by that time, about 350 B.C., men had been taming and domesticating some animals (dogs, and sheep, for example) for more than 6,000 years. And for many thousands of years before that, human survival depended on knowledge about animal habits and habitats; before farming and domestication arose, man survived by hunting and gathering. But the behavioral information that men needed first to hunt and later to domesticate animals did not take on primary scientific significance for many years after the writings of Aristotle and of subsequent hunters, naturalists, explorers, and agriculturalists. It was not until 1859, and then again in 1871, that two books were published which provided the foundation for all subsequent development in the field. The author of both books was Charles Darwin, and the books were *The Origin of Species by Means of Natural Selection* (1859) and *The Descent of Man and Selection in Relation to Sex* (1871).

It is impossible to overemphasize the revolutionary impact of Darwin's ideas on all subsequent biology. The initial effect on behavioral science stemmed from Darwin's assertions about the essential continuity of mental evolution: "There can be no doubt that the difference between the mind of the lowest man and that of the highest animal is immense. . . . Nevertheless, the difference in mind between man and the higher animals, great as it is, certainly is one of degree and not of kind." (*Descent of Man,* Chapter IV.)

Darwin went to great lengths in bringing together all available evidence to bolster this sweeping claim. Virtually all this information consisted of the reports of hunters, travelers, and naturalists (anecdotal evidence) about the activities of animals in natural or artificial environments.

The initial enthusiasm of the anecdotalists waned when it became apparent that genuine understanding of human mentality and behavior in relation to that of nonhuman animals depended on the use of more rigorous scientific procedures. The claims made for the reasoning powers of some animals became so amazing that special commissions were set up to check on them. The best known of these involved a German horse known as "Clever Hans," who, according to his master, could read and do arithmetic. Careful investigations of Hans' behavior under controlled conditions convinced most scientists that his "answers," which he gave by tapping his forefoot on the ground a particular number of times, were terminated whenever the trainer, who stood in front of the horse, made a slight movement of his head. This movement was made unawares by the trainer, but, because both horse and man were rewarded by Hans' stopping at that point (the horse by food or affection and the trainer by another positive example that he was correct about the horse's ability), both maintained their behavior. In this way it was possible to explain Hans' behavior without reference to reading or arithmetical ability (Watson, 1914).

In 1894, the psychologist C. Lloyd Morgan summarized the growing skepticism about "higher mental powers" revealed in nonhuman animals. He said "In no case may we interpret an action as the outcome of the exercise of a higher psychical faculty, if it can be interpreted as the outcome of the exercise of one which stands lower in the psychological scale." This statement, which is commonly referred to as *Lloyd Morgan's Canon,* prescribed simply that for the study of animal behavior, as in all other science, the simplest explanation compatible with the facts is the best one.

Thus by 1900 the groundwork had been laid for the scientific study of animal behavior. During the next 40 years, two lines of theory and experiment developed without much interaction. One, developed primarily by European evolutionary zoologists, concentrated on the analysis of innate or instinctive behavior; this discipline is today called *ethology.* The other stressed the modifiability of behavior, and today is a part of *general experimental psychology.* Each of these fields made fundamental contributions to the theory, method, and data of behavioral science. We will describe the basic concepts of each of the fields in turn, then discuss some special problems that require integration of their separate contributions.

Modern Concepts of Instinct

Development (1900–1940)*

Three ideas coalesced at the beginning of the twentieth century to allow the founding of ethology. They were (1) evolution through natural selection, (2) the gene theory of inheritance, which surfaced at this time because of the rediscovery of Gregor Mendel's earlier work, and (3) the inheritance of behavior patterns as single units. This last idea, which had been originally studied by Darwin but was substantially extended by Whitman and Heinroth, stated that some elements of behavior were inherited in the same way as morphological characteristics, such as bone, muscle, feathers, or fur. Certain behavioral differences between species could be considered "innate" just like differences in size, coloration, and other anatomical characteristics. The significance of this was that a science of comparative behavior could be developed along the lines of comparative anatomy, and Whitman and Heinroth successfully pursued this approach in their studies of the evolution of birds.

Obviously, the behavior patterns that can be used as sure signs of species membership must be quite constant (stereotyped) in their form and must appear in all or most members of the species.† Just as obviously, not everything that animals do fulfills these criteria. In particular, as Wallace Craig pointed out in 1918, much of the behavior of the "higher" animals (by which we will mean birds and mammals) is quite variable from individual to individual, and modifiable within individuals from time to time; yet this relatively "plastic" behavior often places the animal in a position where a stereotyped behavior pattern can occur. Craig therefore made a distinction between *consummatory acts* and *appetitive behavior*. Consummatory acts are the relatively stereotyped patterns that are typical of the species and are involved in basic physiological and social processes: feeding, threatening, fighting, courting, mating, and caring for eggs and offspring. These acts are called consummatory because they complete the longer sequence of behavior which leads up to them. The initial behavior in this longer sequence may be quite variable and dependent on the details of the environment in which it occurs. It is called appetitive behavior because it puts the animal in the position of being able to perform the consummatory act. Consider, for example, a mouse-killing cat who watches you put a mouse inside a box and then put the box underneath a newspaper. In order to get at the mouse, the cat must either crawl under the newspaper or drag it aside, then claw at the box or tip it over or whatever is required to open it. A cat may not get at the mouse exactly the same way at different times, and different cats may use different strategies. However, once the mouse has been loosed, the behavior of the cats becomes more predictable or stereotyped, and different cats will tend to behave similarly. And finally, when the mouse has been killed and carried away, the cats will not return and go through the motions of capturing it again. The relatively stereotyped consummatory acts involved in capturing and killing the mouse complete the sequence begun with the highly variable appetitive behavior of freeing the mouse.

The Unit of Instinctive Behavior

One of the many contributions of Konrad Lorenz to ethology has been his careful analysis of consummatory responses.‡ In 1938 Lorenz and Tinbergen showed that consummatory responses are themselves complexes of two kinds of behavior. They demonstrated this

* Completer histories of ethology may be found in Eibl-Eibesfeldt (1970) and Hess (1962).

† With the exception that species-specific sexual behavior may occur only in the males of the species, for example, and not the females.

‡ Lorenz, who is best known for his semi-popular books *King Solomon's Ring* (1952) and *On Aggression* (1966), has had such an overwhelming impact on the development of ethology that he, like Niko Tinbergen, may be considered a father of the modern field.

principle in experiments with the grey lag goose.

Consider a goose who is sitting on a shallow nest containing a clutch of eggs. If for some reason an egg should roll out of the nest, the goose will retrieve it by stretching out her neck and rolling the egg back under her bill. The facts that the egg is egg-shaped and that the ground over which it must roll may be irregular mean that the goose will probably have to move her neck from side-to-side as well as back and forth. In a series of experiments Lorenz and Tinbergen were able to show that the side-to-side movements were very different from the back-and-forth movements. The details of the side-to-side movements depended quite exactly on the size of the egg and the terrain over which it had to be moved; the goose modified these movements as a consequence of the stimulation she got from the egg moving under her bill. But the back-and-forth movements were not controlled in this way. Once they had begun, they continued with the same frequency and amplitude regardless of what the egg was doing. For example, if the egg slipped completely from under her bill, the goose stopped her sideways movements but continued the back-and-forth movements all the way back to the nest. Moreover, as shown in Figure 33.1, if the goose is presented with an abnormally large egg, she does not modify her back-and-forth motion to accommodate it and hence gets the egg stuck between her neck and breast.

Figure 33-1 Grey lag goose retrieving normal and giant eggs. The fixed pattern of back-and-forth movements is maladaptive with the giant egg. (From Tinbergen, Clarendon Press, Oxford, 1969)

There are many other examples of the principle that consummatory acts contain two components, one which is responsive to feedback from the environment and one which is not (Eibl-Eibesfeldt, 1970; Tinbergen, 1951). A guided missile provides a useful analogy. Once the rocket is fired the engine burns until it runs out of fuel at a rate determined by its internal physical characteristics. But the missile's direction is controlled by its steering mechanism, which can be programmed to account for changes in local conditions.

When we consider which parts of an animal's total behavioral repertoire may legitimately be called "instinctive," we are on safest grounds if we restrict that label to those behaviors that, once released, run their course without further environmental influence. Ethologists call these behaviors *fixed patterns* (or *fixed-action patterns*). These patterns, combined with the more flexible *taxes* (e.g. the sideways movements of the goose's head), together constitute consummatory responses. An important point to remember when thinking about fixed patterns is that they are defined structurally, that is, in terms of the movements of the animal's body; they are not defined functionally, that is, in terms of what the movements accomplish in the environment. This is important for it allows us to say that the back-and-forth movements of the goose's head are instinctive, but *not* that "retrieving the egg" is instinctive. Keeping this distinction in mind makes it easier to come to grips in a meaningful fashion with the concepts of the physical inheritance and evolution of behavior.

The Nature of Social Stimulation

One of ethology's great achievements has been the elegant analysis of stimuli that produce consummatory responding. The principle that emerges from this analysis may be briefly put: *consummatory responses are usually released by a relatively small and quite specific segment of the total information in the environment.* In this brief form the principle perhaps appears more difficult than it is; some examples will clarify it.

Consider a hen tending her chicks. If one chick becomes separated from the rest, it will flap its wings and cheep in a characteristic fashion. The hen will retrieve the chick. The question is, what was it about the chick's behavior that stimulated the hen to retrieve it? From our human perspective we might be tempted to say that the hen was responding to the total situation; that is, "one of my chicks is unhappy." If this is what the hen is doing, then it should not matter whether the hen can see the chick but not hear it, or vice-versa. But, as Brückner showed some years ago, this matters very much. If a chick is restrained behind a barrier, so that the hen can hear its calls but not see it, then the hen reacts vigorously to retrieve the chick. But if the chick is enclosed in a bell jar, so that the hen can see it flapping its wings but not hear its call, then the hen goes on her way without paying further attention to the chick. The auditory stimulation is all-important in releasing the hen's retrieving behavior.

The best-known example of the particular nature of social stimulation involves a small fish called the three-spined stickleback (*Gasterosteus aculeatus*). During the annual breeding season, male sticklebacks develop brilliant coloration, including a bright red belly. At about the same time each male establishes a home base, called a *territory,* which he will defend against other male sticklebacks that might intrude. The question is, how does one male stickleback recognize another? What stimulus releases the stereotyped threat and fighting responses?

In a series of experiments involving wooden models that varied substantially in their resemblance to actual sticklebacks, Tinbergen showed that the key feature in releasing threat or chasing behavior was the sight of the red belly. Very crude models (Figure 33.2) painted red on their undersides released the full complement of aggressive behavior, but a very accurate model lacking only the red underside failed to release this behavior.

Many other examples of this principle have been reported. A special name has been given to those environmental events which are particularly effective in releasing consummatory behavior. They are called *sign stimuli* or *releasers.* The fact that animals can be so easily "fooled" when presented with artificial sign stimuli shows how fixed or "built-in" these

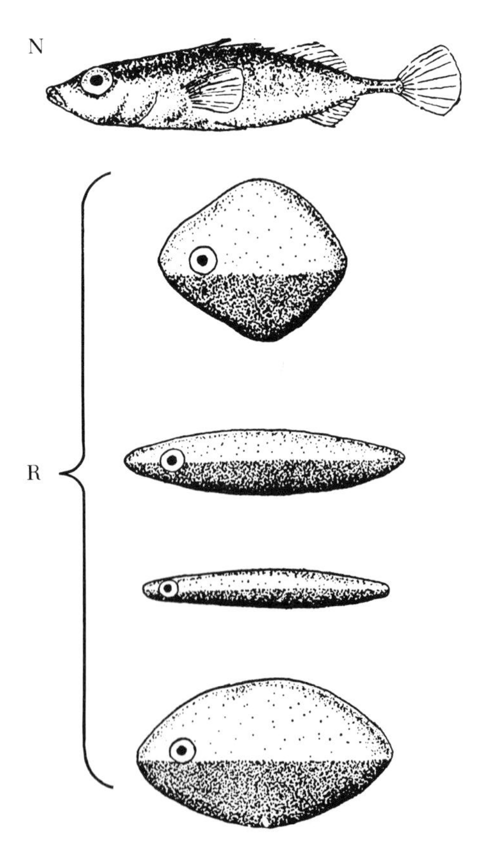

Figure 33-2 Crude dummies with red undersides release fighting in the male stickleback, but an accurate model without the red underside does not. (From Tinbergen, Clarendon Press, Oxford. 1969.)

stimulus-response connections are. Generally speaking, strongly fixed connections between releasers and consummatory behavior are more frequent and obvious in fishes and birds than they are in mammals.

The Innate Releasing Mechanism (IRM)

We have talked about the stimuli that animals take in and the responses that they produce. Quite naturally, we become interested in the mechanisms within the animal that transform the energy or information from the stimulus into the response. Ideally we should be able to specify where and how in the nervous system the relevant activity occurs.

Lorenz originally proposed the concept of the Innate Releasing Mechanism (IRM) to account for the relation between sign stimulus and fixed pattern. His argument was not based on known physiology, but was rather a model that specified the functions of the nervous system in mediating between stimulus and response. Basically, the IRM is a lock or filter which prevents all but the "correct" stimuli from energizing a fixed pattern. For each fixed pattern there is a separate IRM, which is coded or tuned to respond only when stimulated by a particular stimulus. This stimulus signals the presence of an event in the environment which is appropriately dealt with by the fixed pattern associated with the IRM. The "appropriateness" of the fixed pattern to that event has been established over generations by natural selection. Animals inherit their IRM's just as they inherit their skin, heart, and bones.*

The IRM is a functional concept; it would be a mistake to imagine that each IRM exists tucked away in its own part of the brain. For example, some IRMs exist as a result of the filtering characteristics of the sense organs (Eible-Eibesfeldt, 1970; Roeder, 1963). Nevertheless, it has sometimes been possible to localize small brain areas that appear to be critical for the mediation between stimulus and fixed action pattern. For example, Barfield (1969) has shown that capons will exhibit full copulatory behavior when very small pellets of male sex hormone are implanted directly into a particular area of the brain (Figure 33.3). Implants placed in another brain area (not shown in the figure) resulted not in copulation but in stereotyped aggressive behavior. In either case, the behavior was observed only when the capons were in the presence of the appropriate stimulus object: a crouching hen or a cock. Without the implanted hormone, the capons showed neither copulatory nor aggressive behavior. It is as if the IRMs involved in releasing the appropriate behaviors could function only in the presence of the hormone; to use a weak metaphor, the hormone was the fuel to run the IRM.

* Putting the matter just this way leaves open exactly what we are to mean when we say that an animal "inherits" any anatomical structure. Strictly speaking, of course, animals inherit only DNA and a little cytoplasm. When we use "inherit" in any other way, we are speaking loosely, but as long as we are aware and careful, no great harm is done.

The Evolution of Instinctive Behavior

There are two methods by which the evolution of behavior may be studied. The first is to compare the behavior of closely related species which are known to have branched off from their common ancestors at different times. This method was used profitably by the original ethologists, and it has been continued by Lorenz in his detailed studies of ducks and geese. Tinbergen and his students have applied similar methods to the study of gulls. For a number of reasons, this is a very difficult area of study. Hinde (1966) has presented a review of the relevant literature.

The second method is to speed up the course of evolution by bringing it into the laboratory. In this case, natural selection is replaced by "artificial selection," a technique well-known to geneticists and animal husbandmen. As applied to behavior, it consists of choosing the animals which best exemplify the trait of interest (speed in a race horse, aggressiveness in a watch dog, and so on) and breeding them. As a practical matter, the success of the procedure depends solely on the ability of the progeny to equal or surpass their

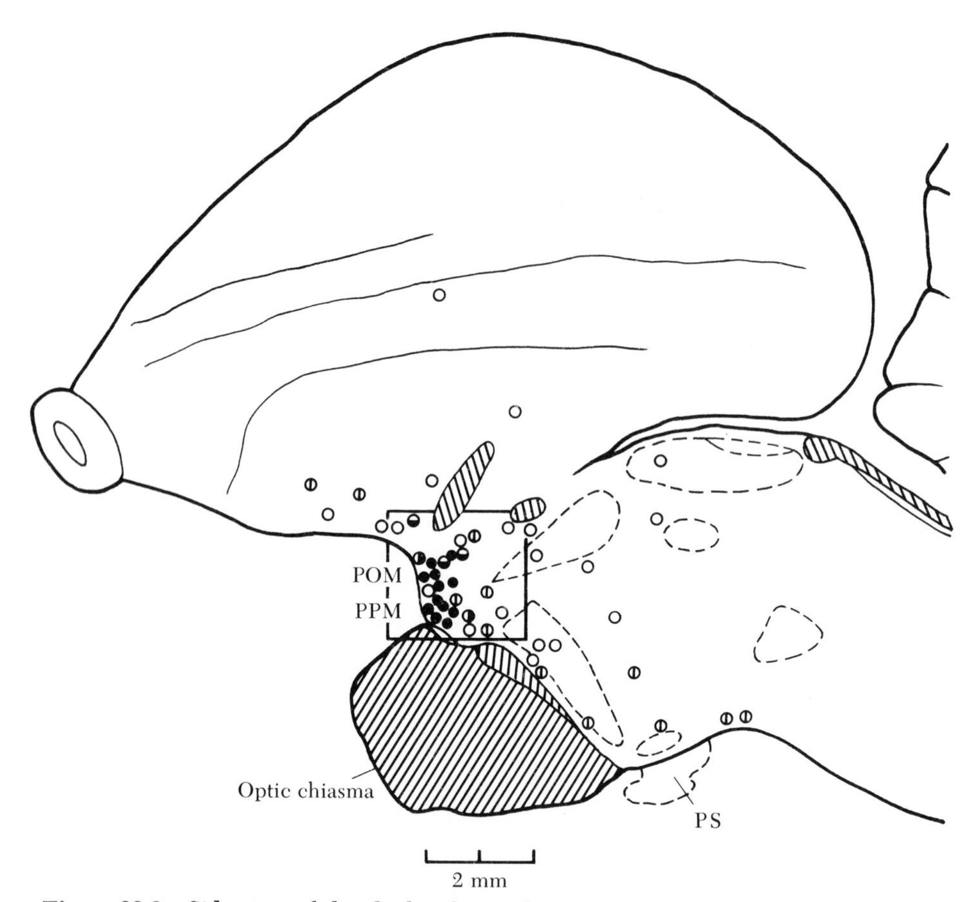

Figure 33-3 Side view of the chicken brain showing where small pellets of male sex hormone allow the release of sexual behavior in capons. Locations shown with filled circles produced full copulation; locations shown with partially filled circles produced mounting behavior only; locations shown with empty circles produced no sexual behavior. POM and PPM represent two sections of what is called the ***preoptic area.*** (From R. J. Barfield, 1969.)

parents in performance. As a scientific matter, great care must be taken before estimates of the behavior's heritability may be made. Full discussion of the relevant problems may be found in Hirsch (1967).

Theory of Learning and Conditioning

Classical Conditioning and Early Behaviorism

American psychology at the turn of the twentieth century was little concerned with the behavior of animals, but rather with trying to understand the contents of the conscious mind (*structural* psychology). This approach to psychology, which had been recently imported from Germany, did not last long in America. A number of scholars, John Dewey and William James in particular, were concerned to understand human behavior from the *functional* point of view, that is, to view human action in terms of its adaptiveness to the demands of the environment. Concern with the adaptiveness of behavior came straight from Darwin's theory of selection, and this, combined with the theory of the continuity of animal species, made animal behavior a sensible topic of interest for psychologists.

By 1920 a revolution had struck psychology,

and its name was *behaviorism.* Led by the forceful John B. Watson, behaviorism in its early radical form maintained that the fundamental principles of human behavior might very well be discovered by careful scrutiny of animal behavior, in particular by the study of animal reflexes. Watson's emphasis on the importance of reflexes was due in part to the success of Ivan Pavlov, the Nobel-Prize-winning Russian physiologist, in demonstrating that dogs could be trained to salivate at the sound of a bell. The importance of this fact lay in its implication for behavior generally: perhaps much of what we do can be traced back through a history or training (or *conditioning*) in which originally "neutral" stimuli gain control over responses by being closely and repetitively placed in time with the stimuli which naturally cause these responses. This is the way that *Pavlovian* or *classical* conditioning works, and Watson stressed its importance, without many data for support, in the control of human behavior (Figure 33.4). Today, psychologists are more cautious in stressing a single form of conditioning as paramount in human or other animal development; nevertheless, classical conditioning still occupies an important place in our theories of how animal and human behavior changes.

The Law of Effect and Instrumental Learning

Radical behaviorism soon gave way to a more sophisticated concern with how events in the

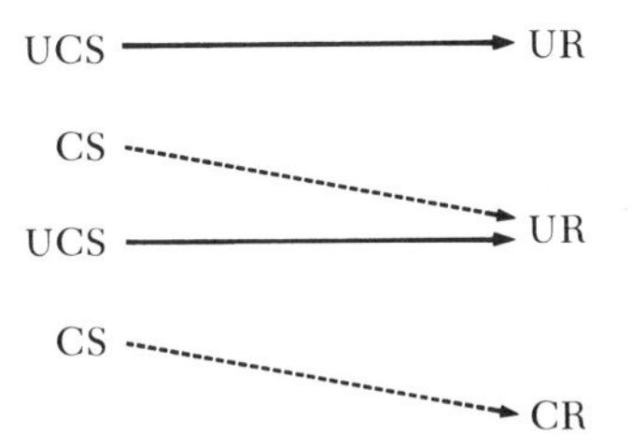

Figure 33-4 The simplest form of classical conditioning. An unconditional stimulus (UCS) naturally elicits an unconditioned response (UR). Repeated pairings of a conditional stimulus (CS) with the UCS lead eventually to the elicitation of a conditioned response (CR) following the CS alone.

environment come to control behavior. The primary principle involved is called the *Law of Effect,* first enunciated in 1911 by Edward L. Thorndike. In simplest terms, this "law" says that organisms will repeat responses that give them satisfaction and will discontinue responses that bring them discomfort. In one form or another, this idea has been central to much of the theory and experimental work in American animal psychology.

The Law of Effect stressed the importance of behavior's *consequences* in determining future behavior. Consider a simple example: if you put a hungry animal in a pen and hide some food under a blanket in one corner, the animal will, eventually, root under the blanket and eat the food. If now you remove the animal and wait 24 hours before replacing it in the pen, again hungry and with food under the blanket, the animal may get to the food more rapidly. And if you repeat this procedure day after day, the animal will soon run right to the blanket. You can measure the time from the animal's entrance to the pen to its getting the food (called the *latency* of response) and therewith have an index of its rate of learning in this situation. A typical learning curve is shown in Figure 33.5.

The important point about this situation and similar situations is that the changes in performance occur because of the animal's success in manipulating its own environment. The animal is instrumental in producing the situation that alters its behavior; e.g., it finds the food. Learning that proceeds in this way is therefore called *instrumental learning.* Another way of stating the Law of Effect is that animals will repeat those responses that operate successfully on the environment and will discontinue responses that do not. However, there usually is a lag between the time a response first becomes unsuccessful and the time the animal stops making the response. In the example above, if you stop putting the food under the blanket but do everything else just the same, the animal will continue to go to the blanket for quite a few days before it quits. The process of getting rid of an unsuccessful response is called *extinction.*

Psychologists use the words *reward* or *reinforcer* to label objects or events in the environment that animals will work to obtain and hence can be used to promote instrumental learning. As we shall see more clearly later, these labels can also legitimately be applied to some behaviors themselves.

Theories of Reinforcement

We described the Innate Releasing Mechanism as the ethologist's concept of the internal process that mediates between sign stimuli and fixed-action patterns. Experimental psychologists have also been concerned with internal mediating processes, but their concern has been with what must go on inside the body during instrumental learning. In particular, they have tried to understand the physiological process of reinforcement.

Initially it was believed that reinforcers worked by reducing physiological needs. For example, animals need food and water, and food and water are very effective reinforcers of instrumental learning. However, there is a problem with this theory, for a number of experiments have shown that reduction of basic "tissue needs" is neither necessary nor sufficient for reinforcement. For example, instrumental learning may be reinforced by allowing animals to ingest saccharin or to copulate, and neither saccharin nor copulation reduces any physiological need for food; animals definitely need a full vitamin complement to survive, but the presentation of vitamins as a potential reward to a deficient animal does not always foster instrumental learning.

Similar difficulties arose with the next major theory of reinforcement, which was called *drive reduction* or *drive-stimulus reduction.* This theory postulated that there were internal states of the animal which it would work to reduce, even though some of these were not related to basic physiological needs.* The generality of this principle was weakened by experimental results which showed that under some circumstances animals will work to *increase* the amount of stimulation they receive, even to the extent that the stimulation produces direct damage to their brains.

* Bolles (1967) provides excellent coverage of the ideas sketched over-simply here.

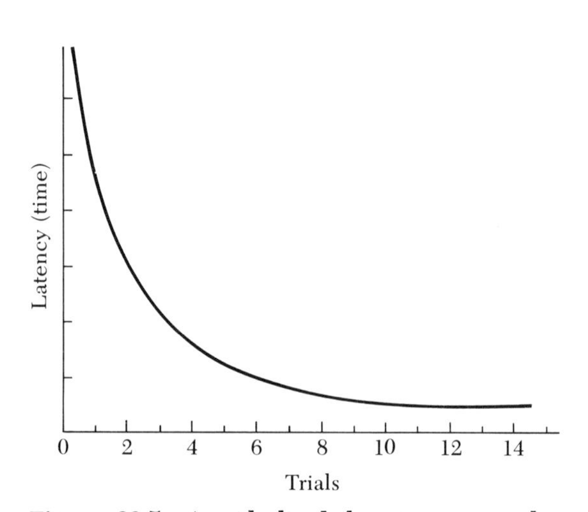

Figure 33-5 An idealized learning curve described by decreasing latencies of response after repeated trials. If the reinforcement were removed the latencies would increase; this process is called extinction.

At the present time, psychologists can offer no single principle that explains the necessary and sufficient conditions of reinforcement. But our understanding of the electrical and chemical properties of the brain is increasing rapidly. This growth of our detailed information will eventually allow us to make meaningful generalizations.

Operant Conditioning: An Engineering Approach to Behavior

Lack of a general theory of reinforcement has not prevented psychologists from learning a great deal about practical principles of behavioral control. Much of this work goes on under the name of *operant conditioning.* The leader of the operant-conditioning movement, B. F. Skinner, is probably the best-known psychologist living in America today. In the 1930's Skinner invented the apparatus which now bears his name, the "Skinner Box." Initially this was just a little box with a lever sticking into it on one side with a food cup beside the lever and a food hopper outside the box connected to the cup by a tube. The apparatus was rigged so that when an animal (a rat) pressed the lever a pellet of food dropped into the cup. Hungry animals soon learned to press the lever and continued to press it at a steady rate, eating after each pellet delivery. It was then possible to change the "rules of the game" so that, for example, the animal would have to press twice for a pellet, or five times, or whatever. Or a clock could be put into the system so that the lever press would produce a pellet only once a minute, or once a minute on the average, and so on. In other words this arrangement of lever and food dispenser allowed the experimenter to manipulate the rules by which food was delivered, and then to see how the animal behaved under those rules. Each rule is called a *schedule of reinforcement* (Ferster and Skinner, 1957).

At the present time the use of these basic ideas is so accepted and widespread in psychology that a "Skinner Box" is now an idea rather than a specific apparatus. The idea is to devise ways to measure the rate of a simple response (pressing a lever, swimming through a hoop, running through a light beam, etc.) as it varies with the presentation of a known reinforcer. The idea is very general and can be applied in many practical ways. During World War II, for example, Skinner trained pigeons to live in the nose cones of torpedo bombs and to guide the flight of the bombs by pecking on keys that controlled the bomb's rudder. Pigeons have also been trained to serve as quality-control inspectors on production lines (Verhave, 1970), and one man has made a successful career using operant-conditioning techniques in training animals for television and other entertainment purposes (Breland and Breland, 1966).

Interactions of Instinctive and Acquired Behavior

Imprinting

Very young precocial birds, such as chicks, ducklings, or goslings, follow their mother very closely as she leads them to food, water, or the nest. One might imagine that the sight or sound of the mother is a sign stimulus which releases some components of the "close following" behavior. But this is not true. A number of experimenters have found that these young birds will follow the first moving object they encounter after hatching. This might be a bird of another species, a human being, or a small ball attached by a string to a slowly rotating arm. And, having once followed this object, the young birds become less likely to follow other objects, even if it should be their mother. The process by which the bird becomes "attached" to the moving object is called *imprinting*. In one form or another, imprinting plays a large role in the behavioral development of a number of animal species (Sluckin, 1965).

Imprinting is a form of learning, but at present we do not know how to relate it precisely to other well-known forms of behavior modification, such as classical conditioning or instrumental learning. There is a question about exactly what it is that the animal has learned. In some species at least, the initial following behavior is only one of the consequences of imprinting, for in adulthood the animals direct their courtship and sexual responses to objects resembling the one they imprinted on. It is as if the imprinted object becomes the sign stimulus for the release of these responses. This feature of imprinting has sometimes created difficulties for zookeepers who have raised orphaned animals by hand. When these animals became adult, they tried to court and mate with their keeper and ignored appropriate sexual partners (Hediger, 1964).

A process similar to imprinting can also occur in adult animals, particularly females at the time of birth. Female sheep and goats, for example, become "imprinted" on their kids very shortly after parturition, with the result that they will later normally accept only their own kids and reject foster kids. Experimental evidence suggests that the mothers identify their young through the sense of smell (Klopfer and Klopfer, 1968).

Instinct-Learning Interlockings

It is useful to consider animal behavior as a chain with links composed of either instinctive or acquired components. Ethological analysis has shown that the instinctive components, the fixed-action patterns, comprise a smaller portion of the chain than we might have imagined. Moreover, they occur when released by the appropriate sign stimulus. Psychological analysis has shown that a good deal of learned behavior (or appetitive behavior) is maintained by reinforcement. In order to be most effective, a reinforcer must be available just after the learned response has been made. When we put these two sets of principles together, we see immediately that fixed-action patterns might serve as reinforcers for the appetitive behaviors that precede them. This idea has been tested experimentally and found to be true in at least some cases.

It is very useful in these experiments to work with fixed patterns which do not themselves produce consequences that could serve as a reinforcer. For example, several investigators have studied the reinforcing value of the "threat display" fixed pattern of the Siamese fighting fish, *Betta splendens.* This response, which consists of extending the membranes covering the gills and adopting a head-on posture to the other animal, is released in a male *Betta* by the sight of another male *Betta;* unless one of the fish swims away following the threat, there will be a serious fight. The question of interest is whether a male *Betta* will learn an unusual instrumental task like swimming through a hoop or down a runway, if completing the task gives him the opportunity to see another *Betta* and perform the threat display. These experiments have been done with positive results (Hogan, Kleist, and Hutchings, 1970); performing the fixed pattern was an effective reinforcer of the instrumental task.

Recent experiments on the behavioral effects of electrical and chemical stimulation of certain brain areas have suggested that there may be substantial overlap in anatomical areas that mediate between sign stimuli and fixed patterns, i.e., IRMs, and the areas that are directly involved in the reinforcement process. Further research along these lines will deepen our physiological understanding of how chains of instinctive and acquired behavior are put together (Glickman and Schiff, 1967).

The Selective Association Principle

Earlier we pointed out that no single theory of reinforcement yet proposed can account for the facts of learning as we know them today. The synthesis of ethology with psychology has produced an important insight into why the earlier theories were inadequate and how a better theory can be developed. The basic idea is simple enough: *as a result of evolutionary history, an animal will learn some things more easily than others.* Students coming on this idea afresh might wonder why it has taken behavioral scientists such a long time to arrive at it. In fact, the idea has been around at least since 1910 or so, but it received only scant attention until recently, primarily because psychologists were trying to find completely general laws of learning and reinforcement, but were basing their experiments on only a few animal species and techniques.

The significance of this "new" idea, which we may call the *Selective Association Principle,* may be seen by considering some of the data which forced it into prominence. Breland and Breland (1961, 1966) report examples of partial failures of operant-conditioning techniques when these were applied to animals not usually found in the operant-conditioning laboratory. These animals, including chickens, raccoons, and pigs, simply did not behave as the "textbook" of operant conditioning said they ought to. Instead of repeating the response which they had been taught to make by operant training, the animals reverted to stereotyped and species-specific responses which were unrelated to the occurrence of reinforcement. The strong natural behavioral tendencies of the animals in these circumstances could not be overcome by the operant-conditioning procedures. Breland and Breland (1961) spoke of this tendency as *instinctive drift*.

Seligman (1970) has reviewed some of the data that bear on this problem and has suggested that we consider animals as (1) prepared, (2) unprepared, or (3) contraprepared for the acquisition or maintenance of a particular learned response. The degree of preparedness for a particular response will vary among species and a combined ethological-physiological approach will be required to understand how the Principle of Selective Association works in particular cases.

We can see this principle operating at a sub-species level in modern breeds of domestic dogs. The most complete study to date of behavioral differences among dogs has been performed by Scott and Fuller (1965) at the Jackson Laboratories in Bar Harbor, Maine. They studied five pure breeds (Basenji, Beagle, Cocker Spaniel, Shetland Sheep Dog, Wire-Haired Fox Terrier) and the F_1, F_2, and backcrossed hybrids of Basenjis and Cockers. Their experiments with adult dogs centered around differences in emotional reactivity, trainability, and problem-solving ability. Results of experiments on reactivity showed a clear difference between the relatively placid Cockers and Shelties and the more excitable Wire-Hairs, Beagles, and Basenjis. Regarding trainability, Scott and Fuller found that over-all generalizations were difficult, because the different breeds varied in response to the different kinds of training tests; no one breed was consistently superior on all tests. Cockers were the most accepting of restraint, a behavioral feature that is related to their history of selection for crouching. This is an example of selective association. Finally, results of problem-solving experiments suggested that Shelties were significantly less successful than the other breeds in over-all speed and accuracy. However, Fuller and Scott emphasize that the Sheltie's performance did not indicate lower over-all intelligence, but rather a mismatch between the breed's natural tendencies and the particular tests employed. Putting the matter this way emphasizes the important point that a single dimension of "intelligence" may not be a very fruitful yardstick to apply to breeds (or species) that have been naturally or artificially selected for special features or capacities.

Implications for Animal Agriculture

Behavioral Characteristics of Domesticated Animals

It should come as no surprise that the behavioral range of domesticated animals is relatively narrow when compared to the rest

Table 33.1.
Behavioral characteristics adapting species to domestication.[a]

I. Group structure
  a. Large social group
  b. Hierarchical group structure
  c. Males affiliated with female group
II. Sexual behavior
  a. Promiscuous matings
  b. Males dominant over females
  c. Sexual signals provided by movements or posture
III. Parent-young interactions
  a. Critical period in development of species-bond (imprinting, etc.)
  b. Female accepts other young soon after parturition or hatching
  c. Precocial development of young
IV. Responses to man
  a. Short flight distance to man
  b. Least disturbed by human ubiquity and extraneous activities
V. Other behavioral characteristics
  a. Catholic dietary habits (including scavengers)
  b. Adaptable to a wide range of environmental conditions
  c. Limited agility

[a] From Hale, 1969.

of the animal kingdom. The origins of domestication were based on a natural compatibility between human needs and the behavior of certain animals. As domestication continued, the fit became closer.

Hale (1962) has listed 14 behavioral traits which characterize a species adaptable to domestication (Table 33.1). Not surprisingly, these traits describe our current domesticated animals rather well. Future attempts to modify behavior, through selective breeding, training, or other procedures, will have to take account of the behavioral baselines that have already evolved. One clear example of how this has been done successfully is the following.

Maximizing Behavioral Potential: An Example in Sexual Behavior

Domesticated animals are generally sexually promiscuous; the formation of monogamous pair-bonds is a rare event. This characteristic is related to the fact that domesticated species seem naturally to arrange themselves in social groupings in which one male tends or guards a number of females. There is now considerable evidence to show that the number of females a bull or ram will inseminate in a given period of time can be greatly increased by some simple technical procedures. The trick is to limit the number of ejaculations with each female to one and to present a number of females to the male in serial order. Under these conditions, the ram's ejaculatory performance shows no decrement over as many as 12 ewes (Beamer, Bermant, and Clegg, 1969). A similar procedure has been used with dairy bulls in combination with an artificial vagina for semen collection. Changing the identity of the teaser female increases the amount of semen collectable in a brief period (Schein and Hale, 1965).

Conclusions

The maximization of behavioral potential by augmenting natural tendencies is one practical application of behavioral science to animal husbandry. A second application is the careful analysis of behavior in a selective breeding program to ensure that disadvantageous traits do not emerge unnecessarily. Third, simple operant conditioning may be utilized to monitor and control food and/or water intake. Although this might be impractical on a large scale, it could be used successfully in pilot projects that would determine optimal feeding or watering schedules for larger groups. Finally, sensitivity to individual or species-specific behavioral tendencies may prevent the use of practices in husbandry which decrease the commercial value of the animals by acting adversely on their food-intake or reproductive capacity. Systematic study and application of the principles presented here ought to pay off in practical terms.

FURTHER READINGS

Bolles, R. 1967. *Theory of Motivation.* New York: Harper and Row.

Eibl-Eibesfeldt, I. 1970. *Ethology: The biology of behavior.* New York: Holt, Rinehart, and Winston.

Ferster, C., and B. F. Skinner. 1957. *Schedules of Reinforcement.* New York: Appleton-Century-Crofts.

Glickman, S., and B. Schiff. 1967. "A biological theory of reinforcement." *Psychol. Rev.*, 74:81–109.

Hale, E. 1969. "Domestication and the evolution of behaviour." In Hafez, 1969.

Hess, E. H. 1962. "Ethology: An approach toward the complete analysis of behavior." In *New Directions in Psychology.* New York: Holt, Rinehart, and Winston. Pp. 157–266.

Hinde, R. 1966. *Animal Behavior*. New York: McGraw-Hill.
Hirsch, J. 1967. *Behavior: Genetic Analysis*. New York: McGraw-Hill.
Lorenz, K. 1966. *On Aggression*. New York: Harcourt, Brace, and World.
Morgan, C. L. 1894. *An Introduction to Comparative Psychology*. London.
Tinbergen, N. 1969. *The Study of Instinct*. London: Oxford Univ. Press.

Section Six

Animal Nutrition

The "Hogamatic" hog-finishing system, in which floor cleaning and feed supply is carried out by automatic machinery. (USDA photo.)

Thirty-Four

General Nutritional Considerations

"The life of animals, then, may be divided into two acts, procreation and feeding; for on these two acts all their interests and life concentrate."

Aristotle
History of Animals, Book VII

Aristotle was correct in asserting that procreation and feeding are the main preoccupations of living animals. The nutritionist may go further in suggesting that of these two preoccupations, feeding is of overriding importance, for without the urge and ability to seek and consume their food, animals would not reach sexual maturity and could not reproduce themselves. Indeed, in some species one of the first signs of an inadequate plane of nutrition is a failure in reproduction.

One of the most fascinating aspects of animal nutrition is the infinite variety of animal types and the wide range of foods they will voluntarily consume. Consider, for example, the quantitative and qualitative differences in diets of a day-old chick requiring a highly nutritious diet constantly available, and that of the Merino sheep grazing the sparsely vegetated salt brush country of arid Australia. A brief glance at Table 34.1 will dramatize the wide variety in both the resources available and some of the animals that use them.

Animal nutrition assumes a special significance when the domesticated farm animal is considered, for here the objectives of production are to maximize growth rates in meat-producing animals, to increase reproductive efficiency, including egg production in poultry, to promote high levels of milk production in females rearing young and in dairy animals, and to sustain high levels of wool growth in sheep. In order to achieve these objectives economically, the right nutrients in adequate amounts must be supplied to the animal under proper feeding regimes.

Superficially, it seems that an infinite number of dietary permutations are possible when the available food resources and the variety of animals which will use them are taken into consideration. However, in spite of the wide range of diets to which animals have become adapted, they seek to derive from their food relatively few common elements and chemical compounds which are now identified as the nutrients. The objectives of much research in the field of nutrition have been: first, to identify what constitutes a nutrient; second, to establish by analytical techniques the contribution each of the available feeds makes in terms of the known nutrients; third, to follow the fate of the nutrients in the body and measure the biological efficiency of the conversion of nutrients into animal products; and, finally, with this knowledge to be able to formulate diets which will meet the specific nutrient requirements of various animals for their various physiological activities. This chapter aims primarily to introduce the reader to the nutrients, and to their classifications and functions.

Nutrient Compounds and Elements

Although the predominant importance of food for survival and procreation has always been assumed, identification of the specific elements and compounds in the food which comprise the nutrients has only been made in relatively recent times. The nutrients fall easily into a six-fold classification, based on their chemical structures and their function in the body. These are carbohydrates, fats, proteins, vitamins, minerals, and water. All foods can be classified according to their content of these six types of nutrients. Table 34.2 illustrates the extremely wide variations in the composition of animal foods in their content of these nutrients. Notice, for example, the wide variation in water content, the very low carbohydrate content of animal products, and the wide variation within foods of both plant and animal origin in their fat content. Vitamins occur only in trace amounts in all foods; their critical importance rests on their physiological function in the body as catalysts to biochemical reactions at the subcellular level, rather than on their presence in large quantities.

Although water is quantitatively and qualitatively the most essential nutrient of the body, it will not be discussed as a nutrient here. It is important as a solvent for the chemical reactions in the body, for thermoregulation, and for excretion, and in many other ways more appropriately discussed in the context of the physiological mechanisms.

Table 34.1.

Classification of animals by their diet	*Examples (with weight in kg)*	*Some available food resources*
Herbivores	Elephant (5000) Sheep (75)	Dry native shrubs and grasses of the arid and semiarid zones
	Field mouse (.01)	Intensively fertilized pasture of the cool temperate zones Wide variety of grains
Carnivores	Large cat (lion) (150)	Game animals
	Domestic dog (20)	Slaughterhouse residues
	Weasle (0.5)	Small fauna species
Insectivores	Spiny ant eater (40)	Terrestrial insects
	Bird (1)	Flying insects
	Fish (2)	Aquatic insects
Omnivores	Pig (150) Man (75) Rat (.3)	Variable diet of animal and plant origin

Carbohydrates

The name "carbohydrate" was derived from the French "hydrate de carbon," a term originally applied to neutral chemical compounds containing carbon, hydrogen, and oxygen, with the last two elements in the same proportions as in water. Glucose, for example, has the empirical formula of $C_6H_{12}O_6$ or $(CH_2O)_6$. This definition is no longer precise, because other compounds which are not true hydrates of carbon are also included in the classification.

Table 34.2.
Approximate percentage nutrient composition of some plant and animal foods.

	Water	*Carbo-hydrates*	*Fat*	*Pro-tein*	*Min-eral*
Green grass	82.0	11.0	1.0	3.5	2.5
Dry grass	13.0	66.0	3.0	10.0	8.0
Cereal grains	13.0	72.0	2.0	11.5	1.5
Peanuts	6.0	20.0	45.0	27.0	2.0
Blood	82.0	Trace	0.5	16.5	1.0
Liver	74.0	1.5	6.5	16.5	1.5
Muscle	72.0	0.5	4.5	21.5	1.5
Milk	87.5	4.5	3.5	3.5	1.0
Pig (whole)	50.0	Trace	35.0	12.5	2.5
Cow (whole)	56.0	Trace	23.0	17.0	4.0

The carbohydrates of nutritional importance may be further classified as follows:

Monosaccharides (simple sugars); examples are glucose, fructose, and galactose.

Disaccharides (two simple sugars condensed); examples are sucrose (glucose and fructose), lactose (glucose and galactose), and maltose (glucose and glucose).

Polysaccharides (long-chain or branched molecules of several disaccharide subunits); examples are cellulose (yields the disaccharide cellobiose on hydrolysis), starch (yields the disaccharide maltose), and amylopectin (yields the disaccharide maltose, and then the monosaccharide glucose).

The structures of such compounds are illustrated in Figure 34.1. There are many other carbohydrates, not shown in this classification, which are perhaps not so important from a nutritional point of view.

Lipids

The lipids are a group of substances which are insoluble in water but quite soluble in common organic solvents. Like the carbohydrates they include a wide range of compounds, of which only fats and oils are of nutritional significance. The five main types can be grouped as follows: fats and oils, plus waxes, are the simple lipids; phospholipids, glycolipids, and lipoproteins are the compound lipids. The fats and oils are important constituents of both plants and animals, and although similar in chemical and structural properties, their physical properties differ, particularly in the melting point.

A fat is a product of the esterification of a glycerol molecule by three fatty acids:

$$\begin{array}{l} CH_2OH \\ | \\ CHOH \quad + 3\,R.COOH \longrightarrow \\ | \\ CH_2OH \\ \text{Glycerol} \qquad \text{Fatty acids} \end{array}$$

$$\begin{array}{l} CH_2.O.CO.R_1 \\ | \\ CH.O.CO.R_2 \qquad + 3H_2O \\ | \\ CH_2.O.CO.R_3 \\ \text{Fat (mixed triglycerides)} \end{array}$$

There are several types of fatty acids. If the same type of fatty acid molecule is attached at each of the three positions on the glycerol molecule, the fat is a simple triglyceride. When more than one type of fatty acid is concerned, then a mixed triglyceride results. Most of the fats and oils that occur naturally are mixtures of different triglyceride forms. R_1, R_2 and R_3 in the above examples represent different fatty acids. There are several fatty acids which oc-

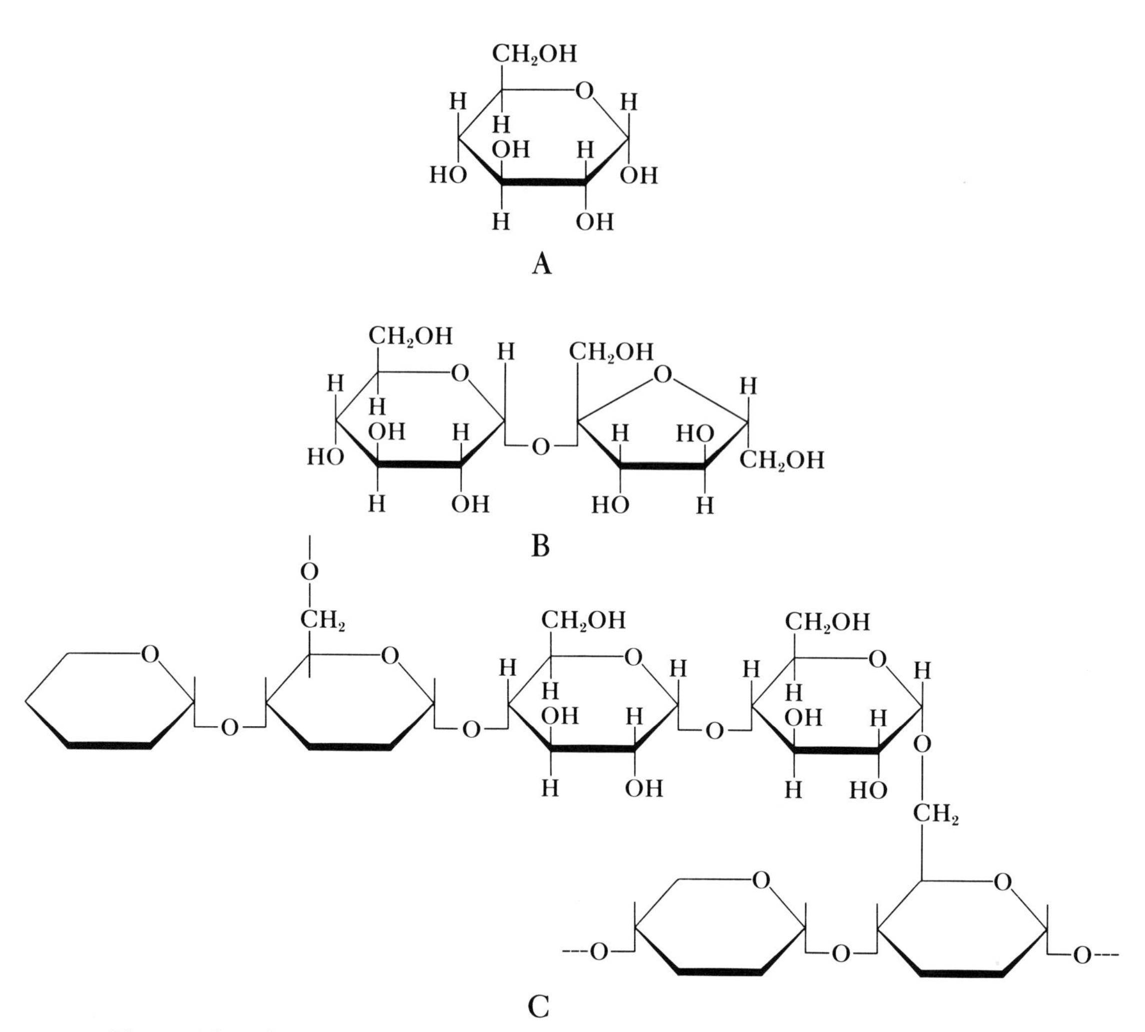

Figure 34-1 Structures of selected carbohydrates. A, glucose, a simple sugar; B, sucrose, a disaccharide; C, amylopectin, a polysaccharide.

cur mostly as long, straight carbon chains, and the features that distinguish them are: (a) the length of the chain (i.e., the number of carbon atoms); (b) whether they are saturated or unsaturated; and (c) if unsaturated, the degree of unsaturation.

The term saturation refers to the lack of double bonds between the carbon atoms. Stearic acid, for example, is completely saturated; it possesses no double bonds, and may be written as:

$$(C_{18}H_{36}O_2) \text{ or } CH_3—CH_2—CH_2—$$
$$—(CH_2—CH_2)_6—CH_2—CH_2—COOH$$

Oleic acid has the same number of carbon atoms, but has one double bond; so it may be written as:

$$(C_{18}H_{36}O_2) \text{ or } CH_3—CH_2—(CH_2)_6—CH=$$
$$=CH—(CH_2)_6—CH_2—COOH$$

There are many fatty acids with a variety of permutations in chain length and saturation.

Proteins

In common with carbohydrates and fats, the proteins contain carbon, hydrogen, and oxygen, but in addition they contain nitrogen and sometimes sulphur. Proteins are highly complex molecules, which, like the carbohydrates and fats, are built up of smaller subunits. The smallest subunits, the building blocks of the proteins, are the amino acids, so-called because they are characterized by having at least one basic amino group (—NH_2) and an acidic carboxyl unit (—COOH) in the molecule. Schematically the amino acid can be represented as:

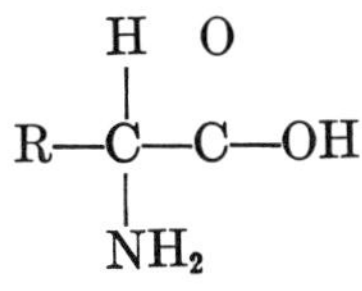

The rest of the molecule (R) may be a chain of varying length or may include ring structures. Although more than one hundred amino acids have been identified, only 25 or so are constituents of proteins. The amino acids can be linked together in long chains in all possible sequences, like railroad cars forming a long train. The linkages between the amino acids, as with a train, are similar between each of the amino acids, and are known as peptide bonds. Two amino acids linked by a peptide bond form a dipeptide (analogous to two sugars or saccharides forming a disaccharide in the carbohydrate classification). Several dipeptides linked together form a polypeptide chain or protein, which may contain several hundreds or thousands of amino acids. The polypeptide or protein may be folded on itself and secondary and tertiary bonds, known as hydrogen bonds (H) and sulphur linkage (S), are formed which hold the large molecule in a fixed spatial arrangement. Notice in Figure 34.2 that only some of the amino acids have sulphur (S) in them; in the folded polypeptide chain, it is when two sulphur-containing amino acids are adjacent to each other that the strong sulphur bond is formed. The hydrogen bonds are more commonly formed between several of the amino acids, and are weaker bonds.

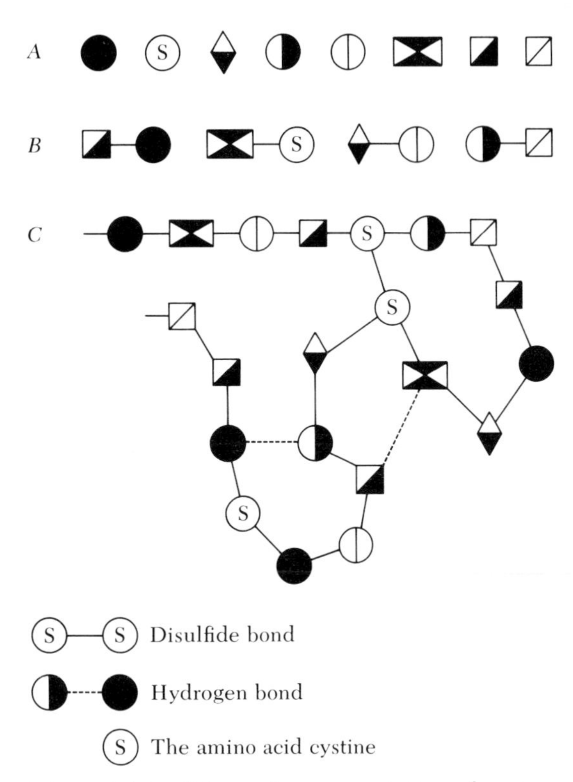

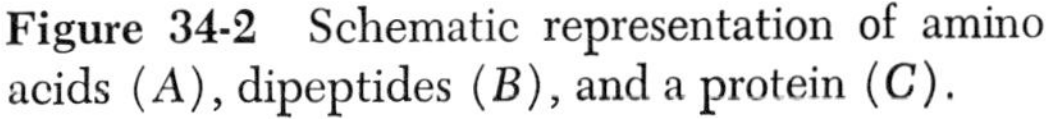

Figure 34-2 Schematic representation of amino acids (*A*), dipeptides (*B*), and a protein (*C*).

Vitamins

A vitamin may be roughly defined as an organic compound required in the diet in minute amounts for normal growth and maintenance of animal life. The vitamins are chemically extremely complex and are often large molecules, but are effective in small amounts, functioning in the transformation of energy, in the regulation of metabolic processes, and in the biosynthesis of diverse compounds in the tissues of the body. However, vitamins are not used as structural units in the body. A list of the known vitamins appears in Table 34.3.

Minerals

In contrast to the vitamins, which are complex molecules, the mineral nutrients are elemental. They are also effective in small amounts, some in minute traces. They constitute approximately 5 per cent of the animal body dry weight. This fraction is concentrated primarily in the bones and teeth, which contain 99 per cent of the body's calcium and 80 per cent of its phosphorus. Other mineral elements present in the body in relatively large amounts are sodium, potassium, magnesium, sulphur, and chlorine. These elements, along with calcium and phosphorus, are often called the major- or macro-mineral elements of nutrition, as contrasted to a second group called micronutrient elements, which are needed only in trace amounts and are present in extremely small amounts in animal tissues. This group includes copper, iron, manganese, zinc and cobalt, iodine, molybdenum, and selenium. Trace amounts are expressed frequently in parts per million (ppm) or in milligrams per pound of ration, whereas calcium, phosphorus, and sodium requirements are expressed as a percentage of the ration.

All feeds contain a wide array of mineral elements, but the levels vary with such factors as forages vs. cereal grains; legumes vs. nonlegumes; and soil-mineral interrelationships. Minerals may be lacking in a soil or rendered unavailable to the plants by factors reducing the solubility, such as the alkalinity of the soil or the presence of mineral-binding complexes. Perhaps in no area of nutrition is there a closer soil-plant-animal relationship than in that of minerals.

Table 34.3.
The known vitamins.

Water-soluble vitamins	*Fat-soluble vitamins*	*Other vitamins*
Thiamine (vitamin B_1)	Vitamin A	Ascorbic acid
Riboflavin (vitamin B_2)	Vitamin D	Inositol
Niacin	Vitamin E	Choline
Vitamin B_6	Vitamin K	
Pantothenic acid		
Biotin		
Folic acid		
Cobalamine (vitamin B_{12})		

Functions of the Nutrients

Life as a Combustion Process

It is a practical impossibility to document the evolution of ideas and discoveries of the nutrients and their functions in a few paragraphs and do justice to a fascinating story, which has been superbly presented at length by McCollum (1957).

Early investigators were concerned to a large extent with whole-animal metabolism, and they were very conscious of the fact that the gaseous exchange involved in respiration had something to do with the ability of the animal to make use of its food. Many became keenly interested in studying the gases themselves. John Mayow, the son of a genteel Cornish family in England, who in 1689 had discovered the formation of an oxide by heating antimony in air with a heated glass rod became one of the first scientists to carry out an animal experiment. He repeated to some extent an experiment carried out by Robert Boyle in 1660, by simply placing an animal and a lighted candle in a closed jar, and noting that the candle went out first, then the animal died. If there was no lighted candle, the animal lived longer. It was not until 1757 that

Black, an English doctor who published his inaugural dissertation at Edinburgh in Latin, discovered CO_2; and not until 1774 that Priestly discovered his "fire air," oxygen. He made the following observations in relation to his newly discovered "fire air," that in a closed space: (1) a flame makes air unfit for a flame; (2) a mouse makes air unfit for a mouse; (3) a flame makes air unfit for a mouse, and vice versa; (4) a flame goes out at about the same time that a mouse dies; (5) the inclusion of a plant in such a confined space improves the air for a mouse (the first demonstration of photosynthesis). Scheele, a Swede, had done the same experiment with bees; moreover, he had produced extra pure "fire air" by heating silver carbonate and absorbing the CO_2 in calcium hydroxide solution. Both Scheele and Priestly were in correspondence with a French aristocrat named Lavoisier who actually gave the name "oxcygene" to the pure "fire air" of Priestly and Scheele. Lavoisier was the first scientist to state that both fire and animals consume oxygen and combine it with organic substances to give off CO_2 and H_2O—a fundamentally valid concept for present-day metabolism. In 1780, Lavoisier and his colleague Laplace expressed a fundamental law of thermochemistry, by stating that the same amount of heat produced by the combustion of glucose to CO_2 and H_2O would be required for the reversal of the process, and noted that the levels of heat produced by animals originated from the combustion of oxygen with organic substances. Lavoisier's entire career was marked with a series of brilliant discoveries in which he elucidated the nature of respiration and oxidation. He was the first scientist to use the thermometer and balancc in metabolic studies. Outside his laboratory, Lavoisier was a public tax gatherer and landed farmer noted for his productive innovations, but he was caught up in the French Revolution and was executed in 1794 at the age of 50. It is said that his close friend Lagrange, who witnessed the execution, said of him, "It took but an instant to cut off his head; yet a hundred years will not suffice to produce one like it."

Meanwhile, in England, Crawford subjected the ideas of Priestly and Lavoisier to animal experimentation by constructing in 1788 the first animal calorimeter for measuring the heat of combustion of hydrogen in oxygen. In 1840, Hess formulated the "law of constant heat sums," which in 1860 Claude Bernard assumed and which in the early 1890's Rubner and Laulanie confirmed, applied to animal metabolism. In the field of bioenergetics, it remained only for more sophistication in the experimental calorimeters to develop before Atwater and Benedict in 1903 were able with confidence to state that the law of conservation of energy applies to living animals as well as physical processes. Advances in bio-energetics since the turn of the century have involved elucidation of details of the individual steps in metabolism by which the animal achieves the oxidation of the organic compounds in its food with strict adherence to the physical laws which were much earlier shown to hold. The result of these oxidations is to make energy available.

Energy

Energy is not a nutrient but rather a property which some of the nutrients possess. In nutrition it is convenient to describe energy in heat units or kilo-calories (kcal), although the chief mechanism for energy storage in the body is through the generation of the high-energy phosphate bonds of ATP (adenosine triphosphate). Reference has already been made in the sections on carbohydrates, fats, and proteins to the fact that these nutrients are sources of energy to the body. Although vitamins and minerals are not sources of energy, their functions are often associated with energy metabolism, in that the minerals are key agents in prompting the activity of vitamins, which in turn assist the enzymes responsible for energy transfer from the substrates (nutrients) into a form the body can use.

To properly understand energy metabolism, a slightly different perspective on the nutrients must be taken. This involves a consideration, not so much of their specific molecular structure, as of their potential energy yield when completely oxidized.

Energy enters the food chain through the well-known reactions of photosynthesis, whereby atmospheric carbon dioxide and water are combined to form carbohydrate with solar energy driving the reaction:

$$\text{Energy} + 6CO_2 + 6H_2O \longrightarrow C_6H_{12}O_6$$

When carbohydrates are metabolized in the body, this reaction is precisely reversed:

$$C_6H_{12}O_6 \longrightarrow \text{Energy} + 6CO_2 + 6H_2O$$

$$1 \text{ gm carbohydrate} \longrightarrow 4.2 \text{ kcal heat}$$

One gram of carbohydrate completely oxidized (burned) will yield 4.2 kcal of heat energy. In a similar way, one gram of fat and protein when completely oxidized will yield 9.3 and 5.8 kcal, respectively. The value of 5.8 kcal of heat from 1 gm protein refers to the yield of energy derived if the protein is completely oxidized to CO_2, H_2O, and NO_2 in a bomb calorimeter under laboratory conditions. In the body, protein is never completely oxidized, because nitrogen is excreted in the form of urea and ammonia in the urine. Under physiological conditions therefore, 1 gm protein yields 4.2 kcal, the same as carbohydrates.

The partition of dietary energy The bomb calorimeter is an instrument devised so that a sample of dry food may be ignited under high pressure of oxygen, ensuring that complete oxidation of all the organic constituents takes place. The energy released gives an estimate of the gross energy (GE) value of the mixture of carbohydrates, fats, and proteins in the sample of food.

Not all the food the animal eats is oxidized in this way, however. In fact, 20 to 50 per cent never passes through the intestinal wall, but appears in the feces. For assessment of the energy value of a feed, the estimated energy loss in the feces must be deducted. Thus:

$$\text{gross energy} - \text{fecal energy} = \text{digestible energy (DE)}$$

The digestible energy is that fraction of the food energy which is absorbed from the digestive tract. Following absorption, there are other losses of energy, particularly in urine, where compounds capable of further oxidation, such as urea and ammonia, are excreted, and also in gaseous losses from the digestive tract. Gaseous loss is important in ruminants, which eruct from the rumen large quantities of methane (CH_4), a highly reduced combustible compound which is an end product of the bacterial fermentation.

If all these energy losses are accounted for, an estimate of the energy available for metabolic processes is derived:

$$\text{digestible energy} - (\text{urine energy loss} + \text{gaseous energy loss}) = \text{metabolizable energy (ME)}$$

As a result of metabolic processes, a considerable amount of heat is generated. Except under very cold conditions this heat represents further loss of energy to the animal. Measurement of the total heat production in the animal is a difficult task involving the use of large calorimeters which will house live animals. However, it is possible to obtain an estimate of total heat production by direct and indirect means, and when this is deducted, the net energy available to the animal can be calculated:

$$\text{metabolizable energy} - \text{heat energy} = \text{net energy (NE)}$$

Since food resources are highly variable, and since no food is completely digestible, it follows that one can ascribe to each feed a value for GE, DE, ME, and ultimately NE. The value of each feed will also vary considerably for different species. Cellulose, for example, has the same GE value for man and a cow, but since man has no mechanism for digesting it, the DE value is very low for man (close, in fact, to zero), but for the cow, which is able to utilize cellulose efficiently because of the microflora in its rumen, the DE value (and consequently the ME and NE values also) are relatively high. The question of the digestion and absorption of food therefore becomes critical in determining value as a source of energy for the animal.

Digestion of Carbohydrates

When carbohydrate enters the digestive tract, the work of breaking down the complex polysaccharide molecules into the simpler disaccharides begins under the action of the digestive enzyme, salivary amylase, secreted in the saliva, and pancreatic amylase, secreted by the pancreas into the duodenum. In the small intestine, the enzymes which attack the disaccharides are secreted from the mucosal cells, and these enzymes bear the name of the disaccharides they attack.

Starch and cellulose are probably the most important carbohydrates to the animal nutritionist, since most animal feeds are rich in one or the other of these compounds. The starches are the storage carbohydrates in seeds and roots. Cellulose is the chief structural carbohydrate component of plants, and it is therefore abundant in grasses and stemy or leafy forages. Cellulose and starch also represent the extreme examples of highly indigestible and highly digestible carbohydrates, respectively; whereas starch is quite readily degraded to glucose, animals possess no enzyme capable of initiating the degradation of cellulose, which is completely resistant to the enzymatic attack from the salivary or pancreatic amylases. Ruminants and nonruminant herbivores depend upon the support of a vast population of microorganisms, which are contained in the rumen or the cecum, respectively. Cellulase, the enzyme that breaks down cellulose, is produced by the microflora present in the rumen of cattle, sheep, and goats, and in the cecum of horses, rabbits and even pigs. Once the initial degradation of cellulose has begun, the microflora utilize the sugars released by fermentation; acids such as acetic, proprionic, and butyric (and to a lesser extent some 4- and 5-carbon acids) are produced as by-products of the process. These volatile fatty acids are absorbed by the host animal, and are metabolized to yield energy or to serve as carbon sources for diverse synthetic processes. Acetic acid, for example, is a precursor for the synthesis of many important compounds, such as cholesterol, nucleic acids, and milk fat.

Further details of the metabolism of sugars and the volatile fatty acids are presented in Chapter 36.

Function of Proteins

Early in the 20th century a considerable body of documented knowledge was available upon energetic processes, and there was wide recognition of the mineral elements and their salts, although their nutritional significance was not well-understood. Although much of the fundamental research in energy metabolism had been completed, there was also a growing consciousness of the antithesis to the generalization that life is solely a combustion process. Justus von Liebig, a proverbial dunce at school but with a passion for chemistry, left his native Germany in 1822 to study in Paris under Alexander von Humboldt. Returning to Geissen, he became Professor of Chemistry at the age of 21 and began a distinguished career as biochemist and physiologist, both there and in Munich, to which he later moved and from where he engaged in violent polemics with Pasteur. Liebig recognized that a considerable proportion of an animal's food did not function as fuel, but as building material for the body, which we now know to be true of proteins and minerals in particular as well as of other substances. Again, as outlined in McCollum's histories, it was Lawes and Gilbert in 1854 who showed for the first time that proteins could differ in their nutritive value, an observation confirmed by Voit in 1897. Voit also established that excreted nitrogen was a useful index of protein metabolism. It was also known that in the digestive tract proteins were broken down to amino acids, which were absorbed in the blood and recombined into body proteins. A great deal of research on the amino-acid composition of foods revealed the marked differences between them in their yield of amino acids. The stage was then set for the pioneering work of Osborne and Mendel, who in 1911 reported on feeding experiments with "isolated food substance"; and of Rose, who demonstrated in the 1930's the essentiality of some of the amino acids, and laid the foundations for a detailed understanding of the species requirements for these.

The amino acids It is now well-known that the features which give the infinite number of proteins found in the plant and animal kingdoms their unique characteristics are (1) their content of the individual amino acids, (2) the size of the protein molecules, and (3) their spatial configuration. Of these three features, the first is of most importance in nutrition. The size and spatial arrangements of the molecules are less important, because upon entering the digestive tract, proteins are attacked by pepsin in the stomach and by a wide range of proteolytic enzymes in the small intestine which split the proteins into their constituent amino acids. The amino acids released by the action of pancreatic and intestinal proteases are rapidly absorbed from the small intestine, and are transported in the blood to the sites of protein synthesis, chiefly the liver and muscles, where they are reconstructed under the control of DNA into the specific structural and functional proteins that the animal needs.

The value of a dietary protein to the animal rests on its ability to supply the amino acids in the correct proportions needed by the animal. Not all proteins are of the same value nutritionally, for their amino-acid content varies. The protein found in corn grain is particularly poor in its lysine content, whereas proteins of animal origin, such as those of fishmeal, are rich in lysine; so the inclusion of a little animal protein in a diet based on corn grain will balance the deficiency. Notice in the polypeptide and protein illustrated in Figure 34.2 that some amino acids occur in much smaller quantities than others. One such amino acid is tryptophan, which is therefore one of the most commonly deficient amino acids in animal feeds.

Amino-acid essentialities Another vitally important characteristic of the amino acids is that some of them *must* be included in the diet. Many of the amino acids can be synthesized by the body if a source of carbon, hydrogen, oxygen, and nitrogen is available in the metabolic pool. There are ten of the amino acids, however, which have a carbon-chain structure that the animal cannot synthesize fast enough for its needs, if at all. It therefore becomes essential to include these amino acids in the diet. The essential, partially essential, and some of the nonessential amino acids are listed in Table 34.4. In practice, most of the essential amino acids are supplied in considerable excess in normal diets. A leucine deficiency, for example, is virtually unknown because its level is high in both plant and animal proteins, but it can be easily demonstrated in the laboratory, using purified diets free of leucine, that this amino acid falls into the essential category. Animals receiving similar purified diets which omit aspartic acid never show any deficiency symptoms because the body can readily synthesize this amino acid; it is therefore a nonessential amino acid.

Amino-acid balance Another important aspect of protein nutrition is the concept of

Table 34.4.
Nutritive classification of the amino acids.

Essential (indispensable)	*Semi-indispensable*	*Nonessential (dispensable)*
Histidine	Arginine[a]	Glutamic acid
Lysine	Tyrosine[b]	Aspartic acid
Tryptophan[b]	Cystine[b]	Alanine
Methionine	Glycine[a]	Proline
Phenylalanine	Serine[a]	Hydroxyproline
Threonine		
Leucine		
Isoleucine		
Valine		

[a] Arginine and glycine are essential for chicks and turkeys. Serine will spare or replace glycine.

[b] Tyrosine will spare but not completely replace phenylalanine. Cystine will spare but not completely replace methionine. Nicotinic acid will spare but not completely replace tryptophan.

amino-acid balance. The consumption, degradation, and digestion of proteins can be visualized as a process in which the animal dismantles the complex proteins in its food to free the amino acids, which it then absorbs and reassembles into proteins specifically required in its own metabolic processes. Clearly the closer the yields of amino acids from dietary protein approximate the proportions in which they are required by the animal, the more efficiently the dietary protein will be utilized for the synthesis of new protein. Amino acids supplied by the diet in excess are deaminated (the NH_2 group is removed), and the carbon chain serves as a source of energy in the same way that a sugar or fatty acid does. This may be rather wasteful from a nutritional point of view, because other energy sources are available to the animal, whereas the essential amino acids as building blocks for protein are unique and cannot be made fast enough or even at all from fats or carbohydrate.

Protein synthesis can only proceed as long as all the amino acids are available. If a dietary protein is fed which is very low in one of the essential amino acids, then protein synthesis may be arrested and all the other valuable amino acids wasted; they are deaminated, and used as an energy source, which may lead to excessive fat deposition and poor growth in the animal. An understanding of this concept is most important now that the essential amino acids are all available in synthetic form. They can be added to the diet in small quantities to refine the over-all amino-acid balance of the diet. Lysine and methionine are currently available commercially, and spectacular improvements have resulted from their inclusion in relatively simple and inexpensive diets for farm livestock.

The special position of ruminants For ruminating animals no specific amino-acid requirements have been demonstrated. Nitrogen supplied from non-protein sources (NPN) such as urea is efficiently used for growth and protein synthesis. The mechanism lies not in the ruminant animal's ability to synthesize the essential amino acids from NPN, but rather in the ability of the rumen microflora to combine NPN with other chemical constituents to form amino acids and subsequently protein.

Vast numbers of the microflora are regularly passed on from the rumen complex to the small intestine and here they in turn are subjected to digestion, releasing their protein and amino acids for absorption in the normal way.

Function of Fats

Fats, as nutrients, have four main characteristics: (1) a high energy content; (2) fatty-acid essentiality; (3) rancidity; and, (4) carriers of fat-soluble vitamins.

Energy content If one gram of any fat is completely oxidized to carbon dioxide and water, approximately 9.0 kilocalories (kcal) of heat will be generated. The energy yield is the same whether this is done in the body by the devious pathways of metabolism or in a calorimeter under high pressures of oxygen. This is 2.25 times the yield of energy which can be derived by the oxidation in the body of 1 g of either carbohydrate or protein. The high-energy yield from fat arises from the fact that fatty acids, as can be seen by their formulas are relatively unoxidized by comparison with carbohydrates or proteins. Clearly if one wishes to increase the energy content of diets, the inclusion of fat is an effective way to do so. Indeed, the newborn, which is often critically short of energy for thermoregulatory purposes soon after birth, is provided for in that the first milk, colostrum, is extremely rich in fat.

Fatty acid essentialities George & Mildred Burr in 1929 demonstrated the essentiality of certain fatty acids. Practically all the fatty acids can be readily broken down and resynthesized within the body, except for linoleic ($C_{18}H_{32}O_2$), linolenic ($C_{18}H_{30}O_2$), and arachidolnic ($C_{20}H_{32}O_2$). In the absence of a synthetic mechanism for making them, it is essential for these three acids to be included in the diet, otherwise severe nutritional disorders will cause scaly skin, cessation of growth, and eventually death. The "essential fatty acids" are required in only minute quantities; so that normally animals eating foods with even small quantities of fat such as grains and pasture grasses will receive adequate quantities.

Rancidity The long fatty-acid chains are readily susceptible to partial oxidation:

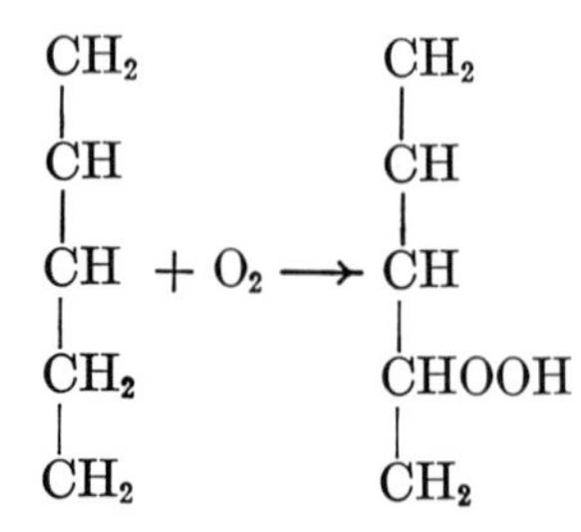

In the partially oxidized state, both the saturated and unsaturated fats develop characteristic rancid taints. These may be regarded as highly unpleasant, as with stale fat, or immensely enjoyable, as with various soft and blue cheeses. The oxidation process tends to be accelerated by heat, exposure to ultraviolet light, and the presence of heavy metals, such as copper and iron, in trace amounts. In nature many of the fats are present in association with anti-oxidants, compounds which prevent this oxidation from taking place, vitamin E being a good example of such a compound. In diets which are formulated to contain fat, anti-oxidants must be incorporated into the diet to prevent oxidative rancidity.

Carriers of fat-soluble vitamins A further important property of fats and oils is their function as solvents for the fat-soluble vitamins, A, D, E, and K. These vitamins are discussed in further detail elsewhere.

In the digestive tract, fats are exposed to two processes which ultimately facilitate their digestion: emulsification, and enzymic attack.

The first contact with a digestive process occurs in the stomach, where gastric lipase attacks certain kinds of fat and splits the triglyceride molecule into its component parts of glycerol and free fatty acids. Most fats, however, reach the duodenum relatively intact, and here come into contact with the bile salts, which are powerful emulsifying and neutralizing agents; that is, these salts break the fat down into smaller droplets and expose a greater surface area for enzyme attack; and they neutralize the HCl produced in the stomach. Pancreatic and intestinal lipases then complete the work of splitting the triglycerides into monoglycerides, diglycerides, glycerol, and free fatty acids, all of which are absorbed through the intestinal wall.

Functions of Minerals

Although minerals were among the first known elements and were included in diets routinely, intensive research on their function in the body did not take place until the 1920's. It was at the University of Wisconsin that so much of the vital early work was carried out, and where men such as Hart, Steenbock, and Elvehjeum were largely responsible for work elucidating the role of iron and copper. More recently the vital role of a wide range of trace minerals as cofactors for the important enzymes in metabolism has been revealed, and thus the interrelationships of minerals, vitamins, and the energy supplying nutrients as we now know them.

Phosphorus is especially important for growth; it is required, along with calcium, for bone development and soft-tissue formation, as well as for metabolic processes. The metabolism of sugar, fat, and protein involves the phosphorylation of metabolic intermediates and the formation of energy-rich phosphate compounds for the storage and transmission of metabolically trapped energy.

Calcium is needed for bone-structure formation. Sodium, potassium, calcium, and magnesium help to maintain the osmotic pressure of the blood and to maintain the proper water balance between circulating fluids and the tissues.

Iron is a part of the hemoglobin molecule; thus, a deficiency of iron can induce an anemic condition in an animal. Copper is also involved in that it facilitates the absorption and utilization of iron.

Cobalt is a part of the vitamin B_{12} molecule. Ruminants can synthesize enough vitamin B_{12} for growth, provided the ration contains at least 0.07 ppm cobalt for cattle or about 0.1 ppm for sheep.

A number of interesting mineral interrelationships have recently been investigated. Swine require zinc, and the apparent requirement is increased as the level of calcium in the ration is increased. Excess molybdenum decreases the utilization of copper. Selenium, which at 5 to 10 ppm in a ration produces toxic symptoms, has recently been found to reduce the vitamin E requirement if present in certain rations at extremely low levels, less than 1 ppm.

Usually classed with the minerals are the halogens (iodine, chlorine, and fluorine), which are important nutritionally. Iodine is needed for the formation of the thyroid hormones, thyroxine and triiodothreonine, and related compounds produced by the thyroid gland. A deficiency induces hyperplasia of the thyroid, called goiter in humans and "big neck" in calves. Fluorine probably should not be included in the list of required minerals, but its presence at low levels (0.75 ppm) in water appears to reduce the number of cavities in human teeth. Too much fluorine can be toxic. Though rarely deficient in normal rations, chlorine is a required mineral. Presumably, it functions in the acid-base relations in the body.

Function of Vitamins

The development of the vitamin concept is a fascinating story. Early investigators found that rations composed of purified nutrients failed to support growth. These investigators correctly concluded that natural foods contained, in addition to the known constituents, minute amounts of unidentified substances that are essential to life. By 1913, there was evidence that two factors existed; one called "fat-soluble A," and another called "water-soluble B." Thiamine (vitamin B_1) was the first vitamin to be synthesized in crystalline form (1935), but today all the vitamins listed in Table 34.4 are available in crystalline form, and their structures have been determined.

Space will not permit an elaboration of specific deficiency symptoms for each of the vitamins or a discussion of the variations among the common classes of farm animals. Reduced food intake, growth rates, and feed efficiency are the common general symptoms of vitamin deficiency in livestock and poultry. Death can occur if the deficiency is acute.

The B-complex vitamins function as enzyme cofactors and catalysts in energy metabolism. In humans, the lack of vitamins causes specific deficiency syndromes. Vitamin B_1 in the diet prevents beri-beri; niacin (nicotinamide) prevents pellagra, and riboflavin prevents cheilosis. Vitamin B_{12} is the "antipernicious-anemia factor."

Fat-soluble vitamins are concerned with specific physiological functions, as are the B-complex vitamins. Vitamin A deficiency causes keratinization of epithelial tissue in the cells and upsets the chemical processes involved in vision. Vitamin D, called the anti-rickets vitamin, or the "sunshine vitamin," functions in activating the processes involved in the utilization of phosphorus and calcium for bone growth and development. Vitamin E is involved in oxidation-reduction (electron transfer) systems, and its deficiency is manifested by damage to the brain in chicks and by "white-muscle" or stiff-lamb disease in young lambs. Vitamin K is needed for normal blood clotting. Vitamin C is the antiscurvy vitamin.

Ruminants again differ from nonruminants in the level of synthesis of B-complex vitamins in the digestive tract. The lower digestive tract of all animals supports the growth of bacteria that synthesize vitamins, which are absorbed and utilized by the animal. The capacity for B-vitamin synthesis is greater in ruminants than in other animals because of the synthetic capability of the microorganisms in the rumen. In horses synthetic capacity is also high because of the microorganisms present in the cecum.

FURTHER READINGS

Maynard, L. A., and J. K. Loosli. 1969. *Animal Nutrition.* New York: McGraw-Hill.

McCollum, E. V. 1957. *A History of Nutrition.* Boston: Houghton-Mifflin.

McDonald, P., R. A. Edward, and J. F. S. Greenhalgh. 1966. *Animal Nutrition.* Edinburgh and London: Oliver & Boyd.

National Research Council publications on nutritional requirements: *Poultry,* no. 1345; *Dairy Cattle,* no. 1349; *Sheep,* no. 1193; *Beef Cattle,* no. 1137; *Horses,* no. 1401; *Swine,* no. 648.

Pike, R. L., and M. L. Brown. 1967. *Nutrition: An Integrated Approach.* New York: Wiley.

Underwood, E. J. 1971. *Trace Elements in Human and Animal Nutrition.* 3d ed. New York: Academic Press.

Thirty-Five

Estimation of the Nutritional Value of Feeds

"Upon what meat doth this our Caesar feed,
That he is grown so great?"

Shakespeare
Julius Caesar, *Act 1, scene 2*

The need for some system to estimate the nutritive value of feeds can be demonstrated by a simple problem frequently confronting the livestock man. "If a ton of corn costs $55, a ton of barley $52, which is the better buy?" To answer this question, the feeder or economist needs information from a nutritionist; i.e., he must find out how much barley will be required to replace one ton of corn with no change in animal performance. For this purpose a relative-value rating of the two grains (barley is worth 97 per cent of corn) could be easily translated into dollars. However, the comparative evaluation of feeds is also important for use in conjunction with feeding standards which give estimates of the nutritional requirements of animals. Used in this manner, nutritive values are important to ensure the formulation of adequate rations and in predicting animal response to a given feeding regime. Nutritive values are also essential for making management decisions concerning cropping systems or for long-range planning and prediction of feed supplies and needs. Thus, some means of expressing feed value in absolute as well as relative terms is necessary.

Chemical Composition

Historical

Our present system of proximate analysis was developed by Henneberg and Stohmann (1864) at the Weende Experiment Station in Germany. Henneberg expected his provisional system of feed analysis to be modified by improved methods ("The present method of fodder analysis greatly needs to be perfected, but in many respects accomplishes more than would be expected from its defectiveness," a quote from Henneberg's lecture notes), but our present procedures are practically the same as those originally published.

Chemical Analysis

Proximate analysis divides feedstuffs into the six heterogenous fractions shown in Table 35.1.

The usefulness of this scheme is the relatively inexpensive and simple procedures of analysis. Much can be inferred concerning the feeding value of feedstuffs by the magnitude of, and the relationships between, the proximate fractions. Examples of analysis of several different feedstuffs are shown in Table 35.2. The higher crude-fiber levels (largely cellulose and hemicellulose) are characteristics of roughages; the very high nitrogen-free extract (largely starch) of corn is typical of most feed grains. Soybean meal (residue after the extraction of oil from the soybean) is high in protein, as is meat and bone meal. The latter feed is also high in ash, which indicates the presence of large amounts of minerals. The proximate analysis, therefore, is useful as an index of a feed's potential value, because it isolates similar compounds into categories which have nutritional significance.

Table 35.1.
The fractions of proximate analysis.

Procedure[a]	*Fraction*	*Major components*
Drying at approximately 100°C to constant weigh	Moisture	Water and any volatile compounds
Ignition at 500–600°C	Ash	Mineral elements
Nitrogen by Kjeldahl sulphuric-acid digestion	Crude protein	Proteins, amino acids, nonprotein nitrogen
Extraction with petroleum ether	Ether extract	Fats, oils, waxes
Residue after boiling with acid and alkali	Crude fiber	Cellulose, hemicellulose, lignin
Remainder; i.e., 100 minus sum of the other fractions	Nitrogen-free extract	Starch, sugars, some cellulose, hemicellulose and lignin

[a] Each procedure is applied to a separate sample, of standard weight, of the feedstuff to be analyzed.

Uses

The proximate-analysis scheme has proven very useful for describing feeds and foods for use in tables of food composition, as a basis for diet formulation, as the starting point for a more specific analysis, and as a basis for food purchasing and merchandising. Commercially prepared feeds are required by law to be labeled with a list of ingredients and a guaranteed analysis showing the minimum amount of crude protein and crude fat, and the maximum amounts of crude fiber and ash, in the feed. These figures are the buyers' assurance that the feed contains at least those minimal amounts of the higher-cost items, protein and fat, and not more than the specified quantities of the lower-cost and potentially less valuable items, the crude fiber and ash.

Table 35.2.
Proximate composition of some common feedstuffs.[a]

Feedstuff	*Moisture*	*Ash*	*Crude protein*	*Ether extract*	*Crude fiber*	*Nitrogen-free extract*
Alfalfa hay	10.8	7.6	15.2	1.8	27.6	37.0
Corn silage	72.1	1.7	2.3	0.8	7.3	15.7
Barley straw	10.0	6.0	3.7	1.6	37.7	41.0
Corn grain	11.0	1.1	8.9	3.9	2.0	73.1
Soybean meal	11.0	5.8	45.8	0.9	6.0	30.5
Meat and bone meal	6.0	29.1	50.6	9.5	2.2	2.6

[a] All figures are percentages.

It is important to recognize that the feed fractions, as separated by a proximate analysis, are all combinations of nutrients, and therefore vary in nutritional significance. Crude protein is determined by analysis for nitrogen, which is converted to protein by multiplying by the factor 6.25 (proteins average 16 per cent nitrogen, so 100/16 = 6.25). The term "crude" is attached in recognition of the fact that all nitrogen present in a feedstuff may not be associated with protein molecules. In some instances, crude protein may not be a good index of true protein content. It is also important to remember that a high content of true protein does not necessarily mean that the feedstuff is a valuable protein source for all animal species, because an animal's requirement is for certain amino-acid molecules rather than for preformed protein. The protein of soybean meal, for example, has a more optimal balance of amino acids than that found in linseed meal, which is low in lysine, isoleucine, histidine, and tryptophan.

The crude-fiber fraction was originally thought to be made up of materials with little nutritive value, and therefore to have a negative relationship with feeding value. This concept is essentially true when only one plant species is concerned, but the crude-fiber fraction from alfalfa, for example, may have proportions of cellulose, hemicellulose, and lignin widely different from the crude-fiber fraction of early cut grass or cereal straw. Crude fiber is of most value when comparing different samples of the same plant species. Meyer and Jones (1962), for example, found the feeding value of alfalfa negatively correlated with its crude-fiber content. This same relationship, however, would not predict the feeding value of oat hay.

The ash fraction of the proximate analysis gives no indication of which minerals are present. Rice straw contains approximately 16 per cent ash, but 85 per cent of this is silica (the chief element in sand) and of no value to the animal. On the other hand, meat and bone meals have about 29 per cent ash and contain large amounts of calcium and phosphorus, both of which are very important components of a well-balanced diet. Thus, the knowledge of the ash content is of limited value unless the elements present in the ash are also known. Ash content is included on the guaranteed-analysis label for commercial feeds chiefly because a value higher than normal is generally associated with soil contamination.

Newer Developments

A rapid method for partitioning the dry matter of feedstuffs into two fractions (cell contents, and cell-wall constituents) based on nutritional availability has been proposed by Van Soest (1967). The basis for this proposal is that the contents of cells are highly digestible and unaffected by lignin and cellulose, whereas the concentrations of these two materials control the digestibility of the cell-wall constituents. The chemical procedure is to use a specific neutral detergent to separate the soluble (cell contents) and insoluble (cell-wall constituents) fractions. Further partitioning of the two fractions is then possible. Studies are continuing along these lines, and the procedures being developed may eventually lead to a simple method of chemically evaluating feedstuffs that will have added nutritional significance.

Digestible Nutrients

Value

Chemical procedures can only give an index of the nutrient content of feeds or foods. In order to have a more precise estimate of nutritive value, the feed must be fed to the particular animal species involved and undergo a biological evaluation. This biological evaluation is necessary to determine the availability of the nutrients to the animal. One common biological procedure is the digestion trial, which involves measuring the feed intake and fecal excretion of animals being fed the feedstuff under evaluation.

Table 35.3.

Procedure for determining the digestion coefficient for protein in alfalfa hay.

Information	*Example*	*How determined*
(1) Feed consumed	8 kg. dry matter	Measure feed consumed
(2) Feed composition	20% protein	Laboratory analysis
(3) Nutrient intake	1.6 kg. protein	8.0 × .20 = 1.6
(4) Fecal excretion	2.9 kg. dry matter	Fecal collection and drying
(5) Fecal composition	13.8% protein	Laboratory analysis
(6) Nutrient excretion	.4 kg. protein	2.9 × .138 = .4
(7) Digestible nutrient	1.2 kg. protein	1.6 − .4 = 1.2
(8) Digestion coefficient	75%	(1.2 ÷ 1.6) × 100 = 75

Methods

The digestibility of any nutrient can be expressed by the relationship: Intake minus fecal excretion, divided by intake. The important measures are the intake of a well-sampled feed or ration, and the total fecal excretion which corresponds to the amount of food consumed. An example procedure is shown in Table 35.3. Several techniques are used to make the fecal collections. One is specifically designed stalls (Figure 35.1), which confine the animal and permit collection of uncontaminated feces. Another is a collection harness and bag (Figure 35.2) which allows the animal more freedom and can be adapted for use under grazing conditions. Chemical analysis of the feed and fecal samples makes it possible to calculate the digestibility of a single nutrient (protein) or a specific fraction of the feed (crude fiber).

Factor Influencing Digestibility

Feed factors Digestibility of a feedstuff is largely the result of three interacting factors: the species of animal (type of digestive tract), the chemical makeup of the feed, and the way the feed was processed.

Generally speaking, the feed grains are highly digestible, and vary in composition within a relatively narrow range. They are well-utilized by all classes of livestock. Barley and particularly oats are more variable in digestibility than the other feed grains, because

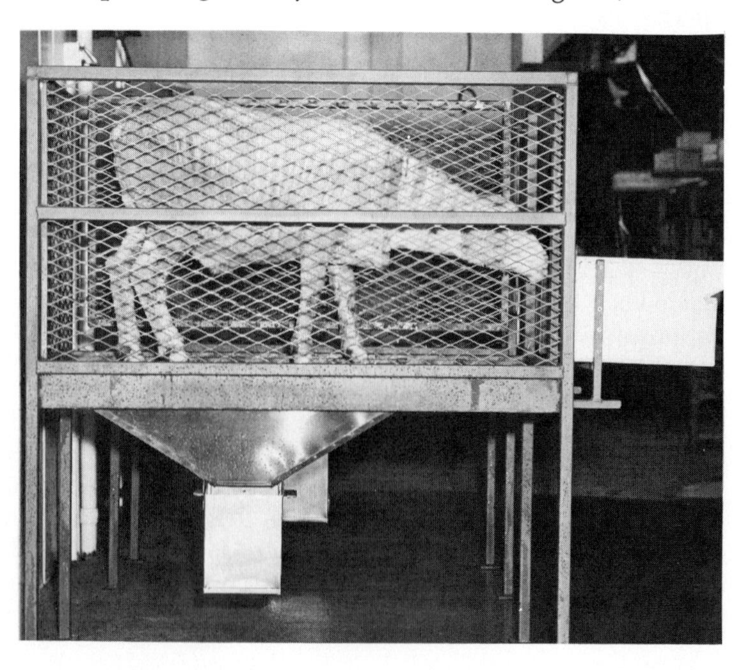

Figure 35-1 A metabolism stall for the separate collection of urine and feces from sheep. The stall is adjustable and can be used by male or female animals. (Courtesy of the Department of Animal Science, University of California, Davis.)

Figure 35-2 A steer with a harness and bag for the collection of feces. Particularly useful in grazing experiments (Courtesy of the Department of Animal Science, University of California, Davis).

the portion of the fibrous hull attached to the kernel or groat depends on the variety and the conditions under which the grain was grown. Even in these grains, the variation is small compared to that found in forage plants.

The predominant factor influencing digestibility of forage crops is the stage of maturity when cut. Thus it has been shown by Reid *et al.* (1959) and Mellin *et al.* (1961) that the dry-matter digestibility for ruminants of first-cut forage in New York state and of timothy hay in Maine could be estimated by the regression equation: percentage of digestible dry matter $= 85 - .48x$, where x was days after April 30 in New York state and days after May 15 in Maine. The major reason for this relationship is that as plants mature, the amount of holocellulose (all cellulose), lignin, and crude fiber increase in a nearly linear fashion up to the time of seed set. Protein usually decreases in a similar manner (Figure 35.3). Relationships between crude fiber, lignin, and digestibility have been used to predict the feeding value of various forage crops (see Sell *et al.*, 1959, and Stallcup and Davis, 1965, for literature reviews).

Digestible crude protein is generally highly correlated with crude protein content, and this relationship has more interspecies significance than the relationship between crude fiber and digestibility. For example, percentage of digestible protein and percentage of crude protein of alfalfa were highly correlated ($r =$ 0.99) in the studies of Meyer and Jones (1962). Their estimating equation, $DP = 0.92\ CP - 3.1$, is quite similar to one established for a variety of forages, $DP = 0.92\ CP - 3.5$ (Reid *et al.*, 1959; Stallcup and Davis, 1965). However, as previously stated, a knowledge of protein quality (amino-acid content and biological availability) is important in addition to protein digestibility, particularly when determining protein value for nonruminant species like swine, chickens, and humans, which have relatively little reliance on intestinal microbes. The ruminant is less dependent on protein quality because the rumen microbial population synthesizes amino acids from dietary nitrogen.

Animal factors The animal also has an important influence on the digestibility of feedstuffs. Within-species variation is present, and some individual animals may have more or less efficient digestive systems than the average for their species. This variation may be important in selecting for efficient production, but is small compared to interspecies differences in digestibility. The ruminant is more efficient in the digestion of high-fiber, low-protein forages; the simple-stomached pig is more efficient in digestion of high-protein, low fiber feedstuffs (Figure 35.4). The reasons for these species differences have been explained in Chapter 31.

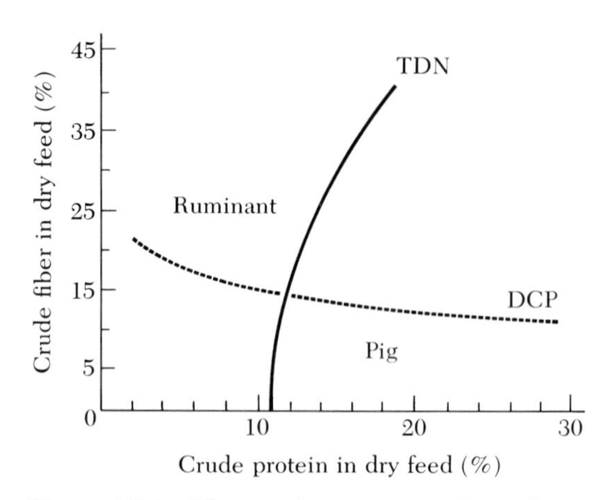

Figure 35-4 The quadrants arising from the intersection of the lines of equal efficiency of digestion of feeds in terms of digestible crude protein (DCP) and total digestible nutrients (TDN). The ruminant is more efficient than the pig in obtaining both DCP and TDN from feeds whose compositions lie in the upper left quadrant while the pig is more efficient than the ruminant in coping with feeds which lie in the lower right quadrant. (From Glover, Comparative efficiency of digestion of feeds by ruminants and pigs. J. Agric. Sci. 56: 113–115. 1961.)

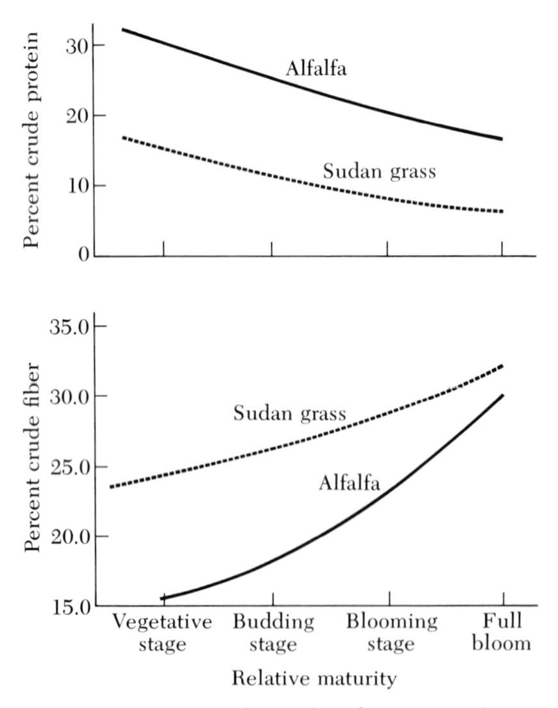

Figure 35-3 The relationship between relative maturity and the crude protein and crude fiber content of alfalfa and sudan grass.

Processing The way a feedstuff is processed can also influence its nutritive value. Fine grinding and pelleting of forages tend to increase rate of passage through the gut, and this lowers crude-fiber digestibility. However, over-all animal response to pelleted forages is usually increased over the same forage fed in chopped form, because the slightly lower digestibility is more than offset by improved efficiencies in other areas. Particularly important is an increased food consumption (Garrett *et al.*, 1961). It is now a common practice, when feed grains are prepared for use in cattle and hog rations, to grind the grain through a hammermill or subject it to steam and then roll it into a flake. These procedures result in small but economically important changes in overall digestibility and efficiency of use for most classes of livestock.

Energy Evaluation

It is not tenable to consider one nutrient more important than another, since all must be available to the animal in adequate amounts if efficient production is to be maintained. For any given feeding situation, the most important nutrient is probably the one which is missing. However, an animal's requirement for energy is, from a quantitative and economic position, the primary consideration. Energy is most often the factor which limits livestock production, and meeting the energy requirements for maintenance and production is the major cost associated with feeding animals. For a great many feeding situations, if the animal is supplied with adequate energy from natural feed sources, the other nutrients may be incidentally supplied. Historically, then, the comparative evaluation of feedstuffs has been on the basis of their ability to supply energy to the animal. Changes have occurred in the methods used to express the nutritive value of feeds as our understanding of energy metabolism has increased.

Energy Units

There has been confusion regarding the nomenclature of energy units. It is presently most common to measure the energy content of feedstuffs in calories. The usual definition of a calorie is the amount of heat required to raise the temperature of one gram of water from 14.5°C to 15.5°C. This is a relatively small unit, and for nutritional purposes it is more convenient to use the kilocalorie (1 kcal. = 1,000 calories) or the megacalorie (1 Mcal. = 1,000 kcal. = 1,000,000 calories). In popular writings, particularly those concerned with human caloric requirements, the term calorie is frequently used erroneously in place of the kcal. The spelling of calorie with a large C, i.e., Calorie, denotes the same thing as kcal., but can lead to confusion and should be considered an archaic means of expression. Similarly, the "therm" is sometimes used as a synonym for megacalorie, but is not as descriptive.

At the present there is an international movement toward the adoption of a single energy unit suitable for expressing mechanical, chemical, or electrical energy, as well as the concept of heat. The unit agreed upon is the Joule (4.184 J = 1 calorie), and sometime in the future we can expect that energy requirements and feed values will also be expressed by this common unit of energy.

Gross energy The gross energy of a feedstuff is the amount of heat evolved when the substance is completely oxidized. This value is obtained by combusting a sample of the feedstuff in an atmosphere of oxygen and measuring heat evolved, usually by the rise in temperature of a known amount of water. The instrument used for this determination is the bomb calorimeter (Figure 35.5). The gross energy of a feedstuff is related to its chemical composition, and, as shown by Nehring (1970), can be estimated from the proximate analysis by means of the following equation:

$$\text{GE (kcal./gm)} = 5.72x_1 + 9.50x_2 + 4.79x_3 + 4.17x_4$$

where

x_1 = percentage of crude protein/100
x_2 = percentage of crude fat/100
x_3 = percentage of crude fiber/100, and
x_4 = percentage of nitrogen-free extract/100

Figure 35-5 A bomb calorimeter, used to measure the heat of combustion of a sample of feed.

This equation indicates that in our common feedstuffs protein has an average caloric value of 5.72 kcal./gm., fat of 9.50 kcal./gm. The carbohydrate components in the crude fiber have an energy content of 4.79 kcal./gm., and the usually more soluble carbohydrates average 4.17 kcal./gm.

Digestible Energy and TDN

The gross-energy content of a material does not give any idea about its value as a feedstuff, since it indicates only energy content without showing how much of the energy is available to the animal. The simplest measure of energy availability is digestible energy (DE) measured in calories, or a comparable measure expressed in weight units, total digestible nutrients. These measures of the energy value of a feedstuff are not strictly comparable (Maynard, 1953), but in general are interconvertible by considering that there are 2.0 Mcal./lb., or 4.4 Mcal./kg, of TDN (Swift, 1957). Both means of expressing the energy value of a feed are obtained by conducting a digestion trial as explained earlier, and subtracting fecal losses from the food consumed. Thus, digestible energy (Mcal./kg.) is determined from the following expression:

$$\frac{\text{Food energy (Mcal.)} - \text{fecal energy (Mcal.)}}{\text{Food intake (kg.)}}$$

TDN is calculated by summing the digestible nutrients, with fat multiplied by 2.25 because of its higher caloric content. The expression is: Percentage of TDN = per cent digestible protein + per cent digestible fiber + per cent digestible nitrogen-free extract + (per cent digestible fat times 2.25).

Since protein has a higher caloric value than carbohydrate, why has no adjustment for protein been made in the digestible-protein figure? Mainly because losses of energy in the urine due to the excretion of nitrogen make digestible protein approximately equivalent to digestible crude fiber and digestible NFE as a source of energy. (For a complete discussion, see Maynard, 1953.)

The TDN and DE systems of feed evaluation have been and continue to be used in much of the Western Hemisphere. These measures are useful as first approximations of a feed's value as a source of energy, and a considerable and valuable volume of knowledge exists concerning the proximate composition and the TDN or DE value of feedstuffs (Morrison, 1956; Schneider, 1947). There are, however, other losses of energy which occur consequent to the utilization of feeds by animals. These other losses, when taken into account, can provide more accurate means of evaluating feedstuffs and expressing the animals' requirement for energy.

Metabolizable Energy

In the complex of physical and chemical changes that occur during the processes involved when an animal consumes food, there are losses of energy besides those accounted for in the feces. Metabolizable energy (ME) represents a more discriminating measure of the value of a feedstuff than DE or TDN, because two sources of energy loss are measured in addition to those which occur in the feces. These losses are in the urine (UE) and in the gaseous products of digestion (GPD), chiefly methane. The latter is of small importance in many species, but is usually 6 to 9 per cent of the gross energy consumed by ruminants. Metabolizable energy, therefore, goes a step beyond DE or TDN, and provides a more accurate measure of the value of a feedstuff.

It has been common to use ME as a measure of feed value for poultry because their feces and urine are excreted through a common orifice; it is actually easier to determine ME than DE for them. However, ME is in general more difficult to determine, especially for ruminants, where the methane production must be measured by sampling the expired air. In these species it is not unusual to measure fecal and urinary energy losses and calculate methane loss from digestible carbohydrate (Swift and French, 1954) or from gross energy (Blaxter, 1962).

In the nutrition of simple-stomached species, where the combustible gaseous losses are usually negligible, ME is generally obtained by considering only fecal and urinary losses. In human nutrition, values thus obtained may be termed physiological feed values, according to the terminology of Atwater and Bryant (1899). It is usual to calculate physiological fuel values from the analysis of a food for carbohydrate, protein, and fat, by considering that protein and carbohydrate have an available energy content of 4 kcal./gm., fat of 9 kcal./gm. Thus fat is worth 2.25 times the caloric value of carbohydrate or protein. This is the origin of the correction factor applied to fat in the calculation of TDN.

For ruminants fed most common rations, it has been found that a relatively constant ratio of ME/DE exists, so that ME can be estimated by multiplying DE by 0.82 (NRC, 1969; ARC, 1965). This ME/DE ratio can be influenced by the nature of the diet, the amount of feed being consumed, the physical characteristic of the feed, and other factors, i.e., by anything which influences the fermentation characteristics of the rumen microbial population. Urinary losses can be related to the protein content of the diet, and with some feedstuffs (like sagebrush) a high content of digestible essential oils (excreted in the urine without utilization) contributes to a ME/DE ratio which differs markedly from 0.82. Therefore, metabolizable energy values which are calculated from digestible energy may not provide much increase in the accuracy of evaluating feedstuffs. An experimentally determined value of ME is, however, a more precise measure of the energy value of a feed than is DE. Furthermore, the concept of metabolizable energy is important in understanding the basis of more advanced methods of feed evaluation.

The Net-Energy Concept

In feed evaluation, the concept of net energy is synonymous with net income in an economic evaluation. Just as net income is gross income minus all expenses and losses, net energy is gross energy intake minus all energy expenses of metabolism and all losses during digestion. Net energy differs from metabolizable energy by the amount of heat lost as a result of the physical and chemical processes which are associated with metabolism. Heat increment is the term generally applied to this heat loss, and its components are the heat of nutrient metabolism and the heat of fermentation. The latter is of more importance in those species which have a considerable reliance on the digestive-tract microbes for aid in preparing food for metabolism, i.e., ruminants and nonruminant herbivores. Heat increment is a loss of heat as a consequence of metabolism, and represents an expense of utilizing metabolizable energy. Thus net energy (NE) is metabolizable energy (ME) minus the heat increment (HI); i.e., $NE = ME - HI$.

For purposes of feed evaluation, knowing each of the individual chemical reactions which occur as an animal utilizes a feed is made unnecessary by the thermodynamic principles of physical chemistry, particularly the law of Hess (Kleiber, 1961), which states that the total amount of heat produced or consumed when a chemical system changes from an initial state to a final state is independent of the way in which this change is brought about. Thus glucose, when an animal uses it as a source of energy, is ultimately transformed into carbon dioxide and water; we do not need precise knowledge of the chemical pathways of oxidation to know that 673 kcal. of heat will be evolved for each mole of glucose oxidized.

The relationship between net energy (NE) and gross energy (GE) of a feed or diet can be expressed as follows:

$$NE = GE - FE - GPD - UE - HI$$

In this expression net energy (NE) includes the amount of energy used to maintain the animal (NE_m) and the amount used for a productive function (NE_p). If the productive function is specified, the commonly accepted (NRC, 1966c) terminology uses a subscript to denote that function, for example, NE_l is net energy for lactation, NE_g is net energy for growth, and NE_f is net energy for fattening.

In order to determine net-energy values, it is necessary to measure, in addition to metabolizable energy, either the energy stored (NE_p) in the animal (or excreted in a product like milk) or the total heat production (HP). When either of these measures are made with animals receiving varying quantities of food (zero to *ad libitum* amounts), it is possible to partition energy used into the components of NE_m, NE_p, or HI. The basic relationships are:

$$ME = NE_m + HI + NE_p$$

This equation says that ME which does not appear in a product has been lost as heat. The heat loss is due to the heat increment and to the energy cost of maintenance.

Determining energy balance There are two general procedures used to partition ME intake into heat production and energy storage (including the energy stored in secreted animal products, like milk and wool). By various techniques of direct or indirect calorimetry (Blaxter, 1962; Kleiber, 1961), the heat produced by an animal is determined and then the energy stored or retained is calculated by difference. The procedures and the equipment necessary for calorimetry with large animals are quite sophisticated, and only a relatively few laboratories are equipped to conduct these experiments. Engineering advances, particularly in the fields of automated analysis, electronic control, and computer science have done much to reduce the labor and time formerly required to conduct this type of investigation, in which accurate measurement and analysis of the expired air for oxygen, carbon dioxide, and methane is necessary.

Another approach to determining energy storage and heat production is by slaughter and analysis techniques, sometimes called the comparative slaughter-feeding trial. This technique was used extensively to study energy metabolism in farm species by Mitchell *et al.* (1926) and later in a modified way by other workers (Garrett *et al.*, 1959; Lofgreen and Garrett, 1968). The technique requires a relatively large number of animals. A random sample is killed at the beginning of an experiment to determine the initial body composition. At the end of a feeding period of sufficient length to result in significant changes in body composition, the remaining animals are slaughtered and analyzed. The difference between the caloric content of the two groups is the energy storage. The modified comparative-slaughter procedure (Garrett *et al.*, 1959) makes use of established relationships between the carcass density and the composition of the animal (Garrett and Hinman, 1969) to estimate the initial-energy and final-energy content without the need to chemically analyze the animal body. The density of the carcass is determined by weighing in water, which does not detract from the carcass's market value; hence the procedure is a relatively inexpensive way to determine net energy values, and has an additional advantage in that the experimental animals can be kept under more natural conditions than those necessary for calorimetric studies. An example of the data obtained in a comparative slaughter-feeding trial is shown in Table 35.4.

Net-energy systems of feed evaluation Recent investigations in various laboratories in the United States and Europe have resulted in several proposals for evaluating feeds on a net energy basis. The British (ARC, 1965) have adopted a procedure proposed by Blaxter (1962) which adjusts the metabolizable-energy value of a feedstuff or diet according to the efficiency with which it is used for a particular purpose. In Rostock, East Germany, many years of energy-balance experiments originating with the classical work of Kellner (1905) have resulted in a proposal by Nehring *et al.* (1969, 1970) that net-energy values be expressed in terms of net energy for fattening (NE_f).

In the United States there has been a slightly different approach to the development of net-energy systems for feed evaluation; however, the theoretical basis is identical to that used by the British and German scientists. The California workers (Garrett *et al.*, 1959; Lofgreen and Garrett, 1968), on the basis of work on the energy metabolism of fattening cattle, have proposed a net-energy system which assigns two net-energy values to each feedstuff. The net energy for maintenance values (NE_m) is assigned depending upon how much of a feed or ration is required to maintain the animal in energy equilibrium, and a net energy for gain value (NE_g) is assigned according to the efficiency with which a feed energy is used for growth and fattening. Since the partial efficiency of energy utilization for maintenance is higher than the partial efficiency of energy used for the production and storage of fat and protein, the NE_m value for a ration or a feedstuff is always higher than an NE_g value. Equations have been derived which give an animal's requirement for maintenance and growth in terms of NE_m and NE_g. This system, like that proposed by the British and German workers, has taken many years to develop, and revisions can be expected as more information accumulates.

A research team working in the USDA laboratories at Beltsville, Maryland, have conducted a large number of energy-balance trials with lactating dairy cows. Their results have also indicated that the efficiency of utili-

Table 35.4.
Some data[a] obtained from a comparative slaughter feeding trial with beef steers.

Item	Feeding level: *Ad libitum*	Feeding level: *Near maintenance*	*How determined*
(1) No. of steers	12	12	—
(2) Days fed	119	119	—
(3) Initial energy, Mcal.	646	646	Calculated from the fat and protein content of an initial slaughter group.
(4) Final energy, Mcal.	1,517	777	Calculated from the fat and protein content of the fed steers.
(5) Daily energy gain, Mcal.	7.3	1.1	$\frac{(4) - (3)}{(2)}$
(6) Daily feed intake, kg.	8.1	3.7	Measured during trial.
(7) Fasting heat production, Mcal.	5.8	5.8	Mathematical extrapolation of the data to zero food intake.
(8) Feed for maintenance, kg.	2.9	2.9	Mathematical extrapolation of the data to zero energy gain.
(9) Net energy for gain, Mcal./kg.	1.4	1.4	$\frac{(5)}{(6) - (8)}$
(10) Net energy for maintenance, Mcal./kg.	2.0	2.0	$\frac{(7)}{(8)}$

[a] Adjusted to equivalent body weights for the two feeding levels. Data source: Animal Science Department, University of California, Davis.

zation of metabolizable energy for maintenance (NE_m) and lactation (NE_l) is influenced by the concentration of ME in the ration. The efficiency of energy use for these two functions is generally similar to, but different from, the efficiency of energy use for fattening. This finding has resulted in the suggestion that the energy requirements and feed values for the lactating cow could be expressed in terms of NE_l (Flatt *et al.*, 1969; Moe *et al.*, 1970).

Net-energy values are definitely the most precise measures of the value of a feed or ration for a particular animal function. There is, at present, no general agreement among scientists working in this area as to which method of expressing net-energy values will be most useful to the livestock industry. The potential advantages of feed evaluation systems based on the principles embodied in the term net energy are, however, universally recognized.

Voluntary Feed Consumption

Food Intake and Nutritive Value

In many, perhaps most, feeding situations it is important to know how much of a particular feed or ration will be consumed. This information is particularly necessary when predicting animal response to a given feeding regime, since a productive response depends to a large extent on how much available energy can be consumed above the maintenance energy re-

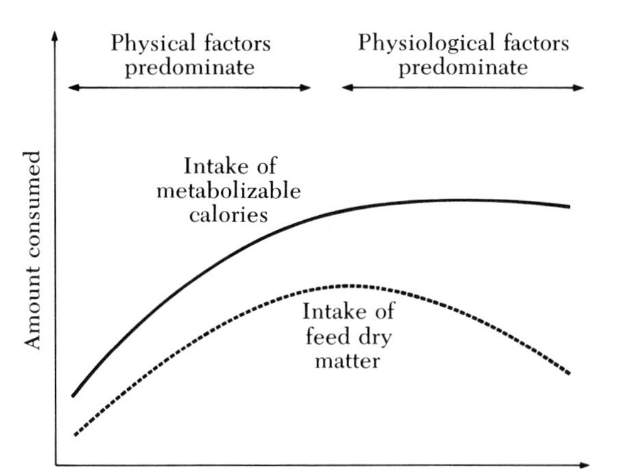

Figure 35-6 A diagrammatic relationship between the metabolizable energy content of a diet and its consumption by the growing ruminant. Physical factors have a major role in appetite regulation until the capacity of the rumen no longer limits energy intake. At some dietary metabolizable energy concentration physiological mechanisms adjust energy intake to a constant level by lowering food consumption.

quirement. Feed intake is, therefore, a critical part of a feed evaluation.

Some of the factors which influence voluntary intake are known, and many of these are related to the digestibility and chemical makeup of the feedstuff. Generally, feed intake increases with increases in digestibility and nutritive value until the physical capacity of the digestive tract no longer exerts a significant control. Further increases in the digestible-nutrient concentration in the feed are accompanied by a decreased intake to adjust the amount of absorbed energy to some level controlled by physiological mechanisms (Figure 35.6).

Feed preparation can also play an important role in feed-intake regulation. For example, the voluntary intake of most high-fiber feedstuffs is increased by pelleting. Decreased digestibility frequently is associated with pelleting. However, the process of grinding and pelleting results in a more rapid rate of passage through the digestive tract and as a consequence the animal is able to eat more.

Feed intake is also influenced by the physiological state of the animal. Heavily lactating cows, as a result of their increased requirements, consume much more food than dry cows. Even in a feedlot situation, the relative feed intake of fat steers is considerably below that of thin animals.

It is important to recognize that a small increase in the digestibility of a forage, say 10 per cent, may be accompanied by a 100 per cent increase in animal production, because the combination of increased digestibility and increased feed intake enables the animal to obtain significant amounts of energy in excess of that required for maintenance.

It should now be understood that the nutritive value of a feedstuff is the result of many interacting factors—some chemical, some physical, some plant, and some animal. The student should not become so involved with the complexities of these interactions that he misses the simplicities and the order that is also present. "Wisdom is the principal thing. Therefore, get wisdom. And with all thy getting get understanding" (*Proverbs* IV, 7).

FURTHER READINGS

Lofgreen, G. P., and W. N. Garrett. 1968. "A system for expressing net energy requirements and feed values for growing finishing beef cattle." *J. Anim. Sci.*, 27:793–806.

Morrison, F. B. 1956. *Feeds and Feeding*. 22d ed. Ithaca, N.Y.: Morrison.

Nehring, K., R. Schiemann, and L. Hoffman. 1969. "A new system of energetic evaluation of food on the basis of net energy for fattening." *In* K. R. Blaxter, J. Kielanowski, and G. Thorbek, eds. *Energy Metabolism of Farm Animals*. EAAP Publ. 12. London: Oriel Press. Pp. 41–50.

NRC. 1966c. *Biological Energy Interrelationships and Glossary of Energy Terms*. Publ. no. 1411. Washington, D.C.: National Acad. Sci.

NRC. 1969. *United States-Canadian Tables of Feed Composition*. Publ. no. 1684. Washington, D.C.: National Acad. Sci.

Schneider, B. H. 1947. *Feeds of the world, their digestibility and composition*. Morgantown, W. Va.: Agr. Exp. Sta.

Thirty-Six

Metabolic Conversions

"The highest aim of science is to see, amid the mass of bewildering facts, the few concepts which are basic, and those from which predictions can be made."

D. E. Green and R. F. Goldberger
Molecular Insights into the Living Process

During the last three decades remarkable advances have been made in our knowledge concerning the chemical reactions involved in supplying the energy for productive purposes: that is, for work performed, for growth (including growth of animal fibers), for lactation, or for fattening. In the 1920's some very simple concepts which involved only two or three chemical reactions were developed to explain how energy was obtained to produce contraction (shortening) of muscle on which our movements depend. As our knowledge increased concerning the enzymes and metabolic products present in muscle, these concepts, of necessity, were modified.

In this chapter an attempt has been made to provide some information on the present state of knowledge of the chemical reactions involved in production by animals. Unfortunately, it is not possible to present *all* such information in a manner which can be readily comprehended by beginning students in biology. Rather the objective here is to provide the student with a preliminary appreciation of the complex series of chemical reactions and of the roles assigned to special structures within cells. An appreciation may also be gained of the function of some of the minor nutrients, such as vitamins, serving as cofactors and catalysts.

The outputs of animals that are used by man and are necessary for the continued existence of the individual and of the species can be described in chemical terminology. For example, we recognize that meat is composed of water, protein, lipid, minerals and other chemical entities. Similarly, the inputs can be described chemically. The inputs include the water, carbohydrate, lipid, protein, and other nutrients consumed, as well as oxygen obtained from the atmosphere. As in any functional physical system, the inputs must provide for the total outputs plus the losses incurred during formation of the products. It is evident that unless the outputs of animals are chemically identical to the inputs (which is not the case), there must be chemical conversions within animals that allow for production of the products from the inputs. The totality of the chemical conversions between the inputs and outputs of animals is known as intermediary metabolism. Therefore it is important to consider the manner in which the chemical inputs are converted in animals to allow for outputs.

Inputs in Animals

The major organic components of the diets and the materials absorbed and available for cellular conversions have been discussed in Chapters 31 and 34. These are the chemical supplies which the cells have to use as inputs for cellular functions.

Outputs of Animals

What the outputs of animals are tend to differ somewhat depending on one's point of view. At this juncture, it may be advantageous to consider the products of animal metabolism to be the outputs, rather than those normally considered in animal production.

Energy

Potential free energy is one of the outputs of animal metabolism. There is an overemphasis on heat units in describing energy metabolism when, in fact, heat of combustion, in itself, has little relevance to occurrences in living cells. It is correct that oxidation of substrates does occur in cells and that the energy derived from oxidation does serve for the energy-requiring functions of cells. However, it must be emphasized that the form in which energy is utilized is not as heat, but as chemical bonds. The common denominator of energy transitions is the "high-energy" phosphate bond (~ P) which is usually encountered as a pyrophosphate ester bond in adenosine-5′-triphosphate, ATP (Figure 36.1). The energy that is potentially useful in metabolism is conserved by forming ~ P by the conversion of adenosine-5′-diphosphate, ADP, and inorganic phosphate to ATP. Biological reactions requiring an energy input are driven by the coupled cleavage of ~ P in which ATP is converted to ADP plus

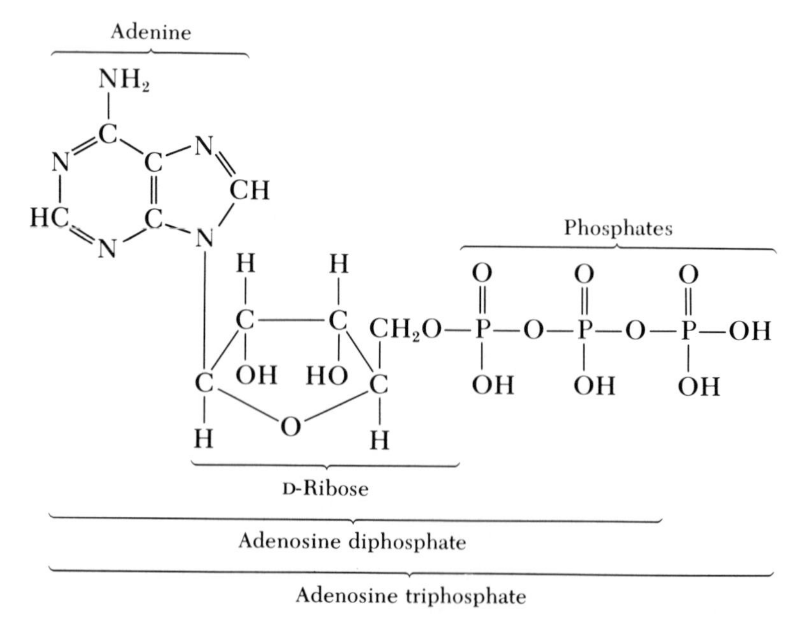

Figure 36-1 Molecular structure of adenosine triphosphate (ATP).

phosphate, making free energy available. Thus the ~ P can be regarded as the "package," or "currency," of energy conversions.

It is pertinent to consider how the "energy package," (~ P) is derived from the chemical materials available for metabolism. Free energy must be available in order to make ~ P. Free energy results from the conversion of a substrate to a product at a lower energy level. Thus, there is a coupling between the energy release and energy conservation. To achieve greatest efficiency of energy conservation, the energy drop during conversion of the substrate should match the energy that may be conserved as ~ P. If there is an excessive energy release, that energy not conserved appears as heat, which is, ordinarily, a waste. The maximum energy that may be derived from nutrients occurs when the carbon of the nutrients is completely oxidized to CO_2. Thus, to obtain maximum energy conservation, there should be maximum energy yield (oxidation), but this must be carried out in steps that match the formation of ~ P. A number of preparative arrangements or conversions are required to achieve such precise steps of degradation, or oxidation, of any molecule. These preparative conversions should result in a minimal energy loss.

Carbohydrate degradation Some of the foregoing relationships of energy conservation in animals are shown in biological degradation of glucose, which is usually the most abundant of the monosaccharides available for metabolism and it occupies a central role in energy metabolism.

The initial, or glycolytic, phase of glucose degradation is represented in Figure 36.2. Initially a phosphate (PO_4) "handle" is attached to the terminal alcohol group of the glucose. This requires the expenditure of a ~ P and is a preparative step. In further preparations, the original glucose is manipulated to give an alcohol group at carbon 1, to which is attached another phosphate handle by expending another ~ P. The hexose diphosphate is then cleaved to two triose phosphates which are readily interconverted by shifting the position of the carboxyl group. The aldotriose is then simultaneously oxidized and phosphorylated to a phosphate anhydride 1,3 diphosphoglycerate. The oxidation is accomplished by the removal of a pair of electrons from the aldotriose phosphate. In animals, electrons derived in oxidation are not turned free (which would generate electricity), but are transferred to electron carriers or acceptors. In this instance the electron acceptor is a compound known as nicotinamide adenine dinucleotide, NAD, a derivative of niacin, one of the B-vitamins. An oxidative step results in a drop in the energy level of the substrate, and part of the energy released is used to phosphorylate the acid group. A phosphate anhydride, when cleaved, provides energy for the formation of a ~ P, and this is what occurs in glycolysis. Thus, from each half of the starting glucose, one ~ P is derived during conversion of diphosphoglycerate to 3-phosphoglycerate. The 3-phosphoglycerate is then taken through two rearrangements to yield phosphoenolpyruvate. The purpose of the rearrangements is to concentrate the energy within the molecule into a bond that, when cleaved, will match the formation of a ~ P. This is achieved during conversion of phosphoenolpyruvate to pyruvate. Under anaerobic conditions, or conditions wherein oxygen is lacking, such as in muscle during sudden periods of strenuous exercise, the electrons derived in the oxidative step of glycolysis are used to reduce pyruvate to lactate, which is the end-product of the scheme. At this point there has been a net yield of two ~ P per hexose.

The glycolytic system serves as a pathway of metabolism of the carbon of all the monosaccharides that may be available for utilization by cells. For example, galactose and fructose, which are components of milk sugar, and sucrose enter the route at the level of hexose phosphate, whereas pentoses may enter as hexose phosphate or triose phosphate. In addition, glycerol, after phosphorylation and oxidation, enters this scheme at the level of the ketotriose phosphate.

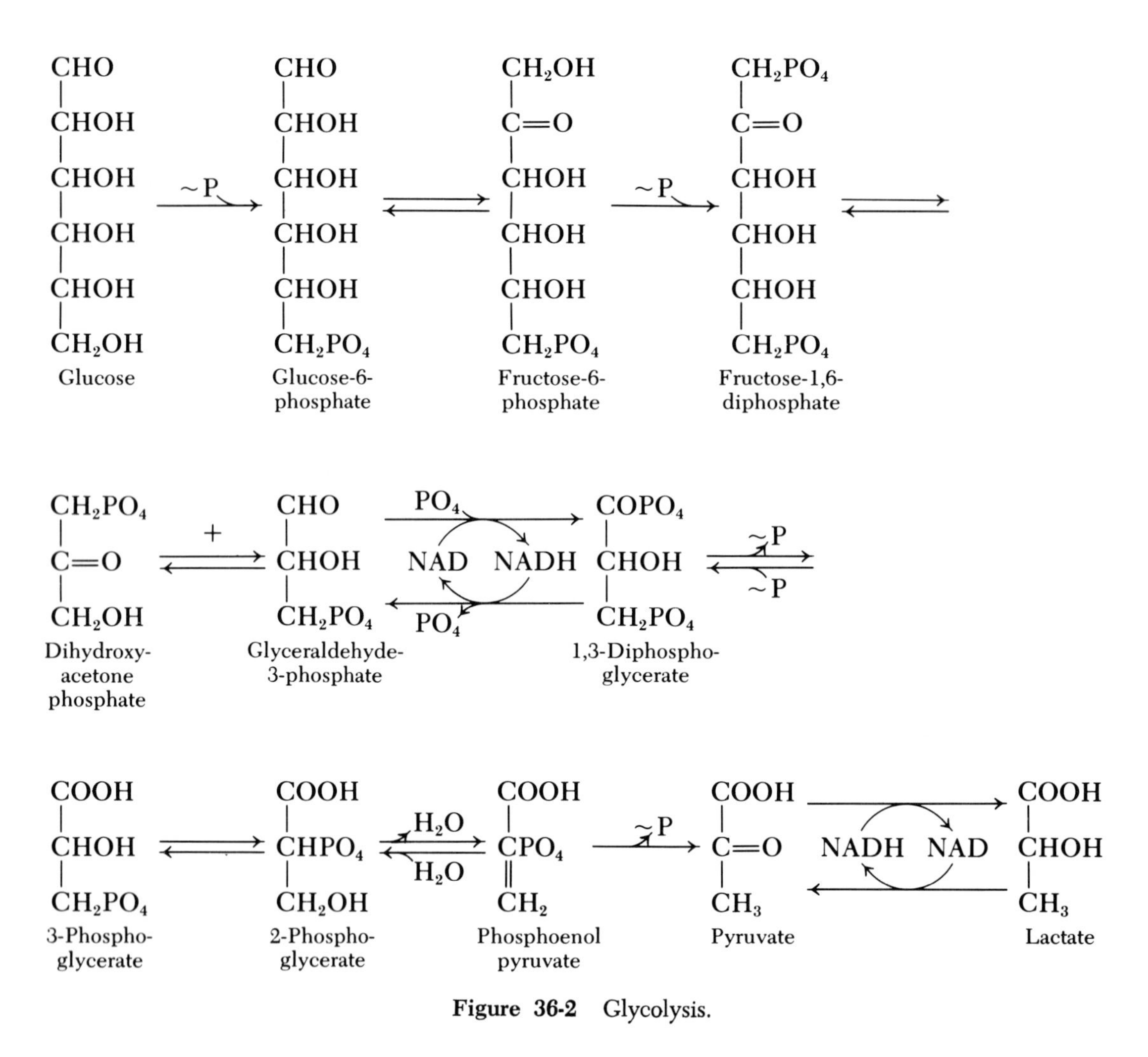

Figure 36-2 Glycolysis.

The glycolytic conversions occur in the soluble portion or cytoplasm of cells (Figure 36.3). However, degradation of pyruvate takes place in the mitochondria, which are membrane-enclosed microbodies within cells.

Pyruvate, formed during glycolysis, must enter the mitochondria, where it is oxidatively decarboxylated in a reaction which requires the participation of lipoic acid and a derivative of thiamine. In this oxidation, part of the energy released from the substrate is used to form an active two-carbon fragment (acetyl coenzyme A) as the product of the reaction. The electrons derived are used to reduce NAD. The acetyl coenzyme A (acetyl CoA) is active because the thioester bond between the

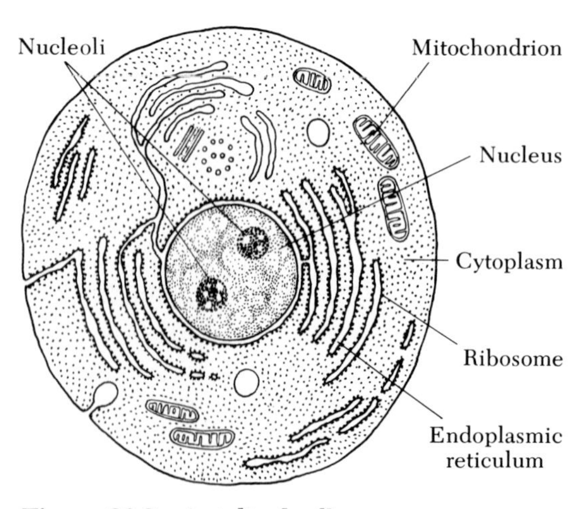

Figure 36-3 A stylized cell.

acetate and CoA, which is a derivative of pantothenic acid, is a high-energy bond. At this point the original glucose has been reduced to two CO_2 and two acetate; it has been partially oxidized, and there has been a net formation of two ~ P and reduction of four NAD. Acetyl CoA represents the carbon that is now prepared for the sequence of reactions leading to complete oxidation. This sequence is known as the tricarboxylic acid (TCA) cycle (Figure 36.4) and was first elucidated by Sir Hans Krebs. The acetyl CoA condenses with oxaloacetate to form citrate. This in turn is rearranged to isocitrate, which is in an appropriate form to undergo oxidative decarboxylation to 2-ketoglutarate. The electrons derived from this oxidative step of the TCA cycle are used to reduce NAD. The 2-ketoglutarate is oxidatively decarboxylated in a fashion analagous to the degradation of pyruvate, yielding CO_2, reduced NAD, and succinyl CoA. The energy of the thioester bond in succinyl CoA is then conserved by formation of a ~ P during cleavage to free succinate. Succinate is oxidized to fumarate; the pair of electrons derived in this oxidative step are used to reduce an electron carrier that is a derivative of riboflavin. The fumarate must be hy-

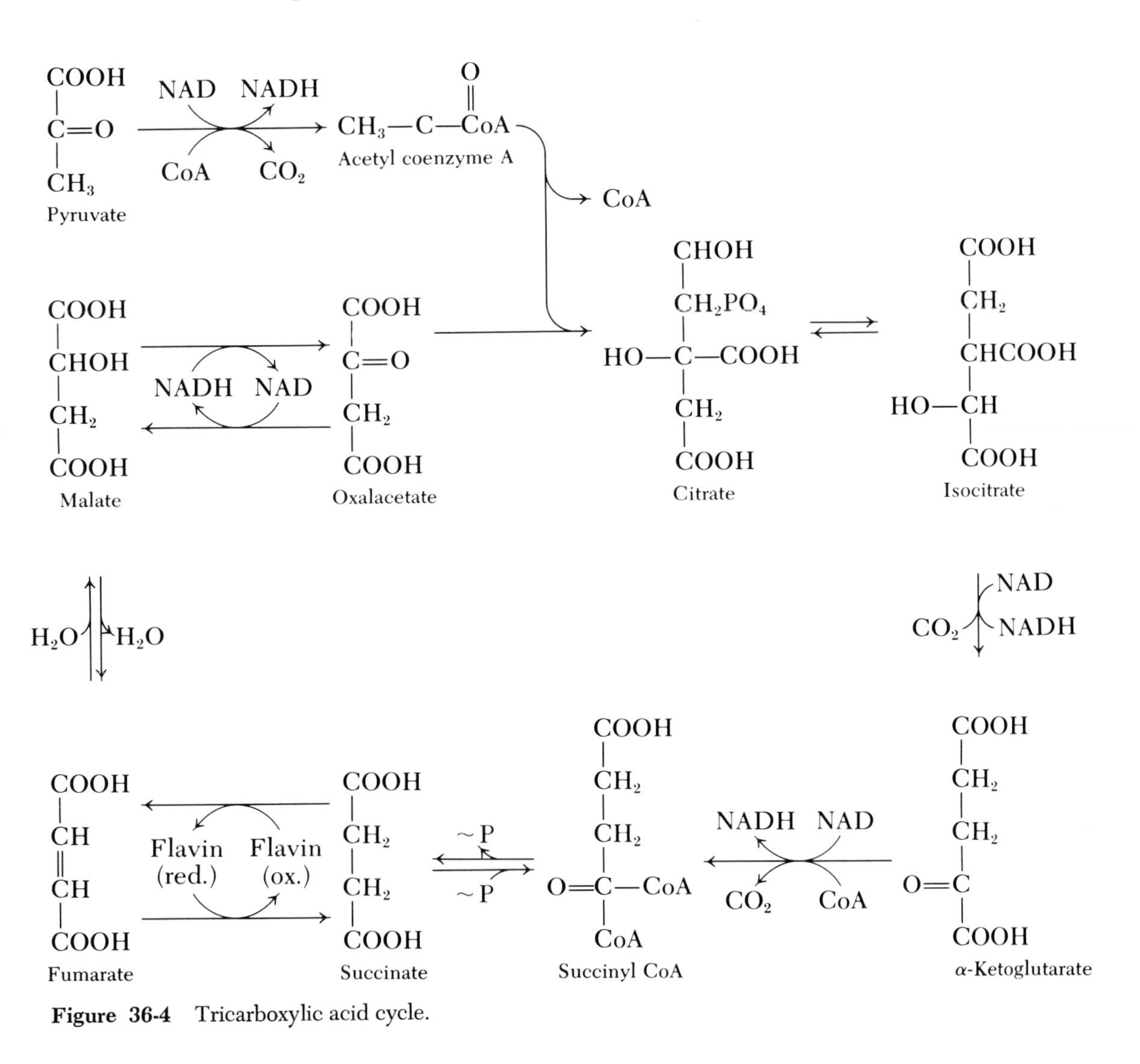

Figure 36-4 Tricarboxylic acid cycle.

drated to malate prior to the next oxidation, which yields oxaloacetate plus reduced NAD. Having regenerated oxaloacetate, it could condense with another active acetate and repeat the sequence. The overall conversion in the TCA cycle, then, is:

$$\text{acetyl CoA} + 2\ \text{NAD} + \text{flavin}_{\text{ox}} \longrightarrow 2\ CO_2 + 3\ \text{NADH} + \text{flavin}_{\text{red}} + \sim P$$

There is an input of six carbons as glucose and an output of six carbons as CO_2 from the complete oxidation of glucose through glycolysis and the TCA cycle. The oxidation of glucose occurred in a very specific fashion in twelve oxidative steps, during which a pair of electrons was released at each step. The electrons were transferred to carriers rather than directly to oxygen as in the combustion through burning in air. In addition, there was net formation of four $\sim$ P.

Fatty acid degradation The oxidation of fatty acids occurs first through a sequence known as beta-oxidation (Figure 36.5). The long-chain fatty acids are first "activated" by expenditure of two $\sim$ P to attach CoA. Then a sequence of reactions follows, entirely in the mitochondria, that results in the sequential cleavage of two carbon units, as acetyl CoA, from the carboxyl end of the fatty acid. Initially, the saturated fatty acyl CoA is oxidized to an unsaturated derivative, which is hydrated and oxidized again to a 2-keto intermediate. This is then subjected to cleavage by CoA to yield a fatty acyl CoA that is two carbons shorter; the two carbons appear as acetyl CoA. The initial oxidation yields a reduced

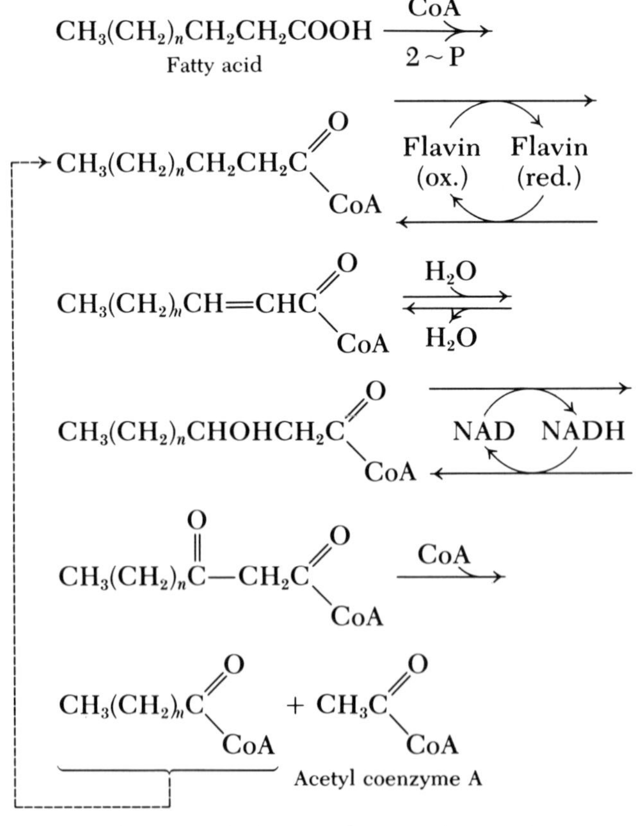

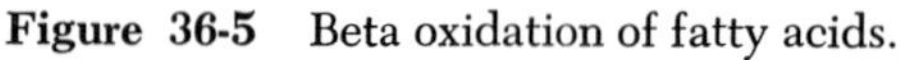

Figure 36-5 Beta oxidation of fatty acids.

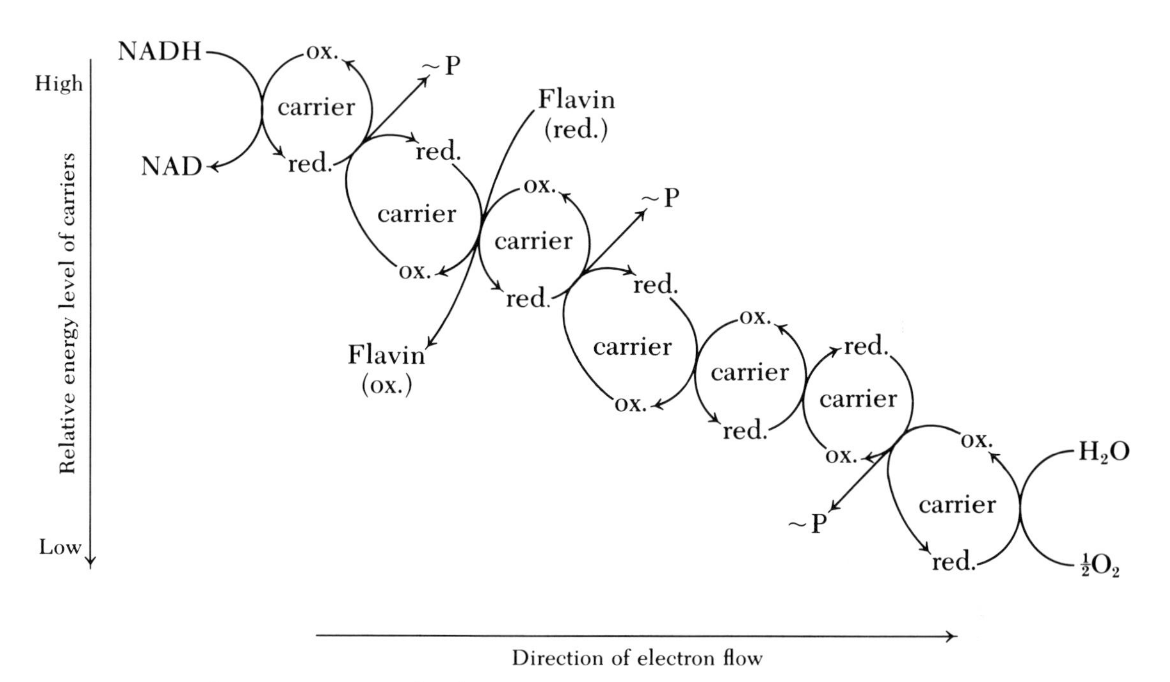

Figure 36-6 Electron transport and P formation.

flavin coenzyme, while the second reduces NAD. The sequence can be repeated until fatty acids containing an even number of carbons are degraded entirely to two carbon fragments. A part of the sequence of conversions in beta-oxidation is analagous to part of the TCA cycle. The essence of manipulation of the fatty acyl CoA involves an oxidation with a flavin coenzyme, followed by hydration in preparation for oxidation with NAD as the coenzyme. The same sequence of chemical changes is seen in going from succinate to oxaloacetate in the TCA cycle (Figure 36.4). The acetyl CoA units produced from fatty acids are indistinguishable from those produced from pyruvate, and enter into and are completely oxidized in the TCA cycle as discussed previously for carbohydrates.

A large variety of fatty acids can be oxidized in the beta-oxidation sequence, because there are virtually no limitations imposed upon the chemical nature of the entering fatty acids.

Electron transport Thus far, in considering the three central routes of metabolism (glycolysis, TCA cycle, and beta-oxidation) as means of derivation of biologically useful energy (~ P), there have been only three steps at which conservation of energy is achieved through the production of ATP. These are the conversions of diphosphoglycerate to 3-phosphoglycerate, of phosphoenolpyruvate to pyruvate, and of succinyl CoA to succinate. However, at each step in which oxidation occurred there was a substantial decrease in the energy content of the substrate. We noted that at each of these oxidations the electrons removed from the substrate were received by a carrier. The reduced carriers are at a higher energy potential than are the oxidized forms of the carriers; hence, at the oxidative steps of metabolism, most of the energy decrease of the substrate is conserved by increasing the potential energy of the electron acceptors. How is the potential energy of the reduced electron carriers converted into energy currency (~ P)?

In the mitochondria, there is a system that exists as a unitized sequence of electron carriers. There are seven individual carriers constituting the chain (Figure 36.6). Electrons are provided, in pairs, to the chain from NADH, or reduced flavin coenzymes, from the oxidative steps where we have seen these compounds to be produced. The electrons flow through the chain down a decreasing energy potential and eventually are accepted by oxygen to form water. The energy gradient between the carriers is arranged in such a way that between NADH and oxygen there are three major energy drops of sufficient magnitude that the free-energy release during transferral of electrons through these sites is harnessed to drive the formation of ~ P. Between the entry from reduced flavin coenzymes and oxygen there are only two sites of ~ P formation. The coupling of electron transferral down an energy gradient to ~ P formation is known as oxidative phosphorylation. The oxidative steps of metabolism giving rise to NAD reduction yield three ~ P; those giving rise to flavin coenzyme may yield two ~ P through oxidative phosphorylation. The great majority of ~ P formation occurs by means of oxidative phosphorylation. For example, during the complete oxidation of a mole of glucose there was formation of four ~ P, two reduced flavin coenzymes, and ten NADH. Therefore, of the 38 ~ P that may be derived, 34 arise from oxidative phosphorylation. This system of oxidative phosphorylation is remarkable, in that the electrons from a great number of different reactions can be channeled into one common sequence for energy conservation.

Amino-acid degradation and the urea cycle The problems posed to the biological system to achieve the degradation of the twenty-odd amino acids are greater than those encountered with either monosaccharides or fatty acids, since chemical differences between individual amino acids are much greater than the differences among the other substrates. Thus, preparation of the carbon skeletons of amino acids for entry into the common schemes of degradation requires virtually individual manipulation of each amino acid, although the basic reactions of oxidation, hydration, cleavage, and isomerization are employed. Space does not allow for discussion of the routes of degradation of each amino acid, but there are some common denominators in-

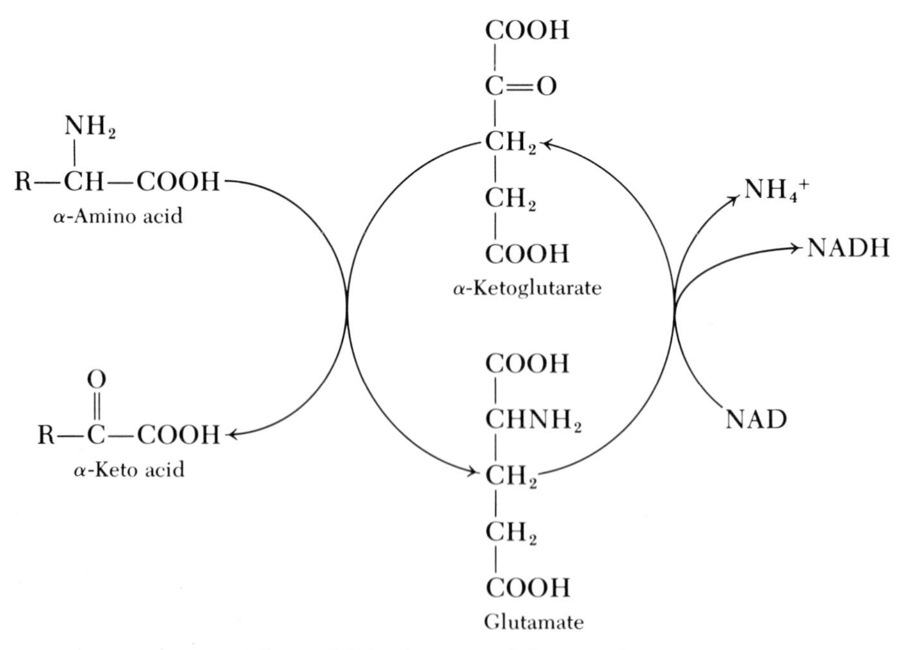

Figure 36-7 Amino acid deamination.

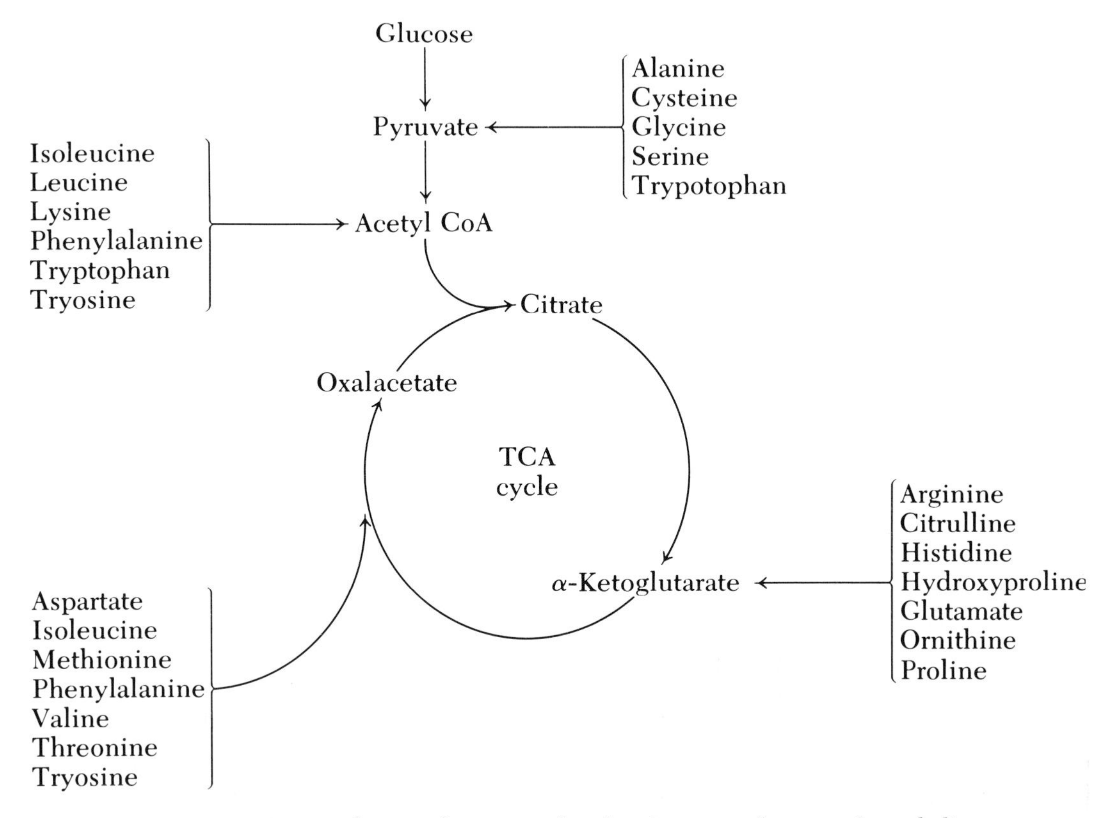

Figure 36-8 Points of entry of amino acid carbon into central routes of metabolism.

volved. Amino acids, by definition, contain amino groups. In order to metabolize amino acids one of the first requisites is the removal of the amino group. The major means of deamination of alpha-amino groups probably involves transamination as the initial step, and α-ketoglutarate, an intermediate of the TCA cycle, is the most common amino group receptor (Figure 36.7). The enzymes catalyzing transamination contain a derivative of pyridoxine as a coenzyme. The glutamic acid formed from transamination is then deaminated to regenerate the α-ketoglutarate in an oxidative reaction resulting in formation of NADH, which can provide electrons for oxidative phosphorylation, plus free ammonia. The carbon skeleton of the amino acids is then carried through preparative reactions until it finally enters the common schemes of complete oxidation at specific points (Figure 36.8). As in other metabolites, about one-third of the ~ P that may be formed from amino acids are derived prior to entry of the amino-acid carbon into the TCA cycle and about two-thirds are formed as a result of the TCA-cycle oxidation.

It was mentioned above that the nitrogen of amino acids was, during their degradation, converted to ammonia. Ammonia is toxic in animals and cannot be allowed to accumulate. The ammonia is detoxified and removed by conversion to urea, which is excreted in the urine. This process appears to occur exclusively in the liver. The first step is the combination of ammonia with CO_2 and phosphate in an energy-expending reaction to form carbamyl phosphate (Figure 36.9). The carbamyl phosphate is condensed with ornithine to form citrulline, which in turn, is combined with aspartate in another energy-expending reaction to form arginosuccinate. The arginosuccinate is then cleaved in such a manner that the carbon derived from aspartate appears as fumarate and the nitrogen of aspartate appears in the guanido group of arginine. Finally, arginine is cleaved to urea and ornithine, thus completing the cycle by generating the starting compound, or "sparker" (ornithine). An addition to the formal urea cycle itself would be the cylic regeneration of aspartate from fumarate. Presumably the fumarate could be

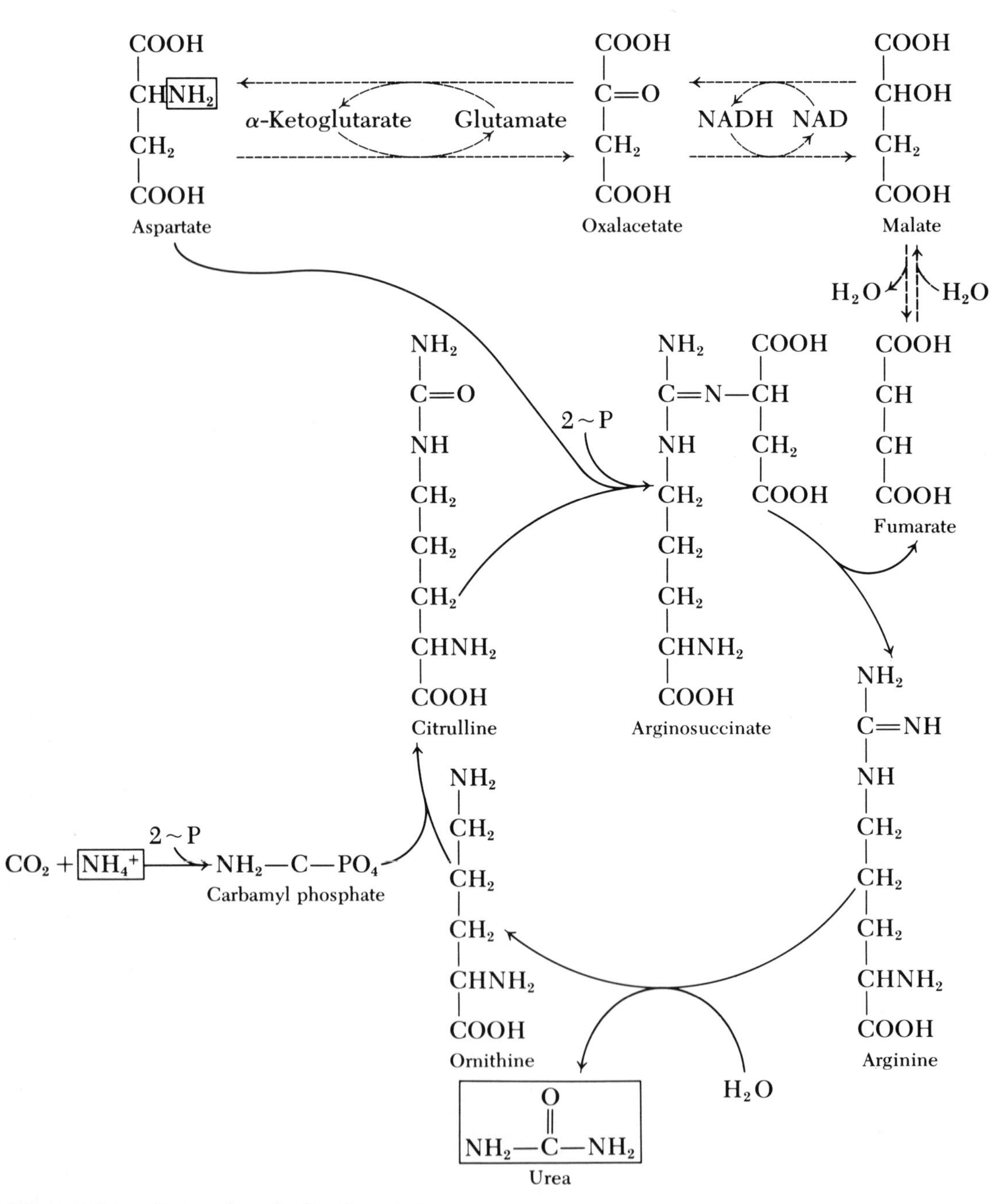

Figure 36-9 Urea cycle and related reactions.

converted to oxaloacetate by means of TCA-cycle reactions, and the oxaloacetate transaminate with glutamate to yield aspartate. The process is then ready to begin again and generate another molecule of urea. The carbon of urea arises from CO_2, whereas one amino group is likely from free ammonia. The other amino group may arise from transaminations rather than ammonia.

Generalities In the foregoing, the economy of systems in intermediary metabolism were emphasized and the major pathways were shown to be able to accommodate inputs from a number of different sources. The pathways were, however, discussed entirely in the light of their functions in degradation of substrates and the formation of ~ P. Such an approach was primarily for the convenience of discussion, as it would be incorrect to infer that ~ P production, or degradation of substrates, can proceed indiscriminately. As was indicated earlier, energy metabolism consists of the formation and utilization of ~ P. The formation must be coordinated with the utilization, since the amount of ADP for conversion to ATP is limited. The animal body requires energy for synthesis of products for maintenance and production. The ATP serves as a reservoir of available energy for these essential functions.

Animal Products

The products of animals result from the expenditure of ~ P and from formation of new molecules from the substrates of metabolism. Individual animal maintenance is one output that requires a rather large, fairly direct expenditure of ~ P for such functions as maintenance of the ionic differences between the interior and exterior of cells, for muscular contraction and relaxation (heart), and for nervous transmission. When physical work is an output of animals, this results from the expenditure of ~ P during muscular contraction and relaxation.

The animal products that are the major energy sources for human consumption and comfort are of course identifiable chemical entities. These involve ~ P expenditure during their formation. This energy is generated through the metabolite degradation reactions. At the same time, products involving synthetic processes are formed from the metabolites available to animal cells. In some instances a great deal of chemical rearrangement is required for synthesis. It is therefore important to consider the means of synthesis of the major chemical outputs of animal metabolism.

Protein Proteins are a particularly important component of meat, milk, fibers, and leather. The ultimate stage of protein synthesis is the chemical joining through peptide bonds of a proper number of amino acids in the right sequence to form a protein. This process occurs largely at the endoplasmic reticulum of cells (Figure 36.3). The amino acids are first attached in an energy-expending reaction to molecules of ribonucleic acid (RNA). There are specific RNA molecules for each amino acid:

$$\text{amino acid} + \text{tRNA} \xrightarrow{2\sim P} \text{amino acid-tRNA}$$

The RNA in this reaction is known as transfer RNA (tRNA). The amino-acid-tRNA complex is then utilized by a particle known as a ribosome (ribosomal RNA) in the presence of messenger RNA (mRNA) for attachment of the amino acid to the carboxyl end of a growing peptide chain:

$$\text{aa-tRNA} + \text{ribosome-mRNA-peptide}_{(n)}$$

$$\xrightarrow{2\sim P} \text{tRNA} + \overset{\text{ribosome}}{\overset{|}{\text{m}}}\text{RNA-peptide}_{(n+1)}$$

The ribosome unit provides the enzymatic machinery for formation of the new peptide bond, whereas the mRNA provides the information that determines the order of addition of each amino-acid-tRNA unit on the basis of specific recognition of the tRNA component. The order of reaction determines the amino acid sequence in the peptide chain. The mRNA also apparently determines the number of additions and hence the size of the protein synthesized. The information in mRNA is probably derived directly from deoxyribonucleic acid (DNA), located in the nucleus of the cell, which carries the genetic information of the cell. The process of protein synthesis is incomparably precise and reproducible, because a "mistake" of only one amino acid in a protein chain can have lethal results.

Before protein synthesis can occur, there must be amino acids available. Amino acids are the absorbed products of digested proteins. Many of these are used directly by animals in the formation of their proteins. The majority of the essential amino acids incorporated into proteins synthesized by animals are derived directly from digestion in the alimentary tract. However, the nonessential amino acids can be synthesized within animal cells. The synthesis of nonessential amino acids is somewhat akin to the degradation of amino acids, in that the carbon compounds employed for synthesis arise from a few major points in metabolism. Glutamate is formed from the TCA-cycle intermediate α-ketoglutarate, simply by the reverse of glutamate deamination. Thus the synthetic reaction is a reductive amination, and the electrons employed for the reduction are from NADH or from reduced nicotinamide adenine dinucleotide phosphate (NADPH):

$$\alpha\text{-ketoglutarate} + \text{NADH, or NADPH} + NH_4^+ \longrightarrow \text{glutamate} + \text{NAD, or NADP}$$

NADP is an electron carrier that differs chemically from NAD only by one phosphate group, but in general its functions in metabolism differ quite markedly from those of NAD. The electrons carried by NAD are often those derived from degradation for ~ P generation, whereas NADP usually transfers electrons that are employed in synthetic reductions.

Glutamate is the precursor of a family of amino acids which includes proline, hydroxyproline, ornithine, citrulline, and arginine. The α-keto acids, pyruvate and oxaloacetate, may readily be transaminated to alanine and aspartate. Serine and glycine can be derived from phosphoglycerate.

Lipids Most of the lipid fraction of animal products is in the form of triglyceride. Usually the three fatty acids making up the triglyceride are fairly long (fourteen or more carbons). The final state of formation of triglycerides is the condensation of three activated fatty acids with a phosphorylated glycerol to form the three-ester bonds:

$$3 \text{ fatty-acyl CoA} + \text{glycerol-phosphate} \longrightarrow \text{triglyceride} + 3 \text{ CoA} + \text{phosphate}$$

The phosphorylated glycerol used for triglyceride synthesis probably arises from glucose by way of reduction of the ketotriose phosphate (dihydroxyacetone phosphate) intermediate of glycolysis:

$$\text{glucose} \longrightarrow \longrightarrow \text{ketotriose phosphate} \xrightarrow{\text{NADH NAD}} \text{glycerol phosphate}$$

The fatty acids used for lipid formation can be those absorbed from the alimentary tract. This is apparent in pigs, which, when fed unsaturated fats in their diets, tend to accumulate unsaturated fats in their lipid stores. Before the absorbed fatty acids are converted to triglycerides it is necessary that they be activated; that is, they must be esterified to CoA. This occurs in a reaction requiring the expenditure of two ~ P per fatty acid.

Fatty acids also can be synthesized in animal cells. The starting point of this synthesis is active acetate, or acetyl CoA (Figure 36.10). The synthesis is catalyzed by a somewhat involved complex of enzymes. Initially one acetyl CoA condenses with the enzyme complex. This acetate unit serves as the foundation stone for the new fatty acid, and its carbon will be found at the methyl end of the product. The acetate-enzyme complex then condenses with malonyl CoA. The malonyl CoA is derived from acetyl CoA in an energy-expending reaction, with biotin as a cofactor, that fixes CO_2 on the methyl group of acetyl CoA:

$$\text{acetyl CoA} + CO_2 \xrightarrow{\sim P} \text{malonyl CoA}$$

The malonate entity is chemically more suitable for later condensation than an acetate unit. The malonate and acetate units, while attached to the fatty-acid synthesizing complex, condense to form a 3-keto intermediate which is reduced using electrons from NADPH. The resultant 3-hydroxy intermediate is dehydrated and reduced again using electrons from NADPH to form a saturated fatty-acyl intermediate that is ready for the addition of another two carbons from malonyl CoA. The addition of two carbon units continues until a fatty acid of the appropriate chain length is formed, at which time it is re-

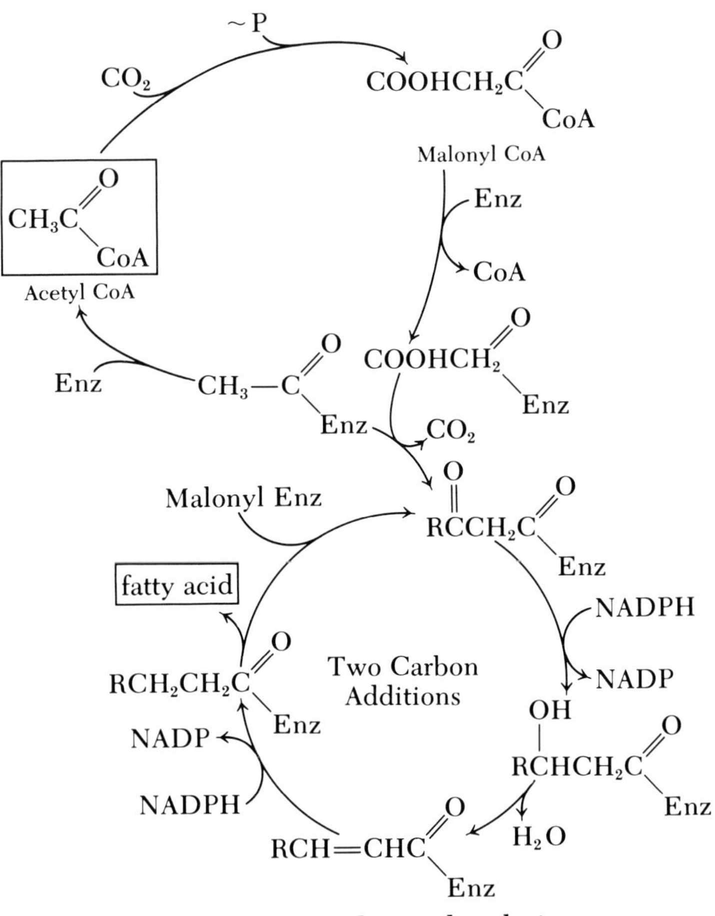

Figure 36-10 Fatty acid synthesis.

leased from the enzyme complex. All the carbon incorporated into fatty acids comes from acetyl CoA; all but the two carbons at the methyl end of the new fatty acid go through malonyl CoA.

Since acetyl CoA, as we have seen previously, can arise from carbohydrate or from amino acids, we have the means of conversion of carbohydrate and protein to fat. In ruminants, acetate and butyrate from rumen fermentation could readily give rise to acetyl CoA. As far as the animal is concerned, the primary reason for formation of fat is to create a store of material that can be drawn upon at some future time for derivation of ~ P through beta-oxidation of the fatty acids produced from hydrolysis of stored triglycerides.

Fat synthesis is another example of the linkage between degradation of substrates and the utilization of ~ P. Since ~ P was expended in the carboxylation of acetyl CoA, the source of ~ P is from degradation of metabolites. In addition, when the acetyl CoA is derived from a monosaccharide by way of glycolysis, it is formed in the mitochondria. However, fatty-acid synthesis occurs in the cytoplasm, and acetyl CoA does not diffuse through the mitochondria membranes. It costs one ~ P to transport an acetate unit out of the mitochondria. Up to one-half or more of the NADPH used in fatty-acid synthesis may arise from a transhydrogenation that can be represented as:

$$\text{NADH} + \text{NADP} \xrightarrow{\sim\text{P}} \text{NAD} + \text{NADPH}$$

The formation of NADPH is another point of direct energy expenditure in fatty-acid synthesis. The NADPH that does not arise from transhydrogenation may be reduced during oxidation of glucose via a route (essentially a loop off the glycolytic scheme) known as the

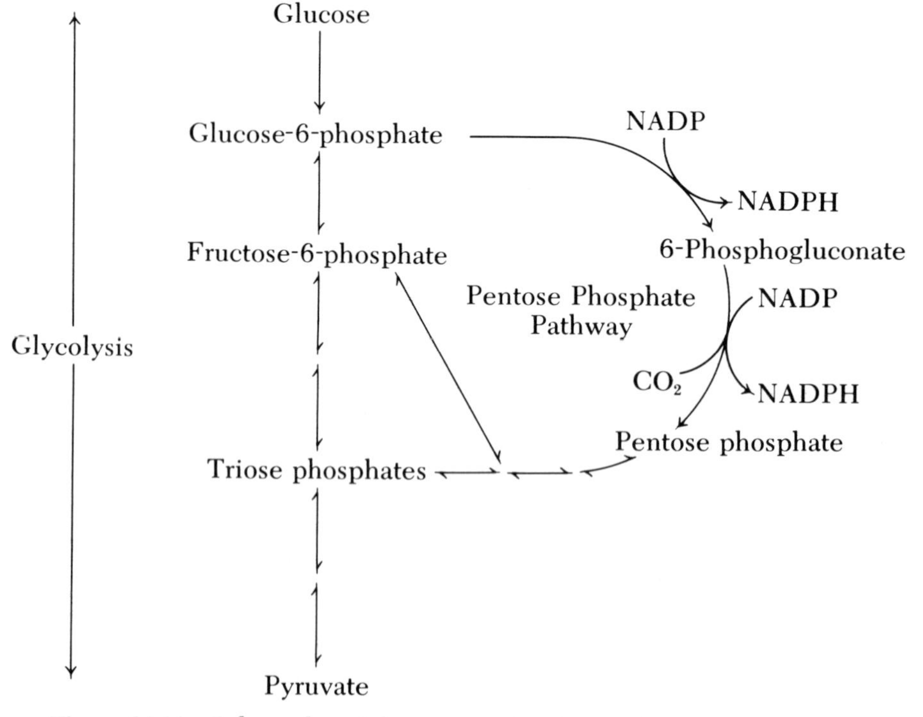

Figure 36-11 Relationship of the pentose phosphate pathway and glycolysis.

pentose phosphate pathway (Figure 36.11). The first two conversions of this route result in the oxidation of the carbon in the first position of glucose phosphate, with NADPH as the electron acceptor. The remaining five carbons of glucose pass through a pool of pentose phosphate that can be used in nucleic-acid synthesis, or they can reenter the glycolytic scheme by way of fructose phosphate and triose phosphate. This alternate route of glucose oxidation generally appears to be more active in those tissues carrying on relatively extensive synthesis, which is logical in view of the observation that NADPH is the more common electron donor in synthesis.

Carbohydrates With the exception of milk, carbohydrates are quantitatively a very minor output of animals. Lactose, or milk sugar, does constitute a substantial portion of the dry matter of milk. Lactose is a disaccharide composed of a glucose and a galactose unit. The final step of lactose synthesis is the condensation of a molecule of free glucose with a ribonucleotide diphosphate derivative of galactose:

$$\text{glucose} + \text{ribonucleotide diphosphate-galactose} \longrightarrow \text{lactose}$$

The glucose can arise from digestion and absorption. The galactose derivative may be formed from the glucose phosphate intermediate of glycolysis:

$$(1)\quad \text{glucose-6-phosphate} \longrightarrow \text{glucose-1-phosphate}$$

$$(2)\quad \text{glucose-1-phosphate} + \text{ribonucleotide} \xrightarrow{\sim P} \text{ribonucleotide diphosphate-glucose}$$

$$(3)\quad \text{ribonucleotide diphosphate-glucose} \longrightarrow \text{ribonucleotide diphosphate-galactose}$$

This then raises the question of where glucose phosphate may come from. An obvious answer would be from glucose that was absorbed. However, in ruminants only small amounts of glucose are absorbed. Glucose phosphate can arise from gluconeogenesis, which is the formation of glucose from noncarbohydrate precursors of dietary origin, notably glycerol, some amino acids, and, particularly in ruminants, propionic acid. Oxaloacetate performs a central role in gluconeogenesis (Figure 36.12). The oxaloacetate can be converted to phosphoenolpyruvate, which is converted to glucose phosphate by the reversal of the conversions involved in glycolysis, with the exception of the conversion of fructose diphosphate to fructose phosphate. In this case the synthetic route is the hydrolytic cleavage of one of the phosphate groups. Similarly, free glucose can be formed from glucose phosphate by hydrolytic cleavage of the phosphate group. This is an example of metabolic economy, in that many of the same conversions are employed in gluconeogenesis and in glycolysis, even though gluconeogenesis probably occurs to a significant extent in only a few tissues (liver, kidney). The oxaloacetate used for gluconeogenesis can be replenished by carboxylation (the addition of CO_2) of pyruvate. Hence, those amino acids that yield pyruvate in their degradation (cysteine, serine, alanine) can provide oxaloacetate. The metabolism of propionate entails conversion to succinyl CoA, which, by way of the TCA-cycle reactions, can yield oxaloacetate. In ruminants, propionate is a very important source for replenishment of oxaloacetate. The amino-acid carbon that, upon degradation, enters the TCA cycle at the level of four (or five) carbon intermediates could also supply oxaloacetate. However (Figure 36.12), if there is entry into the TCA cycle as a two-carbon unit (acetyl CoA), two carbons are lost as CO_2 before there is formation of oxaloacetate. Thus, there is no net formation of oxaloacetate; as a consequence, net gluconeogenesis from acetyl CoA does not occur. The carbon of metabolites (even-numbered fatty acids, some amino acids) that pass through acetyl CoA, by way of the TCA cycle in animals, is not gluconeogenic. This may

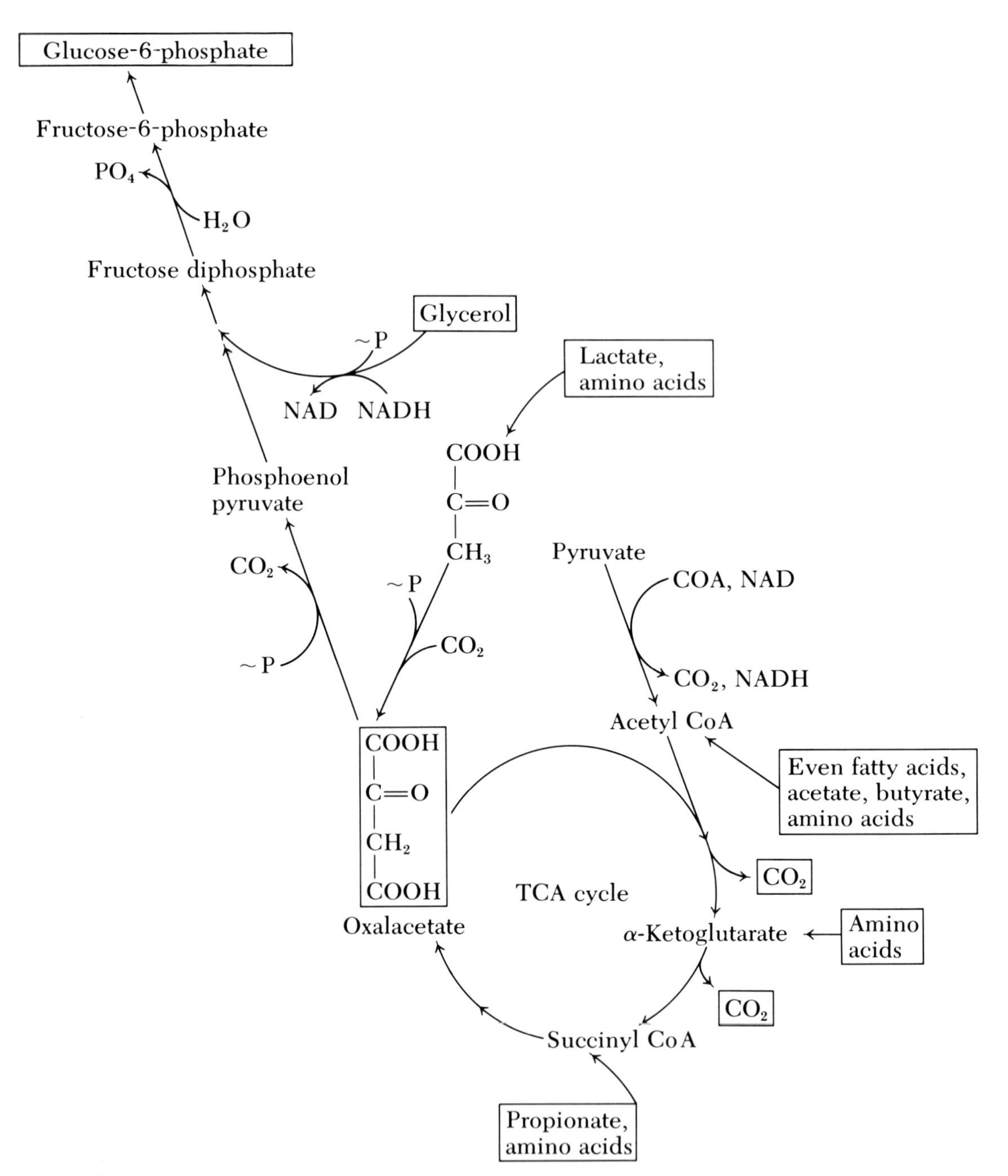

Figure 36-12 Interrelationships in gluconeogenesis.

pose somewhat of a limitation, particularly in ruminants, wherein a very substantial portion of the carbon derived from the diet is absorbed in the form of acetate and butyrate.

This has been a rather fragmentary approach to intermediary metabolism from the point of view of factual presentation. Large areas of metabolism have been skimmed and other areas have not been discussed. Tissue differences in emphasis of the various aspects of metabolism and the metabolism of compounds (nucleic acids, porphyrins, glycogen) that may not be of great quantitative importance in animal products, but which are essential to survival, have not been considered. The object, however, has been to provide a skeleton which, in general terms, shows the relationships between the entities available for chemical utilization and the synthesis of products by animals. There has been substantial emphasis of generalizations, similarities, and economies to demonstrate the logic of the metabolic scene. The metabolic conversions of animals are of two general classifications: (1) degradation of digested and absorbed material with the energy produced conserved as ~ P; and (2) the synthesis of essential products for animal maintenance and production which requires energy in the form of ~ P.

FURTHER READINGS

Baldwin, E. 1967. *The Nature of Biochemistry*. 2d ed. Cambridge, England: Cambridge University Press.

Falconer, I. R. 1969. *Mammalian Biochemistry*. London: Churchill.

Green, D. E., and R. F. Goldberger. 1967. *Molecular Insights into the Living Process*. New York: Academic Press.

Greenberg, D. M., ed. 1967. *Metabolic Pathways*. 3d ed. New York: Academic Press.

Krebs, H. A., and H. L. Kornberg. 1957. "A survey of the energy transformations in living matter." *Ergeb. der Physiol.*, 49:212–298.

McGilvery, R. W. 1970. *Biochemistry: A Functional Approach*. Philadelphia, Pa.: Saunders.

Thirty-Seven

Meeting Nutrient Requirements for Physiological Activities

"The first principle in biology is that there is within living substance a condition of internal pressure tending toward expansion of self or kind . . . the tendency to take in and assimilate everything assimilable . . . to grow, to multiply."

R. E. Coker
Scientific Monthly; 48:61, 121 (1939)

Nutritional costs attached to the various activities carried out by an animal are many and vary over wide limits. An energy charge first comes to mind because energy supplies the driving force necessary for the maintenance of life, growth, lactation, and work. However, consideration must be made also for similar charges for protein, vitamins, and minerals, although initially their functions may not be as obvious.

Since the intent of this chapter is to deal with nutrient requirements in broad terms, quantitative requirements of nutrients for various functions of the different types of livestock will not be detailed, since they can be found in the appropriate published standards, such as the series "Nutrient Requirements of Domestic Animals," published by the NRC, National Academy of Sciences, Washington, D.C.

Requirements for Maintenance

Energy

The first energy requirement of an animal is to have enough energy to keep the animal alive. Generally this is considered to be equal to that required for fasting catabolism. This plus additional energy to provide for expenditure due to eating and walking, but not for production, is equal to the maintenance energy requirement. This level of intake should prevent a net gain or loss of energy from the animal's body.

Basal metabolic rate (fasting catabolism) is the rate of heat production when food is not being absorbed from the intestinal tract. Measurements of the basal metabolic rate are carried out on healthy animals in a good nutritional state under standard conditions which do not stress the animals in any manner which increases their heat production. The environmental temperature must be such that the animal does not increase its heat production to keep warm or increase its metabolic rate trying to keep cool (e.g., as in panting). The animal should be relaxed because an excited animal will produce more heat than one that is at ease. Ideally the animal should be lying down, since standing may increase the metabolic rate 10 to 15 per cent. To establish a post-absorptive state, an overnight fast is sufficient for a single-stomached animal, such as the pig, but one of four or more days may be required for ruminants (Blaxter, 1962). To avoid some of the problems associated with a four-day fast in ruminants, some workers have determined heat production over a specific time period after the last feeding and have referred to this value as standard metabolism.

Larger animals give off more heat per day than smaller ones, but per unit of body weight the smaller animals give off more heat per unit of time. It has been suggested that the rate of heat production of an animal may be dependent on its surface area. However, since this is difficult to determine, others (Kleiber, 1961; Brody, 1945) have developed mathematical formulae that can be used to predict the basal heat production of adult animals from their body weight. Figure 37.1 shows the relationship of daily heat production to body weight. Experimental data were used to calculate the heavy line. It shows that heat production is a function of body weight to the three-fourths power ($W^{3/4}$). The surface-area line fits the data better than does the one for weight ($W^{1.0}$). The daily fasting

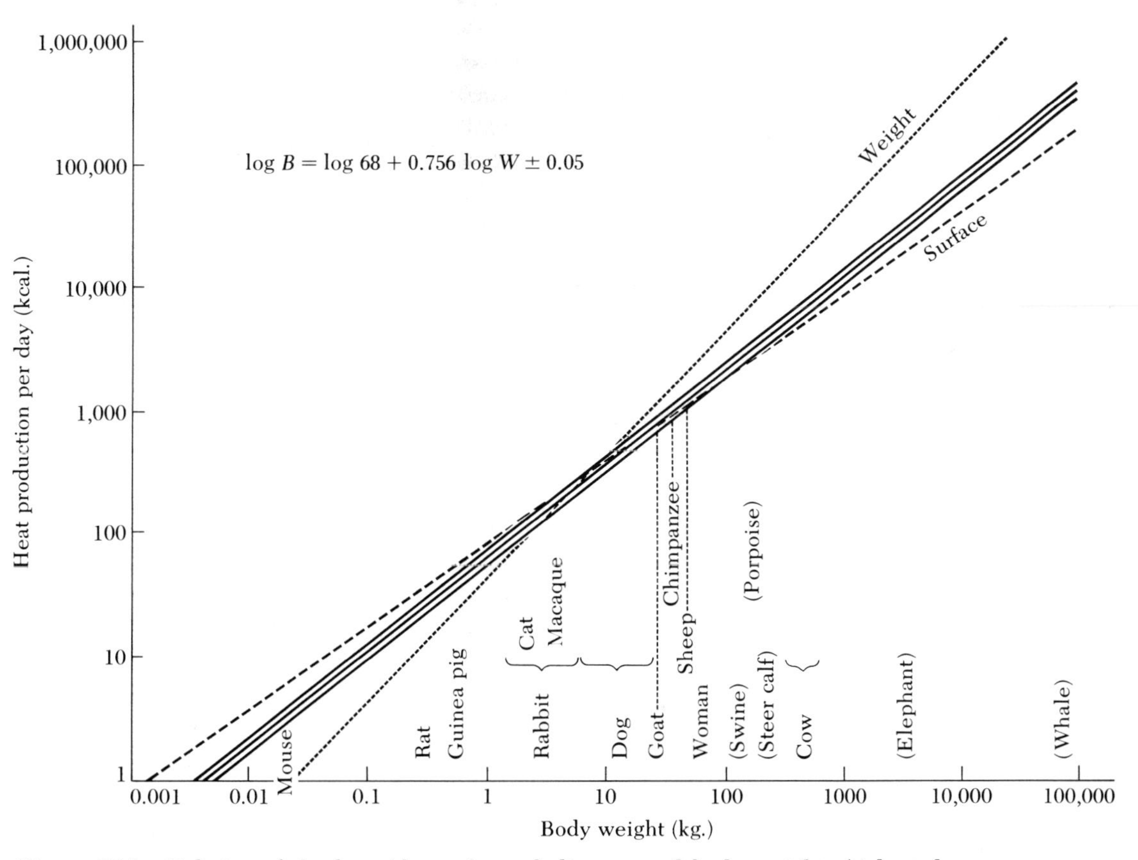

Figure 37-1 Relation of the logarithms of metabolic rate and body weight. (Adapted from Kleiber, 1947.)

heat production can be calculated for animals which have mature body weights of anywhere from 25 g to more than 1,000 kg by use of the generalized formula

$$M = 70\, W^{3/4},$$

where M is basal metabolic rate in kcal per day, and W is the body weight in kg.

Energy required for consumption and digestion of food, that required for the increased respiration and heart rate due to walking and other movements, and that due to low or high temperatures are not accounted for in the determination of basal heat production. Hence, maintenance requirements may vary from 120 to 200 per cent of the basal heat production, depending on the level of activity and on environmental conditions (Blaxter, 1962).

The maintenance energy requirement in terms of total digestible nutrients (TDN), digestible energy (DE), metabolizable energy (ME), and net energy (NE) determined by Garrett *et al.* (1959) from feeding experiments are shown in Table 37.1 for beef cattle and sheep.

Table 37.1.

Equations to estimate the energy requirements of sheep and cattle for maintenance and weight gain.[a]

Sheep	*Steers*
TDN = $0.036W^{3/4}(1 + 2.3G)$[b]	TDN = $0.036W^{3/4}(1 + 0.57G)$
DE = $76W^{3/4}(1 + 2.4G)$	DE = $76W^{3/4}(1 + 0.58G)$
ME = $62W^{3/4}(1 + 2.5G)$	ME = $62W^{3/4}(1 + 0.60G)$
NE = $35W^{3/4}(1 + 1.8G)$	NE = $35W^{3/4}(1 + 0.45G)$

[a] TDN, W, and G are in pounds; DE, ME, and NE are in kilocalories. Adapted from Garrett *et al.* (1959).

[b] The first part of the formulas express maintenance requirement, e.g., TDN = $0.036W^{3/4}$; the factor in parentheses the requirement per unit of gain, e.g., $2.3G$; hence the complete formulas represent requirements for maintenance and gain.

Protein

The reasons why protein is required for maintenance are not as clear as why energy is required, since the animal presumably is neither gaining nor losing weight nor changing in composition. One may wish to use the limited analogy of the engine. At idle, parts wear and ultimately require repair. Wear in the engine may be likened to "protein turnover" in an animal, which is a term used to describe the continuous breakdown and synthesis of the individual body proteins.

The amount of protein (nitrogen times 6.25) lost in the urine and feces of animals, and additional losses, such as hair, skin, and hooves, are thought to represent the amount of nitrogen required for maintenance. In simple-stomached animals, the nitrogen lost in the urine may account for three-quarters of the total. Early work showed that approximately 2 mg of urinary nitrogen was lost per kcal of basal heat production. Urea accounts for the majority of the nitrogen in urine. The nitrogen in the feces is thought to originate from enzymes that were secreted into, but not broken down and reabsorbed from, the gastrointestinal tract and from sloughing of the intestinal linings, in addition to that unabsorbed from foodstuffs.

The maintenance requirement for protein, as with energy, varies with the size of the animal. Crampton and Harris (1969) have suggested that the maintenance requirement of digestible crude protein is 3.4 times the crude protein equivalent of the endogenous urinary nitrogen loss. Results from extensive experiments carried out by Preston (1966) indicate that the maintenance requirements for protein by cattle and lambs are slightly lower than that suggested by Crampton and Harris. A formula developed by Preston can be used to estimate the maintenance requirement for digestible protein (DP) in grams in relation to weight expressed in kilograms.

Vitamins

Little research has been directed strictly at the determination of the maintenance requirement for vitamins. Many of the studies involved in the determination of vitamin requirements use a growth response or one of many biochemical alterations as a means of assessing the requirement. Hence estimates of the vitamin requirement for maintenance unfortunately are confounded with varying degrees of growth.

The requirements for vitamins can logically be divided into two categories: the fat-soluble vitamins in one, the B or water-soluble vitamins in the other. The fat-soluble vitamins appear to be required for the maintenance of tissue and bone, whereas the B vitamins play roles in metabolism as enzyme co-factors and hence may be required in relation to the rate of metabolism rather than the weight of an animal.

Fat-soluble vitamins The requirements for vitamin A and D are fairly well established, whereas those for vitamins E and K are less definite (Mitchell, 1962). The requirements for all the fat-soluble vitamins appear to be a function of body weight, since the requirements per kilogram weight of rats and of larger animals are comparable. For animals that are between rats and cattle in size, the requirement of vitamin A is 5 μg per kilogram body weight. Many animals can convert beta-carotene to supply vitamin A. Requirements expressed as beta-carotene vary somewhat more than those for vitamin A, because different species of animals do not convert beta-carotene with the same efficiency. Cattle, sheep, and swine convert beta-carotene with about 25 per cent of the efficiency of the chicken (Crampton and Harris, 1969).

The requirements for vitamin D are not as well-established as those for vitamin A. One reason is that the biopotencies of vitamins D_2 and D_3 are equal in mammals, but the value of D_3 is 30 times that of D_2 for the chick. The pig and calf appear to require approximately 6.6 I.U. per kilogram of body weight. Poultry seem to require substantially more. Estimates of the requirement for vitamin E are limited, and lie between 0.2 to 0.4 mg per kilogram body weight. The requirements for vitamin K are not well-defined. However, with the exception of the chick, one would not expect to encounter a deficiency of vitamin K because of its wide distribution in feeds and extensive synthesis in the intestinal tract.

Water-soluble vitamins Not all animals require a dietary source of B vitamins. The microorganisms in the rumen synthesize them and hence fulfill the daily requirement of the ruminant for the B vitamins. Although microorganisms in the intestinal tract of the simple-stomached animals do synthesize vitamins, they are apparently of limited benefit, since vitamin deficiencies can be produced in these animals. Small animals, the size of rats or chicks, may require 20 to 30 times more per unit of live weight than that of animals the size of pigs (Brody, 1945; Robinson, 1966). The degree of variability in the requirements for the B vitamins by large and small animals is materially reduced by expressing the requirements per 1,000 kcal of digestible energy (DE) or per unit of food consumed. Two factors contribute here: first, the requirement for B vitamins is apparently related to the metabolic rate of the animal; second, animals eat in relation to their energy requirement.

Minerals

Requirements for minerals can be divided into two broad categories, those for macroelements and those for trace elements. The macroelements, being structural components of the body, and involved in acid-base balance and water balance, are required in larger amounts than the trace elements, which play important roles as co-factors in enzyme reactions and are key elements in compounds like hemoglobin. Estimates of the maintenance requirement of minerals are made more difficult by the mechanisms directed toward increasing the retention and reutilization of body minerals as the dietary source is diminished.

Requirements for calcium and phosphorus seem to be fairly well-established. The requirements for calcium and phosphorus appear to be dependent on the metabolic rate of the animal, since 0.155 and 0.310 mg of calcium and phosphorus, respectively, are required per kcal basal heat production. In NRC (1970), the calcium and phosphorus requirements for beef cattle are stated in relation to the maintenance requirement of protein. The suggestion that 1 g calcium and 2 g phosphorus are required per 100 g protein result in estimates that are slightly higher than those mentioned above.

Although requirements for the other minerals have been investigated, information to date is fragmentary. Interestingly, current information suggests that the requirements for trace minerals are dependent on the metabolic rate of the animal. However, an increase in activity does not materially affect the requirement for minerals—the exception being salt, which is lost in sweat. Daily requirements for some of the minerals published in appropriate standards for the various classes of livestock are not for maintenance, but are estimated for maintenance and production.

Water

Water makes up from 45 to 80 per cent of the body weight, depending upon the proportion of body fat. The primary sources of water are the drinking water, that contained in the feed, and that produced by metabolism. In animals fed at maintenance, water is lost via the urine, respiration, and sweat. However, when an animal lactates, substantial water is lost via milk. Measurement of the half-life of body water of farm animals has revealed that an amount of water equal to that of the total body water enters and leaves the body in from 5 to 12 days, depending on the species considered (Black *et al.*, 1964; Richmond *et al.*, 1962). The magnitude of the water turnover is an indication of the water requirement.

Results from a number of studies suggest that the voluntary consumption of water by animals is equal to 1 ml per kcal of heat production. The ratio of voluntary consumption of water per gram of dry matter consumed may vary from 2.0:2 to 4.0:1, depending on the species of animal being considered (Mitchell, 1962). A figure often used for water requirement is 2 ml per gram of dry matter consumed. The voluntary consumption of water at maintenance and above is affected by many factors, such as temperature, amount of physical activity, stage of lactation, frequency of watering, composition of the ration (especially protein and minerals), and the level of feed consumption.

Requirements for Growth

With the exception of the human, the growth curves of many animals are quite similar (Brody, 1945). When different species are compared, it appears that they attain their adult weight at the same relative rate. Typical growth curves are sigmoid in shape, showing that gains are faster in young animals than in older animals.

Composition of the gains of growing animals change as they become older and larger. In Figure 37.2,A, are shown total body water, fat, protein, and ash in growing pigs weighing from 10 to 100 kg. In Figure 37.2,B, the composition in per cent of gains is shown as calculated from the differences in body composition at the different weights. The water, protein, and ash content of the gain declines steadily, while the fat increases from 20 per cent to in excess of 50 per cent of the gain.

Since neither the rate nor the composition of gain are constant during the growth period, the assessment of requirements for growth is complicated. For example, daily protein requirements are higher for young growing animals to cover needs for more rapid gains and

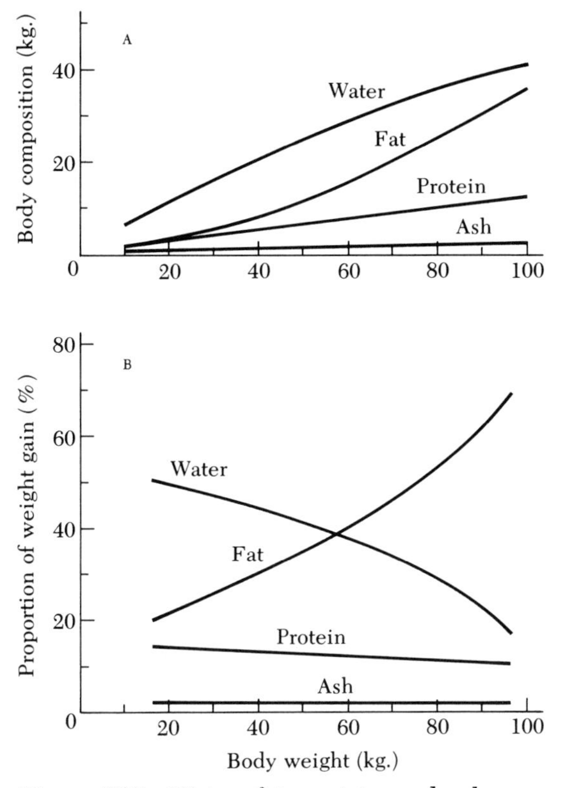

Figure 37-2 Water, fat, protein, and ash content: (A) of growing pigs in kg.; and (B) of gain in growing pigs, as percentage of gain. Figures are for pigs growing from 10 to 100 kg. (Adapted from Doomenbal, 1968.)

relatively greater proportions of protein in the gains than in older animals. As animals become older, requirements change from those related to growth toward those for maintenance of mature animals no longer gaining in weight.

Energy

Estimates, both theoretical and experimental, of the energetic efficiency of the growth process lie between 70 and 78 per cent (Baldwin, 1968). Estimates of the energetic efficiency of fat and protein deposition in the young pig (2.5 to 8.5 kg) are 64 and 80 per cent, respectively. Similar estimates made on older pigs (60 to 90 kg) were 70 and 52 per cent, suggesting a decrease in the energetic efficiency of protein deposition with age (Kielanowski, 1965; Kotarbińska and Kielanowski, 1969).

Extensive studies (Table 37.1) on the energy requirements of sheep and steers for maintenance and growth have been reported by Garrett *et al.* (1959).

Protein

Protein requirements are affected by variations in the composition of gain, by rates of gain, and by alterations in the efficiency of the utilization of protein resulting from changes in the diet. The relative amount of each of the essential amino acids (Chapter 34) in a mixture containing all 20 of the amino acids, as well as the proportion of essential to nonessential amino acids, materially affect the utilization of the mixture. A ratio of essential to nonessential acids close to 1:1 is the most effective. Hence for simple-stomached animals, as the biological value of the protein declines, the amount of protein required to meet the animals' requirements increases, to overcome the poorer balance of amino acids in the diet. The effect of variations in the biological value of the protein fed to ruminants is of less importance, because the rumen microbes break down substantial quantities of dietary protein and use it in the synthesis of amino acids, which are then digested in the lower intestinal tract.

The efficiency of utilization of a protein or a mixture of amino acids is affected also by the proportions of protein and energy consumed. Too little protein per unit of energy will restrict growth, whereas an excessive level of protein will result in a decrease in the utilization of protein, because it will be used to meet the energy needs of the animal. Maximum gains were obtained when the diet contained 28 g of DP per 1,000 kcal of DE in cattle, whereas 22 g of DP per 1,000 kcal of DE was optimal for lambs. Levels of 60 g of DP per 1,000 kcal of DE are required for optimum growth of pigs weighing from 1.5 to 4.5 kg, 52 g for ones between 4.5 and 9.0

kg, and 43 g for ones between 9 and 20 kg (Lucas and Lodge, 1961). Preston (1966) has developed the following formulas to predict the protein requirement for maintenance and gain of growing cattle and lambs:

$$\text{cattle, DP} = 2.79W^{3/4}(1 + 1.905G),$$
$$\text{lambs, DP} = 2.79W^{3/4}(1 + 6.02G),$$

where DP is the digestible protein required in g, W is body weight in kg, and G is the daily gain in kg.

Vitamins

Requirements for some of the vitamins for growing animals, which include maintenance and growth, have been determined.

Fat-soluble vitamins The requirement for vitamin A for growth is not known; however for maintenance plus growth, levels of 5 μg per kg body weight are adequate. Direct sunlight, by converting the vitamin-D precursor in the skin into vitamin D, can materially decrease the amount of vitamin D required in the diet. The vitamin-D requirements per kg of air-dry feed decline with age. Approximately 200 I.U. are required for 5- to 35-kg pigs, whereas ones from 35 to 100 kg require only 125 I.U. Similar declines with weight have been observed in the calf. Although there probably is an additional requirement of vitamin K to cover growth, little is known about the requirement of vitamin K, possibly because animals do not suffer from a deficiency because of its abundance in feeds and extensive synthesis in the intestinal tract. The requirement of vitamin E for growth is not known. Young lambs may develop a muscular dystrophy called "stiff-lamb" disease, which can be prevented or cured by administration of vitamin E and selenium.

Water-soluble vitamins The requirements for the water-soluble vitamins are not known, but since they are required in relation to metabolism, one would expect the dietary requirement to increase in proportion to the increased requirements of energy for growth.

Minerals

Approximately 83 per cent of the ash in the body is found in the skeleton (Mitchell, 1962), and about 85 and 99 per cent of the total body phosphorus and calcium, respectively, are found in the skeleton. As was shown in Chapter 36, much of the phosphorus found in the soft tissues is contained in the compounds intimately involved in the metabolic processes carried out in these tissues. Because of the high concentration of calcium and phosphorus in bone (Table 37.2), the requirements for these two elements for growth are much greater than those of the other elements. The effect of the metabolism of vitamin D on that of calcium and hence phosphorus must be considered when the requirements of calcium and phosphorus are determined, since it acts in the absorption of calcium. For the simple-stomached animals, Ca:P ratios of 1:1 to 1.5:1 are the most effective. Ruminants, on the other hand, can tolerate ratios of 4:1 and, in fact, experience lower levels of urinary calculi formation when fed high ratios than when fed ratios approaching 2:1. One complicating aspect of maintaining the 4:1 ratio during the dry period is that cows are more prone to milk fever upon freshening. Feeding ratios of 2:1 during the dry period has been helpful.

Table 37.2.
Mineral composition of the moisture-free, fat-free bone.[a]

Mineral	*Per cent*[b]
$Ca_3(PO_4)_2$	86.1
$Mg_3(PO_4)_2$	1.0
Ca (combined with CO_2, Cl, F, citrate)	7.3
CO_2	6.2

[a] Mallette *et al.* (1960).
[b] For cattle.

The Ca and P requirements decrease with age because the proportion of ash in the gain (Figure 37.2,B) and the growth of the skeleton decline with age.

An example of how estimates of the requirements of minerals for maintenance plus gain can be determined is shown by the work of Preston and Pfander (1964). By obtaining data based on altering the phosphorus content of the diet, a formula for predicting the phosphorus requirement for lambs gaining at various rates could be obtained:

$$P = 0.0194\ W\ (1 + 0.0171\ G),$$

where P is the phosphorus requirement in g, W is body weight in kg and G is the daily gain in grams. The requirements for other minerals, such as sodium, chloride, potassium, copper, iron, iodine, magnesium, manganese, zinc and selenium, for growth are unknown. However, estimates of the requirements for young animals, which involve maintenance plus growth, can be found in many feeding standards, including the NRC publications.

Requirements for Lactation

Energy

The energy requirement for lactation varies in relation to the amount of milk and its fat content. To compare the production of milks with various fat contents on an equal-energy basis, the concept of fat-corrected milk was introduced. Usually the milks are adjusted to a 4 per cent fat equivalent by the formula:

$$FCM_{kg} = 0.4\ (\text{kg milk}) + 15\ (\text{kg fat}),$$

where FCM = fat-corrected milk. Under normal conditions, the formula can be used with confidence; however, in some cases, where milk-fat percentages are lower than 3 per cent, the FCM value as calculated above overestimates the true energy value of the milk by as much as 15 per cent.

The following calculations illustrate how the energy requirements for the production of one kg of FCM can be estimated. The energy content of 1 kg of FCM is 750 kcal; thus 750 kcal of net energy are required for one kg FCM. Since the average efficiency of conversion of metabolizable energy to the net energy of milk is 62 per cent and that of conversion of digestible energy to metabolizable energy is 82 per cent, then 750 ÷ 0.62 = 1,200 kcal of ME or 1,200 ÷ 0.82 = 1,460 kcal of DE are required to produce one kg of FCM (Flatt *et al.*, 1966). Also, since 4,400 kcal of DE are equal to one kg of TDN, then 1,460 ÷ 4,400 = .330 kg of TDN are required per kg of FCM.

Protein

A complicating factor in the estimation of the requirements for milk production is that a variable proportion of both the protein and energy needs for lactation can be supplied by body tissue. Estimates of the efficiency of utilization of digestible protein for milk production vary from 60 to 70 per cent; hence if one kg of FCM contains 35 g of protein, then 50 to 55 g of DP must be supplied per kg of FCM.

Vitamins

Lactation is impaired in cows that are deficient in vitamins A or D. The specific requirements of the fat-soluble vitamins A, D, E, and K for lactation are not known. Apparently levels of the fat-soluble vitamins which are adequate for growth are also sufficient for the maintenance of lactation. As with the fat-soluble vitamins, requirements of the water-soluble B vitamins for lactation are also unknown, largely because they are synthesized by rumen microorganisms.

Minerals

Calcium and phosphorus make up the bulk of milk ash, and hence are the two minerals which have received the most attention in relation to lactation. Although in experiments with other animals, it has been shown that the efficiency of utilization of both calcium and phosphorus is reduced when the Ca:P ratio deviates greatly from 1:1, ranges from 1:1 to 8:1 are without effect on level of milk production, persistency of lactation, or the total solids, ash, phosphorus, calcium, or vitamin-A content of milk (Smith *et al.*, 1966). Often negative calcium and phosphorus balances are observed early in lactation, while positive balances are observed later in lactation. Apparently under usual conditions the cow has sufficient reserves in her skeleton to withstand the negative balance early in lactation because these deficits are made up in the latter part of lactation and in the dry period.

The requirements for calcium and phosphorus for milk production can be calculated from the following: one kg of FCM contains about 1.3 g of Ca and 1.0 g of P. Dietary Ca and P both are used with an efficiency of approximately 60 per cent. Therefore approximately 2.2 g of calcium and 1.6 g of phosphorus are required per kg of FCM.

Requirements for Reproduction

Mammalian Reproduction

Consumption of excessive quantities of balanced diets, while leading to animals which mature faster, also results in lower levels of production of milk, a shorter productive lifetime, and reduced reproductive performance. Hence, feeding young stock at a high level to bring them into production at an early age may have some drawbacks. Deficiencies may lead to death and reabsorption of the fetus or young that are small and less viable. Nutritional deficiencies that affect the development of the ovum or sperm, the zygote, the embryo, or the fetus may result in infertility or in structural or functional abnormalities, or even the death, of the embryo or fetus (Moustgaard, 1969). An early deficiency results in infertility; deficiencies occurring later result in structural or functional abnormalities.

The major emphasis of this section is on the female. Two major criteria used in determining the requirements for reproduction are that the levels suggested must result in normal offspring and must prevent the maternal tissues from becoming depleted due to the demands of the fetus.

Energy The energy required for reproduction involves that stored in and required for maintenance of the fetus, that stored in the placenta and uterus, and that resulting from the increased heat production of the maternal tissue due to the pregnancy. The daily accumulation of energy in the fetus and surrounding tissues and that due to the increased metabolism of the pregnant cow are shown in Table 37.3. Note that the deposition of energy is moderate during the first two-thirds of gestation and materially increases during the last one-third of the pregnancy. Note also that the heat increment of pregnancy, which is in large part due to the increase in the metabolism of the maternal tissue, accounts for more than three-quarters of the energy used throughout gestation; hence the additional requirement for energy above maintenance is low during the first two-thirds of gestation, and increases materially thereafter.

Bulls can maintain optimal sperm produc-

Table 37.3.
Estimation of the energy used for fetal development in the bovine in kcal per day.[a]

Days after conception	*Energy in the product of conception*	*Heat increment of pregnancy*	*Total energy used*
100	40	575	615
150	100	960	1,060
200	235	1,670	1,905
250	560	2,635	3,195
280	940	3,550	4,490

[a] From Moustgaard (1969).

Table 37.4.
Computed daily deposition and minimum requirement for fetus production in pigs, corrected to 10 fetuses.[a]

	Nitrogen									*"Energy"*		
	Deposited			*Calcium*		*Phosphorus*		*Iron*				
Days after conception	*Uterus*	*Milk gland (g)*	*Requirement (g of digestible protein)*	*Deposited (g)*	*Requirement (g)*	*Deposited (g)*	*Requirement (g)*	*Deposited (mg)*	*Deposited (Kcal)*	*Heat increment (Kcal)*	*Requirement metabolizable energy (Kcal)*
40	1.0	—	10	0.1	0	0.1	0	3	50	350	400
60	1.9	—	20	0.4	1	0.4	1	5	90	750	840
80	3.6	0.8	50	1.2	2	0.9	2	9	170	1,150	1,320
100	6.6	1.8	95	4.0	8	2.2	6	18	320	1,550	1,870
110	9.1	2.7	135	7.2	14	3.7	9	25	440	1,750	2,190
115	10.6	4.7	175	9.8	20	4.7	12	30	510	1,850	2,360

[a] From Moustgaard (1969).

tion with energy intakes as little as 20 per cent above maintenance.

Protein Additional protein is required during the gestational period for the preparation of the uterus, for development of the mammary glands, and, of course, for the developing embryos (Table 37.4). Less than 10 per cent of the total protein required during gestation is used for the preparation of the uterus and mammary glands. The requirement for protein during gestation closely parallels that of energy, being low during the first two-thirds and increasing rapidly during the last one-third of gestation. The protein requirement of dairy cows during the latter portion of gestation may be as high as 80 per cent above that of maintenance. In the bull, optimal sperm production is obtained from animals receiving protein in quantities approximately 70 per cent above that of maintenance.

Vitamins Defiencies of the fat-soluble or water-soluble vitamins can lead to abortion, young that are born small and/or deformed, or young that are born weak and may not survive for more than a few days. An adequate intake of the vitamins is essential if normal reproductive performance is to be attained.

Fat-soluble vitamins Vitamins A and D are the two fat-soluble vitamins of major concern in the reproduction of farm animals. Since vitamin A is stored in the maternal tissues, dietary deficiencies over short periods of time do not result in reproductive failures. Because of the ability of the body to store vitamin A when consumption is in excess of the need, specific increases in the dietary level may not be warranted during gestation, since an increase in the consumption of the diet to meet the increased energy demands results in higher intakes of the vitamin. However, cattle wintered on low-quality roughages may require supplementary vitamin A prior to calving. Since vitamin D is intimately involved in the metabolism of calcium and phosphorus, a deficiency would result in serious consequences; however, normal foodstuffs and summer sunlight provide adequate quantities of vitamin D. Vitamin D should be added to the diet in the winter, however, because of the reduced amount of sunlight and confinement of animals indoors. Although a vitamin-E deficiency produced in sows by use of special diets resulted in lowered reproductive performance, a deficiency under normal conditions is unexpected, since natural feedstuffs contain adequate quantities of vitamin E.

Water-soluble vitamins Quantitative increases in the requirements of the B vitamins during reproduction are not known. However, since the B vitamins function as co-factors in metabolism, their requirement should increase in parallel with that of energy. Therefore a diet which is adequate for rapid growth will be adequate for reproduction, since increases in the consumption of food to cover the increased requirements for energy automatically provide for an increase in the intake of the B vitamins.

Minerals The minerals of major concern are calcium, phosphorus, and iron. It can be seen from Table 37.4 that the pattern of deposition of these minerals is similar to that of energy and protein, in that it is low during the early portions of gestation and increases dramatically during the last one-fifth of pregnancy. The calculated requirements of calcium and phosphorus shown in the tables are based on the assumption that the apparent digestibility of calcium and phosphorus are approximately 50 and 40 per cent, respectively.

The requirements of the pregnant sow or gilt are shown in Table 37.5. They reflect what is known about the pattern of deposition of energy, protein, calcium, and phosphorus. Many standards do not show graded increases in the requirements with the length of gestation, but rather set the requirement high enough so that it meets the requirement late in gestation.

Table 37.5.
Requirement of pregnant sows or gilts.[a]

Days after conception	*Metabolizable energy (Kcal)*	*Digestible protein (g)*	*Calcium (g)*	*Phosphorus (g)*
0–90	6,500	265	9	8
90–105	7,500	350	15	12
105–115	9,000	400	20	15

[a] Adapted from Moustgaard (1962).

Avian Egg Production

The nutritional requirements for egg production are affected by the size and breed of hen, the percentage production (e.g., 75 eggs produced in 100 days is considered 75 per cent production), and the composition of the egg. The composition of an average chicken egg is shown in Chapter 8.

Energy Estimates of the energy requirements for the average hen in production can be made by adding the costs of maintenance to those of producing one egg. For example, let us assume that a 1.8-kg hen lays seven eggs each weighing 54 g each ten days. The requirements can be calculated as follows.

Basic metabolic rate (kcal net energy/day) $= 68W_{kg}^{3/4}$.

Activity increment is estimated at 50 per cent.

Maintenance requirement $= 68W_{kg}^{3/4} \times 1.5 = 68 \times 1.55 \times 1.5 = 159$ kcal of net energy.

Energy content of one egg = 77 kcal of net energy.

Daily energy requirement at 70 per cent production $= 77 \times 0.70 = 54$ kcal of net energy; therefore, maintenance + egg production $= 159 + 54 = 213$ kcal of net energy.

Average efficiency of utilization of metabolizable energy for maintenance and egg production = 68 per cent.

Daily metabolizable energy requirement $= 213 \div .68 = 314$ kcal.

The calculated energy requirement in the example above compares favorably with the daily intake of 300 to 320 kcal of metabolizable energy per day for hens fed well-balanced diets in moderate environments (Scott *et al.*, 1969).

Protein The chicken, unlike the ruminant, requires high-quality protein in its diet. Substantial reductions in the quality of the protein will result in lowered production and hence lowered efficiencies. In considering the protein requirement for egg production, one must consider the quantity deposited in the egg as well as that required for the maintenance of the hen. Scott *et al.* (1969) have estimated the efficiency of utilization of dietary protein for maintenance and production by determining the grams of protein required per day for optimum egg production and calculating that required for maintenance. Table 37.6 shows an example of one of their calculations. The efficiency of protein utilization for maintenance and egg production from the above example at 100 per cent production is 57 per cent. If production falls to 70 per cent, then the daily requirement for egg production is 4.2 g and the efficiency of protein utilization falls to 46 per cent, since an increased proportion of the total protein consumed is used for maintenance. The ratio of energy to protein is important in the efficiency of utilization of the diet. Metabolizable energy to protein ratios (ME/P), calculated by dividing the ME in kcal/kg diet by the per cent of dietary protein, have been assessed in relation to optimum performance. For young growing hens, a ratio of 166 to 170 is suggested; in mature, nongrowing hens, ratios of 193 to 195 are satisfactory. For older hens whose production is declining, a ratio of 196 to 200 is optimal. It is clear from the above that relatively more protein is required per unit of energy in the younger bird.

Table 37.6.
Calculation of Protein Requirement of Chickens.

Need	*Grams of protein required per day*
Production of one egg	6.0
Maintenance of one chicken	3.0
Growth	0.0
Feather growth	0.1
Total	9.1
Daily requirement determined by feeding trial = 16.0	

Vitamins In general, vitamin deficiencies result in lowered production because of their effect on the hen. Excessive consumption of some vitamins results in increased levels in the egg.

Fat-soluble vitamins The requirements for vitamins A, D, E, and K for egg production are not known. However, deficiencies of vitamin A or D result in marked reductions in egg production. A deficiency of vitamin D gives rise to thin-shelled eggs and soft and brittle bones. A deficiency in any of the vitamins A, D, E, or K reduces the number of live, viable chicks.

Water-soluble vitamins Here, again, the requirement for the B vitamins for egg production are not known; however, one would suspect that the requirement for these vitamins would increase in relation to the increase in energy expenditure connected with the production of the egg, since they act as co-factors in metabolism.

Minerals For egg production the minerals required in the greatest quantity are calcium and phosphorus. The requirement for calcium is especially high because it, as calcium carbonate, makes up approximately 98 per cent of the shell. Since the shell contains 2.2 g of calcium and only 50 to 60 per cent of the dietary calcium is absorbed, then approximately 4 g of dietary calcium is required for each egg. The magnitude of the calcium requirement for shell production becomes more apparent when one realizes that this is equal to more than 14 per cent of the hen's total body calcium and that the diet must contain from 3.25 to 4.0 per cent of calcium. Meeting the requirements for phosphorus is complicated by the low availability of organic phosphorus. Diets for laying hens should contain at least 0.6 per cent of phosphorus of which at least 0.5 per cent is in the inorganic form.

Work

The nutritional requirements for work are usually associated with the work performed by horses, whether it be for draft or for pleasure. Since the horse is used mainly for pleasure, this section will tend to emphasize the requirements for light work.

Energy

Work can be measured in foot-pounds or kilogram-meters, both of which can be converted to their caloric equivalents, permitting calculation of the energetic efficiency of work as:

$$\text{gross efficiency} = \frac{\text{Energy equivalent of work done}}{\text{Total energy expended during work}}.$$

Values of 20 to 25 per cent are usually obtained for the gross efficiency of work. Calculation of the net efficiency involves correcting for the contribution of maintenance heat production during the time the work is being performed. This is shown by the formula:

$$\text{net efficiency} = \frac{\text{Energy equivalent of work done}}{\text{Total energy expended less energy expended by the standing animal}}.$$

Values of 30 per cent are normally obtained for net efficiency of work. A specific quantity of work does not require one specific quantity of energy, since the speed at which the work is carried out materially affects the energy requirement (Brody, 1945). As the speed or draft increases, the gross efficiency increases. The increase in efficiency is presumably due to increased work per unit time, and hence a proportionally lower cost of maintenance per unit of work. Above certain levels, however, further increase in speeds or drafts may result in lower efficiencies. Overcoming external (i.e., wind, footing, etc.) and internal (i.e., muscles, etc.) resistances becomes more costly as speed increases. Within limitations of estimating accurately the energy requirements for work, it has been suggested in general that the energy requirements for horses at light (two to three hours per day) and medium (four to five hours per day) work are, respectively, 3.0 and 3.5 times their basal metabolism.

Protein

Numerous studies have shown that there is no increase in urinary nitrogen during work; consequently there is no increase in the protein requirement due to work.

Vitamins

No increases in the fat-soluble vitamins are expected, since they seem to be required in relation to body size; however, increases in the water-soluble vitamins are expected in relation to the increased consumption of energy.

Minerals

Salt requirements increase with work. Horses at moderate work may lose from 50 to 60 g of salt in sweat; an additional 35 g may be lost in the urine. Slight increases in the phosphorus requirements have been suggested due, probably, to something that increases phosphorus turnover. Alterations in the requirements of other minerals in relation to work are not known.

FURTHER READING

Maynard, L. A., and J. K. Loosli. 1969. *Animal Nutrition*. New York: McGraw-Hill.

Section Seven

Livestock Management

Zebu cattle in a primitive corral in Madagascar. (FAO photo.)

Thirty-Eight

Management of Beef Cattle

"Be thou diligent to know the state of thy flocks, and look well to thy herds."

Proverbs 27:23

Beef Cattle Management in the United States

The primary purpose for the existence of beef cattle in the United States is to produce beef. The increase in consumer demand for this product has been dramatic; the per capita consumption in 1970 was 113 pounds, compared to an average of 66 pounds for 1947–1949.

The ultimate goal in a total beef-management program is to produce large quantities of high-quality beef efficiently and profitably. Progress toward this goal, in the beef industry, is spread over the several rather individualized segments listed in Table 38.1, each of which has somewhat different needs that must be met by the beef animal or its products. Certain characteristics of the beef animal may greatly influence economic returns, for one segment, but have little influence on the profitability of another segment. For example, poor milk production may significantly affect the net return of a commercial cow-calf producer, but would have little if any direct effect on the profitability of the packing segment.

Our attention in this chapter will be focused primarily on the first three segments: purebred breeder, commercial cow-calf producer, and feeder. This will permit an evaluation of management practices involving those segments associated with the live beef animal. Correct management decisions should permit each of these segments to produce beef cattle that are profitable for them and the segments that follow. Net profits for the beef producer are influenced primarily by price received at the market place and the efficiency of production. While major attention will be directed toward the latter, some comment should be directed toward the price received for beef.

Historically, cattle numbers and pounds of beef produced has been quite erratic. These peaks and low points of beef supply are referred to as the beef-cattle cycles, and there have been several of these cycles during this century. Prices are highest when beef numbers are at their low point in the cycle. It should be recognized that this over- and under-supply of beef has a greater influence on price and prof-

Table 38.1.

The segments of the beef industry and what each segment needs from beef animals. The flow of animals or products is downward from segment to segment, there being one or more marketing transactions at each step.

Segment	*Characteristics of beef that affect net return or desirability*
Purebred breeder	All the characteristics listed for the other segments (must meet the needs of the entire industry).
Commercial cow-calf producer	Reproductive efficiency Weaning weight Weaning conformation
Feeder	Rate of gain Feed efficiency Live or carcass grade
Packer	Carcass grade Carcass weight Carcass cutability
Retailer	Carcass grade Carcass cutability Product appeal
Consumer	Lean-to-fat ratio Lean-to-bone ratio Tenderness, flavor, and juiciness Consistency of product

itability than beef imports. However, beef imports do depress prices more when beef numbers are at their peak in the United States. As long as the law of supply and demand reflects itself in the beef-cattle cycle, beef producers will have years when market prices will be greatly depressed, and most likely these will be unprofitable years.

Increased costs of producing beef cattle are clearly evident today, and costs continue to rise. The present marketing structure for the beef producer does not permit him to pass on the higher costs of production to the buyer of his product. Certain segments of the beef industry have this ability, and most definitely many other industries in the United States can pass on their higher production costs eventually to the consumer. Successful beef producers must be aware of the market structure for their product.

An increase in the efficiency of beef production has been attained during the past few years. Part of this increase has depended upon defining more specifically what is desired of the beef animal (Table 38.1). Influential industry leaders, representing the various segments, have defined our beef goals and objectives as shown in Table 38.2. It is recognized that these desired levels of productivity do not fit all environments; however, producers can make considerable improvements over present levels of performance in almost all environments. Some cattle in the United States are surpassing the levels of productivity listed, and these will have to be considered as competition to the cattle whose performance is much lower.

Maximum performance in the economically important traits listed does not also imply maximum net income. Costs incurred in increasing productivity must be weighed against the increased return. Each beef-cattle operation is unique in its productivity level of cattle, the environment under which the cattle are raised, and the management decisions imposed on the

Table 38.2.

Desired levels of productivity for the economically important traits of beef cattle.

Trait	*Desired level of productivity*
Calf-crop percentage (weaned)	95 per cent and higher
Seven-month weaning weight	500 pounds and higher
Yearling weight (steer going directly into feedlot after weaning)	1,000 pounds and higher
Feed per hundred pounds of liveweight	600 pounds and lower
Carcass quality grade	Minimum, low choice
Yield or cutability grade	Number 2 (50 per cent of the carcass weight in closely trimmed, boneless, retail cuts from the round, loin, rib, and chuck)

cattle at any one given time. Because each operation is different, there is no one set of management recommendations for improvement that will fit all operations. Each person making the management decisions must decide how specific information on reproduction, nutrition, selection, meats, marketing, disease prevention, and economics will integrate together and apply to his operation. He then must run a test balance on the expected costs and returns to see if a specific management procedure would be feasible to try in his operation.

Feeding Segment

A few decades ago beef was produced primarily from grass-fed animals. A surplus of grain, with the added preference of the consumer for fed beef, has caused the cattle-feeding industry to boom in recent years. Cattle feedlots handle more beef today than the entire cattle industry produced twenty years ago. The number of cattle fed today approaches 25 million head per year, and these cattle are fed by more than 200,000 producers, most of whom have very small operations. Large commercial feedlots, with capacities up to 100,000 head, have developed primarily in the southwest, Colorado, and California. These commercial feedlots, having a capacity of 1,000 head or more, account for only 1 per cent of the number of feedlots, but produce almost half the fed cattle (Figure 38.1).

The cattle feeder who treats his operation as a business must give attention to the following five major areas: (1) investment in facilities; (2) cost of feeder cattle; (3) feed cost per pound of gain; (4) non-feed cost per pound of gain; and (5) total dollars received at the market place. These major areas with some of their component parts are shown in Table 38.3.

The investment costs in facilities vary from $40 per head to over $200. The lower investment costs are found in areas with mild climates, where little or no shelter is provided and open bunk feeding is practiced. The higher investment costs are in areas where extremes in climatic conditions must be counteracted by a more controlled environment. Here we find totally enclosed confinement buildings where temperature and humidity are controlled.

Many commercial feeders rank financing as their number-one management problem. For example, a 20,000 head feedlot filled with 700-pound cattle has approximately 1 million dollars invested in facilities and 4 million dollars invested in cattle. This does not consider the feed, labor, and other costs which must be paid before any return can be obtained.

The kinds of cattle available to the feeder are shown in Figure 38.2. These choices will

Figure 38-1 A large commercial feedlot with feedmill in the background. This feedlot expects to turn over two to two and one-half groups of cattle each year. (Courtesy *Beef*.)

Table 38.3.

The major areas a cattle feeder must consider to evaluate his operation as a business.

Facilities investment	*Cost of feeder cattle*	*Feed cost per pound of gain*	*Non-feed cost per pound of gain*	*Total dollars received*
Land	Grade	Performance of cattle gain	Death loss	Market choice
Feedlot	Weight	feed efficiency	Labor	Transportation
Feedmill	Transportation	Ration	Taxes	Shrink
Equipment	Shrink	ingredients	Interest	Yield
	Gain potential	cost	Insurance	Carcass grade
		form	Utilities	Cutability
		Length of feeding	Veterinary expense	Manure value
			Repairs	

influence not only the cost of the cattle, but also the feed and non-feed costs. Preference has been given by the larger commercial feeder to those cattle showing high compensatory gains. These generally are plainer cattle that have been restricted nutritionally earlier in their life. They have age and body frame in their favor, so when they are put on high-energy feeds, their gain is rapid and efficient. After a 120–150 day feed, they are ready for slaughter. Most calves at weaning are not put directly into the feedlot, since they would be finished at an undesirable slaughter weight of 800–900 pounds. These calves will generally go through a backgrounding (warm-up) operation. Here they are fed, or they graze on high roughage feeds. When they have reached 600–750 pounds they are ready for higher-energy feeds. Some calves weighing 500–600 pounds at weaning are capable of going di-

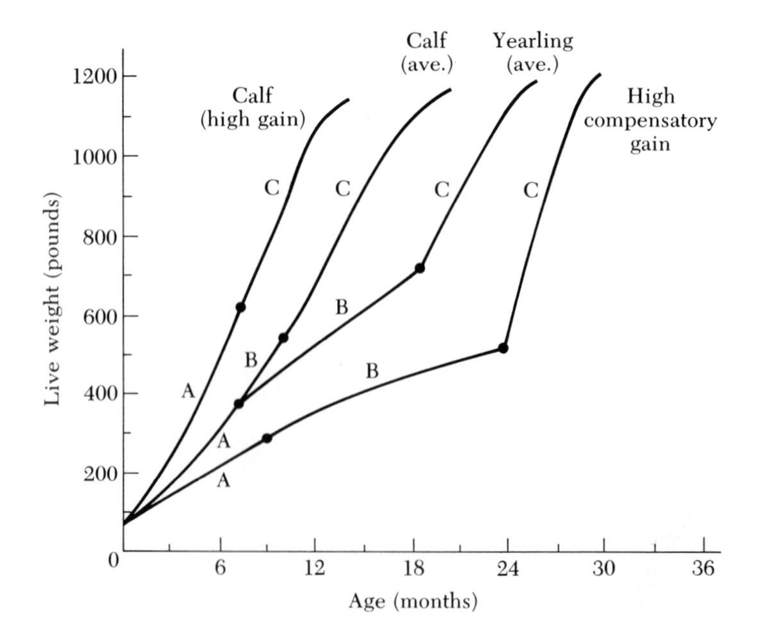

Figure 38-2 Various choices of cattle for the cattle feeder; A, preweaning gain; B, backgrounding (growing) gain, usually on high roughage ration; C, feedlot gain on high-energy ration.

rectly into the feedlot and producing economical gains.

Daily gain, as primarily influenced by nutrition, health, and genetics, is important in determining the economic success of the cattle feeder. Table 38.4 shows the importance of gain as it influences cost per pound of gain. Note that, within the various weight groups, the fast-gaining animals also were the most efficient in their feed conversion.

The rate and efficiency of gain in steers can be increased 7 to 15 per cent by the administration of estrogens. Some estrogens are effective either orally or by implant, others only by implant. Until recently, the feeding of 10 mg daily of diethylstilbestrol (DES) or implanting 24 to 36 mg subcutaneously in the ear at the start of the finishing period were the most commonly used procedures (NRC, 1966b). In 1973, however, the use of DES in the feed or by implant was banned, because, with extremely sensitive analytical procedures, residues were found in the livers of treated steers. According to the provisions of the Delaney Amendment of the Federal Food, Drug, and Cosmetic Act, no residues of administered hormones in the meat of treated animals are allowable. One of the natural estrogens, estradiol-17β, may still be used, as can zearalanol, a synthetic product derived from moldy corn.

Since gains obtained with estrogens are obtained with no appreciable increase in amount of feed consumed, there is an increase in the efficiency of weight gain, because estrogens depress fat deposition in ruminants; it takes about 2.25 times more calories to produce a pound of fat than a pound of lean. A synthetic progestin, Melengestrol acetate (MGA), is as effective as DES in heifers and does not cause undesirable side effects.

Conceivably, the use of all estrogens in finishing steers may eventually be banned. It may then become feasible to feed bulls or short-scrotum bulls (bulls in which the testes are forced up against the body wall by a rubber band around the scrotum thus making them infertile). Untreated or short-scrotum bulls grow more rapidly than steers. Raising of bulls for producing beef is popular in Europe. Presumably, the faster gain of bulls depends on the secretion of the hormone, testosterone, by the testes, but it has not been possible to increase gains of steers to a similar level by injecting or feeding of this hormone. Though feeding bulls for beef introduces some management problems, it appears to be economically feasible and the quality of young bull beef approaches that of steers.

Non-feed costs in many feedlots will fall within the range of 5 to 7 cents per day. Non-feed costs are altered considerably by the animal's gaining ability (Table 38.5).

The farmer feeder may have some advantages over the commercial feeder in feeding cheaper, home-grown feeds and in making fuller use of facilities. At the same time, the

Table 38.4
Comparison of fast- and slow-gaining feedlot cattle.[a]

	Fast 25%			*Slow 25%*		
Purchase weight (lbs)	*Ave. daily gain (lbs)*	*Cost/lb gain*	*Feed/lb gain*	*Ave. daily gain (lbs)*	*Cost/lb gain*	*Feed/lb gain*
Under 500	2.53	16.07	6.47	1.68	21.52	8.57
500–600	2.58	18.36	6.98	1.73	23.95	9.60
600–750	2.78	19.75	7.30	1.84	23.37	9.55
Over 750	2.97	18.74	7.07	1.61	30.51	11.83

[a] Information on 22,000 head of cattle, Iowa Feedlot Business Analysis, Personal communication, Robert C. deBaca, Iowa State University.

commercial feeder has more flexibility in purchasing varied kinds of feed depending on price. Computers not only help the commercial feeders balance rations, but also compute least-cost rations, which helps reduce the feed cost per pound of grain.

Cattle feeding is considered a high-risk enterprise. Price uncertainties pose some of the greatest problems. The margin between the purchase and selling price is important. As was pointed out earlier, there is little control over the yearly prices of beef cattle. Hedging in the future's market provides one means of reducing the risk of loss from price changes. Contract arrangements may also help in reducing uncertainties between the purchase and sale prices.

Table 38.5.
Number of days to market 600 pounds of gain at varying daily gains.

Total gain	*Ave. daily gain (lbs)*	*Number of days*
600	1.5	400
600	2.0	300
600	2.5	240
600	3.0	200
600	3.5	171

Commercial Cow-calf Segment

Beef-cow systems include cow-calf, cow-yearling, and various combinations of the two, with a finishing program sometimes added on the same farm or ranch. In the cow-calf enterprise, calves are usually sold at a weight of less than 500 pounds. In the cow-yearling enterprise, the calves are held longer and grown further on pasture or other crop by-products to weights as high as 800 pounds.

Some operations may be large ranches with thousands of acres of grazing land where cattle get no shelter and a minimum amount of supplemental feed. Other operations may be confinement or semi-confinement facilities, where the cows receive only harvested feeds during most of the year. Most beef-cow enterprises, however, use pasture, much of which is non-tillable land, and salvage crop residues otherwise wasted (Figure 38.3).

The size of operation varies by region, with the western states usually having cow herds from 50 to several thousand head. In the eastern half of the country, beef-cow herds on diversified farms average fewer than 20 cows.

For a cow-calf producer to finance a sizable operation today, he would need an initial investment of $1,000 to $2,500 per cow to cover the cost of the land, cows, buildings, and equipment. An operational unit of 200 to 300 cows requires a large initial investment, and poses problems of paying for itself if the only source of income is from the production of the cattle. Many producers today operate on land that was purchased when land costs were much lower, have grazing rights on government-owned land, or have stayed in business by borrowing money on the anticipated appreciation of their land.

The primary management decisions for a cow-calf producer center around factors which influence per cent of calf crop weaned, annual cow costs, and weight of the calf at weaning time. These three major items are the most influential on net profit (Table 38.6). First and foremost, each producer should know specifically his production level for each of the three categories before he makes any decision on what changes in management decisions should be made. A successful businessman who doesn't have a useful set of records is a rare exception.

The average producer in the United States is weaning approximately 80 to 85 calves per hundred cows, with the calves weighing 250 to 400 pounds, and with an annual cow cost of $100 to 150. Considering the average price received for calves during the past few years, the average cow-calf producer is experiencing difficulty in breaking even with his production costs. Usually the annual cow cost does not include any interest on land investment, which would have to be considered as an ad-

Figure 38-3 A typical pasture.

ditional cost for someone starting out in a cow-calf enterprise.

Some of the more significant management practices in improving per cent of calf crop weaned are given here, with some reasons for considering them.

(1) Replacement heifers should be bred two to three weeks earlier than cows. Heifers take more time to show estrus, after calving, than mature cows.

(2) Proper nutrition, primarily level of energy, is essential prior to calving and during the breeding season. Cows that are gaining weight during this time have a greater chance of conceiving.

(3) Bulls should be semen-evaluated prior to breeding, and under pasture mating the number of cows per bull is usually 20 to 50.

(4) Reproductive diseases, such as brucellosis, vibriosis, leptospirosis, and infectious bovine rhinotracheitis, need to be prevented. Most of the reproductive diseases can be prevented by vaccination or blood test. Calf diseases, such as blackleg, malignant edema, and scours, also need to be prevented. A cow-calf producer should consult his veterinarian in planning an effective herd health program.

(5) Most of the calf losses occur at birth or a few days after birth. It is, therefore, essential to give proper attention to the animals at calving time if a high-percentage calf crop is to be obtained. Assisting cows with difficult births and breeding heifers to bulls known to sire smaller calves has proven to be most beneficial.

(6) Crossbred calves usually are more

vigorous and have a greater survival rate than straightbred calves. Also, crossbred cows will wean a higher percentage of calves than straightbred cows, primarily because of the heterosis for conception rate.

Cow-calf producers attempting to lower annual cow costs will consider some of the following management practices.

(1) Feed costs make up the highest percentage of total costs; therefore, even small changes in feed costs can be beneficial. Feeds can be compared on the basis of cost per unit of energy and of protein. In certain areas, the efficient utilization of surplus crop residues can significantly lower cow costs. The abundance of corn stalks in the midwest is such an example.

(2) Ranges and pastures can be improved to give higher yields of forage. Such practices as brush and tree removal, fertilization, planting of proper species, and controlled grazing systems are some that have been utilized. In the large-range areas, the distribution of water and salt can significantly affect the forage utilization.

(3) Calving heifers at two years of age instead of three is a recommended practice, if heifers can be satisfactorily grown out and observed at calving time. This practice can reduce annual cow costs because the replacement heifer cost will be lower.

(4) The producer should maximize calf weights without maximizing cow size. Large cows have higher maintenance requirements, and a heavier calf must be raised to offset this higher cost. In some environments, it appears that a medium-sized commercial cow with high fertility and heavy milking ability should be preferred. By using a fast-gaining, well-muscled bull, heavy calves can be obtained without using extremely large cows (Figure 38.5).

The following are some of the major management practices in improving weaning weights of the calves.

(1) Increased forage production through improved pastures and ranges will be effective.

Table 38.6.

Necessary selling price, dollars per hundred weight, of calves to break even with varying annual cow costs, calf-crop percentages, and calf weights.

Per cent of calf crop (weaned)	*Annual cow cost*	*Average calf weight at weaning (pounds)*			
		300	*400*	*500*	*600*
95	$140	$49.12	$36.84	$29.47	$24.56
	120	42.10	31.58	25.26	21.05
	100	35.09	26.31	21.05	17.54
85	$140	$54.90	$41.18	$32.94	$27.45
	120	47.06	35.29	28.24	23.53
	100	39.22	29.41	23.53	19.61
75	$140	$62.22	$46.67	$37.33	$31.10
	120	53.33	40.00	32.00	26.67
	100	44.44	33.33	26.67	22.22

(2) Culling cows, selecting replacement heifers, and especially selecting bulls with higher weaning weights, will improve the weaning weights of calves produced. This selection for weaning weight will improve the genetic composition of the herd for growth rate of the calves in the herd and for milk production of their dams.

(3) The implanting of 12 mg of stilbesterol in the ears of three-month-old calves can increase the weaning weight by 10 to 20 pounds without adversely affecting subsequent gains in the feedlot. This management practice appears to give the most consistent results where there is an ample feed supply.

(4) Effective crossbreeding programs will increase weaning weights by permitting heterosis for growth rate in the calf and by increasing the milk production of the crossbred cow.

(5) Calving heifers at two years of age instead of three will increase the total number of pounds of calf weaned over the lifetime of the cow.

Today we have some producers who are producing 500-pound (and over) calves at weaning; these calves go directly into the feedlot and gain over three pounds per day. These

Figure 38-4 Reproductive performance, represented by a live calf each year, ranks as the most important economic trait in beef cattle. (Courtesy American Hereford Association.)

Figure 38-5 These Angus cows with Charolais cross calves are one example of an excellent method to increase pounds of calf weaned per cow in the breeding herd. (Courtesy *Charolais Banner*.)

Figure 38-6 A group of Hereford bulls being developed for other purebred breeders, commercial producers, or artificial insemination studs. (Courtesy American Hereford Association.)

high levels of productivity can be put together at reasonable costs. Logically, producers with this kind of high-gaining cattle should maintain ownership of their cattle through the feedlot.

Purebred Segment

Most of the management considerations for the commercial cow-calf producer will also apply to the purebred breeder. The purebred breeder is in a unique position of being able to make management decisions which can eventually improve the genetic composition of all cattle, which will be of benefit not only to himself, but also to the other segments of the beef industry. The majority of purebred breeders have as their salable product commercial bulls, which are sold to cow-calf producers. A few of the top purebred breeders sell breeding animals only to other purebred breeders, and these top breeders play an important role in directing the progress of a breed.

Table 38.1 shows all the traits of beef cattle that must be considered in developing a management program for beef-cattle improvement. In the past, and to some extent today, breeders have used visual form as the primary method of identifying superior beef cattle. This method alone will not identify those cattle which are superior for the economically important traits. In the past few years, the emphasis has been on performance testing, which utilizes both objective measurements and visual appraisal in selecting superior seedstock.

In a sense, the purebred breeder is a crystal-ball gazer. Each year when he evaluates the direction of his breeding program, he is projecting at least five years into the needs of the future. The economics of beef-cattle traits can change rather drastically and sometimes rapidly; however, cattle cannot change that fast, because of the long generation interval.

At the present time, yearling weight is an important trait for the purebred breeder to improve. Yearling weight reflects both preweaning and post-weaning growth. Pre-weaning growth is influenced primarily by the milk production of the cow and the growth rate of the calf. Post-weaning growth is highly related to feed conversion; so yearling weight actually reflects several economically important beef cattle traits. Along with growth, the composition of this growth (ratio of lean to fat and ratio of lean to bone) is important to the purebred breeder.

Access to scales is an absolute necessity in the management program of a purebred breeder. Weaning, yearling, and mature weights must be taken and put into a usable form so that effective decisions can be made.

Some of the top purebred breeders will own a commercial herd or contract one where bulls can be progeny-tested. The critical test in productivity is when the progeny of several sires can be compared from birth through the carcass under similar environmental conditions. It is most imperative that artificial-insemination studs use progeny-testing, since a bull that is used artificially can sire thousands more calves than a bull used naturally (Figure 38.7). Many of the proven AI bulls cannot be used in purebred herds, because most breed associations do not permit open registrations from AI bulls, which limit the number of bulls which can be sold for natural service. Some restrictions have been lifted, and some breed associations with smaller numbers have an open AI policy.

The purebred breeder can accummulate productivity information from several sources, e.g., within his own herd, on how his animals perform for other breeders and commercial producers, and from central bull testing stations, exhibitions, cooperating feedlots, and others. Once this information begins to accumulate, he has a management decision to make on how to utilize this information in an advertising program. He has a challenge to make other breeders and producers aware of his breeding program and the superiority of his sale animals. Any breeder who misrepresents his animals or their information cannot be assured of success over an extended period of time. The critical test is how his animals produce for other producers who purchase them.

Other segments of the beef industry are dependent on the purebred breeder for the

Figure 38-7 A purebred bull is outstanding when he has proved himself by siring calves that are superior for the economically important traits. Sam 951 is such a bull. (Courtesy Litton Charolais Ranch.)

improvement of certain economically important beef cattle characteristics; thus the breeder should continually be aware of his responsibility.

Some Aspects of Beef Management in Countries Other Than the United States

Different kinds of cattle, climate, feed conditions, topography, level of technology, and desires of people dictate different management decisions throughout the world. Attention will be briefly directed toward some of the different management practices in a few of the twenty countries which rank highest in cattle numbers.

Europe

Most of the cattle in Europe are raised primarily for milk, with beef and veal as by-products. Only in France and the United Kingdom, with few exceptions, are cattle raised solely for slaughter. In Great Britain, cattle raising is becoming more and more tied to dairying, since milk and dual-purpose cattle predominate. Beef bulls are bred to poorer-producing dairy cows, and the calves are fed for slaughter beef. Greater numbers of Holstein calves are left as bulls and fed on high-barley rations. The bulls are slaughtered at approximately a year of age, yielding 450- to 500-pound carcasses, and they produce the so-called "barley beef."

Until recent years very little grain was fed to slaughter animals; however, some feedlots, similar to smaller ones here in the United States, are developing in France, Italy, and Germany. Most of the male calves not slaughtered as veal are grown out as bulls and slaughtered when 20 to 30 months old. Few bulls are needed for breeding purposes because of the widespread use of artificial insemination in many of the European countries.

In Eastern Europe, cattle are raised on state farms or cooperative farms. This form of socialized agriculture is common except in Poland and Yugoslavia, where the small private farmer still exists. In Poland it is a legal requirement that all bulls of the private farmer be castrated before six months of age. This insures better quality bulls to be used at breeding, either by artificial insemination or by natural breeding from sires coming from state farms. The steers are usually stall-fed until they are 1,000 pounds at approximately 18 months of age.

In the Soviet Union, more emphasis than in previous years is now being given to the production of slaughter animals. In many areas the cows are kept in barns the year around, with the barn work commonly being done by the women. Calves are taken from their dams at birth and raised on milk until they are six months old. The steers are fed to slaughter weight, primarily on high-roughage rations. About 85 per cent of the cows in the country are bred artificially.

Asia

Very few of the numerous countries in Asia are considered major beef-producing or beef-consuming countries. The population pressure and low standard of living, in many areas, cannot tolerate the inefficiencies of beef production.

Even though India ranks first among countries in cattle and buffalo numbers, the consumption of beef is low. The Hindu religion prevents the slaughter of cattle for beef; so the only beef produced in India comes primarily from worn-out draft buffaloes.

China also ranks high in the total number of cattle; however, these numbers diminish in importance when the large land area and the number of beef cattle per capita is considered. Details about cattle-management practices in China are not widely known. Generally the cattle are well cared for, since the numbers held by each farmer are small. One or two animals are housed with the farmer or they are located nearby. Considerable emphasis is placed on saving the manure because of the need for fertilizer. One of the biggest problems is winter feed for the cattle, because the land is utilized primarily for human food production.

Japan has had considerable economic growth since World War II, and now is expressing a demand for more beef. Breeding cattle and carcass beef are being imported to fulfill this demand. Extensive expansion of cattle production appears to have serious limitations because of the high concentration of people on the limited land area.

Japan has received worldwide recognition for its production of Kobe beef. This beef is produced by feeding, usually a virgin cow, a high-concentrate ration for over a year. In addition, the animal is exercised and massaged daily, and is given beer every other day (Figure 38.8). This intensive type of beef production is one of national pride, and is demanded by the Japanese for the preparation of the dish sukiyaki.

Africa

Much of the inhabited area is characterized by the nomadic herding of cattle. Tsetse fly and long distances from water prevent large land areas from being grazed.

North of the Sahara, cattle are less intimately involved in the cultural lives of their owners. Cattle are used primarily for draft and milk, with beef coming primarily from animals that are no longer productive.

Figure 38-8 A Kobi cow receiving her supply of beer. (From *World Cattle,* by John E. Rouse, Copyright 1970 by University of Oklahoma Press.)

South of the Sahara, the social and religious significance of cattle dictates how they are handled and managed. These practices vary from tribe to tribe; however, in many respects they are somewhat universal. To many of these people cattle are a sign of wealth and a status symbol (Figure 38.9). Preference is given to keeping this wealth on the hoof with little exchange for money. Size of the herd and its maintenance are the primary management concerns. The flesh of the animals is rarely eaten, except on ceremonial occasions when they are killed for a feast or to salvage an old animal. The milk and, in some instances, the blood are sources of livelihood.

Ranch-type cattle raising by Europeans is of significance primarily in southern Africa. Rhodesia is further advanced in beef production, based on American standards, than any other African country.

South America

Beef production is important in several of the South American countries, and in these countries is found some of the highest per-capita beef consumption in the world. An abundance of feed and cheap labor permits a large supply of beef for both home consumption and export.

The British breeds are most numerous in Argentina; the Criollo (nondescript cattle) are prevalent in Colombia and Uruguay. The Zebu is influential in Brazil and other countries where heat adaptation is important.

In Argentina, as well as some other South American countries, there are large single operators that have thousands of acres of land and cattle under their control. Grass is the primary feed for fattening cattle. In some areas steers are five years old before they are marketed, because of slow growth and fattening; however, improved alfalfa pastures have permitted steers to be marketed at slightly over two years of age.

Figure 38-9 Ankole cattle and a headman in Uganda. (From *World Cattle,* by John E. Rouse, Copyright 1970 by University of Oklahoma Press.)

In many areas, reproductive performance is a serious management problem along with proper pasture utilization. Certain countries, such as Brazil, have a tremendous potential for beef production expansion.

New Zealand and Australia

Per-capita beef consumption is high in these countries, which represent major areas of the world where cattle are produced primarily for beef. The British breeds predominate, but in some areas the Zebu influence is prevalent.

New Zealand's cattle industry is probably secondary to sheep, since the cattle graze after the sheep have taken the best feed. Grass is the primary crop in New Zealand, with some of the best pasture management in the world. The cattle are fattened on pasture, and little grain feeding is done.

Management practices vary widely in Australia depending on the part of the country. Pasture fattening is the dominant practice. Beef is produced in quantities greater than the population can absorb; so the export market is very important to the Australian beef economy. In some areas of Australia the cattle are well managed and under fence. In other areas the human population is scarce and the cattle run over thousands of square miles. In some of these latter situations, the cattle are observed very infrequently and in many cases only once a year at roundup time, when cattle are sorted off for market.

FURTHER READINGS

Gustafson, R. A., and R. N. Van Ardall. 1970. *Cattle Feeding in the United States.* USDA Agric. Econ. Report no. 186.

Neumann, A. L., and R. R. Snapp. 1969. *Beef Cattle.* New York: Wiley.

Rouse, J. E. 1970. *World Cattle.* 2 vols. Norman: Univ. of Oklahoma Press.

Thirty-Nine

Management of Dairy Cattle and Other Milk-Producing Animals

"Will you please tell me which way I ought to go from here?"
"That depends a good deal on where you want to get to," said the cat.
"I don't much care where . . ." said Alice.
"Then it doesn't matter which way you go," said the cat.
". . . so long as I get somewhere," Alice added as an explanation.
"Oh, you're sure to do that," said the cat, "If only you walk long enough."

Lewis Carroll
Alice in Wonderland

Management consists of judgments and decisions based upon the application of the principles of sciences such as nutrition, physiology, and genetics, as well as of economics. The aim is to utilize knowledge of the sciences, to the extent financially feasible, that permits animals to produce at a high level. Thus the successful dairyman must combine good business with the science and art of the husbandman who loves his animals and is willing and able to respond to their needs.

Raising Dairy Calves and Replacements

Care of the Dam

Care of the young dairy calf begins with the dam. Under most feeding regimes for dairy cows in the United States, the nutrient requirements of the fetus will be met if energy and protein are adequate in the dam's diet. Since most of the fetal growth occurs during the last two months of gestation, the nutrient supplies at this time should be increased to about 150 per cent of maintenance.

Physical management of the cow at calving is important for survival of the calf. About the time that noticeable relaxation of muscles and ligaments around the tailhead and inflation in the mammary gland occur, the expectant mother should be separated from other animals and placed in a clean, dry, confined area where assistance can be rendered if difficulty in parturition is experienced.

The navel should be disinfected with an appropriate solution (such as iodine), the calf ear tagged or otherwise positively identified.

Shortly after birth the calf should receive its mother's first milk (colostrum) either by suckling or from the pail.

Care and Feeding of the Calf

Pre-weaning Most dairy calves are allowed to suckle their dam only once, and then are removed so the cow can be milked and managed with the remainder of the herd. The first three weeks appear to be the most critical in the calf's life. Ingredients essential to successful calf raising include regularity and sanitation in feeding, and dry, draft-free housing.

For the first few months, the young calf is basically a monogastric with nutrient requirements similar to those of other young mammals (Huber, 1965). The calf is dependent on milk or a highly digestible liquid food during the first few weeks of life.

Colostrum has a special importance to newborn ruminants because it provides gamma globulin to impart passive immunity; for the first several hours of life, these complex protein molecules are not broken down by digestive enzymes, but enter intact into the bloodstream from the intestinal tract. Nutritionally, colostrum is richer than milk in protein, vitamins A and D, and the minerals iron and cobalt, which are often marginal in the calf's diet during the first several weeks. This colostral yield of 100 to 200 lb. is greatly in excess of the calf's needs. The practice of placing it in cold storage to be fed to all the unweaned calves is a sound management practice.

Normally whole milk or milk replacers are fed to calves at about 8 lb. (one gallon weighs 8.6 lb.) per day during the first three to eight weeks, depending on weaning age. Whereas the calf would consume about 300 lb. of whole milk, it can be well-fed on milk replacer at half the cost. A typical formula for a milk replacer in per cent is: dried skimmilk, 45.0; dried whey, 35.0; emulsified lard oil, 17.5; antibiotic, 0.5; mineral premix (Mg, Fe, Cu, Co, I, Zn), 1.0; vitamin premix (vitamins A and D at 1,000 IU/lb., E at 50 mg/lb., and B-complex, amount not stated), 1.0.

Milk replacers should contain no sucrose and not over 10 per cent starch, because the intestines of young calves contain no sucrase and only limited amounts of amylase and maltase, the enzymes which break down these carbohydrates. This leaves lactose and glucose (dextrose) as the preferred carbohydrate sources.

Satisfactory fats for milk replacers are milkfat, lard, tallow, or coconut oil. Hydrogenation of the unsaturated oils makes them acceptable also.

Calf starter consists primarily of grains and is the concentrate mixture fed for the first three to four months; it, and any hay eaten, go directly into the rumen of the calf and initiate microbial fermentation and other changes necessary for ruminal development and function. The key to early weaning is early starter consumption. Because at weaning the calf is unable to consume and digest large quantities of roughage, the feed must be relatively concentrated in order to provide the energy needed to support life and growth.

The starter should also contain adequate amounts of vitamins A, D, and E, in addition to calcium, phosphorus and trace-mineralized salt. Starters should be offered during the first week of life. Calves can be weaned when consumption reaches 1 to 1.5 lb. of starter per day, which occurs usually at six to eight weeks of age.

A soft leafy hay cut in early maturity can be offered calves at an early age, but little is consumed until about six weeks.

Weaning to puberty At three to four months of age, rumen function is sufficient to synthesize the B-vitamins by microbial action and allow the calf to effectively utilize the usual proteins in cattle rations; a switch to the less expensive herd concentrate can be made at this time. Hay is usually fed free choice, and until puberty (about seven to nine months old) calves can be fed all the concentrate they will eat, although most dairymen limit the concentrate to three or four lb. per day.

Puberty to calving Heifers should be amply fed to "grow out" so that Holsteins will reach breeding size at about 800 lb. when they are 14 to 15 months of age, and thus enter the milking herd when two years old. Figure 39.1 depicts suggested growth rates and breeding sizes for heifers of the different breeds (Hillman *et al.*, 1965). There is an advantage for early breeding in reduced feed costs to first calving, earlier return on investment, and cumulative production per day of age. A disadvantage lies in increased calving difficulties.

The specific feeds used for heifers will vary depending on what is available on the farm. Home-grown forages are generally the most economical nutrient sources. However, grain should supplement any low-quality forages. If heifers are observed to fatten excessively, energy intake should be restricted so that the heifers do not gain over 2 lb./day from seven months of age to calving. This can be achieved by feeding leafy, early cut hay plus limited grain (2 lb./day) or corn silage only. Specific amounts of various nutrients needed for growing dairy heifers as well as lactating cows have been published in NRC, 1971a.

Whether purchased or home-grown, the heifers should be from bulls of proven ability

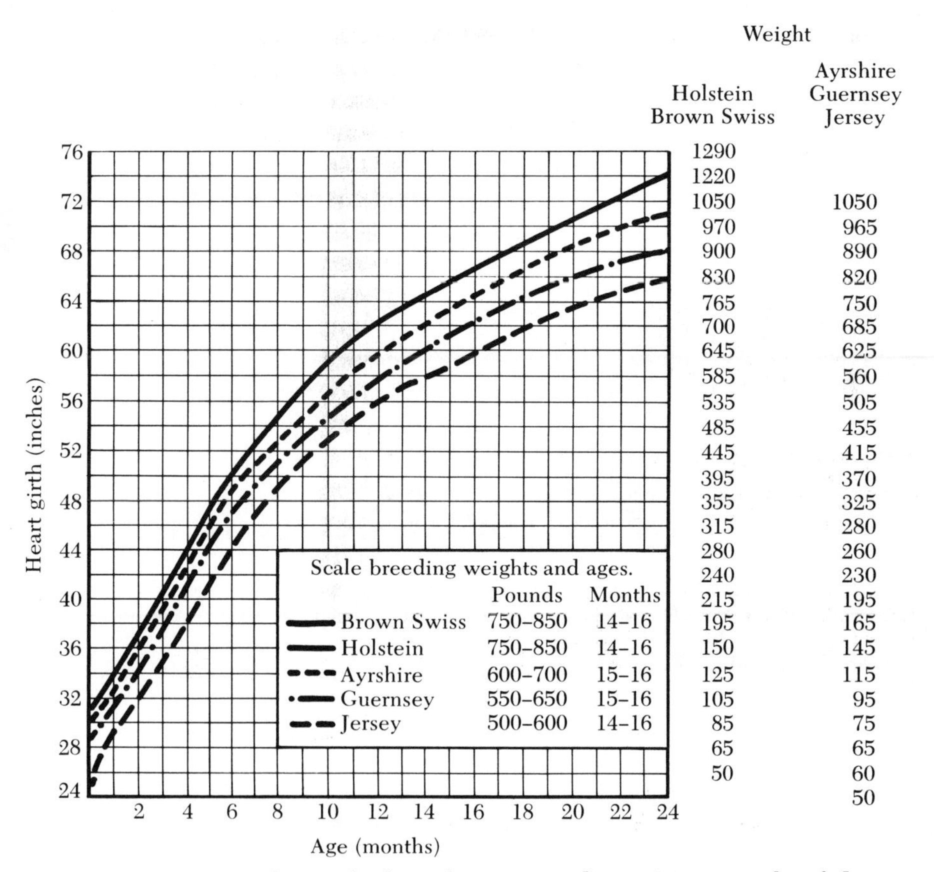

Figure 39-1 Suggested growth chart showing weights and heart girths of dairy heifers of different breeds from birth to 24 months of age. (Adapted from Hillman *et al.*, 1965.)

to sire acceptable type and production as demonstrated by high Predicted Differences (PD) and Repeatability (R). Dams should have produced at or above the breed or herd average.

The number of replacements needed will depend on various management conditions such as culling rate, ability to get cows pregnant, control of mastitis and other health problems, and the desire for improvement in performance. It is not uncommon to have a replacement rate of 33 per cent; so that on the average about every three years a new herd is being milked.

Some dairymen breed heifers to a bull of a small beef breed, hoping to reduce calving difficulties through smaller calves. While this may help to some extent, it should be noted that some plus-proven dairy bulls sire calves with light birth weights; breeding to such bulls can produce not only the desired small calves, but animals that can be used in the breeding herd as well (Boyd and Hafs, 1965).

The Lactating Cow

Feeding for Milk Production

Although proteins, fats, minerals, and vitamins are necessary ingredients in any ration, it is energy that is usually of greatest concern when feeding dairy cattle. Figure 39.2 illustrates the fate of the gross energy consumed by dairy cows and how it might change as the cow eats feed with different ratios of hay and concentrate.

The amount a cow will eat is influenced by many factors, including the acceptability of

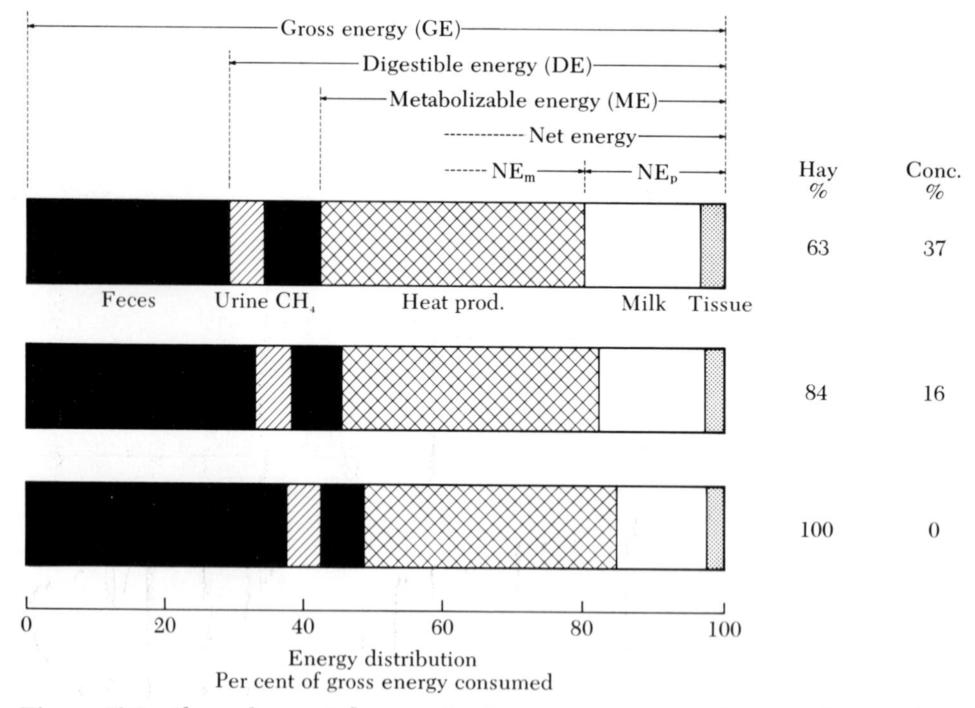

Figure 39-2 Shown here is influence of ration composition on utilization of energy by lactating dairy cows (based on averages of entire lactations of cows used in energy metabolism studies at the Beltsville, Md., Laboratory in 1962–63). This bar chart also illustrates the terminology used to describe energy values of feeds. Total digestible nutrients (TDN) are comparable to the digestible energy (DE) values, because only the losses of energy in the feces are measured to obtain them. (Flatt and Moe, 1967.)

the feed, the capacity of the gastro-intestinal tract, and factors playing upon the hypothalamus (a part of the brain) where two types of appetite regulators operate. Chemostatic-regulating factors in ruminants are blood-volatile fatty acids (especially acetic and propionic) and to a lesser extent glucose and fat; thermostatic regulation results in the tendency of animals to eat to keep warm and to stop eating to prevent overheating.

Baumgardt (1967) has reported that dry matter and digestible energy (DE) intake is maximal when DE is 68 to 70 per cent in the ration fed. This level is achieved when the ration consists of 40 per cent roughage and 60 per cent concentrate.

A cow genetically capable of higher production (beyond 85 to 90 lb.) may not be able to eat enough of even high-energy feed to meet her needs and must withdraw body fat reserves. Effort is made to meet the energy deficit encountered in early lactation by replacing roughages with concentrates; however, the extent to which concentrates can replace roughage is limited. Feeding rations containing more than 60 per cent grain (less than 17 per cent crude fiber) upsets the acetic-propionic acid ratio produced in the rumen and the metabolism of the cow changes essentially to that of a beef animal. First the milk-fat test is drastically reduced; then ultimately the feed energy not needed for maintenance is converted to body fat instead of milk. Further restriction of the roughage can push the cow off feed and into more serious consequences, such as elimination of ruminal protozoa, degenerative erosion of the lining of the rumen, laminitis, abscesses in the liver, and acidosis from the accumulation of D-lactic acid.

A cow normally undergoes a series of dynamic changes in her energy balance during one lactation period. A dairy cow that is not eating enough feed to meet its nutritional requirements continues to produce milk, at the expense of the energy it has stored in its body tissues. Under these circumstances the cow loses body tissue, such as occurs during early lactation with high-producing cows. A cow that is eating enough feed to meet its nutritional requirements is neither gaining nor losing body tissue. This occurs during mid lactation, but rarely can be attained with high-producing cows in early lactation when they are at their peak of lactation. A dairy cow that is eating more feed than it needs to meet its energy needs, regardless of whether the extra energy is coming from concentrates or forages, deposits energy in the form of body fat. This occurs during late lactation if a cow is fed *ad libitum.*

Roughages Legumes and grasses, harvested as hay, silage, or pasture, are the primary forages for dairy cattle in many regions of the United States and throughout much of the world. For best results when fed to cattle, these crops should be harvested in as early a stage of maturity as will allow for acceptable field yields. Hay is usually baled; however, the practice of compressing it into cubes and wafers is rapidly increasing in the Western United States, because these feeds can be stored and handled in bulk with much less space and labor than is required for baled hay.

Corn silage has replaced legume hay (except for possibly 5 to 10 lb. daily) as the main forage fed to dairy cows throughout much of the Central and Eastern United States. This change has mostly occurred during the past 20 years because nutrient yields per acre obtained from corn are much higher than from competitive crops.

The addition of urea and other nonprotein nitrogen compounds to corn (and sorghum) silage at time of ensiling has become a well-accepted practice in many areas. These additives favorably affect silage fermentation and result in a superior feed, in addition to supplying nitrogen for microbial protein synthesis.

Concentrates Corn, barley, wheat, and sorghums are the grains primarily used in concentrates fed to dairy cattle. By-products of various milling and food-production industries are also used. Residues such as soybean oil-meal, resulting from extracting the oil of oil-bearing seeds, supply proteins. Traditionally, cows are offered all the forage they will eat, and a concentrate mixture is formulated to balance nutrient requirements. A simple mixture of rolled barley, molasses (7 per cent) and salt (1 per cent) is adequate with good-quality alfalfa hay. Phosphorus supplementation might be desirable in some areas to provide a 2:1 calcium:phosphorus ratio. If a nonlegume hay is fed, supplemental calcium would be in order.

In the "challenge" system of feeding for maximum production, the cow is offered all the feed she will eat and culled if she fattens instead of producing milk. In practice, at about two weeks prepartum the dry cow is conditioned to high-energy ration by daily feeding her grain at 1 per cent of her body weight. When she begins lactating, the cow is then fed roughage at 1 to 2 per cent of her body weight and as much grain as she will clean up. Such feeding for the first three or four months of lactation will allow the cow to demonstrate her genetic potential for milk production; she is then fed to that level.

An increasing number of dairymen are feeding their cows complete rations prepared by mixing the concentrate and forage into a single feed which can reduce labor costs and provide better nutritional balance for the cows. A complete feeding program has been outlined by California workers (Bath, 1971). Cows may be separated according to age, stage of lactation, and concentrate ratios, depending on desired energy need. The entire ration is fed *ad libitum,* with no grain provided in the milking parlor. Barn-housed cattle can be similarly fed in the stalls. Guidelines for a complete ration should include the following on an air-dry basis: crude protein, minimum, 13.00 per cent; crude fiber, minimum, 17.00; roughage, minimum, 15.00; calcium, minimum, .60; phosphorus, minimum, .40; nonprotein nitrogen, maximum, .75.

In several states, extension services and feed manufacturers have developed programs with the use of computers to guide the dairymen in formulating complete rations tailored to the feeds available and animal needs on specific farms.

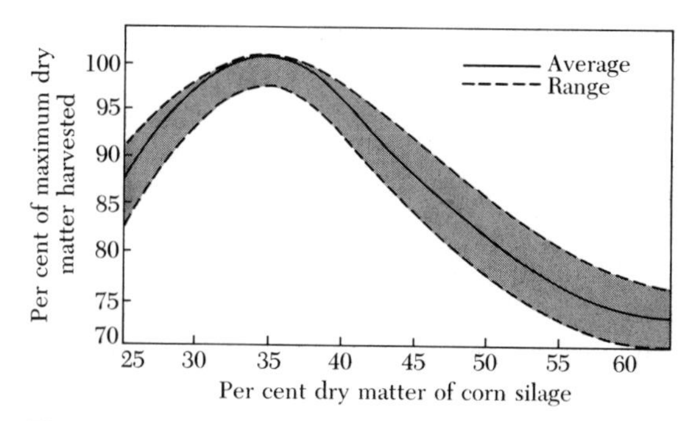

Figure 39-3 Effect of stage of maturity of corn silage on dry matter harvested per acre. Summary of research conducted at Michigan, Indiana and USDA. (From *Mich. State Univ. Ext. Bull.* E-655, 1969.)

Ration calculation Specific amounts of different feedstuffs to supply the needs of a cow can be determined from data such as are included in Table 39.1. For example, what should a cow weighing 1,500 lb. and producing 70 lb. of 3.5 per cent milk be fed each day? For net energy, a cow of this weight requires 11.3 Mcal for maintenance; and since 0.31 Mcal is required per lb. of 3.5 per cent milk, 70 lb. milk will require 21.7 Mcal, making a total daily requirement of 33.0 Mcal. The amount and nature of the ration is determined by this requirement and the economics of the feedstuffs available. Assuming that 50 per cent of the energy is to come from alfalfa hay, as is a common practice in Western United States, and that alfalfa supplies 0.55 Mcal/lb., we will feed $\frac{16.5 \text{ Mcal}}{0.55 \text{ Mcal/lb alfalfa}} =$ 30 lb. of hay (2 per cent of body weight). If we are to provide the remaining energy in barley, containing 0.93 Mcal/lb, $\frac{16.5 \text{ Mcal}}{0.93 \text{ Mcal/lb barley}} =$ 17.7 lb barley should be fed. If the energy needs are met with feeds such as alfalfa and barley, the protein and most other required nutrients are usually met also.

Management of Reproduction

The importance of reproductive efficiency was shown by a 40-year study at the USDA Agricultural Research Center, Beltsville, Maryland, where 41 per cent of the Holsteins and 24 per cent of the Jerseys left the herd because of non-breeding. A failure to get cows settled on time can result in loss of valuable production, because the cow is producing for a longer period of time at the lower level on the lactation curve.

Heifers are more difficult to breed, and conceive less easily than mature animals, because heat periods are shorter and more difficult to detect, and because they are not yet completely mature sexually. For Wisconsin Holsteins, conception was best when bred at 11.4 months old, after two or three heat periods and before reaching 900 lb. (Schultz, 1971). The herdsman should maintain records of all heat periods on every animal in the herd.

Sixty days are usually required for the uterus to return to normal after calving. The gestation period is approximately 279 days, so to calve within 365 days the cow must be pregnant on day 84 postpartum. Consequently, the rule of thumb is to check the health of the reproductive tract 45 days after freshening and breed at the first heat period subsequent to 60 days after freshening. Pregnancy should be checked 45 to 60 days after the last breeding in order to identify conception failures, embryonic deaths, and other abnormalities.

Table 39.1.

Daily requirements for maintenance and production.[a]

A. Maintenance

Body weight (lb.)	*Digestible protein*[b] *(lb.)*	NE_L[c] *(Mcal)*
900	.57	7.7
1,000	.61	8.3
1,100	.66	9.0
1,200	.70	9.6
1,300	.75	10.1
1,400	.79	10.7
1,500	.83	11.3
1,600	.87	11.9
1,700	.92	12.4

B. Production per lb. milk

Fat (%)	*Digestible protein (lb.)*	NE_L[d] *(Mcal)*
2.5	.042	.27
3.0	.045	.29
3.5	.048	.31
4.0	.051	.33
4.5	.054	.35
5.0	.056	.38

[a] From NRC (1971a).

[b] Digestible protein for maintenance = 2.84 gm/$(W_{kg})^{3/4}$.

[c] Energy for maintenance = 0.085 Mcal. $NE_L/(W_{kg})^{3/4}$. Add 20 and 10 per cent for growth for first and second lactations, respectively.

[d] Energy for production = milk energy (Mcal. NE_L/W_{kg}) = 0.353 + 0.096 times the per cent of fat.

Harvesting the Milk Crop

The structure and development of the udder prior to parturition is described in Chapter 29. Parturient edema, the swelling and hardening of the udder and surrounding parts, most frequently seen in heifers before calving, results from leakage of lymph from the capillaries into the interstitial spaces.

The udder is supported by the areolar tissue of the skin, by the sheet-like lateral suspensory ligaments forming the fore and rear udder attachments, and by the important median suspensory ligament between the udder halves. These supports, as well as the balanced arrangement of the teats, are important to ease of milking. Cows with poorly structured and supported udders cannot be tolerated and this characteristic should be considered in sire selection.

Prior to milking the belly, flanks and udder of the cow are usually washed for sanitary and aesthetic reasons. This might be accomplished by spraying the cow with a water jet from a hose. In warm climates, sprinkling devices in the floor of the holding pen wash off the underside of the cow; this water might be accumulated to be reused for flood-washing the alleyways.

Commercial milking procedures simulate nursing by the calf. The cow is prepared by washing the udder vigorously with warm water and drawing a stream of milk from each quarter into a strip cup to check for the abnormal milk of mastitis. Milk "let-down" occurs about 46 to 60 seconds after the oxytocin released from the posterior pituitary gland reaches the udder. Like the suckling of the calf, the modern milking machine reduces the pressure on the outside of the teat sufficiently for the pressure inside to force the milk through the elastic opening in the end of the teat. The teat is subjected to a vacuum of approximately 12 in. of mercury. Alternating vacuum between the flexible liner that surrounds the teat and the hard shell of the teat cut allows the liner to collapse around the teat.

Mastitis (or udder infection) is one of the major problems facing dairymen today. Proper milking, with properly designed and functioning equipment, is imperative. The infection level can further be reduced by dipping the teat ends in a disinfectant solution after milking, and by identifying and treating contaminated cows.

Each year, cows require a 45- to 60-day rest period for alveolar regeneration in the udder. Ordinarily, cows can be dried off by abruptly stopping the milking. Those few highly persistent cows producing over 50 lb. daily may require a concurrent reduced energy and water intake.

Dairy Cattle Selection and Breeding

Selection of dairy cattle is based upon type, performance, and pedigree. The associations for the Holstein, Brown Swiss, Guernsey, Jersey, and Ayrshire breeds have official classification programs in which a trained representative of the association, for a fee, evaluates the conformation of a breeder's animals.

Performance of dairy cattle is usually measured by observing the amount of milk, fat and protein, or solids-not-fat (SNF) produced for two consecutive milkings one day each month. Such records are converted to a mature equivalent, two milkings daily for a 305-day period (abbreviated: ME, 2×305). Sire proofs are calculated from this information. Using these records, the USDA now calculates and publishes data on all bulls with daughters enrolled on DHI and DHIR testing programs. Based on a comparison of the bull's daughters with their herdmates, a Predicted Difference (PD) is calculated, which estimates the sire's ability to transmit milk and fat production to future daughters. In a like manner, genetic capacity of a bull to transmit dairy type to his offspring can be estimated by comparing the classification categories of the offspring to those of their dams. Daughters of bulls with high proof indexes tend not only to produce more milk per lactation but also to generate more lactations.

In selecting the dams for future generations, it is recognized that the best indicator of the cow's productive ability is her own production record; yet a production record of each of four daughters provides as much genetic information as a single record on the cow herself.

Geneticists estimate that 20 to 30 per cent of the differences in milk production and type among cows is due to inheritance, which means that the remaining 70 to 80 per cent is due to environment or to interaction between heredity and environment. This per cent dependent upon inheritance is expressed as a heritability estimate and amounts to 0.2 to 0.3.

Housing of Dairy Cattle

Calves

The young dairy calf can withstand low temperatures, but is very sensitive to drafts and to cold, humid conditions. Winters are severe enough in most of the United States except the south and southwest that some shelter is needed from birth to about four months of age.

In small herds, where only six to eight calves are handled at one time, a cold housing system can work out satisfactorily if it is well-planned. A dry, draft-free cubicle should be constructed. This can be accomplished by boarding up free-stalls with solid sides of plywood at one end of a free-stall barn. In extremely cold or windy weather, a plywood cover placed over the pen will offer added protection to the calf. Water is hand-fed to calves in cold housing because temperatures may drop low enough to freeze water pipes.

Cows

Temperature effects For highest milk production, best use of nutrients, and least stress, the optimum temperature for dairy cows is between 40° and 60°F. Within this range the heat produced by rumen fermentation and nutrient metabolism is efficiently used to maintain the body temperature of the cow. It is estimated that for each 10°F drop in temperature below 40°F, about one Mcal of digestible energy (DE) is required to maintain body temperature. No direct effect on milk yields are usually shown unless daily minimum temperatures drop below 5°F. High temperatures (above 80°F) also depress production, particularly where humidity is high.

Types of housing for cows There are basically three types of sheltered housing systems used for dairy cattle in the United States. The older, conventional type consists of a barn where cows are housed, fed, and milked while confined in stanchions in a ventilated and insulated structure. Many new, warm-enclosed barns with insulated ceilings and walls have, in addition to free stalls, a feeding area and a milking parlor.

A second type of dairy-cattle housing, which is popular in many areas of the United States, particularly for units accommodating 50 to 150 cows, is cold-covered housing. Buildings in these units have no insulation or mechanical ventilation, thus reducing costs of construction and maintenance. As in the warm-enclosed barn, cows are milked, fed, and bedded under one roof. In the cold, northern climates, cows do well in these systems; hence, a large percentage of the new units being constructed are of the cold-covered type.

The third type of housing is the open-lot, free-stall system. Since it adapts so well to larger herds, it is becoming popular both in more moderate climates and in northern areas. This housing usually consists of an open (or partially protected) shed where cows are bedded in free stalls. The feeding bunks are also covered, and milking is done in a parlor. In warmer climates the only housing required may be an open shed.

Dry-lot operation For various reasons, especially with large herds and automated feeding systems, it might be more profitable to mechanically harvest and bring all the feed to the cow rather than to pasture, especially when feeding *ad libitum* complete rations.

Because of the great variation in milk yields among cows, it is desirable to separate them into groups on the basis of their milk production, so the higher producers can be fed what they need without overfeeding the lower producers, which are often later in lactation. Also, the dry cows and pregnant heifers should be fed separately from the milking herd to discourage excessive fattening. Such an arrangement would suggest a minimum of three pens.

Management of Lactating Animals in the Developing Nations

In addition to cattle, the female buffalo, goat, sheep, yak, llama, camel, reindeer, horse, and ass provide significant quantities of milk for human food. Not only the species of animals used for milk production but their management varies widely from place to place. Frequently, in the developing countries of the world, animals have been selected not for milk and meat production, but for draft purposes and survival under rigorous conditions. Great strides in milk-productive capacity have been experienced by first controlling the environment to reduce stress from adverse climate, diseases, parasites, and uncertain nutrient supply, and then introducing genetic improvement through cross-breeding the native stock with improved breeds.

Grazing on pasture is the most common practice, but in some areas where land is especially limited the forage may be cut and brought to the producing animals. More uncommon are situations in which forage is preserved and stored. Such lack of planning and resources lead to heavy losses when the supply of natural pasture is interrupted by flood, drought, or insect infestation.

Joshi *et al.* (1957) describe a practice in Morocco in which the farmer places his cattle in the charge of a herdsman who has the responsibility of finding adequate pasture. His services are paid either in some proportion of increased liveweight of the cattle or in a part of the calf crop.

In the nomadic Fulani tribe of northern Nigeria, cattle provide individual prestige as well as a source of wealth. The tribesman gains aesthetic satisfaction and social approval from his possession of cattle in the same way his western counterpart views his otherwise unproductive flower garden. Consequently, in these situations, cows capable of reproduction are highly prized. The cattle ownership is vested in the men, but the dairy products are claimed by the women. Milk forms the basis of the native diet, and is consumed either as a liquid or as a form of clarified butter (Joshi *et al.*, 1957).

In more populated areas, large numbers of cattle and buffalo are kept in crowded, unsanitary conditions within the city boundaries. To reduce the cattle population within the city, Bombay, India, established a cooperative milk colony 12 miles from the city boundary; the colony included not only facilities for handling the cattle and milk, but such things as housing and shopping centers to make moving to and living there more attractive. Good management of the buffalo by the dairymen is encouraged, and has resulted in a yearly average yield per buffalo of 5,500 lb. (2500 l) of 6.5 per cent milk at the colony, a production three times greater than the national average (Khurody, 1967).

A common system of milking in many parts of the world is practiced twice a day with "calf at foot" (Figure 39.4). The calf, separated from the cow overnight and during the day, is allowed to suckle for a period of time prior to milking. Such a system is deemed necessary with Zebu cattle, many of which let down their milk only when their calves are present (Joshi *et al.*, 1957; Mahadevan, 1966).

Management of the Domestic Buffalo for Dairy Purposes

It has been estimated that 110 million buffalos (*Babalus bubalis*) are to be found in the area stretching from Italy to China and from Nepal to the Philippines. Buffalos are triple-purpose beasts. The buffalo is not particularly tolerant to either heat or cold, and must be provided with shade and water. It survives on very coarse feed, is docile, and can be handled by women and children. Milk yields differ with breed and management, but may vary from less than 10 lb. to more than 50 lb. daily; milk fat is as high as 15 per cent, but 7 per cent is more common, and milk solids-not-fat (SNF) of 9 to 10.5 per cent are common. Ghee, a dehydrated, clarified fat, as well as yoghourt and cheese, are made from the buffalo milk (Cockrill, 1967).

Management of Dairy Goats

Being a ruminant with a highly developed mammary system, the dairy goat is very similar

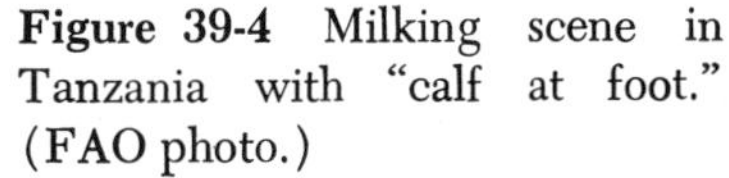

Figure 39-4 Milking scene in Tanzania with "calf at foot." (FAO photo.)

to the dairy cow in nutrition, physiology, and reproduction. Likewise, there are many similarities in management and the adjustment of the environment so that the genetic potential for production can be realized.

In terms of body weight, goats weighing 1/10 that of cows are highly competitive if not superior in milk production. Consider the top 305-day milk and fat production by two Toggenburgs in 1960. Puritan Jon's Jennifer II, T 121022, produced 5750 lb. of milk, and her half-sister Puritan Jon's Janista, T 121820, produced 202.5 lb. of milk fat, which is higher than the top production for cows on a body-weight basis.

Goat milk is characteristically white (the yellow carotene of feed has been converted to colorless vitamin A) with small fat globules. Its protein is sufficiently different from cow's milk that many who have become sensitized against the milk from cows can drink that from goats. The milk flavor characteristically differs from that of other species because of the relatively larger amounts of the "goat" fatty acids: caproic, caprylic, and capric.

Goat-milk production has a relatively high labor requirement, which, together with the seasonal nature of reproduction (and thus milk production), makes necessary a marketing price approximately double that of cow milk.

Greater improvement in the production performance of the dairy goat could come from establishing widely accepted, sound breed goals and deemphasizing nonproductive traits such as colors, ears, and facial characteristics so dear to the hearts of many hobby breeders.

Goats are susceptible to the same diseases and parasites as the cow. Tuberculosis is comparatively rare but more frequently seen are brucellosis, leptospirosis, mastitis, and metabolic diseases such as bloat, ketosis, and milk fever.

Feeding

The structure and function of the digestive system of the goat is very similar to that of the cow. Table 39.3 gives requirements for maintenance and production for lactating goats. The maintenance requirement is higher in the goat than cow because maintenance is a function of metabolic weight (body weight raised to the ¾ power).

Goats will consume dry matter up to 4 per cent of their body weight, depending on energy concentration, acceptability and availability of the feed, and their need for maintenance and production. A crude protein level of 13 per cent in the total ration is generally considered adequate.

Housing

Goats are capable of functioning in a wide variety of climatic conditions. They can withstand low temperatures if they are dry and protected from drafts. The most natural housing might be the loose or shed type, in which the animal is free to move about. In this system, each goat should have about 15 square feet of bedded area and the ceiling height should be at least 10 feet to allow for cleaning. Colby (1966) has provided some details for buildings for goats.

A milk room meeting public health standards and with elevated milking stalls for either hand or machine milking is necessary for convenience and sanitary milk production. These vary from small scale design to units handling several hundred animals (Figures 39.5 and 39.6).

The agility and capriciousness of goats make opportunities for exercise desirable and a fence 4 to 5 ft. high in pens is needed.

Milking

As with cows, does require milking at least twice daily in order to remove the back pressure within the udder. Failure to do so causes hormonal changes resulting in a reduced secretion rate.

The lactation period is usually seven to ten months; however, rebreeding of some does may be delayed so that lactation will be prolonged for up to 20 months in order to provide a more even market-milk supply. Milk yield will increase to a maximum perhaps two months postpartum, and then decline. Six to eight weeks prior to kidding, the doe should be dried off by reducing the energy intake and abruptly stopping milking. Increased yield is observed with each succeeding lactation up to "maturity" at four to eight years.

Table 39.3.

Daily requirements for maintenance and production in the dairy goat.

A. Maintenance

Body weight (lb.)	*Body weight (kg.)*	*Digestible protein*[a] *(lb.)*	*Digestible protein*[a] *(gm.)*	NE_L[b] *(Mcal.)*
80	36	.092	41.74	1.25
90	41	.101	46.01	1.38
100	46	.111	50.27	1.50
110	50	.118	53.39	1.60
130	59	.133	60.49	1.81
140	63	.142	64.18	1.92
150	68	.148	67.31	2.01
160	73	.157	71.00	2.12
170	77	.163	73.84	2.21

B. Milk Production

Fat (%)	*Digestible protein (lb./lb.)*	*Digestible protein (gm./kg.)*	NE_L[c] *Mcal. (per lb.)*	NE_L[c] *Mcal. (per kg.)*
2.5	0.042	42	0.27	0.59
3.0	0.045	45	0.29	0.64
3.5	0.048	48	0.31	0.69
4.0	0.051	51	0.33	0.74
4.5	0.054	54	0.35	0.78
5.0	0.056	56	0.38	0.83
5.5	0.058	58	0.40	0.88
6.0	0.060	60	0.42	0.93

[a] Digestible protein for maintenance = 2.84 gm/$(W_{kg})^{3/4}$.

[b] Energy for maintenance = 0.085 Mcal. $NE_L/(W_{kg})^{3/4}$. Add 20 and 10 per cent for growth for first and second lactations, respectively.

[c] Energy for production = milk energy (Mcal. NE_L/kg) = 0.353 + 0.096 times the per cent of fat.

Figure 39-5 Workers drive 200 milking does to two milking parlors. Each central aisle serves buildings on one side. Does are driven into holding pen just outside each milking parlor. As the does are milked and fed and leave the parlor, they go back to their corral. When all does from one corral are finished, the worker shoos the stragglers home, adjusts the gates, and brings the does from another corral to the parlor. (Photo courtesy Laurelwood Acres, Ripon, Calif.)

Selection and Breeding

The basis of selection consists of the type (conformation), production (milk test), and pedigree. Too often the only information available is the appearance. The desirable goat is one of the appropriate size for its breed and displaying dairy character, viz., sharpness that indicates the ability to convert feed above maintenance into milk rather than body fat. A sound, well-attached udder with medium sized teats, together with strong, straight feet and legs, are desirable. Type classification programs for dairy goats are being developed.

Production records as obtained through the Dairy Herd Improvement Association (DHIA) are based on milk weights and milk fat content measured one day each month as in cattle. Sire proofs based on the performance of daughters as compared with their dams or herdmates are sometimes available for aid in selecting the bucks to use. Such progeny-tested sires can hasten breed improvement.

The pedigree should provide the identities of ancestors, sibs, and offspring, but more important to intelligent selection would be type and production information for each of these animals. Although many other factors are frequently considered in selection (shape of ears color, and friendliness), only those with both economic importance and high heritability should receive selection pressure.

While there are many undesirable genes, usually recessive, in the goat population, of special interest is hermaphroditism, which is relatively common in goats and is thought to be due to a recessive gene closely linked with hornlessness which, in some breeds at least, is partially sex-linked.

Reproduction

Goats are seasonal breeders. The onset of the breeding cycle is triggered primarily by the days becoming shorter. The season onset can be hastened a few days by the presence of a buck with the does and by placing the animals in darkened quarters for longer periods of the day.

In the northern hemisphere, goats breed mainly from September through January, during which time the doe experiences estrus (heat) every 17-21 days. Heat detection and proper record keeping are essential to successful breeding programs. Heat is manifested by easily detected behavioral changes and appearance; frequent urination, nervous bleating, and side-to-side tail quivering are noticeable. These actions are even more evident when other goats are present. There is some riding and standing for other does to mount, and, if a buck is near, the doe will make continuous attempts to familiarize herself with him.

Natural mating is most convenient by taking the doe to the buck for a single service. For artificial insemination, buck semen has been frozen successfully with skimmilk containing 6 to 9 per cent glycerol as diluent (Frazer, 1962). Yearling does are inseminated by passing a pipette blindly but carefully into the vagina and discharging the semen when the tip contacts the cervix which is in the same position in the pelvis as in the cow. With mature does, a lubricated speculum (¾″ diameter, 6 to 8″ long) is used with the tip of the inseminating tube inserted partway into the cervical canal before the semen is expelled.

The length of estrus is variable, but it usually lasts two to three days. It is shorter early and late in the breeding season. Insemination should occur late in estrus whether natural or artificial techniques are applied.

Multiple births are common in goats; they are more frequent at higher levels of nutrient intake and at maturity, and are also influenced by heredity. Triplets are not uncommon among mature, well-fed does.

Kidding is preceded by the development of an enlarged abdomen and udder with varying amounts of mammary congestion. Pelvic ligaments relax a few hours before parturition. Rapid cud chewing, restlessness, pawing at bedding, and a low plaintive bleating are signs of impending parturition. Mucous and fluid discharge associated with straining should precede by about an hour the birth of the kid.

Figure 39-6 Interior of an eight-stall milking parlor. Worker washes udder of doe before applying the milking machine. The milk goes directly into pipeline to a bulk tank. (Photo courtesy Laurelwood Acres, Ripon, Calif.)

Buck Management

A buck may reach puberty at three months and achieve adequate size and maturity for light service at six to ten months. A mature, thrifty buck can serve 30 or more does per season. Because of the strong odor of bucks, they are best kept separated by at least 100 feet from the does except when desired for service. Clipping the hair on or about the head, neck, and belly tends to reduce the strong odor.

Bucks should be kept in thrifty condition, which is possible with *ad libitum* feeding of a good quality roughage, steamed bone meal, and trace mineralized salt. An additional one or two pounds of grain daily may be required during the breeding season.

Care of the Young

Kids should have their navels painted with iodine upon birth. They should be positively identified by tattooing. Ear tags tend to tear out more frequently in kids than in calves.

Kids should receive colostrum four or five times daily for three days, whether they are left with their dams or separated at birth. Colostrum provides passive immunity as well as being a rich source of protein, minerals, and vitamins. Feed milk not to exceed a daily total of 8 per cent of body weight in two or three portions by bottle or pan. As with calves, kids can be successfully raised on a suitable milk replacer. From the beginning, kids should be encouraged to eat a grain-based calf starter available free choice. Good-quality hay is fed after a few weeks of age. Since puberty is more a function of body size than age, kids should be adequately fed to reach breeding size (85 to 90 lb.) at an early age (ten to twelve months), and thus come into early production. Buck kids should be separated from doelings at two to four months of age to prevent premature pregnancy.

Goat hooves tend to overgrow on animals maintained on other than abrasive surfaces, thus requiring regular trimming. Sharp pruning shears are effective tools for this task.

Management of Sheep for Dairy Purposes

Although rarely milked in the United States, in many Balkan and Mediterranean countries, ewes supply milk for commercial production.

In recent years in the Near East trucks have been especially equipped to commence processing the ewe milk into cheese. France and Israel have large flocks of dairy ewes milked by machine in intensive farming systems. Morag *et al.* (1967) have described a parlor with a 4-abreast arrangement providing for machine attachment from the rear between the hind legs. Teat cups are applied as soon as the ewe is in the milking position and without udder washing or fore-milking; milk flow starts 5 to 15 seconds later and lasts from 4 to 50 seconds. Machine stripping accompanied by massage for 10 to 15 seconds accounts for 10 per cent of the total yield. In a simple milking routine, two men can milk and hand-strip 80 ewes per hour. Under more primitive conditions, one shepherd, with the help of his family, is expected to be able to milk from 150 to 200 ewes per day (Betrone, 1966).

Sheep milk, with its 5 to 9 per cent fat and 4 to 7 per cent protein, is not usually consumed as a beverage but as yogurt and cheese. The famous Roquefort blue cheese is an example of its widespread acceptance.

FURTHER READINGS

Bines, J. A., S. Suzuki, and C. C. Balch. 1969. "The significance of long-term regulation of food intake in the cow." *Brit. J. Nutr.*, 23:695.

Boyd, J. S. 1969. "Calf and youngstock housing." *MSU Ext. Bull.* 619.

Hoglund, C. R., J. S. Boyd, and J. A. Speicher. 1969. "Free-stall dairy-housing systems." *MSU Exp. Sta. Bull.* 91.

Huber, J. T. 1969. "Development of the digestive and metabolic apparatus of the calf." *J. Dairy Sci.*, 52:1303.

Huber, J. T., J. W. Thomas, and R. S. Emery. 1969. "Response of lactating cows fed urea-treated corn silage harvested at varying stages of maturity." *J. Dairy Sci.*, 51:1806.

Joshi, N. R., E. A. McLaughlin, and R. W. Phillips. 1957. *Types and Breeds of African Cattle*. Rome: FAO.
Khurody, D. M. 1967. "Milk colonies." *World Rev. of Anim. Prod.*, 3:46.
Lindahl, I. L. 1968. *The Digestive System of the Goat: Structure, Function and Dysfunction*. Spinedale, N.C.: American Dairy Goat Association.
Lunca, N. 1965. "The present state of artificial insemination in sheep and goats." *World Rev. of Anim. Prod.*, 1:73.
Moe, P. W., and W. P. Flatt. 1969. "The new energy value of feeds for lactation." USDA-DCRB pub. no. 69.

Forty

The Science and Husbandry of Swine Management

"Patriot and benefactor, cosmopolite and bon vivant—*no matter how you slice ham,* sus scrofa domesticus *merits our affectionate salute."*

R. C. Davids

The pig has enjoyed a prominent role in the development of American animal agriculture. Since swine are prolific, gain rapidly and efficiently, and utilize feedstuffs not directly consumed by or acceptable to humans, they early proved profitable to the American farmer. Historically, they have been recognized as "mortgage-lifters." The versatility and adaptability of the pig have permitted changes appropriate to consumer needs. For example, through selection programs, and the advantage from the relatively short generation interval, swine breeders have been able to change the body type—short and fat, long and rangy, intermediate or modern meat type—to meet market demands. But in addition to synthesizing those body tissues which ultimately we enjoy as barbecued pork chops, ham steaks, and other delicious meats, the pig has contributed much to medical science. His physiology is remarkably similar to the human; furthermore, he sometimes develops ulcers and is susceptible to certain diseases and allergies common to man.

The modern pig leads a rather short, but pampered, life. Man has removed him from his natural habitat to a man-made environment. In this confined state, pigs can individually be docile, stubborn, or aggressive. But given appropriate management, housing, and nutrition, swine prove to be among the most profitable of meat-producing animals.

Modern Swine Production

Modern swine production is a blend of business acumen, superior husbandry, and technological know-how. Since World War II, the trend toward specialization has resulted in fewer, but larger, production units. For example, the number of farms from which hogs are sold has declined dramatically in Illinois and the entire United States Corn Belt area (Figure 40.1). As individual enterprises increased in productive capacity, the traditional field-rearing (Figure 40.2) systems have been replaced with confinement and intensive swine-rearing systems (Figure 40.3). The change to confinement (defined as any system that confines animals to non-dirt-floor housing units) has been justified on the following factors. (1) Rising per-unit cost of a decreasing supply of manual labor. (2) Increased land values and economically competitive alternative uses: as price of land increased and tillage practices permitted more complete utilization of crop land, use of such land for hog areas became less economically justifiable. (3) Advantage potential for mechanization and automation: innovative equipment and housing designs and arrangements permitted labor-saving techniques not possible in field systems. (4) Volume production: moving to confinement, with its mechanization and automation, meant raising the same number of hogs with less labor, or raising more with the same available labor. (5) Environmental control: modification of seasonal weather extremes can be more effectively accomplished in confinement. (6) Nutritional knowledge: the vast amount of research information of the late 1940's and early 1950's provided release from the need for forage and soil for certain nutrients.

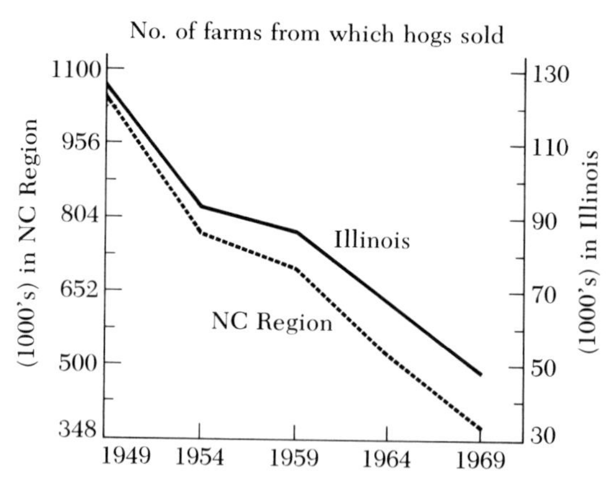

Figure 40-1 The number of farms from which hogs are sold in the North Central United States, designated as the corn belt area, is less than half the number in 1949. A similar trend has occurred in the state of Illinois where abundant quantities of corn and soybeans are produced.

The importance of each of these factors is probably related to geographical and localized circumstances. Since we are primarily concerned with management in our discussions

Figure 40-2 A field rearing system of swine production. Portable houses and shelters permit rotation of pen or field use to reduce disease and parasite buildup.

Figure 40-3 A confinement swine production system utilizing such labor saving factors as slotted floors, pneumatic conveyance of feed from mixing center (round units) to storage tanks by each building, automatic distribution of diets to feeders in the buildings and a lagoon for waste disposal.

here, we should also recognize that confinement has in most cases provided more comfort and convenience for management and labor. This is an asset of high value, but difficult to measure in dollars.

Challenges of confinement production include (1) high investments, (2) manure handling and disposal, (3) concentration of disease problems and (4) exercise of excellent management, which includes all those decisions necessary for best possible environment and greatest operational efficiency, and which determines total productive efficiency.

Historically, in the United States, swine production has been associated with corn production. Currently, because of new and rejuvenated grain-production areas, such as irrigated acreages or expansion of sorghum grain production, particularly in the southwestern states, the Corn Belt does not enjoy the privileged position of years past. However, the bulk of swine production is, and will continue to be for the foreseeable future, in this area.

Kinds of Production

In the United States the two major kinds of production are (1) rearing and selling of breeding stock and (2) rearing and selling of animals as feeders or for slaughter.

Production of Breeding Stock

Quality of pork product and the efficiency with which it is produced are largely determined by the genetic potential of the parent stock. The seed stock producer, whether he supplies purebred, crossbred or hybrid animals, is entrusted with the responsibility of providing genetic material for the industry (see Chapter 21).

Production of Market Animals

The fraternity of commercial swine producers includes: (1) feeder-pig producers, those who produce pigs to be sold at about 40 lb. in weight; (2) feeder-pig finishers, those who buy the feeder pigs and feed them to a market weight of at least 200 lb.; and (3) those operating a total enterprise, raising pigs from birth to sell at market weight.

Feeder-pig production requires the least total feed supply and is common in areas of marginal grain production. Excellent management is required to ensure the maximum number of pigs weaned per sow per year.

Feeder-pig finishing, on the other hand, requires much greater quantities of feed, and is usually concentrated in areas or localities of abundant grain production and availability. Mechanization and automation of feed handling and distribution, along with building designs to eliminate daily removal of manure, minimize need for manual labor. Management and husbandry requirements are less demanding than for feeder-pig production.

The total enterprise is a combination of the other two producers and is the most common system. Excellent management, abundant feed supplies, and modern facilities are utilized to greatest economic advantage. Programs resulting in 2,500 to 3,000 market hogs per year per man are not uncommon.

Detailed records covering several years show that profit is greatest for the total enterprise system, least for the feeder-pig finisher, and in between for the feeder-pig producer.

Housing Considerations

The rapid growth of the pig emphasizes the need for a specific environment for each segment of growth and development.

Climatic Factors

Air temperature and humidity are the two factors most obvious and most readily detected. Thus, in deciding on building designs, the producer chooses the degree of environmental control which is economically feasible in his area. Housing suited to Thailand (Figure 40.4), for example, would be unsuited to Canada.

Physiological Factors

The newborn pig is extremely sensitive to cold. He will have a body temperature of 102°F, but his thermoregulatory capacity is inadequate to successfully withstand cold stress the first few hours after birth. This pig's mother (or dam), on the other hand, will weigh 300 pounds or more and have sufficient thermal insulative protection from subcutaneous fat and other tissues to withstand considerable cold. She will, however, be quite sensitive to heat, since, unlike man she does not perspire when hot.

Figure 40-4 A slotted floor pig unit peculiarly adapted to environmental conditions and production practices in Thailand. Manure drops into water, which may be associated with rice or fish farming practices.

Production Efficiency

Housing can provide an environment more conducive than natural environments to efficient use of feed. Responses of growing pigs to different housing environments during midwestern United States winter seasons are shown in Table 40.1. Table 40.2 shows performance data recorded during a warm season.

Building and Equipment Design

Slotted floors These consist of slats with openings between them through which manure may drop or be trampled. They were not generally used in the United States until the late 1950's and early 1960's. The slotted-floor principle, however, is certainly not new. It is reported that slotted floors were in use in housing for sheep as early as 1760 (Hammer, 1960). A partial-slotted-floor piggery used on a farm in Rhode Island in 1790 has been described (Harris, 1870). But the reintroduction and rapid adoption of slotted floors in swine housing has been given much credit for accelerating the trend toward confinement (Jensen and Becker, 1961). Previously, manure had been (and in many units still is) removed mechanically from confinement units by scraping (by hand or tractor scrapers), water pressure, gutter-cleaner equipment, and various combinations of these. The labor required and the volume of material handled proved discouraging in many early attempts at confinement.

Table 40.1.
Response of growing swine to different housing units during cold weather.

	Housing		
	Enclosed[a]		
	Heated[b]	Not heated[c]	Open-front[d]
Av. initial wt., lb.	30	30	30
Av. final wt., lb.	74	74	69
Av. air temperature, °F	71	50	20
Av. daily gain, lb.	1.76	1.76	1.65
Av. daily feed, lb.	3.33	3.61	3.96
Gain per 1,000 lb. feed	522	488	416

[a] Insulated windowless wood-frame building, partial slotted floors.
[b] Gas-fired space heater used to maintain minimum air temperature of 70°F.
[c] No supplementary heat or bedding used.
[d] Concrete walls, open front on southern exposure, solid concrete floor with wooden overlay, plus straw provided in sleeping area.

Table 40.2.
Performance of finishing pigs in open front building and in dirt lots during warm weather.

	Open front	Dirt lot
No. of pigs	60[a]	60[b]
Space per pig, sq. ft.	10	180
Av. initial wt., lb.	99	96
Av. final wt., lb.	200	194
Av. daily gain, lb.	1.56	1.41
Av. daily feed, lb.	5.15	5.21
Gain per 1,000 lb. feed, lb.	303	271

[a] Four pens of 15 pigs each.
[b] Two lots of 30 pigs each.

The main advantages of slotted floors are elimination of daily cleaning of pens, elimination of use of bedding (this assumes necessary warmth provided), and increased sanitation potential due to cleaner pens and pigs.

Materials or designs that cause discomfort or injury to the animal are undesirable. In general, space between slats should be in proportion to slat width.

Typical designs for partial and total slotted floors are shown in Figures 40.5 and 40.6. Minimum labor and maximum cleanliness can be obtained with use of total slotted floors. Cleanliness is not always maintained on partial slotted floors. Manure on the solid portion is common during warm weather, since the pigs benefit from evaporative cooling by laying on the wet surface. In contrast, on total slotted floors pigs do not have moist-floor sleeping

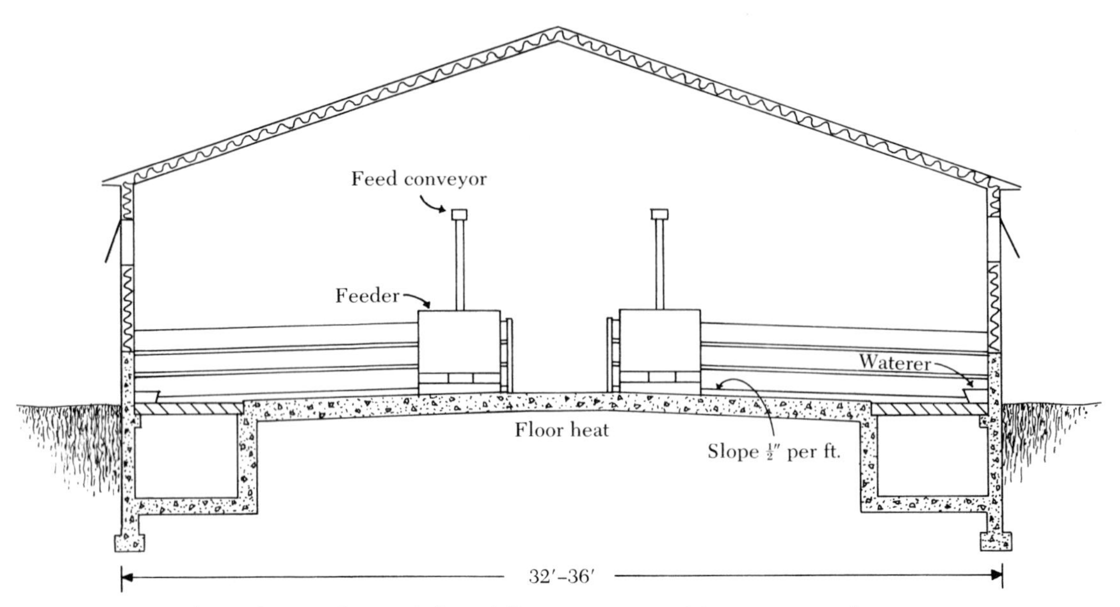

Figure 40-5 A basic design of partial slotted floor swine unit. Manure is stored in gutters along the walls and removed periodically. Heat elements, either electric or tubing for hot water circulation, keep sleeping area dry and warm.

areas, and supplementary cooling—mist spray, zone air cooling, etc.—may be desirable. When provided equally comfortable environments, growing-finishing pigs perform similarly on floors of varying degrees of slotted surface area. Excellent management and greater environmental control are required with total slotted than with partial slotted or solid floors.

When open-front building units are used during cold weather, comfortable sleeping areas can be provided by use of heating elements in the solid floor portion, radiant heat brooders, or bedding.

Environmental control In our considerations here, environment includes all known factors external to the pig, which, singly or in combination, affect behavior and performance. It is recognized that all effects of these factors have not been delineated, and factors not yet identified may be the stimuli for certain of the currently observed behavior.

Air temperatures above the pig's critical temperature significantly reduce voluntary feed intake, with subsequent lower rate of gain and decreased feed efficiency. Conversely, cold air temperature stimulates feed intake, but this compensatory consumption is largely for energy demands of increased metabolic activity for thermoregulation.

The percentage of mated females that farrow varies with season, and tends to be lowest when mating occurs during warm weather. This may reflect lowered fertility of the boar, abnormal cycling and behavior of the females, or both.

The 24 to 48 hours immediately prepartum is a particularly heat-susceptible period. Increasing air movement over the animal, wetting the animal, or providing cooler air for breathing can modify the stress of heat. Postpartum response to high air temperature is reflected in minimal activity, increased respiration rate, and decrease in voluntary feed consumption. Snout-cooling (directing cooled air to the head of the animal) or zone cooling effectively relieves air temperature stress.

Since the pig has a low rate of evaporative heat loss, humidity *per se* has less effect than with other species. Relative humidity appears to have little direct effect until it reaches above 80 per cent in conjunction with high air temperatures.

There is little evidence to indicate that variations in light intensity or length of exposure period significantly affect performance of growing-finishing pigs.

Sudden movements or sounds cause an excited, but usually momentary, response from swine. However, pigs appear to adapt readily to sounds of different intensities and duration.

The velocity and microcontent of the air surrounding the pig are important parts of the effective environment. Velocity sufficient to produce "draft" conditions causes increased discomfort of young pigs in cold temperatures. Oxygen consumption rate is increased and feed efficiency markedly decreased. Growing-finishing pigs are similarly affected.

Atmospheric gas contaminants are also of concern. The excreta collected and stored under slat floors provide an effective medium for anaerobic organisms, which produce hydrogen sulfide, ammonia, and other gases. These are odorous, but more importantly can, at specified levels of concentration, be irritating, and at higher concentrations can prove toxic (Taiganides and White, 1968). Insidious and/or toxic effects that may accrue from prolonged, continuous exposure to low gaseous levels have not been determined. It is obvious that building design must incorporate ventilation systems that maintain minimal odor and contaminant conditions.

Effects of dust *per se* have not been determined, but dust particles may be stressor agents and predispose animals to respiratory infections and aberrations.

Unit housing systems Confinement systems are composed of various arrangements of individual segments designed specifically for most efficient performance of a given age or

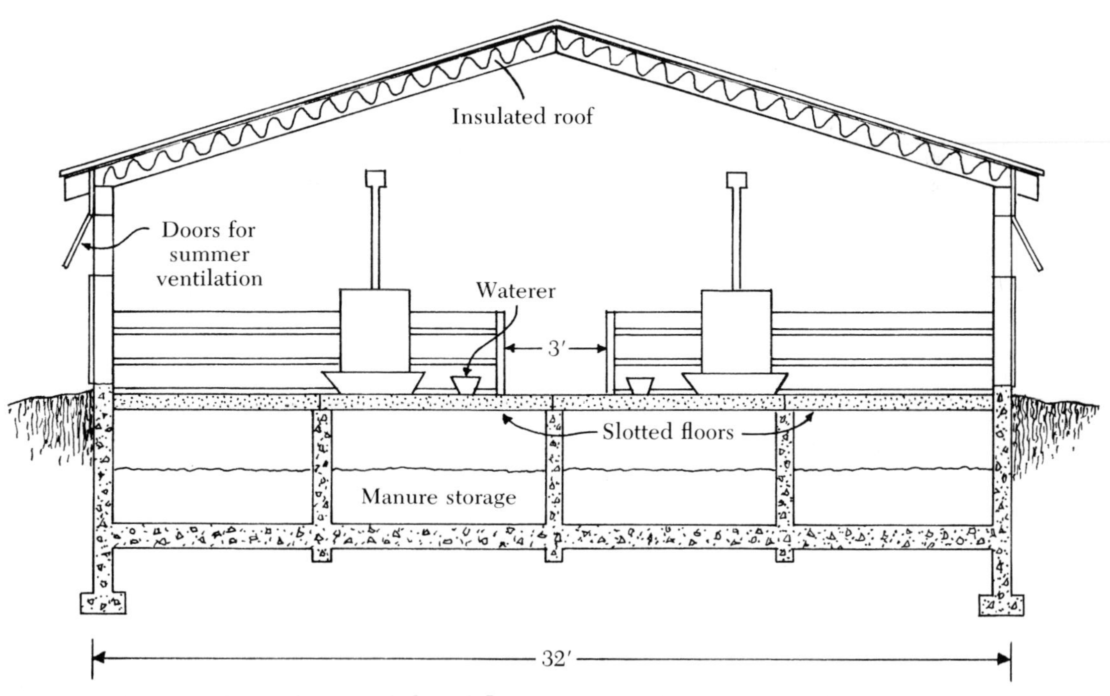

Figure 40-6 Basic design for a total slotted floor swine unit. Various slat materials and designs can be used. Sufficient environmental control to maintain comfortable temperatures is essential.

physiological function of the animals. Units may be separate buildings or separate areas within a building complex.

The farrowing unit should be so designed that both the baby pig and the dam can be comfortable. A 50 to 55°F environment is suitable for the sow, but a newborn pig requires 90 to 95°F. Supplemental heat for the baby pigs, such as heat lamp or infrared radiant heater suspended over the sleeping area, is frequently provided. Heat—electrical elements or water pipe—in the floor also keeps the sleeping area dry and warms the pigs.

Farrowing stalls, or crates (Figure 40.7) are commonly used to restrict movement of the sow. This reduces loss of baby pigs due to overlaying by the sow, since the pigs can have isolated sleeping and activity areas.

Floor surface should be smooth to minimize skin abrasion to knees of nursing pigs, but not so smooth that the sows have difficulty in standing. Solid floors should be sloped about ½ inch per foot from the pig sleeping area both to front and rear of the farrowing crate to drain liquids away from pigs.

The nursery unit is designed for pigs from the day of weaning until they reach 50 to 60 pounds. Usually this will be for pigs weaned at about 10 to 12 pounds in weight. This weight pig needs an environmental temperature of about 80–85°F to be comfortable—that is, when he has minimal metabolic rate and energy expenditure to maintain body temperature. Rough hair coats and excessive huddling are indicative of environmental stress. The nursery design must facilitate provision of such warmth, either through entire-building heat or through heated localized areas in each pen. Room temperature can be somewhat lower in a partial slotted floor unit than in a totally slotted floor unit since the pigs will have a solid floor sleeping area (Figure 40.8).

The growing-finishing unit is for pigs from about 50 pounds in weight to market. A 70°F temperature for pigs from 50 to 100 pounds, and 60°F from 100 to market weight, appears to be about optimum for maximum performance. Given equal comfort, pigs perform similarly in a wide variety of unit designs and arrangements. Mechanization and automation of feed handling is emphasized in these units because of the large volume of feed handled.

Breeding animals have traditionally been allowed to graze forage areas or at least have considerable acreage of ground and freedom of movement. When you move these animals to confinement, you dictate their environment. Less experience and research data are available for specific recommendations for the breeding and gestation unit than for other phases of production. It is especially critical that floor material and design are conducive to "comfort" for the animals. For example, floors that are slick when moist or have rough edges or other characteristics injurious to feet and legs discourage breeding aggressiveness of the male, cause reluctant behavior of the female, and result in unsatisfactory performance.

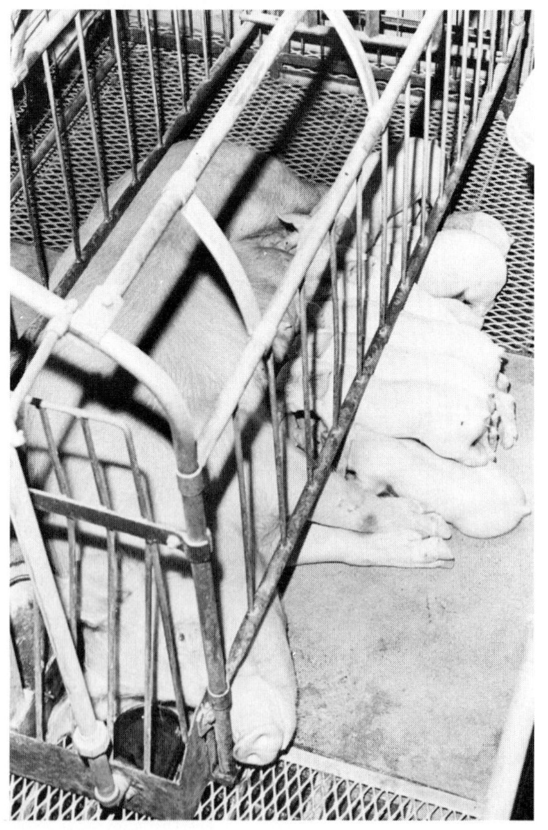

Figure 40-7 A farrow unit with expanded metal across the front and rear of the pen, concrete in between.

Figure 40-8 Partial slotted and expanded metal floors in nursery units. The wooden overlay and heat lamp provide a sleeping area protected from drafts and cool room temperatures. The notches in the ears represent individual number identification.

Separate pen space should be provided for the boars, the gilts, and the older sows.

The holding unit is a multi-purpose unit that should be suited to take care of "overflow" situations and provide an isolation section for disease control. A total slotted floor is recommended.

Feed and water supplies and distribution Combinations of grains and supplemental ingredients, including amino acids (protein), minerals, and vitamins, permit precise formulation of diets to provide the nutrients the pig needs. Modern systems permit automatic arrangements for grinding, mixing, and dispersion (auger conveyers, pneumatic tubing, etc.) of diets to desired storage or feeder location, eliminating traditional manual labor. Watering systems provide animals with ready access to fresh water. They are usually designed so that soluble nutrients and medicants can be administered via the water. Provision for water pressure for cleaning facilities and equipment is essential.

Waste handling and disposal Provisions for storage, removal, and disposal of wastes must be integral parts of any confinement system. Daily excreta (feces and urine) for a 100-pound pig will average about 7.7 pounds, which is equivalent to 1.0 gallon volume or ⅛ cubic foot of space. Waste-disposal systems will usually involve one or more of the following: (a) regular cleaning, hauling, and disposal of waste on land; (b) retention of wastes in storage basins under slotted floor, with hauling and disposal on land as needed, or when uncropped land is available; (c) same as b, plus lagoon for waste that cannot be removed to land; and (d) lagoon as a total and final disposal area (Muehling, 1969).

Waste management also includes methods of handling and treatment. When liquid manure has been anaerobic for some time, vigorous mixing or agitation for the sake of efficiency of removal can release dangerous, if not lethal, concentrations of H_2S. Thus, adequate building ventilation is essential.

The system of aerobic liquid-manure storage in the oxidation ditch essentially eliminates odor from swine waste and can increase biological oxidation of the organic matter (Jones *et al.*, 1970).

The system of manure handling, storage, and disposal must be such that pollution of air and water sources will not result. State regulatory agencies should be consulted to ensure proper guidelines and approval of design prior to construction of new facilities.

Management and Husbandry Practices

Production Schedules and Facility Use

The high investment for confinement facilities dictates a schedule of production that will guarantee maximum use efficiency. The number of farrowing groups per year or the frequency of feeder pig acquisition are examples of programs which must be rigidly scheduled to avoid periods when buildings are overcrowded or empty. Basically, each facet of the program must be calendarized so that specific actions are accomplished at scheduled times. But a program that looks clear-cut and effective on paper, regardless of managerial effectiveness, is seldom completely adhered to, since the biological variations encountered defy rigid compliance.

Selection of Breeding Stock

Selection of females Gilts (females that have not produced a litter) should be selected at a weight of from 175 to 220 lb. They should have rapid growth rate, moderate backfat thickness, evidences of desirable meat quality, such as full ham and loin and firm fleshing, at least 12 well-spaced fully formed teats, strong feet and legs, and be free of disease.

Selection of males The genetic influence of the boar on the next generation is greater than that of the individual gilt, since one boar will service several gilts (or sows). The boar should be selected from proven herds, have capacity for rapid growth (at least 200 lb. in weight at 150 days of age), show evidence of excellent meat quality (not more than 1.0 inch backfat thickness at 200 lb., electronic estimate of loin-eye area of at least 4.5 sq. in.), have good mammary development, and be free of disease. Further, performance and carcass data on littermates or other close relatives should meet certification standards.

Breeding Systems

For maximum marketable product per litter farrowed, a systematic crossbreeding program should be followed. Primary production advantage of crossbreeding is use of the crossbred dam—she farrows and weans more pigs. And on the average, crossbred pigs gain more rapidly and more efficiently than noncrossbred pigs.

The Breeding Herd

Physiological and social behavior patterns of breeding animals can be affected by confinement. Care of and comfort for animals will affect breeding and reproductive efficiency.

The boar He should be at least 7.5 months of age or 250 pounds in weight when first used for breeding. He should be acquired about 30 days prior to use to get adjusted to the new environment. A boar may be a poor worker if overfat, sick, under stress of high air temperature, overworked, or mishandled (an impatient herdsman can ruin a good boar). Under pen-mating management, a young and a mature boar, respectively, should effectively service 12 and 20 females. With individual mating management, up to twice as many females could be served.

The gilt and sow Except in very specialized programs, gilts should not be bred prior to 7.5 months of age or 250 pounds in weight. The advantage of older ages at first breeding is that the number of eggs ovulated increases up through the third or fourth estrous period.

Overt symptoms of estrus in young gilts may be less obvious and of shorter duration in confinement than in field systems.

Sows can be successfully rebred the first estrus following weaning. If litters are weaned under three weeks of age, conception rate at first estrus will be less than if weaning occurred at four weeks or later.

Number of gilts or sows per pen should be small, for example, 8 to 15, to minimize competition and irritation.

Artificial insemination and estrus control These would allow precise scheduling of breeding and farrowing. Both natural and artificial methods can be used. For lactating sows, estrus will occur within four to seven days after the litter is weaned; thus, several litters could be weaned on the same day, and the sows bred at the earliest opportunity.

Induced estrus and ovulation can be accomplished by judicious administration of certain hormonal or hormone-like substances. In one test, in which 27 sows were injected 39 or more days after parturition with 1,000 I.U. of pregnant mare serum, 26 came into heat, and 20 of the 23 on which information was available gave birth to normal-sized litters (Cole and Hughes, 1946).

Sows in which estrus is induced during lactation do not show as strong signs of estrus as do those coming into estrus normally; therefore, more careful observations for estrus must be made.

Artificial insemination offers several potential advantages to the producers and will be used much more extensively when procedures are perfected.

The gestation period During gestation the sows can be kept in small groups, individually penned or tethered. In groups, gilts and sows, respectively, should have a minimum of 12 and 15 square feet of floor space each.

The Sow and Litter

The pregnant animals should be in the farrowing unit by four or five days prior to expected farrowing to allow for adjustment to their environment. Management is especially critical, since most baby-pig losses occur during the first 72 hours after birth (Bauman, Kadlac, and Powlen, 1966). With special care and attention to dam and litter, up to 97 per cent survival to eight weeks of age has been realized (England and Chapman, 1962). In contrast, the commerical production average is closer to 70 per cent (ARS, 1965). Many losses are precipitated by stress due to chilling (Curtis, 1970).

Within twenty-four hours of birth, individual pigs are identified for record and future reference. Ear-notching is the most reliable system. It is also desirable to clip needle teeth to reduce potential injury from fighting each other and damage to the udder of the dam. Tying off the navel cord and daubing it with tincture of iodine or comparable solution reduces possibility of navel infections.

Since sow milk is very low in iron, oral or injectable iron preparations are administered to the pigs in confinement to prevent development of iron-deficiency anemia. Injections are made prior to five days of age, and oral preparations to the pigs usually are continuously accessible in soluble or textured forms.

Male pigs not retained for breeding should be castrated prior to two weeks of age. A creep ration (formulated especially for the small pigs and placed in an area not accessible to the sow) should be available to the pigs by the time they are two weeks of age.

Weaning is usually accomplished when pigs are from three to six weeks of age. Weight and condition are better criteria than age *per se*. Currently, in a very few specialized programs, pigs are removed from the dam two or three days postpartum (or after piglets have suckled and obtained colostrum) and housed in units (Figure 40.9) that allow automated feeding of a liquid diet in specified amounts at regular intervals to individually caged pigs (Danielson, 1968). Colostrum-deprived pigs, given adequate isolation and care, can also be successfully reared and have been used to establish Specific Pathogen Free, or Minimal Disease, herds (Woods *et al.*, 1962). It must be emphasized that superior management, rigid environmental control, and continuous attention to minute detail are essential for success with these programs.

Growing and Finishing Pigs

Continued emphasis on management and health is necessary if maximum rate and efficiency of gain from weaning to market weight are to be realized.

Figure 40-9 These individually caged pigs were removed from their dams about eight hours after birth. They thus had opportunity to suckle and ingest colostral milk. Here they are fed a milk replacer at hourly intervals for the first few days. By three weeks of age they will be on dry feed and can be moved to nursery units.

Space allowance Weight, number of pigs per group, air temperature, method of feeding, ventilation, and floor design can affect space needs. Minimum floor space allowances have been suggested by Gehlbach *et al.* (1966) as follows: 25 to 50 lb., 3 sq. ft.; 50 to 100 lb., 4 sq. ft.; 100 to 150 lb., 6 sq. ft.; and 150 to 210 lb., 8 sq. ft.

Number per group Optimum number differs with size of pig and environment. Small groups tolerate high ambient temperatures better than larger groups, but are less tolerant of cold. Heat dissipation is more efficient in small groups. Group size of 10 to 30 animals each is desirable during the period from weaning to market.

Cannibalism Most commonly seen as tail-biting, it can decrease level of performance of affected pigs. Cause is not known, but discomfort and "boredom" perhaps increase tendency to this vice. Incidence and intensity increase with size of group and variation in pig sizes. Removal of the offender(s) prevents spread of the activity, but surgical removal of tails at birth and comfortable environment are at the moment the best preventatives.

Mixing This is least stressful when pigs are small. Resorting and mixing groups of growing pigs usually result in fighting, and reduced performance. Use of sedation, odor-masking compounds, and other agents at time of mixing has been variably effective in reducing fighting.

Crowding Stress of crowding has been implicated in frequency of esophagogastric ulcers. Lack of exercise does not significantly affect carcass quality.

Management and Feeding Programs

Omnivorous by nature, the pig has historically been extolled for scavenging ability. When placed in confinement housing, the animal became completely dependent upon feed provided by man. Although not necessarily in anticipation of intensive production systems, most available data on nutrient requirements fortunately have been obtained under laboratory conditions which can be equated to confinement units (NRC, 1968). Feed cost may represent up to 75 per cent of the total cost of production. Any housing, breeding, management, or nutritional decisions which improve feed utilization will increase profit. And with the knowledge available there is little excuse for poor nutrition.

Methods of Feeding

Age and productive function of the animal frequently dictate method of feeding. Management selects one of several methods that might be available.

Full feeding For maximum daily gain, it is usually necessary to allow the pig access to feed at all times. Many different designs of feeder can be used, but to minimize feed wastage constant attention to feeder adjustment is required. One feeder hole should be provided for each four to six pigs.

In the United States, growing-finishing pigs are self-fed to encourage maximum growth and marketing at a young age. In Canada, Great Britain, and certain European countries, however, a controlled feeding program is followed to ensure production of a uniform quality of carcass.

On-floor feeding This is suited particularly to controlled feeding of finishing swine or the breeding herd. Feeding in the sleeping area, on solid floors, encourages cleanliness, since pigs are less inclined to dung in the eating and sleeping area.

Liquid feeding This usually involves mixing predetermined amounts of feed and water prior to or at the time of feeding. This method properly used can essentially eliminate feed dust in the feeding area and minimize wastage.

Interval feeding Allowing breeding herd animals, especially during gestation, access to a self-feeder every third day is a labor-saving technique. The number of hours the animals should have access to the feeder will depend on condition and gain of the animals. Regularity of schedule is important to prevent undue concern and commotion of animals.

Quality and Quantity of Feeds

Knowledge of the nutritional value of grains and other feedstuffs enables the producer to selectively evaluate each on the basis of both nutritional and economic worth. A few processing techniques, such as pelleting and cooking, improve utilization of nutrients in certain feedstuffs. Finishing pigs can utilize feedstuffs that would be inefficiently used by the young pig.

Levels of feed for breeding animals are determined by size and condition of the animals. Gestating animals should receive that quantity which would allow for maximum reproductive efficiency. In the United States, guideline weight gains for gilts and sows during gestation are 75 and 40 pounds, respectively. During lactation, feeding programs are usually either *ad libitum* or a specified amount per day related to sow weight, or a basal allowance for the sow plus an additional increment for each pig in the litter.

Management and Disease Control

Swine diseases, specific and nonspecific, clinical and subclinical, present continuing challenges.

Sanitation

This is critical in confinement, since swine concentrated in small areas carry a greater "respiratory load" of organisms than animals in field environments. Cleaning, disinfecting, and fumigating confinement facilities between crops of pigs prevent the progressive microbial buildup that can occur during continuous prolonged use of a given facility. Complete confinement does allow isolation from birds, rodents, and wildlife that may be disease carriers.

Internal and external parasites can be problems. A clean, sanitary environment provides the best prevention. Confinement affords isolation from contaminated fields and dirt lots. Anthelmintics and other drugs when properly used aid in elimination of parasites and prevent deleterious parasitism.

Specific Pathogen Free Program (SPF)

"Specific Pathogen Free" replaces the previously used "disease-free" term. It refers to pigs that have been delivered surgically from the dam two or three days prior to normal expected farrowing (Primary SPF) and reared in strict isolation until about four weeks of age. This program was first used by Young, Underdahl, and Hintz (1955) as a means of breaking the chain of disease transmission from the dam to her offspring. The viral disease atropic rhinitis (AR) and mycoplasmic pneumonia were of particular concern. They are not fatal, but are estimated to cost the swine producers several million dollars annually due to decreased growth rate and poor feed efficiency. Secondary SPF pigs are the offspring of Primary SPF.

Disease Control by Drugs and Biologics

Antibiotics and other drugs have provided protection against disease proliferation. They also enable producers to overcome specific disease outbreaks. Certain of these drugs, when included in diets at low levels, such as 20 to 50 grams per ton of diet, are spoken of as growth promoters. Levels of 300 to 500 grams per ton are used for therapeutic purposes. Vaccines are used to prevent occurrence of many diseases.

Disease Eradication Programs

The swine industry in the United States, under a Federal-State Eradication Program Plan initiated in 1961, has made tremendous progress toward elimination of swine cholera, a disease which has cost the industry millions of dollars from death losses each year for the past half century.

Management and Future Production Systems

The trend toward more intensive swine-rearing systems will continue. As specialization increases, enterprises will become larger and more sophisticated, with a greater demand for more precise control of all those factors implicated in affecting production efficiency. Management teams will include nutritionists, geneticists, engineers, microbiologists, veterinarians, and others as needed. Thus, continuing efforts must be expended to identify the management, environmental, nutritional, and facilities factors that affect biological and behavioral patterns. Feasible methods for control of these factors must then be established. Multidiscipline research efforts will be essential to integrate these physical and biological principles into programs of economic advantage.

FURTHER READINGS

Curtis, S. 1970. "Environmental-thermoregulatory interactions and neonatal piglet survival." *J. Animal Sci.*, 31:576.

Danielson, D. M. 1968. "Utilizing automation in rearing baby pigs." *J. Animal Sci.*, 27:1132.

England, D. C., and V. M. Chapman. 1962. "Some environmental factors related to survival of newborn pigs." *Oregon Agr. Exp. Sta. Special Report,* 137.

Jensen, A. H., and D. E. Becker. 1961. "Floor design and materials in housing for growing-finishing swine." *Illinois Agr. Exp. Sta. Mimeo.* AS-534.

Muehling, A. J. 1969. "Swine housing and waste management: A research review." *Illinois Agr. Exp. Sta. Bul.* A Eng-873.

Forty-One

Management of Sheep and Other Fiber-Producing Animals

"Sir, I am a true laborer: I earn that I eat, get that I wear, owe no man hate, envy no man's happiness, glad of other men's good, content with my farm, and the greatest of my pride is to see my ewes graze and my lambs suck."

Shakespeare,
As You Like It

Besides the fiber obtained from sheep, large quantities of fibers are obtained from goats and some members of the camelid family. The first section of this chapter is concerned with the management and feeding of sheep, and the second section with that of goats and *Camelidae*. Little is known about management needs of the latter.

Sheep

Sheep are unique among domestic animals in the great variety and types that have evolved and that are adaptable to a wide range of environmental conditions. They can convert diverse types of forage, on marginal agricultural land, mountains, hills, moorlands, plains, and deserts, into valuable products for mankind: wool, pelts, lamb, mutton, and milk.

Latitude, altitude, topography, soil, water, vegetation and climatic conditions influence sheep farming. Methods of sheep husbandry may be different on farms or ranches even within the same district. Sheep breeders of the past partly solved their problems by developing breeds to suit the divergent conditions of the various areas in which they lived. Each of the many different breeds found today was developed to suit a certain type of environment. Merinos and Rambouillets, some carpet wool breeds, and hair sheep that do not grow wool are favored for their hardiness in hot desert areas with sparse vegetation and for their gregarious traits. Mutton breeds, such as Suffolk and Hampshire, are popular in cooler climates with abundant vegetation. The increasing importance of lamb as a desirable dinner meat has resulted in the development of Columbias and Targhees in the U.S., which cross well with Suffolk and Hampshire rams to produce excellent carcasses and heavy fleece.

Today the importance of wise management, selection within a breed, and crossbreeding, in addition to correct feeding, is widely recognized. One thing is certain, however: breed alone cannot substitute for feed and management. Economical maintenance of breeding animals, a large lamb crop, continuous and rapid growth of lambs, heavy weaning weights, and heavy clean-fleece weights all require adequate nutrition and management for their complete expression.

Four general systems of sheep production fit into the agricultural pattern: (1) range, (2) farm, (3) purebred breeding, and (4) lamb feeding. Each represents unique management and feeding problems.

Range-Sheep Production

In range operations, general management cycles are similar, but calendars of operation vary substantially, depending upon geographical and climatic characteristics of given areas. A common objective is to plan the time of lambing to avoid severe weather and yet to coincide with optimum range forage production between lambing and weaning.

Typical range-sheep management systems A summary of the management system widely applicable in the intermountain areas of the western United States, such as in Utah, may exemplify major factors involved in range-sheep management. In the United States, 70 per cent of the sheep are located in the 17 westernmost states, which constitute 40 per cent of the continental land area. The area can be divided roughly into range regions according to vegetation type as shown in Figure 41.1.

Most of the ranges in the mountain, desert, and foothill areas are best adapted to grazing during only one season: winter range (Figure 41.2) at lower elevations, where rainfall commonly is less than 10 inches, and desert shrubs, sagebrush, and grass forage types are predominant; summer range in the mountains, where precipitation is greatest (18 inches or more), temperatures low, growing season short, and topography steep, with the prominent vegetation being ponderosa pine, spruce, fir, aspen, and alpine grasses and sages; and the spring-fall ranges intermediate between the two in topography and growing conditions and with vegetation consisting of sagebrush, juniper, mountain brush, some forbs, but usually very little grass.

Typically, Rambouillet, Columbia, and Targhee purebred rams are used to produce lambs to be replacement ewes for the range flocks or to be sold as foundation breeding ewes. Mainly Suffolk and Hampshire rams are used to produce market lambs.

In the intermountain areas of the western United States, lambs are weaned in late August, September, or early October, at which time the ewes are on the high summer range. The heavy lambs are sent directly to market, and the lighter and thinner ones are sent to fattening yards or pastures. About October 15, the ewes are trailed to the fall foothill range, where they remain for about two weeks before being trailed to the winter range. This trail may be 50 to 250 miles long. In recent years, many operators have begun to ship their animals by truck or rail.

The winter-range forage is dormant, limited in amount, and in most areas low in protein and phosphorus. The sheep, therefore, do not consume sufficient nutrients from this range forage to keep them gaining; so supplemental feed usually must be provided. The midwinter months are the most critical, requiring about 0.25 lb. daily supplement, containing 12 to 36 per cent protein, depending on specific conditions; if ewes become snowbound, even more feed, including hay, may be needed. Adequate feeding and good management may improve lamb crops by 15 to 30 per cent and minimize death losses of ewes.

For range lambing, the rams are usually put with the ewes during November or December. For shed lambing, the rams are put with the ewes after the lambs are weaned in August.

In April the ewes are usually sheared with portable shearing machines (Figure 41.3).

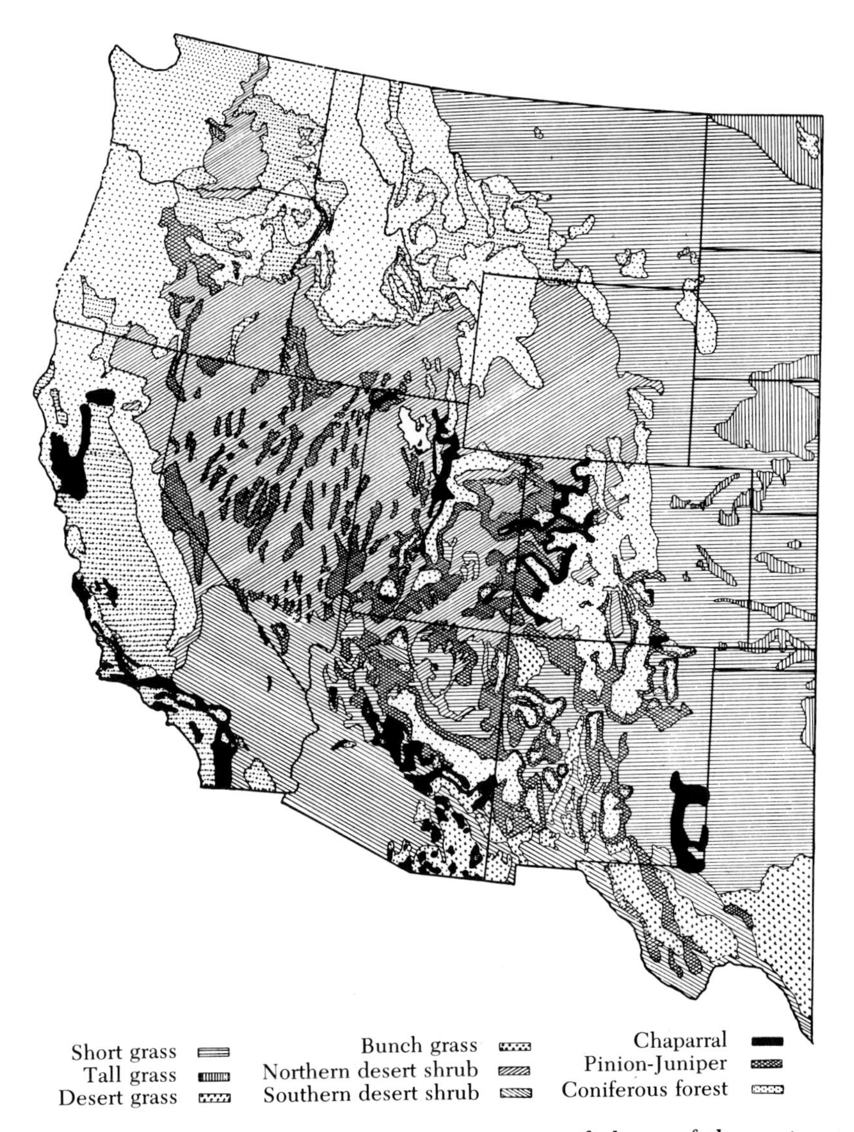

Figure 41-1 Natural range vegetation regions of eleven of the western range states. (From Stoddart and Smith, 1953.)

While black wool is generally separated from the rest of the wool and tags are removed and packed separately, sorting of the wool into grades is most often done in warehouses in the West or in Boston.

After shearing the ewes are trailed to the spring range, where they lamb. There are two types of lambing procedures, range lambing and range-shed lambing.

For range lambing an area is selected that has protection from the weather provided by brush or trees on south slopes where there is sunshine, water, and grass. Groups of about 500 ewes are bedded in separate areas, and each morning those that have not lambed are moved to new areas.

Sometimes small portable tents are placed over the lambs and ewes at the time of lambing; these are especially useful when it is storming. The most progressive operators feed

Figure 41-2 Sheep on the winter range in southern Utah. This range is typical of much of the intermountain region. The valleys are broad with low mountains. Note the playa in the background.

the ewes after lambing a high-protein pellet fortified with minerals.

As lambing proceeds, the animals are separated into three groups: those that have lambed (Figure 41.4), those that are about to lamb, and those in which lambing is not imminent.

When lambing is done under sheds on the range, feeding yards or pasture may be provided. For one month before lambing, ewes are commonly kept on improved pasture or are confined to yards and fed hay and concentrates with free access to salt and a mineral mixture.

Soon after lambing is completed, the ewes and lambs are trailed to the summer range, where they are divided into bands of about 1,000 ewes with their lambs. On many summer ranges, two men look after one or two bands; one of the men tends camp while the other does the herding.

The shepherd, using a horse and dogs to control the sheep, guides them so they will graze the forage to the best advantage. In late summer or fall, when the lambs are weaned, the ewes are combined into bands of 2,000 to 4,000 to be taken to the fall and winter range.

On parts of the western great plains, such as in west-central Montana, more modest operations are maintained the year around on improved rangelands equipped with sheep-tight fencing. In such operations, ewe bands are smaller and are not under direct control of shepherds. Bands are rotated for grazing, often seasonally, among sub-fenced range pastures, each with water supplies and feeding facilities as appropriate. The animals are gathered at central locations for both lambing and shearing operations under a permanent shed. In Arizona, northern New Mexico,, and the Sierra Mountains of California, sheep are commonly herded on unfenced rangelands, whereas in other parts of New Mexico, Texas, and the coastal hills of California they are kept in fenced range pastures. Fencing rather than herding is commonly the method of choice where year-round rather than seasonal pastures prevail.

Low rainfall areas in New Zealand, Australia, South Africa, and South America, especially in Uruguay, southern Argentina, and Chile, are well-noted for huge range operations developed in the nineteenth century for production of wool and lamb for British markets. Operations with wool breeds on ranches called stations may involve bands of 5,000 or more ewes kept year-round in permanently fenced range pastures. Both breeding and lambing are conducted in the open with minimum attention and little or no supplementation or shelter; so lambing percentages are relatively low. The sheep are "mustered" or gathered for marketing or culling and shearing at highly specialized centers where the clip is graded, baled, and readied for marketing. In higher rainfall areas, other breeds are utilized for fat-lamb production. Under these more favorable feed conditions, operations tend to approach the mixed-crop farming and farm-flock combination characteristic of better feed areas of the world.

Another unique example of what may be classed as range operations can be found in the desert areas of North Africa and the Near and Middle East. Here small flocks are in the care of nomadic shepherds and shepherd families. They move with the light rainfall to glean what sparse vegetation is available. Sheep are largely of carpet wool, hair, fat-tailed, and fat-rumped types. Much of the production is to supply the shepherd and his family with fiber, meat, and milk. Substantial effort is being expended in some of these areas toward improvements in sheep types and development of management systems unique to these environmental conditions.

Nutrition and Management of Range Sheep

Under favorable conditions and on well-managed ranges, forage consisting of browse, grass, and forbs frequently supplies all the nutrients necessary. At times this may not be true. The condition of the sheep, the amount and kind of forage on the range, climatic conditions, and the time of year determine what kinds of supplements to feed and when to feed them.

Aside from a shortage of water and salt, the deficiencies common among sheep grazing on winter-range forage, particularly on mature, dried forage, are in phosphorus, protein, and energy. Calcium deficiency is rare. Many areas are deficient in trace minerals. When mineral

Figure 41-3 Sheep being sheared with portable equipment. This procedure is used frequently in the west, and protects the range from being trampled and overgrazed, since only one band of sheep is sheared in a given area. Blades are being used here. Sheep are now sheared with machines. (Photo by U.S. Forest Service.)

Figure 41-4 Ewes that have just lambed. When ewes are lambed on the range, they are distributed over a large area, allotting each ewe and offspring a larger area for grazing. (Courtesy J. L. Van Horn.)

deficiencies exist, iodine, cobalt, and copper may be supplied by use of trace-mineralized salt.

On some ranges, carotene (provitamin A) may be deficient, especially on a grass-type range or when there are long periods between growing seasons. However, sheep store vitamin A in their livers when grazing on succulent green forage, and if periods of inactive plant growth are not unduly long, animals usually will not suffer from a shortage. Since vitamin D is synthesized in the skin by the ultraviolet rays from the sun, its supplementation is not recommended in areas of adequate sunlight. Sheep are able to synthesize in their rumens all the known B-vitamins; therefore, supplementary sources of these vitamins are not needed.

Before one can formulate a supplement for range ewes, one must determine the composition of the forage on which they graze, and estimate the amount they are consuming. Then feeds are combined to furnish those nutrients which are deficient. Supplements usually are pelleted so they can be fed on the ground. The rate of feeding is important. As the amount of feed is increased, the percentage of protein, phosphorus, and vitamin A may be decreased, making it possible to use more of the less-expensive energy feeds. For spring-lambing ewes that are in good condition when they go to the winter range, it is usually not economical to feed supplements at levels which keep them gaining or maintaining their weights throughout the winter months.

In recent years high levels of salt have been mixed with protein meals to regulate their consumption. This method of feeding saves labor but is recommended only where it is not possible or practical to hand-feed the animals.

Care must be taken to prevent overconsumption of salt, which may be fatal. Salt-meal mixtures greatly increase water consumption; so where possible such mixtures should be fed near water.

In many range areas, suitable water is the most limiting nutrient. For best production on partially or completely dry feed, range sheep should be watered once each day, but if green lush feed is available, watering every other day is sufficient. When soft snow is available, range sheep do not need additional water unless dry feed supplements are used, in which case they may not be able to consume sufficient snow.

Farm-Sheep Production

The main objective of farm-flock production is to grow fat lambs that can be marketed as soon as they are weaned; consequently most of the farm flocks are of the mutton type. Farm-flock management differs from that of range sheep in that it is not nearly so directed by climatic conditions or forage growing seasons. The over-all management is much more intensified, especially for feeding. Farm sheep are fed largely on pasture, crop aftermath, and harvested forages. At times they need to be given protein supplements, phosphorus, iodine, copper, cobalt, and the vitamins A, D, and E. Sheep do best if provided with a succulent, fertilized pasture of legumes and grasses. A good irrigated pasture, grazed in rotation, will carry six to eight ewes and their lambs per acre. If the pasture is part of a crop-rotation system, the control of sheep parasites is simplified, because the animals do not graze the same areas continuously. Sheep prefer alfalfa, clover, or lespedeza hay. In the western United States, the standard grain is barley; in the Midwest and East, it is corn.

Sheep should have free access to crushed salt in one side of a self-feeder, and to one part dicalcium phosphate and one part salt in the other side. If trace minerals are needed, a trace-mineralized salt mixture may be used in place of plain salt. One lb. of salt or trace-mineralized salt and 1 lb. of dicalcium phosphate are often added to every 100 lb. of grain or pellet mixture. When sheep are fed entirely on corn silage and cereal grains, 0.02 to 0.03 lb. of limestone (38 per cent calcium) should be added to the daily ration to supply required amounts of calcium.

Management During the Breeding Season

The sexual season and its control Ewes have a period of anestrus and a sexual season during which estrus occurs at 17-day intervals. The period of anestrus varies among breeds, but usually occurs between midwinter and late spring. Merino and Rambouillet ewes, as well as those with considerable Dorset or Tunis blood, may come into estrus in late spring to early summer. Other breeds usually come into heat in late summer or fall.

Estrus in ewes ranges from 20 to 42 hours, with an average of 30 hours. Ovulation occurs about 24 to 30 hours after the onset of estrus. If the ewe is not bred, or if she fails to conceive, estrus recurs after an interval of about 17 days.

Research is being directed at certain reproductive processes so that estrus can be synchronized to occur at approximately the same time in all females of a breeding group. This synchronization is made possible primarily by the use of progesterone, a female sex hormone, and synthetic progesterone-like hormones. Continued research in this field will doubtless solve present problems so that these hormones can be used for practical and economical control of reproductive processes of farm animals. Such control, combined with the use of artificial insemination, would result in greater uniformity in age of offspring; it could also result in more efficient utilization of labor, facilities, and markets.

Puberty Rams reach puberty at the age of 4 to 7 months, but there is considerable variability. Ram lambs that have developed well may be used with 15 to 20 ewes the first breeding season, but usually they are not used until after they are one year old. Many young ewes will breed at the age of 6 to 10 months, depending partly on growth attained. Ewe lambs that have been born in January and

February and that are well-matured can be bred in the first fall under farm conditions. If ewe lambs are born later than February, however, they should not be bred until the next breeding season unless extremely well-grown.

Time of year to breed The time of year to breed ewes depends on the method of sheep raising, climatic conditions, and the feed and equipment available at lambing time. If range ewes lamb without protective sheds, it is more desirable to have the lambs born late in the season when the weather has moderated. If sheds are available, however, range and farm lambs may be born earlier. For hothouse, Easter, and early spring lambs, the ewes are bred so that the lambs are born in fall or winter.

Flushing and breeding Flushing consists of supplying the ewe with an abundance of feed to make her gain weight just before and during the breeding season. It is believed that this process increases the number of eggs ovulated.

After the lambs are weaned the ewes are usually kept on pasture. Sometimes they may be fed roughage. They should be managed so they are in thrifty condition—not too thin or too fat. This usually requires reduced feed intake to reduce or at least hold down body weight. Two weeks before the breeding season, the amount of feed is increased to permit the ewes to gain weight. Breeding is begun two weeks after the ewes have started to gain. The supplementary feeding is continued during the breeding season, which is usually six weeks in length.

In warm climates the flock is sheared and a cool place is provided for the sheep. In cool climates the ewes are tagged and most of the belly of the ram is sheared. If early lambs are desired for special markets the flock may be put in a dark building and held there to decrease the amount of light. Decreasing the amount of light causes the ewes to come into heat sooner.

Number of ewes per ram The average number of ewes bred per ram is as follows: a ram lamb will breed 20 ewes on the farm, 15 on the range, a yearling ram 35 and 30, and a mature ram 40 and 35. If the ewes are hand-bred with special attention and procedures, one ram can serve up to 100 ewes.

Rams should be checked for live, viable sperm, blindness, normal testicles, inflamed penis, lameness, and crooked legs before being used for breeding. It pays to use sound high-quality purebred rams.

The pregnant ewe The length of the gestation period in ewes is approximately five months. This period is somewhat variable by breed, mostly ranging from 142–153 days. Most of the fetal lamb growth occurs during the last six weeks. From the standpoint of nutritional need in the pregnant ewe, the last month is the most critical.

In warm climates, pregnant ewes can be kept on pasture throughout the winter, but in cool climates they are kept on fall pasture or crop aftermath until snow comes. They are then fed 0.5 to 0.75 lb. of a concentrate mixture tailored to supplement the forage being fed.

Management During Lambing

Sheds A shed for sheep should provide a dry place where the sheep may lie, free from drafts but with adequate ventilation, with well-drained yards on the south side or away from the prevailing wind. Some lambing sheds consist of a wooden frame with a canvas top. Early lambing necessitates warm quarters, which can be insured by insulating a small area of the shed to be closed off and used for lambing pens. In New Zealand and Australia, most of the sheds are equipped with slatted floors to help keep animals clean and reduce labor costs.

Shed lambing Lambing pens are usually set up with panels in a portion of the sheep shed, or a special lambing shed with lambing pens is provided. If the weather is cold, some of the pens are equipped with brooders and heat lamps, and should receive dry bedding daily; if newborn lambs are kept dry and free of drafts, they can stand considerable cold. Hay and water are provided in each pen.

Unless the ewes have already been sheared, they should be tagged and the wool removed from around the udder about four weeks before lambing.

When lambing time approaches, the vulva swells, the udder fills, and the ewe is uneasy. A hollow appears between the ribs and the hip, and the ewe begins to strain. The water bag that has cushioned the unborn lamb bursts with a gush of fluid. Usually the ewe will lamb without help. In a normal birth, both front feet appear first, with the head lying snugly between them (Figure 41.5). The ewe should not be helped unless the lamb is in an abnormal position or is excessively large, or unless the ewe has strained for an hour or more and no part of the lamb has appeared.

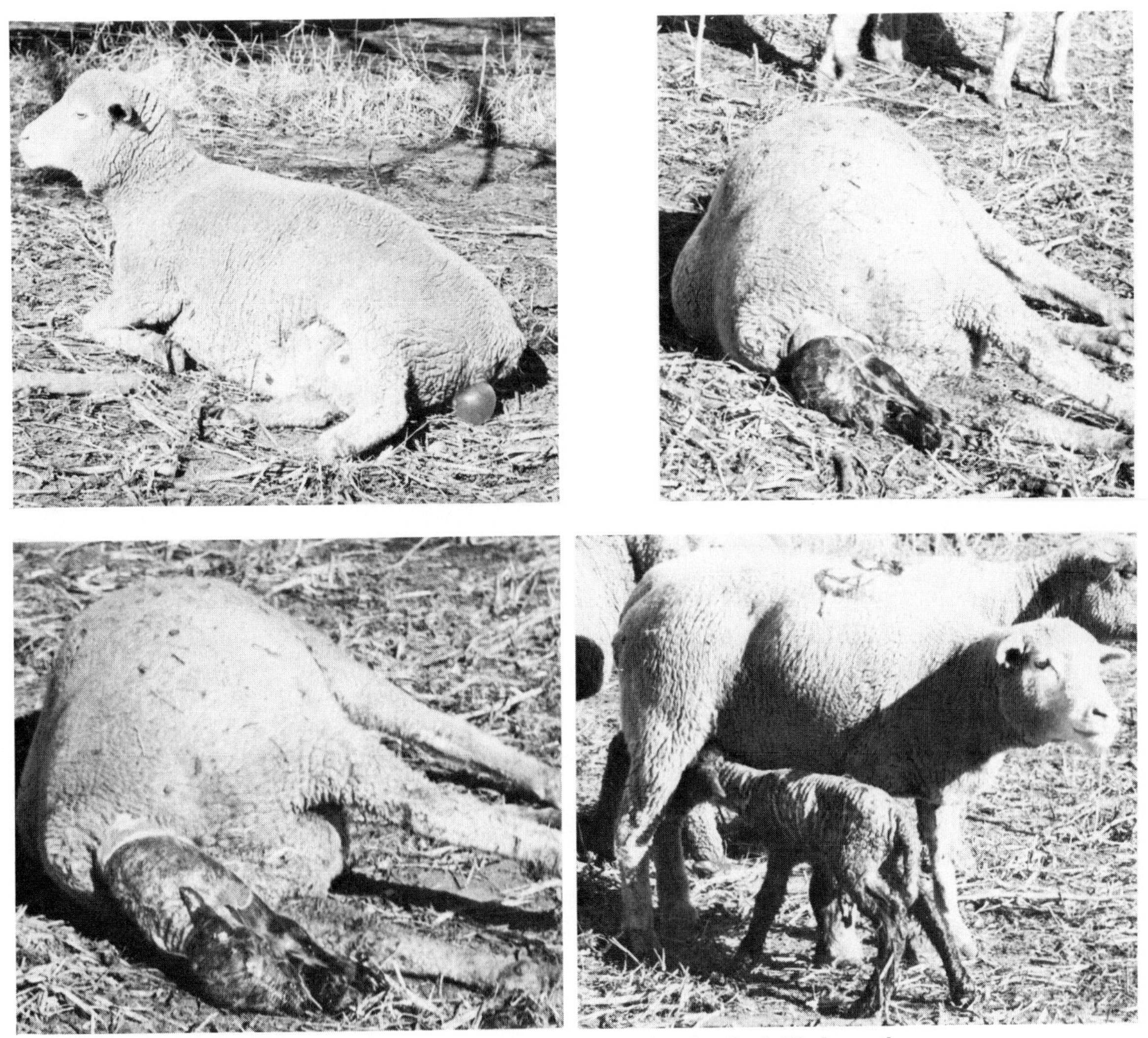

Figure 41-5 A lamb is born. From upper left to lower right, the fluid-filled membranes (water bag) become visible, the head and feet appear, the lamb slides out, and finally the lamb begins to nurse.

After the birth the ewe should be examined to make certain there are no other unborn lambs. The attendant inspects the ewe's udder and milks it a little to see that the milk canal is open. If the lamb is weak, he places the ewe's teat in its mouth and milks a small amount of milk into it. In doing this, care must be taken that the milk does not get into the lamb's lungs.

In cold weather a lamb brooder is used to keep the lambs warm. Single lambs are left in the lambing pen for about 12 hours. Twin lambs are left for about 24 to 48 hours or until the ewe comes to know them intimately.

As each ewe lambs, a number should be marked in paint on her side and the same number marked on the side of her lamb or lambs. This permits immediate identification of lambs that may have become lost or that are not receiving proper care from their mothers. Only paint that will scour out of the wool should be used. In purebred flocks, ear tags are used showing the name of the owner and the number of the sheep. Sometimes permanent numbers are tattooed in the ears.

Feeding lambs and ewes during lactation After ewes lamb, their rations are gradually increased. Ewes may be fed any number of rations consisting of hay and concentrate mixtures compounded to supplement the hay. In the western United States, for example, where excellent-quality alfalfa hay is available, it may not be necessary to feed any grain. If succulent pasture is available, ewes usually do not need any grain, and in the best areas may not need any hay.

When it is desired to market the lambs early, and for twin lambs, extra hay and grain may be fed in a lamb creep.

Care of newborn lambs Chilled lambs should be rubbed dry and wrapped in a blanket or put under a heat lamp or other warming devices such as jugs of hot water. A stiff, cold lamb can often be revived by immersing in warm water for 2 to 10 minutes. It may be well to inject such lambs with penicillin and streptomycin to prevent secondary complications such as pneumonia. Ewes, however, usually will not accept a lamb that has been immersed when it is returned to her.

Ewes sometimes do not willingly accept their newborn lambs. They should be confined and forced to accept the lamb with the aid of tricks, such as rubbing the ewe's milk on the lamb, arousing protective instincts through presence of a dog, or administering tranquilizers. An unwanted lamb may be "grafted" on a ewe that has lost her own, by inducements such as smearing the lamb with the ewe's afterbirth or covering it with the skin of the dead lamb.

Orphaned lambs can be fed from a nippled bottle. They should receive colostrum (first milk) drawn from a ewe or frozen cow's colostrum kept for this purpose. Thereafter, warmed cow's milk or a milk replacer of high fat content can be fed, frequently in small amounts at first followed by gradually increasing amounts and decreasing feedings.

Lamb Feeding

Feeder lambs are usually produced from matings of white-faced ewes (Rambouillet, Columbia, or Targhee) to purebred black-faced rams (Suffolk or Hampshire). They are the smaller lambs, weighing between 65 and 85 lb., remaining after fat lambs, weighing more than 85 lb., have been topped out at weaning time.

After the lambs have been received by shipment from the range, they are rested and given free access to water, hay, and salt for a few days. Preferably during this time, they are vaccinated for sore mouth and a disorder from overeating (enterotoxemia) and treated for internal parasites if needed. If crop aftermath is available, lambs may be turned in to eat the grain stubble, corn, or beet tops while this is available. They are then put in dry lot and fed grain and roughage or placed on pasture to be finished for market.

Finishing in dry lot Lambs are separated into weight groups and started on about 0.25 lb. of concentrate per head per day and given hay free choice. These concentrates are gradually increased and the hay decreased according to a schedule, such as is shown in Table 41.1, for both hand feeding and self feeding. Toward the end of the feeding period, bulky concentrates are decreased and high-energy concentrates increased.

If lambs scour, they should be checked for infectious diseases and internal parasites, including coccidiosis. The amount of grain allowed should be reduced, and about one-third of the hay replaced with straw or oat hay when available.

Many large commercial feeders are pelleting the entire diet for fattening lambs. Several experiments have been conducted which have shown that lambs on pelleted rations eat 6 per cent more feed, gain 23 per cent more in weight, and require 19 per cent less feed per lb. of gain than those fed nonpelleted rations.

Table 41.1.
Feeding schedule for finishing lambs.

	Hand-feeding		*Self-feeding*[a]	
Days on feed	*Concentrate mixture*[b] (%)	*Hay* (%)	*Concentrate mixture*[b] (%)	*Hay* (%)
7 to 14	25	75	25	75
15 to 28	35	65	45	55
29 to 56	45	55	55	45
57 to 100	55	45	55	45

Suggested grain mixtures for the above feeding schedule

Grain	%
Maize or wheat, whole or rolled[a]	65.0
Beet, pulp with molasses, dried; or oats, grain	33.0
Salt or trace-mineralized salt	1.0
Dicalcium phosphate	1.0
	100.0

[a] If hay is of poor quality, substitute 5 to 10 per cent of linseed, cottonseed, or soybean meal, solvent-extracted, for part of these grains.

[b] The hay is ground and mixed with concentrates to be fed as a meal or pelleted.

When the entire diet is pelleted, more hay can be fed in proportion to concentrates than under other feeding systems.

Finishing lambs on pasture In the southern U.S., Arizona, and California, many feeder lambs are fattened on pasture. In these areas, alfalfa fields are also used, particularly in the early spring. Later in the season, when losses from bloat may become severe, sheepmen generally prefer native feed, such as bur clover and alfilaria, for finishing spring lambs. Sudan grass is an excellent summer pasture crop under irrigation, and it is a favorite green forage crop among purebred breeders. In the early-lamb districts, the lambs are always marketed by late May, before the green feed becomes dry.

Rotation of pastures improves availability of fresh green feed to develop the lambs quickly. This can be provided for by dividing pastures into small units by cross fencing.

Market lambs also can be produced by running both ewes and lambs in a pasture-rotation system. This practice insures continual fresh feed supply for good milk flow in the ewes. The lambs develop more rapidly and attain heavier weights, and more sheep can be grazed on a given area than in non-rotational systems.

Control of Parasites

The more important parasites of sheep include stomach worms, nodular worms, bankrupt worms, lungworms, tapeworms, liver flukes, coccidia, ticks, and lice.

Phenothiazine, thiabendazole, loxon, and tramisol are some of the products that may be used to control stomach and nodular worms. These may be alternated in use to avoid the development of resistant strains of parasites. Each year, the ewe flock and rams should receive one of these worming medicines: two weeks before the breeding season; before beginning any dry-lot feeding in cold climates; in the fall in warm climates where sheep are kept on pasture; and within the week before ewes and lambs go to spring pasture. At the beginning of summer, all ewes and those lambs weighing over 40 lb. should be treated, and all ewes and lambs at weaning time. Ticks (keds) can be controlled by dipping or spraying with toxaphene, malathion, methoxychlor, or by dusting with rotenone.

Handling Sheep

A sheep should be caught by the nose, flank or by the hind leg (least-preferred method). Never catch or hold a sheep by the wool since this will bruise the flesh just under the skin. If the left hand is under the jaw and the right hand on the sheep's rump, the animal can easily be controlled or "nosed" over at will. To mouth a sheep, as in examining the teeth, straddle the neck, raise the head with the left hand under the jaw, and part the lips with the right hand. To examine the conformation of the sheep, the fingers should be kept close together.

Predators

On western ranges, large losses of sheep occur from predators. This has resulted in widespread control programs, consisting of the use of both government trapping and private trapping and hunting, much of it organized by livestock associations. The coyote, a species of wolf, is by far the most important of the predators, which include bobcat, bear, and cougar (puma or mountain lion).

Animals Producing Specialty Hair Fibers

There are many animals that produce specialty hair fibers. The fibers are used with wool and synthetic fibers to give additional beauty, luster, color, or softness. Most of these fibers are obtained from species of the goat and camel families (Figure 41.6). Fibers are also obtained from horses, swine, cows, and various fur-bearing animals, including the Karakul lamb and Angora rabbit.

Angora Goats

Angora goats are animals best suited to dry, mild climates providing an ample supply of browse, weeds, and grass (Figure 41.7). These goats are sensitive, intelligent animals that adjust easily to different management practices. They are compatible with other classes

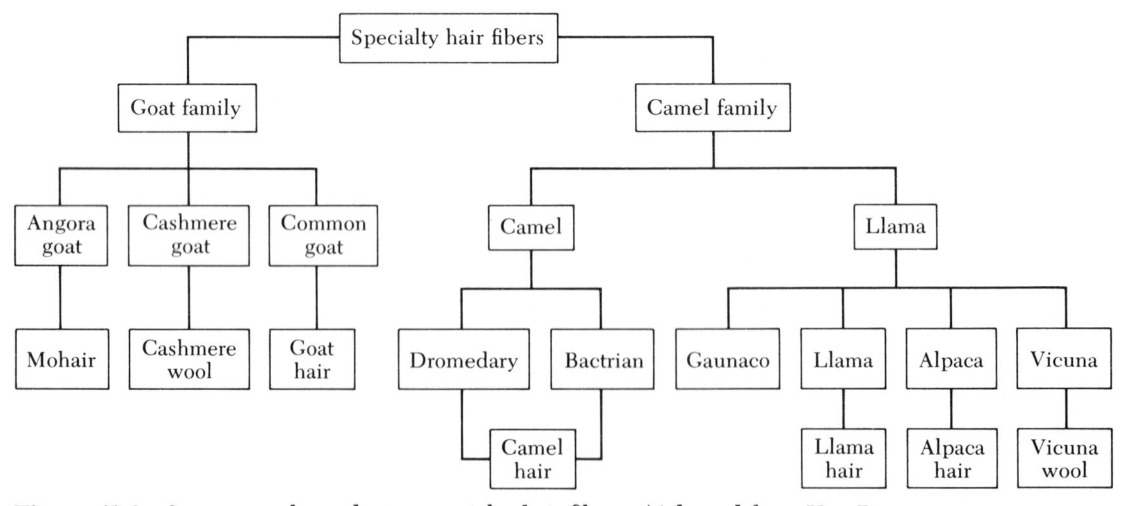

Figure 41-6 Some animals producing specialty hair fibers. (Adapted from Von Bergen, 1963.)

Figure 41-7 Angora goats on western range. Sometimes they are used to clear the land. (Courtesy USDA.)

of livestock in their grazing habits and adapt readily to farm or range conditions.

Angora goats, being excellent climbers and good travelers, are well-suited to high mountainous areas. They are helpful in clearing some types of brush from lands and in controlling sprouts from brush on cleared lands. Angora goats relish many species of brush and will graze on this type of forage during the entire year.

Angora goats are maintained either in small farm flocks of 25 to 30 head for the primary purpose of controlling weeds and brush along roads, ditches, and fences or in large flocks of several thousand head on open-range country. These large flocks are often used in a dual capacity to eradicate or control brush as well as to produce meat and fiber.

Types of fleece Angora goats produce fiber of three distinct types of fleece associated primarily with the locks. These are ringlet or C type, flat lock or B type, and the web lock.

The production of mohair varies with feed and range conditions from 3.5 to 5 pounds for the doe and kid in bands being managed under range conditions. Most often mohair is taken in two clips per year, spring and fall. Angora goats should produce about one inch growth of fine-textured mohair per month.

Equipment Angora goats require very little specialized equipment. To protect animals from cold, wet weather after spring shearing, a shed of some type should be provided. This will provide protection from the cold weather until the goats grow sufficient mohair to afford self-protection from the elements.

For farm flocks, good tight fences are required. Goats will seldom jump over a fence, but wherever a hole is found they will crawl through. Net wire makes a suitable enclosure, and barbed wire can be used if the lower strands are six inches apart and vertical stays are used every few feet. Feed troughs, mineral feeders and corrals used for sheep are adequate for goats.

Breeding

Angora goats have seasonal breeding habits; they come into estrus in the fall, with the kids born about five months later. The gestation period is from 147 to 155 days. Because the seasonal breeding characteristics are so well-established, bucks often are left with the does during all the year.

The number of does per buck will vary depending on the terrain on which they graze, but the normal is usually three to four bucks for each 100 does. Bucks should be well past the age of yearlings prior to their use in the breeding flock.

It is recommended that does be "flushed" with a good grain supplement. Good green forage is very important during the breeding season to insure a good kid crop and heavy clip of mohair. Bucks should be supplemented with a good grain mixture three to four weeks prior to the breeding season.

Care should be taken to keep the doe in good condition during the gestation period to prevent abortion. Angora goats abort much more easily than do most other species of animals. A doe that aborts from a low plane of nutrition one season may continue to abort on subsequent years.

Kidding Most ranchers raising Angora goats place does in pasture for kidding. Care must be taken to make sure does are in good condition; if they are not, oftentimes they will not claim their kid. When good thrifty does are maintained in the breeding flock, good kidding results can be expected. Kids are weaned following fall shearing, at which time they are placed in pastures remote from the does.

Feeding When a wide variety of brush, weeds, and grasses are available, goats do not need any supplemental feeding. When drought restricts plant growth or when there is a lack of green feed during the winter months, supplemental feeding similar to that of sheep is advised.

Occasionally brush and weeds of various types are cut and fed to goats. These may consist of singed pricklypear, live oak brush, or Russian thistle.

Angora goats will consume from one quart to more than one gallon of water daily, depending on the moisture content of the forage being eaten and the amount of dew formed on this vegetation. Weather conditions will also influence water intake.

Kid goats should be protected from open watering troughs, because their climbing habits sometimes cause them to fall into the trough and drown. Stepping stones or other means of climbing out will prevent drowning.

When the weather is hot, shade, either natural or artificial, should be provided.

Parasites Goats are infested with several types of lice. These include both blood-sucking varieties and biting lice. Because the lice seem to congregate on the under parts of the body, dipping is more effective than spraying because the solution covers the complete body surface, whereas in spraying it is very difficult to reach many areas of the body. The preferred time for treating for lice control is after shearing, both in the spring and fall. Lice live on the goat all year round and can live off the goat for periods of three to eighteen days. Infestation is usually transmitted from one herd to another, or from one ranging area that has been used recently by a herd of goats to those currently occupying this area. Concentration of lice is usually heaviest during the winter months, and during this time the greatest damage to the mohair occurs. A heavy infestation of lice will cause goats to shear fleece of light weight lacking luster, strength, and quality.

Internal parasites are present and will oftentimes be a serious problem during wet years. Symptoms of infestation are not easily observed except in an extreme infestation of stomach worms when scouring and anemia will usually occur. Eating of dirt and other gestures may suggest a depraved appetite, alerting the rancher to this problem. Treatment can be made by drenching with various worm medicines as described for sheep. Rotation of pasture is helpful in controlling serious infestation. Pasture lands grazed heavily over extended periods of time will become heavily infested with the larvae of various genera of internal parasites and be a prolific source of infestation.

Cashmere Goat

Cashmere is one of the finest and softest of animal fibers. This is the down fiber obtained by combing from the domesticated goat, *Capra hircus*. This goat is indigenous to Asian countries extending from Asia Minor to the Himalayas and Outer Mongolia. Much cashmere comes from the northwestern provinces of China.

There are several varieties of Cashmere goats, varying in size from animals three feet high at the shoulder and weighing 150 pounds to small ones two feet high and weighing only 50 pounds. As with all true goats, it has two coats of fur; the undercoat or down, and the outercoat. With Cashmere goats, the undercoat is very well-developed, giving a good supply of soft, fine fibers. Annual production per goat is one ounce to one pound, with half a pound considered to be an average annual clip.

Figure 41-8 Vicuna in Peru. (Courtesy Darrell Matthews, Southern Utah State College, Cedar City, Utah.)

Importation of Cashmere goats was attempted by several countries, including the United States, during the nineteenth century, but all efforts resulted in failures.

The Camel Family

Several species of the *Camelidae* are prized for the fiber they produce. Both the Arabian (*Camelus dromedarius*) and the Asiatic (*Camelus bactrianus*) camels produce fine hair fibers used in fine brushes and for fabrics. There are apparently no special management systems used for this exploitation; the hair is simply harvested, since it is shed by the camels more or less continually.

The vicuña, a wild specie of the *Lama* genus (*Lama vicuña*) in South America is well known for its dense, silky fleece and the luxurious, velvety touch of fabrics produced from its wool (Figure 41.8). At one time this was reserved for the exclusive use of Inca royalty. These animals are quite intractable and apparently not well-disposed to domestication. However, with the threat of extinction and loss of a valuable resource, there is renewed effort among Andean countries to improve vicuña production through application of proper wildlife management practices.

Another wild South American specie of the *Camelidae*, the guanaco (*Lama guanicoe*), is of interest not for itself but because it was the progenitor of two domestic varieties, the llama (*Lama guanicoe glama*) and the alpaca (*Lama guanicoe paca*). Both of these varieties have for centuries been of socioeconomic importance to the Andean cultures as beasts of burden and sources of meat (often the only source of dietary protein) and of fiber (Figures 41.9 and 41.10). Early Spanish conquistadores described these animals as the "sheep of Peru." Their management, however, has been rather primitive and apparently based largely on tradition, so that productivity seems to have been well below the potential.

There has been renewed effort, especially in Peru, and primarily with the alpaca, to study the basic physiology of these animals in order to develop appropriate management programs. From information available, unique physiological and behavioral characteristics indicate that their management requirements are quite different from those for sheep, goats, or cattle.

Figure 41-9 Alpaca in Peru. (Courtesy, Darrell Matthews.)

Adaptability The *Lama* species usually are associated with the high altitudes of the Andean mountains, where indeed they do thrive. This is often cited as the reason for the unique, dense characteristics of the fleece, an adaptation for protection against the extreme temperature fluctuations at altitudes of 12,000 to 16,000 feet. However, the guanaco, the wild progenitor of alpacas and llamas, ranges widely in western South America as far south as Patagonia. Many ranchers have small numbers on their land at low elevations in the central valleys, and remains have been found in archeological excavations along the coast. The latter may of course be remains from members of pack caravans, for which these animals have been used extensively.

Lamas are classified as ruminants, although their digestive tract is different from that of cattle and sheep, both anatomically and functionally. Of ecological importance is the apparent ability of llamas to utilize more efficiently than sheep the vegetation at high altitudes, even though these plants are high in lignin.

Figure 41-10 Llama on the Alto Plano in Bolivia. (Courtesy, Warren Foote, Utah State University, Logan, Utah.)

Reproduction More physiological studies of the alpaca have been reported, but presumably the llama is similar because of the common origin. Sexual maturity is attained at about one year of age. The female does not have recurrent estrus cycles, but remains continually in heat until about five days after copulation, which triggers ovulation. Approximately 20 per cent of apparently normal females, however, fail to ovulate and remain in heat, a condition which may be influenced by the plane of nutrition.

Figure 41-11 Shearing alpaca in Peru. (Courtesy Darrell Matthews.)

The gestation period of the alpaca is relatively long, being on the order of 342 to 345 days. Young are born in relatively advanced stage of maturity, and birth occurs only in the morning hours. The birth rate seldom surpasses 50 per cent and multiple births are not known. Infant mortality may be as high as 50 to 60 per cent.

The male is vigorous in his mating behavior, having been observed to mate five to six times with a single female and as many as 18 times when first turned in with a flock of females. Libido appears to decrease rapidly in two to three days; accordingly, it appears that about one breeding male may be needed per five females. In mating the female is mounted while "kneeling" on all four legs.

Behavior *Lama* are quite intractible animals, as is learned especially from the llama when it is used as a beast of burden. When feeling overloaded or overworked, they protest by lying down, kicking, and spitting a characteristically stinging saliva in the face of unwary handlers. All *Lama* have the characteristic of using common dunging areas which is advantageous to the control of internal parasites. They also utilize common dusting beds.

Fleece *Lama* hair ranges in color from black to brown to white, with varying intermixing. Alpaca (Figure 41.11) is finer and more desirable than llama, and white fleeces from either is much preferred. Average yearly production is about 8 lb. per animal.

Development of unique management systems needed for efficient exploitation of *Lama* presents a great challenge.

FURTHER READINGS

Belschner, H. G. 1956. *Sheep Management and Diseases.* 4th ed. London: Angus and Robertson.

Harris, Lorin E. 1968. "Range nutrition in an arid region." Honor Lecture Series, Utah State University, Logan, Utah.

Harris, L. E., C. W. Cook, and L. A. Stoddart. 1956. "Feeding phosphorus, protein and energy supplements to ewes on winter ranges of Utah." *Utah Agric. Expt. Sta. Bull.* 398, pp. 1–28.

Kammlade, W. G., Sr., and W. G. Kammlade, Jr. 1955. *Sheep Science.* Rev. ed. New York: Lippincott.

Kunkel, H. D. 1971. "Nutritional requirements of the angora goat." Circular B-1105. College Station, Texas: Texas A&M University Agric. Expt. Sta.

McKinney, J. 1959. *The Sheep Book.* New York: Wiley.

NRC. 1964. *Nutrient Requirements of Sheep.* Pub. 504. Washington, D.C.: NRC.

Weir, W. C., and R. Albaugh. 1954. "California sheep production." *Calif. Agric. Expt. Sta. Manual* 16.

Forty-Two

Horse Management

Wherever man has left his footprints in the long ascent from barbarism to civilization, we will find the hoofprint of the horse beside it.

John Trotwood Moore

The world horse population has declined about 20 per cent during the last two decades (Table 42.1). The decline has been attributed to mechanization of agriculture in certain areas of the world, since there is a close correlation between rising tractor numbers and declining horse numbers. In many areas of the world, such as Latin America and Africa, the horse is still used for draft and transportation as well as for recreation purposes. In these areas the horse population has been stable or slightly increasing. According to the FAO, the number of horses kept on farms in the United States and Canada has been stable for the last ten years at slightly over three million head. However, the United States' total horse population has been growing rapidly. The growth from 3 million in 1960 to approximately 7 million head in 1970 has been attributed to affluence, increased recreation time, and a certain longing for a return to nature. Even though there is no precise way to measure the extent of the economic impact resulting from the rapid growth, the American Horse Council has estimated that the nation's horse industry has an income of about seven billion dollars annually.

Who is today's horseman in the United States? About 75 per cent are pleasure-horse owners; the other 25 per cent are ranchers, horse breeders, and race-horse owners. The pleasure-horse owner is found in suburban areas, and seven out of eight prefer western over English riding. About 70 per cent of all horsemen are males, and 60 per cent are over 20 years of age.

Removal of horses from their natural habitat of free roaming and grazing to one of close confinement and hand feeding has led to many problems in horse management. With increased confinement, management practices are based on procedures to insure that the horse has adequate exercise, proper feet and leg care, parasite and disease prevention, and a balanced diet. Since stallions are no longer run with the mares, management practices must allow for estrus detection and breeding at the proper time. These factors and many others offer unique management problems to the horseman in the United States.

Table 42.1.
World horse population.[a]

Area	1947–48/ 1951–52	1951–52/ 1955–56	1962–63	1963–64	1964–65	1965–66	1966–67
Europe	16,524	15,378	10,568	9,923	9,351	8,998	8,817
North America[b]	9,337	5,886	3,351	3,327	3,209	3,197	3,231
Latin America	22,797	23,758	22,813	23,334	23,814	24,297	24,287
Near East	2,106	2,290	2,576	2,506	2,569	2,554	2,553
Far East	4,179	4,180	3,626	3,462	3,382	3,208	3,255
Africa	2,994	3,006	3,319	3,337	3,572	3,578	3,644
Oceania	1,298	1,062	683	668	650	631	612
World Total	77,328	76,913	63,636	62,657	62,049	62,040	61,989

[a] Figures are in thousands. From FAO (1968). [b] Horses kept on farms.

Nutrient Requirements of Horses

In the United States, most equine research was discontinued with the passing of the draft horse. Many of the nutrient requirement recommendations for today's light horses are based on beef-cattle and draft-horse research. Even though many comparisons can be made between light horses and draft horses or cattle, the similarity is far from complete. Draft horses work at slow speeds and have a docile temperament compared with that of light horses. The caecum of the horse, like the rumen of the cow, serves as a fermentation vat for large numbers of microorganisms. In both species, volatile fatty acids are the major products of carbohydrate fermentation. In the rumen, however, fermentation precedes gastric and intestinal digestion, whereas cecal microorganisms act upon the residues of gastric and intestinal digestion; consequently, the nutritional significance of microbial activity may be quite different in these species. Only recently have the nutrient requirements of light horses come under investigation. Slade and Robinson (1970) have established the digestible protein maintenance requirement of adult Thoroughbred and Quarter Horse mares to be 58 g per 100 kg of body weight. This value confirms the level of 60 g recommended by the National Research Council. Other studies have established that the horse can use urea as a source of dietary nitrogen. Upon completion of these and other studies, horse nutrition will have a scientific basis and will no longer be based upon fads, fables, or trade secrets.

The National Research Council recommendations shown in Table 42.2 are merely rough guides to the nutrient requirements for horses in various physiological states, such as maintenance, growth, work, gestation, and lactation. Many factors must be taken into account in considering the nutrition of an individual horse. Horses working at a slow steady pace require less feed than those working with short, quick bursts of speed. The recommended allowances that are adequate for the horse during the summer are apt to be inadequate during the winter. Nervous horses that weave or pace in their stalls probably have a higher requirement than more placid ones.

Under most circumstances, horses are flexible enough to eat and prosper on practically any food. In the United States, most horse rations are composed of cereal grains and a roughage, such as alfalfa or timothy hay. Other countries may or may not use similar rations. In the Sahara desert, dates may be used rather than barley in the ration of the Arabian horse. Oftentimes Arabians are given sheep or camel's milk. In Denmark, corn is sometimes replaced with a mixture of beet molasses and dried swamp kelp. The availability of local feeds will periodically influence the composition of the ration. Utah farm horses have been known to consume as much as 20 pounds of wet sugar-beet pulp daily without any ill effects. Many other substitutes, such as cooked potatoes, carrots, turnips, prunes, apples, sugar, bread and beans, have been used in the rations for horses. Oats, corn, and barley, however, are universally considered to be standard grains for horses.

The requirement for bulk in the horse's ration has always been of concern to armies, farmers, and pleasure-horse owners. Recently, it has been shown that the bulk may be reduced by feeding a completely pelleted ration. Pellets offer several advantages, such as labor economy, less storage space, convenience in feeding, and opportunity for increased utilization of by-product feeds. If proper constituents are used in correct proportions for preparing the pellets, the horse is assured a balanced ration. However, some horsemen believe that pellet feeding leads to an increased tendency to chew wood.

In the past, there has been considerable discussion concerning the sequence of feeding hay and grain and watering. It is generally recognized, however, that horses may be watered before, during, or after feeding without

Table 42.2
Daily nutrient requirements of horses (adapted from NRC, 1973).

Class of animal	*Body weight (kg)*	*Daily feed[a] (kg)*	*Digestible energy (Mcal)*	*Protein (g)*	*Digestible protein (g)*	*Vitamin A (thousands IU)[b]*	*Ca (g)*	*P (g)*
Mature horses at rest (maintenance)	200	3.00	8.24	300	160	5.0	8.0	6.0
	400	5.04	13.86	505	268	10.0	16.0	12.0
	500	5.96	16.39	597	317	12.5	20.0	15.0
	600	6.83	18.79	684	364	15.0	24.0	18.0
Mature horses at light work (2 hr/day)	200	3.80	10.44	383	202	5.0	8.0	6.0
	400	6.68	18.36	672	355	10.0	16.0	12.0
	500	7.96	21.89	803	424	12.5	20.0	15.0
	600	9.23	25.39	930	491	15.0	24.0	18.0
Mature horses at medium work (2 hr/day)	200	4.79	13.16	483	255	5.0	9.2	7.0
	400	8.65	23.80	871	460	10.0	17.2	13.0
	500	10.43	28.69	1,047	553	12.5	21.2	16.0
	600	12.22	33.55	1,229	649	15.0	25.2	19.0
Mares, last 90 days of pregnancy	200	3.16	8.70	364	216	10.0	10.4	8.0
	400	5.41	14.88	613	375	20.0	19.5	15.0
	500	6.31	17.35	725	434	25.0	24.0	18.0
	600	7.25	19.95	837	502	30.0	28.0	21.0
Mares, peak of lactation	200	5.54	15.24	750	480	10.0	34.0	23.4
	400	8.91	24.39	1,181	748	20.0	42.0	35.6
	500	10.04	27.62	1,317	829	25.0	47.0	38.6
	600	10.92	30.02	1,404	876	30.0	64.0	43.4
Growing horses, 600 kg mature weight	140	5.15	14.15	958	705	5.6	52.0	32.2
	265	6.26	17.21	870	582	10.6	51.2	32.0
	385	6.86	18.86	837	524	15.4	32.9	20.6
	480	6.98	19.20	775	458	19.2	31.3	19.6
	600	6.83	18.79	684	364	15.0	24.0	18.0

[a] Assume 2.75 Mcal of digestible energy per kg of 100 per cent dry feed.
[b] One mg of beta-carotene equals 400 IU of vitamin A.

any ill effects. Convenience of the caretaker usually determines the watering time if water is not available *ad libitum.* It has been shown that maximum feed utilization is obtained when the hay is fed before the grain. A certain amount of feed remains in the stomach at all times. Therefore, if the hay is fed first, it passes into the caecum for fermentation and the grain will remain in the stomach longer for maximum digestion, resulting in maximum feed efficiency.

Common Digestive Ailments

There are several common digestive ailments which result from feeding dirty or moldy feed, abrupt changes in diet, or other improper management practices. The horse is particularly susceptible to "colic" or stomachache, since he is unable to vomit. It is a symptom of abdominal pain that can be caused by several conditions. Frequently it is caused by indigestion, poisons, or an impacted caecum, but it can be caused by improper watering or bloodworms. Azoturia occurs when a horse is exercised vigorously after a period of idleness in which the feed intake is high. It is associated with a fault in carbohydrate metabolism that is characterized by a sudden onset of a semi-paralysis of the horse's legs. Founder is an inflammation of the laminae of the foot which may be caused by overeating, overwork, drinking cold water while overheated, or inflammation of the uterus following foaling, but most commonly is due to overeating grain. Grain contains a toxin, histidine, that is formed during digestion and when decarboxylated to histamine causes the laminitis. Heaves or chronic alveolar emphysema is a result of continued feeding of dusty or moldy feed. The horse can inhale normally but cannot exhale properly because of the breakdown and loss of elasticity in some parts of the lungs.

In addition to the common digestive ailments, there are common vitamin-deficiency symptoms. Periodic ophthalmia, commonly called moon blindness, is characterized by periods of cloudy vision in one or both eyes. It may be caused by riboflavin deficiency, as well as by leptospirosis, parasites, or reaction to repeated streptococcal infections. Vitamin A deficiency is characterized by faulty vision and is especially noticeable when the horse moves at night.

Reproduction

Problems associated with reproduction are most critical in the horse business. Hutton and Mescham (1968) reported an analysis of reproductive efficiency from 14 farms and six breeds involving 1,876 mare years. They observed conception, foaling, and weaning efficiency means of 80.1 per cent, 73.8 per cent, and 70.8 per cent, respectively. Considerable variation in efficiency between years and farms was observed. The reason for the low reproductive efficiency is of utmost concern to horsemen, since it often determines success or failure of the breeding farm.

Reproduction in the mare is characterized by so many variations that it is difficult to rely on normal values. Most horsemen consider the pubertal age of the filly to be 15 to 18 months. However, estrous cycles may start as early as 12 or as late as 24 months. Once puberty is attained, they are seasonally polyestrus; i.e., they have regular estrous cycles during the sexual season. In the western United States, mares cycle regularly from April through June. During the remainder of the year, some mares will continue to cycle while others experience anestrus, a period of sexual inactivity. The degree of anestrum varies considerably. Some anestrous mares continue to have ovarian cycles but fail to display behavioral signs of estrus. Cessation of ovarian activity occurs in other mares until the following breeding season is approached. The wild horse is reported to have a more restricted sexual season than the domesticated horse; the longer sexual season is a genetically determined trait developed by selection.

As daylight hours increase, mares begin to come into estrus. This was first established by Burkhardt and later confirmed by Nishikawa

(1959) when they succeeded in inducing estrus in anestrous mares during the fall and winter by lighting their stalls. Only recently have horsemen in the United States recognized the value of Burkhardt's and Nishikawa's work and started to provide artificial lighting for barren or open mares. Since estrous cycles are usually long and irregular during the early part of the breeding season, artificial lighting induces the mares to cycle regularly during this normally irregular part of the sexual season. Once established, the estrous cycles are 21 or 22 days in length, but only about 60 per cent of mares will display the so-called normal cycle. Not only does the length vary with the season, but it varies from cycle to cycle. The duration of estrus or heat also varies between individuals as well as between cycles. About 50 per cent of the mares display a "normal" five- to six-day estrous period. Ovulation normally occurs on day 3 of estrus, but varies from day 1 to day 5. The only constant feature of the mare's reproductive cycle is that within 48 hours after ovulation, she will cease displaying estrous behavior. Because of the individual variations in estrous cycle length, estrous periods, and time of ovulation, most stud-farm managers tease the mares daily with a stallion to determine the proper breeding time. Maximum reproductive efficiency is achieved by farms that determine the time of ovulation by rectal palpation and that breed just prior to and just after ovulation. Based on the facts that ovum survival is approximately 24 to 48 hours, that spermatozoa survival is approximately 72 hours, and that it requires about eight hours for sufficient number of spermatozoa to reach the ovum for fertilization, this practice allows sufficient live spermatozoa to be present for fertilization at all times. In many instances, rectal palpation is not feasible so that mares displaying short three-day estrus periods are bred on day 1 and alternate days. The mares displaying normal or long estrous periods are bred on day 3 and alternate days until cessation of estrus. These methods also insure that adequate numbers of live spermatozoa are present to fertilize the ovum. In mares with a prolonged estrous period, it is becoming common practice to induce ovulation with exogenous hormones. They are given intravenously 1,000 to 2,000 I.U. of human chorionic gonadotropin (HCG), an amount which is sufficient to stimulate ovulation within 20 to 60 hours. Assuming eight hours for spermatozoa to reach the uterine horn, mares are bred 12 hours after injection.

Investigations of the hormonal control of the mare's estrous cycle have lagged behind those for other species. Only recently, the pattern of progesterone secretion during the cycle has been determined by Stabenfelt *et al.* (1972). The plasma levels increase quite rapidly for five to six days after ovulation, are maintained for eight days, and then decline to a minimal level five to six days prior to ovulation (Figure 42.1).

One of the reasons for low reproduction efficiency on many farms is the failure to determine pregnancy. Often mares that are shy breeders will be left open or pregnant mares will show signs of heat and be rebred if a pregnancy test is not made. There are several biological tests for pregnancy, but the simplest test is palpation of the fetus through the rectal wall. An experienced technician can palpate the fetus at 40 days and sometimes as early as at 35 days of development. The method has the advantage in that the result is available at once, but has the disadvantage in that it is accurate only in the hands of a skilled and experienced person.

The biological tests are based on two unusual hormonal conditions in the pregnant mare. In early pregnancy, the test is based upon the presence of a gonadotropin, pregnant-mare serum gonadotropin (PMSG), in the blood that has considerable follicle-stimulating activity. The gonadotropin appears between 35 and 42 days, reaches a maximal concentration by the 60th to 90th day and disappears by the 110th to 180th day (Figure 42.2). Therefore it can be used between the 45th and 140th day after breeding for pregnancy diagnosis. Serum is assayed for presence of PMSG by injection into immature mice or rats. The presence of ripe follicles or corpora lutea in the ovaries, or an increase in ovarian weight, as well as an enlarged uterus, is a positive test. In late pregnancy, the test is based on the presence of estrogen in the urine. At 140 days after breeding, sufficient estrogen is being voided in the urine of pregnant mares to allow detection of pregnancy (Figure 42.2). Injection of the urine in spayed rats stimulates the accessory sex glands. The estrogen can also be detected by a chemical test.

Artificial control of estrus and ovulation in the mare would be beneficial to breeders and showmen. Prevention of estrous behavior and ovulation is desirable in fillies and mares during race training. Loy and Swan were successful in blocking estrus and ovulation by daily intramuscular injection of 100 mg progesterone in oil solution. The mares returned to normal cycling after treatment. A successful procedure for the induction of estrus and ovulation has not been developed. Its development would allow breeding farms to have their foals born at a specific time of the year. The induction of estrus in anestrum mares has been demonstrated by Nishikawa with estrogen injections. However, the ovaries are not stimulated, and there is a failure to produce and ovulate follicles.

The length of gestation in the mare is 340 days, but it varies considerably. Mares have been known to carry their fetuses over one year, but many others foal 2 to 3 weeks before the 340-day "due" date. During pregnancy, mares require a minimum of special care. Most stud farms keep their pregnant mares on pasture where they receive adequate nutrition and exercise. If confined, mild but regular exercise to maintain muscle tone is recommended. Many pleasure mares can be ridden up to the day of foaling without any harm. Approaching parturition is indicated by distention of the udder two to six weeks before parturition, marked shrinkage of muscles near the tail head at seven to ten days preceding parturition, and presence of wax on the ends of the teats four to six days before foaling. At this time, the mare's feed is reduced since birth seems to be easier if the mare is as empty as possible. Feed reduction also decreases her milk flow, which prevents the foal from getting diarrhea as a result of drinking too much milk. Immediately prior to foaling, mares become quite nervous, urinate frequently, lie down and get up frequently, and break out in a sweat. Once started, normal foaling is accomplished in 15 to 30 minutes. Many stud farms keep a man on duty in the foaling barn during the foaling season. He observes the mare from a place where he cannot be seen to prevent disturbing her, and assists only if trouble occurs during foaling.

Immediately following foaling, bacterial invasion of the umbilical cord is prevented by tincture of iodine. Otherwise, "navel ill" may occur in many foals that are born under unsanitary conditions. The foal's bowels are filled with excrement accumulated during fetal life. If the bowels aren't voided within 4 to 12 hours after birth, colic is observed, and an enema should be given. Soon after foaling, the

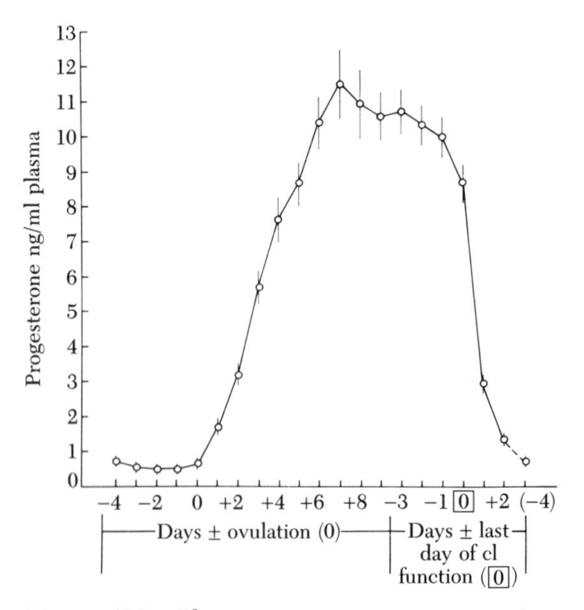

Figure 42-1 Plasma progesterone concentration during estrous cycle. (From Stahenfeldt *et al.*, 1972.)

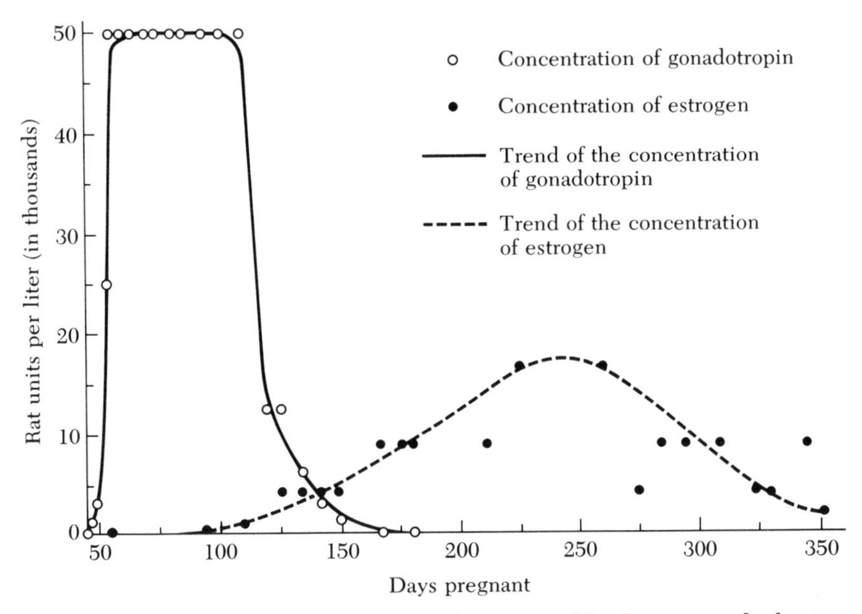

Figure 42-2 The concentration of gonadotropin in blood serum and of estrogen in the urine of a mare throughout pregnancy. (Cole and Hart, 1942.)

mare and foal are usually turned out in a clean pasture where they get adequate exercise and nutrition.

Stallion management practices vary from farm to farm and often depend on the value of the stallion. In any event, the stallion must be considered as an individual in regard to feeding, exercise, and handling. The feed requirements have been listed in Table 42.2. To keep the stallion in a thrifty condition, he should receive daily exercise. Most horsemen feel that regular exercise is a necessity for good libido and maintenance of quality semen. The kind of exercise depends on his training, soundness, and temperament.

Stallions are considered to be serviceable on a limited basis at two years of age and at four years can be used heavily. The number of services in a season varies with the individual, but 80 to 100 services are common for a mature stallion. When the breeding schedule becomes heavy, it is a good practice to evaluate periodically the semen to assure maximum efficiency. Actual breeding practices vary from pasture breeding to highly supervised mating with a veterinarian in attendance. As the value of the stallion increases, the greater his care. The practice of artificial insemination has been limited in the United States, but is used extensively in Russia and China. Rules established by most horse breeding organizations either severely restrict or prohibit its exclusive use. It does have a definite advantage in that several mares can be bred to the same stallion in one day.

The major concern at breeding is sanitation and safety. Unsanitary breeding practices lead to bacterial invasions and result in abortion or failure to conceive. Safety procedures such as hobbling the mare and keeping the stallion under strict control serve to prevent injury to the attendants and horses.

Health Management

Light horses are used in many ways that require maximum performance and endurance. To achieve a continued high level of performance, the horse must be in excellent health and relatively free of parasites. One of the major managerial responsibilities of the horse owner in cooperation with a veterinarian is to evaluate and formulate an effective parasite and disease prevention program for his particular situation.

There are at least 150 kinds of parasites that infect horses, internally and externally. Parasites can cause an unthrifty appearance, colic, damaged blood vessels and lungs, and otherwise reduce the strength and efficiency of performance. A successful parasite-prevention program is based on sanitation, manure management, and chemical control practices. Many of the internal parasites, of which the most common are bots, ascarids, strongyles, and pinworms, go through a life cycle such that they are passed out of the horse in the feces, become infective, and reenter by way of the mouth. The aim of treatment and preventive measures is to break the life cycle and prevent infestation. Proper manure removal to prevent coprophagia, and maintenance of the horse in clean surroundings free from stagnant surface ponds, sloughs, or poorly drained corrals will help break parasite life cycles during the period spent outside the horse. Chemical treatment will break the life cycle while in the host, but does not prevent tissue damage resulting from their migration or feeding habits. It should be pointed out that strict adherence to prevention and control programs will result in a low level of infection, but not a parasite-free horse.

Horses are subject to attack by a variety of external parasites. Their damage ranges from mere annoyance to loss of "condition" and anemia. Some of the external parasites serve as vectors of diseases. Mosquitos transmit equine infectious anemia (swamp fever) and equine encephalomyelitis (sleeping sickness), whereas a specific tick transmits equine piroplasmosis. External parasites, such as lice, mites, ticks, and flies, can be controlled or prevented by the use of proper insecticides as a supplement to sanitation and good manure-management practices. Control of the external parasites with insecticides is based upon proper formulation and timing of application.

Respiratory diseases are the most common infectious disease problems in the horse. Three of the most common, equine viral rhinopneumonitis, equine viral arteritis, and equine influenza, are caused by viruses; while the fourth, "strangles," is caused by a specific bacterium. These infections are potentially contagious; so they present management problems. Since transmission is usually through direct contact by rubbing noses and indirect contact by airborne secretions or contaminated feed and water troughs, infected animals must be isolated to prevent outbreaks.

Unethical Management Practices

Since the early history of horse-performance events, means for improving their performance have been eagerly sought. Drugs, hormones, and chemicals have been used in a variety of ways to gain an unfair advantage over other contestants. "Doping" of race horses has been a problem, and cases such as Dancer's Image, winner of 1968 Kentucky Derby, receive wide publicity. The rules governing the administration of drugs to race horses are basically similar throughout the world, but certain differences do exist. In California, phenylbutazone can be administered on the day of the race whereas in other countries, such as Great Britain, it cannot be used within 48 hours of a race. The use of tranquilizers and other drugs by trainers in gaited and pleasure horses has become a problem, and is being studied by The American Horse Show Association. A law has just been passed to prohibit "soring," the practice of applying caustics to wounds, of walking horses. The pain causes the horse to move his center of gravity backward, thus causing a longer stride behind and light-footed high action in front.

General Care

Teeth

The horse, like a human, has two sets of teeth during its life. The eruption of the "milk" and permanent teeth and their subsequent wear offer means of determining the age of a horse. Generally, the deciduous incisor teeth appear in pairs at eight days, eight weeks, and eight months. When the horse is 30 months old, the deciduous incisors begin to be replaced by permanent teeth. At five years the replacement is complete, and the horse has a full set of permanent teeth. The permanent teeth continue to erupt all during life. As a result of the horse grinding the feed as he eats, the teeth are worn off in a characteristic manner which enables age determination from 5 to 12 years of age (Figure 42.3). This method of age determination is not exact, since horses consuming sandy or dirty feed wear their teeth more rapidly than others. However, there is nothing mysterious about this method, and it has been used for many years to determine the period of probable usefulness of a horse.

The upper-cheek teeth are wider than the lower; therefore, as the teeth wear off, there is a tendency for sharp edges or "hooks" to be produced on the outer edge of the upper and inner edge of the lower teeth (Figure 42.4).

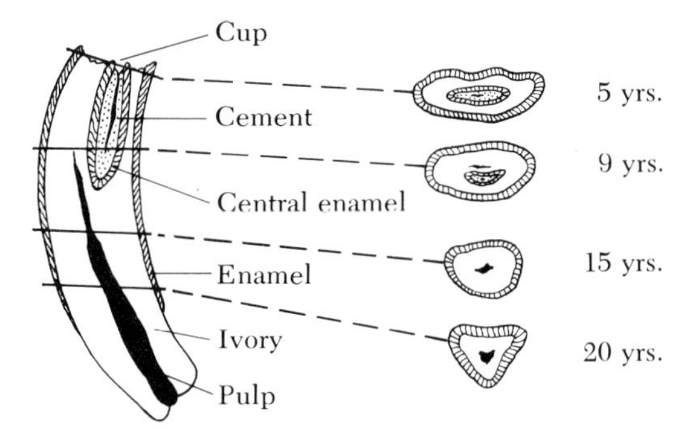

Figure 42-3 The horses tooth. Longitudinal section of a permanent lower middle incisor tooth; and cross-sections of permanent lower middle incisor teeth at different age levels. (Ensminger, 1969.)

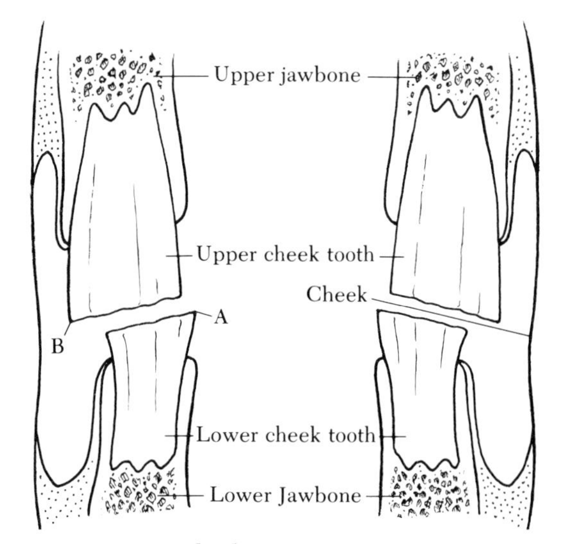

Figure 42-4 Cheek teeth of horse, viewed in cross section from the front. Note how upper teeth are wider. Constant wear produces sharp points at A and B. (Fowler, 1970.)

These hooks must be floated (rasped off) periodically to prevent the teeth from cutting into the cheek or the tongue. Failure to remove hooks will result in improper chewing of the feed, and "colic" may develop. Trotting-horse trainers round off the rough edges of young colts' teeth before "hooks" develop. If the cheeks are cut by the edges, the colt will start carrying his head off to one side to escape the pain of the bit pressing the cheek against the teeth.

In some horses, small pointed teeth grow in front of the molars on each side and are called wolf teeth (Figure 42.5). They serve no useful purpose and are a prime source of mouth trouble. If a wolf tooth gets broken off, it acts as a sliver and can cause trouble for years. The constant irritation causes the horse to carry its head to one side or the other and to fuss with its mouth.

Feet

The rapidity and/or preciseness in which a horse can move its feet and legs determines the value of most horses. Management practices must include proper care of the feet and legs if the horse is to remain serviceably sound.

The foot of the horse (Figure 42.6) primarily consists of the pedal bone and elastic structures encased in a hard outer shell or hoof wall. The hoof wall grows continuously, like the human fingernail. Excessive or disproportionate growth tends to keep the horse off balance and places an excessive stress on the supporting tendons and ligaments. If neglected for long periods of time, lameness will usually result. Horses traveling on rocky or sandy terrain continuously wear off the hoof wall at approximately the same rate as growth so that the length of the hoof does not become excessive. Actually, excessive wear must be prevented for most horses that are used for work or pleasure. Shoeing prevents the wear, and protects the feet from injury. The hoof wall grows at a rate which requires reshoeing approximately every six to eight weeks.

Foot problems are usually encountered in horses standing in bedding or areas covered with manure or soaked with urine. One of the most common problems is thrush. It is caused by a necrotic fungus that causes deterioration of tissues in the cleft of the frog or in the junction between the frog and bars.

Many foot and leg problems can be avoided by careful inspection of horses prior to purchase, since poor conformation of the feet and legs either directly or indirectly result in unsoundness or lameness. In viewing the horse from the front, a line drawn from the point of the shoulder perpendicular to the ground should equally bisect the forearm, knee, cannon bone, fetlock, and foot. Any deviation of the leg from this line is a fault in conformation. The "splay-footed" condition (toes turned outward) prevents the front foot from breaking over the toe, causing the striding foot to

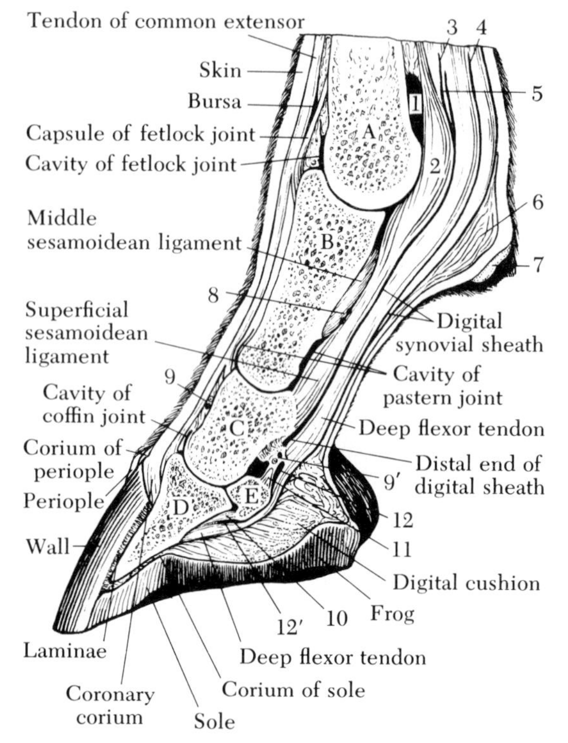

Figure 42-6 Sagittal section of digit and distal part of metacarpus of horse. A, Metacarpal bone; B, first phalanx; C, second phalanx; D, third phalanx; E, distal sesamoid bone. 1, volar pouch of capsule of fetlock joint; 2, intersesamoidean ligament; 3,4, proximal end of digital synovial sheath; 5, ring formed by superficial flexor tendon; 6, fibrous tissue underlying ergot; 7, ergot; 8,9,9′, branches of digital vessels; 10, distal ligament of distal sesamoid bone; 11, suspensory ligament of distal sesamoid bone; 12,12′, proximal and distal ends of navicular bursa. The superficial flexor tendon (behind 4) is not marked. (Sisson and Grossman, *The Anatomy of Domestic Animals*, W. B. Saunders, 1953.)

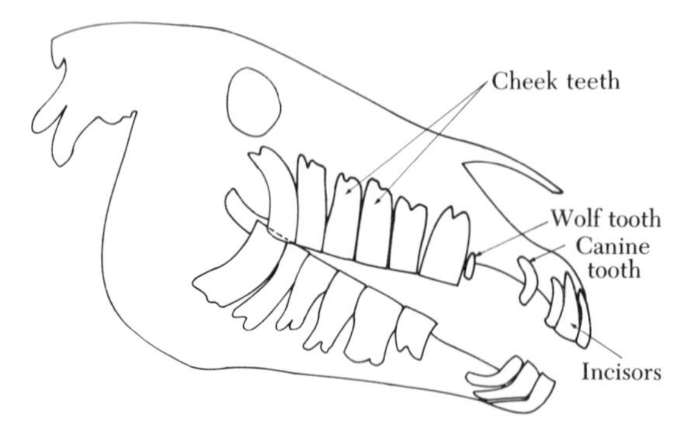

Figure 42-5 Side View. Horses' head, showing arrangement of teeth. (Fowler, 1970.)

swing inward. In such case, interference (striking opposite front leg) is inevitable and may cause splints. As the foot strikes the ground, increased strain is placed on the inside hoof wall and medial sidebones may develop. The opposite condition, pigeon-toed or toed-in, causes an outward deviation in the flight of the foot known as paddling. As it strikes the ground, increased strain is placed on the outside hoof wall and may result in lateral sidebones. Lateral deviation of the cannon bone (bench knees) causes increased concussion on the inside splint bone. The bone periosteum is disturbed, resulting in new bone growth or splints and lameness. From these few examples, one can quickly conclude that faulty foot and leg conformation causes excess strain to be placed on bone structures, resulting in unsoundnesses and lameness. Avoidance of horses with faulty conformation will often save time, money, and disappointment.

Housing

Generally speaking, horses require little or no shelter. They are not delicate animals, and get along fine with the protection provided by trees and hillsides. In severe cold, they require some shelter from the wind, and in severe heat it is advisable to furnish a shade. However, most of today's horses are kept in confinement and are provided box stalls. Many confined horses still prefer to be outside in the rain or sun, and will do so if they have access to an outside paddock.

Grooming

Most horsemen take pride in ownership. An anonymous expression is, "There is nothing as good for the inside of man as the outside of a horse." Frequent grooming cleans the hair and stimulates the pores of the skin to produce natural oils that bring a shine to the coat. Rubbing and brushing massages the skin, which stimulates blood flow and helps keep it healthy.

Exercise

One of the most common mismanagement practices of today's horseman is infrequent exercise for pleasure horses. Until recently horses received daily exercise, since they were used for work and transportation or were kept in pastures where they had to travel for food and water. Lack of proper exercise for the confined horse is one reason for poor health in many horses. A vigorous workout for about 30 minutes per day is necessary for the confined horse to maintain proper muscle tone. Another major concern is that lack of exercise causes the horse to become bored. In an effort to break his boredom, he acquires vices, such as wood chewing, cribbing, stall kicking, and stall weaving or pacing. Once a vice is acquired, it is difficult or impossible to break the habit.

FURTHER READINGS

Adams, O. R. 1966. *Lameness in Horses.* Philadelphia, Pa.: Lea and Febiger.

Ensminger, M. E. 1969. *Horses and Horsemanship.* 4th ed. Danville, Ill.: Interstate.

Harrison, J. C. 1968. *Care and Training of the Trotter and Pacer.* Columbus, Oh.: United States Trotting Association.

Kays, D. J. 1969. *The Horse.* Cranberry, N.J.: W. S. Barnes.

Lyles, L. L. 1969. *Horseman's Handbook.* Santa Rosa: California State Horseman's Association.

Forty-Three

Poultry Management

Raising poultry requires the acumen of a sane man.

An old Indian proverb

Poultry production is widespread, as Table 43.1 shows. Similar management practices may be adapted the world over, partly because returns are obtained in about 3 to 12 months. Poultry can be raised even in areas where domestic cattle would not survive because of diseases such as trypanosomiasis, spread by the tsetse fly, and foot-and-mouth disease. Since poultry can be maintained under conditions of intensive confinement, it is a practical operation in areas which lack land mass for grazing other animals.

Poultry production formerly was a small farmyard operation of minor economic importance in the U.S. Usually, the chickens and other fowl were kept to meet the need of the family, and the excess produce was sold, as is still the practice in many developing countries. Eventually, specialized poultry farms made their appearance and became economically feasible operations. This specialization is continuing, and poultry production is emerging as a primary industry, with many flock populations in the hundreds of thousands. With the same trend, the sight of the chickenyard in the countryside may even become a rarity in the developing countries in the near future.

Since 1930, egg production per hen has nearly doubled. Today's hen is a product of the efforts of many breeders and geneticists, and receives the most nutritious foods and the comforts which a superior management can provide.

Domestication of Chickens

It is generally assumed that the domestic chickens originated from one or all four species of Jungle Fowl, namely, *Gallus bankiva, G. lafayette, G. varius,* and *G. sonnerati.* Jungle fowl are still found wild in India, Pakistan, Burma, Ceylon, southwestern China, Thailand, the Philippines, Indonesia, Malaysia, and New Guinea, and on many other islands. The exact date of domestication cannot be determined with certainty, but it appears to be somewhere in the Bronze age. The archaeological discoveries indicate that domestic

Table 43.1
The distribution of poultry in the world.[a]

Poultry species	*Europe*	*North and Central America*	*South America*	*Africa*	*Asia*	*Oceania*	*World*
Chickens	1,508.2	654.4	394.0	303.9	1,068.3	28.8	3,597.6
Ducks	19.9	0.9	9.2	4.7	64.3	1.2	100.2
Geese	12.0	0.2	0.2	2.8	2.3	0.2	17.7
Turkeys	11.2	8.4	4.7	0.9	0.9	0.7	26.8

[a] All figures in millions. Based on FAO (1969a). The accuracy of the estimates in this table is doubtful, because the U.S.A. alone produced 104 million turkeys in 1968. Many countries, like the U.S.S.R., China, and India, lumped all species under chickens.

chickens were familiar to the inhabitants of Mohenjo Daro in the Indo-Pakistan subcontinent by 3250 B.C. This predates the finding in Egypt, where a portrayal of chicken is found in the tomb of Rekhmara (about 1500 B.C.). The Egyptian tombs constructed before 2000 B.C. depict bas-reliefs of about 30 birds, and the Assyrians mention pelican, heron, stork, crane, cormorant, gull, duck, swan, blackbird, bunting, pigeon, turtledove, eagle, hawk, and sparrow hawk, but no chickens.

The Chinese mention the introduction of chickens from the west during the dynasty which ruled about 1400 B.C. The depiction of chickens on coins and vases in Greece also starts about 700 B.C. It appears reasonable to assume that the domestic chickens must have dispersed from Asia, southeastern Europe, and Egypt to the other parts of the world.

In Greece, chickens were used as sacrificial animals and might have been specially bred for this purpose. Bantam breeds were well portrayed in sculptures decorating the Lycian Tombs (ca. 600 B.C.). The sound and duration of crowing also impressed ancients and people may have selected birds for this trait.

The sport of cockfighting, which was very popular in the world, also contributed to the wide distribution of chickens and the development of game cocks. It is believed that the interest in this sport was the reason for the domestication of the chicken. Cockfighting was very popular in England and there was a special cockpit in Parliament for cockfights up to the beginning of the 19th century. The fighting is still popular in many parts of the world, including the U.S.A.

The ornamental value of the chickens was responsible for development of Sumatras in Indonesia and of the Longtailed fowl, Onagadori, of Japan.

Chickens were brought to the Americas by the Spanish explorers in the 1500's and to the U.S.A. later from Europe by the early settlers (1607). By the early 1800's, the U.S.A. was importing birds directly from China and the East Indies.

The objective of today's breeders is not a fancy-looking bird, but one capable of maximum meat or egg production.

Other Domestic Poultry

The birds which are also classified under domestic poultry (besides the chicken and turkey) are guinea fowl, ducks, geese, swans, pigeons, pea fowl, and game birds. Economically, chickens and turkeys are the most important birds in North America, but ducks and geese are of great economic importance elsewhere.

The turkey (*Meleagris gallopovo*) is indigenous to America and was raised by the Indians long before the arrival of the Europeans. According to archaeological evidence, Indians in Mexico domesticated the bird thousands of years ago. Turkeys were taken to Spain in 1519, and appeared in England five years later.

Most breeds of domesticated ducks probably originated from wild mallards (*Anas platyrhynchos*). Mallards were domesticated independently in different countries and are reported to lose their migratory instincts in about three to four generations. The Muscovy ducks (*Cairina moschata*) were domesticated by the Indian natives of South America before Columbus. In this breed the drake's tail feathers have no curl.

Duck meat must have been esteemed by early man, because duck bones were found among the remains of cave dwellers of the Stone age. The catching and killing of ducks is also represented on Egyptian monuments as far back as 3000 B.C. The Chinese probably have raised domesticated ducks longer than other countries. The high fat content of duck meat prevents it from being as popular as chicken in some parts of the world.

Khaki Campbell and Indian Runner ducks bred for egg production can produce annually 300 eggs, each weighing 80 to 85 grams. Ducks bred for meat, such as the Pekin, grow very rapidly and may reach a body weight of 2.7 to 3.2 kg (6 to 7 pounds) at seven to eight weeks of age. Feed efficiency for this age is about 2.6 units of feed for each unit of gain. Duck eggs are not very popular in the U.S.A. and Canada, but they are in Europe and Asia. Diets may be designed to produce moderate yolk color and mild flavor.

The date of domestication of the wild goose has not been established with any certainty, but it is generally assumed that the wild Grey-lag goose (*Anser anser*) is the ancestor of all the domestic breeds of geese, with the exception of the Canada goose, which originated from *Branta canadensis*, and the Egyptian goose, from the wild *Chenalopex aegyptiacus*.

The goose was regarded as a sacred bird in Egypt about 2000 B.C. White domestic geese are described in the *Iliad* and the *Odyssey* (twelfth to seventh centuries B.C.). The use of feathers for filling cushions and mattresses probably originated with the Celts and the Germans. Force-feeding of geese was practiced under the Pharaohs in Egypt, and specialized force-feeding to produce an overgrown liver was known to the Romans of the first century A.D. It is still the most practical way of producing *pâté de foie gras*.

The guinea fowl originated from the wild species *Numidia meleagris* in West Africa, where it still exists. It was reintroduced into Europe by the Portuguese about 1486–1500. Its economic potential still awaits exploitation.

Pigeons are represented in bas-reliefs of Egypt of at least 2700 B.C. According to Darwin, pigeons originated from the wild rock pigeon (*Columba livia*) still found in many places. The production of squabs is of very limited economic importance.

Genetic Improvements

It is difficult to assess the time when emphasis started on the food-producing potential of poultry. Specialized breeding of chickens for meat or eggs had been practiced by the ancients, but was lost until the nineteenth century A.D.

Poultry breeding has passed successively through the following stages for selection: (1) cockfighting, (2) fancy feathers and other exhibition traits, and (3) selection for meat and egg production.

Cornish birds were kept solely for exhibitions until less than a century ago, when it found its place as the male parent in the breeding of birds for fryer production. Egg production is generally low in strains bred for meat production.

Breeding chickens for high egg production turned out to be more tedious than selection for meat. Egg production is greatly influenced by environmental factors and is less heritable than body weight. A satisfactory device for measuring egg production, such as a trapnest, was only invented about 1898 by Silberstein as well as Gowell.

Although more attention might be paid to the breeding of stock adapted for local climates and diseases, birds can tolerate considerable variations in the weather, and disease resistance can be supplied by proper vaccination. When proper feed and management are available, the improved strains of birds may be exported and raised all over the world.

The present day Broadbreasted turkey can be traced back to superior stock developed in England by Cattell. This stock was exported to British Columbia in 1927, and reached the U.S.A. in 1937. Both egg and meat breeds of ducks have been developed by selection. Pigeons and geese have relatively low reproductive rates, but in spite of this both meat and fancy-feathered breeds of pigeons have been developed. Behavioral traits in pigeons form the basis of some breeds of homers, rollers, tumblers, and fantails.

Disease Prevention and Control

The high morbidity and mortality from diseases, with the consequent loss of production in poultry, from day-old to adult life was a serious early drawback in the development for many species of poultry as an industry. Disease prevention is much more economical than treatment for disease.

Newcastle disease virus was isolated by Doyle in England in 1927, and a successful method for the immunization of fowl against this disease was developed there by Iyer and Dobson (1940) and in the U.S.A. by Beach (1942). However, Britain refused to permit an immunization program until very recently.

A serious problem associated with a microorganism, *Mycoplasma gallisepticum,* occurs in chicken and turkey broilers causing "air sac disease." The mycoplasma almost always provides favorable conditions for many bacteria to become active. The infection is egg-transmitted, and can be eradicated from commercial turkey and broiler stocks in most parts of the world by dipping the eggs in and treating chicks with antibiotic solutions.

Marek's disease, considered part of the avian leukosis, today is the poultry industry's most serious disease. The causative virus has been isolated and identified and its highly infectious nature demonstrated. The results of research activity suggests that control measures seem imminent. The disease affects chickens, turkeys, and ducks.

Sanitation is always important in disease prevention, regardless of the species. Sanitation has been defined as keeping the bird and everything the bird comes in contact with *clean.* The work of Van Es and Olney (1940) demonstrated the influence of environments on the incidence of poultry diseases and the value of wire floors. Proper sanitation measures prevent Salmonella and parasites such as round worms, tape worm, mites, fleas, and lice. The depopulation procedure of controlling poultry diseases is very costly unless it is planned as a part of normal management procedure. One age flock on the entire ranch, called the all-in all-out system of fryer or pullet raising, is widely used and has the inherent advantage of helping to control certain poultry diseases.

Vaccines are available to protect chicken flocks against the following diseases: infectious bronchitis, infectious lâryngotracheitis (also pheasants), Newcastle disease (also turkeys and pheasants), and fowl pox (also other fowl). Since fowl pox is spread by such insects as mosquitoes, it is generally advisable to vaccinate against this disease. For the other diseases the decision to vaccinate depends on the local situation. Only a flock in good health should be vaccinated and, preferably, with only one vaccine at a time. The vaccination program should be completed before the females start to lay. It should be done by, or under the supervision of, a qualified specialist, since most vaccines are live viruses, capable of causing disease if improperly used.

The poultry industry now has remarkable sophistication in handling mass populations. For example, it is possible to vaccinate 10,000 to 30,000 or more individuals by mass-method techniques (aerosol, drinking water, dust). A high and durable immunity is produced in most instances. If this procedure were adapted for other large populations (swine, cattle, or even humans) it would provide interesting results, and many ecological benefits. The possibility of producing geographic areas relatively immune to one or more devastating diseases has much to offer this overcrowded world.

Hatching of Chicks

Today, the hen is too valuable an egg machine to be allowed to waste many weeks of her life in hatching and brooding of chicks; and too fickle to be entrusted with this job on a mass scale.

The artificial incubation of eggs was practiced in Egypt long before the time of Aristotle (Figure 43.1). The immense Egyptian hatching ovens were constructed of sun-dried bricks and mortar covered with soil to form several chambers, each capable of holding several thousand eggs. The present-day recommendations for incubation of some species of poultry are given in Table 43.2. The recommendations are for forced-air incubators;

Figure 43-1 Native Egyptian hatcheries, according to Ghany *et al.* (1967), average over 60 per cent hatches and produce over 95 per cent of baby chicks in Egypt. These hatching ovens of mudbrick have been in use for many centuries.

Table 43.2.
Recommended dry- and wet-bulb temperatures for eggs of poultry species during (forced air) incubation. (See text for recommendation for hatching.)

	Incubation temperatures				
Species	*Dry bulb* °C (°F)	*Wet bulb* °C (°F)	*Incubation period (days)*	*Egg weights (g)*	*Mating ratio (M:F)*
Chicken	37.5 (99.5)	29.4–30.0 (85–86)	21	56–60	1:15
Turkey	37.5 (99.5)	30.0–30.6 (86–87)	28	75–100	1:10
Coturnix	37.5 (99.5)	30.0–30.6 (86–87)	17–18	9–11	1:2
Geese, except Canada	37.5 (99.5)	31.1–31.7 (88–89)	30	215	1:3
Ducks, except Muscovy	37.5 (99.5)	31.1–31.7 (88–89)	28	80–85	1:5
Chukar Partridge	37.5 (99.5)	30.0–30.6 (86–87)	23	22–23	1:2

in still-air incubators, the temperature should be higher when measured at the top of the egg. At hatching the dry-bulb temperature is reduced 1°F for chickens, turkeys, and coturnix eggs, but only 0.5°F for ducks and geese. The wet-bulb temperature at hatching is increased 3° or 4°F for chickens and turkeys, and 2° or 3°F for ducks and geese. For coturnix, no change is necessary.

Brooding and Rearing of Chicks

Parallel with the development of the mammoth incubator has been the evolution of brooders which supply warmth to the young. The source of fuel may be wood, coal, kerosene, gas, or electricity. Since then many modifications have been made to improve the performance of brooders. The transfer of heat can be by convection (hot air), conduction (hot water pipes), or radiation (heated slab or heat lamps). Proper insulation of buildings will reduce the cost of heating. The general principles of brooding applies to all poultry except pigeons, which feed and warm their young.

In climates where the room temperature does not fall below 21°C (about 70°F), no supplementary heat is necessary if provisions can be made to conserve the body heat as in feather-brooders. Otherwise enough heat is needed to maintain a temperature of 35°C (95°F) under the brooders for the first week. As the birds develop feathers, the need for extra heat decreases, and the temperature is decreased by 2.78°C (5°F) per week until they are three weeks old, when in most climates supplemental heat is no longer needed.

When brooded on litter, chicks are confined to an area around the brooders by chick guards about 30 cm. (12″) high and placed at a suitable distance around the brooding hover for the first few weeks. The litter on the floor is usually covered by paper for the first few days. When chicks are brooded on wire floors and wire pens, higher livability may be anticipated. Food and water is made available to the birds at all times (Figure 43.2).

Chicks are often debeaked as early as one

Figure 43-2 A deep-litter fryer house with mechanized feeding and watering devices in Lebanon. The gas brooders, used earlier, are on the right. The birds have the necessary space to grow. (FAO photo.)

week old to prevent feather picking or cannibalism. Debeaking is usually done with a hot-blade guillotine-type shearing and cauterizing device. The portion of the upper beak that is removed is about one-half the distance from the nares to the tip of the beak. After birds are debeaked, the depth of feed and water in the troughs may need to be increased.

Young chicks require a changing environment during the growing period, since the rates for production of heat and moisture increase as they grow older. In general, the following information is useful for brooding the young of all poultry. The relative humidity should be about 50 to 80 per cent. The ventilation rate is based on body weights. The rate varies from 0.03 to 0.06 m^3/min./kg. (0.5 to 1 ft^3/min./lb.) in winter to 0.12 m^3/min./kg. (2 ft^3/min./lb.) in summer. The floor space changes with age and season from 5.6 to 9.3 dm^2/bird (0.6 to 1.0 ft^2) and the space allotment under brooder from 45 to 58 cm^2/bird (7 to 9 $in.^2$). For wire floors of hardware cloth or welded wire, the spacing of the wire may vary from 0.5 to 2 inches, depending on the size of the bird.

Rearing of Chicken Replacement Flock

Chicks are generally removed from the brooder by the time they are about six weeks old, and may be reared in any suitable way in wire cages or in deep-litter houses, taking into consideration the disease prevention measures. Under intensive confinement conditions, as the birds grow, their need for more space increases.

The industry is becoming so specialized that even the raising of replacement pullets has become a specialty. The pullet-raising enterprise has certain advantages for the production of a uniform quality product and for disease prevention. An egg producer can replace all his old hens by buying ready-to-lay pullets, in order to make the most efficient use of his housing and equipment and to keep egg production on a high level all through the year.

Cages for Poultry

The majority of commercial layers are now kept in cages, notwithstanding that the system floundered in the 1920's in the U.S.A., partly because of poor ventilation. The ventilation problem was solved two decades later. The system was developed as a means of keeping the birds from fouling the feed and water with droppings, by using cages made of welded wire or hardware cloth.

With certain modifications, mainly in size, ducks, turkeys, and quail may be kept successfully in cages.

Houses for Poultry

The housing for poultry may be one of the following types: (a) open sheds, (b) semi-enclosed, or (c) enclosed houses with light and temperature control (LTC).

Open sheds are generally suited for birds to be reared as breeders, especially ducks, geese, turkeys, and game birds. The birds may be provided with some green forage or pasture in fenced range pens.

Semi-enclosed houses give more protection for the birds, and provide more environmental control. These houses are usually associated with the semiintensive method of poultry management.

The intensity of light required for poultry (about 1 foot-candle) is much less than is required for plant growth or for man's everyday activities such as reading. The recommendations for lighting regimens for chickens are given in Table 43.3.

The general concepts that were used in formulating these recommendations for chickens are as follows.

The physiological requirement for light by the birds is less than that generally needed by the caretaker; hence a compromise is needed.

During the rearing period, the birds should not be exposed to increasing amounts of light prior to 21 weeks of age if best egg production is to be attained. Short or decreasing daylength during the growing period is desirable. Long or increasing daylength (the reverse) is required for layers. The best laying-period light regimen is dependent on what the regimen was during the growing period. An increase in the light regimen in excess of 18 hours is of doubtful value.

A sudden decrease in the amount of light hours during the laying period for more than one week usually results in lowered egg production. This is especially true in cool weather. This reduction in egg numbers may be temporary and in another seven weeks unnoticeable.

Light regimens for broilers are much more simple than those for layers or breeders. Optimal growth can be expected from either a continuous light or with intermittent light at night to supplement daylight. Turkeys, pheasants, and chukars appear to have a short day requirement. Coturnix are unique, for they may be sexually mature by six weeks if exposed to daylengths in excess of 12 hours. This compares with 20 and 30 weeks for chickens and turkeys, respectively. Mature ducks may be brought into production in winter by providing a 14-hour day.

Other Environmental Factors

Under conditions of low ventilation in intensive-type housing, excess ammonia production from decomposition of waste from birds may be harmful. Chickens exposed continuously to 20 ppm (mg./liter) of ammonia for six weeks show gross or histopathological damage to

Table 43.3.
Recommended methods of lighting chickens.

Age of birds (weeks)	Minimum light (foot-candles)	Light-to-dark ratio (hours): Windowless houses, Method 1	Windowless houses, Method 2	Open house (Method 3)
Replacement pullets for egg production				
0–3	1.0	20:4 hours	20:4 hours	20:4 hours
3–12	0.1 to 0.5	Seasonal daylight	16:8 hours	Decreasing at rate of 15–30 minutes/wk or natural light for March–Sept. hatches.
12–21	0.5	Seasonal daylight, if decreasing; otherwise, short daylengths 6:18 or 8:16 hours.	Continue 16:8 to 16 weeks of age, then give 8:16 until 21 weeks.	Continue with simulated decreasing daylengths.
Commercial chicken layers				
21 or over	1.0	Seasonal light if increasing, or 16:8, 15:9 or 14:10 hours, depending on latitude	16:8 hours	Increasing light at rate of 15 minutes/wk until 18 hours then maintain.
Parent stock chicken breeders				
24–26 or over	1.0	As above.	16:8 hours	As above.

their respiratory tracts. Resistance to Newcastle disease virus is reduced by exposure to ammonia at this concentration. Air concentrations of 25 to 35 ppm ammonia will cause irritation to the eyes of most people. Concentrations of 105 ppm or even less ammonia result in lowered egg production and some eye injury (keratoconjunctivitis) in chickens.

Dusty conditions are increased by low humidity and have been associated with increased incidence of respiratory diseases. The problem is greatest in houses with litter floors. In litter houses as much as 1.8 kg (4 pounds) of dust per 1,000 birds per day may be collected on filters for air conditioners. Caged birds will produce about one-fourth this amount of dust. Poultry dust is composed of particles of litter, feed, feather, and skin debris, and is a means of transporting etiological agents for certain diseases.

Miscellaneous environmental factors that are known to influence poultry production include noise and space allotments. Sounds of 120-decibel intensity should be avoided if possible, because sudden noises frighten poultry and may cause mortality or injury due to stampeding. A number of the recommendations for space allotments are based on man's benefit rather than the birds. For example, the suggested requirement for feeder space using shallow feeders is related to the number of times the feeders would have to be refilled every day. For hoppers or self-filling feeders the space required is related to the geometry of the chicken. More birds can feed from a round hopper than from a linear trough. The area of floor space allotted to each bird in a littered floor house is related to the wetness of the litter. A crowded house or a house during times of high humidity will have a moist litter, which would reduce the chance of maintaining cleanliness of eggs for the operator.

However, poultry behavior under crowded conditions may result in feather picking, cannibalism, mortality, and reduced production.

Biological rhythms are entrained to certain environmental factors. The most obvious ones are feeding and time of oviposition which are linked to the onset of light and the onset of darkness, respectively.

Poultry laying houses have two systems of management, one with the birds on built-up litter or slatted floors and the other with birds held in cages (from one or two birds per cage up to 25 or 30 birds in wire-floor colony pens). The floor space allowed per layer is .05 to .06 m^2 (0.5 to 0.7 ft^2) in cages and 0.18 to 0.28 m^2 (2.0 to 3.0 ft^2) on the floor.

Nutritional Improvements

Poultry when growing very fast need a high level of dietary protein. As the growth levels off, the need for dietary protein decreases. The protein level of the diets of poultry depends upon the species, age, and reproduction state.

The daily feed consumption of poultry is influenced by the energy content and the other nutrients in the feed, age and body weight, the environmental temperature, and level of production of the species. The addition of fat increases the energy level of the diet and decreases feed intake by the birds. Unless an adjustment is made in the dietary protein level to keep a constant ratio of energy to protein, the reduction in total intake due to the addition of fat will depress the growth of young birds due to a decrease in protein intake.

Poultry diets look very similar and contain the same ingredients but in different proportions. When any alterations are desired because of relative changes in price of the ingredients, then preferably a cereal grain should replace another cereal grain, a protein source another protein source, while still keeping the amino acids in proper balanced ratios and retaining the necessary amounts of minerals and vitamins. In some underdeveloped countries, if the cereals are more expensive than the plant proteins, a relatively higher level of protein may be substituted for some of the cereal.

Poultry diets are formulated to meet certain standards based on information on the nutrient requirements of poultry. Most of this information has been developed in temperate climates and a close agreement exists on the approximate dietary protein levels. However, the requirements suggested for vitamins and amino acids differ considerably in the U.S.A. and the British publication of standards (Table 43.4). Based on these studies, educated guesses are used to formulate recommendations for poultry diets. A generalized formulation of poultry diets is given in Table 43.5.

Because of the small amounts required, the trace elements (minerals) and vitamins are usually added at the required levels with little consideration of the amounts present in the other ingredients. In some underdeveloped countries, it may be difficult to supplement all the vitamins at reasonable cost. In such places, one should consider the contributions of the rest of the ingredients and supplement those which tend to be deficient. The nutrient requirements can also be expressed in terms of units per 100 kilocalories metabolizable energy rather than total weight of the diet. This system is more helpful than percentages when adjustments in the composition of the diets are necessary due to differences in feed intake.

Some of the nutritional goals for other species of poultry are also given in Table 43.4. No definite information is yet available on the amino-acid requirements of ducks and geese, and the tabulated values are educated guesses. Species like ducks and geese can use some forages and water plants. These can be raised very easily under range conditions in the developing countries. The protein requirements of ducks appear to be, in general, about 17 per cent, lower than those of other poultry (chickens, 20 per cent).

Certain natural feedstuffs may be found to contain inhibitors or deleterious substances

Table 43.4.
Suggested levels of nutrients in diets of different aged poultry.

Nutrient	*Turkeys, 0–8 weeks*	*Coturnix, starting*	*Ducks, starting, 0–4 wks.*	*Geese, starting*	*Chickens* 0–8 weeks	*Chickens* Layers and breeders
Protein, %	28	25	20	20	20	15
Metabolizable energy, kcal/kg.	2,750	2,800	2,800	2,350	2,800	2,850
Lysine, %	1.5	1.3	0.7	0.9	1.2	0.5
Methionine plus cystine, %	0.9	0.8	0.6	0.6	0.8	0.55
Tryptophan, %	0.26	0.22	0.18	0.18	0.2	0.13
Calcium, %	1.2	1.2	0.8	1.0	1.0	2.75
Phosphorus, inorganic, %	0.6	0.6	0.4	0.4	0.6	0.6
NaCl, %	0.37	0.37	0.37	0.37	0.33	0.33
Mn, %	0.006	0.006	0.006	0.006	0.006	0.006
Zn, %	0.007	0.007	0.007	0.007	0.005	0.007
Vitamin supplements per kg.						
Vitamin A, I.U.	6,000	8,000	8,000	4,400	1,500	4,000
Vitamin D_3, I.C.U.	1,600	1,500	1,100	1,100	200	500
Vitamin E, mg.	10	20	5	5	10	—
Biotin, mg.	0.2	0.2	0.2	0.2	0.09	0.15
Choline, mg.	1,900	1,000	800	1,000	1,300	1,100
Folacin, mg.	1	1	1	1	0.55	0.35
Niacin, mg.	30	40	55	30	30	10
Pantothenic acid, mg.	15	10	15	4	10	10
Riboflavin, mg.	4	4	4	4	4	4
Thiamin, mg.	2	2	2	2	2	1
Vitamin B_{12}, mg.	0.004	0.004	0.004	0.002	0.009	0.003

that limit their use in poultry rations. The intermixing of certain feedstuffs to provide nutritional balance must be done carefully. The high requirement of the laying hen for calcium is well-known, and laying rations must contain at least 2.75 per cent calcium or supplemental oyster shell or limestone feed. Chicks, on the other hand, have a much lower requirement for calcium (1.0 per cent) which needs to be in balance with the content of phosphorus, manganese, and zinc. A deficiency of manganese or an excess of phosphorus may cause perosis (slipped tendon) in growing chicks and poults.

Guides for feed requirements and growth to specified ages for growing poultry are given in Table 43.6.

The egg producer is interested in a small hen that lays many large eggs. The feed requirements of chickens of different live weight as related to various levels of egg production are given in Figure 43.2. The feed efficiency is a measure of the conversion of grams of feed into grams of produce, such as eggs or meat. It is improved by the following: moderate high temperature, small inherited body size, adequate diet with high energy, and high rate of growth or egg production. The proper composition of the ration is of prime importance to get optimum performance.

Feeding System for Poultry

The system used for feeding of poultry is governed by economic and managemental considerations. The feeding may be done manually if the labor cost is reasonable, or be completely mechanized as it is in most industrially advanced countries under intensive management practices. It is practical and economically feasible to raise poultry on the range or under conditions of partial confinement in certain areas of the world. This system requires less out-of-pocket cost when plenty of good range is available. Some grain and minerals still have to be provided in addition to the food available on the range. Good range may not be possible the year round, and for this reason it is not the most reliable method of raising poultry. The alternate method is to provide balanced diets in a premixed form to confined birds.

The balanced diets can be used in the following forms: (a) mash and grain, (b) mash, (c) pellets, and (d) crumbles. The mash consists of a mixture of ground grains with protein, minerals, and vitamin concentrates. Mash is the base material for pellets and crumbles. The pelleting of diet increases the density and palatability, reduces dustiness and feed wastage, and increases efficiency of feed utilization in young poultry. In general, economic factors govern the choice of mash, crumbles, or pellets, and the processing of mash may be worthwhile under certain conditions.

It may be advantageous for a poultryman to purchase specially compounded premixed feeds containing a higher level of protein,

Table 43.5.
Generalized formulae patterns.[a]

	Chickens				
Class of feedstuffs	*Starter, 0–8 wks.*	*Grower, 8–maturity*	*Broiler, 0–6 wks.*	*Broiler, 6–10 wks.*	*Layer, breeder*
Proteins					
Vegetable: soybean meal, sesame cake, peanut cake, cottonseed cake, and other oil cakes.	30	16	22	22	15
Animal: fish meal, meat meal, etc.	5	5	10	5	3
Cereals and by-products					
High energy: corn (maize), sorghum, wheat, rice, rice polish, etc.	46	50	58	64	54
Low energy: barley, wheat bran, rice bran, etc.	10	18	—	—	17
Minerals: limestone, bone meal, phosphates, salt, manganese, and trace elements.	3	4	3	3	5
Vitamins and additives: synthetic vitamins, stabilized vitamin supplements, yeast, antibiotics, drugs, molasses, alfalfa grass, or other dried ground greens.	6	6	7	6	6

[a] Figures in each column show percentages for the classes of feedstuffs in making up the formula.

Table 43.6.

Guides for cumulative feed consumption (cum.) and growth of broilers, turkeys, ducks, and pheasants of different ages and sex.[a]

Age in weeks	*Broiler chickens*				*Turkeys (med. to large)*				*Ducks*				*Pheasants*			
	Male		*Female*		*Male*		*Female*		*Male*		*Female*		*Male*		*Female*	
	Cum. feed	*Body wt.*	*Cum. feed*	*Body wt.*	*Cum. feed*	*Body wt.*	*Cum. feed*	*Body wt.*	*Cum. feed*	*Body wt.*	*Cum. feed*	*Body wt.*	*Cum. feed*	*Body wt.*	*Cum. feed*	*Body wt.*
0	—	0.04	—	.04	0.054	—	—	0.054	—	0.05	—	0.05	—	0.02	—	0.02
1	.07	0.09	.07	0.09	0.09	0.12	0.09	0.12	0.16	0.18	0.16	0.18	0.06	0.04	0.06	0.04
3	0.57	0.36	0.55	0.35	0.55	0.43	0.5	0.36	2.0	1.02	2.0	1.02	0.29	0.14	0.29	0.14
5	1.41	0.77	1.3	0.7	1.55	0.86	1.32	0.73	5.2	2.01	5.2	2.0	0.64	0.27	0.59	0.25
7	2.7	1.32	2.36	1.11	3.05	1.41	2.55	1.23	9.2	3.14	9.2	2.96	1.3	0.45	1.1	0.42
9	4.32	1.91	3.59	1.55	5.5	2.23	5.2	1.95					1.9	0.66	1.6	0.54
12					11.36	3.73	9.55	3.18					3.1	0.96	2.8	0.75
14					15.59	5.0	12.73	4.1					4.1	1.15	3.6	0.86
16					19.91	6.36	15.5	5.0					5.1	1.30	4.6	0.90
18					24.0	8.95	22.27	6.73								
24					39.68	11.45	29.1	7.86								

[a] All figures are in kg.

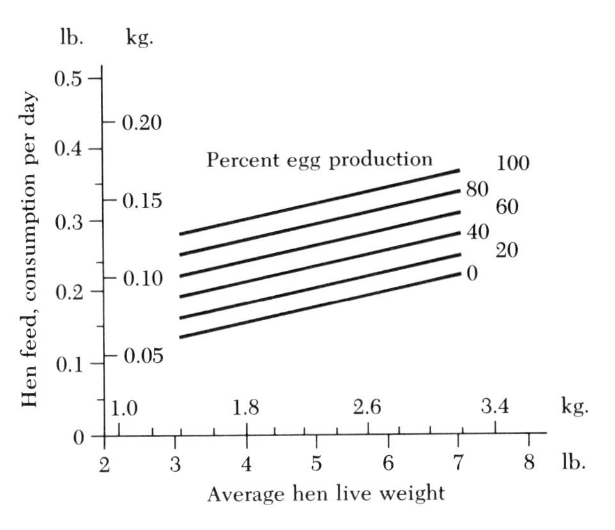

Figure 43-3 Daily feed requirement in pounds or kilograms of chicken hens of varying body weight and egg production rates. (Longhouse *et al.,* 1960). Developed from equation by Byerly (1941). Estimates are most accurate between 4.4–24°C (40–70°F). Above this temperature range, the feed requirements are overestimated and below are underestimated from calorimetric data.

minerals, and vitamins, which he could dilute with some locally available ingredient such as cereal grains (ground or whole), and still feed a well-balanced diet.

Restricted Feeding

It is possible to keep a parent stock of meat birds from getting overweight and at the same time improve their egg production by diet control. To achieve this purpose, the intake of nutrients is restricted by such practices as skipping a day's feeding each week, limiting the amount fed daily, or increasing the fiber content of the diet.

Broiler chickens can be given a high-protein diet during the initial period of rapid growth, and then the level of protein can be reduced as the rate of growth decreases. Feeding based on the stage of egg production (phase feeding) can be economical in the hands of an experienced poultryman.

Poultry Manure

The disposal and utilization of manures from barnyard poultry flocks formerly was not a problem because it was scattered or diluted and constantly subjected to microbial action and drying. A laying bird each day will consume about 115 g (0.25 lb.) of feed and void about 160 g (0.35 lb.) of fresh droppings. On this basis, ten layers provide about as much waste as one human being.

With the disappearance of barnyard flocks and the concentration of poultry operations into units containing more than 100,000 layers, the handling of manure is becoming a problem, especially when one realizes that this size of operation creates as much waste per day for disposal as the sewage from a town of 10,000 people.

The owner of a large poultry unit often does not have enough land to utilize this amount of manure. Poultry ranches are generally not far away from human populations, and therefore complaint of flies and odors is common. Dry manure is an excellent source of nitrogen, but storage of wet manure for as few as five days may cause about 85 per cent of the nitrogen to be lost. The final composition and the value of the manure depends upon the management practices. The manure from caged layers is more liquid and problematic than that from deep litter houses, where it is more dry through dilution with litter material (30–40 per cent moisture).

The methods which are most suitable for

the disposal and utilization of poultry manure in an area will depend on the labor costs, the state of local technology, the pressure from surrounding residents and conservation groups, and the relative economic value of manure as compared to the synthetic fertilizers.

At the present time, the following methods are used: anaerobic decomposition, city sewers, compositing, dehydration, feed supplement for ruminants, incineration, lagoons, and spread on land as fertilizer. In some countries recycling exists as another possibility in the form of feeding poultry litter to ruminants, but consideration must be given to the possibility that drug residues in the litter may contaminate the meat of animals consuming the litter.

FURTHER READINGS

Ewing, W. R. 1963. *Poultry Nutrition*. 5th ed. Pasadena, Calif.: Ray Ewing.

Ghany, M. A., M. A. Kheir-Eldin, and W. W. Rizk. 1967. "The native Egyptian hatchery: structure and operation." *World's Poultry Sci. J.*, 23:336–345.

Ministry of Agriculture, Fisheries, and Food. 1964. *The Rearing of Chickens*. Bull. no. 154. London: H.M.S.O.

NRC. 1971. *Nutrient Requirements of Domestic Animals, No. 1: Nutrient Requirements of Poultry*. 6th ed. Pub. no. 1345. Washington, D.C.: NRC.

Norris, L. C. 1958. "The significant advances of the past fifty years in poultry nutrition." *Poultry Sci.*, 37:256–274.

Romanoff, A. L., and A. J. Romanoff. 1967. *Biochemistry of the Avian Embryo*. New York: Wiley.

Termohlen, W. D. 1968. "Past history and future developments." *Poultry Sci.*, 47:6–22.

Wood-Gush, D. G. M. 1959. "A history of the domestic chicken from antiquity to the 12th century." *Poultry Sci.*, 39:321–326.

Forty-Four

Management and Disease

"Less thick and fast the whirlwind scours the main with tempest in its wake, than swarm the plagues of cattle; Nor seize they single lives alone, but sudden clear whole feeding grounds, the flock with all its promise, and extirpate the breed."

Virgil
Georgics

Disease prevention is a *prerequisite* to, as well as an essential part of good livestock-management practice. Efforts to improve animal production, by genetic upgrading or other means, are likely to end in failure unless they are carried out in animal populations which enjoy at least minimal standards of good health, and from which major disease threats to survival have been eliminated or brought under reasonable control.

Many aspects of management may be important parts of primary disease prevention and control efforts. Properly designed breeding programs, for example, may be the key to the prevention or reduction not only of diseases which are actually genetic in origin, but also of other diseases for which some lines of animals may show either some genetically determined predisposition or resistance.

Nutrition is another aspect of management closely related to primary programs for the promotion of animal health. Not only are the more gross forms of malnutrition an important class of diseases in themselves, but, more commonly, less obvious or even marginal nutritional deficiencies may alter an animal's resistance to diseases of other origins. Conversely, many infectious diseases cause an animal to eat less or stop eating entirely, and thus they may precipitate complicating nutritional imbalances.

Diseases of livestock are among the most serious impediments to a significant worldwide expansion in production of human foods and other useful products of animal origin, and primary programs for disease prevention and control* should not be undertaken in isolation, but should be integrated as closely

* Disease *prevention* means all measures to exclude disease from an *unaffected* population of animals; *control* means measures to reduce the level of existing disease, ultimately to a level of little or no consequence. *Eradication* is the complete elimination of an existing *infectious* disease (and is a term most aptly applied to entire countries or to relatively large regions).

as possible in general programs for improvement of all aspects of management.

What Causes Disease?

Disease is simply a state of ill health, of less than optimum vitality or efficiency. It may result from *any* structural or biochemical abnormality or injury which reflects itself in the malfunctioning of an animal's body. If the malfunction is severe enough, the animal may die. If it is less severe, the animal becomes unthrifty, with a lowering of its full genetic potential for production. Thus, the relative state of *health* of individual animals may be measurable in such familiar terms as their milk production, their growth rates, their efficiencies of feed conversion, or their rates of producing offspring. These types of health parameters are often very sensitive ones, and they are ones which any livestock producer can readily observe. Sometimes, however, disease or prodromal states leading to disease may only be detected, or their seriousness estimated, by less obvious signs, symptoms, or other bodily changes, recognition of which may require specially trained and conditioned senses on the part of the observer, or even special amplifying instruments and highly sophisticated tests.

The most important questions to be answered initially when an individual animal's body malfunctions, or is diseased, are: (1) What is the nature of the disease? and (2) What has caused it? The first of these questions concerns clinical and laboratory *diagnosis,* which is a technical aspect of veterinary medicine outside the scope of this book. The second question of disease *causes* is one, however, which should be understood by every livestock producer, for it is necessary to understand the causes of specific diseases in order to be able to take rational steps to prevent their repeated occurrence or their possible spread.

An Ecological View

Even though it is common practice to speak of a certain disease as being "caused" by some specific *agent*—for example, by a particular virus or some particular poison—the real *causes* or determinants of diseases in livestock are rarely quite this simple. Usually the presence of such a specific agent is only a part of a more complicated picture in which certain host factors or environmental conditions, or both, may also play important "causal" roles (Figure 44.1).

What this means is that, if certain diseases are to occur in a herd or flock, it is often necessary that some specific living or nonliving agent be present in or upon the bodies of the affected animals. These possible agents of disease are varied. In one instance it might be a virus or a bacterium, in another a type of worm or an insect larva. It might also be some organic or inorganic chemical, or even some physical agent, such as excessive heat or cold or a defective milking machine. Whatever it may be, the *presence* of this specific agent in the immediate surroundings of an animal, or in or on its body, depends on some particular combination of circumstances. Moreover, for the disease process to actually *manifest* itself, the agent may have to occur in or act upon animals in a certain age group, or of a particular sex or breed or genetic predisposition, or only in animals in a certain physiological state.

Now let us take a brief look at a few diseases in order to see something of the complicated variety of such causal interrelationships.

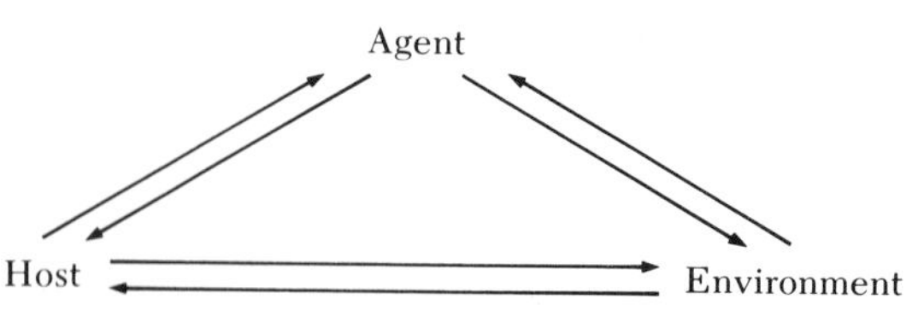

Figure 44-1 Livestock diseases are caused by various types of interactions between agent, host, and environmental variables.

Pendulous crop Pendulous crop is a disease of turkeys. It was first described in birds raised in the Central Valley of California. About one-third of the affected birds die. Those who observed the disease noted that it occurred only during very hot, dry summer months and only in turkeys of the Bronze breed. Turkeys of the Bourbon Red breed did not develop pendulous crop even when raised under equally severe environmental conditions, nor did Bronze turkeys develop it when raised in a cooler summer climate, such as that on the sea coast of California.

The *agent* of pendulous crop could simply be regarded as water, the excessive consumption of which "causes" an irreversible stretching of the crop wall of Bronze turkeys, with resultant food stasis, fermentation, and toxin formation in that organ. However, a real understanding of the determinants of this disease, and its prevention, must also take into account a genetic predisposition in one particular breed of turkeys, which may manifest itself as a disease under one set of environmental circumstances, yet fail to reveal itself as such in another (Figure 44.2).

Shipping fever Shipping fever is the name given to a clinical syndrome which is most frequently seen in beef cattle in North America. It usually occurs epidemically and manifests itself as an acute debilitating bronchopneumonia with a mortality rate of about 5 per cent.

Bacteria of the genus *Pasteurella* can be isolated from the nasal membranes of about 60 per cent of affected cattle (and about 3 per cent of normal cattle). A specific parainfluenza virus may also be isolated from many affected cattle, and serological surveys of slaughter cattle indicate that about 70 per cent of them have been infected with this particular virus sometime during their lives.

Shipping fever almost invariably follows some stressful episode, most frequently the shipping of animals to a new location (e.g., combined effects of fear, crowding, change in feed, inadequacy of feed or water, exhaustion, and exposure to cold) or their weaning (e.g., change in feed, dehorning, castration, branding, and fear).

Pasteurellae alone or in combination with stress will not produce this clinical disease in experimental cattle, and inoculation of parainfluenza-3 virus alone results only in a mild respiratory disease. However, the two agents given together to animals which are stressed will produce a disease identical to the naturally occurring one.

The most plausible current explanation for the "cause" of shipping fever requires a stressful experience (reflecting environmental-management factors) which lowers the resistance of the animal to infection, partially by stimulating the adrenal glands to produce an excess of adreno-cortical hormone. This hormone then depresses the cellular systems which normally produce antibodies and cells protective against infectious agents. As a result, the parainfluenza virus may damage the lining of the respiratory tract, thus providing conditions favorable to multiplication of the *Pasteurella* bacterium and resultant severe pneumonic lesions and symptoms.

The "cause" of epidemic losses in the calf scours syndrome is probably also very similar to this one.

Carcinoma of the eye Cancer of the eye in cattle occurs far more commonly in Hereford cattle than in those of any other breed. Most of the animals which are affected lack pigmentation of the eyelids and associated structures. These patterns of pigmentation are genetically controlled. Other factors which have been associated with the development of this disease are chronic exposure to environmental influences, such as intense sunlight and dust.

These few examples should be adequate to make the point that the determinants or causes of diseases are usually fairly complex. The key to prevention or control of any particular

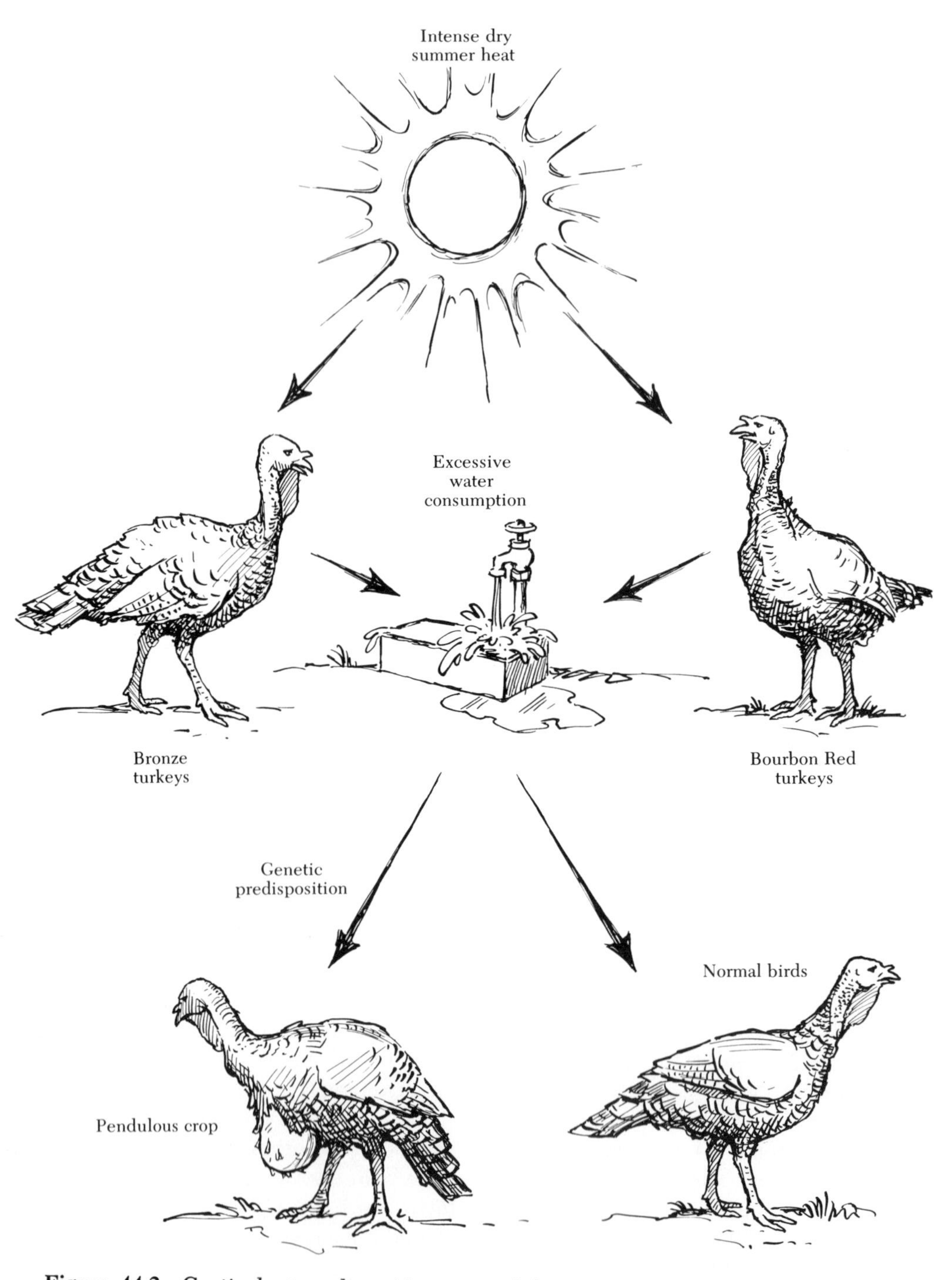

Figure 44-2 Gentic *host* predisposition to pendulous crop occurs in Bronze turkeys. They develop the disease, however, only when subjected to a severe *environmental* stress.

disease may reside in *any* of its contributing factors or circumstances.

A Word About Infections and their Manner of Spread

One of the most important groups of livestock diseases is that associated with infectious agents. Infection merely means that the body of a *host* animal has been invaded by some other living organism. There are many kinds of invading organisms, or infectious agents. They include viruses, certain bacteria, various types of organisms intermediate between viruses and bacteria (e.g. rickettsiae and mycoplasms), certain molds and related fungi, various protozoa, worms or helminths of many different kinds, and certain insect larvae (Table 44.1).

Some of these agents (e.g., viruses, bacteria, protozoa) are capable of multiplying within the bodies of their hosts; other invaders merely undergo a period of development there (e.g., certain of the helminths). Invasion of a susceptible animal's body is an obligatory part of the life cycle for most infectious agents, while for some (e.g., certain molds) it is only an optional feature. In fact, some infectious agents can survive only briefly outside of an animal's body (e.g., many viruses), whereas others (e.g., the spores of the anthrax bacterium or the eggs of certain helminths) may be able to survive in a free-living state for long periods of time. Whether an agent is an obligatory invader, how well it survives outside of an animal's body, and how it passes from one infected animal to another susceptible animal, are all important questions bearing on the prevention or control of the disease with which that agent may be associated.

Infection without disease Infection, however, does not necessarily imply that a state of disease actually exists, because some infectious agents may live within the bodies of their hosts without producing symptoms or signs of illness. We speak of these infections as being *subclinical* or *inapparent*.

Some infections may be inapparent throughout their entire course or even throughout the entire life of the host animal. For example, a rickettsia called *Coxiella burnetii* readily infects both cattle and sheep and is transmissible from one animal to another. These infected hosts, however, do not develop any symptoms or show other outward evidence of the infection. This infection is a very important one, nonetheless, because it can also be transmitted from sheep or cattle to man. And in people this agent *does* produce a serious disease which is called Q fever.

As with the shipping-fever complex, some other inapparent infections are what is called *latent;* that is, they may precipitate a disease process at any time the host animal's general resistance is appreciably lowered through *any* cause. The latent infection of most people with the herpes simplex virus which sometimes "causes" cold sores is a good example of this type of relationship. Latent infections are particularly insidious.

Inapparent infections, in general, are a very important complicating factor in the prevention and control of many livestock diseases, particularly when the infectious agent is capable of being spread from an inapparently infected animal to other susceptible animals in the herd or flock. Such "silent" transmitters of infectious agents are called *carriers*. The carrier state may only occur during certain stages of the infection or only in certain individual animals.

How infections are transmitted Different infectious agents may be transmitted from one animal to another in a variety of ways. Some require *direct contact* and can only pass from an infected host to a susceptible one when the animals actually touch one another. Venereal infections are in this class. Other infections are transmitted by means of *indirect contact,* as when animals touch one anothers' excretions or secretions, or such things as their placentae

Table 44.1.
Principal types of infectious agents of disease.

Type of agent	*Properties*	*Some important livestock infections with which associated*
Viruses	Tiny acellular organisms, most of which are visible only with the electron microscope; obligate intracellular parasites; generally not susceptible to antibiotics or other chemotherapeutic agents.	Hog cholera, vesicular exanthema, vesicular stomatitis, foot-and-mouth disease, rinderpest, rabies, scrapie, encephalomyelitides, bluetongue, and Newcastle disease.
Bedsoniae (psittacoid organisms; chlamydiae).	Microscopic organisms, sometimes considered as large viruses; obligate intracellular parasites, but can be grown outside of cells; selectively susceptible to antibiotics and other chemotherapeutic agents.	Psittacosis, ornithosis, sporadic bovine encephalomyelitis, polyarthritis.
Rickettsiae	Microscopic organisms intermediate between viruses and bacteria, which they resemble; obligate intracellular parasites; produce endotoxins; usually transmitted by arthropod vectors (one species is known to be transmitted by a helminth); selectively susceptible to antibiotics.	Q fever.
Mycoplasmas (PPLO organisms)	Microscopic organisms intermediate between viruses and bacteria; multiply within and outside of cells; most are obligate parasites; susceptible to relatively few antibiotics.	Bovine and goat pleuropneumonias, infectious synovitis and airsacculitis of turkeys, respiratory infections of chickens.
Bacteria	Microscopic, unicellular, plant-like organisms; spherical, rod-shaped, curved, or spiral; many are free-living, others are facultative or obligate parasites; some produce highly resistant spores; some produce ecto- or endotoxins; occur intracellularly or extracellularly; except for spores, are generally susceptible to wide range of germicides and disinfectants; selectively susceptible to chemotherapeutic agents, such as sulfonamides and antibiotics.	Anthrax, clostridial infections and intoxications, erysipelas, salmonellosis, coliform infections, vibriosis, listeriosis, pasteurelloses, infectious keratitis, tuberculosis, pseudotuberculosis, Johne's disease, brucellosis, leptospirosis, mastitis complex.
Mold-like bacteria (actinomycetes)	Microscopic, filamentous organisms which may fragment into spherical or rod-like forms resembling bacteria; facultative and obligate parasites; selective susceptibility to antibiotics and other chemotherapeutic agents.	Actinomycosis, nocardiosis.

Table 44.1. (Continued)

Type of agent	*Properties*	*Some important livestock infections with which associated*
Yeastlike fungi	Unicellular fungal organisms; reproduce by budding; facultative parasites with free-living cycles; selectively susceptible to chemotherapeutic agents.	Cryptococcosis, blastomycoses, candidiasis.
Molds	Fungal organisms, usually multicellular, but some are unicellular and yeastlike in tissues; cells nucleated; generally reproduce sexually and asexually; many produce resistant spores; most are facultative parasites with free-living cycles; selectively susceptible to chemotherapeutic agents.	Ringworm and other dermatophytoses, aspergillosis, coccidioidomycosis, histoplasmosis.
Protozoa	Unicellular animal organisms; various shapes and means of locomotion; many are free-living, others are obligate parasites during at least a portion of their life cycles; occur intracellularly and extracellularly; some produce resistant cysts; many are transmitted by arthropods; selectively susceptible to chemotherpeutic agents.	Coccidiosis, histomoniasis, trypanosomiases, piroplasmosis, anaplasmosis, toxoplasmosis.
Cestodes (tapeworms)	Multicellular elongate, bandlike worms; adults, a few millimeters to several meters long, consist of segments and head (with suckers and sometimes hooks); sexes not separate; produce resistant eggs; all adults are obligate parasites of intestinal tract, or its tributaries, and larvae of vertebrate or invertebrate tissues; some larvae produce cyst-like bladderworms; adults selectively susceptible to chemotherapeutic agents; therapy of larval infections generally unknown.	Cysticercosis, gid, hydatid disease, many types of intestinal infections.
Trematodes (flukes)	Multicellular worms, often ovoid or lancet-shaped; adults few millimeters to about 1 1/2 inches long; nonsegmented; sucker on anterior end; sexes not separate; adults found in many organs; produce eggs; complicated larval cycle; transmitted by snails; most adults selectively susceptible to chemotherapeutic agents.	Fascioliasis, schistosomiasis, a number of others.

(continued)

Table 44.1. (Continued)

Type of agent	*Properties*	*Some important livestock infections with which associated*
Nematodes (roundworms)	Multicellular, long, cylindrical worms; separate males and females; adults, a few millimeters to many inches long, are nonsegmented; parasitize many organs; many produce resistant eggs; many are totally free-living or plant parasites, many others are parasites of animals; many produce free-living larval stages; some are transmitted by arthropods; selectively susceptible to chemotherapeutic agents.	Hookworm disease, mixed trichostrongyle infections, lungworm diseases, ascariasis, strongyle infections, pinworm infections, many others.
Insect larvae	The larvae or maggots of certain flies are facultative or obligate parasites.	Screwworm infection, ox warbles, other myiases.

or aborted fetuses. Indirect contact also includes transmission of an agent within relatively confined surroundings by exhaled droplets from the nose or mouth of an infected animal or by airborne droplet nuclei. Sometimes airborne transmission of infectious agents even takes place over relatively great distances.

Some infectious agents are also transmitted by *vehicles,* which are various nonliving substances, such as food or water, or contaminated articles, such as unclean syringes or common cloths used to wash cows' udders. Gastrointestinal infections, such as salmonellosis, and bovine mastitis are transmitted in these ways.

The fourth principal mode of transmission of infections is by *vectors,* that is, by living invertebrate animals, such as insects, mites, ticks, or snails. In some instances, this may merely be the mechanical transfer of the agent by the vector, for example, by a biting fly, which acts something like a "flying needle"; in many other instances a period of development and/or multiplication in the vector is actually required by the infectious agent before it may be passed on to a new host. For both vehicle and vector transmission, it is not necessary that infected and susceptible animals be near one another for transmission to take place.

The rapidity of spread of different infections, their seasonal or geographical distributions, and the relative ease or difficulty of their prevention or control may all depend, at least in part, upon their possible means of transmission.

Livestock diseases may also be very *rapidly* spread by the transport of livestock through a succession of public markets and stockyards. A remarkable instance of rapid spread by this means was provided in the case of vesicular exanthema, a viral infection of swine which was first reported in California in 1932. For the next 20 years it was not diagnosed outside of that state (California being an importer rather than an exporter of swine). Then, on June 11, 1952, vesicular exanthema was discovered in a herd of swine in Grand Island, Nebraska. This outbreak was traced to the feeding of uncooked garbage containing pork scraps obtained from a California-to-Chicago dining car. Within one month this disease had spread to 13 states: California, Nebraska, Wyoming, Utah, Washington, Oregon, Idaho, Arizona, New Mexico, Texas, South Dakota, Alabama, and New Jersey! One year later vesicular exanthema was present in 42 states and the District of Columbia. Fortunately, this serious disease of swine has since been completely eradicated from the United States.

The language of epidemiology A last point we must mention about how disease spreads concerns the resulting patterns of occurrence of diseases in different herds or flocks. In carrying out an over-all program of disease prevention, it is important that as accurate a record of disease occurrence and death be kept as is possible. These records enable the veterinarian to define the disease experience of the particular herd or population, and to accurately observe and describe significant changes in the disease patterns. Among the terms used for these purposes are *endemic,* which means a pattern of disease in which there is little fluctuation in the rate of occurrence of new cases of this disease over relatively long periods of time. A disease may be *endemic* at high, low, or medium levels.

The *sporadic* occurrence of disease means, on the other hand, that cases occur only occasionally and irregularly, and not at any more or less constant level. Lastly, the *epidemic* occurrence of disease means the development of new cases at a rate "clearly in excess of that expected." Medically speaking, epidemic does not therefore necessarily mean a *large* number of cases of a disease. It is a purely relative term implying an unusually high incidence of disease. An *outbreak* is an approximate synonym for an epidemic.

Disease as a Barrier to Increased Production

The literally hundreds of infectious and noninfectious diseases which affect livestock are major barriers in every country to increasing present levels of animal production. In the United States, some of the major infectious plagues of livestock have been eradicated completely or brought under effective control by past efforts on a national scale. The prosperous animal agriculture which we now enjoy would not have been at all possible had such livestock diseases as Texas fever (bovine piroplasmosis), foot-and-mouth disease, hog cholera, bovine pleuropneumonia, fowl plague, Newcastle disease, and scabies remained unstudied and been permitted to rage unchecked, as they once did, in the United States.

Fights Against Disease Yet to Be Won in America

If we cannot afford complacency about animal plagues not now of immediate importance to us, neither can we afford it about the many battles against animal diseases in the United States which we have not yet won. Each year in this country about 1.5 million cattle and at least 2.25 million calves still die on our farms. Our current annual mastitis losses in dairy cattle have been carefully estimated at about $70 per cow. Diseases also cost us a 20 per cent loss in the total value of our swine production and a 15 per cent loss in our poultry production each year in the United States. As the result of these still unchecked animal diseases the total cost to the American farmer and the national economy is probably more than $4 billion annually.

The Global Importance of Livestock Diseases

In a number of areas of the world which are potentially productive for animal agriculture, little or no development of modern livestock industries has yet occurred. Frequently this is because of the presence in these areas of major endemic and epidemic diseases, which, either by making animals unthrifty or by causing their deaths, make developments in animal agriculture uneconomical or impossible.

Little development of modern animal agriculture has occurred, for instance, in most of the vast grasslands areas of sub-Saharan Africa, some 4.5 million square miles of which are potentially capable of supporting immense livestock populations. The political analyst, Eric Sevareid, admonished the new leaders of Africa to "make room in [their] social pecking order for scientific farmers . . . [and] veterinarians [for they] need them badly." What can be done in animal agriculture in Africa has been dramatically illustrated in the far south of that continent, where in the late nineteenth and early twentieth centuries Sir Arnold Theiler and other veterinary pioneers began to set up innovative programs for livestock disease research and control which have since received world acclaim.

In an animal-disease control and management program being carried out in Colombia by the Rockefeller Foundation, after only 2 years effort the beef-calving rate in one area was increased 62 per cent, the death losses of calves from birth to weaning were reduced from 25 per cent to less than 5 per cent.

The threat of foreign diseases There are important reasons why animal diseases in distant places must interest us today. The increasing rapidity of international travel strains more and more the quarantines and other defenses which our official veterinary services have established to keep foreign or exotic diseases out of the United States. As a consequence, at no time in our history has our livestock population been more vulnerable to the risk of potentially catastrophic introductions of new diseases. The legal, and sometimes illegal, movement from one corner of the world to another of domestic and wild animals, of animal products, of semen, and of animal caretakers may now be literally a matter of hours.

Among the major livestock plagues which could enter the United States from abroad at any time are foot-and-mouth disease, rinderpest, African swine fever, fowl plague, contagious pleuropneumonia, African horse sickness, East Coast fever, and bovine piroplasmosis. We already have had experience with several of them. For instance, foot-and-mouth disease, which affects cattle, swine, and other species, including human beings, entered the U.S. nine times between 1870 and 1952. In each instance it was stamped out, although, in 1914, it had spread into 3,556 herds in 22 states and the District of Columbia before it was eventually eradicated. That immense effort required the slaughter of 172,000 infected and exposed animals.

Although a period of quarantine is required of most animals imported into the United States, the present danger is suggested by the fact that the tick which transmits East Coast fever, one of the most important cattle diseases in Africa, and which is also a vector of bovine piroplasmosis and several other important diseases, entered this country in 1960 and passed through quarantine on a shipment of zebras destined for wild animal gardens in Florida and New York. In a dramatic and successful multistate effort, federal and state veterinary authorities were mobilized, and they stamped out this invader before it could establish itself.

Despite such present ability to respond quickly to disease invaders and their vectors, important livestock infections, such as bluetongue and scrapie, have successfully entered the United States and established themselves in the recent past. *Dermacentor nitens,* a tick which is the vector of equine piroplasmosis, also became established in Florida in 1960, and piroplasmosis was first diagnosed in Florida horses less than a year later.

How Animal Diseases Also Affect Man's Health

Diseases of domestic animals are of much more than economic consequence to us, because they may directly affect the health of people in several ways other than by simply causing a reduction in potentially available supplies of high-quality protein foods.

Infections Shared by Animals and People

There are approximately 250 recognized infectious diseases of man. A vast majority of them, some 200, are shared by various other vertebrate animals. We refer to this whole group of shared infections as *zoonoses*. The zoonoses include such important human infections as rabies, tuberculosis, African sleeping sickness, cysticercosis, brucellosis, and trichinosis, as well as several types of hemorrhagic fevers and encephalitides. Many zoonotic infections are transmitted to people through meat, milk, or eggs of animals; others are transmissible only through some form of close contact with an animal or its excreta; still others may only be carried from animals to a man by insect, tick, mite, or snail vectors. The livestock producer and his family are in particular danger from some of these diseases.

It is important that we realize that *many zoonotic diseases cannot be prevented or controlled in people except through their control in animals.* Until 1947, for example, over 6,000 cases of human brucellosis (undulant fever) were diagnosed in the United States every year. Most of the individuals who suffered from this painful and recurring disease had acquired their infections directly from infected cattle or by drinking milk from infected cows. At that time approximately 10 per cent of adult cattle in the United States had this disease. Now, because of a tremendous cooperative effort by veterinarians and producers (Figures 44.3, 44.4), only about 200 new cases of this disease are diagnosed in people each year. Most of these few remaining infections are acquired from swine.

Research Leads

Finally, animal diseases may influence man's health when they become subjects for research. Virtually all diseases of people have one or more counterparts in spontaneously occurring diseases of livestock and other animals. Research on animal diseases may thus not only directly benefit the economy through improved animal health, but also have other far-reaching implications for human health. For example, research on the serious and baffling disease of sheep called *scrapie*, now known to be an infectious disease for which there is probably some form of genetic predisposition, led directly to discovery of the cause of *kuru*, a previously mysterious and fatal disease of people in New Guinea.

Victory over tuberculosis Something of the broad over-all impact of veterinary medicine and animal disease on human health can be appreciated in a few illustrations from the history of a single disease, tuberculosis. At the beginning of the century, tuberculosis was the

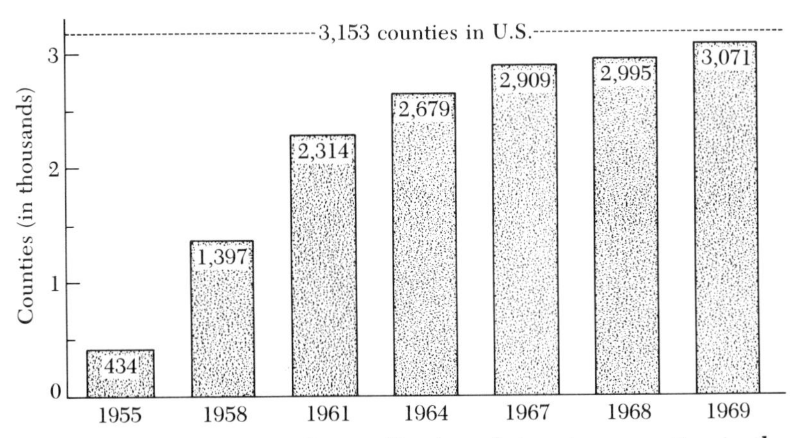

Figure 44-3 Progress in the certification of American counties in the cooperative state–federal brucellosis eradication program for cattle.

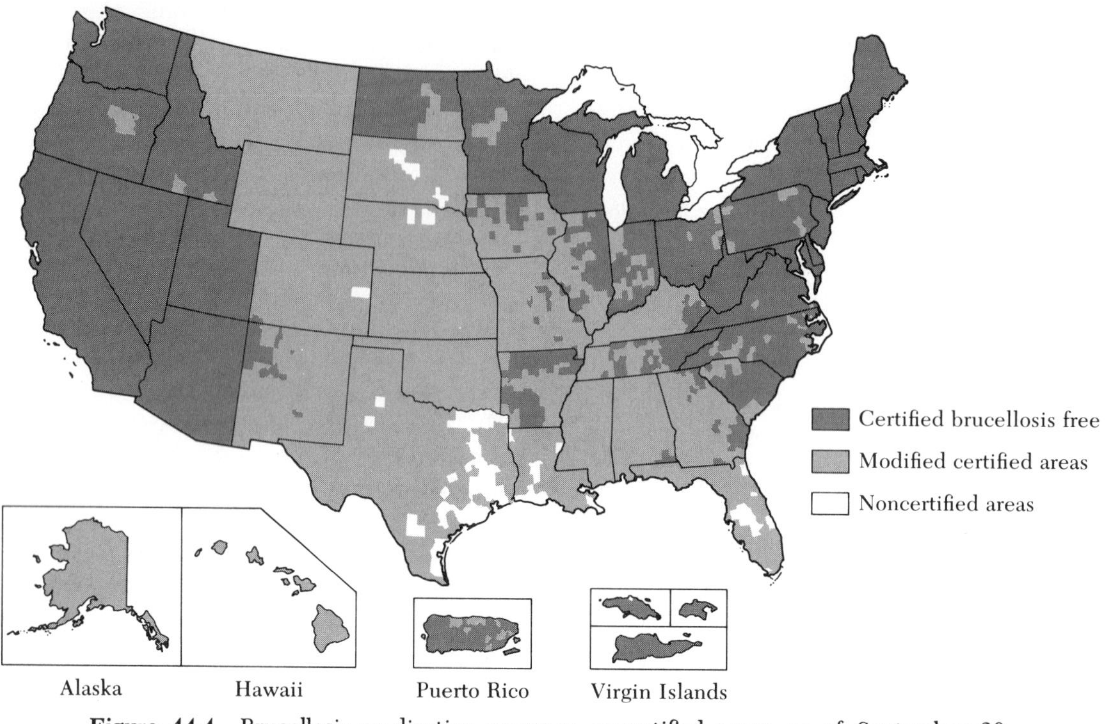

Figure 44-4 Brucellosis eradication program, uncertified areas, as of September 30, 1969. (Courtesy USDA.)

leading killer of people in the United States. The bovine form of this infection once accounted for up to 10 per cent of pulmonary tuberculosis in man, as well as for almost all tuberculosis of the bones and joints, the lymph glands, the digestive system, and all other organs except the lungs. Today, non-pulmonary tuberculosis, which caused so much human death and deformity, has virtually disappeared in this country. A major reason for this remarkable change has been the near-elimination by veterinarians of tuberculosis from cattle in the United States.

Basic to all direct efforts to control tuberculosis in man has been the availability of a simple test to apply to people. This is the tuberculin test, first developed by the Danish veterinarian Bernard Bang as a test for tuberculosis in cattle and now also used as the basis for its control in people.

In addition to tuberculin, the worldwide effort of the World Health Organization against tuberculosis in man depends also on a tuberculosis vaccine, BCG. This vaccine was developed by a French physician-veterinarian team from the bovine-type organism for use in cattle. Ironically, BCG is of no value as a cattle vaccine, but it has protected almost countless millions of people.

A final example of how veterinary medicine has helped reduce the toll of tuberculosis in people concerns the treatment of the disease. Up until a few years ago, there were no drugs available to treat human tuberculosis, and patients were often subjected to prolonged sanitarium rest and radical surgery, both often unsuccessful. That picture is now completely changed, with many tubercular patients effectively treated with drugs at home. This remarkable advance owes much to the research of William Feldman, a veterinary pathologist, who first showed in animal experiments that drugs could be effective in the treatment of the disease.

Today's Veterinary Medicine

Veterinary medicine contributes to societal needs in many ways and not solely through its vital and important role as a protector of animal agriculture. It is concerned with the diseases and health of all animals other than man. Veterinarians thus engage in a wide variety of activities, most of them related to the study, alleviation, prevention, and control of animal diseases, or concerned with the maintenance of the health of animals. Zoologists, wildlife biologists, parasitologists, animal husbandmen, microbiologists, psychologists, and physicians also contribute to over-all veterinary knowledge, just as many veterinarians work with, and contribute directly to knowledge of, diseases of man or other aspects of biology.

Keeping Livestock Healthy

The most important charge of the veterinarian is keeping livestock healthy. This is an activity which *requires* an active partnership between the veterinarian and the livestock producer. These efforts are frequently difficult ones, because health and disease prevention, as factors influencing livestock production and management, are perhaps the most difficult to visualize in conventional cost-benefit terms. *This is partially because many farmers have become accustomed to "living with" a certain level of disease, and therefore do not consider the possibility that new efforts for prevention might reduce what they may falsely regard as "fixed" operational costs.* It is also partially because the benefits of preventive efforts appear to manifest themselves in negative terms —the *absence* of disease—which are sometimes psychologically difficult to grasp.

Most infections of livestock which still exist in the United States do so because a reservoir* of infection is being perpetuated in some livestock species, or perhaps in some species of wildlife.

Since the real causes of a disease are often complex, the objective of all preventive and control programs is to identify the weakest links in, or the most vulnerable points for an attack against, the disease in question. Whether an infectious disease can be effectively prevented or controlled, and how readily, depends in part on its epidemiological complexity. For example, if a particular infection is limited to swine and is transmissible from infected swine to susceptible swine only by direct contact or by exhaled droplets, then the conceivable points of attack are practically limited to what can be done to swine themselves or to their surroundings. However, if we are dealing with a mosquito-transmitted infection of swine, an attack on the disease could be made *either* through the swine population *or* through the mosquito population.

Disease Prevention as Part of Good Management in Our Changing Animal Agriculture

Until comparatively recently, much of the animal agriculture in the United States was in the hands of a relatively large number of small-scale private producers. Changing patterns of livestock ownership and management would now favor the concentration of larger numbers of animals in fewer hands, the development of more intensive and mechanized

* An infection's *reservoir* host is the species, or group of species, in which the infection is biologically maintained in nature and upon which that particular infectious agent depends for survival. For infections of livestock, this host could be the particular livestock species with which we are directly concerned (e.g., swine are the reservoir for hog cholera), or some different domestic animal species, or wildlife (e.g., foxes, skunks or bats for cattle rabies), or even man (e.g., the human-type tuberculosis transmitted to cattle).

systems of husbandry, and an increasing trend toward integrated or coordinated production and marketing operations.

Although most of these changes favor progress in the more economic production of animals and animal products, at the same time they substantially heighten the risk to the producer of serious losses from disease, because larger and larger numbers of animals are being raised under conditions of closer and closer contact. To meet these increased risks, new approaches to disease prevention are being evolved, many of which are now economically feasible for the first time. Instead of scarce veterinary talent being partially dissipated in time-consuming responses to medical emergencies or for the treatment of individual sick animals in small herds or flocks, as often happened in the past, private veterinarians are now increasingly able to concentrate their time and efforts on the problems of either a single large producer or several moderately large producers.

Several different types of producer-veterinarian relationships have developed. Frequently one or more veterinarians are employed fulltime by an individual agricultural enterprise. Another arrangement is for one veterinarian or a group practice of veterinarians (e.g., sometimes including specialists in areas such as reproduction, epidemiology, and clinical pathology) to enter into contractual arrangements with several agricultural enterprises to provide continuous clinical and preventive services. Still another arrangement involves the veterinarian as a consultant for specific areas of disease prevention and control, but not in the delivery of specific clinical and preventive services. All these patterns for producer-veterinarian relationships promote maximum emphasis on prevention of disease before it occurs rather than on attempts to repair damage already done.

The Strategy of Disease Prevention

Among the general approaches to the prevention of livestock diseases common to all these newer patterns of producer-veterinarian cooperation are (1) education, (2) epidemiological investigation and analysis, (3) accurate diagnosis, (4) early detection, and (5) environmental measures. Some of the more specific types of preventive tools employed include such things as (1) quarantine, (2) mass immunization, and (3) chemoprophylaxis.

Education Programs for the prevention of diseases of livestock, no matter how scientifically and economically sound, will risk failure if they are not fully understood by all the persons who must be trusted to carry them out. In the design of a specific disease prevention program, this means that the producer must work closely with his veterinarian so that he and his employees become adequately acquainted with the principal features of the epidemiology of the diseases to be guarded against, and with the basis of and requirements for the program being designed.

Epidemiological study and analysis Epidemiology is the discipline which provides an important part of the broad scientific basis for preventive herd medicine. The epidemiological or ecological approach to disease problems is dependent in considerable measure on ability to enumerate and identify individual animals, or in some instances specific groups of animals, and on development of an accurate system of record-keeping for such things as births, sicknesses and deaths. These prerequisites permit the calculation of descriptive rates and indices, and the application of techniques of biostatistical and computer analysis to the identification of disease problems, the testing of disease-related associations, and the design of programs for disease prevention and control.

Properly compiled epidemiological data also provide a basis for predicting the occurrence of certain diseases, for indicating the optimum time for application of specific preventive measures, such as vaccination, drug administration, or vector control, and for the evaluation of the effects of specific preventive measures (Figure 44.5).

Accurate diagnosis An absolutely essential prerequisite to rational disease prevention and control programs is a capacity for rapid and *definitive* diagnosis of diseases. Clinical medicine, pathology, and epidemiology serve as the complementary sequence of diagnostic disciplines in preventive veterinary medicine. To be utilized fully, they must be backstopped by suitable laboratory facilities for a variety of microbiological, serological, pathological, and chemical tests and examinations.

Early detection Early detection is based on the routine application of some relatively simple test (or tests) to the individuals in an animal population with the objective of diagnosing an illness in an early preclinical stage, to diagnose the presence of a disease *before* a real problem arises, to determine the state of immunity of the population, or to detect abnormalities likely to predispose individual animals to a particular illness. Tools for early detection include: (1) all clinical procedures for the routine physical examination or health examination of animals; (2) immunodiagnostic screening tests, such as skin tests for tuberculosis and serological tests for brucellosis; (3) chemical tests on blood, urine, feces, milk, saliva, etc.; (4) tests for infectious agents, such as stool examinations, skin scrapings, or milk cultures; (5) cytological or superficial biopsy tests for abnormal or specific types of cells; (6) strip cup and other simple physical tests on milk; (7) radiographic screening tests; and (8) pregnancy examinations.

Controlling the environment Environmental approaches may be extremely varied,

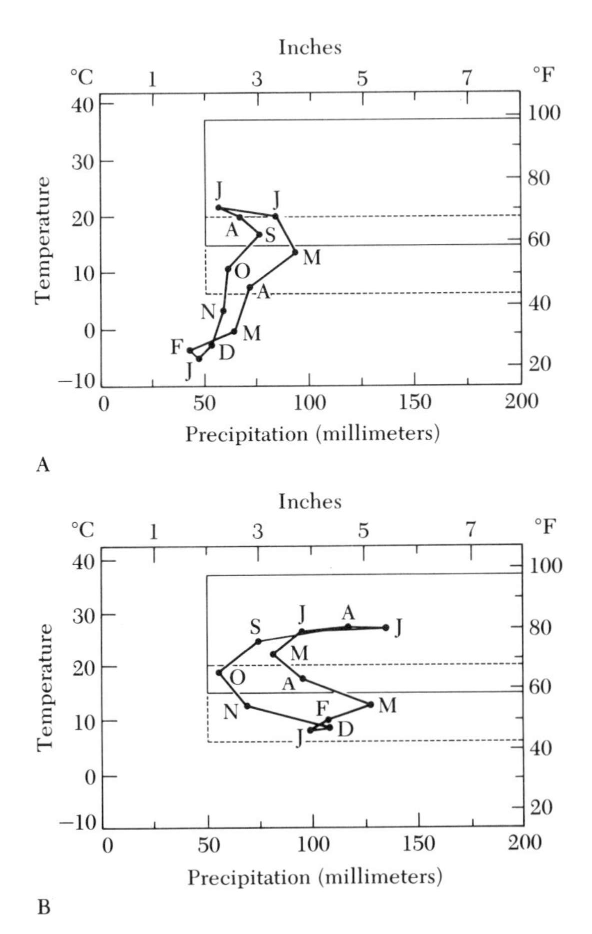

Figure 44-5 One way to analyze epidemiological data in order to help accurately determine the optimum time for implementation of specific control measures, here against gastrointestinal nematode parasites of sheep. These *bioclimatograms* were constructed to show the actual periods during which these parasites can be transmitted locally. The area of the figure enclosed by a solid line represents climatic conditions for optimum transmission of *Haemonchus* (i.e., 5 cm. or more of total monthly precipitation, and 15 to 37°C mean monthly mean temperature), and that by broken lines for *Trichostrongylus* and *Ostertagia* (i.e., 5 cm. or more of total monthly precipitation and 6 to 20°C mean monthly mean temperature). The bioclimatogram for Lansing, Michigan (A), shows, for example, that optimum conditions for transmission of *Ostertagia* and *Trichostrongylus* occur from the end of March until the beginning of June and again from the beginning of August through the middle of October. In Columbus, Georgia (B), however, optimum transmission of these parasites can take place from the end of September through the middle of April. (From Levine, N.D., *Advances in Veterinary Science* 2:234–235, 1966. Academic Press, Inc.)

depending on the particular problems, diseases, or programs. They could include such things as: excrement or litter treatment or disposal; provisions for safe water supplies; systems of moveable pens, which can be moved frequently to interrupt the life cycles of parasites, or other improvements in housing, ventilation, lighting, or drainage; regimens for pest and vector control; cleaning and disinfection; provision of sanitary feeders or special diets; and measures for climate control.

Quarantine Quarantine is the isolation or separation of sick (and possibly exposed) animals from well animals in an effort to prevent the spread of diseases. It is a basic approach to disease prevention which can be used at many different levels, international, interstate, intrastate, or local (on the farm).

Mass immunization For many infectious diseases of livestock, specific vaccines have been developed which, when used properly, can prevent serious or catastrophic occurrences of that particular disease. This development of efficient vaccines has been one of the most significant areas of veterinary advance.

Vaccines are all preparations of antigens* of either infectious agents or the metabolic products of infectious agents. Most of the diseases against which vaccines have been developed are "caused" by viruses or bacteria (or the several intermediary groups of infectious agents). These vaccines are of three principal types. *Killed vaccines* are preparations in which the infectious agents are killed by formalin (or some other means) in such a way that the antigen molecules are not significantly affected. Such a vaccine obviously cannot *infect* the immunized animal, because it is nonliving. The total number of antigenic molecules available to stimulate antibody production are those actually injected. Examples of killed vaccines are blackleg and swine erysipelas bacterins.

A second type of vaccine consists of metabolic products of the infectious agent (toxins) treated in such a way as to make them harmless without destroying their antigenic properties. The best known vaccines of this type are the toxoids which were developed against tetanus (and against human diphtheria) by a French veterinarian, Gaston Ramon.

In the third principal type of vaccine, the living infectious agent has been altered in the laboratory in such a way as to be still alive but harmless, and thus able to infect and multiply in the vaccinated or immunized animal without producing disease. These are called *live, attenuated vaccines,* and have an advantage over the previous two types by simulating natural infection and thus permitting multiplication of the antigenic material and magnification of the stimulus for antibody formation. Such vaccines are usually attenuated by growing the infectious organism in embryonated hen's eggs, in tissue cultures, or in some abnormal host animals. Examples of living, attenuated vaccines are spore preparations against anthrax, strain 19 brucellosis vaccine, and some types of rabies vaccines.

The object of mass immunization as a preventive measure against disease in animal populations is to achieve a high level of *herd immunity,* that is, a high ratio of resistant animals to susceptible animals in the population. Usually if about 80 per cent or more of the animals in a population are resistant to infection (protected), relatively few cases of the disease will occur if the population is subsequently exposed to the infection (Figure 44.6). That is, epidemic disasters can be prevented. This is important to know, because most vaccines will not protect 100 per cent of *individual* animals, even when they are correctly administered.

* An antigen is usually a large protein or carbohydrate molecule which, when injected into an animal, will stimulate the animal's lymph nodes and other tissues to produce other specific proteins called antibodies. These antibodies can unite with or bind to the antigen which stimulated their formation. They act on the specific virus or bacterium which contains that antigen, killing it or preventing it from invading the host's cells.

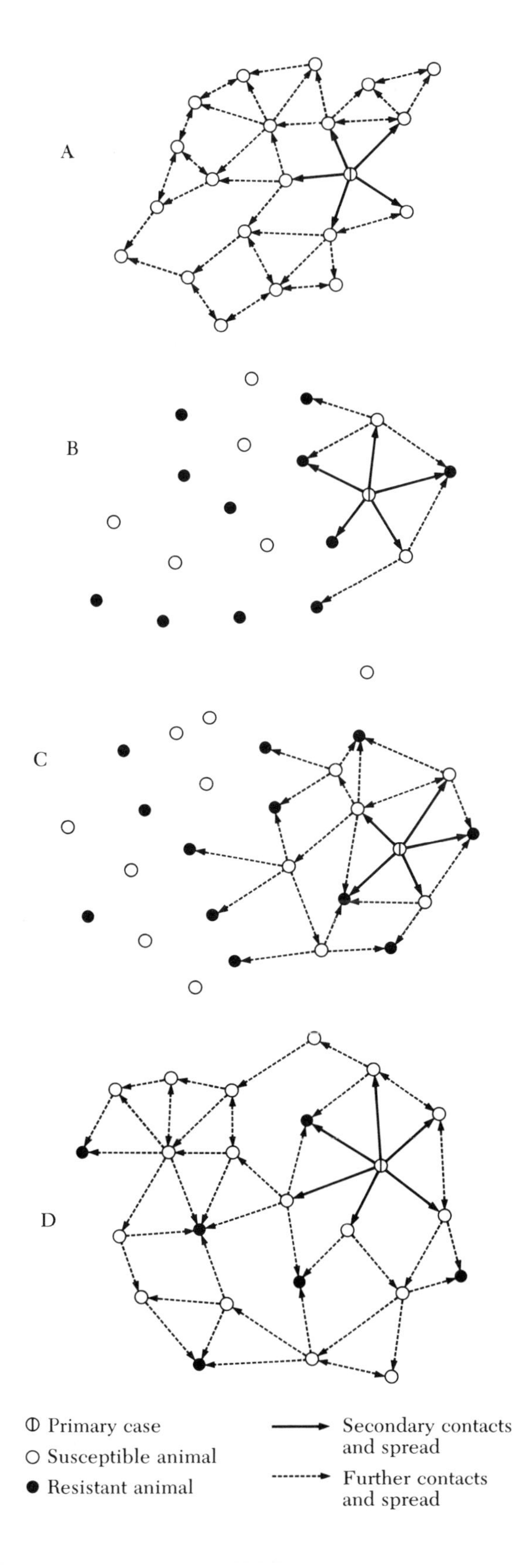

Chemoprophylaxis This approach involves the routine administration of drugs or other chemicals to animals in order to prevent illnesses or their severe manifestations, rather than to treat the full-blown disease. Broadly considered, chemoprophylaxis includes use of dietary supplements, as with specific vitamins and minerals, and use of insecticides and other pesticides, as well as the addition of particular drugs to the animal's salt, feed, or water, application to the surfaces of their bodies, or injection into their bodies.

Controlling and Eliminating Animal Diseases

Diseases and the circumstances which cause them are often not limited to single premises. Many diseases have important regional and national implications, and must be approached accordingly. Therefore, in addition to such measures as a livestock producer and his veterinarian may privately undertake to promote

Figure 44-6 The effects of different levels of herd immunity (ratio of resistant to susceptible animals) on the ability of a hypothetical contact or droplet-transmitted infection to spread within an animal population. (A) Introduction of infection into a herd with no herd immunity (i.e., all animals susceptible). These animals all become infected and either recover or die. (In this hypothetical instance, we will assume that all recovered animals are then resistant for life). (B) introduction of infection into the same herd one year later. (The susceptible animals represent new-born and purchased replacements). The herd immunity is moderately high and only a few of the susceptible animals are affected. (C) one year later, the ratio of resistant to susceptible animals (herd immunity) is reduced, and slightly greater spread of infection occurs. (D) after one additional year, the level of herd immunity is so low as to be completely without effect, and all susceptible animals eventually contract the infection. (Adapted from Schwabe, 1969. Courtesy Williams and Wilkins Co.)

and maintain the health of his own animals, the livestock producer often has the opportunity and the responsibility to participate in cooperative federal-state programs designed to control or eradicate on a national scale certain designated diseases of livestock. Present programs of this type in the United States include those directed against brucellosis in cattle and swine, tuberculosis, hog cholera, screwworm infection, and scabies.

As more and more of the most dramatic plagues of livestock are brought under effective control in this country, attention will be directed increasingly to more complex and less dramatic, but often even more costly diseases, such as bovine mastitis, gastrointestinal parasitism in all species, the shipping-fever complex, reproductive failures, and the various causes of prenatal and neonatal mortality in all livestock species.

FURTHER READINGS

Pan American Health Organization. 1969. *International Symposium on Health Aspects of the International Movement of Animals.* PAHO Sci. Publ. 182. Washington, D.C.: PAHO.

Schwabe, C. W. 1969. *Veterinary Medicine and Human Health.* 2d ed. Baltimore, Md.: Williams and Wilkins.

USDA. 1965b. *Losses in Agriculture.* Agric. Handbook no. 291. Washington, D.C.: USDA.

Section Eight

Classification, Grading, and Marketing of Livestock and Their Products

A Colombian villager delivering fiber to the farmers' co-op for storage. (FAO photo.)

Forty-Five

Classification, Grading, and Marketing of Livestock and Meat

Pray, butcher, spare yon tender calf!
Accept my plan on his behalf;
Let clover tops and grassy banks
Fill out those childish ribs and flanks
Then may we, at some future meal
Pitch into beef, instead of veal.

Ogden Nash

The marketing of both livestock and meat is a complex process. The reflection of consumer preference for meat back through the marketing system to the original producer is highly dependent upon a national system for classifying and grading livestock and meat. The USDA system for classifying and grading provides the national terminology or vocabulary for describing supplies and prices of livestock and meat for day-to-day and hour-to-hour functioning of the national market.

The basic principles of classifying and grading livestock and meat in this country date back to colonial days. As soon as the early settlers began to produce more than enough livestock for their own needs, they started to trade for other materials and commodities. Consequently, local terminology evolved to describe differences in supplies and preferences in each community.

With the development of large urban centers and improved refrigeration and transportation, large slaughter facilities developed around such centers as Chicago, Cincinnati, and St. Louis, and later at many western points. The need to compare prices being paid at competitive markets resulted in the development of Federal grade standards for livestock and meat.

In 1926, a producer group known as the Better Beef Association was formed with the primary objective of establishing a national program for grading the higher qualities of beef. It was the contention of this group that if a reliable identification of excellence were placed on these better grades of beef, consumers would buy them with greater confidence, and that this would indirectly encourage the production of better beef cattle. Accordingly, in May 1927, at the request of this group and with cooperative support of the National Livestock and Meat Board, the U.S. Department of Agriculture inaugurated the Federal grading of beef on a voluntary, experimental basis for one year. At the end of this first year, the program was continued on the same voluntary basis, but the users of the service were charged fees to cover the operating costs of the service. The program was later expanded to cover veal, calf, lamb, and mutton.

Standards for grades of pork carcasses have been available for many years. Their use has consisted primarily of providing information to producers to guide in the production of more desirable hogs. Since practically all the pork wholesale trade is in the form of wholesale cuts, grades for pork carcasses are used very little by the meat trade.

Although Federal class and grade standards for livestock and meat are published and available to all segments of the industry, their use has been strictly voluntary except for two periods of national emergency, during World War II and the Korean Conflict, when Federal grading of beef, veal, calf, lamb, and mutton was made compulsory in conjunction with price control programs. In 1940, just prior to World War II, the volume of beef being graded was about 8 per cent of total production. However, following each of the periods of compulsory grading, the volume of grading remained at a greatly increased level. Presently, more than 65 per cent of the total beef production in the U.S. is federally graded.

The introduction of yield grades for beef in 1965 was also on a voluntary basis. Provisions were made to permit beef to be graded for quality grade, for yield grade, or for both. The use of yield grades has increased steadily until at the end of 1972 more than 50 per cent of the beef that was quality graded was also yield graded.

While the U.S. Department of Agriculture does not officially grade livestock in the same manner that it does meat, Federal grade standards for livestock are used extensively throughout the industry. These standards are used as the basis for the nationwide USDA market news service. They are also used by most of the private market news services throughout the country. Some states do offer an official livestock grading service. This grading by state employees usually involves feeder animals, and is done when the animals are offered for sale at special auctions or sales. Federal grades for livestock became the basis for trade on livestock futures contracts in the mid 1960's. When deliveries are called for on these contracts for cattle or hogs, they are officially graded by USDA market reporters to determine compliance with the futures contracts.

With increasing emphasis on direct trading in livestock, many producers are assuming a greater responsibility for determining the price of their livestock than when sold through an agent. Therefore, it becomes increasingly important for the producer to be as well informed as possible in the application of livestock grades both for slaughter and feeder animals.

Hence, from its earliest beginning when marketing and market terminology consisted only of local practices and terms, livestock and meat production and marketing has become a highly specialized, complex business. Although many private companies have developed private brands to merchandize their products, Federal grades have become the nationwide standard for buying and selling of livestock and meat. The following discussion is limited to the classification and grading based on Federal standards.

Classification and Grading of Meat

Class, as used in a broad sense, is a designation that identifies—or groups together—carcasses or cuts that have a similar commercial use and which come from the same species or kind of animal. For example, lamb and mutton are both derived from the ovine species. However, since they are not interchangeable from the standpoint of their commercial use, lamb and mutton are considered as two distinct classes of meat. Subclasses of carcasses within a specific class or kind of meat usually refer to the sex condition of the animal from which the meat was derived.

Grade is a designation which identifies carcasses or cuts of the same class on the basis of certain utility or value-determining characteristics. A grade includes a sufficiently narrow range of grade-determining factors such that carcasses or cuts of the same grade have a high degree of interchangeability.

The two basic considerations utilized in grading meat are: (1) to reflect differences in the proportion of the more desirable to the less desirable parts of the carcass and in the ratio of meat to bone; and (2) to evaluate the characteristics of the lean that are associated with its ultimate consumer acceptability. In the case of beef and lamb and mutton, separate standards have been developed for quality grades and yield grades. Since these two grading systems are independent of each other, beef and lamb carcasses can be graded for quality only, yield only, or both. The standards for grades of veal and calf are primarily "quality" grades, whereas the standards for grades of pork are essentially yield grades but also give limited consideration to differences in quality.

Quality Grades

Quality grades—as distinguished from yield grades—are primarily intended to reflect differences in eating quality. However, as presently developed, quality grades give consideration not only to "quality" factors, but also to conformation.

Conformation Conformation refers to the proportionate development of the various parts of the carcass and to the ratio of meat to bone. It is primarily a function of the relative development of the muscular and skeletal systems. However, since over-all thickness and fullness of a carcass is a part of the conformation evaluation, the quantity of external fat on a carcass is also a factor that must be considered. Superior conformation implies a high proportion of meat to bone and a high proportion of the weight of the carcass in the more valuable parts. It is reflected in carcasses which are very thickly muscled, very full and thick in relation to their length, and which have a very plump, full, and well-rounded appearance.

Quality Quality, as used in grading meat, refers to its expected palatability: its tenderness, juiciness, and flavor. The most important factors considered in evaluating the quality of a carcass are: (1) the firmness of the lean and its degree of marbling or intramuscular fat (interspersion of fat within the muscles), and (2) the indications of the maturity of the animal from which the meat was produced. Differences in maturity are evaluated by consideration of the color and texture of the lean; the color, size, and shape of the rib bones; the ossification of the cartilages, particularly those on the ends of the split thoracic and lumbar vertebrae, and the cartilaginous connections of the sacral vertebrae.

Since increasing firmness and marbling and advancing maturity have opposite effects on quality, the Federal standards permit increased marbling and firmness to compensate, within certain limits, for advancing maturity to maintain the same degree of quality. Excellent quality in meat, as evidenced in the cut surface, usually implies a full, well-developed, firm muscle of fine texture and bright color containing a liberal amount of marbling and a minimum of connective tissue.

Illustrations of the degrees of marbling used in the Federal grade standards for beef are available from the U.S. Department of Agriculture.

The revised USDA beef grade standards which became effective in June 1965 required, for the first time, that all carcasses be ribbed, i.e., required that the hindquarter be separated from the forequarter between the twelfth and thirteenth ribs, to expose the large rib-eye muscle which lies adjacent to the backbone. This permits a direct observation of the cut surface of the largest muscle in the carcass. However, most ovine and veal and calf carcasses are graded as intact (unribbed) carcasses, without the benefit of observation of a cut surface of the lean. Therefore, when grading these carcasses, it is necessary to evaluate their quality-indicating characteristics indirectly, on the basis of the development of other closely related characteristics. The development of certain interior fats—feathering between the ribs, fat streaking in the inside flank muscles, fat covering over the diaphragm, and overflow fat over the ribs—are some of the most important indications of quality in unribbed carcasses. Color of fat is a price-determining factor in some instances, but since it has not proven to be directly related to quality, it is not used as a factor in grading. However, the character of fat is a factor associated with quality in grading unribbed carcasses. Firm, brittle fats indicate a higher degree of quality than do soft, oily, or powdery fats. Although the quantity of external fat is positively associated with quality, its use in conjunction with other grade factors adds little to this determination.

Yield Grades

Yield grades are predicted on an entirely different basis than are the quality grades. Yield grades reflect differences in yields of boneless, closely trimmed, major retail cuts. Yield grades for beef were adopted in June 1965, and for lamb and mutton in March 1969. These will be discussed in more detail in subsequent sections of this chapter.

USDA meat grades Table 45.1 summarizes the applicable grades for the various kinds of meat for which the USDA has issued official standards. (These are listed as references at the end of this chapter.)

Table 45.1.
Classes and grades of meat.

Class and subclass	*Quality grades*
Beef[a]	
Steer, heifer, and cow[b]	Prime, Choice, Good, Standard, Commercial, Utility, Cutter, and Canner
Bull[c]	
Calf and veal	Prime, Choice, Good, Standard, and Utility
Lamb[a]	Prime, Choice, Good, Utility, and Cull
Yearling mutton[a]	Prime, Choice, Good, Utility, and Cull
Mutton[a]	Choice, Good, Utility, and Cull
Pork	
Barrows and gilts	U.S. no. 1, U.S. no. 2, U.S. no. 3, U.S. no. 4, U.S. Utility
Sows	U.S. no. 1, U.S. no. 2, U.S. no. 3, Medium, Cull

[a] Yield grades for this class are 1, 2, 3, 4, 5.
[b] Cow carcasses are not eligible for the Prime grade.
[c] After this chapter was written, the beef standards were revised to provide separate quality grades for beef from young bulls, which are now identified as "Bullocks." In general, the quality grades of bullock beef are comparable to those for steers and heifers. Carcasses from older bulls are now yield-graded only, and grades for stag beef have been eliminated.

Classes and Grades of Bovine Carcasses

Classification

Carcasses and cuts produced from animals of the bovine species are grouped into three classes based essentially on the evidences of maturity present in the carcass. These three classes—from youngest to oldest—are veal, calf, and beef. The primary carcass characteristics used to differentiate between these classes include: the color and texture of the lean; the character of the fat; the size, shape, and color of the bones; and the degree of ossification of the cartilages. Although typical carcasses of each class have distinctive characteristics, carcasses near the borderline between two classes seldom show an equal development of each characteristic. Hence, there may be some overlapping of these characteristics between classes, and the final determination must represent a composite evaluation of all characteristics.

Veal The color of lean is the most important characteristic used in differentiating between veal and calf. The lean of typical veal carcasses is grayish-pink in color and is very smooth and velvety in texture. Such carcasses also have slightly soft, pliable fat, and round, red rib bones. Most veal carcasses weigh less than 150 pounds, but some exceed that weight, particularly in the higher grades. Historically, most veal carcasses have been produced from vealers less than 3 months in age and which have been fed on milk. In recent years, however, feeding regimes based on the use of carefully formulated rations to replace milk permit the production of very high-quality veal from considerably older and heavier vealers.

Calf The lean of typical calf carcasses is grayish-red in color and is usually somewhat firmer than that of veal. The fat is drier and flakier, and the rib bones are flatter and lack some of the redness characteristic of veal carcasses. Calf carcasses usually weigh from 150 to 275 pounds and rarely exceed 350 pounds even in the higher grades. (Most calf carcasses will be produced from calves three to eight months in age.)

Beef Carcasses with evidences of more advanced maturity than included in the calf class are classed as beef. (Most beef carcasses

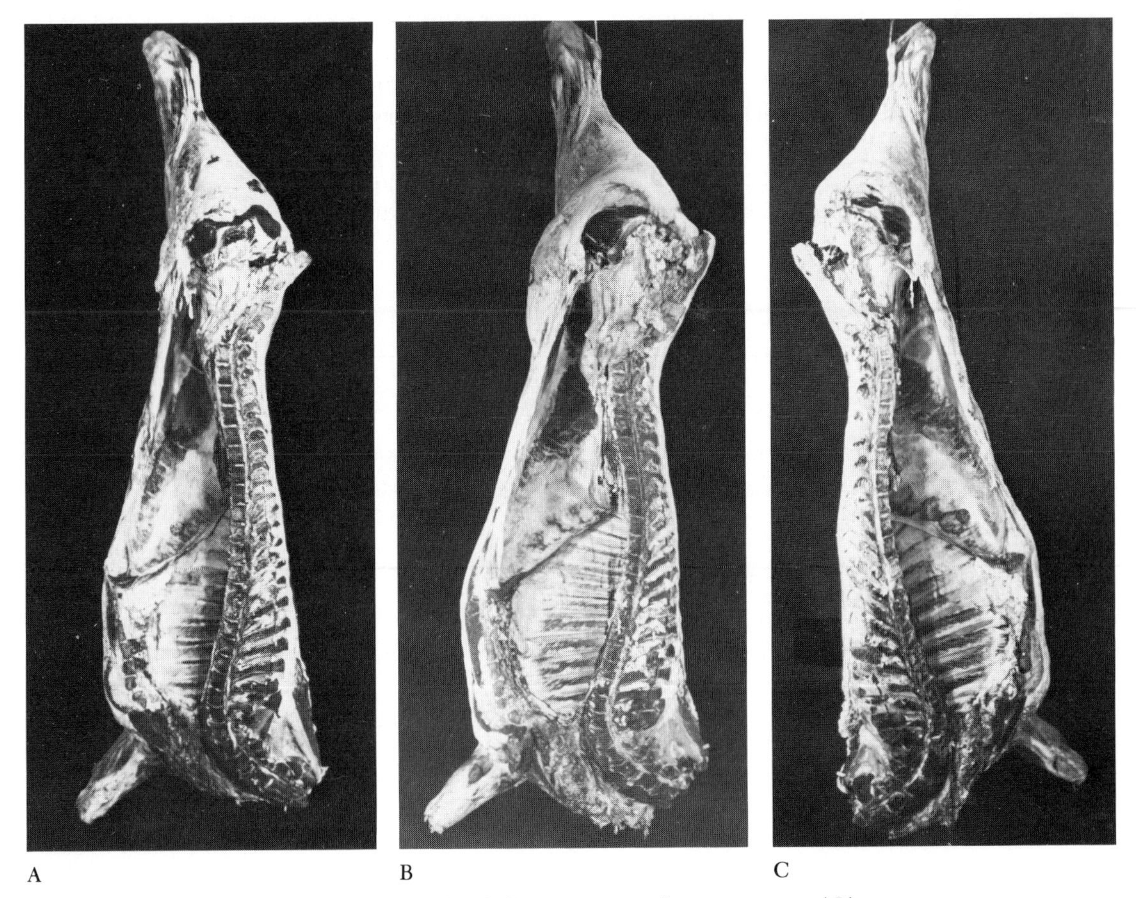

A B C

Figure 45-1 Young steer (A) and heifer (B) carcasses and mature cow (C) carcass.

will be produced from animals nine months of age or older.)

In beef carcasses the sex condition of the animal at time of slaughter is an important economic consideration and affects the trade use of the beef. Therefore, five subclasses of beef are recognized: steers, heifers, cows, bulls, and stags.

Steer carcasses are recognized by the rough, rather irregular fat in the region of the cod, the small pelvic cavity, the small "pizzle eye," the curved aitchbone, and the small area of lean posterior to the aitchbone (Figure 45.1, A).

Heifer carcasses are identified by the smooth udder fat, the slightly larger pelvic cavity and straighter aitchbone than in steers, and the much larger area of lean posterior to the aitchbone (Figure 45.1, B).

Cow carcasses are characterized by a large pelvic cavity and a nearly straight aitchbone. The udder is usually removed, but in some cows that are not lactating at the time of slaughter, the udder will be left on the carcass. Cow carcasses usually have at least slightly prominent hips, and since most are rather advanced in age when marketed, the bones and cartilages are usually hard and white (Figure 45.1, C).

Bull carcasses are recognized by their disproportionately heavy muscling in the round and shoulder; the heavy, crested neck; and the large, prominent "pizzle eye." The cut surface of the meat is usually dark and coarse. Stag carcasses exhibit characteristics somewhat intermediate between those of steers and bulls.

In the Federal grading system, steer, heifer, and cow carcasses are graded on the same standards, and the grade stamp does not include an identity of the sex condition.

Bull and stag carcasses are graded on separate standards, and the word "bull" or "stag" is included with the grade identification when such carcasses are graded by a Federal grader. Most bulls marketed are rather advanced in age, and usually have very little fat and quite dark and coarse lean. Consequently bull beef is generally considered inferior to beef of the other sexes for sale as fresh beef cuts through retail outlets, and such beef is primarily used in processed meat—bologna and similar sausage-type products. However, in the past few

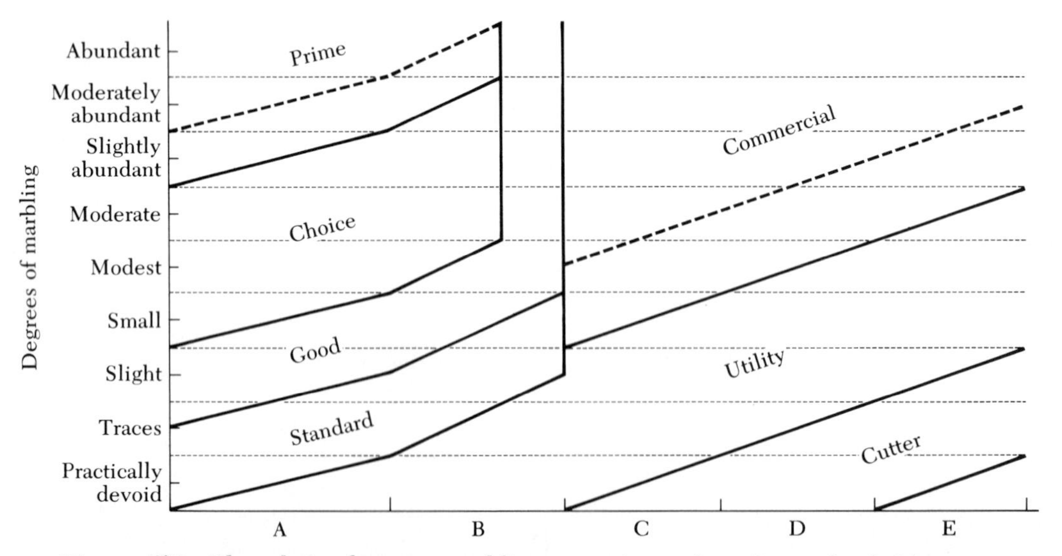

Figure 45-2 The relation between marbling, maturity, and quality in beef. Maturity increases from left to right (*A–E*); the middle ranges of the prime and commercial grades are indicated by dotted lines.

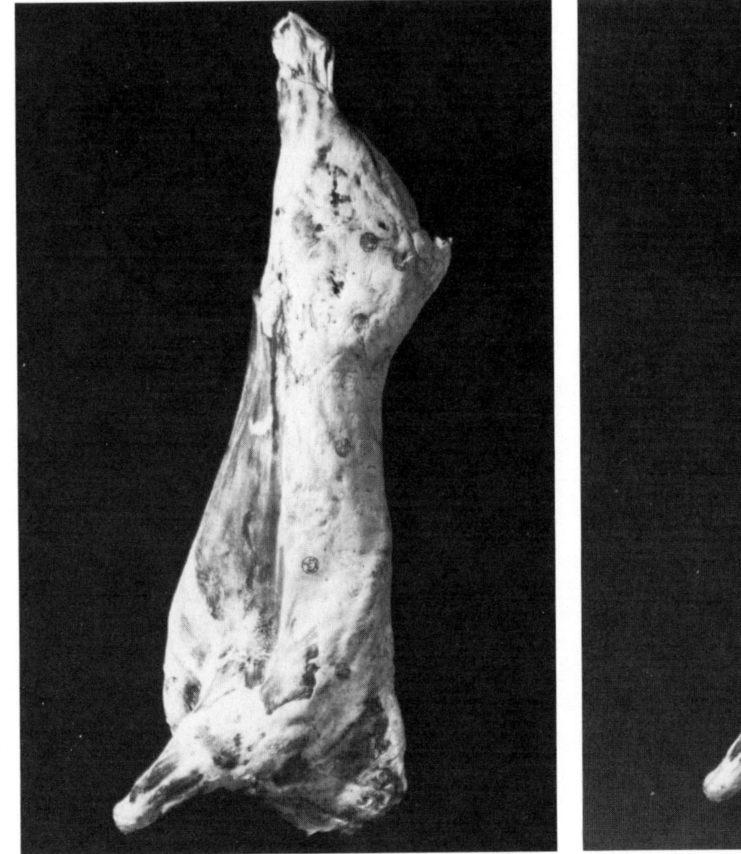

A. High conformation

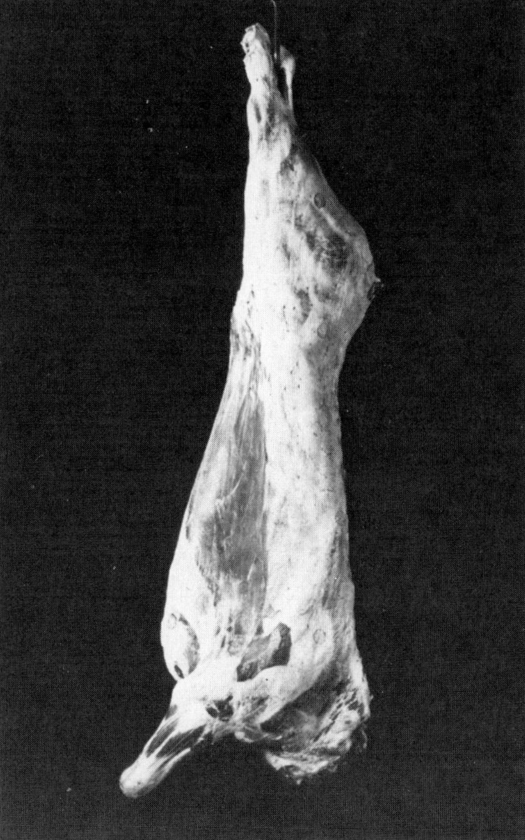

B. Low conformation

Figure 45-3 High (A) and low (B) conformation. The carcass on the left is much thicker and fuller in relation to its length and therefore has a higher degree of conformation. Note, in particular, its much plumper, fuller round.

years, an increasing number of young bulls have been fed and are now marketed as bullock beef.

In veal and calf carcasses, the sex condition does not materially affect the quality of the flesh. Therefore, all sexes are graded on the same standards.

Quality Grades

Although the Federal standards for beef, calf, and veal are contained in three separate standards, it is intended that they be considered a continuous series. Therefore, the grade of a carcass that is near the borderline in maturity between two of these groups would be essentially the same regardless of the standard under which it was graded.

The quality grade standards for beef describe the quality requirements in terms of firmness and marbling of the rib-eye muscle for different degrees of maturity. The Prime, Choice, Good, and Standard grades are restricted to carcasses from young cattle, and involve two maturity groups, A and B. Although no references are made in the standards to chronological age, the maturity groups described are designed to correspond, in gen-

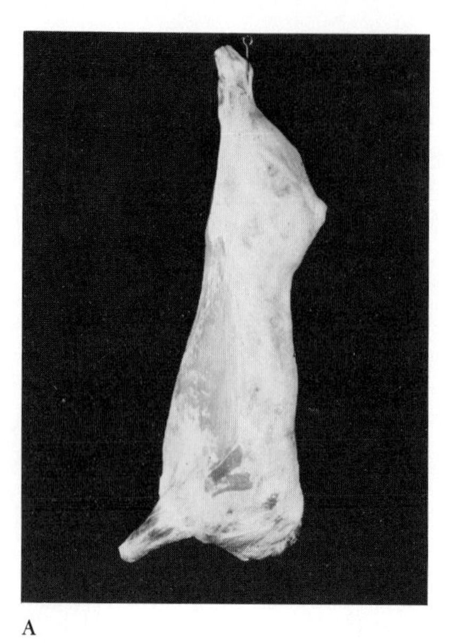

A

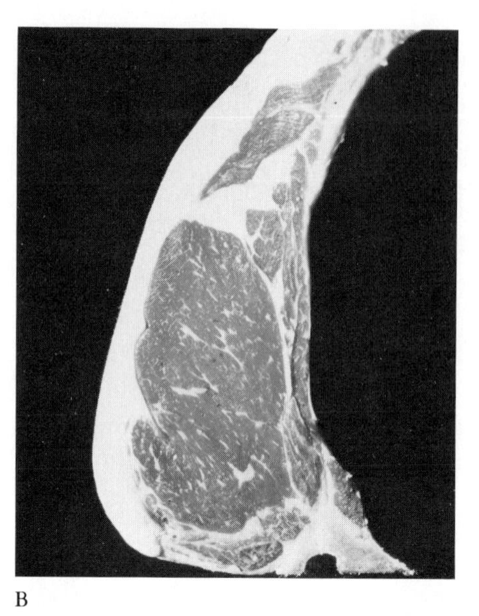

B

Figure 45-4 U.S. Choice grade beef carcass, 614 pounds. The rib-eye muscle (B) of this typical A maturity carcass displays average Choice quality. It has a weak moderate amount of marbling and slightly firm lean. This degree of quality, combined with average Choice conformation—evidenced by a moderately plump round and a moderately thick loin, rib, and chuck—results in an average Choice grade carcass.

eral, to certain chronological ages. In all four grades, the "younger" group is intended to include beef from cattle about nine to 30 months of age. The "older" group for Prime and Choice grades is designed to include beef from animals about 30 to 42 months old, for Good and Standard grades 30 to 48 months of age. The Commercial grade is restricted to carcasses older than those eligible for Good and Standard. Carcasses within the full range of maturity may qualify for the Utility, Cutter, and Canner grades; in these grades, therefore, five maturity groups (designated A through E) are recognized. For each grade, the degree of marbling and firmness of lean are specified for each maturity group. Since marbling and maturity are the most important factors used in the determination of quality, Figure 45.2 shows how these two factors are combined into a quality evaluation, provided the firmness of lean is proportionately developed with the marbling.

The standards for each quality grade also give detailed descriptions of the conformation requirements. By evaluating the conformation of the various parts, an over-all evaluation of the conformation of the carcass can be made (Figure 45.3).

To arrive at the final quality grade, the quality and conformation evaluations must be combined. The relative importance of each of these factors varies somewhat with grade. For example, in the Prime, Choice, and Commercial grades, a superior development of conformation is not permitted to compensate for a deficient development of quality. In the other grades, this type of compensation is permitted, but is limited to one-third of a grade of deficient quality. This means that in the Prime, Choice, and Commercial grades, carcasses must have the quality requirements specified regardless of how superior their conformation may be, whereas, in the other grades, carcasses with conformation at least one-third of a grade superior to that specified for the particular grade may have quality equivalent to the upper one-third of the next lower grade. In all grades, however, a superior development of quality is permitted to compensate for an inferior development of conformation on an equal basis and to an unlimited extent through the upper limit of quality. For example, a carcass with average Choice quality would qualify for the Choice grade with only average Good conformation. By the same token, a carcass with low Prime quality and low Good conformation would also qualify for Choice.

The theory behind these differences in rate and extent of compensation between different developments of conformation and quality stems from the use that is made of the different grades. Since the eating quality of the lean is the primary consideration in grading beef, the standards for all grades permit a superior development of quality to compensate, without limit, for a deficient development of conformation. The reverse type of compensation, i.e., a superior development of conformation for a deficient development of quality is not permitted in the higher grades, such as Prime or Choice, because beef of these grades is utilized largely by hotels, restaurants, and individual consumers whose major interest is the quality of the lean. For such users, any extra increase in the ratio of edible to inedible portions that might result from a superior development of conformation would not compensate for a lowering of the eating quality of the lean. On the other hand, to consumer buyers who normally use the lower grades of beef, the ratio of edible to inedible portions is considered to be of greater relative importance; so in these grades superior conformation is permitted to compensate for deficient quality, but only to the extent of one-third of a grade.

Figures 45.4 and 45.5 illustrate Choice and Standard grade carcasses.

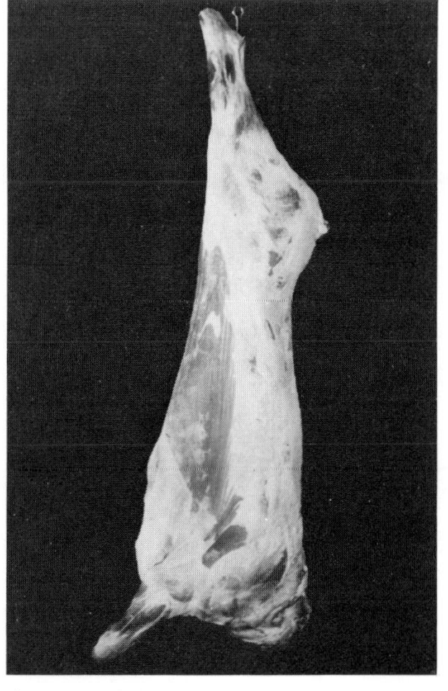

A

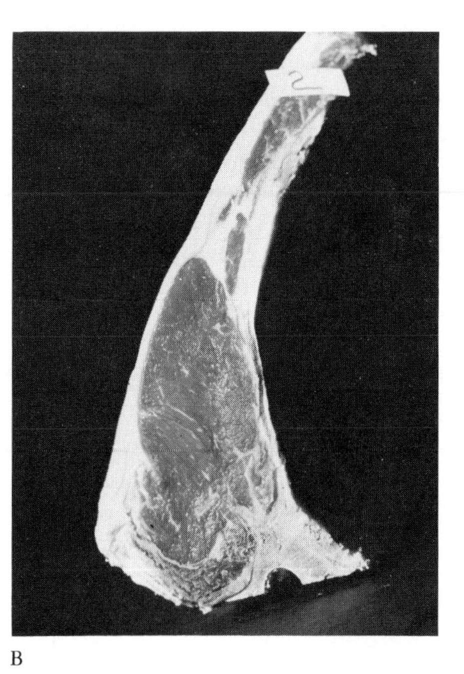

B

Figure 45-5 U.S. Standard grade beef carcass, 620 pounds. This carcass is approaching the maximum maturity for the youngest maturity group (A plus). The rib-eye muscle (B), which is moderately soft and has traces of marbling, is typical of average Standard grade quality. The combination of conformation and quality, both typical of average Standard, qualifies the carcass for average Standard grade.

Yield Grades

In the early 1950's, the USDA recognized the need for a change in the beef grade standards which would precisely identify carcasses within the same quality grade for differences that existed in their yields of usable meat. Many retail buyers were making an effort to select for cutability through personal selection. During the next ten years, studies involving measurements, evaluations, and cutting data on hundreds of carcasses served as the basis for the development of such standards. These studies also indicated that variations in yields of usable meat among carcasses probably have a greater effect on value than do variations in the quality of their lean.

The adoption of yield-grade standards in June 1965 was perhaps the most significant change made in the grading of beef since its inception in 1927, because: (1) the yield grades are based largely on factors which can be measured objectively; (2) they were formulated on more well-substantiated research results than any previous change; and (3) they identified important differences in value among carcasses of the same quality grade, differences that had previously received only token recognition in trading. Price differentials based on yield grades and which fully reflect these value differences could provide the entire industry with the much-needed guide and incentive to increase the production of truly meat-type cattle, those that combine thick muscling with a high quality of lean and a minimum of excess fat.

Specifically, yield grades identify carcasses for differences in "cutability" or yield of boneless, closely trimmed retail cuts from the round, loin, rib, and chuck. In a more general sense, however, they identify differences in yields of all retail cuts from a carcass. There are five yield grades, numbered from 1 to 5. Carcasses qualifying for Yield Grade 1 have the highest degree of cutability, those qualifying for Yield Grade 5 the lowest. Among steer, heifer, and cow carcasses, each yield grade reflects a specific range in carcass yields of retail cuts regardless of sex, weight, quality grade, or other consideration. These yields are shown in Table 45.2.

A limited amount of research indicates that bull carcasses of a specific yield grade will have a higher degree of cutability than a steer, heifer, or cow carcass with the same development of yield grade factors. The same is likely true for stag carcasses.

The yield grade of a carcass is determined by considering four characteristics: (1) the amount of external fat; (2) the amount of kidney, pelvic, and heart fat; (3) the area of the rib-eye muscle; and (4) the carcass weight.

The amount of external fat on a carcass is evaluated in terms of its thickness over the rib-eye muscle at a point three-fourths of the length of the cross section of this muscle from its chine bone end, where this muscle is exposed by ribbing. This measurement may be adjusted, as necessary, to reflect unusual amounts of fat on other parts of the carcass. The quantity of external fat is the most important factor affecting cutability. As the amount of external fat increases, the cutability decreases—each four-tenths of an inch more fat thickness over the rib-eye changes the yield grade by one full grade.

The amount of kidney, pelvic, and heart fat considered in determining the yield grade is expressed as a percentage of the carcass weight. It is evaluated subjectively. As the amount of these fats increases, the cutability

Table 45.2.
Per cent of boneless retail cuts from round, loin, rib, and chuck by yield grades.

Yield Grade	*Per cent of cuts*
1	52.4 and higher
2	50.1 to 52.3
3	47.8 to 50.0
4	45.5 to 47.7
5	45.4 and lower

decreases; a change of 5 per cent of the carcass weight in these fats is required to change the yield grade by one full grade.

The area of the rib-eye muscle is determined where this muscle is exposed by ribbing between the 12th and 13th ribs. In normal grading operations, the area usually is estimated subjectively; however, it may be measured. When making rib-eye measurements, Federal graders use transparent grids calibrated in tenths of a square inch (Figure 45.6). An increase in the area of rib eye increases the cutability; an increase of three square inches in area of rib eye changes the yield grade by approximately one full grade.

The hot carcass weight is the weight used in determining the yield grade. As carcass weight increases, the cutability decreases slightly; an increase of 250 pounds in hot carcass weight will decrease the yield grade by approximately one full grade.

The yield grade standards actually are based on a mathematical equation derived from research results involving the very detailed measurement and cutting of a large number of carcasses (Murphey *et al.*, 1960). That equation is as follows:

$$
\begin{aligned}
\text{Yield grade} = {} & 2.50 \\
& + (2.50 \times \text{adjusted fat thickness, inches}) \\
& + (0.20 \times \text{per cent of kidney, pelvic, and heart fat}) \\
& + (0.0038 \times \text{hot carcass weight in pounds}) \\
& - (0.32 \times \text{area of rib eye in square inches}).
\end{aligned}
$$

In addition to the equation, the yield grade standards for each grade also describe the development of the yield grade factors for two weights of carcasses, 500 and 800 pounds. These descriptions are not specific requirements, but are included as general illustrations of the development of the yield grade factors in carcasses of these weights. An adaptation of the yield-grade equation to provide a simplified, short-cut method of determining yield

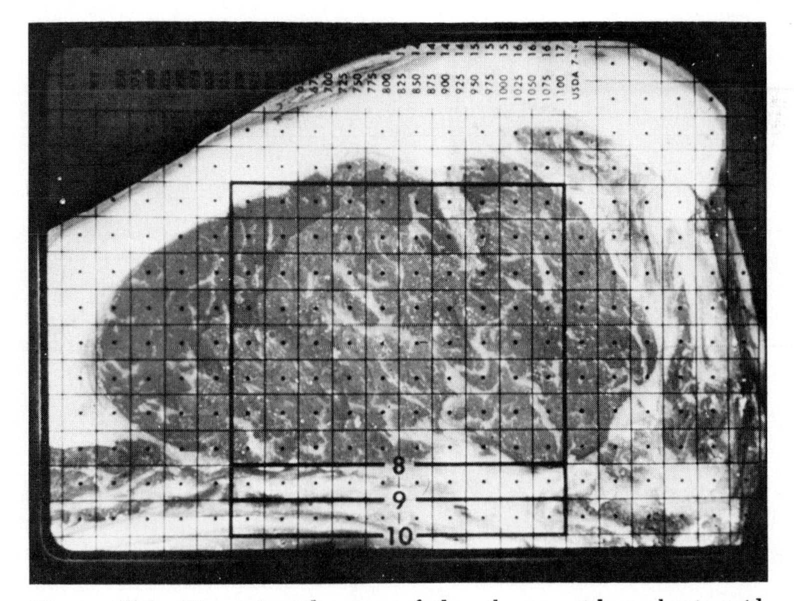

Figure 45-6 Measuring the area of the rib eye with a plastic grid. Each square is 0.10 of a square inch. The areas outlined in dark lines include 8, 9, and 10 square inches. This rib eye measures 11.2 square inches.

grade also has been developed by the USDA, and copies are available from the Livestock Division, Agricultural Marketing Service, USDA.

Figures 45.7 and 45.8 illustrate Choice grade carcasses from yield grades 1 and 5. Such carcasses have a difference in value to a retailer in excess of $17.00 per hundredweight when the average carcass price is about $50.00 per hundredweight. A much more comprehensive discussion of the yield grades for beef is contained in USDA Marketing Bulletin 41, *U.S. Yield Grades for Beef.*

Under the USDA grades, beef carcasses may now be identified for quality grade, yield grade, or both, at the option of the packer. The quality grade is applied in a ribbon-like manner so that the grade name appears on most retail cuts. The yield grade is applied once on each major wholesale cut.

USDA meat graders become highly skilled in applying the beef-carcass yield grades based on a rapid, over-all comparison of the carcass to be graded with mental images which each grader has of different kinds of carcasses that qualify for each yield grade. Through intensive subjective training and experience, graders are able to recognize with confidence carcasses which are in the middle portion of each of the yield grades. By the same token, they are able to recognize carcasses which are near the limits of each of the grades. Such carcasses are then analyzed more carefully to determine their correct yield grade. This may include measuring the fat thickness over the rib eye, measuring the area of the rib eye, or both.

Classes and Grades of Ovine Carcasses

Classification

Carcasses and cuts produced from animals of the ovine species are grouped into three classes, essentially on the basis of evidences of maturity present in the carcass. These three

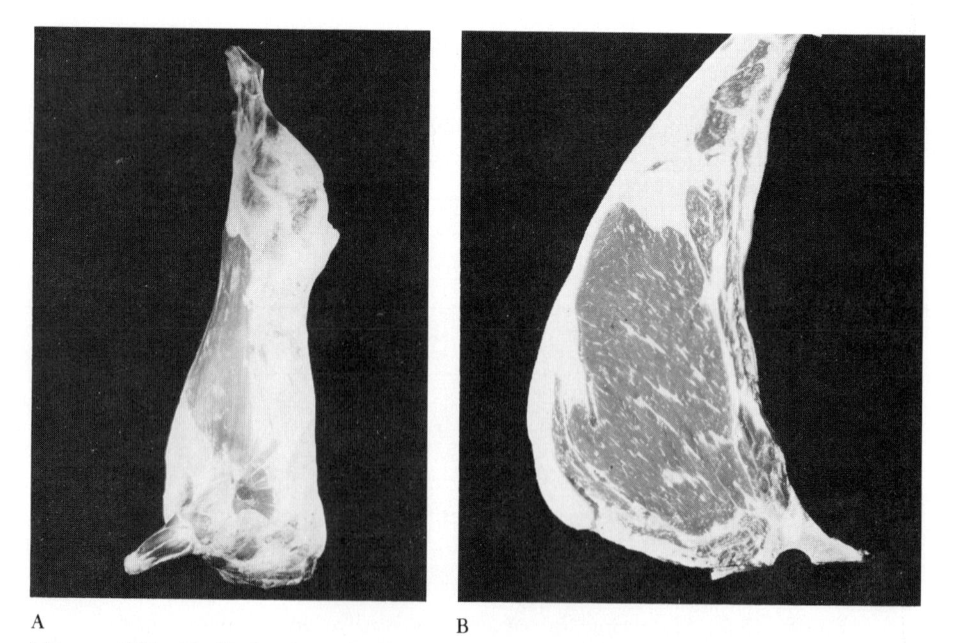

A B

Figure 45-7 Yield Grade 1 beef carcass. Carcass weight, 645 pounds; fat thickness over rib eye, 0.2 inch; rib-eye area (B), 13.9 square inches; kidney, pelvic, and heart fat, 2.5 per cent; yield grade, 1.5; quality grade, High Choice.

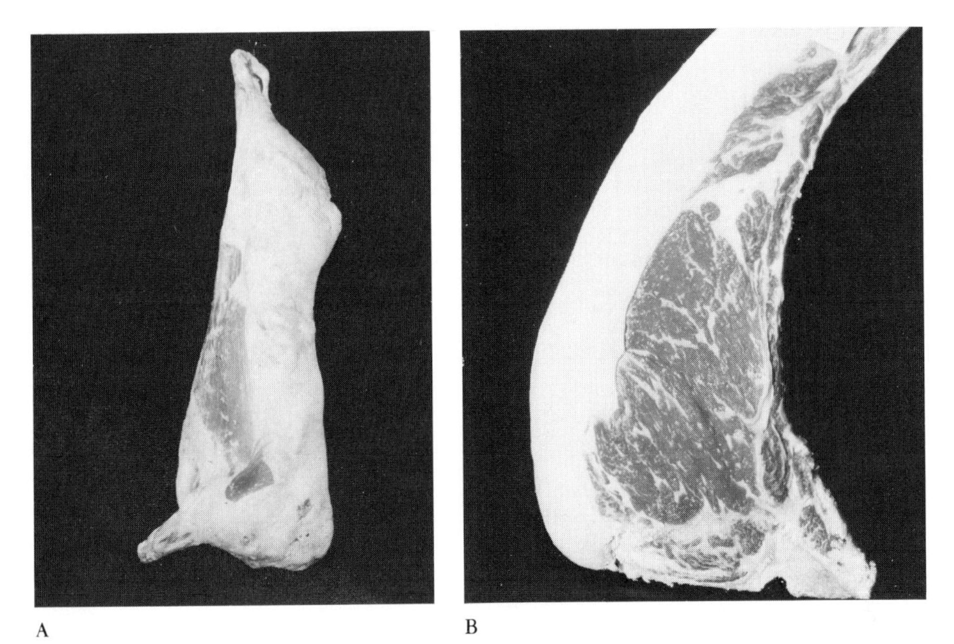

Figure 45-8 Yield Grade 5 beef carcass. Carcass weight, 750 pounds; fat thickness over rib eye, 1.1 inches; rib-eye area (B), 10.9 square inches; kidney, pelvic, and heart fat, 5.0 per cent; yield grade, 5.6; quality grade, High Choice.

classes, from youngest to oldest, are lamb, yearling mutton, and mutton.

Lamb carcasses always have "break joints" on their front shanks; these are moist and fairly red and show well-defined ridges. The rib bones are moderately flat and slightly wide; the lean is light red, and has a fine texture.

Yearling mutton carcasses may have either break joints or spool joints on their front shanks; the rib bones tend to be flat and moderately wide; and the lean is slightly dark red and slightly coarse in texture.

Mutton carcasses always have spool joints on the front shanks and have wide, flat rib bones. The lean is dark red in color and coarse in texture. Figure 45.9 illustrates a break joint and a spool joint.

In the United States, there is a very strong preference for lamb over yearling mutton or mutton because of the progressively stronger flavor that develops in ovine carcasses as they advance in maturity. Price differentials between these classes of carcasses are substantial. For this reason, a high percentage of ovines are marketed as lambs. Mutton is used largely in manufactured meat products.

Grading

As with beef, the grades for ovine carcasses include separate grades for both quality and yield. Also as with beef, there are separate quality grade standards for the three different maturity groupings, lambs, yearling mutton, and mutton, and these are intended to be a continuous series. This means, for example, that the grade of a carcass which is borderline in maturity between lamb and yearling mutton will be essentially the same regardless of whether it is classed as lamb or yearling mutton. The quality grades also apply to all carcasses without regard to their sex. However, carcasses that show the heavy shoulders and thick necks typical of uncastrated males are discounted anywhere from less than half a grade to two full grades, depending on the extent to which these secondary sexual charac-

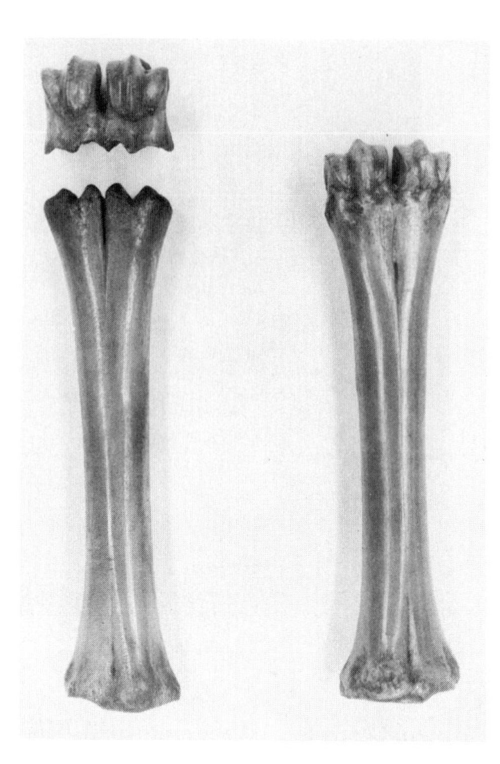

Figure 45-9 Ovine break joint (left) and spool joint (right). In young ovines, up to about 1 year of age, the foreshank can be broken at the break joint. In older ovines this joint becomes ossified and cannot be broken.

teristics are developed. Grade is determined on the same general basis as for beef, by considering variations in conformation and quality.

Except for one important difference, the combination of quality grade and conformation grade into a final grade is much the same as for beef. Lamb and mutton carcasses that have at least midpoint Choice conformation are permitted to have a development of quality equivalent to the minimum of the upper third of the Good grade and still be graded Choice. This compensation is not permitted in Choice beef.

Another difference between the standards for these two species relates to the extent to which a superior development of quality can compensate for a deficiency in conformation. The lamb and mutton standards specify that, regardless of how high a degree of quality a carcass may have, it will not be considered for a given grade if it does not have a development of conformation equivalent to at least that of the next lower grade. For example, a carcass could not be graded Choice if it had less than Good grade conformation. There is no such minimum conformation specified for beef. Also, the lamb and mutton standards provide for a minimum external fat covering for carcasses in the Prime and Choice grade. Figures 45.10 and 45.11 illustrate differences in two grades of lamb carcasses.

Yield grades (Figures 45.12 and 45.13) applicable to all ovine carcasses and which reflect differences in yield of boneless, closely trimmed retail cuts from the leg, loin, hotel rack, and shoulder were adopted in March 1969. Under these standards, carcasses may be graded for quality alone, yield alone, or both.

Classes and Grades of Pork Carcasses

Classification

Carcasses produced from swine are grouped into five classes—barrows, gilts, sows, boars, and stags—based on their sex condition.

Barrow carcasses have the typical pocket in the split edge of the belly (where the sheath has been removed) and have a small "pizzle eye." *Gilt* carcasses have a smooth split edge of the belly, but do not show any development of mammary tissue. *Sow* carcasses have a smooth belly edge similar to gilts, but show a rather pronounced development of mammary tissue as a result of advanced pregnancy or lactation. *Boar* carcasses have a somewhat larger belly pocket than barrows, and have a large, coarse "pizzle eye." The shoulders are heavy, the skin and joints coarse, the lean dark red and coarse. *Stag* carcasses—depending on the age at which the stag was castrated—have characteristics somewhat intermediate between barrows and boars.

Barrow and gilt carcasses are considered interchangeable in trading, since they produce pork of similar quality, and are graded on the same standards. Sow carcasses are graded on separate standards, but some young, lightweight sows also produce cuts that are considered interchangeable with those from barrows and gilts.

Carcasses from most boars have a strong "sex" odor; those that do are not passed for use as food. There are no Federal grade standards for boar or stag carcasses.

Grading

The standards for grades of barrow and gilt carcasses were last revised in April 1968. The discussion which follows relates only to these classes of carcasses.

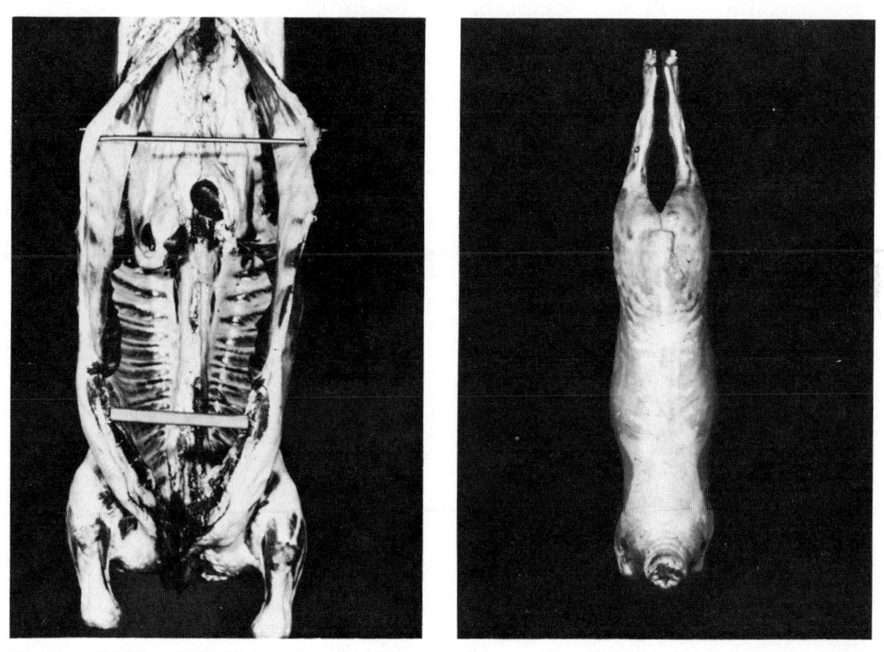

Figure 45-10 U.S. Prime lamb carcass, 48 pounds. This lamb carcass has typical A maturity as indicated by the moderately narrow, slightly flat rib bones and slightly dark pink color of the inside flank muscles. Its plump, full legs, wide, thick back, and thick, full shoulders qualify its conformation for average Prime. The moderate amount of feathering between the ribs, the modest streakings of fat in the inside flank muscles, and the moderately full and firm flanks combine to indicate average Prime quality and average Prime conformation qualifies the carcass for average Prime.

Grades for barrow and gilt carcasses are based on two general considerations: (1) quality-indicating characteristics of the lean; and (2) the expected combined yields of the four lean cuts, ham, loin, picnic shoulder, and Boston butt. Two general levels of quality are recognized, acceptable and unacceptable, for the lean in the four lean cuts. Quality of lean is best evaluated by a direct observation of its characteristics in a cut surface. However, in grading carcasses when a lean cut surface is not available for observation, quality is evaluated indirectly from quality-indicating characteristics evident in the carcass. These include firmness of fat and lean, feathering between the ribs, and color of lean. Thickness of back fat, as such, is not a factor in determining quality. Carcasses having evidence of unacceptable lean quality or which have bellies that are too thin to be suitable for bacon production are graded U.S. Utility. Soft and oily carcasses are also graded U.S. Utility regardless of their other characteristics.

Four other grades—U.S. no. 1, U.S. no. 2, U.S. no. 3, and U.S. no. 4—are provided for carcasses which have acceptable lean quality and acceptable belly thickness. These grades are based entirely on the following expected carcass yields of the four lean cuts: U.S. no. 1, 53.0 per cent and over; U.S. no. 2, 50.0 to 52.9; U.S. no. 3, 47.0 to 49.9; U.S. no. 4, less than 47.0. These yields are based on cutting and trimming methods used by the USDA in de-

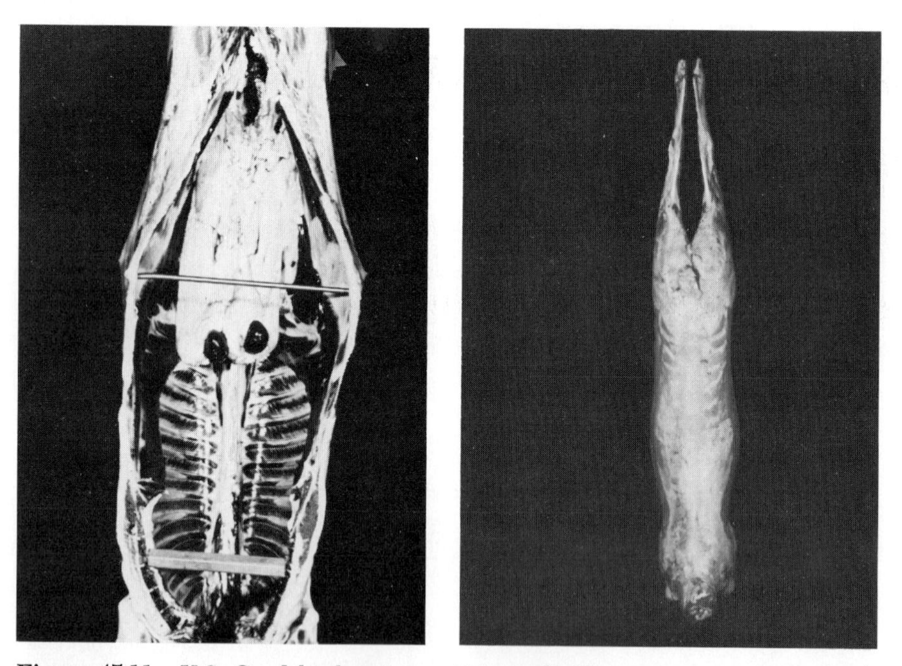

Figure 45-11 U.S. Good lamb carcass, 41 pounds. This is a more mature lamb carcass (B minus), as indicated by the slightly wide, moderately flat rib bones and light red color of the inside flank muscles. It has average Good conformation in that the legs tend to be slightly thin and tapering, and the back and shoulders tend to be slightly narrow and thin. Likewise, average Good quality is indicated by the weak slight amount of feathering between the ribs, the strong traces of streakings of fat in the inside flank muscles, and the slightly full and firm flanks. Average Good conformation and average Good quality combine to qualify this carcass for average Good.

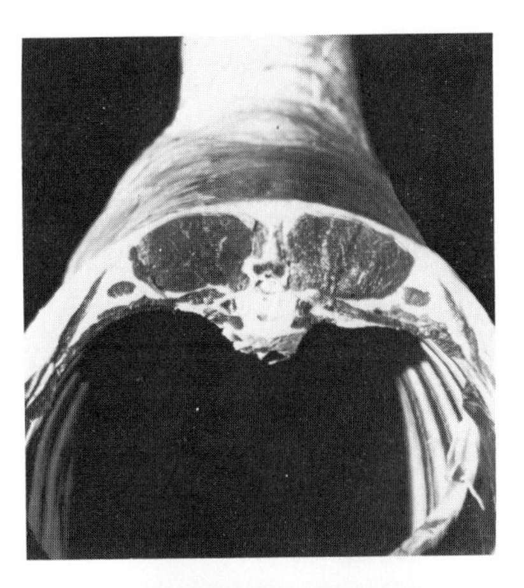

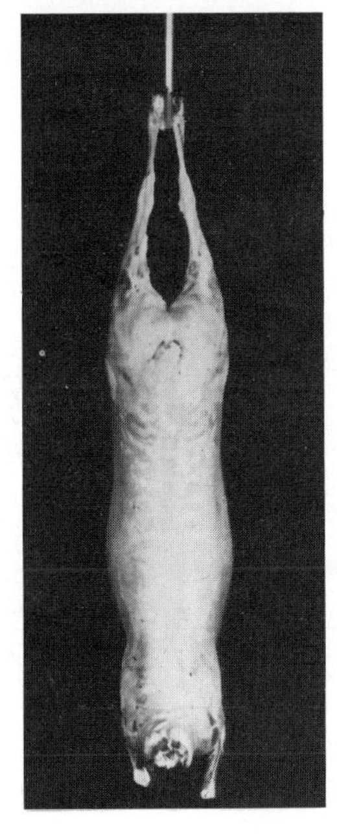

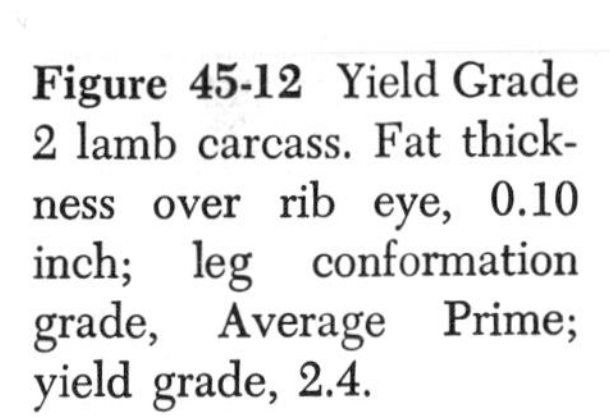

Figure 45-12 Yield Grade 2 lamb carcass. Fat thickness over rib eye, 0.10 inch; leg conformation grade, Average Prime; yield grade, 2.4.

veloping the standards. Other cutting and trimming methods may result in different yields.

Carcasses vary in their yields of the four lean cuts because of variations in their degree of fatness and in their degree of muscling (thickness of muscling in relation to skeletal size). The standards for the four numbered grades are based on these two general factors; see Figures 45.14 and 45.15. A detailed explanation of the pork carcass grades and how they are determined is contained in USDA Marketing Bulletin 49, *USDA Grades for Pork Carcasses.*

Sow carcasses are graded on much the same basis as barrow and gilt carcasses, except that no consideration is given to either carcass length or weight. Guides to grades are included in the standards based entirely on average thickness of backfat. However, the standards do permit this grade based on backfat thickness to be changed—up to half a grade—based on such factors as meatiness, conformation, and fat distribution.

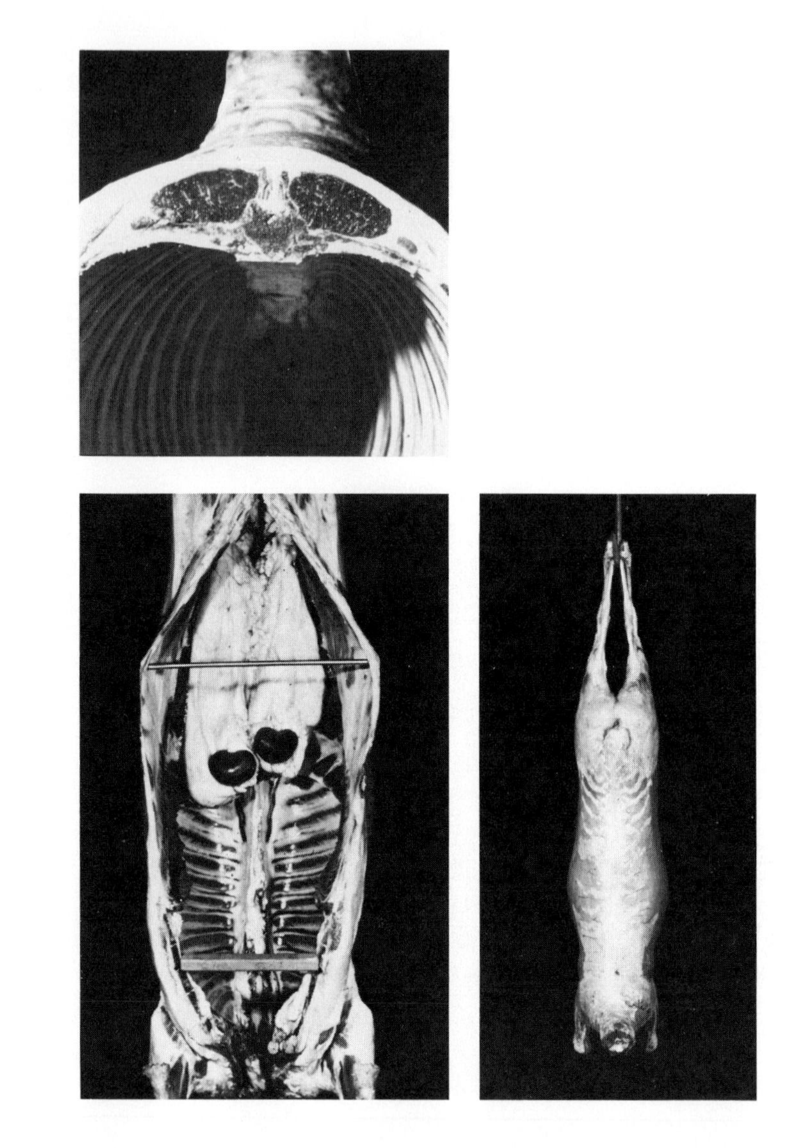

Figure 45-13 Yield Grade 4 lamb carcass. Fat thickness over rib eye, 0.35 inch; leg conformation grade, Low Choice; yield grade, 4.6.

Classification and Grading of Livestock

Basis for Class

The class and grade standards for slaughter livestock are based on the class and grade of carcass they will produce. The class and grade standards of feeder animals are based on the class and grade of slaughter animals they will logically produce under normal feeding and management practices. This closely interrelated system of standards thus provides a means of communication whereby consumer's wants and preferences for meat can be relayed back through the entire marketing channel to the producers. This enables producers to plan their production and marketing program more intelligently.

Class, as used in its broader sense, refers to the segregation of animals into groups on the basis of their commercial use. Thus, there are two general classes of livestock, slaughter and feeder. Subclasses of livestock are usually based on their sex condition. Thus, steers, heifers, cows, bulls, and stags are considered subclasses of cattle.

In formulating grade standards for the various classes of livestock, some of the subclasses may be grouped together. For instance, grade standards for slaughter steers, heifers, and cows are combined as are standards for grades of slaughter barrows and gilts.

Basis for Grade

The grade of a slaughter animal is the estimated grade of carcass it will produce. Therefore, the basic considerations involved in grading are the same for livestock and meat. Reduced to their most simple form, these considerations relate to (1) the proportions of lean, fat, and bone, and (2) the factors that are indicative of differences in the quality or palatability of the lean.

The evaluation of an animal's proportions of lean, fat, and bone must necessarily involve a consideration of its width (thickness) and depth in relation to its skeletal size (length and height). Animals which are wide, thick, and plump in relation to their height and length obviously will have a low proportion of bone in relation to their combined proportions of lean and fat. Although there is a considerable variation in the proportion of bone between animals, variations in their proportions of lean and fat are much greater. Since variations in proportions of lean and fat are very important value-determining factors, they likewise become important factors in determining grade.

An animal's over-all thickness and plumpness in relation to its skeletal size is a function of both its fatness and muscling, and this is relatively easy to evaluate with a high degree of accuracy. Although it is much more difficult to make individual estimates of muscling and fatness, these too can be made with considerable accuracy. This is best accomplished by making the muscling and fatness evaluations simultaneously, and by giving careful consideration to the development of the various parts, based on an understanding of how

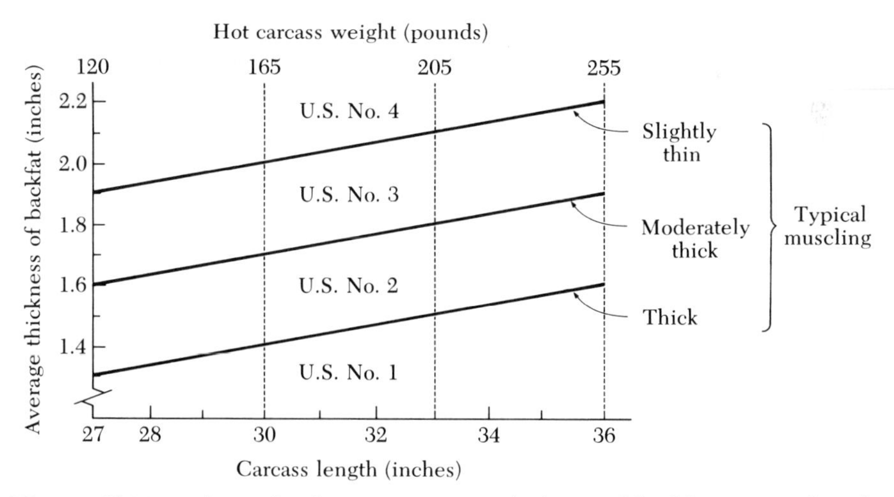

Figure 45-14 Relationship between average thickness of backfat, carcass length or weight, and grade for carcasses with muscling typical of their degree of fatness. The thickness of the backfat is an average of three measurements, including the skin, made opposite the first and last ribs and the last lumbar vertebra. It also reflects adjustments, as appropriate to compensate for variations from normal fat distribution. The carcass weight is the weight of a hot packer-style carcass. The carcass length is measured from the anterior point of the aitch bone to the anterior edge of the first rib.

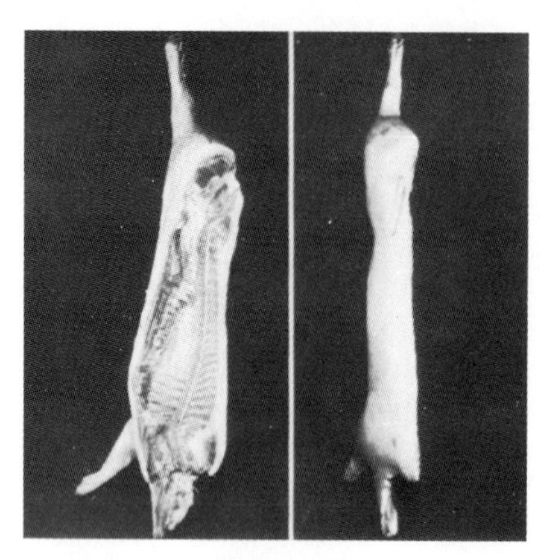

U.S. No. 1
Carcass length, 30.4 inches
Average backfat thickness, 1.3 inches
Degree of muscling, thick

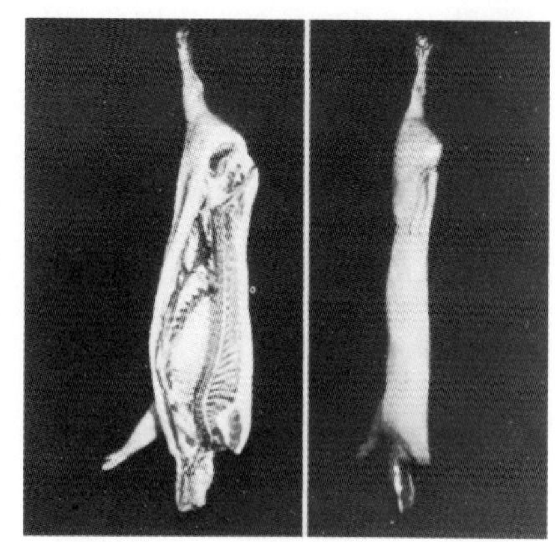

U.S. No. 2
Carcass length, 29.5 inches
Average backfat thickness, 1.6 inches
Degree of muscling, moderately thick

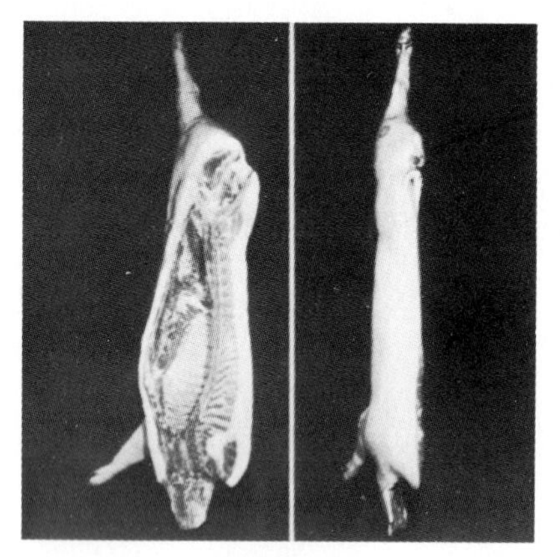

U.S. No. 3
Carcass length, 30.0 inches
Average backfat thickness, 1.9 inches
Degree of muscling, slightly thin

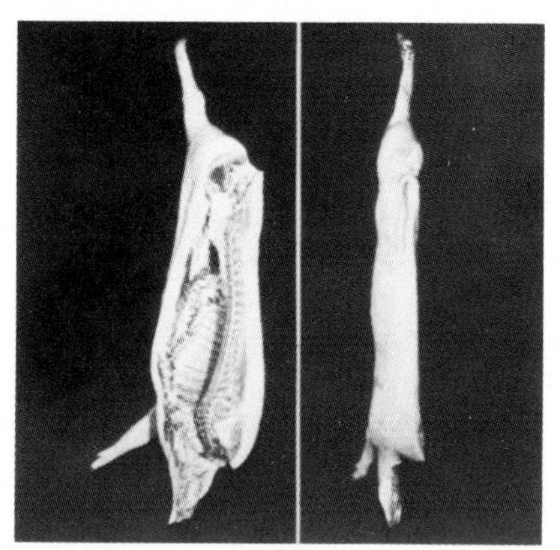

U.S. No. 4
Carcass length, 28.5 inches
Average backfat thickness, 2.2 inches
Degree of muscling, thin

Figure 45-15 U.S. Grades of Pork Carcasses. Utility grade is omitted.

the appearance of each is affected by variations in muscling and fatness.

Muscling, thickness of muscles in relation to skeletal size, tends to develop rather uniformly in all parts of the animal. On the other hand, fat is normally deposited at a considerably faster rate on some parts than on others. Therefore, muscling can be appraised best by giving primary consideration to the parts on which the least fat is deposited, such as the ham in hogs, the round and foreshank in cattle, and the leg in sheep. Other parts should not be ignored, but consideration of them—particularly in highly finished animals—should be made with appropriate adjustments for the fat that is likely present. Conversely, the overall fatness of an animal is best appraised by giving primary consideration to those parts on which fat is normally deposited at the most rapid rate. In cattle and sheep, these parts include the back, loin, rump, hips, ribs, brisket, and fore and rear flanks. In hogs, the jowl is another area where fat is deposited at a faster-than-average rate. As animals become fatter, these parts usually appear fuller, thicker, and more distended in relation to the thickness and fullness of the other parts, especially the thickness through the rear quarter.

In thinly muscled animals with a low degree of fatness, the width of the back usually will be greater than the width through the center of the rear quarter and the back on either side of the backbone will appear sloping and flat or even sunken. Conversely, in thickly muscled animals with a low degree of fatness, the thickness through the rear quarter and shoulder will be greater than through the back and the back will be full and well-rounded.

Very fat animals, especially swine, will be wider through the back and belly than through the rear quarter and shoulder. This difference will be greater in thinly muscled animals than in those that are thickly muscled. A very fat animal—especially if it is thin-muscled—also may be nearly flat across the top of its back and loin. As animals increase in fatness, they also become deeper bodied because of the deposits of fat in the brisket, flanks, and around the middle.

An accurate appraisal of a live animal for its quality of lean is usually more difficult to make than an appraisal of its proportion of fat, lean, and bone, since the criteria that are useful for estimating marbling and other quality characteristics of the lean are, at best, only moderately accurate indicators of these characteristics. The amount of finish carried by an animal is, of course, the best single criterion of the quality of the meat in its carcass. However, the distribution and firmness of the finish are probably nearly as important. Maturity is also a very important consideration in grading slaughter livestock. Among the criteria that are useful in evaluating maturity are: (1) the size of the animal; (2) the width of the muzzle, which reflects differences in the number of small, temporary teeth and of the larger, permanent teeth; (3) the length of the tail, since in vealers and calves the hair on the tail has not had time to grow long; (4) the general symmetry and smoothness of outline, since, as animals advance in maturity, they frequently develop more prominent hips and become irregular in contour; and (5) in horned cattle, the size of the horns. As animals advance in maturity, they are required to have greater amounts of finish to qualify for a given grade.

USDA Grades for Slaughter Livestock

The quality and yield grades for the various kinds of slaughter livestock are the same as are listed in Table 43.1. Examples of some of them are shown in Figures 45.16 through 45.19. Detailed descriptions of the grades of slaughter livestock are provided in the USDA publications listed at the end of this chapter.

Grades of Feeder Livestock

Whether a particular animal is considered a feeder or slaughter animal is determined not by his characteristics, but rather by the use for which he is purchased. If he is purchased for immediate slaughter, he is a slaughter animal. If he is purchased for further feeding, he is considered a feeder animal. However, under certain economic conditions, some livestock, especially cattle and lambs, may be bought by packers for slaughter or by feeders for further feeding. Livestock whose value is almost the same to either slaughterers or feeders are frequently referred to as "two-way" or "warmed-up" animals.

Detailed descriptions of the grades of feeder livestock can be found in the USDA publications listed at the end of this chapter.

Changes in Livestock and Meat Marketing

Only a few of the significant developments influencing the complex job of marketing are mentioned here. However, the role of the national system for classifying and grading livestock and meat has become increasingly important with the changes in marketing.

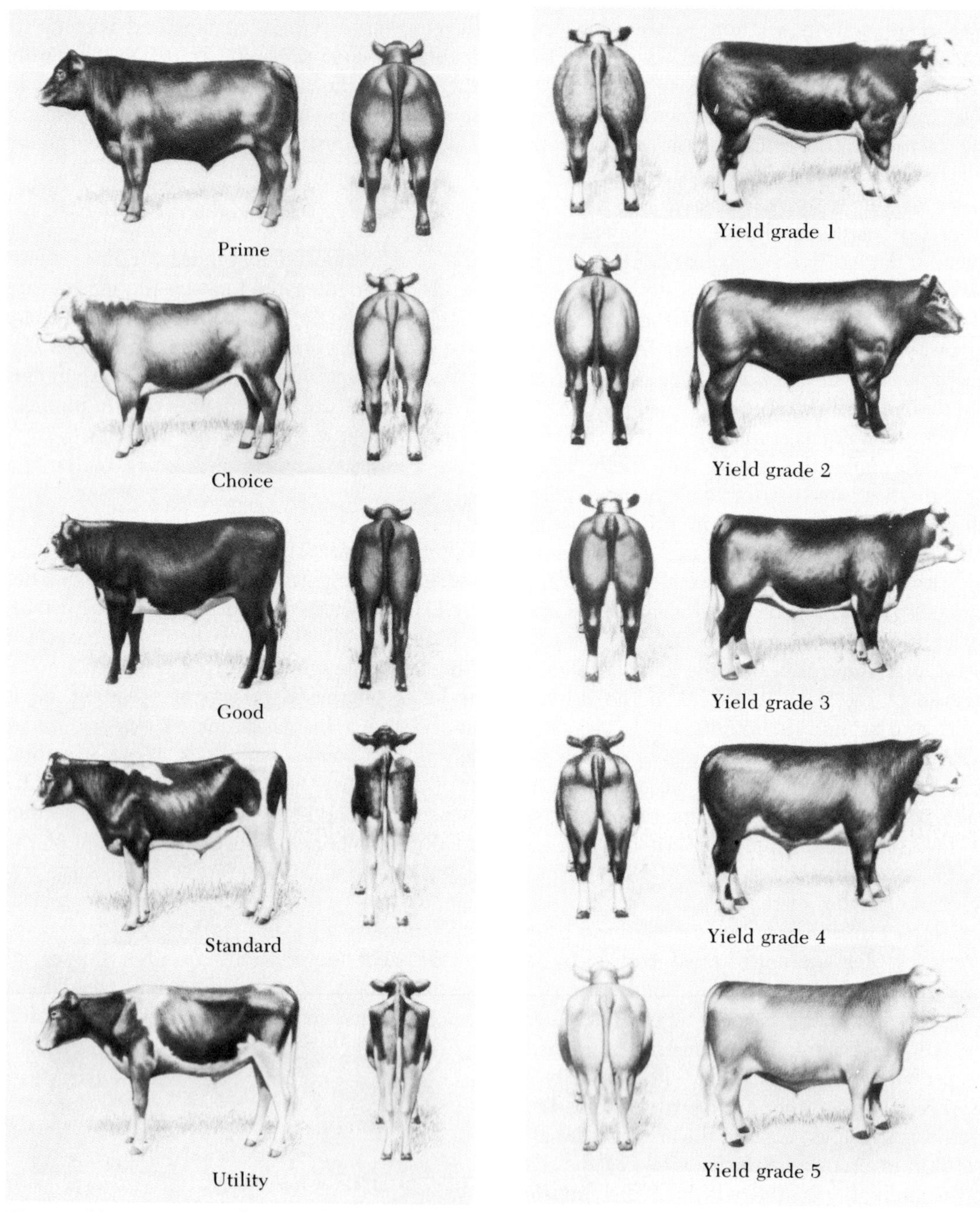

Figure 45-16 U.S. quality grades of slaughter steers. Commercial, Cutter, and Canner grades are omitted.

Figure 45-17 U.S. yield grades of slaughter steers.

Figure 45-18 U.S. grades of slaughter lambs. Cull grade is omitted.

Cattle Feeding and Meat Packing

The growth and changes in the feeding and slaughtering of cattle are typical of the developments that have had a strong influence in shaping our modern marketing system. During the late 1950's and through the 1960's, there was a significant shift in cattle feeding from small-scale, farm-feeding operations to large-scale, year-round, professionally managed, and highly automated feedlots. In 1970, 1 per cent of the feedlots in the United States produced more than half of the fed cattle. In general, the concentration in feeding occurred in areas where feed, cattle, weather, and other natural advantages prevailed. Concurrent with

U.S. No. 1

U.S. No. 2

U.S. No. 3

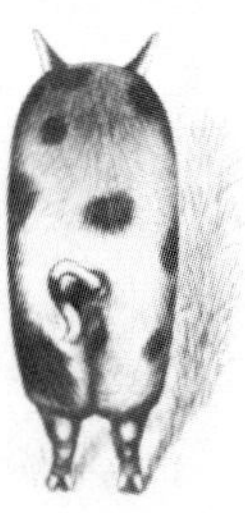

U.S. No.4

U.S. Utility

Figure 45-19 U.S. grades of slaughter swine.

this increase in size and number of very large feedlots, there has been an increase in the number of cattle fed on high energy rations, from about 50 per cent of the total beef produced in 1960, to 75 per cent in 1970. During this period, per capita annual beef consumption increased from 85 to 113 pounds.

The meat-packing industry, on the other hand, has tended to become more decentralized and also more specialized in the last 15 years. Old plants located at central markets or in other metropolitan areas have been replaced by new, modern plants located in production areas. Many of these new plants also have specialized and slaughter only one species of livestock. These new plants have utilized improved slaughtering and handling techniques, and these have resulted in greatly increased productivity and efficiency.

Methods of Marketing Livestock

Until recent years, the predominant method of marketing livestock was through terminal or central markets. These markets developed prior to or about the turn of the twentieth century. Their development was strongly influenced by the development of the refrigerated railcar and the nearby location of large packing plants. The central markets afforded a full line of services and assured producers that their livestock would have a ready sale. The stockyards company provided facilities and personnel for receiving and caring for livestock; commission firms were available to serve as expert agents for selling producers' livestock; buyers from nearby packing plants actively competed for livestock to slaughter; and order buyers were available to purchase feeder livestock for the farmer or slaughter livestock for shipment to packers at other locations.

In recent years, however, marketings through central markets have declined significantly. This has paralleled the development of the large-scale commercial livestock-feeding operations, the decentralization of the packing industry, and a better understanding and use by producers of available market information.

Selling Livestock by Auction

Selling livestock by auction is one of the oldest methods of selling, and is still a very important method of selling certain classes. For example, in 1969, nearly 45 per cent of the cows and bulls, and more than 50 per cent of the vealers and slaughter calves, were sold through auction markets. In some areas, substantial numbers of feeder cattle and feeder pigs are marketed through auctions.

Direct Marketing

The movement of livestock from country points directly to the packer has increased greatly in recent years. In 1969, over 60 per cent of the cattle and over two-thirds of the hogs and sheep were marketed direct.

Increasing numbers of livestock are also being sold direct to packers on a carcass basis. In 1969, nearly 20 per cent of the cattle, about 4 per cent of the calves and hogs, and over 6 per cent of the sheep and lambs were purchased by packers on the basis of their carcass grade and/or weight.

Meat Marketing

The list of changes in meat marketing are impressive, but only a few will be mentioned. Many of the newer plants slaughter only one species, and sell carcasses only on a carlot basis. Many of these shipments of beef and lamb go directly to retail chain-store warehouses. These developments have been quite important in the establishment of a national market for meat, in which prices in different parts of the country primarily reflect differences in delivery costs.

Fabrication of Cuts

Another recent innovation in the marketing of meat has been the large-scale fabrication of beef carcasses by some packers into trimmed, boneless or semi-boneless cuts which are "saw or knife ready" for conversion into retail cuts. Also, many of these cuts now move directly to hotels, restaurants, and other large-scale users of meat without further fabrication. Prior to this innovation, the "hotel and restaurant" trade was based on the sale of carcasses to fabrication plants in large metropolitan areas, where they were converted into oven-ready and portion-control items. Some of this is still done, but in many instances the former fabricator is now essentially a distributor, in that he only buys and sells fabricated cuts. Many retail chain stores that have large, central warehouses for receiving, handling, and distributing meat to many stores in an area, have further extended these operations to the breaking of carcasses into "saw or knife-ready" cuts prior to delivery to the retail store.

Self-Service Meats

The use of "self-service meats" in the nation's supermarkets has become the rule rather than the exception in the last 25 years. To date, these cuts are almost always prepared and packaged in the individual retail store. However, it appears likely that, in the foreseeable future, much of this cutting and packaging will be done in central locations for subsequent delivery—either fresh-chilled or frozen—to retail stores.

Market Information

Most of the foregoing changes in marketing livestock and meat could not have occurred without the national market concept. A national market requires the use of a nationally uniform method of classifying and grading, and the system of USDA grades for livestock and meat is the only such system of grades available. The widespread official use of Federal meat grades as the basis for trading in meats, and the corresponding understanding and use of Federal grades as the basis for trading in livestock, has been a requisite of our very successful national market.

The existence and success of our national market are also dependent upon a well-established national system of providing almost continuous market information. Although private-industry sources of market information are available and used by many, the USDA's livestock and meat market-reporting service makes a significant contribution to the dissemination of timely, accurate, market information. In 1970, Federal and Federal-State cooperative livestock market reports covered some 22 terminal markets, 23 direct marketing areas, and about 180 auction markets. Meat market reports also were issued for 14 producing and consuming areas. The dissemination of this information is primarily by teletype, telephone, radio, television, and newspapers.

Futures Trading

A futures market was opened for slaughter cattle in 1964 and for slaughter hogs in 1966. These are likely being used by an increasing number of livestock feeders and producers to "hedge" their operations. By "hedging," a cattle feeder, for example, can assure himself of a certain price for his cattle some months ahead of when he expects to have them finished for sale. He does this by selling cattle at a specified price for delivery at a specified future date. On the delivery date he can deliver his own cattle; he can buy other cattle (usually at the delivery point) and deliver these in fulfillment of his contract; or, more often, as the time to deliver approaches, he simply buys a contract for the same delivery period. This eliminates making an actual delivery. Thus, for the cost of the fees involved, he is assured, months in advance, of the price he will receive for his cattle. Many "speculators" also trade in livestock futures. Speculators profit from such trading only to the extent that they are better-than-average in predicting future levels of price.

If and when actual deliveries of livestock are made in connection with futures contracts, the animals in each acceptable lot are certified as meeting certain requirements for grade, class, weight range, and dressing percentage by USDA graders—the same personnel who issue the USDA livestock-market reports. Pork bellies and boneless beef are two meat commodities on which there also is futures trading. When deliveries of these are made, the examination of the delivery unit to determine whether or not it meets the contract specifications is made by official USDA meat graders. Under most conditions, practically all meat contracts also are "traded out" and relatively few actual deliveries are made.

Integration

Integration has had some impact in recent years on the livestock and meat industry, but it has not become a major influence involving livestock production. Two types of integration have been and are taking place. *Vertical integration* is when two or more phases of production, processing, or distribution are controlled by the same ownership or management, for example, cattle feeding and meat packing, or meat packing and retailing. *Horizontal integration* involves the combination of two or more enterprises of the same type into a larger unit under the same management. Both types of integration have taken place to a limited degree in the livestock and meat industry.

While we have seen many dramatic changes in the marketing of livestock and meat recently, likely we are seeing only the beginning of many more. More and more products are competing for the consumer's meat dollar, including nonmeat sources of protein. If livestock and meat are to remain competitive for the consumer's food dollar, improved technology undoubtedly will need to develop in both production and marketing. Efficiency and innovation may well be the key to the future. The Federal classification and grading system for livestock and meat also will need to change if it is to continue to meet the needs of all segments of the industry: to provide guidance for consumers; to provide a common language for marketing; and to provide the means for communicating consumer demands back to producers. It must be a dynamic tool to do a job in a dynamic industry.

FURTHER READINGS

Doty, D. M., and J. C. Pierce, 1961. *Beef Muscle Characteristics as Related to Carcass Grade, Carcass Weight, and Degree of Aging.* USDA Technical Bulletin No. 1231.

National Livestock and Meat Board. 1969. *Meat Evaluation Handbook.*

USDA standards:

- *Slaughter Cattle.* (Steers, heifers, cows, bulls, and stags.) Effective April 23, 1966.
- *Slaughter Vealers and Calves.* Effective October 1, 1956.
- *Feeder Cattle.* Effective September 25, 1964.
- *Slaughter Lambs and Sheep.* Quality, effective June 18, 1960. Yield, effective March 1, 1969.
- *Slaughter Barrows and Gilts.* Effective July 1, 1968.
- *Slaughter Sows.* Effective September 1, 1969.
- *Feeder Pigs.* Effective April 1, 1969.
- *Carcass Beef.* (Steer, heifer, cow, bull, and stag.) Effective June 1, 1965.
- *Carcass Beef.* (Bullock.) Effective July 1, 1973.
- *Veal and Calf Carcasses.* Effective October 1, 1956.
- *Lamb, Yearling Mutton, and Mutton Carcasses.* Quality, effective March 1, 1960. Yield, effective March 1, 1969.
- *Barrow and Gilt Carcasses.* Effective April 1, 1968.
- *Sow Carcasses.* Effective September 1, 1956.

Yield Grades for Beef. USDA Marketing Bulletin No. 45, July 1968.

Yield Grades for Lamb. USDA Marketing Bulletin No. 52, July 1970.

Forty-Six

The Production, Marketing, and Classification of Wool and Other Animal Fibers

Sheep, goats, and other fiber-bearing animals are raised throughout the world under a variety of production and marketing conditions. The breeds are diverse and so is the quality of fibers they produce. As one would expect, the fiber-marketing systems of the countries differ, and so do their fiber classification methods. Those countries which produce fiber for export, and thus must compete with other producers in the world market, generally have the more sophisticated marketing systems, including fiber classification.

Trends in World Wool Production

The trend in world wool production has been steadily upward, from slightly more than four billion pounds, grease basis, in 1950–1951 to more than six billion pounds in 1968–1969. Production seems to be holding at about this level, with increases in Australia, New Zealand, and Eastern Europe offsetting continuing declines in the United States, Uruguay, France, and the United Kingdom (see Chapter 49 for additional information on wool importing and exporting countries).

United States shorn wool production has declined rather steadily from nearly 400 million grease pounds at the beginning of World War II to less than 166 million pounds in 1969 (USDA, 1970e).

Trends in Mohair and Other Animal-Fiber Production*

Apart from wool, a number of other types of fine and coarse animal fibers or hair are regularly used in textile manufacturing. The chief producers of fine hair—known as specialty fibers—are the Angora goat (mohair), Auchenidaes (alpaca, llama, and vicuña), the Cashmere goat (cashmere), the Bactrian camel, and the Angora rabbit. The coarse hairs used in textiles consist primarily of common goat hair and horse hair. A fair amount of in-

* Adapted largely from *Industrial Fibers—A Review*, published by the Commonwealth Secretariat, London, S.W. 1.

formation is available on the production of and international trade in mohair and the Auchenidae group, but statistics for other types of hair are nearly nonexistent. The following sections relate to the more important types of fine hair and what information is available on their production.

Mohair

Mohair is the main specialty fiber used by the textile industry. It is largely used in the worsted sector of the wool-textile industry, and demand for it is greatly influenced by fashion changes. The finest types are manufactured into apparel fabrics and knitting yarns; other qualities go into the manufacture of upholstery and blankets, and the coarser types into carpets and rugs.

The production of mohair is concentrated almost entirely in the United States, Turkey, South Africa, and Lesotho. Total production increased steadily from 31 million pounds in 1950 to a peak of over 66 million pounds in 1965, but has subsequently declined each year in response to lower prices and generally less favorable conditions (Table 46.1).

The United States produces nearly half the world's mohair supply, and more than 97 per cent of that production is in Texas.

In South Africa, goat numbers in 1969 were expected to total 1.5 million head, continuing their small annual decline since 1965. The reduction since 1965 has been characterized by the slaughter of crossbred and older goats, with hopes of raising the general quality of mohair production in the future.

In the United States, during recent years, the average clip per goat has increased from 4 pounds to over 6 pounds; in South Africa, Turkey, and Lesotho, the average is about 7.6 pounds, 3.3 pounds, and 3.2 pounds, respectively. Angora goats are usually clipped twice a year in the United States and South Africa, the fiber length being about 4 to 6 inches for each half-year's growth. The first three shearings of young goats are usually marketed separately from those of adult goats because of its

Table 46.1.
Angora goat numbers and Mohair production in the chief producing countries.[a]

Countries	*Average 1951–1955*	*Average 1956–1960*	*1964*	*1965*	*1966*	*1967*	*1968*
			Angora goat numbers (million)				
Turkey	4.8	5.8	5.6	5.5	5.6	5.5	5.5
United States[b]	2.5	3.5	4.6	4.8	4.7	4.1	4.0
South Africa	0.6	0.9	1.7	2.0	1.9	1.7	1.6
Lesotho	0.6	0.6	0.7	0.7	0.7	0.7	0.6
Total	8.5	10.8	12.6	13.0	12.9	12.0	11.7
			Mohair production (million lb., grease basis)				
United States	13.9	21.2	29.7	32.4	29.6	27.1	26.0
Turkey	16.4	21.2	18.7	18.3	19.4	18.5	18.7
South Africa	4.9	5.9	12.5	13.4	14.0	11.7	11.2
Lesotho	1.2	1.1	2.1	2.1	2.2	2.2	2.5
Total	36.4	49.4	63.0	66.2	65.2	59.5	58.4

[a] Adapted from *Industrial Fibres—A Review*, published by the Commonwealth Secretariat, London, S.W. 1.
[b] Number of goats clipped.

higher quality. The clean yield of mohair is considerably higher than wool, and averages about 83 per cent in South Africa, 82 per cent in the United States, and 78 per cent in Turkey.

Auchenidae Group—Alpaca, Llama, and Vicuña

The animals in this group, which inhabit the high mountain regions of South America, are found chiefly in Peru, southern Equador, Bolivia, and northwestern Argentina. Good information is lacking, but the world production of auchenidae hair is believed to be around 10 million pounds annually, with Peru being by far the largest producer (see Chapter 13 for description of these fibers). Wholesale slaughter of the vicuña in recent years is reported to have reduced their numbers to around 20 thousand, and has placed them in danger of extinction. The Peruvian Government has banned the export of vicuña hair or skins, and Bolivia has imposed similar restrictions.

Cashmere

Cashmere is obtained from the cashmere goat, found in Tibet, Mongolia, China, northern India, Iran, and Afghanistan (see Chapter 13 for a description of cashmere fibers).

Camel Hair

Camel hair is of two types, the soft undercoat ranging from 1 to 5 inches in length, and a coarse outer coat with fibers measuring up to 15 inches in length. The fleece of the camel is neither sheared nor pulled, but is shed naturally, each animal yielding 5 to 10 pounds annually. The better types of hair used in the textile industry come mainly from the Bactrian camel, found in Mongolia, China, and the Soviet Union. Iran and other Middle Eastern countries are mainly producers of Arabian camel hair. Fine camel hair is of worsted spinning quality, and because of its high insulating and weaving properties finds its greatest use in high-quality overcoats and in fine knitted garments. Coarse camel hair is used for belting.

Angora Rabbit Hair

The Angora rabbit is kept in captivity in many European and Asian countries. It produces one of the finest fleeces used in the production of hand knitting yards, knitted garments, and outerwear. While sometimes spun alone, it is usually blended with wool and other fibers to provide additional softness and warmth. Each rabbit produces about 8 ounces of spinnable hair annually, of up to 5 inches in length. The main producing countries are France, China, and Japan.

Trends in Trade of Wool and Other Animal Fibers*

Mohair

World trade in mohair has rapidly expanded, from 14.7 million pounds average for 1951–1955, to 31.3 million pounds in 1956–1960, and to more than 37 million pounds in 1968. Increased production and trade in the United States and South Africa were primarily responsible. The United States first became a net exporting nation in 1953, and exports increased from less than 2 million pounds in 1951–55 to nearly 16 million pounds in 1968. Comparable figures for South Africa were 5.3 million pounds and 13.5 million pounds. Turkey's mohair exports during this period have remained around 8 million pounds, but were slightly higher in the late 1950's. Detailed distribution data for the 1950's are not available.

The United Kingdom is by far the largest importer of mohair in the world, and is the leading market for the United States and

* Adapted largely from *Industrial Fibres—A Review*, published by the Commonwealth Secretariat, London, S.W. 1.

South African exports. Italy and Japan are also very important importers of these countries' mohair. Spain has recently gained importance for South African exports. The Communist countries have met their sharply increasing mohair demands through imports from Turkey, with the Soviet Union being the largest importer. The United Kingdom is also an important market for Turkish mohair, although the trade is declining.

Auchenidae

Peru and Bolivia are the primary exporters of llama and alpaca hair, mainly to the United States and the United Kingdom, and are the only countries for which export information is available (see Table 46.2). Export of vicuña hair or skins has been banned by these countries. Bolivia reserves some 75,000 pounds of its production for a local processing plant. In the late 1960's the United States was becoming the major importer from this declining industry.

Cashmere

Complete information is not available on world trade in cashmere, but is adequate for the major trading nations (see Table 46.3). The trade relationships shown in Table 46.3 have held fairly constant over the years, as has the quantity of cashmere exported.

Table 46.2.
Exports of llama and alpaca hair from Peru and Bolivia. Figures are in thousands of pounds.

	1964	*1965*	*1966*
Peru:			
Alpaca	7,520	7,160	5,300
Llama	460	420	230
Bolivia:			
Alpaca	197	86	127
Llama	70	92	136
Total	8,247	7,758	5,793

Table 46.3.
Imports of cashmere, in thousands of pounds.

Country importing	*1964*	*1965*	*1966*	*1967*	*1968*
United Kingdom[a]	3,350	3,560	2,100	2,710	2,230
United States[b]	4,250	4,230	3,040	3,220	3,320
Japan[c]	990	1,130	1,170	1,020	1,660
Total	8,590	9,920	6,310	6,950	7,210

[a] Primarily from China and the Soviet Union.
[b] Primarily from Iran and Mongolia.
[c] Primarily from China and Mongolia.

Camel and Angora Rabbit Hair

The United Kingdom and Japan are the dominant importers of these specialty fibers. The United Kingdom is the main processor of camel hair; its imports in recent years have been under 4 million pounds, and dropped to less than 2 million pounds in 1967 and 1968. These imports came primarily from China, Czechoslovakia, and the Soviet Union.

Japan has traditionally been a notable exporter of Angora rabbit hair, but in recent years substantial quantities have been imported, primarily from China, for local processing together with its domestic production. The United Kingdom relies chiefly on China and France for its supply. Imports to these major consuming countries average about 600,000 pounds annually, with Japan accounting for three-fourths of the total.

Major Value Characteristics of Wool and Mohair

The more uniform fiber requirements of increasingly specialized textile-manufacturing equipment, and ever-changing domestic and foreign demands for wool and mohair, have created a strain in traditional methods of fiber classification. In the more advanced manufacturing countries, the day is quickly passing when companies can afford to send representatives to personally inspect and appraise each lot of wool or mohair. They are increasingly relying on laboratory test results and more objective fiber-classification systems, in addition to wanting information on an increasing number of fiber characteristics.

The physical and chemical properties of wool and mohair determine their value as textile fibers (see Chapter 7 for additional information). Their major physical characteristics, which are about all that one can expect to measure under current marketing conditions, include the fineness of the fiber, the length of the fiber, and the yield of clean fibers. Other physical factors relating to their commercial value include crimp, strength, color, character, and elasticity. Within recent years, fiber technologists have developed scientific means of measuring some of the more important physical characteristics. They have devised sampling plans and laboratory tests for the determination of such factors as fineness, length, and yield based on small quantities or samples from commercial-size lots.

Yield

Yield—or, conversely, shrinkage—is probably the most important factor in determining the value of grease (or shorn) wool, especially in the United States. For mohair yields are much higher, and the variation in yields is less. In evaluating a clip of wool, the first factors a buyer will check are its grade and yield. Wool, as shorn from the sheep, contains varying amounts of natural grease, dried perspiration, dirt, and vegetable matter, such as grass, burrs, and straw. These extraneous substances largely are removed from the wool through "scouring" or washing. Their removal results in a considerable loss in the original weight of the wool. This loss, known in the trade as "shrinkage," may range as high as 80 per cent and as low as 35 per cent. The per cent of clean wool fiber remaining is referred to as the "yield."

Experienced wool buyers are known to estimate very closely the actual shrinkage, but wide variations exist between buyers and during the season. To overcome the hazards of estimates of shrinkage, the U.S. Treasury Department, Bureau of Customs, has developed a scientific means of determining the amount of clean wool present in a lot. This development is known as "core testing," and involves drawing representative samples from bags or bales of wool using a power-driven metal tube equipped with a sharp cutting edge (Figure 46.1,A). Research results showed that estimates of yield based on core-test results were much more accurate than estimates made in the routine manner followed in traditional wool buying (Pohle *et al.*, 1958; USDA, 1949).

Increasing proportions of the domestic wool clip are baled each year, and a recent development for rapidly drawing core samples from bales is the "post-punch," designed and tested by the USDA (Figure 46.1,B).

The yield of mohair, or proportion of clean fiber, is much greater and more uniform than for wool. The average yield is about 80 per cent, but as with wool, it varies with grade and growing conditions. The yields generally decrease from around 85 to 90 per cent for "kid hair" (numerical counts 30s to 40s) to around 70 per cent for "grown" or adult hair (18s to 30s). Currently, some mohair is core tested for yield like wool, as a basis for pricing, and buyers now visually appraise most lots rather than assume the yields.

Grade

The word "grade," when referring to wool or mohair, relates to the fineness or diameter of fiber. There are two systems of wool and mohair grade terminology. For wool there are the Blood or American system and the numerical or count system. The two systems are used interchangeably in the trade, although the count system is preferred and is more specific in describing the average fiber diameter of a particular lot of wool.

A

B

Figure 46-1 Core sampling bagged wool.

The Blood or American system originally specified wool types grown on sheep with fractional quantities of Merino blood. Merino wool was called Fine. Other wools were grouped according to their relative degree of coarseness as compared with the Merino. In the Blood system the following terms are generally used: Fine, Half Blood, Three-eighths Blood, Quarter Blood, Low Quarter Blood, Common, and Braid.

The wool grade terms "numerical" and "count" are synonymous. They are associated with yarn-spinning capability, and originally designated the number of hanks of yarn that could be spun from one pound of clean wool of that particular grade. A hank is 560 yards of yarn. For example, one pound of a 64s grade of wool, when processed and spun to capacity, theoretically produced 64 hanks of yarn.

Table 46.4 shows the relationship between wool grades of the numerical or count system and of Blood system, along with the appropriate length designations by grade. The nu-

Table 46.4.
Relationship of blood and numerical wool grades and length classes by grade.[a]

	Length Classes[b]			
Grade	*Staple*	*Good French*	*Average French*	*Short French and clothing*
Fine: 64/70/80s	2.75	2.25	1.25	Under 1.25
Half Blood: 60/62s	3.0	2.5	1.5	Under 1.5
Three-eighths Blood: 56/58s	3.25	2.25	—	Under 2.25
Quarter Blood: 50/54s	3.50	2.5	—	Under 2.5
Low Quarter Blood: 46/48s	4.0	—	—	—
Common and Braid: 36/40/44s	5.0	—	—	—

[a] Adapted from von Bergen (1963) by permission of John Wiley & Sons, Inc.

[b] Minimum average requirement, except for the shortest length class, in inches.

merical system is generally used in international trade and is becoming increasingly popular in the domestic United States market.

In 1966 average fiber-diameter specifications and a maximum variability in fiber diameter in terms of microns were established for each of the 14 grades (Table 46.5). A micron is 1/25,000 inch. The standards also provided that, if the variability in fiber diameter of wool being graded exceeded that specified as maximum for wools of that average fiber diameter, the wool was assigned the next coarser grade.

The first official standards for grades of grease mohair became final in 1971. The currently proposed grease mohair standards are related to differences in processing performance, and like the wool standards are based on the numerical or count system. The proposed standards provide specifications in terms of average fiber diameter for 12 grades: finer than 40s, 40s, 36s, 32s, 30s, 28s, 26s, 24s, 22s, 20s, 18s, and coarser than 18s. The mohair specifications cover a fineness range from 23 microns for a 40s grade to 43 microns for an 18s grade (USDA, 1970c).

Table 46.5.

Measurement schedule for designating grades of wool.[a]

Grade	*Limits for average fiber diameter (microns)*	*Limit for standard deviation, maximum (microns)*
Finer than 80s	Under 17.70	3.59
80s	17.70 to 19.14	4.09
70s	19.15 to 20.59	4.59
64s	20.60 to 22.04	5.19
62s	22.05 to 23.49	5.89
60s	23.50 to 24.94	6.49
58s	24.95 to 26.39	7.09
56s	26.40 to 27.84	7.59
54s	27.85 to 29.29	8.19
50s	29.30 to 30.99	8.69
48s	31.00 to 32.69	9.09
46s	32.70 to 34.39	9.59
44s	34.40 to 36.19	10.09
40s	36.20 to 38.09	10.69
36s	38.10 to 40.20	11.19
Coarser than 36s	Over 40.20	—

[a] From USDA (1966b).

Staple Length

Staple length is one of the more important physical properties of wool and mohair relating to value. There are no official length standards for domestic wools; however, staple length classes based on objective measurement have been suggested for grades of grease wool (Pohle *et al.*, 1953).

To describe wool length, such terms as staple, good French combing, average French combing, short French combing, clothing, and stubby are used in wool trading. These terms mean that within a grade, staple wools are the longest, good French combing comes next, and so forth down to stubby. Generally, longer wools of the same grade are more valuable than shorter wools. Methods of testing have been developed which will give very reliable estimates of the average staple length and range in staple lengths within fairly uniform lots (ASTM, 1957; Johnston *et al.*, 1951). In the absence of official wool length standards, the USDA prepared some general guides that are workable and generally accepted in domestic wool trading (Table 46.4).

There are no official length standards, or even guidelines, for mohair. Fiber length is more variable in mohair than wool. An angora goat will grow a staple of 8 to 12 inches in a year. However, mohair is usually shorn twice a year and the differences in age of kids at shearing and the time between shearings add to the variation. The mohair is separated into lots of uniform lengths prior to processing and a fiber length of 4 to 6 inches for a half-year's growth is greater than the length of most other fibers with which it is blended for manufacturing.

Noilage or Wastiness

The term "noilage" or "wastiness" is a percentage figure representing the relationship between the weight of noils combed out of a given lot of wool or mohair and the sum of the weight of the top obtained and the noils. Noils are short fibers separated from the longer ones in the combing process preliminary to producing worsted yarn and fabric.

In evaluating grease wool or mohair, noilage estimates are used as a measure of wastiness. Generally, a lot is thought of as being wasty or having a potentially high noilage when the fibers are short, weak, or tangled, or show an excessive amount of weathered, brittle tips. An unsound, weak wool is made up of fibers that are weak throughout, or of fibers with a tender area or distinct break. Environmental and nutritional conditions and health of the sheep are the main factors which may cause weak wool. These characteristics are objectionable in processing since they will result in a higher percentage of wastes and noils, these items being of less value than the main product. The estimation of wastiness of raw wool or mohair is entirely subjective, and the accuracy of estimation is dependent on one's knowledge of this characteristic. After the lot has been combed into top, the percentage of top, noils, and waste can be calculated from the combing reports, but a number of lots are usually blended prior to combing, and performance of a specific lot is impossible to identify.

If the wool is processed into a woolen item instead of combed into top for worsted use, fiber weakness is relatively unimportant because far less stress is placed on the fibers in the woolen processing system. However, the woolen market is generally a lower-value market, and the raw wools and mohair are priced accordingly.

Color, Character, and Crimp

Color in raw wool and mohair is generally considered to be the non-scourable discoloration of otherwise white or cream-colored fibers. It does not include heavily pigmented wool fibers, which grew in their colored form on the sheep. Color is attributed to many factors; the major ones are urine and fecal material, bacteria, and fungi. A combination of heat and mineral or vegetable contaminants may also produce discoloration, as well as other harmful effects on the value of the fibers.

Color is an important first-impression factor to buyers (Poats and Fong, 1957). Color is associated with lower value and a reason for discounting price. Recent trends toward white and pastel shades in wool and mohair garments is the major reason for increasing emphasis on discounting colored fibers.

There is no precise meaning of the term "character," but it encompasses all those undefined characteristics that make a lot of wool or mohair attractive. In wool, generally, a major factor is the distinctiveness and uniformity of staple crimp which in turn makes the wool more appealing. Staple crimp is the natural waviness or curl occurring in the fibers. The number of waves or crimps per inch is used by wool buyers and graders as an indication of fineness or grade. Generally, the more crimps per inch, the finer the wool. Crimp also seems to have a beneficial effect on processing performance and fabric-dying properties (Pohle *et al.*, 1958).

In mohair, the same fiber characteristic as crimp is referred to as "lock-style." There are three primary categories of fleeces based on the type of lock formation; the tight lock, the flat lock, and the fluffy fleece. Angora breeders generally prefer a well-developed tight lock or ringlet. The tight lock is ringleted throughout most of its entire staple. It is the type most highly associated with fiber fineness. The flat lock is usually wavey and forms a bulky fleece. The flat lock is normally associated with heavy shearing weight and a coarser but satisfactory quality mohair. The fluffy or open fleece is low in character, often harsh, and subject to greater snagging and damage by brush as the goats graze. Buyers still object to the presence of kemp, short and brittle hairlike fibers, in mohair, but much has been accomplished through improved breeding to eliminate this problem (Chelton Research Services, 1968).

Wool and Mohair Marketing Systems

The major raw wool or mohair marketing countries are Australia, New Zealand, Argentina, South Africa, and the United States. These five countries account for nearly 60 per cent of the world's wool production, and about 65 per cent of its mohair production. Highlights of the marketing systems for the United States and for a few major wool-producing countries are briefly discussed.

The United States

The more important domestic wool and mohair processing plants are in the southeastern states, moving out of New England after World War II to gain some relief from relatively high taxes and labor costs. Today the large integrated mills are buying directly from producers or their agents. Additionally, many textile companies have merged, with the result that a few large integrated companies dominate the industry, and they are increasingly relying on manmade fibers. Thus the dominant power in the market is a few large integrated mills, whose plants are located primarily in the South, and who are relying less and less on wool and mohair.

Wool warehouses In terms of volume of wool handled, warehouses are the most important single type of marketing agency within the domestic wool-marketing system. Each year the bulk of the U.S. wool clip passes through these warehouses, either on consignment or warehouse account (purchased) or for special handling. It is estimated that perhaps 125 warehouses in the United States handle wool each year (Figure 46.2), but that only 25 to 30 of them may account for as much as 70 per cent of the wool marketed through all warehouses (O'Dell, 1969).

Most warehouses handle some portion of their wool volume on the warehouse account. About two-thirds purchase all of the wool they handle. About one-third of the warehouses, and often the larger ones, handle 50 per cent or more of their total volume on con-

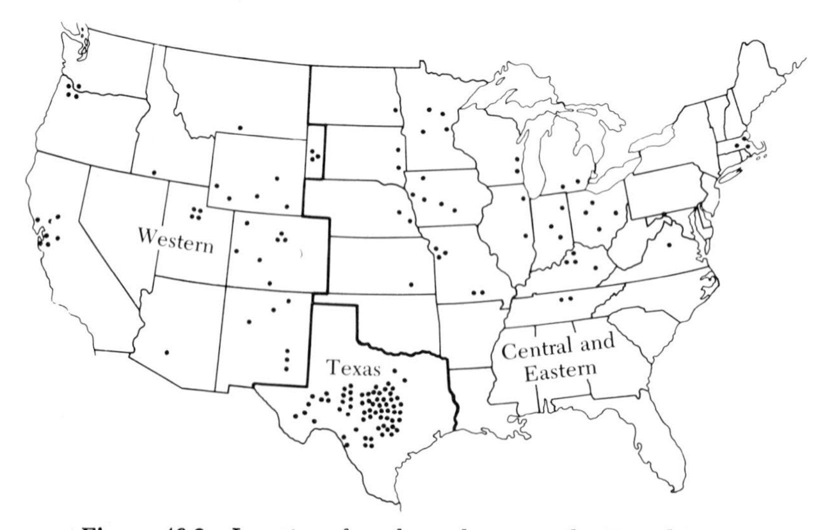

Figure 46-2 Location of wool warehouses in the United States.

signment. Operators of warehouses that consign wool prefer to do so for two major reasons: (1) the warehouse needs less cash to do business; and (2) the producer maintains the risk of market fluctuations. Over-all, warehouses in the west, north central, and southern regions purchase about 95 per cent of the total wool volume they handle.

Depending on market conditions and other factors, warehouses purchase wool at basically three stages of marketing. Wool may be purchased prior to shearing (often referred to as contracting), at the farm or ranch after shearing, or at the warehouse. The proportion of wool contracted has been declining each year. For most warehouses, contracting is the last method of procurement used in response to competitive forces in the market.

Quality and yield differences of individual lots are more likely to be considered when the wool is purchased at the warehouse rather than at the farm; there, the operator or other experienced wool and mohair appraisers are usually available, along with the potential for objective measurement of quality and yield.

To strengthen their competitive position, some warehouse operators offer a number of services to producers in addition to handling and selling wool (Jones, 1961). Approximately 20 per cent of the operators give assistance to producers in grading-up ewe flocks and selecting rams.

Services related directly to the preparation of the wool for marketing include sampling, core testing, grading (the classification of whole fleeces), sorting (the segregation of different grades within a fleece), scouring, and baling. Both the proportion of wool receiving these services and the charges made for them vary from year to year and from one region of the country to another.

Drawing samples by warehouse operators, for inspection by prospective buyers, is a form of aid, regardless of whether the wool is owned by the producer or the warehouse. Frequently these small hand samples, in conjunction with the operator's description of the lot and possibly a core-test report, are a sufficient basis for sale, even though the purchaser may not have seen the lot from which the sample was drawn.

Core testing is much more prevalent in the larger warehouses. It is estimated that more than one-third of the domestic clip is core tested annually, primarily at warehouses handling 1.5 million pounds or more.

Before 1920, wool was seldom graded at the grower level; however, since then, this practice has become more common. This grading is usually done at the warehouse where the wool is sold, or, in some instances, at the shearing pens. This service provides some quality information to buyers and sellers, and also supplies a more uniform product to the buyer. A uniform lot reduces the consideration the buyer must give to resale values of those wools in the clip which he cannot use. Grading is particularly desirable in the central and eastern states, where a variety of breeds are raised and crossbreeding is common.

Grading mohair is an operation very similar to sorting wool. The mohair fleece is not tied before bagging, and consequently the grader or sorter must work with bits and pieces of individual fleeces (sorting) instead of classing whole fleeces (grading). Until recently very little mohair was graded at warehouses. The mohair was sold "original bag" with a broad designation as "adult" or "kid" hair. Recently, however, buyers have become more exacting in their requirements, and more Texas warehouses are starting to grade mohair for fiber diameter.

The wool-sorting operation has traditionally been carried out by the topmakers and manufacturers. They maintain that sorting must be guided by the specific yarn or cloth types to be produced: the higher the quality of goods to be manufactured, the more careful the sorting that is required. But observations of mill graders and sorters in operation suggests that generally sorting requirements are not so rigid as is sometimes claimed.

Topmakers buy and use grease wool almost exclusively because of the more rigid requirements of the combing operation. To arrive at suitable blends, topmakers like to prepare and blend grease wool to meet specifications before it is scoured. Once wools are scoured, it is extremely difficult to determine how well they will comb, as large amounts of undetected short fibers may be present in them. Also wools scoured at different times may take dyes differently.

The baling of grease wool became a popular practice in larger western warehouses during the 1950's. The main object of baling by marketing firms has been to reduce transportation costs. Baling has increased the competitive position of both trucks and water carriers, with truck lines leading the way. (Few wool warehouses have steamship facilities available.)

Assembling and preparing wool and mohair for market is only part of the marketing process. The warehouse operators' objective is to sell the wool or mohair at the best price. They rely on one of two basic methods of sale: sealed bid, and private treaty. In most major wool-producing and exporting countries, the auction method is used, but it has not developed in the United States.

Mohair sales are nearly always private treaty, with a few large Texas dealers and three or four national and foreign processing firms furnishing the competition. The demand for mohair is so sporadic that sealed-bid sales generally have been unsuccessful.

Thirty-eight warehousemen surveyed in 1964 reported that nearly 85 per cent of the 52 million pounds of wool they marketed was by private treaty. Buyers purchased more than 60 per cent of this volume based only on visual appraisals. An additional 27 per cent was sold on the basis of the warehousemen's description and core tests.

Topmakers and manufacturers are the major buyers of domestic wool. Topmakers purchased about half the wool sold at warehouses in 1964, manufacturers about a third, and dealers the remainder. Except in Texas, where they acquired about a fourth of the wool, dealers were relatively unimportant buyers (O'Dell, 1969).

Local wool pools Local pools usually result from insufficient market outlets and producer dissatisfaction with local marketing conditions. They are concentrated largely in the eastern and southern United States, and in the northern regions of the Rocky Mountains, the areas with the fewest number of wool warehouses (Figure 46.3). More than 200 local wool pools exist throughout the United

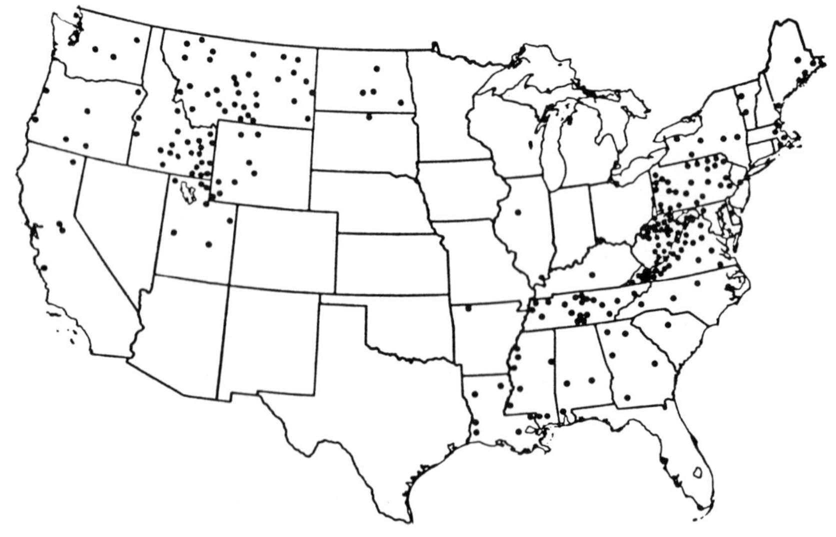

Figure 46-3 Location of wool pools in the United States.

States, handling from 5 to 10 per cent of the annual clip (O'Dell, 1969; Wilson, 1965).

In several States, local pools have attempted to overcome common problems by coordinating their marketing activities, and in some instances have organized pool associations. In Virginia, North Carolina, Georgia, and Montana, statewide associations assist local pools by arranging for personnel with some experience to help them in preparing and marketing their wool.

The ultimate success of a pool's operation is measured in terms of its ability to market wool to each grower's advantage. However, it is not uncommon for pools to penalize members with higher-quality wool by selling all wool for an average price, regardless of grade or quality differences.

Local pools could overcome their lack of services by consigning the assembled wool to a qualified marketing agency for preparation and sale, but less than 20 per cent of all pools usually employ this procedure. They prefer instead to make immediate cash sales and hold out-of-pocket costs to a minimum.

Income support program The United States government has supported the domestic market for wool and mohair for some time through tariffs, purchase and loan programs, and differential or "incentive" payments. Imports of wool and mohair textiles and apparel also are dutiable. Under the National Wool Act of 1954, import duties on wool and mohair could be used to finance incentive payments to wool and mohair producers. The purpose of this act was to encourage 300 million pounds of domestic wool production for national defense purposes. Incentive payments have been as high as 80 per cent of the market price of grease wool. The act also authorizes payment on unshorn lambs, based on a fixed amount per hundred weight of animals sold to reflect the approximate number of pounds of raw wool on the animals. This act also provided the mechanics for a wool and mohair promotion program supported by deductions from the incentive payments subject to the approval of the growers; growers have consistently approved this option.

Other Countries

Australia The Australian wool producers do the bulk of grading or classing at their ranches or "stations," and then usually consign their clips to wool brokers in one of about 15 selling centers. The shearing shed is supervised by a head classer, who separates a fleece into the various classes or grades, placing similar "sorts" together. Nearly all the warehouses supply or recommend classers to their clients, feeling that this provides a more uniform level of preparation and one that meets the Australian Wool Commission's standards.

The wool is packed into bales at the sheds by presses. The bales contain an average of 40 fleeces, and weigh about 300 pounds. Heavy jute bagging is used for baling and each bale is marked with the owner's name or brand, the classer's name, and the grade or type of wool it contains.

The grower is free to consign his clip to any broker, including cooperatives, in the center. He also may negotiate a direct private sale and bypass the brokers and auctions, but it is not a common practice. Sydney is the largest wool center, handling over 20 per cent of the clip, with Brisbane, Melbourne, Adelaide, and Geelong also ranking as significant wool markets. In these markets, the brokers handle the wool for the growers, and it usually is sold at public auctions during a ten-month period, beginning in late August and continuing through June.

Increased wool production and lower prices have created some income and marketing problems. The increased production has placed a physical strain on the facilities and on the buyer's ability to appraise the large volume of offerings at the centers. The recent and severe price declines have created problems both for the individual grower and for Australia, which depends on wool exports for a large percentage of its foreign exchange.

The current problems and the importance of wool to Australia resulted in the government's revamping the existing price-support measures and establishing the Australian Wool Commission in November 1970.

The Commission has broad powers and is responsible for operating a price-support program for wool sold at auction and for performing other functions, relating to the entire Australian clip, aimed at improving wool marketing. Specifically, the major functions and powers of the Commission are to: (1) operate a flexible reserve price system for all wool presented for sale at auctions; (2) develop and enforce standards of clip preparation of all wool, whether or not it is sold at auction; (3) develop the terms and conditions governing the sale of wool at auction and control wool auction rosters and offerings; (4) eliminate small lots from sale at auctions to the extent desirable; (5) pay advances to growers whose wool sale is delayed because of Commission requirements; (6) sell outside the auction system or have wool processed if it is being passed over at auction; (7) participate in negotiations on all service charges associated with wool marketing, including freight; and (8) operate a market intelligence unit (IWS, 1970c).

New Zealand New Zealand's wool clip is prepared for market and marketed in much the same manner as Australia's. Because of the smaller average size of clips, however, there are some differences. Classing, similar to Australia's, is usually done only on the larger clips. About all that is done at the ranch with the smaller clips is remove some of the off-sorts, such as belly wool, the seedy and burry portions, and the crutching or tags.

Beginning with the 1968–69 season, the New Zealand Wool Commission set minimum prices for the various qualities of wools, and if a particular lot on the market failed to reach the appropriate minimum price, the Commission either paid the grower a supplement up to the minimum price, or purchased the wool, or both.

Union of South Africa The classing of wool in the Union of South Africa is primarily done at the warehouses, many of which are grower-owned cooperatives. It is not as elaborate or thorough a process as in Australia or New Zealand. Only about 40 per cent is considered well-classed.

South Africa has a price-stabilization plan operated by the South African Wool Commission, comprised of growers, brokers, and government representatives. Until 1970, the Commission announced an average price at the beginning of each season to which reserve prices for more than 200 types or classes of wool were adjusted.

The South African mohair clip traditionally is sold in small lots at biweekly auctions, the summer clip marketed from March to May, the winter clip from September to November. However, low prices in 1969 and 1970, along with increasingly reduced competition at the auctions, has caused the South African Mohair Board to suggest significant changes in the system. The Board has proposed to the government that it be authorized to operate the auction sales through the brokers on a pool system, incorporating advance partial payments to the growers. The advance payment would be based on the Board's estimated value of the mohair. All mohair would be delivered to the brokers, acting as agents on behalf of the Board, to be binned or grouped into some 250 types for sale at auction. The Board would have the power to accept or reject any bid. If the auction system should fail altogether, the Board has requested authorization to act as the growers' sale agent and deal directly with processors and other buyers. This back-up proposal would also empower the Board to prepair mohair for sale by processing it up to and including top. Thus, it could sell mohair in the grease, scoured, or top form (*Wool Digest,* 1971b).

Argentina There is relatively little preparation of wool by growers in Argentina, and less than one-third of the clip is consigned to the central market warehouses in Buenos

Aires, Bahia Blanca, and Rosario, where limited grading services are offered.

Uruguay In Uruguay, wool is far more important to the national economy than in Argentina, and more emphasis is placed in its preparation and marketing. The bellies are separated from the fleece at shearing, and the wool is usually skirted at the warehouses in Montevideo, which handle about 80 per cent of the clip.

Trends in Prices and Consumption of Wool and Other Fibers in the United States

Until the late 1940's, the United States and the world depended basically upon two fibers for clothing and other textile products: cotton and wool. Following World War II, the civilian demand for apparel goods was very large. Old wardrobes needed replacing, and large numbers of servicemen returning to civilian life, were adding to the demand. The stockpiles of wool throughout the world were released to meet this demand. Then the Korean War added further demand, and wool prices increased to a peak that has never been reached before or since. In 1951, the national average price for shorn wool was 97 cents grease basis, nearly 3 times the 1970 average price. Mohair prices averaged $1.18 compared to 39 cents in 1970. They priced themselves out of the reach of the ordinary consumer and into the realm of specialty fibers. The chemical companies, with this incentive, increased their fiber-research activities, and by 1953 a larger number of new and improved manmade fibers were on the market. Three major competitors entered the market that year, and at a price very competitive with wool (Figure 46.4). Their introduction, along with subsequent manmade fibers, was accompanied by extensive advertising to create a consumer awareness of their desirable properties, particularly wash-wear. As new and improved manmade fibers became available in large supplies, fiber producers started reducing prices and competing for the growing textile market. The only way for wool prices to go was down, and they did, farm prices reaching a 30-year low in 1970. Prices in Australia, New Zealand, and the other major countries responded similarly.

Per capita mill consumption of apparel wool in the United States dropped from 2.24 pounds in 1953 to 1.08 pounds in 1969, as manmade fibers, primarily noncellulosics, captured an increasing share of the expanding apparel market. Carpet wool consumption

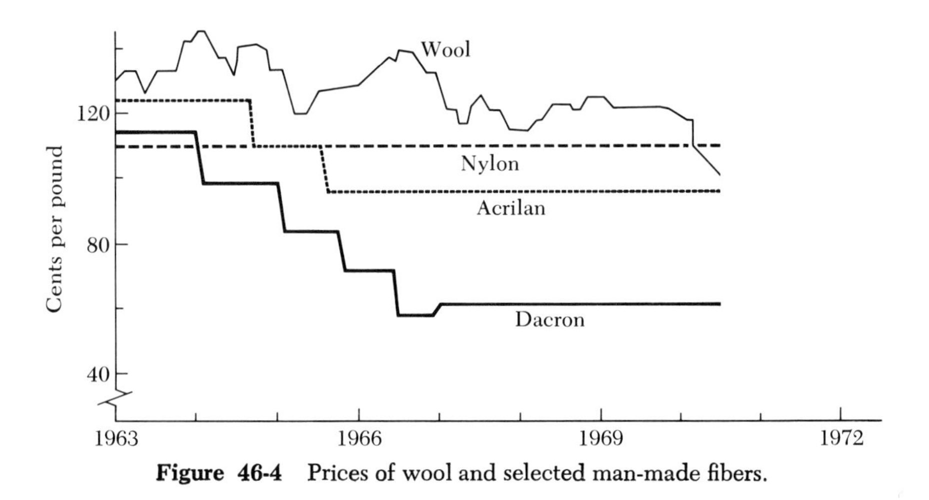

Figure 46-4 Prices of wool and selected man-made fibers.

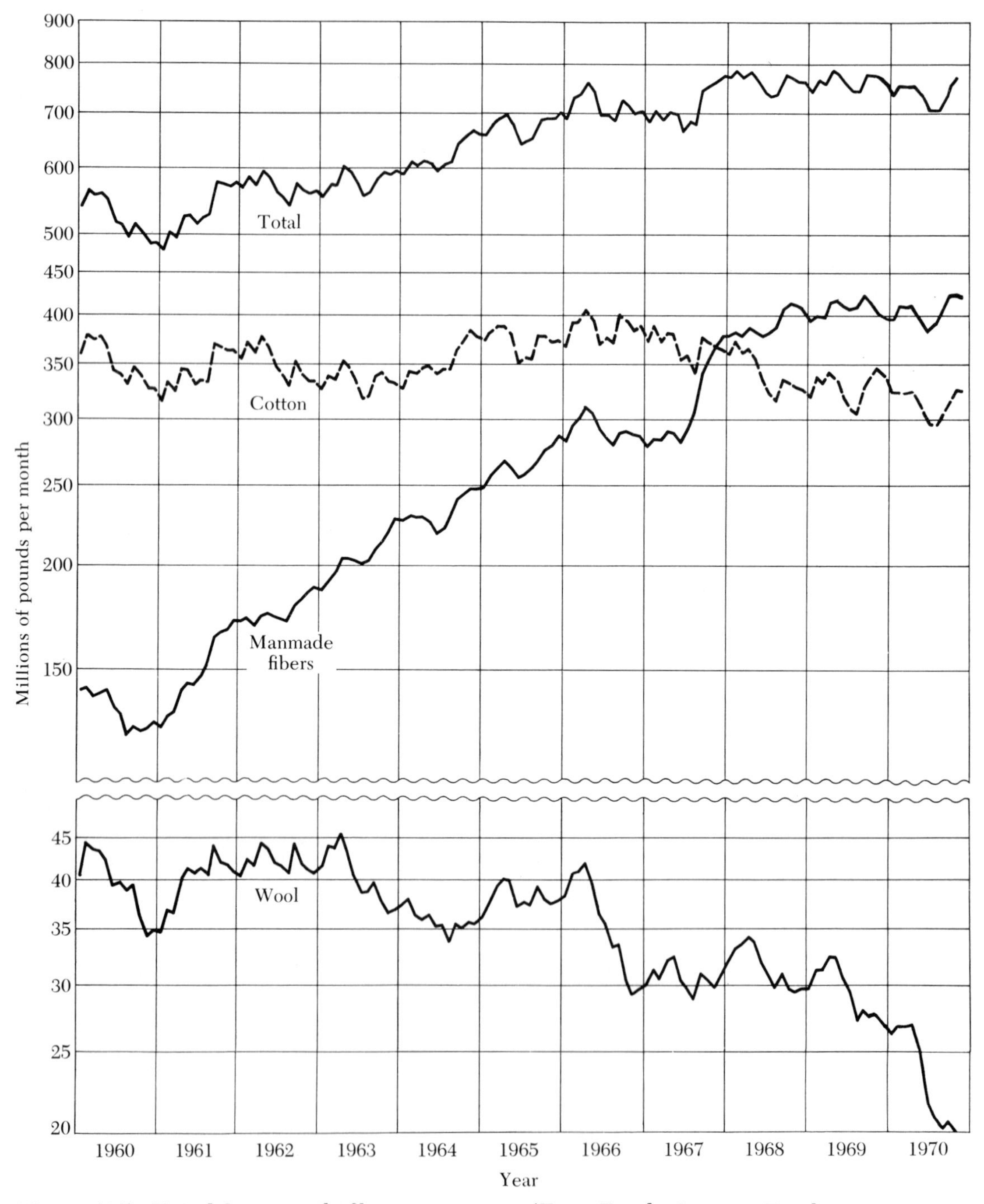

Figure 46-5 United States textile fiber consumption. (From *Textile Organon.* Textile Economics Bureau, Inc.)

dropped from 0.85 pound per capita to 0.46 pound, although total carpet production expanded nearly 40 times its 1953 level. The per-capita consumption of rayon and acetate remained nearly steady during this period, but that of the other manmade fibers increased from 1.75 to 18.51 pounds in 1969. The United States textile-mill fiber-consumption trends (Figure 46.5) clearly indicate the impact of manmade fibers on wool (the raw-wool content of imports semiprocessed and manufactured wool textiles held fairly constant during the 1960's). Mohair has also been affected, with consumption being erratic, and prices dropping more than 50 per cent between 1953 and 1970.

The major impact of manmade fibers on wool and mohair may now be over, since prices have adjusted to very competitive levels and increased usage of these fibers in blends with manmades can be expected. The market for items made entirely of wool and mohair will probably continue to decline, and per-capita consumption of these fibers may decline a little further, but population growth and growing incomes should result in an expanding textile market that will more than offset these losses.

FURTHER READINGS

American Society of Testing Materials. 1956. *Standard Method of Sampling and Testing Staple Length of Wool in the Grease.* New York: ASTM.

Clayton, L. B. 1970. "Wool in the United States: Major Trends and Prospects." *Wool Situation,* USDA, ERS, TWS-92, August 1970.

Jones, A. D. 1961. *Wool Warehouses—Practices, Facilities, Services, Charges, Problems.* Economic Research Service, Tech. Bull. 1259. Washington, D.C.: USDA.

Pohle, E. M.; *et al.* 1958. *Value-Determining Physical Properties and Characteristics of Domestic Wools.* Consumer and Marketing Service, Mktg. Res. Rpt. no. 211. Washington, D.C.: USDA.

USDA. 1966b. *Official Standards of the United States for Grades of Wool.* Consumer and Marketing Service, Service and Regulatory Announcement no. 135.

USDA. 1970c. *Grease Mohair Standards—Proposed Standards for Grade.* Consumer and Marketing Service, 7 CFR, Part 32, 1970c (Processed).

von Bergen, W. 1963. *Wool Handbook.* 3d ed., vol. I. New York: Interscience.

Wilson, D. 1965. *Wool Pools—Organization, Practices, Services, and Problems.* Farmer Cooperative Service, General Rept. 127. Washington, D.C.: USDA.

Wool Digest, published by International Wool Secretariat and the Wool Bureau, Incorporated, London. Volume XXI: no. 15/16, August 1970a; no. 19/20, October 1970b; no. 21/22, November 1970c. Volume XXII: no. 1, January 1971a; no. 2, February 1971b.

Forty-Seven

Grading and Marketing of Milk

The production of milk and the processing of fluid milk and manufactured milk products have been discussed in previous chapters. Here we will consider the marketing of fluid milk and manufactured dairy products, such as cheese, butter, and ice cream, the evolution of the modern dairy industry, and the influence of pricing plans as related to marketing.

Of all food industries, the dairy industry is the most regulated by ordinances, codes, permits, and regulations. Inspections by many jurisdictions, Federal, State, county, city, and military, are often made of a single plant. The Conference of Interstate Milk Shipments has sponsored reciprocal inspections which tend to reduce the number of interstate milk inspections and thus permit the free flow of milk between the states. Another development to reduce duplicate inspections has been the trend of states to take over milk control rather than to permit counties and cities to inspect the same plants.

Reasons for Grading

Milk must be produced, transported, and processed in a manner that will prevent its deterioration due to bacterial growth or to chemical changes. The composition of milk makes it an ideal medium for the growth of bacteria and other microorganism when time and temperature conditions are favorable for their growth. Factors influencing milk quality are discussed in Chapter 6.

In order to secure a safe, good-quality product, milk supplies must be inspected and the products tested to make certain that code and ordinance standards are being met. In addition to the inspection and testing by the health authority, most dairy-processing companies maintain quality-control laboratories to monitor the bacterial quality and composition of the milk they buy and the pasteurized milk and manufactured milk products processed in the plant.

Grading of Milk

In order to operate a fluid-milk plant, a processor must secure a permit from a local health authority. The permit is issued after it has been demonstrated that the milk supply, the buildings, and the equipment meet the required standards. The permit may be revoked if violations of grade standards are found. The most usual standard for Grade A milk is that if three out of the last five tests do not meet standards for temperature, bacterial counts, and certain other requirements, the permit could be withheld until further testing indicated that the plant could meet the standards. Milk must also meet minimum standards for fat and solids-not-fat.

The Grade A raw-milk supply must be derived from herds that meet USDA requirements for programs designed to establish tuberculosis-free and brucellosis-free areas. State programs are usually in force to reduce the incidence of mastitis, pesticides, and antibiotics in milk. Manufacturing milk standards are not as restrictive as Grade A requirements for bacterial counts and building and equipment facilities. State dairy codes are usually the controlling authority when the areas have not been preempted by Federal authorities, such as the USDA, the FDA, or the Department of Defense. Congress has set the legal requirements for butter composition and for the definition of nonfat dry milk. The FDA has established standards of identity for many cheeses and for ice cream. State codes frequently govern the composition and other properties of dairy products shipped intrastate.

The grading of milk and milk products involves essentially the supervision of production and processing facilities and procedures, and the testing of raw milk and pasteurized products to determine if code and ordinance standards are being adhered to. Modern milk production and processing has made milk a safe, uniform product in which the customer has confidence. Food-borne diseases due to pasteurized milk are extremely rare, and when they do occur the cause can usually be traced to improper pasteurization or to contamination after pasteurization.

Raw milk should not be considered safe for human consumption. Before the extensive use of pasteurization, Certified raw milk was the only milk supply that could be depended upon to be reasonably safe and free from pathogenic bacteria of either human or bovine origin. The development of tuberculosis-free and brucellosis-free herds and pasteurization has resulted in the decline of Certified milk. Most of the Certified milk now sold is pasteurized.

Nutrients Supplied by Milk

The dairy industry in the United States and in most of the developed countries of the world supplies a significant proportion of the total nutrients consumed.

About 76 per cent of the calcium, 43 per cent of the riboflavin (vitamin B_2), 37 per cent of the phosphorus, and 23 per cent of the protein in the American diet are derived from dairy products. Dairy products also provide important amounts of vitamin A, thiamine (vitamin B_1), niacin, vitamin D (from fortification), and magnesium. In certain countries of the world, the percentages of nutrients supplied by dairy products are even higher. Table 47.1 shows the total amount of milk products consumed in 17 of the countries that have developed dairying as an important source of food. As the table shows, dairying is highly developed in western Europe, in North American countries, and in Australia and New Zealand.

Japan is rapidly becoming a large consumer of all types of dairy products. Dairying is being advanced not only in Asia but also in South America. In Asia many dairy products are used in the form of nonfat dry milk re-

Table 47.1.

Consumption of fluid milk and cream, butter, cheese, and whole-milk equivalent in selected countries.[a]

Whole Milk Equivalent		*Fluid Milk and Cream*		*Butter*		*Cheese*	
Country	*Pounds*	*Country*	*Pounds*	*Country*	*Pounds*	*Country*	*Pounds*
1. Finland	1,363	1. Finland	593	1. New Zealand	40.6	1. France	28.8
2. Ireland	1,307	2. Norway	547	2. Finland	35.0	2. Switzerland	22.1
3. New Zealand	1,219	3. Ireland	487	3. Ireland	29.6	3. Denmark	20.6
4. Norway	1,024	4. New Zealand	398	4. Australia	22.9	4. Italy	19.9
5. France	1,016	5. Sweden	374	5. Denmark	20.8	5. Netherlands	19.5
6. Switzerland	968	6. Switzerland	370	6. France	19.9	6. Norway	18.9
7. Denmark	929	7. Denmark	361	7. United Kingdom	19.7	7. Sweden	18.3
8. Australia	911	8. United Kingdom	350	8. Belgium	18.7	8. Belgium	15.2
9. United Kingdom	878	9. Netherlands	337	9. West Germany	18.7	9. United Kingdom	11.0
10. Sweden	873	10. Austria	327	10. Switzerland	16.4	10. UNITED STATES	10.6
11. Belgium	843	11. Australia	304	11. Sweden	16.3	11. West Germany	9.3
12. Canada	789	12. Canada	288	12. Canada	16.2	12. Canada	9.0
13. West Germany	763	13. UNITED STATES	273	13. Austria	13.2	13. Austria	8.4
14. Austria	742	14. France	230	14. Norway	13.1	14. New Zealand	7.9
15. Netherlands	738	15. West Germany	213	15. Netherlands	5.7	15. Australia	7.8
16. UNITED STATES	565	16. Belgium	192	16. UNITED STATES	4.7	16. Finland	7.3
17. Italy	411	17. Italy	137	17. Italy	4.0	17. Ireland	5.5

[a] Figures are pounds consumed per person per year. Data for the United States is for 1969, for other countries the latest available (usually 1968). Data is from the National Dairy Council (1970).

combined with milkfat or with vegetable oil. Whole milk, cottage cheese, ice cream, and other dairy products can be made with recombined dairy products. Yogurt and other cultured dairy products have long been a staple food item in Turkey and other countries bordering upon the Mediterranean.

Surprisingly, the United States is well down the list in the consumption of both fluid milk and other dairy products. The milk consumption in the 17 countries in Table 47.1 refers chiefly to cow's milk. In other parts of the world, milk of the goat, sheep, and water buffalo is used more extensively for fluid milk and for making certain types of cheese.

Evolution of the Modern Dairy Industry

Dairy farming and dairy-product processing as we know them today in the United States and in some of the developed countries are of fairly recent origin. Many inventions that have made possible today's practices were developed during 1870–1890. The major inventions were the centrifugal cream separator (1878), the glass milk bottle (1886), and the Babcock milkfat test (1890). The work of Pasteur (1864), which resulted in the pasteurization process, the development of mechanical refrigeration (1834) and the start of tuberculin testing of cows (1890) all contributed to quality control of products, as well as to the growth of dairying as an industry.

Improvements that caused rapid changes in the fluid-milk segment of the industry came a little later, principally during 1920–1940. The major events were the wax-coated fiber carton, the high-temperature-short-time (HTST) plate pasteurizer, the homogenizer, and the use of stainless steel equipment. Important developments since the 1950's that have further accelerated changes in fluid-milk processing include: welded sanitary stainless steel pipelines; the cleaning and sanitization in place (CIP) of pipelines and equipment; and the remote control of air-actuated valves for automation of all processes.

Major developments in dairy farming are also of recent origin. The nearly universal use of milking machines, pipeline milkers, milking parlors, CIP cleaning and sanitization of equipment, and the bulk farm cooling/storage tanks with tanker pickup all contributed to making dairying a business rather than a sideline operation. Milk is picked up from farm bulk-cooling storage tanks by tanker trucks, some with a capacity of 4,000 gallons. This system has largely displaced the use of ten-gallon cans at the fluid-milk tank, and has made the hauler the receiver of the milk; he checks the milk for flavor at the farm, records the temperature, and takes samples for fat, nonfat milk solids, and bacterial analysis. Better blood lines, due largely to artificial insemination, better feeding, and culling on the basis of production, have greatly increased the production of milk. On large Grade A dairies the average production per year per cow may average from 10,000 to 14,000 pounds of 3.5 per cent fat milk. The changes in the period since 1950 have resulted in marked decreases in the numbers of dairy farms and processing plants. The increase in the size of dairy-farm operations in the fluid or Grade A segment is shown by the daily average delivery of milk per producer in the Federal Market Order markets, which in 1950 was 325 pounds, in 1968 1,098 pounds.

There are two dairy industries in the United States, the fluid-milk industry and the manufacturing-milk industry. Although they are distinct in the type of products processed and in the degree of public supervision involved, they are somewhat dependent on each other in price structure. Table 47.2 shows the per cent of the total milk supply used for fluid and manufacturing milk purposes. The fluid-milk industry utilized nearly half of the total (47.4 per cent). The remainder (52.6 per cent) was used for manufacturing products, such as butter, cheese, evaporated milk, and cottage cheese. The total yearly milk production in the United States is approximately 114 billion pounds.

Table 47.2.
Use of total United States milk supply, 1969.[a]

Dairy food	*Per cent*
Fluid milk and fluid cream	47.4
Butter	20.7
Cheese	16.4
Ice cream	9.5
Evaporated and condensed milk	3.4
Other: eggnog, yogurt, sour cream, etc.	2.4
Total	100.0

[a] National Dairy Council (1970).

The Pricing of Milk

The pricing of milk is one of the most complex problems in the dairy industry, and the establishment of equitable prices is related to the stability of the industry, to the amount of milk available for fluid consumption, and to the surplus that can be used for manufacturing dairy products. Historically, milk used for manufactured dairy products brings a lower price than does milk for fluid consumption. The price is thus related to the surplus problem and to its effect in reducing the blend price to Grade A producers. The blended price refers to the weighted average of milk sold as Class I for bottling and the portion as surplus that is utilized as manufacturing milk.

In the past 50 years, many price plans have been developed in an attempt to secure a stable market and to secure equitable prices for producers, processors, and consumers (Henderson, 1971). The problems relative to pricing have resulted in two significant developments: (1) the rise of dairymen's cooperative bargaining associations; and (2) federal and state programs for setting minimum producer prices. The Federal Milk Marketing Order program was started in 1938 as an attempt to stabilize milk marketing. At the present time there are 62 Market Orders that cover most of the metropolitan milk markets and over half of the fluid milk sold. In some areas that do not have Federal Markets, State programs apply.

The primary function of the Federal Milk Marketing Order is to define, in specific marketing areas, the terms under which handlers engaged in milk distribution purchase milk from dairy farmers. The Orders do not control production, guarantee a market for any producer, or set price ceilings to producers, but they do establish price floors consistent with local and general economic conditions affecting the supply and demand for milk. The producer, through his cooperative association, is still able to negotiate prices higher than the Class I price set as a minimum by the Order. The Order in a marketing area is administered by a local Market Administrator who is a representative of the Dairy Division of the USDA.

If the price of Class I milk is set too high in relation to that available from other areas, production will be stimulated and this results in unnecessary surpluses that must be used for manufacturing milk. This, in effect, defeats the purpose of the Order, which generally sets the price after investigating the supply-demand factors for milk. Surpluses are generally due to negotiated premiums or the use of unrealistic "formulas" based on economic factors that are not related to the farmers' decisions to increase or decrease milk production.

Following World War II, a second program was established by the federal government to aid in maintaining prices of manufacturing milk (including Grade A surpluses). The "support" program was designed to maintain milk prices at not less than 75 per cent or more than 90 per cent of parity (Public Law 439, 1949). Parity refers to a percentage of the price the Secretary of Agriculture determines as necessary in order to assure an adequate supply. The need for the program was the result of increased milk production encouraged by the government during World War II, and of an imbalance between Class I and manufacturing milk prices.

Manufacturing milk prices are supported by the USDA through purchases of butter, cheese, and nonfat dry milk at the price that will equal the announced per cent of parity. The products are stored and later disposed of in various ways; welfare and aid programs, school lunch programs, and assistance to developing countries are the major beneficiaries. Stored products may be sold when the market equals the support price plus storage costs. Each year the Secretary of Agriculture reviews the status of milk prices and probable production and announces the new price. The support program could be greatly reduced if the price of Class I milk in certain marketing areas did not stimulate unnecessary surpluses.

Productivity in the Dairy Industry

Increases in the productivity of labor in the dairy industry, including production on the farm and in the processing plants, are important from the standpoint of dairy-product marketing. These developments have resulted in slower escalation of prices of dairy products when compared with many other foods and services, and has resulted in continuation of milk and dairy products as good values in the food budget.

The number of minutes of factory labor required to earn a half gallon of milk illustrates the effect of productivity of labor and facilities in the industry: in 1947 it was 18 minutes; in 1961 it was 12.5 minutes; and in 1969 it was 10 minutes. Dairy-farm labor output, related to milk production, has more than doubled during the period 1960 to 1968 (USDA, 1969a), largely because of the factors discussed relative to the evolution of the modern dairy industry.

Processing plants also have increased the productivity of labor due largely to developments introduced during the last 15 years. In Figure 47.1 the pasteurizing department of a modern milk plant illustrates some of the developments that have made possible the high productivity of labor. The high-temperature-short-time (HTST) pasteurizing system, with its control mechanisms, such as the sealed timing pump, flow-diversion valve, air-actuated valves, and temperature controls, have resulted in continuous operation at a high rate. The cleaning in place (CIP) systems with digital programmers have further reduced labor in the plant. The sciences of metallurgy, refrigeration, heat transfer, and electronics have made possible the development of the modern processing plant. Industry associations, the U.S. Public Health Service, and dairy engineers have guided the developments in dairy equipment improvement.

Fluid Milk Marketing

The United States Public Health Service Grade A Pasteurized Milk Ordinance (1965) forms the basis for the production, processing, and distribution of fluid milk. Most of the states and many counties and cities have adopted the Ordinance as the basis for their milk codes. Other jurisdictions that have not formally adopted the Ordinance do use it as the guide in forming their codes.

Fluid-milk marketing can be divided into two activities: home delivery and wholesale distribution.

Home Delivery

In the evolution of the fluid-milk industry, home delivery was the first method used. Before about 1900, raw milk was dipped from vessels and deposited in the customer's bowl, pan, or other container. The invention of the glass milk bottle and pasteurization, per-

Figure 47-1 A modern fluid-milk processing department: Control panel (A), homogenizer (B), holding tube (C) (minimum pasteurizing conditions, 161°F for 15 seconds), high-temperature-short-time (HTST) press plates for heating and cooling of milk (D), constant-level supply tank (E), milk storage tank (F), centrifugal booster pump (G) (the homogenizer is used as a timing pump), and flow diversion valve (H). If milk is not up to temperature, it is returned to the constant-level tank and recirculated through the press. (Courtesy, Cherry-Burrel Corp, Cedar Rapids, Iowa.)

mitting longer shelf life, made possible the present-day home-delivery system.

The home-delivery system in recent years has been declining as a major factor in milk distribution. In 1940 approximately 70 per cent of the milk was distributed by home delivery. Currently 15 to 40 per cent is distributed by this method. Areas vary in the amount of home-delivered milk. The productivity of labor on a home route cannot equal that of a wholesale route, since a driver is limited in the amount of product he can distribute in one day. The retail driver may make 100 stops to deliver 500 units (quarts) of milk, whereas the wholesale driver may not only deliver 5,000 units at one trip, but make several trips to supermarkets. Some methods that have been tried to increase productivity on wholesale routes include reduced delivery days, quantity discounts, elimination of 1 to 3 quart customers, and consolidation of routes. Increases in wages unaccompanied by proportionate increases in productivity makes it imperative to improve home-delivery procedures if it is to continue as an important part of the fluid-milk distribution system.

Wholesale Distribution

Currently, wholesale distribution of milk in the United States accounts for about 80 per cent of the total. Wholesale methods vary depending on the extent of the service requirements. The most common and the most expensive method is full service. With this method the driver deposits the milk in the store cabinets and often marks the store price on the cartons. A second and cheaper method is the drop-shipment service, by which the driver deposits the milk on the receiving dock of the store. Because of pricing arrangements the shipment is usually an entire load of milk. Store personnel stock the cases, mark cartons, and place the excess milk in storage cabinets. The least-expensive method is dock delivery. With this procedure, the customer himself uses either his own truck or that of a contract hauler to take delivery of the milk at the dock of the processing plant. This procedure is often used by large supermarkets; the company truck may deliver the milk directly to the individual stores or to a central warehouse, where it is reloaded on other trucks that also deliver eggs and other products.

The trends in milk marketing are being influenced by the changes in the structure of the industry: control of raw milk supplies and handling of surpluses by regional dairy cooperatives; the merger of Federal Milk Marketing areas into larger units; the increased market power of supermarkets and independent buying chains; and the growth in size and the decrease in numbers of the dairy-processing companies. In addition to the above changes, the trend of large supermarket

Figure 47-2 Delivery of milk and dairy products to Thais who live on Bangkok's countless waterways. Similar methods are used in Hong Kong. (Courtesy, Foremost-McKesson, Inc., San Francisco, Calif.)

Figure 47-3 Milk delivery truck used in Bangkok, Thailand, to serve stores and other sales outlets. This is the type of wholesale truck used in the United States, and indicates the modernization of the dairy industry in Southeast Asia. (Courtesy, Foremost-McKesson, Inc., San Francisco, Calif.)

chains to establish their own processing plants and the entry of some large cooperatives as processors has decreased the market power of the dairy companies. Since the processor now has less control of his raw milk supply and less control of marketing, he has tended to grow in size as a processor to reduce his costs and to serve a larger geographical area.

Marketing of Manufactured Milk Products

The milk for this segment of the dairy industry comes from two sources, as previously mentioned. The volume of milk available for butter, cheese, and other products varies considerably, being greater in the spring and summer than in the fall and winter.

Table 47.3 shows the amounts of whole milk equivalent required to produce a given amount of major dairy products. For example, 21.7 pounds of whole milk are required to make one pound of butter. Skimmilk and buttermilk are by-products of this operation; they are concentrated and then dried for human food.

Butter

Prior to the invention of the centrifugal cream separator, cream was separated by gravity, by placing the milk in shallow pans, allowing the cream to rise, and then skimming it off. This was a very inefficient method, and did not re-

Table 47.3.
Pounds of whole milk required to yield some manufactured milk products.

To make one pound of	*Pounds of whole milk needed*[a]
Butter	21.7
Cheese	9.9
Evaporated milk	2.1
Condensed milk	2.9
Dry whole milk	7.9
Dry cream	19.0
Ice cream[b]	15.0
Nonfat dry milk	11.2[c]
Cottage cheese	6.3[c]

[a] Composition: 9.1 per cent milk-solids-not-fat; 3.7 per cent milkfat.
[b] One gallon.
[c] Nonfat milk.

sult in cream with satisfactory characteristics for commercial production. Most of the butter was made at home and either used at home or sold to neighbors. The invention of the separator made it possible to furnish cream with a satisfactory fat percentage for commercial churning. At this stage in the development of the industry, the cream was held on the farm until sufficient amounts were collected to ship to the creamery. The skimmilk was retained on the farm and fed to hogs, calves, or chickens. The butter made in the "centralizer" creameries was not of very good quality, since the cream was often received at the creamery in poor condition. The butter often scored 90 or lower, whereas good-quality butter scores 92 or 93.

The next development was design of systems to process skimmilk for use as human food. The invention of the milk dryer and the condenser made it possible to dry the nonfat milk and use it for human consumption where it was a more valuable product than if used for animal feed. The dairyman shipped whole milk to the creamery instead of cream. This procedure resulted in better-quality butter. During World War II, the demand for food resulted in the establishment of many milk-drying plants. Butter was also a major product of the new plants. Most of the new plants were built by dairymen's cooperatives with some government assistance. After the war the continued high rate of milk production resulted in depressed prices. At this point the Federal "support" programs previously discussed were established.

The per-capita consumption of butter has been declining in the United States due to the competition from oleomargarine. The improved quality and variety of oleomargarine, coupled with the lower price, has resulted in the loss of 65 per cent of the prewar butter market to this competitor. In 1950 the per-capita consumption of butter in the United States was 9.1 pounds, compared with 4.7 pounds in 1969. As Table 47.2 shows, the United States is 16th in the list of 17 countries reported. From the data available, it appears likely that butter consumption will become stabilized at approximately its current level.

The composition of butter was established by Congress to consist of not less than 80 per cent milkfat. Most of the butter made in the United States is churned by large cooperative creameries. It is generally packaged in 60-pound cubes and sent to special packaging plants in urban areas, where it is "printed" as ¼, 1, or 2 pound cubes. Butter is graded by a USDA grader who may be in residence at the churning plant or the packaging plant. The top grades of butter are 92 or 93, and are the ones most commonly sold in grocery stores. Butter with grades of 90 or 91 are lower in quality and command a lower price.

The USDA Marketing Services publishes the price of butter daily in the market-news page of most large newspapers. The price is that paid for most of the butter sold in a particular market. The support price of manufacturing milk establishes a floor price below which butter will not be sold. The butter and cheese prices are of importance to the fluid-milk producer, since the Class I prices of most of the Federal Market Orders are based on a premium over the price of milk sold in Minnesota and Wisconsin as manufacturing milk.

Cheese

Many types of hard cheese are produced in the major dairying countries; Swiss, Edam, Guda, Romano, and Cheddar, to name a few of the most important kinds. In contrast to butter consumption in the United States, the per-capita consumption of cheese has continued to rise, from 7.6 pounds in 1950 to 10.6 pounds in 1969. In 1968 (Table 47.2) the United States was 10th in cheese consumption of the 17 countries reported; France was first with 28.8 pounds, and Switzerland was second with 22.1 pounds.

The cheese factories in the past were generally small operations. The trend as with other parts of the industry has been to larger plants. Mechanization of the cheesemaking process and development of equipment with large capacity has resulted in more large plants and the closing of many smaller ones.

Whey is produced from the cheesemaking process. The whey-disposal problem grew in importance with the increased size of the cheese plants. When the plants were small, the dairyman delivered his milk to the cheese plant, which was usually only a few miles from the farm. The whey-disposal problem was simple: the dairyman took it home in 10-gallon cans to feed to hogs. As whey volume increased in the larger plants, its disposal became a problem. Disposal by putting whey in public sewers or on land-settling ponds or in streams became environmentally unacceptable. Whey is better utilized as human food. When dried and graded, USDA Extra Grade, dried whey can be used in many food products. Whey is also a valuable source of lactose (milk sugar) and the whey proteins.

Although cheese is made in many plants, only a few marketers handle most of the product. Basic prices are based on the Plymouth, Wisconsin, auction, from which the price is determined each week.

Evaporated and Condensed Milk

The development of the evaporated-milk plant was one of the first "factory-type" operations of a manufactured dairy product. The large investment in plant and equipment and the bulkiness of the product requires that plants be located near milk production areas and not too far distant from marketing centers. The market for evaporated milk in the United States has steadily decreased, from 18.5 pounds per capita in 1950 to 6 pounds in 1969. The decrease has been attributed largely to the quality and availability of bottled fluid milk and to the development of coffee whiteners. Much of the evaporated milk now sold is used for infant feeding or for cooking. The market for evaporated milk may be expected to continue to decline at a slow rate.

Condensed milk (more correctly, sweetened condensed milk) is not now an important dairy product in the United States. The consumption remains fairly constant at 2 pounds per capita. Sweetened condensed milk is an important item in certain Asian countries.

Evaporated and condensed milks are usually marketed through wholesale grocers, who in turn supply the retail stores. The prices are set by demand-supply factors. Private-labeled products for wholesale grocers or for supermarkets comprise a large share of the market. A few large companies process almost all of the evaporated milk in the country.

Ice Cream

The per-capita consumption of ice cream in the United States has remained essentially static for the last 20 years, from 14.8 quarts in 1950 as compared with 15.2 quarts in 1969. The amounts of all frozen desserts, that is ice cream, ice milk, and sherberts, have increased from 16.4 quarts in 1950 to 21.2 quarts in

1969. A number of factors have contributed to this result: concern of consumers for weight control; the saturated-animal-fat, cholesterol controversy; competition from imitation (mellorine) ice cream and other desserts, such as gelatine and puddings.

The manufacture of ice cream in the United States is being concentrated in large plants that distribute over wide areas, frequently in a number of adjacent states. The cost of labor has made it necessary to use large efficient equipment in the production of frozen desserts. The number of small ice-cream plants is rapidly declining.

All states have composition and quality requirements for ice cream and other frozen desserts made within their jurisdictions, and federal standards control the composition of products shipped interstate. Military specifications cover the composition and production of products that are shipped to military installations. A number of states (12) have standards for imitation ice cream and ice milk. There are no Federal standards for these products.

Twenty years ago a significant portion of ice cream was sold in "ice cream parlors" or drugstore fountains. Most of the product was packaged in 2, 3, or 5 gallon bulk containers. The ice cream was "dipped" for cones or sundaes or for filling small cartons. A smaller portion of the total production was packaged in the plant in pint and quart containers, and the novelty business was largely confined to the chocolate-coated bar. Today the ice-cream store or drugstore fountain is becoming a thing of the past, except for small specialty shops. The bulk-can sales have become relatively minor, and the largest part of the ice cream is sold in half-gallon rectangle or round containers. The novelty business, involving bars of all kinds and shapes, is a significant part of the total production of an ice-cream plant.

Thirty years ago the grocery-store supermarket seldom sold packaged ice cream, whereas now it is a major distributor. Often a chain of supermarkets will own and operate an ice-cream plant to supply their stores. There has been a tendency to make ice cream a loss leader in the stores, and as a result the competition has made it difficult to manufacture a really high-quality product. The federal standards have tended to control the minimum composition and weight of ice cream.

Miscellaneous Manufactured Milk Products

Cottage cheese, yogurt, sour cream, sour half and half, and toppings are products that are increasing in importance as consumer items. For example, in 1950 the per capita consumption of cottage cheese was 3.1 pounds, compared with 4.8 pounds in 1969. When a good-quality product has been made, the consumption has been approximately 10 pounds in some areas.

The Impact of Filled and Imitation Products on the Dairy Industry

As previously discussed, oleomargarine was the first important imitation dairy product. In recent years other imitation and filled dairy products have been developed. The price motive has been the major factor for the development of imitation products. Filled milk usually sells for five cents less per quart than milk. It is reported that coffee whiteners have taken 35 per cent of the market for products used to whiten coffee (half and half and table cream). Whipped toppings have an even larger percentage of the whipping-cream market. The imitation fluid-milk market potential can become serious competition to the fluid-milk industry.

Filled Products

Filled milk products are those to which has been added or into which has been incorporated any fat or oil other than milkfat to make them resemble milk products (Weike,

1969). Products of this class cannot be shipped in interstate commerce since this would be a violation of the Congressional Filled Milk Act. A product that does not contain milk products and is properly labeled can be shipped interstate. About 30 states also have laws that prevent their shipment intrastate. Filled milks have been introduced into a few markets where they are permitted. Sales at first have often captured a small percentage of the total fluid milk market but then have declined. Filled milks are generally made from nonfat dry milk powder and hydrogenated coconut oil. Research has not demonstrated that filled milks are nutritionally equal to milk, especially for infants and children.

Imitation Dairy Products

The imitation or synthetic dairy products are made with ingredients that are not identical with those found in milk. The protein, however, is usually furnished by sodium caseinate, a manufactured dairy product. Technological developments are improving the flavor and other characteristics of soy-bean protein, and it eventually may replace casein in the synthetic products. Coffee whiteners and whipped toppings are usually synthetic rather than filled products.

Imitation milk has not been a very successful product because of its flavor characteristics, but improvements in the ingredients could make it more acceptable. To date the FDA has not been able to develop standards for the imitation products that will improve their nutritional characteristics.

Future of Milk Marketing

The dairy industry is an important segment of the food industry and is essential for the nutrition of infants and children. Dairy products are also desirable for the good nutrition of adults, especially for pregnant and lactating women. The industry will continue to grow and expand as population increases and as educational programs of the National Dairy Council and the American Dairy Association further acquaint the public with the unique qualities of dairy products. In areas where the National Dairy Council maintains affiliates, the per-capita consumption of milk is 43 pints or about 18 per cent over that in areas where the programs are not available. The current per-capita consumption of milk is approximately 220 pints, and this has been declining in recent years. The structure of the dairy industry will continue to change; the urban character of the population will favor larger processing plants covering wider areas, possibly to distances of 300 to 500 miles. The dairymen's cooperatives will control more of the Class I milk supply and maintain country plants to process the surpluses. Milk-pricing plans will be adjusted to make dairy products more competitive with the imitations. A greater percentage of the total milk supply will be Class I as the pressure of population and the demands for high fat products decreases.

Home delivery may remain at 20 to 25 per cent of the total milk market if proper adjustments are made in the costs of this type of distribution. Wholesale distribution will continue to become more concentrated in the hands of supermarkets and other sales outlets that are growing in importance, such as gallon jug stores, independent distributors, vending machines, and convenience stores.

The industry has advanced with the growth of sciences: bacteriology, chemistry, engineering, biochemistry, and animal husbandry being the most important. These sciences will continue to alter the industry through improvements and development of dairy products that may replace as much as 25 per cent of the fluid-milk market. Sterile milk and sterile-milk concentrates, frozen concentrates, and instant whole-milk powders are among the most likely candidates. The low-fat dairy products also will become more important; ice milk, 2 per cent fat milk with added milk solids, partially creamed cottage cheese (2 per cent fat), and low-fat yogurt will increase their share of the market. In the manufacturing field, butter spreads, low-fat cheese, and other products may be expected to be developed as a result of research in dairy technology.

FURTHER READINGS

Arbuckle, W. S. 1966. *Ice Cream.* Westport, Conn.: AVI.

Henderson, J. L. 1971. *The Fluid-milk Industry,* 3d ed. Westport, Conn.: AVI.

USDA. 1966. "Changes in farm productivity and efficiency; A summary report." *Statistical Bull.* 233, Washington, D.C.: USDA.

USDA. 1969. *Summary of Major Provisions of Federal Milk Marketing Orders.* Washington, D.C.: USDA.

Webb, B. H. and E. O. Whittier. 1970. *Byproducts from Milk.* 2d ed. Westport, Conn.: AVI.

Forty-Eight

Processing and Marketing of Poultry and Eggs

"Ignorance of fundamentals is a greater obstacle to progress than lack of popular appreciation of known facts."

Whitman J. Jordan, 1851–1931

One purpose of this chapter will be to describe how living domestic birds are converted into human food by poultry processing. The many steps involved are based on necessity, tradition, government regulations, art, science, and technology. The pathways by which eggs move from the hen to the consumer are not at all similar to those which apply to poultry meat.

Poultry and eggs were among the last of the agricultural products to be produced industrially in the United States. Until well into this century, most poultry and eggs moved directly from producer to consumer. In the 1930's backyard flocks were disappearing rapidly, but World War II encouraged them again, along with "Victory" gardens, to increase food production. Following World War II, small flocks decreased, and now probably provide less than 5 per cent of the total poultry production of the country.

Originally poultry was delivered alive to the city and sold to special poultry shops. Typically birds were selected by the purchaser, then slaughtered, dressed, and prepared for cooking by the poultry shopkeeper. Long before the final closure of the live poultry markets, poultry was slaughtered and New York dressed (blood and feathers only removed) at processing plants located in the areas of production. It was not until about 1950 that full preparation of the birds at the processing plant became a common practice; not until refrigeration, sanitation, and handling practices generally were greatly improved did ready-to-cook poultry become a reality in retail stores.

In comparison, the changes over the years in the marketing of eggs have been small. Typically eggs are taken to an egg packing plant, sorted for size and quality, packaged, distributed through a varying number of "middle men," and sold to consumers. Today packaging material of many kinds is employed at all stages of the marketing process. The other change is that handling, transporting, grading, sizing, and all other facets of distribution have been mechanized.

The three major types of poultry are broiler-fryers, turkeys, and fowl (cull hens). Generally speaking, young chickens are called broilers east of the Rocky Mountains and fryers in the west. The term broiler is sometimes used for slightly smaller birds than fryers, but even this difference is rapidly disappearing. Fowl refers to hens that have completed profitable egg production. The USDA defines poultry as follows: Poultry means any domesticated bird (chickens, turkeys, ducks, geese or guineas), whether live or dead (USDA, 1971e). Other birds used for food, such as squabs, pheasants and domestically grown game birds, are not covered by present regulations. A voluntary inspection service is available on a fee basis from the USDA, and some states have regulations covering the processing of these other birds. Each class is divided by age and sometimes sex into several groups as shown in Table 48.1.

The name Rock Cornish game hen is an anomaly. These are small, young birds usually sold frozen, whole, and stuffed, to be cooked for an individual serving. Legally, they are supposed to be derived from stock in which there is Plymouth Rock and Cornish Game blood. In practice, there is no way of determining whether this is true; so any small bird of either sex, if slaughtered and processed at 1 to 1.5 pounds live weight, is commonly labeled Rock Cornish game hen.

Table 48.1.
Market classes of poultry.[a]

Species	*Class*	*Sex*	*Age*
Chicken	Cornish game hen	Either	5–7 weeks
Chicken	Broiler or fryer	Either	9–12 weeks
Chicken	Roaster	Either	3–5 months
Chicken	Hen or stewing	Female	Over 10 months
Turkey	Fryer-roaster	Either	Under 16 weeks
Turkey	Young hen	Female	5–7 months
Turkey	Young tom	Male	5–7 months
Duck	Roaster duckling	Either	Under 16 weeks
Goose	Young	Either	

[a] USDA, 1971e.

Processing Poultry

There are few differences in the manner in which any of the classes of poultry are processed. Special crews rather than producers take care of the collection and hauling of most poultry from the farm to the processing plant. These crews usually are not employees of either the producer or the processor, but operate independently. The collection and hauling of poultry is responsible for considerable damage to the birds (Shackleford *et al.*, 1969). Catching crews tend to handle the birds roughly in their efforts to perform the job quickly. Much of the loading is done at night in order to meet the processing plant's customary early morning opening.

In the West most poultry moves from the farm to the processing plant in trucks with specially built bodies. In all other parts of the country, the birds are put in coops made of wooden rods or plastic and carefully stacked and tied onto flat-bed trucks.

Before unloading at the processing plant, all poultry is subjected to antemortem inspection. Only a small sample of the birds is examined to determine whether the entire load is in generally good health. If not the load may be returned to the source. In 1970 only 0.44 per cent of all poultry failed to meet antemortem requirements (USDA, 1971e).

At the processing plant, coops are lowered to roller conveyors passing in front of a "hanging" crew standing at ground level. The hangers open the crates, pull out the birds, and place the feet in metal shackles, so that the birds are suspended head downward from the continuously moving overhead chain conveyor (Figure 48.1).

The birds are dispatched by having the

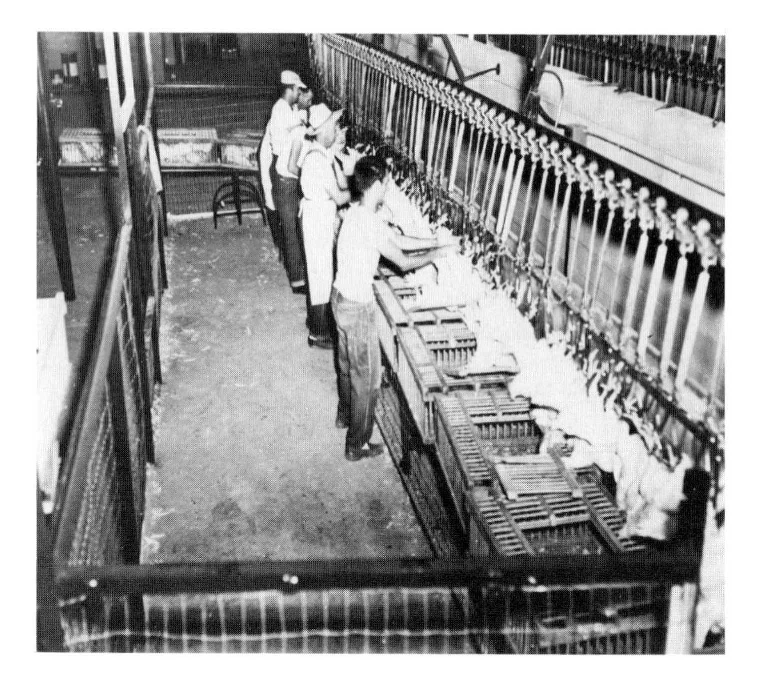

Figure 48-1 Hanging live poultry on processing plant conveyor line. (U.S.D.A.)

main blood vessels cut in the throat area, where the head joins the neck. Mechanical devices are now available for severing the blood vessels for bleeding. It requires 60 to 90 seconds for a satisfactory "bleed" or drainage of blood from the bird. Therefore, the conveyor line passes through a long tunnel to provide for this lapse of time. This "bleeding tunnel" also serves to collect the blood for later use and to keep it out of the plant waste water.

In order to remove the feathers, the skin of the bird must be heated sufficiently, usually by a hot water scald, to cause the feathers to be released. New developments, however, may eliminate the scald tank. In one version, heat is applied to the birds by a combination of live steam and hot water sprays. There is no reuse of the water, and thus no opportunity for the birds to inhale contaminated scald water.

Feathers are removed in machines with rapidly moving rubber fingers that hit the carcass from all sides. The force of the fingers is a compromise, because if it is too great, the underlying meat may be toughened, if too light, not all the feathers, particularly pin feathers, will be removed, thus requiring hand labor to finish the job. With broiler-fryers, the force must be limited in order to prevent the removal of the outer layer of skin, which contains the yellow pigment.

Ducks and geese present an unusual problem, in that scalding and rubber-finger-picking do not remove all feathers and down. Therefore, following the rubber-finger-picking operation, ducks and geese are dipped in a special "wax," and then sprayed with cold water to solidify the wax. The wax is stripped off, carrying away nearly all remaining feathers and pin feathers.

The last stages of the defeathering process consist of passing the bird through a gas flame. The jets are usually operating at low pressures, so that the gas flame is not too hot. The purpose of this treatment is to "singe" the hairs (filoplumes) from the skin of the bird.

For reasons of sanitation, USDA regulations require the birds to be transferred to a new conveyor in the eviscerating room. First the abdomen must be opened without puncturing the intestinal tract. If any fecal contamination occurs in this operation which cannot be adequately removed, the bird must be discarded. The viscera are drawn out of the carcass, but are left attached so that the inspector may examine them with the carcass. After inspection the viscera are removed, the heart, gizzard, and liver are separated for cleaning, and the inedible viscera dropped into a trough and flushed away with water. The head, and finally the crop, esophagus, and trachea, are also removed. The interior of the carcass is then given a cleaning by a vacuum device, which removes any remnants of lungs, intestines, kidneys, reproductive organs, and other unwanted matter not previously removed. The carcass is then conveyed to a washing area, where it is thoroughly sprayed with clean water under high pressure both inside and out.

As soon as the carcass has been thoroughly washed, the chilling operation begins. The carcasses are gently tumbled in cold water or ice slush until the carcass has reached 40°F or lower. After leaving the chilling equipment, the carcasses are hung on a conveyor line for several minutes to drain off any excess moisture.

Frequently the final conveyor line is also an automatic weighing system. The bird passes over a series of bins and tanks and depending on size, drops into one of the bins from the automatic weighing shackles on the conveyor line.

The temperature of the carcasses having been reduced to 40°F in the chiller, it is essential that refrigeration be maintained afterward to prevent spoilage or loss of quality. Usually, broiler-fryers move through distribution channels without freezing. Crushed ice or dry ice is placed inside the shipping container just before closure. Most turkeys are packaged individually in plastic film bags from which all air is removed, then heat shrunk to give a skin-tight wrap.

Most turkeys, practically all fowl, and some broiler-fryers are frozen. Freezing should be performed rapidly for best color and quality preservation. Turkey carcasses are usually "crusted" in a blast-freezing tunnel, where air at −10° to −40°F is blown at high velocity over the packaged birds, or in a liquid-freezing tunnel, where the birds are sprayed with refrigerated brines or propylene glycol at a −20 to a −30°F, until a crust of about an inch is formed. The remainder of the freezing process may then be carried out at a slower rate (at ca. 0°F).

Postmortem Inspection

All the carcasses must be inspected for wholesomeness after slaughter. Postmortem inspection is performed after the body cavity of the bird has been opened and the viscera withdrawn, but not separated from the carcass. Both antemortem and postmortem inspections are performed by a State or Federal employee, who is either a veterinarian or a trained "lay" inspector working under veterinary supervision. The inspector looks for signs of disease, contamination, or any other condition which might render the bird unfit for human consumption. Poultry presented for slaughter is generally quite healthy. In 1970 only 3.2 per cent of young chickens (broiler-fryers), 5.6 per cent of mature chickens (fowl), and 2.5 per cent of turkeys were condemned as unwholesome by inspectors.

Inspection is not confined to the poultry; the entire processing plant must also meet minimum sanitation requirements. The other major function of inspection is concerned with labeling. Requirements specify that the product must be labeled with an appropriate identification, such as fryer, stewing hen (fowl), or young turkey. If the product has been further processed, appropriate designations such as turkey roll, drumsticks, or chicken weiners must be used. If anything has been added, as in the making of various special products, all ingredients must be listed on the label.

In addition to inspection, most poultry is graded for quality using either USDA standards or brand-name standards of the processing firm. It is not mandatory that poultry be graded. There are only three USDA grades for poultry: A, B, and C. Since a high percentage of all poultry qualifies for USDA Grade A, this is the only grade usually used on the labels.

Further Processing

If poultry is prepared beyond the eviscerated "ready-to-cook" stage, it is referred to as "further processed." In 1970, 30.2 per cent of broiler-fryers, 42.8 per cent of turkeys, and 77.7 per cent of fowl were used in further processing. The quantity of poultry used in this manner is increasing rapidly, and the trend is expected to continue.

The most common form of further processing for broiler-fryers is to cut up the carcass. Probably the most popular style of cutting in the United States is the nine-piece cut, which is two legs (drumsticks), two thighs (with attached backbones), three pieces of breast, and two wings. After cutting-up, the chickens may be packaged in trays, with one bird per tray. They may be bulk-packed in 30-pound cans or boxes. They may be frozen or distributed chilled. The popular trend has been toward the "Chill-Pack," in which the chicken is cut into pieces, placed in a tray, overwrapped with plastic film, then chilled to 28° to 30°F. If kept at this temperature, the carcass will retain its freshness for two weeks or more.

Many popular items are made from poultry meat which has been removed from the carcass. Turkeys are usually deboned raw; most fowl are deboned after cooking; and broiler-fryers are deboned both ways. The uses for the deboned meat are many and varied. Many turkeys are used to make fully-cooked turkey rolls, which are popular with hotels, restaurants, and other institutional users. Cooked, deboned chicken meat is diced and frozen for such uses as chicken salad, chicken pies, or casserole dishes. The major use for deboned fowl meat is in soups. Specialty items, such as deboned breasts or legs of broiler-fryers stuffed with wild rice, ham and cheese, or chestnut dressing, are popular in restaurants and on air lines.

Comminuted, deboned poultry meat is used in such products as frankfurters, bologna, luncheon meat, meat loaves, poultry rolls, and poultry roasts. Present regulations permit the use of up to 15 per cent of this product in cooked sausage products without changing the traditional name. It is necessary, however, to include the type of poultry meat used in the list of ingredients on the label.

The major canned poultry products are chicken and turkey meat, both of which may be made up of light or dark meat or a combination of the two. Canned chicken or turkey spread is a comminuted product of meat plus skin and broth.

Plant Sanitation

The needs for good plant sanitation are: (1) safety, to prevent contamination of poultry meat products with disease-producing microorganisms; (2) economic, to promote good shelf-life for these products, by preventing spoilage by contaminating microbes; and (3) esthetic, to assure consumers that their foods have been prepared in a clean environment. State and federal inspection services have plant sanitation as one of their major functions.

The plant should be designed and built in such a way and use such materials as to provide for easy cleaning. Concrete or tile should be used for walls and floors. Equipment should be made of easily-cleaned material and designed so that all parts and surfaces can be reached and cleaned easily. Stainless steel is commonly used for all equipment which comes in contact with the poultry, and other non-rusting or non-corroding metal is used for all other parts. In addition, the equipment should be arranged so that it can be conveniently cleaned frequently and thoroughly.

Also important to a sanitation program is

assigning specific individuals to the job so there will be no question as to who has the responsibility and authority. Every program should be monitored for its effectiveness. Daily microbiological examination of walls, floors, equipment surfaces, and the products should be made.

Quality Control

Physical Defects

The processor usually has no control over bruises, broken bones, torn skin, and other injuries to the birds before they arrive at the plant. The hanging crews are under his control and can therefore be supervised. If an injury occurs, the affected part must be removed unless the damage is so extensive as to cause the entire carcass to be discarded. Since there is now a good market for parts or deboned meat from both turkeys and broiler-fryers, the major loss is the weight of the damaged parts removed, plus trimming costs.

Sensory Defects

Quality control of flavor in poultry is mainly preventing the development of off-flavors. If the feed contains too much oil from fish meal or too much of the highly unsaturated fatty acids from any other source, the poultry meat may have a "fishy" off-flavor. Off-flavors may be caused also by the use of highly odorous cleaning materials in the plant sanitation program. Such products should not be used in processing plants. The third source of odor and flavor problems is from microbiological growth. Prevention of this particular problem will be discussed later.

The development of optimum tenderness can be influenced by processing procedures. More tender meat is produced if the birds are allowed to go through the onset and resolution of *rigor mortis.* Broiler-fryers require about five hours for aging after slaughter and usually present no serious problem. However, if they are cut-up prior to chilling the parts will be less tender than otherwise.

Turkeys develop and resolve rigor in approximately 16 to 24 hours. For maximum tenderness, the carcass should be aged about 24 hours after processing before they are frozen or further processed. Other factors influence tenderness, such as the scalding temperatures and times. If the temperature is above 140°F and the scald time is longer than about two minutes, the tenderness may be reduced. Feather-removing machines utilizing rubber fingers may also reduce tenderness if they beat the bird too vigorously or for too long a time.

Proper skin color requires control in broiler-fryer operations. U.S. processors, wholesale buyers, and supermarket operators are convinced that broiler-fryers should have a bright yellow skin to get highest consumer acceptance. Because of the trade preference, broiler-fryers are fed pigmented (oxycarotenoids) feeds to produce a yellow-skinned bird. Processing must be such as not to remove the color. Since the yellow pigment is present only in the very outer layer of the skin, scald times and temperatures are regulated very carefully so as not to remove the pigmented layer.

Microbial Quality

Most poultry processors try to produce a finished carcass contaminated with the lowest possible number of microorganisms. There are several key points to be observed for accomplishing this objective. The scald water contains many microorganisms, but actually contributes little to carcass surface counts, if proper washing is used afterwards. A more serious problem comes from the scald water which is inhaled by birds still struggling (though unconscious) when they enter the scald tank.

High pressure spray washing follows the picking operations. If the sprays are well-designed and properly operated, surface counts on poultry carcasses are usually quite low following the washing operation.

Removing the viscera is an operation providing many opportunities for microbial contamination. If the intestinal tract is cut or ruptured, if there is any leakage from the vent, or if the hands of the worker become contaminated, microbial counts on the outer and inner surfaces of the bird can be high. Actually, those handling the birds during evisceration should wear rubber gloves, and these should be rinsed thoroughly and frequently in clean, sanitary water.

As mentioned in the section on plant sanitation, the microbial quality of finished products should be monitored frequently to detect any contamination. Refrigerated rooms should be carefully checked also to make sure that proper temperatures are maintained.

Utilization of Poultry Meat

The manner in which poultry finally reaches the consumer is changing rapidly. Although most poultry is still cooked and served in the home in the traditional manner, many millions of pounds are being used in other ways. Perhaps the single greatest change has been the phenomenal rise in popularity of the "carry-out" fried-chicken establishments typified by Kentucky Fried Chicken. In 1970, this chain marketed 10 per cent of the total U.S. production of broiler-fryers. The total volume sold by all such firms probably accounted for 20 per cent or more.

Americans are now consuming about one-third of their meals away from home. Schools, air lines, hotels, restaurants, company lunch rooms, hospitals, nursing homes, and many, many other institutions are using large quantities of poultry and poultry products.

Turkey is still the traditional holiday bird, particularly at Thanksgiving, Christmas, and New Year's. The frozen, whole-body bird is the most popular for these festive occasions, but turkey rolls and roasts now account for sizable portions of the total market. Turkey consumption is gradually increasing at other times of the year. Turkey processing plants in the past have operated only a few months of the year; now many operate most of the year.

The popularity of ducks, geese, and other forms of poultry appears to change very little, and institutional use for these products probably exceeds home use.

Marketing Poultry

Poultry marketing has undergone many changes since the days when processors sent their trucks through the countryside to gather up chickens or turkeys from the local farmers. In the last 25 years, the poultry industry has gone from a rather primitive marketing system to one of the most highly sophisticated in all of agriculture. Broiler-fryer market channels are diagrammed in Figure 48.2.

Production Arrangements

Most producers of broiler-fryers or turkeys make arrangements in advance for selling their birds. The following are examples of current practices. Nearly all independent producers have either a written or a verbal agreement with a processor to buy their birds when ready for market. The provisions of the U.S. Packers and Stockyards Act have been extended to cover poultry buying operations, so that producers are given full information on price, weights, culls, and other deductions, and are assured prompt payment.

A highly popular arrangement for producing broiler-fryers and turkeys is called contract growing. The grower contract may be with a processor, a feed company, a breeder, or some combination of these. If a feed company is the contractor, it will make arrangements with a processor to process the birds. Contract arrangements are usually based on a certain amount per bird, plus bonuses for above-aver-

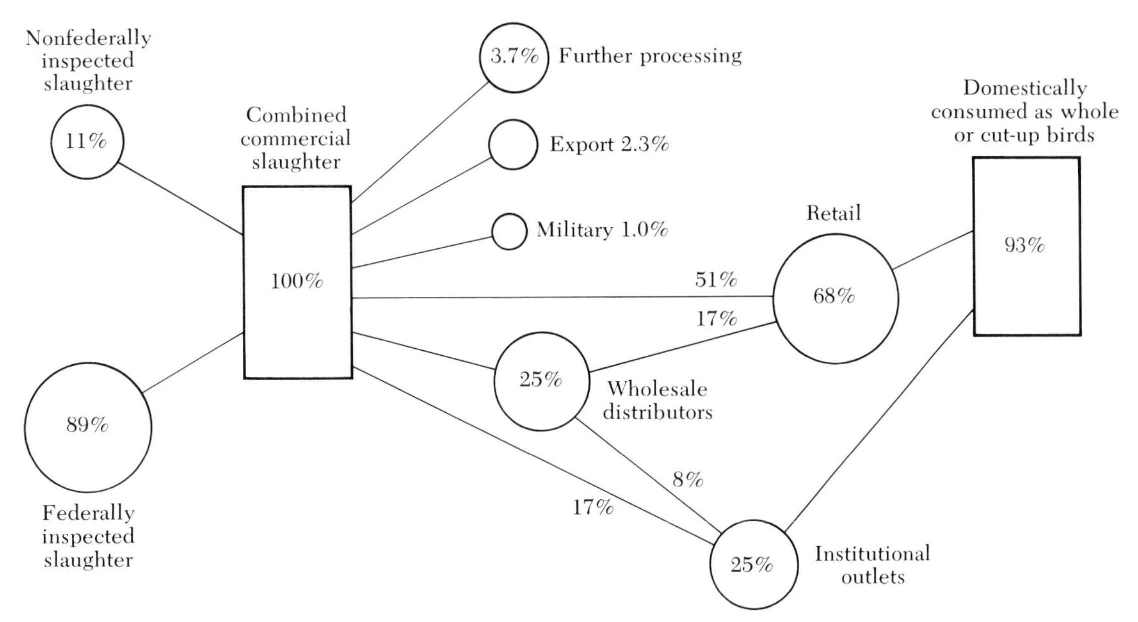

Figure 48-2 Major marketing channels for ready-to-cook broilers, 1969. (USDA)

age weight gains and feed conversion and below-average losses.

If there is ownership of the birds or the facilities at more than one stage of the production and processing cycle, it is called vertical integration. There is a great deal of vertical integration in the broiler-fryer industry. The turkey industry has some, and it is increasing. In the duck industry, vertical integration is almost total. Some integrated firms now engage in wholesaling, brokering, and production of further-processed retail items. These items may be marketed under brand names, and be accompanied by extensive advertising and promotion programs.

Selling to Institutions

Institutions are served in three major ways. One is the wholesaler buying the product and reselling it to the institution. A larger volume is handled through brokers, who do not actually ever take possession of the product, but perform the function of locating the institutional buyer, finding out his requirements, agreeing on a price, and arranging for delivery. Finally, some poultry-processing firms are now selling directly to institutions, using their own sales personnel.

Selling to Retailers

Retailers purchase their poultry in much the same manner as do institutions, except for supermarket chains, which buy in such large quantities that they may deal directly with processors. One of the major differences in selling to retailers is the planning which must go into preparing for special sales. Preparations must be made many weeks in advance to have adequate supplies to accommodate such sales. Prices may be agreed upon weeks or months in advance at an exact figure or one related to a quoted-market price.

Processing Eggs

Processing implies a change in or conversion of a raw agricultural product into food. Eggs in the shell (or "shell eggs" as they are usually called) are similar to fresh fruits and vegetables. Processing consists merely of procuring the product, sorting it for size and quality, cleaning, and packaging. Producing egg products is quite a different thing; it is truly a conversion of a raw agricultural product into food.

Shell Eggs

In 1970, nearly 20 per cent of all laying hens were in flocks of over 100,000. Because large numbers of eggs are available in one location, processing operations are located at or near the production units.

Egg-packing operations Most egg-packing plants are now located in the country and are associated with production units. In addition to the large size of the production unit, other things making this possible are the availability of good roads and high-speed refrigerated trucks, which can deliver eggs to the city users quickly.

Most eggs are first gathered by hand, although a strong trend toward mechanical gathering is under way. In laying cage operations, the gatherer usually rides on an electrical cart, and gathers eggs on to filler-flat trays for delivery to the packing room. Stacks of filler flats of eggs are usually accumulated on a pallet or some other means for transporting to the packing room. If the eggs are to go to a packing plant some distance away, they may be loaded into racks or placed on flat-bed trucks on pallets.

Cracked and broken eggs are a problem in gathering and assembling eggs for packing. It is estimated that 70 per cent of all cracked and broken eggs occur before eggs ever reach the packing plant.

A refrigerated holding room is a necessity in order to cool and hold the eggs for further handling. The ideal holding environment is about 55°F and 80 per cent relative humidity. Lower temperatures cool the eggs faster and preserve quality more effectively, but the eggs may "sweat" when taken to the packing room. Sweating is the condensation of moisture on the shell when the egg is cooler than the dew point of the surrounding air. In warm, humid climates, the dew point can be quite high, and even eggs kept at 55°F may sweat when taken from the holding room. Higher humidities in the holding room reduce evaporation from the eggs but also encourage mold growth on packing supplies, walls, fixtures, and equipment. Therefore 55°F and 80 per cent relative humidity is a reasonable compromise considering all factors.

Nearly all egg packing plants use the same basic equipment. The eggs are first loaded from filler flats to a conveyor. Vacuum rubber cups are used to lift the eggs. Next is the washing operation. In larger operations, cleaning machines convey the eggs either six or twelve abreast through sprays of water and brushes for the removal of dirt and stains. The wash water usually contains detergents and sanitizers.

The next operation is candling. The eggs are conveyed over a light source so that the operator (in a darkened enclosure) can look down on the eggs. Light is transmitted through the egg, giving the "candler" an opportunity to make judgments concerning the egg contents. The shadow of the egg yolk can be seen, but the egg white, if it is normal and clear, will show nothing. The condition of the shell can be observed as well as the size and condition of the air cell. We will discuss in a later section how these observations are used in judging the quality of the eggs.

After examination by candling, the eggs are usually weighed, in order to sort them into weight ranges so the eggs in a given pack will weigh the proper amount and be uniform in size.

It is a common practice, although not universal, for eggs to be given a coating of oil before being packed into cartons or cases. Oil prevents the evaporation of water which reduces egg weight and increases the air-cell size. If the egg is less than 24 hours old, oiling will also preserve interior quality. The *p*H of the newly laid egg is about 7.0, and begins to rise as the naturally present CO_2 escapes. After oiling, the CO_2 cannot escape, and this prevents a rise in *p*H. If the *p*H of the envelope of thick white surrounding the yolk does not go above 7.5 to 8.5, it will become thin at a much slower rate and thus retain its high quality longer (see Table 48.2). Refrigeration also slows this thinning process.

Eggs are packaged in cartons. There are a number of machines in use today which perform this operation mechanically. In smaller plants the eggs are placed in cartons by hand. The filled cartons are conveyed to closing machines and onto benches or turntables where they are packed into cases or wire baskets for delivery to retailers.

Detecting defects The two most common defects are eggs with cracked or broken shells and with blood spots or bloody albumen. During the candling operation, the severe cracks and the larger blood spots are not difficult to detect. Unfortunately, however, the smaller cracks and small blood spots are frequently missed during visual inspection. For this reason, special detection devices have been developed. The blood-spot detector operates on the principle that blood absorbs certain wavelengths of light which accounts for its red color (Figure 48.3). It is possible with photoelectric sensing devices to determine whether an egg has absorbed this light. If so, the egg containing blood is rejected.

To detect cracks, the egg shell is vibrated gently at a high frequency. If the egg shell is intact, it will vibrate at the same frequency as the impulse vibrator. If there is a crack in the shell, extra vibrations or "noise" will be created.

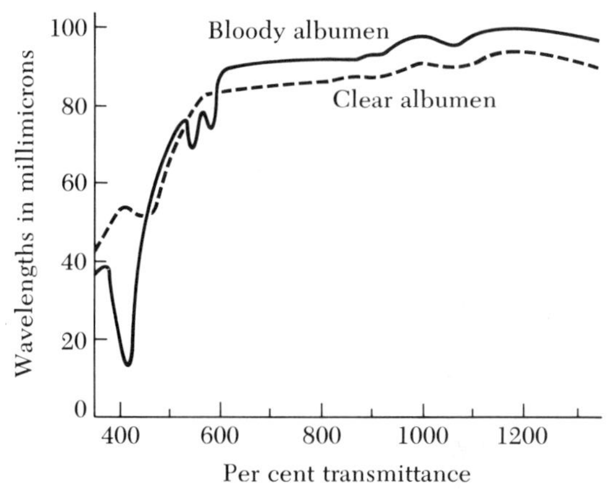

Figure 48-3 Light absorption curves for clear and bloody egg albumen. (A. W. Brant.)

Holding and delivery Eggs are usually held for very short periods of time prior to delivery at 55°F and 80 per cent relative humidity. Cold storage of eggs for a long period of time is no longer practiced, but eggs may be stock-piled for periods of 30 to 60 days in anticipation of special markets, such as Easter, when the demand for eggs is quite high. Such rooms are usually kept at 35°F and 80 per cent relative humidity. Before delivery, the eggs are usually taken to a holding room at about 50°F and "tempered" to avoid sweating.

Quality standards and grades As mentioned earlier, the candler has an opportunity to observe external quality and to some extent the interior quality of each egg. The USDA has organized the various factors into a set of official standards and grades. Minimum requirements are specified for various qualities, as well as tolerances to compensate for human judgment and human error. There are also weight standards. (See Tables 48.2 and 48.3.)

Egg Products

Approximately 10 per cent of the annual production is converted into egg products. Most of the eggs used are not suitable for table use, though they are quite wholesome and nutritious.

Most eggs are broken by machine except for the salvage operations in small plants. Breaking machines open the eggs with mechanical devices resembling human hands and drop the contents onto separating cups (Figure 48.4). An operator then inspects the egg for defects. If the egg is not satisfactory, it is tripped into the discard container. If the yolk is broken but otherwise normal, the egg is tripped into the whole egg (mixed yolk and white) container. If the yolk is unbroken, the

Table 48.2.
Summary of USDA standards for quality of individual shell eggs.

Quality factor	*AA quality*	*A quality*	*B quality*	*C quality*
Shell[a]	Clean. Unbroken. Practically normal.	Clean. Unbroken. Practically normal.	Clean; to very slightly stained. Unbroken. May be slightly abnormal.	Clean; to moderately stained. Unbroken. May be abnormal.
Air cell	1/8 inch or less in depth.	3/16 inch or less in depth.	3/8 inch or less in depth.	May be over 3/8 inch in depth. May be free or bubbly.
White	Clear. Firm. (72 Haugh units or higher.)	Clear. May be reasonably firm. (60 to 72 Haugh units.)	Clear. May be slightly weak. (31 to 60 Haugh units.)	May be weak and watery. Small blood clots or spots may be present.[b] (Less than 31 Haugh units.)
Yolk	Outline slightly defined. Practically free from defects.	Outline may be fairly well defined. Practically free from defects.	Outline may be well defined. May be slightly enlarged and flattened. May show definite but not serious defects.	Outline may be plainly visible. May be enlarged and flattened. May show clearly visible germ development but no blood. May show other serious defects.

[a] Eggs with dirty or broken shells may also be classed as: *Dirty*, if dirty but unbroken; *Check*, if checked or cracked, but not leaking; and *Leaker*, if broken so contents are leaking. Such eggs, or C quality, are not sold at retail.
[b] If they are small (aggregating not more than 1/8 inch in diameter).

machine then lifts the yolk on one part of the cup until the albumen separates and falls into a separate container. The yolk travels to another station and is tripped into another container. In most countries outside the U.S., egg-breaking machines do not separate yolk and albumen, and produce only whole egg. The U.S. market uses considerable quantities of the separated products.

Any number of egg products are produced, depending on the specifications of the customer (Figure 48.5). Common types are: egg white; egg yolk; whole egg; whole egg blends (contain other ingredients); fortified whole egg (contains extra yolk); sugared yolk (sucrose or corn sugar added); salted yolk (table salt added).

Pasteurization In order to reduce the number of microorganisms present in the liquid egg and to kill any human pathogens such as Salmonella, Federal law requires that all egg products be pasteurized. Whole eggs must be heated to 140°F and held for three minutes.

Table 48.3.
USDA weight classes for consumer grades for shell eggs.

Size or weight class	*Minimum net weight per dozen (oz.)*	*Minimum net weight per 30 dozen (lbs.)*
Jumbo	30	56
Extra large	27	50 ½
Large	24	45
Medium	21	39 ½
Small	18	34
Peewee	15	28

Figure 48-4 Egg breaking machine. (USDA)

Other products may require longer or shorter heating to achieve the same effect. Egg white, if it is to be used for angel-food cake, must be pasteurized very gently, otherwise foaming and beating properties are impaired.

Preservation Egg products are highly perishable and must be handled with extreme care. Unless refrigerated, frozen, or dried, they soon spoil. Egg products are so rich in nutrients that almost any microorganism can multiply in them readily.

Freezing is usually accomplished by air blast at temperatures as low as −40°F directed at staggered piles of 30-pound cans of product. Freezing time is from 16 to 72 hours. Smaller containers and other methods of freezing have been tried, but few are in common use.

The largest quantity of egg products is available in the frozen form. There are many reasons for this. Most users of egg products are accustomed to receiving the product in this form and know exactly how to use it; their formulas and mixing methods are developed for the use of frozen-egg products; so they prefer not to change. Since frozen storage is widely available, frozen eggs present few problems to their users. Properly frozen and stored egg products are quite stable, and freezing is less costly than drying.

A little less than 30 per cent of all egg products produced are dried. Eggs are dehydrated by pan, foam, or spray drying. Dried eggs of satisfactory quality are difficult to produce. The many functional properties for which eggs are used can be severely damaged by improper processing. These properties are usually disturbed very little by freezing; therefore, users who have had bad experiences with dried eggs tend to prefer the frozen product. On the other hand, dried eggs present many advantages. Shipping costs are greatly reduced. Frozen transport and storage are not required; so costs are lower. Thawing is not necessary. The exact amount of dried product needed can be measured from a large container with ease thus avoiding refreezing or chances of spoilage. Dried eggs can be used in systems which blend powdered ingredients for such items as cookies or noodles. Properly packaged dried eggs are very stable in relatively inexpensive storage facilities.

Inspection for wholesomeness On July 1, 1971 it became mandatory that all egg products be produced under continuous inspection by the U.S. government or an approved agency, such as a state or local government. Inspection is concentrated on four major aspects of egg products processing: (1) sanitation; (2) approval of raw materials; (3) mon-

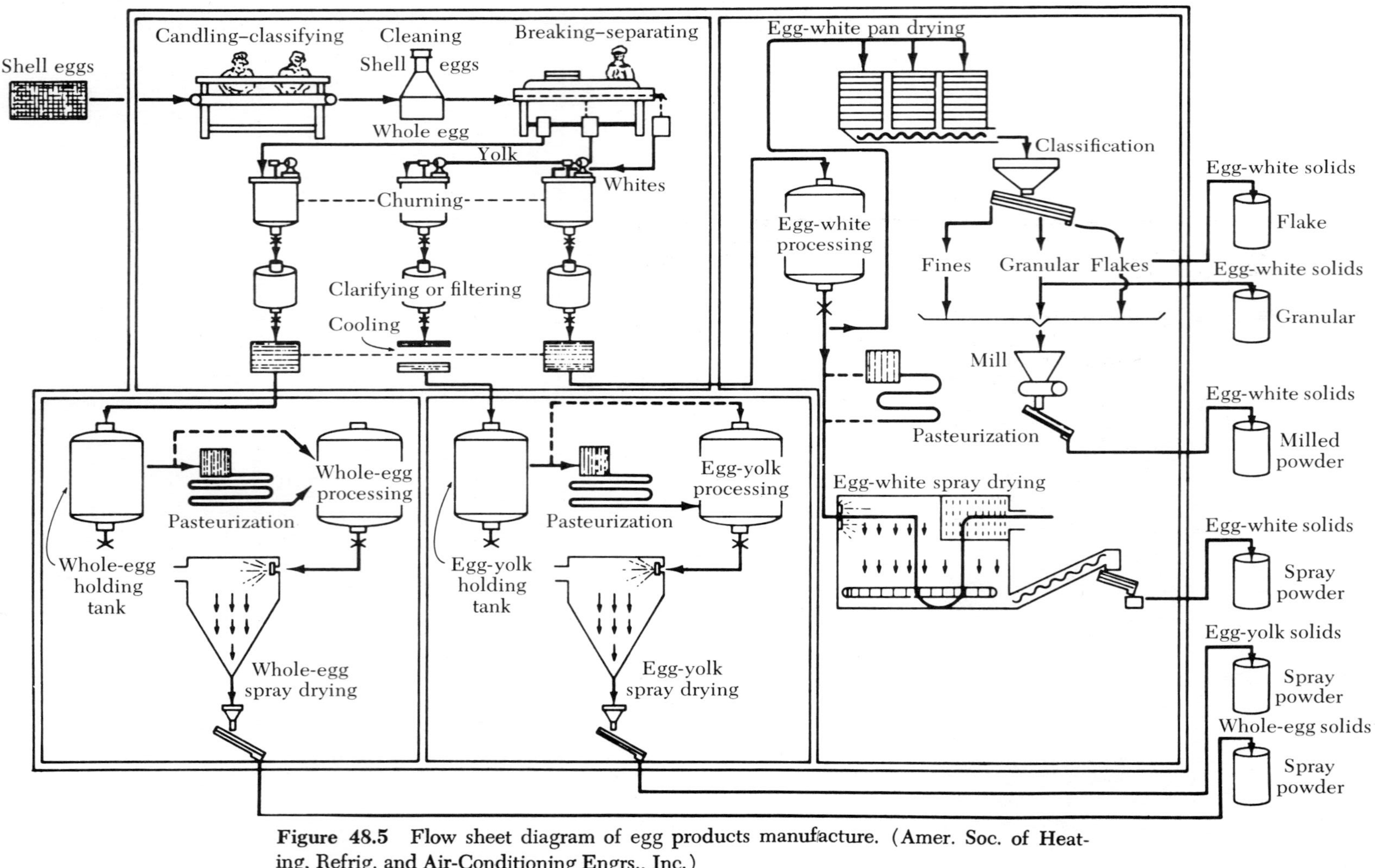

Figure 48.5 Flow sheet diagram of egg products manufacture. (Amer. Soc. of Heating, Refrig. and Air-Conditioning Engrs., Inc.)

itoring of processes; and (4) labeling of products. The sanitation requirements are essentially the same as those for processing poultry already described.

The inspection program certifies that the product is wholesome at the time it leaves the processing plant. Many users, of course, inspect the product on arrival to make sure it is wholesome when they receive it, but there is no mandatory program requiring this be done to all egg products. The U.S. Food and Drug Administration, however, examines commercial egg products from time to time. If not satisfactory, the product is destroyed, diverted to nonhuman uses, or reprocessed. Some quality attributes of egg products are listed in Table 48.4.

Table 48.4.
Quality attributes of egg products.

Quality attribute	*Characteristics*
Solids	Solids content usually required by users: whole egg, 24.75–25%; egg yolk, 43–45%; egg white, 10–12%.
Fat	Whole egg, 10–12%; egg yolk, 30–32%.
Color	Super yellow preferred for many uses of egg products containing yolk. Color intensity measured in colorimeter using beta carotene as standard.
Viscosity	Processing procedures such as homogenizing, pasteurization, freezing may alter viscosity. Viscosity may determine performance of egg products for certain uses.
Microbial count	Egg products should be Salmonella-negative, and total count should be below 100,000 per gm.
Performance tests	Volume of angel-food cake per gm of batter is one criterion. Foam stability is test for egg white. For whole egg, emulsion test may be made.
Yolk content of white	Traces of egg yolk in egg white depress foaming power. Egg yolk content should not exceed 0.03%.

Marketing Eggs

Recent years have seen important changes in the ways eggs are marketed. These changes are continuing, and the statements made here may already be out of date. Figure 48.6 diagrams the egg-marketing channels and illustrates the extent of recent changes.

Shell Eggs

Since about 1920, egg prices have been dominated by the Commodity Exchanges in New York City and Chicago. Eggs are bought and sold on the Commodity Exchanges by buyers, sellers, and speculators who conduct business under carefully prescribed rules. These Exchanges were considered for many years to be the true expression of a free market, where the forces of supply and demand were allowed to interplay and establish fair values for the product. Now the volume of eggs traded on the Exchanges is quite limited, causing many market experts to feel that their influence exceeds their importance.

Another influence on market prices is market-news reporting. The USDA and one commercial firm provide price-reporting services. Reporters maintain daily contact with numerous buyers and sellers of eggs. They obtain information on prices offered and asked, sales made, volume of sales, supplies available, and the "tone" of the market. Market tone usually signals a change in prices as supplies begin to exceed demand or vice versa. The sales recorded plus the reporter's experienced judgment are condensed into a daily report of the egg market. The reports are widely distributed by all means of communication—radio, television, teletype, telephone, newspaper, and direct mail. The reports of one day's market activity sets the stage for the following day.

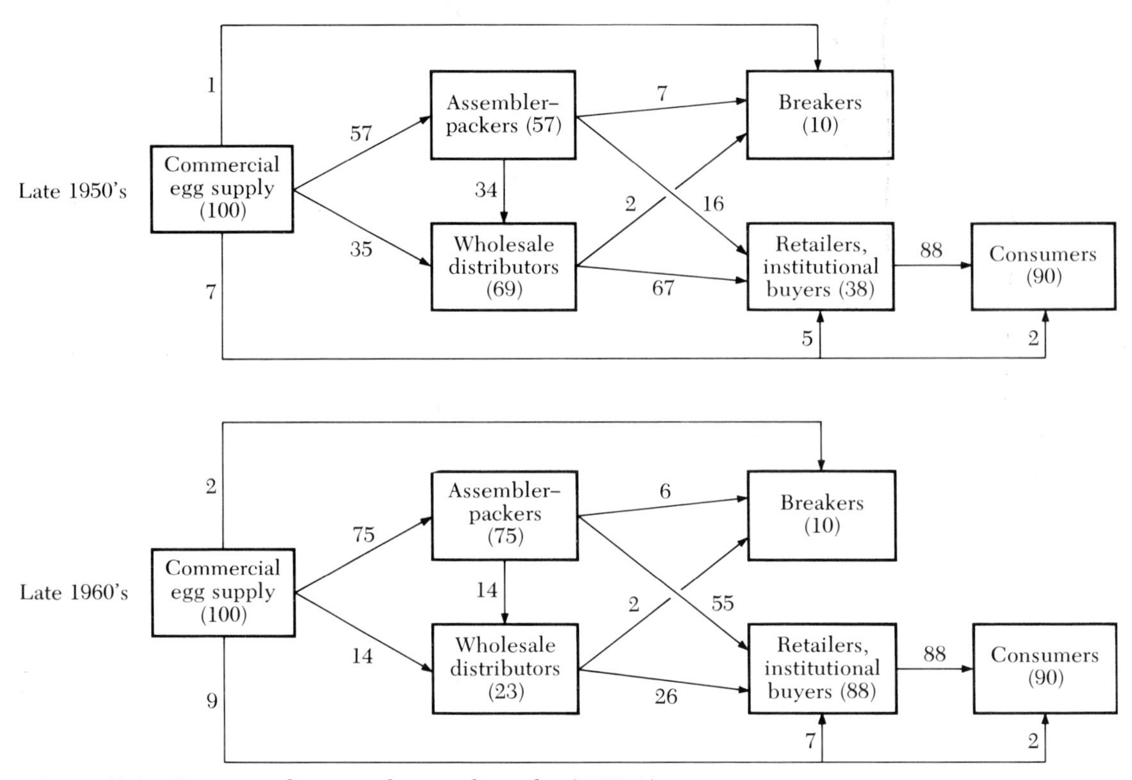

Figure 48-6 Commercial egg marketing channels. (USDA)

Producer pricing When large numbers of eggs were traded on the New York and Chicago Exchanges, prices to producers in all parts of the country were established in reference to these prices. Packing and shipping charges and egg quality were, of course, taken into account.

As the industry changed, fewer and fewer eggs were bought and sold on the Exchanges. Instead, direct selling between packer and retailer became the standard practice, using the Exchange figures as a basis for establishing price. Today, the buying and selling of a few hundred cases on the Exchanges establishes the reference prices for millions of eggs throughout the country. The weakness in this system has now become apparent to everyone. There is, however, no general agreement on another pricing system.

Selling to retailers and institutions Most retailers purchase their eggs from one or two suppliers over long periods of time. Some of the large retail chains pack their own eggs, but there is very little retailer-owned production. When a packer agrees to supply a retailer, a fixed price is usually not set, but is negotiated for each delivery or for a week's deliveries at most.

The U.S. military establishments, both here and abroad, are large users of shell eggs. Their influence in the marketplace is considerable. The purchases are made on the basis of bids. There is one very interesting feature about military purchases, which relates to the size of the eggs. All offers and bids are made on the basis of Grade A Large Eggs. If the supplier cannot provide Large but can provide Mediums, he is permitted to substitute them to whatever extent he desires at a previously agreed-upon difference in price. This practice has the net effect of preventing the accumulation of small surpluses of either Medium or Large in a given market area.

Futures market and hedging By the late 1800's, transportation and communications in the U.S. had developed to the point where markets for many agricultural commodities, including eggs, became national rather than local. Buying and selling between individuals on street corners gradually changed to trading in private clubs and later into more highly organized meetingplaces, such as the Commodity Exchanges. The Exchanges have their own rules for fair and open trading, and are also regulated by the federal government under the Commodity Exchange Act. The regulations prevent attempts to manipulate prices on the cash market or the futures market. The New York Mercantile Exchange was organized in 1872, and the Chicago Mercantile Exchange in 1898. These were the only two Commodity Exchanges which dealt in eggs.* The New York Mercantile Exchange deals only in cash or "spot" trading which means the eggs bought are due at the buyer's place of business by 2 p.m. of the same day and must be paid for in cash. The Chicago Mercantile Exchange also trades in eggs on the spot, and in addition provides the only "futures" market for eggs. As the name suggests, the futures market means that eggs can be bought or sold for delivery at a future time. The Exchange lays the ground rules for these transactions and provides severe penalties for any infractions. Thus, trading may be carried out with complete confidence.

Hedging started as the practice of selling contracts to deliver stored inventories at a future date. Grains and cotton are classical examples because harvest is once a year. Selling contracts for future delivery protects the owner of the grain or cotton from price changes until the next harvest. Until about 15 or 20 years ago, egg production was highly seasonal and large quantities of eggs were placed in cold storage in the spring to be withdrawn in the fall and winter. Eggs, therefore, somewhat resembled the once-a-year crops and futures trading or "hedging" against stored inventories was common. Egg production is no longer seasonal in nature, and cold storage of eggs is seldom longer than 30 days. On the Commodity Exchanges eggs held no longer than 29 days may be called "fresh."

If a producer does not wish to take chances on the price fluctuations on eggs, he may sell a futures contract for the number of eggs he expects to produce in a given time, say, one month. He may sell his eggs in regular market channels and receive the going price. He must, however, *buy* a futures contract for that month to offset the contract he sold. If the price has gone down, he receives less for the production he delivered at the going prices, but it costs less to buy a futures contract to offset the one he sold. If the price for eggs has gone below the cost of production, the "profit" from the futures transaction would tend to offset the loss. If the price goes up after selling a futures contract, buying the offsetting contract will cause a loss on that transaction. The actual eggs, however, have been sold at higher prices, thus making a profit from which the futures trading loss may be substracted. The purpose of the hedge, therefore, is to make it possible for producers, merchandisers, and users of eggs to avoid the risks of price fluctuations over periods of several months if they wish to do so.

* In 1973 the Pacific Coast Exchange authorized trading in egg futures.

Retail sales and promotion The standard method of selling eggs at retail is in the one-dozen carton. The same practice is used in most English-speaking countries. Most other countries sell eggs by the piece, by weight, or by a combination of the two.

The major method by which eggs are promoted or advertised is in the newspaper advertisement of the retail food stores. Most eggs are sold under the brand name of the retailer. There are some exceptions, however, where eggs are packed under the brand name of the distributor.

There is some nationwide commodity promotion of eggs by the American Egg Board. This organization receives its funds from producers and other segments of the egg industry. Its programs include dissemination of information, development of new products and recipes, and the advertising of eggs in order to encourage their increased consumption. It is virtually impossible to measure the effectiveness of commodity promotion activities.

Egg Products

Approximately 10 per cent of the eggs used for food (some are used for hatching) are broken for egg products. Because of increased total production of eggs, the production of egg products has about doubled since the early 1950's. The north central states, between the Rockies and the Appalachians, continue to supply most of the eggs for breaking. The percentage, however, has declined from nearly 90 per cent in the late 50's to a little over 50 per cent in 1970.

New sources of eggs for breaking stock are the southeastern and western states. These eggs are primarily undergrades from shell egg operations, or sizes and qualities that may be in temporary oversupply. It is expected that these trends will continue into the future. The middle western states will probably continue to decline as sources of breaking stock, and the other areas will continue to increase.

Bakeries are the largest users of egg products. About 80 per cent of their eggs are purchased frozen, about 15 per cent dried, and the remainder in liquid form. Premix firms which manufacture cake mixes, doughnut mixes, cookie mixes, and the like use dried eggs almost exclusively. Confectioners use egg white in all three forms, liquid, frozen, and dried. The other users include processors of baby food, meat and fish products, noodles, macaroni, ravioli, mayonnaise, salad dressing, and a variety of other specialty products. These firms purchase mostly frozen eggs, some liquid and dried, and a surprisingly large portion (30 per cent) as shell eggs.

Many users of egg products have developed sets of specifications for purchasing their supplies. These specifications enable them to procure egg products which will provide the quality characteristics and functional properties they desire and to provide more uniformity in the product from purchase to purchase. Producers of high-quality egg products will frequently provide the users with assistance in developing these specifications, since it provides useful guidelines for the egg breakers as well.

FURTHER READINGS

Benjamin, E. W., J. M. Gwin, F. L. Faber, and W. D. Termoblen. 1960. *Marketing Poultry Products.* 5th ed. New York: Wiley.

Faber, F. L. 1971. *The Egg-Products Industry: Structure, Practices, and Costs, 1951–69.* MRR 917. Washington, D.C.: USDA.

Faber, F. L., and R. J. Irvin. 1971. *The Chicken-Broiler Industry: Structure, Practices, and Costs.* MRR 930. Washington, D.C.: USDA.

Forsythe, R. H. 1968. "The science and technology of egg products manufactured in the United States." In Carter, T. C. (ed.), *Egg Quality.* Edinburgh: Oliver and Boyd. (1968a), pp. 262–304.

Institute of American Poultry Industries. 1971. *Current Poultry Trends.* Chicago: IAPI.

Lenz, R. B. 1971. *Future Trading in Fresh Shell Eggs.* MSR 1. Washington, D.C.: USDA.

Rogers, G. B., and L. A. Voss. 1969. *Pricing Systems for Eggs.* MRR 850. Washington, D.C.: USDA.

Shackelford, A. D., R. E. Childs, and I. A. Hamann. 1969. *Determination of Bruise Rates on Broilers Before and After Handling by Live-bird Pick-up Crews.* ARS 52-47. Washington, D.C.: USDA.

Stadelman, W. J., and C. G. Haugh. 1971. Poultry Products. In *ASHRAE Guide and Data Book—Applications* (New York: American Society of Heating, Refrigerating and Air-Conditioning Engineers), pp. 311–322.

Stewart, G. F., and J. C. Abbott. 1961. *Marketing Eggs and Poultry.* Rome, Italy: FAO.

USDA. 1969b. *Egg Grading Manual.* Agriculture Handbook 75. Washington, D.C.: USDA.

Forty-Nine

World Marketing and Distribution of Animal Products

"The white man knows how to make everything, but does not know how to distribute it."

Sitting Bull
quoted in Vestal (1957)

Much of the world's food, whether it is of plant or of animal origin, is consumed in the countries where it is produced. Most nonfood animal products, such as hides, skins, and wool, are also used in the countries where they are produced and harvested. However, some countries produce more than they need, or they must export some to earn foreign exchange, even though those animal products might have been used to good advantage at home. Other countries do not produce enough to meet their domestic requirements, and some of them earn sufficient foreign exchange by exporting other products to enable them to import part or all of the animal products they need. It is with these products of animal origin that move in international trade that this chapter is concerned.

The circumstances under which countries may be in a position to export animal products are quite varied.

Countries like Argentina, Australia, New Zealand, and South Africa have extensive grassland areas suited to livestock production. They have access to markets for those products in the United Kingdom and other European countries. Their own populations are not large; so they have surpluses to export.

Countries like the Netherlands, France, the Federal Republic of Germany, and others in northwestern Europe have intensive agricultural systems in which large amounts of both animal and plant products are produced to meet the needs of their industrialized economies. Since both the climate and the economic environment are favorable to animal production, several of these countries produce in excess of their own needs and have considerable amounts of animal products for export.

Still other countries, which are in early stages of development and which have substantial livestock numbers, have some products in excess of domestic demand. From nutritional and other standpoints these countries might, in most instances, make good use of these excess products at home, but their peoples do not have the economic resources to be able to afford them. Too, the countries need the foreign exchange. So countries like Syria and Ethiopia export live animals and China exports carpet wool. There are other countries that are much more developed, but which are in somewhat similar conditions for political or economic reasons. Thus, some of the satellite countries of eastern Europe that export considerable quantities of animal products are relatively low consumers of animal protein, but the political and economic demands for exports are such that they must remain in short supply at home.

The main importing countries are for the most part those that have industrialized economies and fairly dense populations. They are of three broad types. First, there are highly industrialized countries such as Japan, with little room for animal production, and highly commercialized but limited areas like Hong Kong, into which substantial imports are made to meet local demands. Then there are the countries like the United Kingdom and Italy that have considerable animal production of their own but which have demands well in excess of that production. In addition, there are countries like the Netherlands, the Federal Republic of Germany, and France that were mentioned earlier as exporters, but which are also major importers. The United States falls in this third group too, since it has considerable importance both as an importer and an exporter.

There is a fourth type of importing country that should also be mentioned, into which the total imports of animal products are not large, but where significant amounts of certain products may be imported to help meet nutritional or other needs. For example, although not important importers of other animal products, both Malaysia and the Philippines import notable amounts of dairy products and eggs.

These general introductory remarks are intended to indicate both the considerable variability and the complexity of trade among countries in animal products. The extent of that variability and complexity will become more evident in subsequent sections, where the level of international trade, the main sources of exports and the destinations of imports, and some of the reasons for contrasts in the world market scene are examined.

Level of International Trade and Its Relation to Production

Trade in Live Animals and Meat and Meat Products

The world exports of meat in all its forms, including live animals, accounted in 1959–61 (FAO, 1965) for 13 per cent of all world exports of food and beverages, compared with 21 per cent for grains, 16.5 per cent for coffee, cocoa, and tea, and 5.5 per cent for dairy products.

The discussion in this section is based on exports, since, in theory, data on imports should equal those for exports; so it is simpler to grasp the volume of trade if only one set of figures is used. It should be noted, however, that because of variability in precision of data, international statistical accounts seldom balance neatly.

World production of meat in terms of dressed carcass weights, not including poultry, was 78,404,000 metric tons in 1968 (FAO, 1970a). Of this total, 48.2 per cent was beef, veal, and water-buffalo meat; 43.4 per cent pork; and 8.4 per cent mutton, lamb, and goat meat.

An exact comparison of meat production with the amount of meat moving into international trade is hardly possible because of the variety of forms in which meat moves across national borders. Some of it moves as live animals; some as fresh, chilled, or frozen meat; some in dried, salted or smoked form; and some in airtight containers and in other forms. However, leaving aside live animals, poultry, and horse meat, we can see from Table 49.1 that approximately 4.5 million metric tons of meat and meat products move in international trade each year. This is about 5.8 per cent of the carcass weight of meat produced. Since

Table 49.1.
Annual volume of international trade in live animals, meat, and meat products.[a]

Type of animal or product	*Value of exports, in units of $1 million*[b]
Cattle, including water buffaloes	673
Sheep, lambs, and goats	108
Swine	148
Fresh, chilled or frozen meat	2,619
Bovine	1,311
Sheep and goat	286
Swine	320
Poultry	266
Horse, donkey, and mule	46
Edible offals, and other	192
Dried, salted, or smoked meat[c]	360
Meat in airtight containers and meat preparations[d]	783
Lard and other rendered pig and poultry fat	96
Other animal oils, fats, and greases	216

[a] From FAO (1970c).
[b] Data are rounded to nearest whole million.
[c] Includes all such meats, whether or not in airtight containers.
[d] Includes meat in airtight containers not elsewhere stated, and meat preparations whether or not in airtight containers.

some of the meat and meat products moving in international trade are in forms more concentrated than dressed carcass weight, it is estimated that perhaps 7 to 7.5 per cent of the meat produced enters international trade channels. FAO (1965) estimated the amount at about 7 per cent.

The categories and values of meat and meat products set out in Table 49.1 are the major categories normally used in the tabulation of international statistics. It will be seen that fresh, chilled, and frozen meat have a dominant place in international trade.

It is even more difficult to estimate the proportion of marketable live animals moving into international trade channels than it is to estimate the proportion of meat and meat products that enters these channels. However, using values rather than numbers, and assuming that the proportions of the available supplies of live animals which move into international trade are roughly the same as for meat and meat products, it appears that up to about 5 per cent enter international trade. Limited numbers of these animals are, of course, maintained alive for breeding purposes.

Trade in Dairy Products and Eggs

World production of milk in 1968 amounted to 395,316,000 metric tons (FAO, 1970a), of which 92 per cent (363,793,000 metric tons) was from cows, 4.9 per cent from buffaloes, 1.5 per cent from sheep, and 1.6 per cent from goats.

Here, also, a direct comparison of the amount produced with the amount moving in international trade is hardly possible, since most of it moves in the form of evaporated, condensed, or dried milk, or as butter, cheese, or curd. The data summarized in Table 49.2 relate to approximately 3.3 million metric tons of these concentrated products which move in international trade. This is a small part of the total production; so it is evident that most of

Table 49.2.
Annual volume of international trade in dairy products and eggs.[a]

Type of product	*Value of exports, in units of $1 million*[b]
Milk and cream; evaporated, condensed, dried, and fresh	676
Butter	570
Cheese and curd	603
Eggs in the shell	192
Eggs not in shell: liquid, frozen, or dried	41

[a] From FAO (1970c).
[b] Data are rounded to nearest whole million.

the milk produced is used in one form or another in the countries where it is produced.

World production of hens' eggs in 1968 totaled 1,699,000 metric tons (FAO, 1970a). Something like 2 per cent of the production moved into international trade as eggs in the shell and somewhere between 41 and 68 thousand metric tons as liquid, frozen, or dried eggs. Thus, while the monetary and nutritional value of eggs moving in international trade is considerable, such eggs represent only a small fraction of the total production.

Trade in Wool

World production of greasy wool in 1968 totaled 2,803,000 metric tons (FAO, 1970a). Of this, approximately 1.5 million metric tons moved into international trade (Table 49.3).

Main Sources of Exports and Destinations of Imports

The pattern of movements of animals and animal products is extremely complex. In some cases all that may be involved is the driving of live animals across a national border, or shipping them across by rail or road. In others, live animals may be moved into a country, and reshipped to another, or they may be slaughtered and carcasses, cuts, or processed products in turn exported. The data available internationally are not sufficiently complete to enable one to map precisely these and other kinds of movements. However, examination of available information on the value of animals, and on products exported from and imported into many countries, gives a reasonably good picture of their flow in international trade.

Table 49.3.
Annual volume of international trade in wool.[a]

Type of wool	*Value of exports, in units of $1 million*[b]
Greasy and degreased[c]	1,520
Greasy or fleece-washed	1,212
Degreased[d]	308

[a] From FAO (1970c).
[b] Data are rounded to nearest whole million.
[c] Includes sheep and lamb's wool.
[d] Includes all degreased wool, whether or not bleached or dyed.

Table 49.4.
Leading countries in international trade in live animals, meat and meat products, and animal fats.[a]

Country	*Value, in units of $1 million*[b] *Exports*	*Imports*
Netherlands	573	112
Argentina	380	2
United States	351	864
Australia	331	5
New Zealand	291	1
Ireland	273	31
France	268	377
Belgium/Luxembourg	194	130
Poland	171	45
Hungary	143	17
Yugoslavia	143	12
Germany, Fed. Rep.	136	514
Canada	128	78
U.S.S.R.	111	59
United Kingdom	69	1,134
Italy	30	633
Japan	10	164
Spain	5	102
Hong Kong	2	101

[a] From FAO (1970c).
[b] Data are rounded to nearest whole million.

Leading Traders in Live Animals, Meat and Meat Products, and Animal Fats

Data on the value of exports and imports for 19 countries are given in Table 49.4. These countries were selected on the arbitrary basis that each had exports or imports valued at $100 million or more per year. Some of them for various reasons have volumes of this magnitude in two-way flows in animals and their products. For example, the United States is a net importer of meat, but a net exporter of animal fats. Dairy calves may be exported to Italy for growth and fattening to meet the demand for veal, while beef steers may be imported from Canada for fattening. Broilers may be exported to Europe while beef is being imported from Australia.

Some other interesting points can be observed from this tabulation. Leaving aside Hong Kong because of its limited size and its unique commercial and political status all of these countries except Argentina are in the developed category.

The Netherlands, with the largest amount of exports, outranks such major livestock-producing countries as Argentina, Australia, New Zealand, and the United States, yet ranks eighth as an importer. This reflects both a highly intensive system of mixed agriculture, and an industrialized economy in which there is a large demand for animal products.

The United States ranks third among the exporters and second among the importers. For meat and meat preparations, the United States ranks sixth as an exporter and second as an importer; in this respect the United States imports over 4.5 times as much as it exports. For animal fats, the United States substantially outranks all other countries as an exporter, while its imports are quite limited. For live animals, the United States ranks tenth as an exporter, and third as an importer.

Japan, with its highly intensive agriculture built around the production of rice, and with its highly industrialized economy, has been building up its imports of animal products and now ranks sixth among the importers. Concurrently Japan has been building up its imports of feed grains and its animal production.

Apart from Japan and Hong Kong, none of the leading trading nations are found in the otherwise developing regions of Asia and the Far East, the Near East, and Africa. On the other hand, three of the Latin American countries have exports in excess of $50 million per year, and Brazil falls only a little below this level, with exports valued at $48.5 million. All these four Latin American countries have extensive grazing areas, and although two of them (Argentina, Uruguay) consume large quantities of meat at home, all need the foreign exchange that can be gained from the export of animal products (Figure 49.1).

On the other hand, developing countries appear to feel a greater need for animal fats than for other animal products. For example, Cuba, Bolivia, Korea, Haiti, and Malaysia rank 2nd, 6th, 11th, 13th, and 14th among the importers of lard and other rendered pork and poultry fat. For other types of animal oils, fats, and greases, India, Pakistan, U.A.R., Colombia, Korea, and Brazil rank 3rd, 7th, 9th, 12th, 13th, and 15th among the importers.

Over 16 million live animals move across national borders in international trade each year. Using value as the criterion, eight of the ten leading exporter nations are in Europe, the other two being Canada and the United States. Within the next ten are six developing countries: Argentina, Mexico, Somalia, Turkey, Sudan, and Bolivia, all of which have some market for live animals in their neighboring countries.

On the importing side, six of the leading ten countries are in Europe, the other four being the United States, Hong Kong, Lebanon, and Chile. The United States draws on adjacent sources in Canada and Mexico for live animals. Hong Kong, with very limited production capacity of its own, but with considerable demand, draws on nearby countries, particularly for swine and cattle. Large numbers of cattle, sheep, and goats are driven into Lebanon from neighboring countries to meet demands in large cities such as Beirut. In a similar way, Chile brings in live animals from neighboring Argentina to meet demands for meat in Santiago.

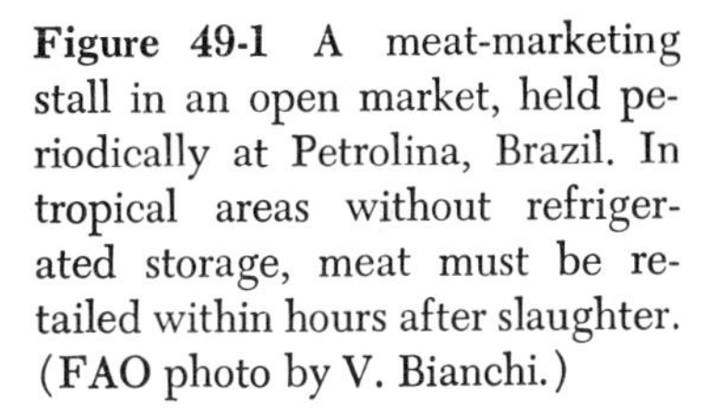

Figure 49-1 A meat-marketing stall in an open market, held periodically at Petrolina, Brazil. In tropical areas without refrigerated storage, meat must be retailed within hours after slaughter. (FAO photo by V. Bianchi.)

Leading Traders in Dairy Products and Eggs

Listed in Table 49.5 are the 20 countries having either exports or imports valued at $25 million or more per year. Nine of them have both export and import volumes of this magnitude, and all of these, except for the United States, are in Europe. As with live animals, meats, and meat products, the Netherlands is the leading exporter of dairy products and eggs, the United Kingdom the leading importer.

Four Asian countries appear in Table 49.5 among the major importers. Some other Asian countries, such as Singapore, Burma, India, and Vietnam, also import moderate amounts of these products as one means of making modest additions to their animal-protein supplies (Figures 49.2 and 49.3).

Leading Traders in Hides and Skins

Hides and skins constitute one of the major nonfood products of the animal industry, and substantial numbers of them move in international trade. Those countries that have either exports or imports valued at $25 million or more are listed in Table 49.6. Among the eighteen countries included in this table, fourteen also appear among the leading traders in live animals, meat and meat products, and animal fats (Table 49.4). The others all have substantial livestock industries from which hides and skins are harvested.

Among the eleven countries having imports in excess of $25 million, only three (United States, Canada, Japan) are outside Europe. It is evident that the tanning and leather-processing industries are concentrated in the countries that are also otherwise highly industrialized.

Leading Traders in Wool

Wool moves in international trade in greasy, fleece-washed, and degreased forms. The leading trading countries, i.e., those having either exports or imports valued at $25 million or more annually, are listed in Table 49.7. Only

Figure 49-2 Milk from small producers is delivered to a milk cooperative in Anand, India. The Anand Milk Union was developed as a means of getting the milking cows and buffaloes out of Bombay and to provide a way in which some 100,000 small farmers in the Kaira district could effectively combine their milk-producing and marketing efforts. (FAO photo by P. N. Sharma.)

three countries qualify at this arbitrary level on both counts.

Quite naturally, the leading exporters are the major sheep-producing countries, such as Australia, New Zealand, South Africa, Argentina, and Uruguay; the main importers are the industrialized countries with large weaving industries and with climates suited to use of woolens.

Although the amounts are not large in relation to total trade in wool, there are substantial amounts of carpet wool exported from some of the developing countries, such as Mongolia, China, Pakistan, India, and Afghanistan. Mongolia's wool exports, for example, were valued at $14,220,000 in 1968.

Some Reasons for Contrasts in the World Market Scene

Thus far we have focused the volume and value of world trade in animals and animal products. Let us now consider some of the reasons why countries differ in the supplies of animals and their products that may move into international trade channels, and in their capacity to import animals and/or their products.

The major exporters are found, of course, among those countries that have substantial feed supplies, as productive range lands or pastures, or as such grasslands in combination with substantial supplies of grain for animal feeding, or in some cases as feed grains that are imported. Such countries are for the most part developed countries where the domestic demand for livestock products is also high. Thus, there is a good economic base for their livestock industries as a result of both domestic and export demands. In Denmark, for example, which is an important exporter of pork, abundant feed supplies allow strong domestic needs for animal products to be met, and still leave a surplus which makes up a substantial portion of Denmark's export trade (Figure 49.5).

At the other extreme, there are the developing countries, in which the populations are very dense, where there is little land or feed supply for use in commercial livestock production, and where economic levels are too low to permit the use of substantial amounts of livestock products, whether they be produced domestically or imported. Such countries have few if any animals or animal products to export, even though many of them may in fact have relatively large populations of animals (Phillips, 1967), and many of them lack the foreign exchange with which to finance imports.

In between these two extremes there are many variations. Some countries that are considered as developing, but which have large and generally thriving cities, have rather substantial demands and may therefore appear among the major importers. Others may maintain considerable numbers of livestock, and may even be basically pastoral, but for various reasons they may produce little or nothing for export. Masai tribesmen in Eastern Africa lead an almost totally pastoral life, but the main livestock product harvested is blood, taken from the jugular veins of their cattle and used immediately for food. In India, which has the largest cattle population of any country, the animals are used primarily for draft, and to a more limited extent for milk production, since Hindu religious beliefs limit the slaughter of cattle and the eating of beef and veal. Still others, such as those in the area of Africa infested with the tsetse fly, have few cattle, or

Figure 49-3 Eggs ready for marketing, at a demonstration and training center at Fort Archambault, Chad. Training is provided to farmers, selected primary and secondary school students, and teachers, as part of an effort to increase the supply of eggs. (FAO photo by Aide Suisse a l'Etrangeres.)

Table 49.5.
Leading countries in international trade in dairy products and eggs.[a]

	Value in units of $1 million[b]	
Country	*Exports*	*Imports*
Netherlands	369	61
France	296	50
New Zealand	213	—
United States	141	78
Germany, Fed. Rep.	115	248
Australia	104	4
Belgium/Luxembourg	86	83
Switzerland	74	35
U.S.S.R.	69	37
Ireland	52	1
Italy	38	206
Finland	37	1
United Kingdom	31	480
Canada	29	23
Poland	25	9
Japan	6	44
Malaysia	6	28
Hong Kong	3	33
Philippines	—	35
Spain	—	25

[a] From FAO (1970c).
[b] Data are rounded to nearest whole million.

other types of livestock for that matter, and obtain their food largely from plant sources.

Some other countries have considerable potentials for livestock production and for export that have not been realized, because of a lack of adequate disease control or of attention to the development of feed supplies and other good management practices, or because the genetic potential of their animals is low, or because of the lack of transportation and processing facilities, or because of combinations of these factors. Many of the countries of Africa, the Near East and Asia, and the Far East are affected by these kinds of limitations, and so are quite a few of those in the Latin American and Caribbean areas.

Table 49.6.

Leading countries in international trade in hides and skins.[a]

	Value in units of $1 million[b]	
Country	*Exports*	*Imports*
United States	169	181
Australia	72	3
U.S.S.R.	64	52
Argentina	57	1
France	52	105
Canada	51	31
South Africa	47	3
New Zealand	44	—
Norway	38	4
Germany, Fed. Rep.	36	192
Netherlands	34	30
Finland	33	6
Sweden	25	18
United Kingdom	20	151
Belgium/Luxembourg	13	30
Italy	13	141
Spain	4	29
Japan	2	83

[a] From FAO (1970c).
[b] Data are rounded to nearest whole million.

Table 49.7.

Leading countries in international trade in wool.[a]

	Value, in units of $1 million[b]	
Country	*Exports*	*Imports*
Australia	776	4
New Zealand	174	1
South Africa	134	4
Argentina	100	—
Uruguay	58	—
France	43	136
U.S.S.R.	40	109
United Kingdom	35	237
Belgium/Luxembourg	22	76
Germany, Fed. Rep.	6	120
Italy	1	150
Japan	1	361
United States	1	149
Czechoslovakia	—	30
Germany, Dem. Rep.	—	36
Poland	—	29

[a] From FAO (1970c).
[b] Data are rounded to nearest whole million.

On the other hand, some such countries are rather important sources of live animals or animal products. For example, as was mentioned earlier, considerable numbers of livestock, particularly cattle and sheep, may be driven across national borders in the normal process of moving grazing animals to markets.

Japan provides a further example of the many variations in capacity to produce and export or import animals or their products. Japan has perhaps the most intensive agriculture of any country in the world. With rice as its major crop and source of calories, it is making substantial efforts to increase its supplies of animal protein both by developing dairy, poultry, and pork production at home and by importing considerable quantities of animal products.

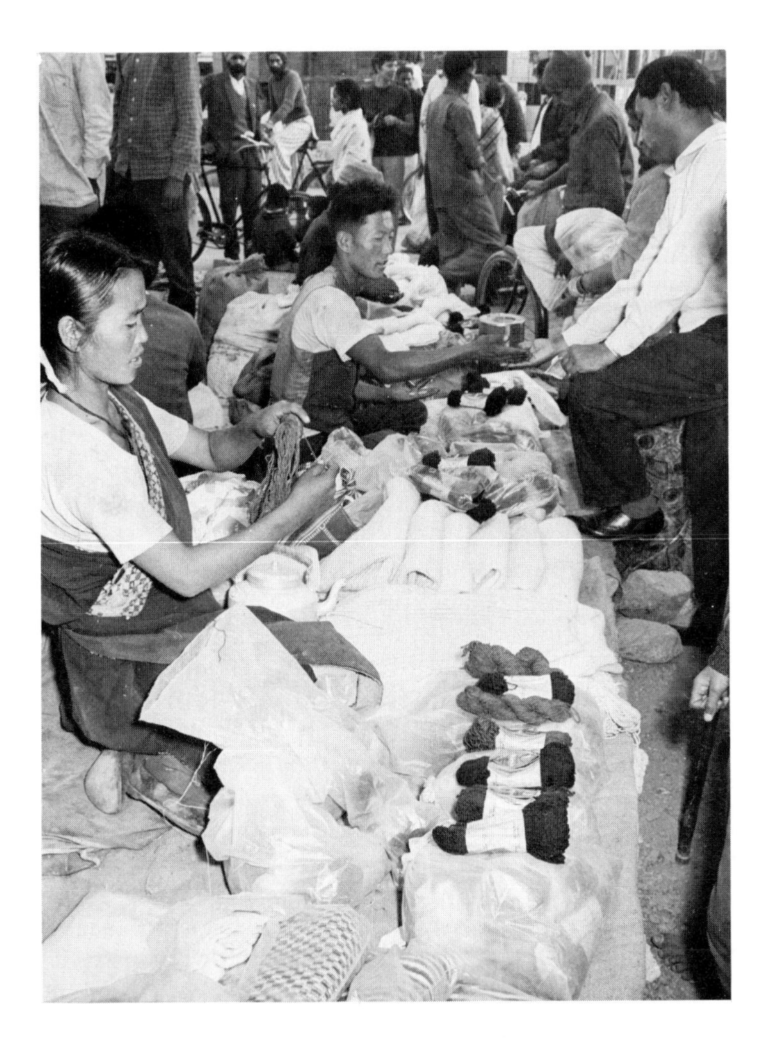

Figure 49-4 Tibetans market wool products at Dehra Dun, Uttar Pradesh, India. (FAO photo by P. Pittel.)

Some limitations or constraints on livestock production and on international trade in animals and animal products may be imposed by governments for political reasons to achieve various objectives (Pryor, 1970). In order to assure the availability of meat to low income groups, official controls may be placed on both producer and consumer prices. Such a policy has been followed in some Latin American countries. This practice may, however, reduce incentives for production and consequently limit amounts of animal products available for consumption or export.

Both national and local governments may impose high and discriminatory taxes on livestock. National governments may, in order to bolster their incomes, place unduly high taxes on exports of animals and animal products, thus reducing sales and serving to discourage production-improvement programs. Local governments may exact taxes merely for the movement of animals from one jurisdiction to another. All these practices have been followed to some degree, for example, in a number of Latin American countries.

Controls may also be imposed by major importing countries. Such controls may be designed to protect producers against the bringing in of diseases or against competition from imports. Actions of the latter type by major importers tend both to maintain high prices in the consuming countries, and to inhibit the development of forward-looking investment and production policies in countries that may have major potentials and advantages in the output of animal products which could move into international trade.

International Standards for Animal Products

The classification and grading of livestock and poultry, and of various animal products, was discussed in Chapters 45 to 48, with particular emphasis on the systems of classification and grading used in the United States. Many countries have developed standards, and it is only natural that these should vary owing to the differing circumstances in which they evolved and the various needs they were designed to meet.

An effort is being made to develop and establish international standards for food, including various animal products. These standards are intended to serve two main purposes: to protect the health of consumers, and to facilitate world trade. This is being done under an international Codex Alimentarius Commission, established in 1963 jointly by the Food and Agriculture Organization of the United Nations (FAO) and the World Health Organization (WHO). The two parent bodies are independent intergovernmental organizations in the so-called family of United Nations organizations. Each has its own member countries, budget, program of work, and governing bodies. The governing bodies of FAO and WHO were responsible for the establishment of the Codex Alimentarius Commission. The Secretariat of the Commission is based at FAO headquarters in Rome, Italy. Any government that is a member of either FAO or WHO may send representatives to meetings and otherwise participate in the work of the Commission and its Committees.

The intent is for the Codex Alimentarius standards eventually to include all principal foods, whether processed or raw, that are distributed to consumers. A Codex standard is essentially a combined standard of identity and a standard of quality. The standards deal with composition, minimum quality, adulterants, additives, labeling, and related factors. The standards, once they are drafted, are considered at a number of stages in Codex Committees and by governments before being accepted by the Commission. Once the Commission accepts a standard, member countries are in turn invited to accept it for application to domestic products and imported products. Acceptance is voluntary, and in practice, means modifying a country's own standards, if neces-

Figure 49-5 Although Denmark is not among the leading exporters of live animals, meat and meat products, and animal fats as a whole, it is an important exporter of pork. In this picture, halves of carcasses are being moved and stored on racks in a modern slaughterhouse. (Photo by courtesy of The Corporation of Danish Bacon Factories.)

sary, to correspond to the Codex Alimentarius standard.

Each set of standards is developed by a Codex Committee established for the purpose. Among the many such committees thus far established, the following deal specifically with animal products: Codex Committee on Meat; Codex Committee on Processed Meat Products; and Joint FAO/WHO Committee of Government Experts on the Code of Principles Concerning Milk and Milk Products. The Codex Committee on Meat has a subcommittee on carcasses and cuts of carcasses. The Committee dealing with milk and milk products differs somewhat from the others, because it was set up several years prior to the establishment of the Codex Alimentarius Commission, and was later brought within the framework of the Commission.

Other Codex Committees deal with Fats and Oils, Food Hygiene, Food Labeling, Pesticide Residues, and Food Additives.

Since Codex Alimentarius standards move through a total of eight formal steps before they are submitted to governments for acceptance, and since many differing and often quite divergent national views must be taken into account and harmonized, the process is time-consuming. Thus, although the Commission began its work in 1963, it was late 1970 before some standards began to emerge in final form to be considered for acceptance by governments.

FURTHER READINGS

Phillips, Ralph W. 1964. "Animal agriculture in the emerging nations." In A. H. Moseman, ed., *Agricultural Sciences for the Developing Nations.* Washington, D.C.: AAAS.

Phillips, Ralph W. 1968. "The use of animals in providing foods: Factors favoring animal production." *Proceedings, Second World Conference on Animal Production,* pp. 15–23.

Bibliography

The following abbreviations are used in the references and bibliography.

AAAS. American Association for the Advancement of Science.
AMIF. American Meat Institute Foundation.
AMSA. American Meat Science Association.
ARC. Agricultural Research Council, London.
FAO. Food and Agriculture Organization of the United Nations.
NRC. National Research Council of the National Academy of Sciences.
USDA. United States Department of Agriculture.

Adams, O. R. 1966. *Lameness in Horses.* Philadelphia, Pa.: Lea & Febiger.

Alexander, F. 1963. "Digestion in the horse." In D. P. Cuthbertson, ed., *Progress in Nutrition and Allied Sciences* (Edinburgh, Scotland: Oliver and Boyd), Chap. 23, pp. 259–268.

Alexander, P., and R. F. Hudson. 1963. *Wool: Its Chemistry and Physics.* 2d ed., rev. by C. Earland. London: Chapman and Hall.

Altman, P. L., and Dorothy S. Dittmer, eds. 1962. *Growth, Including Reproduction and Morphological Development.* Washington, D.C.: Federation of American Societies for Experimental Biology.

Amann, R. P. 1970. "Sperm production rates." In A. D. Johnson, W. R. Gomes and N. L. Van Demark, eds, *The Testis: Development, Anatomy, and Physiology,* Vol. 1 (New York: Academic Press).

American Poultry Association. 1958. *The American Standard of Perfection.* Great Falls, Mon.: APA.

American Society of Testing Materials. 1956. Committee D-13 on Textiles. *Standard Method of Sampling and Testing Staple Length of Wool in the Grease.* D1234-54. New York: ASTM.

AMIF. 1960. *The Science of Meat and Meat Products.* San Francisco, California: W. H. Freeman and Co.

Amoroso, E. C. 1952. "Placentation." In Parkes (1952), II, 127–311.

Amoroso, E. C., and F. H. A. Marshal. 1960. "External factors in sexual periodicity." In Parkes (1960), I, 707–831.

AMSA. 1967. *Recommended Guides for Carcass Evaluation and Contests.* Chicago, Ill.: AMSA.

Anonymous, 1971. "Frozen boar semen." *An. Nutr. Health,* 26:3.

Arbuckle, W. S. 1966. *Ice Cream.* Westport: Conn.: Avi Publishing.

ARC. 1965. *The Nutrient Requirements of Farm Livestock, No. 2: Ruminants.* London: ARC.

Arey, L. B. 1965. *Developmental Anatomy.* 7th ed. Philadelphia, Pa.: Saunders.

Arthaud, V. H., C. H. Adams, D. R. Jacobs, and R. M. Koch. 1969. "Comparison of carcass traits of bulls and steers." *J. Animal Sci.,* 28(6):742–745.

Asdell, S. A. 1964. *Patterns of Mammalian Reproduction.* 2d ed. Ithaca, N.Y.: Cornell Univ. Press.

Astbury, W. T., and A. Street. 1931. "X-ray studies of the structure of hair, wool, and related fibers." *Trans. Roy. Soc. London,* Ser. A, 230:75–101.

Astbury, W. T., and H. J. Woods. 1933. "X-ray studies of the structure of hair, wool, and related fibers; II: The molecular structure and elastic properties of hair keratin." *Trans. Roy. Soc. London,* Ser. A. 232:333–394.

Atwater, W. O., and A. P. Bryant, 1899. "The availability and fuel value of food materials." *Storrs Agric. Exper. Stat. Ann. Rept.,* pp. 73–100.

Austin, C. R. 1969. "Fertilization and development of the egg." In Cole and Cupps, 1969, pp. 355–384.

Bailey, M. E. 1964. "Meat tenderness." *Agric. Sci. Rev.,* 2(3):29–34.

Baldwin, E. 1967. *The Nature of Biochemistry.* 2d ed. Cambridge, Eng.: Cambridge Univ. Press.

Baldwin, R. L. 1968. "Estimation of theoretic calorofic relationships, as a teaching technique: A review." *J. Dairy Sci.,* 51:104.

Banks, E. M., ed. 1967. "Ecology and behavior of the wolf." *Amer. Zool.,* 7:220–381.

Barcroft, J., 1952. "Foetal respiration and circulation." In Parkes (1952, II, 398–441.

Barfield, R. 1969. "Activation of copulatory behavior by androgen implanted into the preoptic area of the male fowl." *Horm. Behav.,* 1:37.

Barnawell, E. B. 1965. "A comparative study of the responses of mammary tissues from several mammalian species to hormones *in vitro.*" *J. Exper. Zool.,* 160:189.

Bateson, W. 1902. "Experiments with poultry." *Repts. Evol. Comm. Roy. Soc.,* 1:87–124.

Bath, D. L. 1971. "Rations and feeding methods for tomorrow's cows." 10th Annual Dairy Cattle Day, University of California, p. 33.

Bauman, R. H., J. E. Kadlac, and P. A. Powlen. 1966. *Some Factors Affecting Death Loss in Baby Pigs. Indiana Agric. Exper. Stat. Bull.,* no. 810.

Baumgardt, B. R. 1967. "Efficiency of nutrient utilization for milk production: Nutritional and physiological aspects." *J. Animal Sci.,* 26:1186–1194.

Beadle, G. W., and E. L. Tatum. 1941. "Genetic control of biochemical reactions in *Neurospora.*" *Proc. Nat. Acad. Sci.,* 27:499–506.

Beamer, W., G. Bermant, and M. Clegg. 1969. "Copulatory behavior of the ram, *Ovis aries,* II: Factors affecting copulatory satiation." *Anim. Behav.,* 17:706.

Bearse, G. E. 1966. "Pesticide residues and poultry products: U.S.A. legislation and pertinent research." *World's Poultry Sci. J.,* 22:194–232.

Becker, R. B. 1971. *Dairy Cattle Breeds: Origin and Development.* Gainesville: Univ. of Florida Press.

Becker, W. A. 1964. *Manual of Prodedures in Quantitative Genetics.* Washington State. Univ. booklet no. 1. Pullman, Wash.: University Bookstore.

Bell, A. E., C. H. Moore, and D. C. Warren. 1955. "The evaluation of new methods for the improvement of quantitative characteristics." *Cold Spring Harbor Symposia,* 20:197–212.

Belschner, H. G. 1965. *Sheep Management and Diseases.* 8th rev. ed. Sydney, Australia: Angus and Robertson.

Bendall, J. R. 1960. "Post-mortem changes in muscle." Bourne, 1960, III, 227–274.

Benjamin, E. W., J. M. Gwin, F. L. Faber, and W. D. Termoblen. 1960. *Marketing Poultry Products.* 5th ed. New York: Wiley.

Bennett, J. A., D. H. Matthews, and M. A. Madsen. 1963. *Range Sheep Breeding Studies in Southern Utah. Utah Agric. Exper. Stat. Bull.,* no. 442.

Bennett, S. C. J., E. R. John, and J. W. Hewison. 1948. "Animal husbandry." In J. D. Tothill, ed., *Agriculture in the Sudan* (London: Oxford Univ. Press), pp. 633–667.

Betrone, E. B. 1969. "Modern sheep production methods in the developing countries." *World Rev. of Anim. Prod.*, 5:76.

Bhasin, N. R., and R. N. Desai. 1967. "The influence of crossbreeding on the performance of Indian cattle." *Indian Vet. J.*, 44:405–412.

Bhimasena Rao, M., R. C. Gupta, and N. N. Dastur. 1970. "Camels' milk and milk products." *Indian J. Dairy Sci.*, 23:71–78.

Bianca, W. 1965. "Reviews in the progress of dairy science, Sect. A., Physiology: Cattle in a hot environment." *J. Dairy Res.*, 32:291.

Bindernagel, J. A. 1968. *Game Cropping in Uganda.* Uganda: Ministry of Agriculture, Game, and Fisheries.

Bines, J. A., S. Suzuki, and C. C. Balch. 1969. "The significance of long-term regulation of food intake in the cow." *Brit. J. Nutr.*, 23:695.

Biondini, P. E., T. M. Sutherland, and L. H. Haverland. 1968. "Body composition of mice selected for rapid growth rate." *J. Animal Sci.*, 27:5–12.

Bird, H. R. Undated. *Nutritive Value of Eggs and Poultry Meat. Nutr. Res. Bull.*, no. 5. Chicago, Ill.: Poultry and Egg National Board.

Black, A. L., N. F. Baker, J. C. Bartley, T. E. Chapman, and R. W. Phillips. 1964. "Water turnover in cattle." *Science,* 144:876.

Blackwell, R. L., D. E. Anderson, and J. H. Knox. 1956. "Age, incidence and heritability of eye cancer in Hereford cattle." *J. Animal Sci.*, 15: 943–951.

Blaxter, K. L. 1962. *The Energy Metabolism of Ruminants.* London: Hutchinson.

Blaxter, K. L. 1970. "Domesticated ruminants as a source of human food." *Proc. Nutr. Soc.*, 29: 244.

Blaxter, K. L., J. Kielanowski, and Greta Thorbek, eds. 1969. *Energy Metabolism of Farm Animals.* Newcastle, England: Oriel Press.

Blumer, T. N. 1963. "Relationship of marbling to the palatability of beef. *J. Animal Sci.*, 22(3): 771–778.

Bodenheimer, F. S. 1951. *Insects as Human Food.* The Hague, Netherlands: W. Junk.

Boessneck, J. 1953. "Die Haustiere in Altägyten." *Veröff. Zool. Staatssamml. (Munich)*, 3:1–50.

Bolin, Bert. "The carbon cycle." *Scientific American,* 223 (Sept. 1970):125. Available as Offprint no. 1193 from W. H. Freeman and Company.

Bolles, R. 1967. *Theory of Motivation.* New York: Harper and Row.

Bond, T. E. 1967. "Microclimate and livestock performance in hot climates." In Shaw (1967), pp. 207–220.

Bonner, James. 1961. "The world's increasing population." *Fed. Proc. Am. Soc. Exp. Biol.*, 20(1): 111.

Borgstrom, George. 1969. *Too Many: A Study of the Earth's Biological Limitations.* London: Collier-Macmillan.

Borland, R. 1964. "The chromosomes of domestic sheep." *J. Hered.*, 55:61.

Botkin, M. P., and Leon Paules. 1965. "Crossbred ewes compared with ewes of parent breeds for wool and lamb production." *J. Animal Sci.*, 24: 1111–1116.

Bourne, G. H., ed. 1960. *The Structure and Function of Striated Muscle.* New York: Academic Press.

Boyd, J. D. 1960. "Development of striated muscle." In Bourne, 1960.

Boyd, J. S. 1969. *Calf and Young Stock Housing. Michigan State Univ. Ext. Bull.*, no. 619.

Boyd, L. J., and D. H. Hafs. 1965. "Body size of calves from Holstein dams and sired by Holstein or Angus bulls." *J. Dairy Sci.*, 48:1236.

Bradley, B. P. 1964. "Comparison of three breeds of sheep and their general and specific combining ability." M.S. thesis. University of Wisconsin, Madison.

Braidwood, R. J., H. Çambel, C. L. Redman, and P. J. Watson. 1971. "Beginnings of village-farming communities in southeastern Turkey." *Proc. Nat. Acad. Sci.*, 68:1236–1240.

Brant, A. W., A. W. Otte, and K. H. Norris. 1951. "Recommended standards for scoring and measing opened egg quality." *Food Technol.* 9:356–361.

Bray, R. W. 1968. "Variation of quality and quantity factors within and between breeds." In Topel (1968), pp. 136–144.

Breidenstein, B. B., C. C. Cooper, R. G. Cassens, G. Evans, and R. W. Bray. 1968. "Influence of marbling and maturity on the palatability of beef muscle, I: Chemical and organoleptic considerations." *J. Animal Sci.*, 27(6):1532–1541.

Breland, K., and M. Breland. 1961. "The misbehavior of organisms." *Amer. Psychologist,* 16: 681.

Breland, K., and M. Breland. 1966. *Animal Behavior.* New York: Macmillan.

Briggs, H. M. 1969. *Modern Breeds of Livestock.* 3rd ed. London: Collier-Macmillan.

Briskey, E. J. 1963. "Influence of ante- and postmortem handling practices on properties of muscle which are related to tenderness." In *Proc. Meat Tenderness Symp.* (Camden, N.J.: Campbell Soup Co., pp. 195–221.

Briskey, E. J. 1964. "Etiological status and associated studies of pale, soft, exudative porcine musculature." *Advances Food Res.,* 13:89–178.

Briskey, E. J., R. G. Cassens, and B. B. Marsh, eds. 1970. *Physiology and Biochemistry of Muscle as Food,* Vol. II. Madison: Univ. of Wisconsin Press.

Briskey, E. J., R. G. Cassens, and J. C. Trautman, eds. 1966. *Physiology and Biochemistry of Muscle as Food.* Madison: Univ. of Wisconsin Press.

Briskey, E. J., and D. Lister. 1968. "Influence of stress syndrome on chemical and physical characteristics of muscle postmortem." In Topel (1968), pp. 177–186.

Brocklesby, D. W., and B. P. Vidler. 1966. "Haematozoa found in wild members of the order *Artiodactyla* in East Africa." *Bull. Epizoot. Dis. Afr.,* 14:285.

Brody, Samuel. 1956. "Climatic physiology of cattle." *J. Dairy Sci.,* 39,6:715.

Brody, Samuel. 1945. *Bioenergetics and Growth.* New York: Reinhold.

Brown, E. 1906. *Races of Poultry.* London: Arnold.

Brown, Lester R. 1963. *Man, Land, and Food.* Foreign Agricultural Economic Report no. 11. Washington, D.C.: USDA.

Brown, Lester R. 1970. "Human food production as a process in the biosphere." *Scientific American,* 223 (Sept. 1970), 160. Available as Offprint no. 1196 from W. H. Freeman and Company.

Bruner, W. H. 1958. *Breed Comparisons. Nat. Hog Farmer Swine Information Serv. Bull.,* no. C2.

Brungardt, V. H., and R. W. Bray. 1963. "Estimate of retail yield of the four major cuts in the beef carcass." *J. Animal Sci.,* 22(1):177–182.

Bull, S., and H. P. Rusk. 1942. *Effect of Exercise on Quality of Beef. Illinois Agric. Exper. Stat. Bull.,* no. 488.

Burns, Marca, and Margaret N. Fraser. 1966. *Genetics of the Dog.* 2d ed. Philadelphia, Pa.: Lippincott.

Butcher, D. F., and A. E. Freeman. 1968. "Heritabilities and repeatabilities of milk and milk fat production by lactations." *J. Dairy. Sci.,* 51: 1387.

Buzzati-Traversa, A. 1952. "Heterosis in population genetics." In Gowen, 1952, pp. 149–160.

Campbell, John R., and John F. Lasley. 1969. *The Science of Animals that Serve Mankind.* New York: McGraw-Hill.

Carmen, Ezra A., H. A. Heath, and J. Minto. 1892. *Special Report on the History and Present Condition of the Sheep Industry of the United States.* Washington, D.C.: USDA.

Carpenter, Z. L., R. G. Kauffman, R. W. Bray, and K. G. Weckel. "Interrelationships of muscle color and other pork quality traits." *Food Technol.,* 19(9):115–118.

Carter, T. C., ed. 1968a. *Egg Quality: A Study of the Hen's Egg.* Edinburgh: Oliver and Boyd.

Carter, T. C. 1968b. *Environmental Control in Poultry Production.* Edinburgh: Oliver and Boyd.

Cartwright, T. C. 1970. "Selection criteria for beef cattle for the future." *J. Animal Sci.,* 30:706–711.

Cartwright, T. C., O. D. Butler, and S. Cover. 1958. "Influence of sires on tenderness of beef." *Proc. 10th Res. Conf., AMIF, Chicago, Ill.,* pp. 75–79.

Cassens, R. G. 1966. "General aspects of postmortem changes." In Briskey *et al.,* 1966, pp. 181–196.

Catchpole, H. R. 1969. "Hormonal mechanisms during pregnancy and parturition." Cole and Cupps, 1969, pp. 415–440.

Chamberlain, F. W. 1943. *Atlas of Avian Anatomy Mich. Agric. Exper. Stat. Memoir Bull.,* no. 5. East Lansing: Michigan State College.

Chapman, A. B. 1969. "Genetic improvement of swine." *Agric. Sci. Rev.,* 7:23–28.

Chilton Research Services. 1968. *The Strategy of Marketing Mohair: A Report to the Mohair Council of America,* San Angelo, Texas: Chilton Research Services.

Christensen, Raymond P., William E. Hindrix, and Robert D. Stevens. 1964. *How the United States Improved Its Agriculture.* Economic Research Service Foreign Report, no. 76. Washington, D.C.: USDA.

Church, D. C. 1969. *Digestive Physiology and Nutrition of Ruminants.* 2 vols. Cornvallis: Oregon State Univ. Bookstores.

Clausen, H., R. Nortoft Thomsen, and O. K. Pedersen. 1970. "Sammenlignende forsøg med svin fra statsanerkendte avilsrentre, 1967–68 og 1968–69." *Beretning fra forsøgslaboratoriet,* 379.

Clayton, L. B. 1970. *Wool in the United States: Major Trends and Prospects.* Reprint from *Wool Situation,* TWS-92. Washington, D.C.: USDA.

Clegg, M. T., and W. F. Ganong. 1969. "Environmental factors affecting reproductions." In Cole and Cupps, 1969, pp. 473–488.

Colby, B. E. 1966. *Dairy Goats: Breeding, Feeding, Management. Massachusetts Agric. Exper. Pub.,* no. 439.

Cole, H. H., ed. 1966. *Introduction to Livestock Production.* 2d ed. San Francisco, Ca.: W. H. Freeman and Co.

Cole, H. H., and P. T. Cupps. 1969. *Reproduction in Domestic Animals.* 2d ed. London: Academic Press.

Cole, H. H., and H. Goss. 1943. "The source of equine gonadotropin." In H. H. Cole and H. Goss, eds., *Essays in Biology,* (Berkeley: Univ. of Calif. Press), pp. 107–119.

Cole, H. H., and G. H. Hart. 1930. "The potency of blood serum in progressive stages of pregnancy in effecting the sexual maturity of the immature rat." *Amer. J. Physiol.,* 93:57–68.

Cole, H. H., and G. H. Hart. 1942. "Diagnosis of pregnancy in the mare by hormonal means." *J. Amer. Vet. Med. Assn.,* 785:124–128.

Cole, H. H., and E. H. Hughes. 1946. "Induction of estrus in lactating sows with equine gonadotropin." *J. Animal Sci.,* 5:25.

Cole, J. W., C. B. Ramsey, and R. H. Epley. 1962. "Simplified method for predicting pounds of lean in beef carcasses." *J. Animal Sci.,* 21(2): 355–361.

Colvin, H. W., Jr., and L. B. Daniels. 1965. "Rumen motility as influenced by physical form of oat hay." *J. Dairy Sci.,* 48:935.

Commonwealth Economic Committee. 1970. *Industrial Fibres—A Review.* London: HMSO.

Connor, J. J. 1918. "A brief history of the sheep industry in the United States." In *Ann. Report of the Amer. Hist. Assn.*

Cook, R. 1937. "A chronology of genetics." In *Yearbook of Agriculture* (Washington, D.C.: USDA, pp. 1457–1477.

Cory, L. 1950. *Meat and Man.* New York: Viking.

Craft, W. A. 1958. "Fifty years of progress in swine breeding." *J. Animal Sci.,* 17:960–980.

Cramer, D. A., J. B. Pruett, and W. C. Schwartz. 1970. "Comparing breeds of sheep, I: Flavor differences." *J. Animal Sci.,* 30:1031. (Abstract)

Crampton, E. W., and L. E. Harris. 1969. *Applied Animal Nutrition.* 2d ed. San Francisco, Ca.: W. H. Freeman and Co.

Crawford, M. A., ed. 1968. *Comparative Nutrition of Wild Animals.* Symp. Zool. Soc. London, no. 21. London and New York: Academic Press.

Crawford, M. A., and S. M. Crawford. 1973. *What We Eat Today.* London: Neville Spearman.

Crawford, M. A. 1969. "Dietary prevention of atherosclerosis." *Lancet,* II, 1429.

Crawford, M. A., M. M. Gale, M. H. Woodford, and N. M. Casperd. 1970. "Comparative studies on fatty-acid composition of wild and domestic meats." *Int. J. Biochem.,* 1:295.

Crawford, S. M. 1970. "Wild protein: a vital role for Africa." *Animals,* 12:540.

Crewther, W. G., R. D. B. Fraser, F. G. Lennox, and H. Lindley. 1965. "The chemistry of keratins." *Adv. Protein Chem.,* 20:191–346.

Crocker, E. C. 1948. "Flavor of meat." *Food Res.* 13(3):179–183.

Cundiff, L. V. 1970. "Experimental results on crossbreeding cattle for beef production." *J. Animal Sci.,* 30:694–705.

Cundiff, L. V., and K. E. Gregory. 1968. "Improvement of beef cattle through breeding methods." *No. Central Regional Publ.* no. 120. *Nebraska Res. Bull.,* no. 196.

Cundiff, L. V., K. E. Gregory, R. M. Koch, and G. E. Dickerson. 1969. "Genetic variation in total and differential growth of carcass components in beef cattle." *J. Animal Sci.,* 29:333–344.

Cunningham, F. E. 1971. "Eggs and egg products." In *ASHRAE Guide and Data Book Applications.* (New York: American Society of Heating, Refrigerating, and Air-Conditioning Engineers), pp. 441–446.

Cunningham, M. P. 1968. "Trypanosomiasis in African wild animals." *E. Afr. Agric. For. J.,* 33: 264.

Curtis, S. 1970. "Environmental-thermoregulatory

interactions and neonatal piglet survival." *J. Animal Sci.*, 31:576.

Cuthbertson, D. 1970. "Role of the ruminant in world food supply." *World Rev. Nutr. Diet.*, 24:414.

Dam, R., G. W. Froning, and J. H. Skala. 1970. "Recommended methods for the analysis of eggs and poultry meat." *Nebraska Agric. Exper. Stat. Bull.*, no. 509.

Danielson, D. M. 1968. "Utilizing automation in rearing baby pigs." *J. Animal Sci.*, 27:1132.

Dart, R. A. 1965. "Ahla, the female Baboon goatherd." *S. Afr. J. Sci.*, 61:319–324.

Davidson, E. H. 1965. "Hormones and genes." *Scientific American*, 212(6):36. Available as Offprint no. 1013 from W. H. Freeman and Company.

Davidson, S., and R. Passmore. 1969. *Human Nutrition and Dietetics.* Baltimore, Md.: Williams and Wilkins.

Davies, R. E. 1967. "Recent theories on the mechanism of muscle contraction and rigor mortis." *Proc. Meat Industry Res. Conf., AMIF, Chicago, Ill.*, pp. 39–53.

Dawson, W. M., R. W. Phillips, and E. B. Krantz. 1950. *Performance of Morgan Horses under Saddle.* Circ. no. 824. Washington, D.C.: USDA.

Degerbol, M. 1961. "On a find of a preboreal domestic dog (*Canis familiaris L.*) from Starr Carr, Yorkshire, with remarks on other Mesolithic dogs." *Proc. Prehist. Soc.*, 27:35–55.

Dettmers, A. E., W. E. Rempel, and R. E. Comstock. 1965. "Selection for small size in swine." *J. Animal Sci.*, 24:216.

Dickerson, G. E. 1952. "Inbred lines for heterosis tests?" In Gowen (1952), pp. 330–351.

Dickerson, G. E. 1969. "Experimental approaches in utilising breed resources." *Animal Breeding Abs.*, 37:191–202.

Dickerson, Gordon. 1970. "Efficiency of animal production: Molding the biological components." *J. Animal Sci.*, 30:849–859.

Dinsmore, Wayne. 1935. *Judging Horses and Mules.* Book no. 219. Chicago, Ill.: Horse and Mule Assoc. of America.

Dobb, M. G., F. R. Johnston, J. A. Nott, L. Oster, J. Sikorski, and W. S. Simpson. 1961. "Morphology of the cuticle layer in wool fibers and other animal hairs." *J. Textile Inst.*, 52:T153–160.

Doornenbal, H. 1968. *Relationship to Body Composition of Subcutaneous Backfat, Blood Volume, and Total Red-Cell Mass in Body Composition in Animals and Man.* Publication 1598. Washington, D.C.: National Academy of Sciences.

Doty, D. M., and J. C. Pierce. 1961. *Beef-Muscle Characteristics as Related to Carcass Grade, Carcass Weight, and Degree of Aging.* USDA Technical Bulletin no. 1231.

Dougherty, R. W., ed. 1965. *Physiology of Digestion in the Ruminant.* Washington, D.C.: Butterworths.

Downs, J. F. 1960. "Domestication: An examination of the changing social relationships between man and animals." *Pap. Kroeber Anthropol. Soc.*, 22:18–67.

Eccles, J. 1965. "The synapse." *Scientific American*, 212(1):56. Available as Offprint no. 1001 from W. H. Freeman and Company.

Eckstein, P. and S. Zuckerman. 1956. "The oestrus cycle in the mammalia." In Parkes, (1956), I, 226–396.

Ehrlich, P. R., and A. H. Ehrlich. 1972. *Population, Resources, Environment: Issues in Human Ecology.* 2d ed. San Francisco, Ca.: W. H. Freeman and Co.

Eibl-Eibesfeldt, I. 1970. *Ethology: The Biology of Behavior.* New York: Holt, Rinehart, and Winston.

Elson, C. E., W. A. Fuller, E. A. Kline, and L. N. Hazel. 1963. "Effect of age on the growth of porcine muscle." *J. Animal Sci.*, 22:946.

England, D. C., and V. M. Chapman. 1962. "Some environmental factors related to survival of newborn pigs." *Oregon Agric. Exper. Stat. Special Report*, no. 137.

Ensminger, M. E. 1969. *Horses and Horsemanship.* 4th ed. Danville, Ill.: Interstate.

Epstein, H. 1969. *Domestic Animals of China.* Farnham Royal, Bucks, England: Commonwealth Agric. Bureaux.

Epstein, H. 1971. *The Origin of the Domestic Animals of Africa.* 2 vols. New York: Africana Publishing.

Estes, J. A. 1958. *Pedigrees.* Stud Manager's Handbook, no. 82–101. Lexington, Ky.

Estes, J. A. 1967. *Thoroughbred Pedigrees.* The Stud Manager's Course, 1–8. P.O. Box 4218, Lexington, Ky. 40504.

Everson, Gladys J., and Helen J. Souders. 1957.

"Composition and nutritive importance of eggs." *J. Amer. Dietetic Assn.*, 33:1244–1254.

Ewing, W. R. 1963. *Poultry Nutrition.* 5th ed. Pasadena, Ca.: Ray Ewing.

Faber, F. L. 1971. *The Egg Products Industry: Structure, Practices, and Costs, 1951–69.* MRR 917. Washington, D.C.: USDA.

Faber, F. L., and R. J. Irvin. 1971. *The Chicken-Broiler Industry: Structure, Practices, and Costs.* MRR 930. Washington, D.C.: USDA.

Falconer, I. R. 1969. *Mammalian Biochemistry.* London: J. and A. Churchill.

FAO. 1963. *Third World Food Survey.* FFHC Basic Study, no. 11. Rome, Italy: FAO.

FAO. 1964. *The State of Food and Agriculture.* Rome, Italy: FAO.

FAO. 1965. *The World Meat Economy.* Commodity Bull. Series, no. 40. Rome, Italy: FAO.

FAO. 1969a. *Production Yearbook, 1968.* Rome, Italy: FAO.

FAO. 1969b. *The State of Food and Agriculture.* Rome, Italy: FAO.

FAO. 1970a. *Production Yearbook, 1969.* Rome, Italy: FAO.

FAO. 1970b. *The State of Food and Agriculture.* Rome, Italy: FAO.

FAO. 1970c. *Trade Yearbook, 1969.* Rome, Italy: FAO.

Farner, D. S., F. E. Wilson, and A. Oksche, 1967. "Neuroendocrine mechanisms in birds." In L. Martini and W. F. Ganong, eds., *Neuroendocrinology* (New York: Acadamic Press), vol. II, chap. 30.

Feeney, R. F. 1964. "Egg proteins." In H. E. Schultz and A. F. Angelmier, eds., *Proteins and Their Reactions.* (Westport, Conn.: Avi), pp. 209–224.

Fell, H. B. 1964. "Some factors in the regulation of cell physiology in skeletal tissue." In H. M. Frost, ed. *Bone Biodynamics.* Henry Ford Hospital Symposia. Boston: Little, Brown.

Ferster, C., and B. F. Skinner. 1957. *Schedules of Reinforcement.* New York: Appleton-Century-Crofts.

Field, C. R. 1968. "A comparative study of the food habits of some wild ungulates in the Queen Elizabeth National Park." In Crawford, 1968a, pp. 367–389.

Field, R. A. 1971. "Effect of castration on meat quality and quantity." *J. Animal Sci.*, 32(5): 849–858.

Finci, M. 1957. "The improvement of the Awassi breed of sheep in Israel." *Bull. Res. Council Israel, B, Biol. Geol.*, 68 (no. 1–2):106.

Findlay, J. D., and D. Robertshaw. 1965. "The role of the sympatho-adrenal system in the control of sweating in the ox (*Bos taurus*)." *J. Physiol. (Lond.)*, 179:285.

Fisher, Tadd. 1969. *Our Overcrowded World.* New York: Parents' Magazine.

Flatt, W. P., and P. W. Moe. 1967. "High Producing Cows—How Should They Be Fed?" *Dairy Herd Management*, October.

Flatt, W. P., P. W. Moe, L. A. Moore, N. W. Hooven, R. P. Lehman, and R. W. Hemken. 1966. "Energy requirements for milk production." *Distillers Feed Research Council Proceedings*, 21:36.

Flatt, W. P., P. W. Moe, A. W. Munson, and T. Cooper. 1969. "Energy utilization by high-producing cows, II: Summary of energy-balance experiments with lactating Holstein cows." In Blaxter *et al.*, 1969, pp. 235–251.

Foote, R. H. 1969. "Physiological aspects of artificial insemination." In Cole and Cupps, 1969, pp. 313–355.

Forsythe, R. H. 1968. "The science and technology of eggs products manufactured in the United States." In Carter, 1968a, pp. 262–304.

Foster, E. M., F. E. Nelson, M. L. Speck, R. N. Doetsch, and J. C. Olson. 1957. *Dairy Microbiology.* Englewood Cliffs, N.J.: Prentice-Hall.

Fowler, M. E. 1970. "Dental problems." *The Quarter Horse of the Pacific Coast,* 7:5.

Fraps, R. M. 1955. "Egg production and fertility in poultry." In A. McLaren, ed., *Progress in the Physiology of Farm Animals* (London: Butterworth), vol. II, chap. 15.

Fraser, R. D. B., and G. E. Rogers. 1955. "The bromine allwörden reaction." *Biochem. et Biophys. Acta,* 16:307–316.

Fraser-Darling, F. 1960a. "Wildlife husbandry in Africa." *Scientific American,* 203:123.

Fraser-Darling, F. 1960b. *Wildlife in an African Territory.* London: Oxford University Press.

Frazer, A. F. 1962. "A technique for freezing goat semen and results of a small breeding trial." *Canad. Vet. J.*, 3:133.

Fredeen, H. T. 1958. "The genetic improvement of swine." *Animal Breeding Abstracts,* 24:314–326.

French, M. H. 1970. *Observations on the Goat.* Rome, Italy: FAO.

French, M. H., I. Johansson, N. R. Joshi, and E. A. McLaughlin. 1966. *European Breeds of Cattle.* 2 vols. FAO Agr. Studies, no. 67. Rome, Italy: FAO.

Galal, E. S. E., L. N. Hazel, G. M. Sidwell, and C. E. Terrill. 1970. "Correlation between purebred and crossbred half-sibs in sheep." *J. Animal Sci.,* 30:475.

Garrett, W. N., and N. Hinman. 1969. "Re-evaluation of the relationship between carcass density and body composition of beef steers." *J. Anim. Sci.,* 28:1–5.

Garrett, W. N., J. H. Meyer, and G. P. Lofgreen. 1959. "The comparative energy requirements of sheep and cattle for maintenance and gain." *J. Animal Sci.,* 18:528–547.

Garrett, W. N., J. H. Meyer, G. P. Lofgreen, and J. B. Dobie. 1961. "Effect of pellet size and composition on feedlot performance, carcass, characteristic, and rumen parakeratosis of fattening steers." *J. Anim. Sci.,* 20:833–838.

Gehlbach, G. D., D. E. Becker, J. L. Cox, B. G. Harmon, and A. H. Jensen. 1966. "Effects of floor-space allowance and number per group on performance of growing-finishing swine." *J. Animal Sci.,* 25:386.

Giffee, J. W., M. C. Urbin, J. B. Fox, W. A. Landmann, A. J. Siedler, and R. A. Sliwinski. 1960. "Chemistry of animal tissues: proteins." In AMIF (1960), pp. 56–110.

Gilbert, A. B. 1967. "Formation of the egg in the domestic chicken." In A. McLaren, ed., *Advances in Reproductive Physiology.* (New York: Academic Press), vol. II, chap. 3.

Gilbert, A. B. 1969. "A reassessment of certain factors which affect egg production in the domestic fowl." *World's Poultry Sci.,* 25(3):239–258.

Gilmore, Robert O. 1947. "Statistics on sires and dams." *The Thoroughbred of California,* vol. 6, no. 8, p. 24, and no. 9, p. 22.

Glass, G. B. J. 1968. *Introduction to Gastrointestinal Physiology.* Englewood Cliffs, N.J.: Prentice-Hall.

Glickman, S., and B. Schiff. 1967. "A biological theory of reinforcement." *Psychol. Rev.,* 74:81.

Glover, J. 1961. "Comparative efficiency of digestion of feeds by ruminants and pigs." *J. Agric. Sci.,* 56:113.

Goll, D. E. 1968. "The resolution of rigor mortis." *Proc. 21st Annual Reciprocal Meat Conf. AMSA National Livestock and Meat Board, Chicago, Ill.,* pp. 16–46.

Goll, D. E., N. Arakawa, M. H. Stromer, W. A. Busch, and R. M. Robson. 1970. "The chemistry of muscle proteins as a food." In Briskey *et al.,* 1970, pp. 755–800.

Goodall, Daphne M. 1965. *Horses of the World.* New York: Macmillan.

Gowen, John, ed. 1952. *Heterosis.* Ames: Iowa State University Press.

Graf, G. C., and R. W. Engel. 1965. "Effect of maturity of the nutritive value of corn silage for lactating cows." *J. Dairy Sci.,* 48:1121.

Graham, E. F. 1966. "Comments on freezing spermatozoa." *Proc. First Technical Conf. Artificial Insemination Bovine Reprod., National Assn. Animal Breeders,* pp. 61–3.

Gray, James A. 1959. "Selecting Angora goats for increased mohair and kid production." *Texas Agric. Exper. Stat.,* MP-385.

Gray, James A., and Jack L. Groff. 1959. *Texas Angora Goat Production.* Texas Agricultural Extension Service Circular B-926.

Grayson, J. 1960. *Nerve, Brain, and Man.* New York: Taplinger.

Green, D. E., and R. F. Goldberger. 1967. *Molecular Insights into the Living Process.* New York: Academic Press.

Greenberg, D. M., ed. 1967. *Metabolic Pathways.* 3d ed. New York: Academic Press.

Gregory, Keith E. 1969. *Beef Cattle Breeding. USDA Agric. Res. Serv. Info. Bull.,* no. 286.

Guenther, J. J., J. A. Stuedemann, S. A. Ewing, and R. D. Morrison. 1967. "Determination of beef-carcass fat content from carcass specific-gravity measurements." *J. Animal Sci.,* 26(1): 210. (Abstr.)

Gustafson, R. A., and R. N. Van Ardall. 1970. *Cattle Feeding in the United States. USDA Agric. Econ. Rpt.,* no. 186.

Hafez, E. S. E., ed. 1968. *Adaptation of Domestic Animals.* Philadelphia, Pa.: Lea and Febiger.

Hafez, E. S. E., ed. 1969a. *The Behaviour of Domestic Animals.* 2d ed. London: Bailliere, Tindall, and Cassell.

Hafez, E. S. E. 1969b. "Prenatal growth." In E. S. E. Hafez and I. A. Dyer, eds., *Animal Growth and Nutrition.* Philadelphia, Pa.: Lea and Febiger.

Hafs, H. D., L. J. Boyd, S. Cameron, and F. Dombroske. 1969. "Fertility of bull semen with amylase and polymixin." *A.I. Digest*, 27:8.

Hale, E. 1962. Domestication and the evolution of behaviour. In Hafez, 1969a, pp. 21–56.

Hall, C. W., and T. I. Hedrick. 1966. *Drying Milk and Milk Products*. Westport, Conn.: Avi.

Ham, A. W. 1969. *Histology*. 6th ed. Philadelphia, Pa.: Lippincott.

Hammer, Wilfried. 1960. "European experience with slatted floors for livestock." Paper presented at the Winter Meeting, Am. Soc. Agr. Eng., Memphis. Tennessee.

Hammond, J., Jr. 1927. *The Physiology of Reproduction in the Cow*. London and New York: Cambridge University Press.

Hammond, J. 1961. "Growth in size and body proportions in farm animals." In: M. X. Zarrow, ed., *Growth in Living Systems*. New York: Basic Books.

Hammond, J., Jr., and F. H. A. Marshall. 1952. "The life cycle." In Parkes (1952), II, 793–848.

Hammond, J., Jr., J. T. Robinson, and I. L. Mason. 1971. *Hammond's Farm Animals*. London: Arnolds.

Hankins, O. G., and P. E. Howe. 1946. "Estimation of the composition of beef carcasses and cuts." *USDA Tech. Bull.*, no. 926.

Harris, J. 1870. *Harris on the Pig*. New York: Orange Judd.

Harris, Lorin E. 1968. "Range nutrition in an arid region." Honor Lecture Series, Utah State University, Logan, Utah.

Harris, L. E., C. W. Cook, and L. A. Stoddart. 1956. "Feeding phosphorous, protein, and energy supplements to ewes on winter ranges of Utah." *Utah Agric. Exper. Stat. Bull.*, no. 398, pp. 1–28.

Harris, M. 1954. *Handbook of Textile Fibers*. Washington, D.C.: Harris Research Laboratories.

Harris, M. 1965. "The myth of the sacred cow." In Leeds and Vayda, 1965, pp. 217–224.

Harrison, D. L., R. Visser, and L. Shirmer. 1959. "Meat tenderness: A resumé of the literature related to factors affecting the tenderness of certain beef muscles." *Kansas Agric. Exper. Stat. Bull.*, no. 208.

Harrison, J. C. 1968. *Care and Training of the Trotter and Pacer*. Columbus, Ohio: The United States Trotting Association.

Harvey, E. B., H. E. Kaiser, and L. E. Rosenberg. 1948. *Atlas of the Domestic Turkey*. Washington, D.C.: U.S. Atomic Energy Commission.

Hazel, L. N. 1943. "The genetic basis for constructing selection indexes." *Genetics*, 28:476–490.

Hazel, L. N. 1963. "Influence of breeding on production efficiency of pigs." *Proc. World Conf. on Anim. Prod.*, Rome.

Hazel, L. N., and J. L. Lush. 1942. "The efficiency of three methods of selection." *J. Hered.*, 33: 393–399.

Hazel, L. N., and C. E. Terrill. 1945. "Heritability of weaning weight and staple length in range Rambouillet lambs." *J. Animal Sci.*, 4(4):347–358.

Hearle, J. W. S., and R. H. Peters, eds. 1963. *Fibre Structure*. London: Butterworth.

Hediger, H. 1964. *Wild Animals in Captivity*. New York: Dover.

Hedrick, H. B., J. B. Boillot, D. E. Brady, and H. D. Naumann. 1959. "Etiology of dark-cutting beef." *Missouri Agric. Exper. Stat. Res. Bull.*, no. 717.

Hedrick, H. B., G. B. Thompson, and G. F. Krause. 1969. "Comparison of feedlot performance and carcass characteristics of half-sib bulls, steers, and heifers." *J. Animal Sci.*, 29(5): 687–694.

Hemingway, Ernest. 1932. *Death in the Afternoon*. New York: Scribners.

Henderson, J. L. 1971. *The Fluid-Milk Industry*. 3d ed. Westport, Conn.: Avi.

Henneberg, N. and F. Stohmann. 1864. *Beiträge zur Begrundung einer rationellen Fütterung der Wiederkäuer, Zweites Heft: Ueber die Ausnutzung der Futterstoffe durch das volljährige Rind und über Fleischbildung im Körper desselben*. Braunschweig: C. A. Schwetschke und Sohn.

Henrickson, R. L., and R. E. Moore. 1965. "Effects of animal age on the palatability of beef." *Oklahoma Agric. Exper. Stat. Tech. Bull.*, no. T-115.

Heptner, W. G., and A. A. Nasimowitsch. 1967. *Der Elch*. Wittenberg Lutherstadt: H. Ziemsen.

Herre, W. 1952. "Studien über die wilden und domestizierten Tylopoden Sudamerikas." *Zool. Gart.*, 19(20):98.

Herre, W. 1955. *Das Ren als Haustiere: Eine Zoologische Monographie*. Leipzig: Geest and Portig.

Herre, W. 1969. "The science and history of domestic animals." In D. R. Brothwell and E. S. Higgs, eds., *Science and Archaeology: A Survey of Progress and Research* (London: Thames and Hudson, 2d ed.), pp. 257–272.

Herz, K. O., and S. S. Chang. 1970. "Meat flavor." *Advances Food Res.,* 18:2–83.

Hess, E. H. 1962. "Ethology: An approach toward the complete analysis of behavior." In R. Brown *et al., New directions in psychology* (New York: Holt, Rinehart, and Winston), pp. 157–266.

Hess, E. A., H. S. Teague, T. M. Ludwick, and R. C. Martig. 1957. "Swine can be bred with frozen semen." *Ohio Farm Home Res.,* 42:100.

Hetzer, H. O., and W. R. Harvey. 1967. "Selection for high and low fatness in swine." *J. Animal Sci.,* 26:1244–1251.

Higgs, E. S. 1967. "Early domesticated animals in Libya." In W. W. Bishop and J. D. Clark, eds., *Background to Evolution in Africa* (Chicago, Ill.: Univ. of Chicago Press), pp. 165–173.

Higgs, E. S., and M. R. Jarman. 1969. "The origins of agriculture: A reconsideration," *Antiquity,* 43:31–41.

Higham, C. F. W., and B. F. Leach. 1971. "An early center of bovine husbandry in southeast Asia." *Science,* 172:54–56.

Hildebrand, Milton. 1959. "Motions of the running cheetah and horse." *J. Mammalogy,* 40: 481.

Hildebrand, Milton. 1965. "Symmetrical gaits of horses." *Science,* 150:701.

Hillman, D. H., D. V. Armstrong, C. C. Beck, B. F. Cargill, and D. J. Ellis. 1965. "Raising calves." *MSU Ext. Bull.,* no. 412.

Hinde, R. 1966. *Animal Behavior.* New York: McGraw-Hill.

Hinman, R. B., and R. B. Harris. 1939. *The Story of Meat.* Chicago, Ill.: Swift and Co.

Hirsch, J. 1967. *Behavior-genetic Analysis.* New York: McGraw-Hill.

Ho, P. 1969. "The loess and the origin of Chinese agriculture." *Amer. Hist. Rev.,* 75:1–36.

Hick, R. J. 1970. "The physiology of high altitude." *Scientific American,* 222:52.

Hodgson, R. E., ed. 1961. *Germ Plasm Resources.* Publ. no. 66. Washington, D.C.: AAAS.

Hoffman, H. H. 1959. "Experiments in the deep freezing of boar semen." *Animal Breed. Abstr.,* 29:350.

Hoffman, Paul G. 1960. *One Hundred Countries, One and One Quarter Billion People.* Washington, D.C.: Albert D. and Mary Lasky Foundation.

Hogan, J., S. Kleist, and C. Hutchings. "Display and food as reinforcers in the Siamese fighting fish (*Betta splendens*)." *J. comp. physiol. psychol.,* 70:351.

Hoglund, C. R., J. S. Boyd, and J. A. Speicher. 1969. "Free-stall dairy housing systems." *MSU Exper. Stat. Bull.,* no. 91.

Holman, R. T. 1968. "Essential fatty-acid deficiency." *Prog. Chem. Fats Lipids,* 9:123.

Holmes, W. 1970. "Animals for food." *Proc. Nutr. Soc.,* 29:237.

Hornstein, I. 1967. "Flavor of red meats." In H. W. Schultz, E. A. Day, and L. M. Libbey (Westport, Conn.: Avi), pp. 228–250.

Howell, F. C. 1972. "Our earliest ancestors." In *Science Year: The World Book Science Annual* (Chicago, Ill.: Field), pp. 224–227.

Huber, J. T. 1969. "Development of the digestive and metabolic apparatus of the calf." *J. Dairy Sci.,* 52:1303.

Huber, J. T., G. C. Graf, and R. W. Engel. 1965. "Effect of maturity on the nutritive value of corn silage for lactating cows." *J. Dairy Sci.,* 48:1121.

Huber, J. T., J. W. Thomas, and R. S. Emery. 1969. "Response of lactating cows fed urea-treated corn silage harvested at varying stages of maturity." *J. Dairy Sci.,* 51:1806.

Huggins, M. L. 1943. "The structure of fibrous proteins." *Chem. Rev.,* 32:195–218.

Humphrey, Elliott, and Lucien Warner. 1934. *Working Dogs: An Attempt to Produce a Strain of German Shepherds which Combine Working Ability and Beauty of Conformation.* Baltimore, Md.: Johns Hopkins Press.

Hungate, R. E. 1968. *The Rumen and its Microbes.* New York and London: Academic Press.

Hunziker, O. F. 1940. *The Butter Industry.* LaGrange, Ill.: Hunziker.

Hutt, F. B. 1949. *Genetics of the Fowl.* New York: McGraw-Hill.

Hutt, F. B. 1964. *Animal Genetics.* New York: Ronald.

Hutton, C. A., and T. N. Meacham. 1968. "Reproductive efficiency on fourteen horse farms." *J. Anim. Sci.,* 27:434–438.

Institute of American Poultry Industries. 1971. *Current Poultry Trends.* Chicago, Ill.: IAPI.

International Wool Secretariat and the Wool Bureau Incorporated. *Wool Digest*: vol. 21, no. 15/16 (August 1970), no. 19/20 (Oct. 1970), no. 21/22 (Nov. 1970), vol. 22, no. 1 (Jan. 1971), no. 2 (Feb. 1971). London.

Isaac, E. 1962. "On the domestication of cattle." *Science,* 137:195–204.

Isaac, E. 1970. *Geography of Domestication.* Englewood Cliffs, N.J.: Prentice-Hall.

Ives, Paul. 1947. *Domestic Geese and Ducks.* New York: Orange Judd.

Jarman, M. R. 1969. "The prehistory of upper Pleistocene and Recent cattle, part I: East Mediterranean, with reference to northwest Europe." *Proc. Prehist. Soc.,* 35:236–266.

Jasper, W. 1953. "Some highlights from consumer egg studies." Agric. Inf. Bull., no. 110, Washington, D.C.: USDA.

Jenness, R., and S. Patton. 1959. *Principles of Dairy Chemistry.* New York: Wiley.

Jenness, R., and R. E. Sloan. 1970. "The composition of milks of various species: A review." *Dairy Sci. Abstr.,* 32:599–612.

Jensen, A. H., and D. E. Becker. 1961. "Floor design and materials in housing for growing-finishing swine." *Illinois Agric. Exper. Stat. Mimeo.* AS-534, Urbana, Ill.

Jensen, P., H. B. Craig, and O. W. Robinson. 1967. "Phenotypic and genetic associations among carcass traits of swine." *J. Animal Sci.,* 26:1252.

Jewell, P. A. 1969. "Wild mammals and their potential for new domestication. In Ucko and Dimbleby (1969), pp. 101–109.

Johansson, I., and J. Rendel. 1968. *Genetics and Animal Breeding.* (English transl.) San Francisco, Ca.: W. H. Freeman and Co.

Johnston, D. D., H. D. Ray, W. J. Manning, and E. M. Pohle. 1951. *Relationship of Staple Length in Grease Wool and Resultant Top.* Washington, D.C.: USDA.

Jones, A. D. 1961. *Wool Warehouses—Practices, Facilities, Services, Charges, Problems.* ERS Tech. Bull. no. 1259. Washington, D.C.: USDA.

Jones, A. D., and H. A. Richards. 1965. *Scouring, Baling, and Transporting Western Wools—Practices, Problems, Possibilities.* ERS Mktg. Res. Rpt. no. 723. Washington, D.C.: USDA.

Jones, D. D., D. L. Day, and A. C. Dale. 1970. "Aerobic treatment of livestock wastes." *Illinois Agric. Exper. Stat. Bull.,* no. 737. Urbana, Ill.

Joshi, N. R., E. A. McLaughlin, and R. W. Phillips. 1967. *Types and Breeds of African Cattle.* FAO Agr. Studies, no. 37. Rome, Italy: FAO.

Joshi, N. R., and R. W. Phillips. 1953. *Zebu Cattle of India and Pakistan.* FAO Agr. Studies, no. 19. Rome, Italy: FAO.

Joubert, D. M. 1968. "An appraisal of game production in South Africa." *Trop. Sci.,* 10:200.

Judge, M. D. 1969. "Environmental stress and meat quality." *J. Animal Sci.,* 28(6):755–760.

Jull, M. A. 1927. "The races of domestic fowl." *National Geographic Magazine,* 51:379–452.

Kammlade, W. G., Sr., and W. G. Kammlade, Jr. 1955. *Sheep Science.* Rev. ed. New York: Lippincott.

Kassenbeck, P. 1959. "Kinetics of the process of keratinization and the morphogenesis of keratin fibers." *Bull. Institut Textile de France,* no. 83, pp. 25–40.

Kassenbeck, P. 1961. "Structure de la laine." In *Textes et discussions du colloque* (Paris: Institut Textile de France), pp. 51–75.

Katz, B. 1961. "How Cells Communicate." *Scientific American* 205(3):209. Available as Offprint no. 98 from W. H. Freeman and Company.

Kays, D. J. 1969. *The Horse.* Cranberry, N.J.: W. S. Barnes.

Kellner, O. 1905. *Die Ernährung der landwirtschaflichen Nutztiere.* Berlin: P. Parey.

Kelly, R. B. 1949. *Sheep Dogs.* 3d ed. London and Sydney: Angus and Robertson.

Kelsall, J. P. 1968. *The Caribou.* Ottawa, Canada: Queen's Printer.

Khurody, D. M. 1967. "Milk colonies." *The World Rev. of Anim. Prod.,* 3:46.

Kielanowski, J. 1962. "Estimates of the energy cost of protein deposition in growing animals." In Blaxter, 1962, p. 13.

Kiesselbach, T. A. 1951. "A half-century of corn research." *Amer. Scientist,* 39:629–655.

Killebrew, J. B. 1880. *Sheep Husbandry.* Nashville, Tenn.: Travel, Eastman and Howell.

King, G. J., F. N. Dickinson, C. A. Rampendahl, J. J. Corbin, and A. H. Kienast. 1971. "DHIA participation report." *DHI Letter,* ARS-44-229.

King, G. T., and Z. L. Carpenter. 1967. "Cutability of bull, steer, and heifer carcasses: Beef cattle research in Texas 1967." *Texas Agric. Exper. Stat. Prog. Rept.,* no. 2497.

Kleiber, M. 1961. *The Fire of Life: An Introduction to Animal Energetics.* New York: Wiley.

Klopfer, P., and M. Klopfer. 1968. "Maternal 'imprinting' in goats: fostering of alien young." *Zeitschr. f. Tierpsychol.,* 25:862–866.

Koch, R. M., L. A. Swiger, Doyle Chambers, and

K. E. Gregory. 1963. "Efficiency of food use in beef cattle." *J. Animal Sci.*, 22:486–494.

Kon, S. K., and A. T. Cowie. 1961. *Milk: The Mammary Gland and its Secretion.* 2 vols. New York and London: Academic Press.

Kosikowski, F. 1966. *Cheese and Fermented Milk Products.* Ithaca, N.Y.: Kosikowski.

Kotarbinska, M., and J. Kielanowski. 1969. "Energy-balance studies with growing pigs by the comparative-slaughter technique." In Blaxter *et al.*, 1969, p. 299.

Krainov, D. A. 1960. "Pestichernaya stoyanka tash-ai-r." *Mater. Issled. Arkeol. SSSR*, vol. 41. Cited by Higgs and Jarman, 1969.

Krebs, H. A., and H. L. Kornberg. 1957. "A survey of the energy transformations in living matter." *Ergeb. der Physiol.*, 49:212.

Kunkel, H. D. 1971. *Nutritional Requirements of the Angora Goat.* Texas Agric. Exper. Stat. Circular B-1105.

Lambrecht, F. L. 1966. "Principles of tsetse control and land use with emphasis on wildlife husbandry." *E. Afr. Wildl. J.*, 1:63.

Lamprey, H. E. 1963. "Ecological separation of the large mammal species in the Tarangire Game Reserve, Tanganyika." *E. Afr. Wildl. J.*, 1:63.

Lanning, E. P. 1967. *Peru Before the Incas.* Englewood Cliffs, N.J.: Prentice-Hall.

Lasley, John F. 1963. *Genetics of Livestock Improvement.* Englewood Cliffs, N.J.: Prentice-Hall.

Lawrence, B. 1968. "Antiquity of large dogs in North America." *Tebiwa,* 11(2):43–49.

Lawrie, R. A. 1966. *Meat Science.* Oxford, Eng.: Pergamon Press.

Laws, R. M., and I. G. C. Parker. 1968. "Recent studies on elephant populations in East Africa." In Crawford, 1968a, p. 319.

Laxminarayana, H., and N. N. Dastur. 1968. "Buffaloes' milk and milk products." *Dairy Sci. Abstr.*, 30:177–186, 231–241.

Leakey, L. S. B. 1965. *Olduvai Gorge, 1951–1961.* Cambridge, Eng.: Cambridge Univ. Press.

Leeds, A., and A. P. Vayda, eds. 1965. *Man, Culture, and Animals: The Role of Animals in Human Ecological Adjustments.* Washington, D.C.: AAAS.

Legates, J. E. 1966. "Crossbreeding of cattle." *World Rev. Animal Prod.*, 3:69–74.

Leggett, William F. 1947. *The Story of Wool.* Brooklyn, N.Y.: Chemical Publishing.

Lennon, H. D., Jr., and J. P. Mixner. 1958. "Relation of lactation milk production in dairy cows to maximum initial milk yeild and persistency of lactation." *J. Dairy Sci.*, 969.

Lenz, R. B. 1971. *Future Trading in Fresh Shell Eggs.* MSR 1. Washington, D.C.: USDA.

Lerner, I. 1950. *Population Genetics and Animal Improvement.* New York: Cambridge Univ. Press.

Lerner, I. M., and H. P. Donald. 1966. *Modern Developments in Animal Breeding.* New York: Academic Press.

Leupold, J. 1968. "The camel: An important domestic animal of the subtropics." *Blue Book of the Veterinary Profession,* 15:1–6.

Liebelt, R. A., and H. L. Eastlick. 1952. "The subcutaneous fat organs of the chick embryo." *Anat. Rec.*, 112:422.

Lindahl, I. L. 1968. *The Digestive System of the Goat: Structure, Function, and Dysfunction.* Spinedale, N.C.: American Dairy Goat Assn.

Lindholm, H. B., and H. H. Stonaker. 1957. "Economic importance of traits and selection indexes for beef cattle." *J. Animal Sci.*, 16:998–1006.

Little, Clarence C. 1957. *The Inheritance of Coat Color in Dogs.* Ithaca, N.Y.: Cornell Univ. Press.

Lofgreen, G. P., and W. N. Garrett. 1968. "A system for expressing net energy requirements and feed values for growing finishing beef cattle." *J. Anim. Sci.*, 27:793–806.

Long, Robert A. 1970. *The Ankony Scoring System—Its Uses in Herd Improvement.* Rhimebeck, N.Y.: Ankony Angus Corp.

Longhouse, A. D., H. Ota, R. E. Emerson, and J. O. Heishman. 1968. "Heat and moisture design data for broiler houses." *Trans. of A.S.A.E.*, 11:695–700.

Lorenz, K. Z. 1952. *King Solomon's Ring.* New York: Crowell.

Lorenz, K. Z. 1955. *Man Meets Dog.* Boston: Houghton-Mifflin.

Lorenz, K. Z. 1966. *On Agression.* New York: Harcourt, Brace, and World.

Lowenberg, M. E., E. N. Todhunter, E. D. Wilson, M. C. Feeney, and J. R. Savage. 1968. *Food and Man.* New York: Wiley.

Loy, R. G., and S. M. Swan. 1966. "Effects of exogenous progestogens on reproductive phenomena in mares." *J. Animal Sci.*, 25:821–826.

Lucas, I. A. M., and G. A. Lodge. 1961. *The*

Nutrition of the Young Pig: A Review. Tech. Comm. No. 22, Commonwealth Bureau of Animal Nutrition, Commonwealth Agricultural Bureau, England.

Luick, J. R. 1971. "Progress Report." U.S.A.E.C. Reindeer Research Sta., Univ. Alaska.

Lunca, N. 1965. "The present state of artificial insemination in sheep and goats." *World Rev. of Anim. Prod.*, 1:73.

Lundgren, H. P., and W. H. Ward. 1962. "Levels of molecular organization in α-keratins." *Arch. Biochem. Biophys. Suppl.*, 1:78–111.

Lundgren, H. P., and W. H. Ward. 1963. "Keratins." In R. Borasky, ed., *Ultrastructure of Protein Fibers* (New York: Academic Press), pp. 39–122.

Lush, J. L. 1931. "The number of daughters necessary to prove a sire." *J. Dairy Sci.*, 14:209–220.

Lush, Jay L. 1945. *Animal Breeding Plans*. 3d ed. Ames: Iowa State Univ. Press.

Lush, J. L. 1948. *Genetics of Populations*. Iowa State Univ. Mimeo.

Lush, J. L., and L. N. Hazel. N.d. *Dwarfism in Beef Cattle*. Ames: Iowa State Univ. Mimeo.

Lyles, L. L. 1969. *Horseman's Handbook*. Santa Rosa: California State Horseman's Association.

Lyne, A. G., and B. F. Short, eds. 1965. *Biology of the Skin and Hair Growth*. New York: Elsevier.

Mackenzie, David. 1967. *Goat Husbandry*. Rev. ed. London: Faber and Faber.

Mahadevan, P. 1966. *Breeding for Milk Production in Tropical Cattle*. Farnham Royal, Bucks, England: Commonwealth Agric. Bureaux.

Maijala, Kalle. 1969. "Need and methods of gene conservation in animal breeding." *European Assoc. for Animal Prod., Commission on Animal Gen., Helsinki, June 23–26, 1969.*

Makino, S. 1951. *An Atlas of the Chromosome Numbers in Animals*. 2d ed. Ames: Iowa State College Press.

Mallette, M. F., P. M. Althouse, and C. O. Clagett. 1960. *Biochemistry of Plants and Animals*. New York: Wiley.

Maloney, M. A., J. C. Gilbreath, J. F. Tierce, and R. D. Morrison. 1967. "Divergent selection for twelve-week body weight in the domestic fowl." *Poultry Sci.*, 46:1116–1127.

Mann, T. 1964. *Biochemistry of Semen and of the Male Reproductive Tract*. New York: Wiley.

Marsden, S. J. 1971. *Turkey Production*. Agric. Handbook No. 393. Washington, D.C.: U.S. Govt. Printing Office.

Marsh, B. B. 1970. "Muscle as food." In Briskey *et al.*, 1970, pp. 3–10.

Marshall, F. H. A. 1956. "The breeding season." In Parkes (1956), I, 1–42.

Martin, E. L., L. E. Walters, and J. V. Whiteman. 1966. "Association of beef carcass conformation with thick and thin muscle yields." *J. Animal Sci.*, 25:682–687.

Mason, I. L. 1967. *Sheep Breeds of the Mediterranean*. Farnham Royal, Bucks, England: Commonwealth Agric. Bureaux.

Mason, I. L. 1969. *A World Dictionary of Breeds, Types, and Varieties of Livestock*. Technical Communication No. 8 (rev.) of the Commonwealth Bureau of Animal Breeding and Genetics, Edinburgh, and the Commonwealth Agricultural Bureaux, Farnham Royal, Bucks, England. London: Morrison and Gibb.

Mason, I. L., and J. P. Maule. 1960. *The Indigenous Livestock of Eastern and South Africa*. Edinburgh: Commonwealth Bureau of Animal Breeding and Genetics.

Matoltsy, A. G. 1958. "The Chemistry of Keratinization." In Montagna and Ellis (1958), pp. 135–165.

Mavimow, A. A., and W. Bloom. 1948. *A Textbook of Histology*. 5th ed. Philadelphia, Pa.: Saunders.

Maynard, L. A. 1953. "Total digestible nutrients as a measure of feed energy." *J. Nutr.*, 51:15–22.

Maynard, L. A., and J. K. Loosli. 1969. *Animal Nutrition*. New York: McGraw-Hill.

McBee, J. L., and J. A. Wiles. 1967. "Influence of marbling and carcass grade on the physical and chemical characteristics of beef." *J. Animal Sci.*, 26(4):701–704.

McCarthy, J. F., and C. G. King. 1942. "Some chemical changes accompanying tenderization of beef." *Food Res.*, 7(4):295–299.

McCauley, W. J. 1971. *Vertebrate Physiology*. Philadelphia, Pa.: Saunders.

McCollum, E. V. 1957. *A History of Nutrition*. Boston: Houghton-Mifflin.

McDaniel, B. T., and J. E. Legates. 1965. "Associations between body weight predicted from heart girth and production." *J. Dairy Sci.*, 947.

McDaniel, B. T., R. H. Miller, and E. L. Corley. 1965. "DHIA factors for projecting incomplete records to 305 days." *DHI Letter* ARS-44-164.

McDonald, P., R. A. Edward, and J. F. S. Greenhalgh. 1966. *Animal Nutrition.* Edinburgh and London: Oliver and Boyd.

McDowell, R. E. 1966. "The role of physiology in animal production for tropical and subtropical areas." *World Rev. of Animal Prod.,* 1:39.

McDowell, R. E. 1968. "Climate versus man and his animals." *Nature,* 218:641.

McDowell, R. E. 1972. *Improvement of Livestock in Warm Climates.* San Francisco: W. H. Freeman and Co.

McDowell, R. E., G. V. Richardson, B. E. Mackey, and B. T. McDaniel. 1970. "Interbreed matings in dairy cattle, V: Reproductive performance." *J. Dairy Sci.,* 757.

McGilvery, R. W. 1970. *Biochemistry: A Functional Approach.* Philadelphia, Pa.: Saunders.

McKinney, J. 1959. *The Sheep Book.* New York: Wiley.

McLean, F. C., and M. R. Urist. 1968. *Bone: Fundamentals of the Physiology of Skeletal Tissue.* Chicago, Ill.: Univ. of Chicago Press.

McMeekan, C. P. 1940. "Growth and development in the pig, with special reference to carcass quality characters." *J. Agr. Sci.,* 30:276.

Mecchi, E. P., M. F. Pool, G. A. Behman, M. Hamachi, and A. A. Klose. 1956. "Role of tocopherol content in the comparative stability of chicken and turkey fat." *Poultry Sci.,* 35:1238–1251.

Meggitt, M. J. 1965. "The association between Australian aborigines and dingoes." In Leeds and Vayda (1965), pp. 7–26.

Mellaart, J. 1967. *Catal Huyuk: A Neolithic Town in Anatolia.* New York: McGraw-Hill.

Mellen, I. M. 1952. *The Natural History of the Pig.* New York: Exposition Press.

Mellin, T. N., B. R. Poulton, and M. J. Anderson. 1962. "Nutritive value of timothy hay as affected by date of harvest." *J. Anim. Sci.,* 21:123–126.

Mendel, G. 1866. "Experiments in plant hybridization." *Proc. Brunn Nat. Hist. Soc.* Translation in J. A. Peters, ed. *Classic Papers in Genetics* (Englewood Cliffs, N.J.: Prentice-Hall), 1959.

Mercer, E. H. 1961. *Keratin and Keratinization: An Essay in Molecular Biology.* London: Pergamon Press.

Meredith, R., ed. 1956. *The Mechanical Properties of Textile Fibers.* New York: Interscience.

Meredith, R., and J. W. S. Hearle, eds. 1959. *Physical Methods of Investigating Textiles.* New York: Interscience.

Meyer, J. H., and L. G. Jones. 1962. "Controlling alfalfa quality." *Calif. Agric. Exper. Stat. Bull.,* no. 784.

Midgley, A. R., and R. B. Jaffe, 1968. *J. Clin. Endocrinol. Metab.,* 28:1699.

Miller, C., L. D. Sanborn, H. Abplanalp, and G. F. Stewart. 1960. "Consumer reaction to egg flavor." *Poultry Sci.,* 39:3–7.

Miller, M. W., and G. G. Berg, eds. 1969. *Chemical Fallout.* Springfield, Ill.: Thomas.

Miller, W. H., F. Ratliff, and H. K. Hartline. 1961. "How cells receive stimuli." *Scientific American,* 205(3):223. Available as Offprint no. 99 from W. H. Freeman and Company.

Ministry of Agriculture, Fisheries and Food. 1964. *The Rearing of Chickens.* Bull. no. 154. London: HMSO.

Mitchell, H. H. 1962. *Comparative Nutrition of Man and Domestic Animals.* New York: Academic Press.

Mitchell, H. H., L. E. Card, and T. S. Hamilton. 1926. *The Growth of White Plymouth Rock Chickens.* Bull. no. 278. Urbana: Illinois Agric. Exper. Stat.

Mitchell, H. H., L. E. Card, and T. S. Hamilton. 1931. *Technical Study of the Growth of White Leghorn Chickens.* Bull. no. 367. Urbana: Illinois Agric. Exper. Stat.

Mitchell, H. H., W. G. Kammlade, and T. S. Hamilton. 1926. *A Technical Study of the Maintenance and Fattening of Sheep and their Utilization of Alfalfa Hay.* Urbana: Illinois Agric. Exper. Stat. Bull. no. 283.

Moe, P. W., and W. P. Flatt. 1969. *The New Energy Value of Feeds for Lactation.* DCRB Pub. no. 69. Washington, D.C.: USDA.

Moe, P. W., H. F. Tyrrell, and W. P. Flatt. 1970. "Partial efficiency of energy use for maintenance, lactation, body gain, and gestation in the dairy cow." In Schurch and Wenk (1970), pp. 65–68.

Moe, P. W., H. F. Tyrrell, and W. P. Flatt. 1971. "Energetics of body tissue mobilization." *J. Dairy Sci.,* 54:548.

Montagna, W., and R. A. Ellis, eds. 1958. *The Biology of Hair Growth.* New York: Academic Press.

Moody, W. G., J. A. Jacobs, and J. D. Kemp. 1970. "Influence of marbling texture on beef rib palatability." *J. Animal Sci.,* 31(6):1074–1077.

Morag, M., J. A. C. Gibb, and S. Fox. 1967. "A milking parlour for experimental work with lactating ewes." *J. Dairy Res.*, 34:215.

Morgan, C. L. 1894. *An Introduction to Comparative Psychology*. London: Scott.

Morley, F. H. W. 1968. "The use of animals in producing foods: Challenges to animal production." *Proceedings, Second World Congress on Animal Production,* pages 23–30.

Morrison, F. B. 1956. *Feeds and Feeding*. 22d ed. Ithaca, N.Y.: Morrison.

Morrison, S. R., H. Heitman, Jr., T. E. Bond, and P. Finn-Kelcey. 1966. "The influence of humidity on growth rate and feed utilization of swine." *Int. J. Biometeor.*, 10:163.

Mountney, G. J. 1966. *Poultry Products Technology*. Westport, Conn.: Avi.

Moustgaard, J. 1962. "Foetal nutrition in the pig." In J. T. Morgan and D. Lewis, eds., *Nutrition of Pigs and Poultry* (Washington, D.C.: Butterworth).

Moustgaard, J. 1969. "Nutritive influences upon reproduction." In Cole and Cupps (1969), p. 489.

Muehling, A. J. 1969. "Swine housing and waste management: A research review." *Illinois Agric. Exper. Stat. Bull.*, A Eng-873.

Muller, H. J. 1961. "Human evolution by voluntary choice of germ plasm." *Science,* 134:643.

Murphey, C. E., D. K. Hallett, J. C. Pierce, and W. E. Tyler. 1960. ""Estimating yields of retail cuts from beef carcasses." *J. Animal Sci.*, 19(4): 1240 (Abstract).

Murray, J. 1970. *The First European Agriculture: A Study of the Osteological and Botanical Evidence until 2000* B.C. Edinburgh: Edinburgh Univ. Press.

Nalbandov, A. V. 1964. *Reproductive Physiology*. 2d ed. San Francisco, Ca.: W. H. Freeman and Co.

National Dairy Council. 1970. *How Americans Use Their Dairy Foods*. Chicago, Ill.: NDC.

National Livestock and Meat Board. 1969. *Meat Evaluation Handbook*. Chicago: NLMB.

Neale, P. E., G. M. Sidwell, and J. L. Ruttle. 1958. "A mechanical method for estimating clean fleece weight." *New Mexico Agric. Exper. Stat. Bull.* no. 417.

Nehring, K. 1970. "Tabellen zur Bewertung der Futterstoffe auf energetischer Grundlage." In Schurch and Wenk (1970), pp. 257–259.

Nehring, K., R. Schiemann, and L. Hoffman. 1969. "A new system of energetic evaluation of food on the basis of net energy for fattening." In Blaxter *et al.*, 1969, pp. 41–50.

Nelson, J. A., and G. M. Trout. *Judging Dairy Products*. Milwaukee, Wis.: Olsen.

Neumann, A. L., and R. R. Snapp. 1969. *Beef Cattle*. New York: Wiley.

Newbold, R. P. 1966. "Changes associated with rigor mortis." In Briskey *et al.*, 1966, pp. 213–224.

Nishikawa, Y. 1959. *Studies on Reproduction in Horses*. Tokyo: Japan Racing Association.

Nishikawa, Y. 1964. "History and development of artificial insemination in the world." *Proc. Fifth Int. Congr. Reprod. Artificial Insemination, Trento, Italy,* pp. 163–256.

Norris, L. C., 1958. "The significant advances of the past fifty years in poultry nutrition." *Poultry Sci.*, 37:256–274.

NRC. 1964. *Nutrient Requirements of Sheep*. Pub. no. 504. Washington, D.C.: NRC.

NRC. 1966a. *Nutrient Requirements of Domestic Animals, No. 1: Nutrient Requirements of Poultry*. Pub. no. 1345. Washington, D.C.: NRC.

NRC. 1966b. *Biological Energy Interrelationships and Glossary of Energy Terms*. Publ. no. 1411. Washington, D.C.: NRC.

NRC. 1966c. *Hormonal Relationships and Applications in the Production of Meats, Milk, and Eggs*. Pub. no. 1415. Washington, D.C.: NRC.

NRC. 1968a. *Nutrient Requirements of Domestic Animals, No. 2: Nutrient Requirements of Swine*. Pub. no. 648. Washington, D.C.: NRC.

NRC. 1968b. *Nutrient Requirements of Sheep*. Pub. no. 1193. Washington, D. C.: NRC.

NRC. 1969. *United States-Canadian Tables of Feed Composition*. Publ. no. 1684. Washington, D.C.: NRC.

NRC. 1970. *Nutrient Requirements of Beef Cattle, No. 4*. Pub. no. 1137. Washington, D.C.: NRC.

NRC. 1971a. *Animal Environmental Research and Facilities Guide*. Washington, D.C.: NRC.

NRC. 1971b. *Nutrient Requirements of Domestic Animals, No. 3: Nutrient Requirements of Dairy Cattle*. 4th rev. ed. Pub. no. 1349. Washington, D.C.: NRC.

NRC. 1973. *Nutrient Requirements of Horses*. 3d rev. ed. Pub. no. 1401. Washington, D.C.: NRC.

O'Dell, C. A. 1969. *The Domestic Wool Marketing System.* ERS Rpt. no. 400. Washington, D.C.: USDA.

Odum, H. T. 1971. *Science, Power, and Society.* New York: Wiley-Interscience.

Omtvedt, I. T. 1968. "Some heritability characteristics and their importance in a selection program." In Topel (1968).

Onions, W. J. 1962. *Wool: An Introduction to its Properties, Varieties, Uses, and Production.* London: Ernest Benn.

Opel, H., and A. V. Nalbandov. 1961a. "Follicular growth and ovulation in hypophysectomized hens." *Endocrinol.*, 69:1016–1028.

Opel, H., and A. V. Nalbandov. 1961b. "Ovuliability of ovarian follicles in the hypophysectomized hen." *Endocrinol.*, 69:1029–1035.

Osman, A. H. 1966. "Animal breeding in the Sudan." *Sudan J. Vet. Sci. and Animal Husb.*, 7: 102–112.

Owen, C. 1969. "The domestication of the ferret." In Ucko and Dimbleby (1969).

Owen, J. B. 1971. *Performance Recording in Sheep.* Technical Communication no. 20. Edinburgh: Commonwealth Bureau of Animal Breeding and Genetics.

Palmer, A. Z. 1963. "Relation of age, breed, sex, and feeding practices on beef and pork tenderness." *Proc. Meat Tenderness Symp., Campbell Soup Co., Camden, N.J.*, pp. 161–169.

Pan American Health Organization. 1969. *International Symposium on Health Aspects of the International Movement of Animals.* Publ. no. 182. Washington, D.C.: PAHO.

Parkes, A. S., ed. 1952, 1956, 1960. *Marshall's Physiology of Reproduction.* 3d. ed. Vol. I, part 1, 1956; vol. I, part 2, 1960; vol. II, 1952. London: Longmans, Green.

Patten, B. M. 1964. *Foundations of Embryology.* 2d ed. New York: McGraw-Hill.

Patterson, R. L. S. 1968. "5-a-androst-16-ene-3-one: Compound responsible for taint in boar fat." *J. Sci. Food. Agr.*, 19(1):31–38.

Pauling, L., and R. B. Corey. 1951. "Configurations of polypeptide chains with favored orientations around simple bonds: Two new pleated sheets." *Proc. Natl. Acad. Sci. U.S.*, 37:729.

Pauling, L., and R. B. Corey. 1953. "Two pleated-sheet configurations and two rippled-sheet configurations of polypeptide chains." *Proc. Natl. Acad. Sci. U.S.*, 39:253–256.

Pauling, L., R. B. Corey, and H. R. Branson. 1951. "The structure of proteins: Two hydrogen-bonded helical configurations of the polypeptide chain." *Proc. Natl. Acad. Sci. U.S.*, 37:205–211.

Penionzhkevich, E. E. 1962. *Poultry Science and Practice,* Vol. 1: *Biology, Breeds, and Breeding.* Moscow. Translated from Russian by Israel Program for Sci. Transl., 1968.

Perkins, D., Jr. 1964. "Prehistoric fauna from Shanidar, Iraq." *Science,* 144:1565–1566.

Perkins, D., Jr., 1969. "Fauna of Catal Huyuk: Evidence for early cattle domestication in Anatolia." *Science,* 164:177–179.

Perry, E. J., ed. 1968. *The Artificial Insemination of Farm Animals.* 4th, rev. ed. New Brunswick, N.J.: Rutgers Univ. Press.

Peters, R. H. 1962. *Textile Chemistry, Vol. I: The Chemistry of Fibers.* New York: Elsevier.

Petty, R. R., and T. C. Cartwright. 1966. *A Summary of Genetic and Environmental Statistics for Growth and Conformation Traits of Young Beef Cattle.* Dept. Anim. Sci. Tech. Rep. no. 5. College Station: Texas A&M University.

Pfaffenberger, Clarence. 1963. *The New Knowledge of Dog Behavior.* New York: Howell.

Phillips, Ralph W. 1949. *Breeding Livestock Adapted to Unfavorable Environments.* FAO Agric. Studies. no. 1. Rome, Italy: FAO.

Phillips, Ralph W. 1961. "Untapped Sources of Animal Germ Plasm." In Hodgson (1961), pp. 43–75.

Phillips, Ralph W. 1963a. "Animal products in the diets of present and future world populations." *J. Animal Sci.*, 22:251.

Phillips, Ralph W. 1963b. "The necessity of defining needs and establishing priorities for the solution of animal production problems, taking into consideration the needs of human nutrition. *Proceedings, World Conference on Animal Production,* vol. 1, main reports, pp. 7–45.

Phillips, Ralph W. 1964. "Animal agriculture in the emerging nations." In A. H. Moseman, ed., *Agricultural Sciences for the Developing Nations* (Washington, D.C.: AAAS), pp. 15–32.

Phillips, Ralph W. 1967. "Animal agriculture in Asia: Present and future." *Proc. Fifteenth Annual Meeting, Agricultural Research Institute,* pp. 147–178. Washington, D.C.: NRC.

Phillips, Ralph W. 1968a. "National, regional and international societies serving animal science." *J. Animal Sci.*, 27:251.

Phillips, Ralph W. 1968b. "The use of animals in

providing foods: Factors favoring animal production." *Proc. Second World Congress on Animal Production,* pp. 15–23.

Phillipson, A. T., ed. 1970. *Physiology of Digestion and Metabolism in the Ruminant.* Newcastle upon Tyne, England: Oriel.

Pike, R. L., and M. L. Brown. 1967. *Nutrition: An Integrated Approach.* New York: Wiley.

Pirchner, F. 1968. *Population Genetics in Animal Breeding.* (English trans.) San Francisco: W. H. Freeman and Co.

Poats, F. J., and W. Fong. 1957. *Economic Evaluation of Color in Domestic Wool.* ERS Mktg. Res. Rpt. no. 204. Washington, D.C.: USDA.

Pohle, E. M., D. D. Johnston, H. R. Keller, W. A. Mueller, H. D. Ray, and H. C. Reals. 1958. *Value-Determining Physical Properties and Characteristics of Domestic Wools.* CMS Mktg. Res. Rpt. no. 211. Washington, D.C.: USDA.

Pohle, E. M., D. C. Johnston, H. D. Ray, and W. J. Manning. 1953. *Suggested Staple Lengths for Grades of Grease Wool.* Washington, D.C.: USDA.

Polge, C. 1968. "Frozen semen and the A.I. programme in England." *Proc. Second Conf. Artificial Insemination and Reprod,* pp. 46–51.

Porter, A. R., J. A. Sims, and D. F. Foreman. 1964. *Dairy Cattle in American Agriculture.* Ames: Iowa State Univ. Press.

Posselt, J. 1963. "The domestication of the Eland." *Rhodesian J. Agr. Res.,* 1:81–87.

Powell, W. E. and D. L. Huffman. 1968. "An evaluation of quantitative estimates on beef carcass composition." *J. Animal Sci.,* 27(6): 1554–1558.

Preston, R. L. 1966. "Protein requirements of growing-finishing cattle and lambs." *J. Nutr.,* 90:157–160.

Preston, R. L. 1971. "Effects of nutrition on the body composition of cattle and sheep." *Proc. Georgia Nutrition Conference, University of Georgia, Athens,* p. 26.

Preston, R. L., and W. H. Pfander. 1964. "Phosphorous metabolism in lambs fed varying phosphorous intakes." *J. Nutr.,* 83:369–378.

Proceedings of the Fourth International Textile Research Conference, Berkeley, California, 1971. New York: Interscience.

Protsch, R. 1970. "Radiocarbon dates from some of the earliest domesticated mammals in Europe." M.A. thesis, University of California at Los Angeles.

Pryor, Donald. 1970. "The road to market." *Finance and Development,* 7(4):22.

Quinlivan, T. D., and T. J. Robinson, 1967. "The number of spermatozoa in the fallopian tubes of ewes at intervals after artificial insemination following withdrawal of SC-9880 impregnated intravaginal sponges." In T. J. Robinson, ed., *The Control of the Ovarian Cycle in the Sheep* (Sydney: Sydney Univ. Press).

Rae, A. L. 1956. "The genetics of the sheep." *Advances in Genetics,* 8:189.

Rasmussen, H. 1961. "The parathyroid hormone. *Scientific American,* 204(4):59. Available as Offprint no. 86 from W. H. Freeman and Company.

Raspe, G., ed. 1969. *Advances in the Biosciences.* New York: Permagon.

Reed, Charles A. 1959. "Animal domestication in the prehistoric Near East." *Science,* 130:1629–1639.

Reed, C. A. 1969. "The pattern of animal domestication in the prehistoric Near East." In Ucko and Dimbleby (1969), pp. 361–380.

Reed, Charles A., and William M. Schaffer, 1972. "How to tell the sheep from the goats." *Field Museum of Natur. Hist. Bull.,* 43:2–7.

Reid, H. W., M. G. Burridge, N. B. Pullan, R. W. Sutherst, and E. B. Wain. 1966. "A survey of trypanosomiasis in the domestic cattle of South Busoga." *Int. Scient. Comm. Trypanosom.,* no. 11, p. 31.

Reid, J. T., W. K. Kennedy, K. L. Turk, S. T. Slack, G. W. Trimberger, and R. P. Murphy. 1959. "Effect of growth stage, chemical composition, and physical properties upon the nutritive value of forages." *J. Dairy Sci.,* 42:567–571.

Retief, G. P. 1971. "The potential of game domestication in Africa with special reference to Botswana." *J. So. Afr. Vet. Med. Assn.,* 42:119–127.

Reynolds, M., and S. J. Folley. 1969. *Lactogenesis: The Initiation of Milk Secretion at Parturition.* Philadelphia: Univ. of Pennsylvania Press.

Rhoad, A. O., ed. 1955. *Breeding Beef Cattle for Unfavorable Environments.* Austin: Univ. of Texas Press.

Rice, V. A., F. N. Andrews, E. J. Warwick, and J. E. Legates. 1967. *Breeding and Improvement of Farm Animals.* 6th ed. New York: McGraw-Hill.

Richardson, K. C. 1949. "Contractile tissues in the mammary gland, with special reference to myoepithelial in the goat." *Proc. Roy. Soc. London, B*, 136:30–45.

Richmond, C. R., W. H. Langham, and T. T. Trujillo. 1962. "The comparative metabolism of tritiated water by mammals." *J. Cellular Comp. Physiol.*, 59:45–53.

Richter, C. P. 1952. "Domestication of the Norway rat and its implication for the study of genetics in man." *Amer. J. Hum. Genet.*, 4:273–285.

Ridgeway, W. 1905. *The Origin and Influence of the Thoroughbred Horse.* Cambridge, Eng.: Cambridge Univ. Press.

Riemann, Hans, ed. 1970. *Food-Borne Infections and Intoxications.* New York: Academic Press.

Riggs, E. S. 1936. *The Geological History and Evolution of the Horse.* Leaflet no. 13. Chicago, Ill.: Field Museum of Natural History.

Robinson, F. A. 1966. *The Vitamin Co-Factors of Enzyme Systems.* New York: Pergamon.

Robinson, T. J. 1951a. "The control of fertility in sheep, II." *J. Agric. Sci.*, 41:6.

Robinson, T. J. 1951b. "The necessity for progesterone with estrogen for the induction of recurrent estrus in the ovariotimized ewe." *J. Agric. Sci., Camb.*, 41:6.

Robinson, T. J. 1954. Quantitative studies of the hormonal induction of oestrus in spayed ewes." *Endocrinol.*, 55: 403.

Roeder, K. 1963. *Nerve Cells and Insect Behavior.* Cambridge, Mass.: Harvard Univ. Press.

Rogers, G. B. 1971. *Vertical and Horizontal Integration in the Market Egg Industry, 1955–69.* Washington, D.C.: USDA.

Rogers, G. B., and R. M. Conlogue. 1970. *Economic Characteristics of and Changes in the Market Egg Industry.* Washington, D.C.: USDA.

Rogers, G. B., and L. A. Voss. 1969. *Pricing Systems for Eggs.* Washington, D.C.: USDA.

Rogerson, A. 1968. "Energy utilisation by the eland and the wildebeest." In Crawford (1968a), p. 153.

Rollins, W. C. 1966. "Genetic improvement in horses." In Cole (1966), pp. 330–332.

Roman, J., C. J. Wilcox, and F. G. Martin. 1970. "Milk production of tested Holsteins in Ecuador." *J. Dairy Sci.*, 53:673.

Romanoff, A. L., and A. J. Romanoff. 1949. *The Avian Egg.* New York: Wiley.

Romanoff, A. L., and A. J. Romanoff. 1967. *Biochemistry of the Avian Embryo.* New York: Wiley.

Roth, H. H., and R. Osterberg. 1971. "Studies on the agricultural utilisation of semi-domesticated eland. (*Taurotragus oryx*) in Rhodesia, IV." *Rhod. J. Agric. Res.*, 9:45.

Rouse, G. H., D. G. Topel, R. L. Vetter, R. E. Rust, and T. W. Wickersham. 1970. "Carcass composition of lambs at different stages of development." *J. Animal Sci.*, 31:846.

Rouse, J. E. 1970. *World Cattle.* 2 vols. Norman: Univ. of Oklahoma Press.

Rowson, L. E., and C. E. Adams. 1957. "An egg transfer experiment on sheep." *Vet. Rec.*, 69: 849.

Royal Veterinary College. 1970. *East African Research Team Report.* London: RVC.

Ruhweza, S. 1968. "Game management practices in Uganda." *E. Afr. Agric. For. J.*, 33:275.

Rusoff, L. L. 1964. "The role of milk in modern nutrition." *Borden's Rev. Nutr. Res.*, 25:17.

Saacke, R. G. 1970. "Morphology of the sperm and its relationship to fertility." *Proc. Third Conf. Artificial Insemination and Reprod.*, pp. 17–30.

Saeger, S. W. J. 1969. "Successful pregnancies utilizing frozen dog semen." *A.I. Digest,* 27:6.

Salisbury, G. W. 1968. "Aging of spermatozoa during storage and DNA." *Proc. Second Conf. Artificial Insemination and Reprod.*, pp. 27–32.

Salisbury, G. W., and N. L. VanDemark. 1961. *Physiology of Reproduction and Artificial Insemination of Cattle.* San Francisco, Ca.: W. H. Freeman and Co.

Sauer, C. O. 1969. *Agricultural Origins and Dispersals. The Domestication of Animals and Foodstuffs.* 2d ed. Cambridge, Mass.: MIT Press.

Saunders, A. H. 1925. "The Taurine World." *Nat. Geographic Mag.*, 48:591–710.

Scheele, C. W. 1780. *Kongl. Vetenskaps Akadamiens nya Handlingbar,* tom I, 116, 269.

Schein, M., and E. Hale. 1965. "Stimuli eliciting sexual behavior." In F. Beach, ed., *Sex and Behavior* (New York: Wiley), pp. 440–482.

Schley, P. 1967. "Der Yak und seine Kreuzang mit dem Rind in der Sowjetunion." *Oesteuropastudien der Hochschulen des Landes Hessen Reiche,* I, 44. Wiesbaden: Otto Harrassowitz.

Schmidt-Nielsen, K. 1964. *Desert Animals: Physi-*

ological Problems of Heat and Water. New York and Oxford: Oxford Univ. Press.

Schneider, B. H. 1947. *Feeds of the World, Their Digestibility and Composition*. Agr. Exp. Sta., Morgantown: West Virginia Agric. Exper. Stat.

Scholander, P. F., R. Hock, V. Walters, F. Johnson, and L. Irving. 1950. "Heat regulation in some artic and tropical mammals and birds." *Biol. Bull.*, 99:237.

Schulthess, W. 1967. "Yak and Tsauri in Nepal." *World Rev. of Animal Production*, 3:88–97.

Schurch, A., and C. Wenk, eds. 1970. *Energy Metabolism of Farm Animals*. EEAP publ. no. 13. Zürich: EEAP.

Schwabe, C. W. 1969. *Veterinary Medicine and Human Health*. 2d ed. Baltimore, Md.: Williams and Wilkins.

Scott, George E., ed. 1970. *Sheepman's Production Handbook*. Denver, Col.: Industry Development Program.

Scott, John Paul, and John L. Fuller. 1965. *Genetics and the Social Behavior of the Dog*. Chicago, Ill.: Univ. of Chicago Press.

Scott, M. L. 1956. "Composition of turkey meat." *J. Amer. Dietetic Assn.*, 32:941–944.

Scott, M. L., M. C. Nesheim, and R. J. Young. 1969. *Nutrition of the Chicken*. New York: M. L. Scott.

Searle, A. G. 1968. *Comparative Genetics of Coat Color in Mammals*. New York: Academic Press.

Sebastian, L., V. D. Mudgal, and G. P. Nair. 1970. "Comparative efficiency of milk production by Sahiwal cattle and Murrah buffalo." *J. Animal Sci.*, 30:253–256.

Self, H. L. 1959. *Traits of Major Breeds*. National Hog Farmer Swine Information Service Bulletin no. B3.

Seligman, M. 1970. "On the generality of the laws of learning." *Psychol. Rev.*, 77:406.

Sell, O. E., J. T. Reid, P. G. Woolfolk, and R. E. Williams. 1959. "Intersociety forage evaluation symposium." *Agronomy J.*, 51:212–245.

Shackelford, A. D., R. E. Childs, and I. A. Hamann. 1969. *Determination of Bruise Rates on Broilers Before and After Handling by Live-Bird Pickup Crews*. Washington, D.C.: USDA.

Shaw, R. H., ed. 1967. *Ground-Level Climatology*. Washington, D.C.: AAAS.

Shelton, M. 1960. "The relation of face covering to fleece weight, body weight, and kid production of Angora does." *J. Animal Sci.*, 19:302–308.

Shelton, M., and J. W. Bassett. 1970. *Estimate of Certain Genetic Parameters Relating to Angora Goats*. Texas A. and M. Univ. *Agric. Exper. Stat.* R-2750.

Shikama, T., and G. Okafuji. 1958. "Quarternary cave and fissure deposits and their fossils in Akiyosi District, Yamaguti Prefecture." *Sci. Rep. Yokohama Nat. Univ., Sect. II: Biol. Geol. Sci.*, (7):43–103.

Shull, G. H. 1909. "A pure-line method in corn breeding." *J. Am. Breeders' Assn.*, 5:51–59.

Sidwell, G. M. 1956. "Some aspects of twin versus single lambs of Navajo and Navajo crossbred ewes." *J. Animal Sci.*, 15:202.

Sidwell, G. M., D. O. Everson, and C. E. Terrill. 1962. "Fertility, prolificacy, and lamb livability of some pure breeds and their crosses." *J. Animal Sci.*, 21:875–879.

Sidwell, G. M., D. O. Everson, and C. E. Terrill. 1964. "Lamb weights in some pure breeds and crosses." *J. Animal Sci.*, 23:105–110.

Sidwell, G. M., and L. R. Miller. 1971. "Production in some pure breeds of sheep and their crosses." *J. Animal Sci.*, 32:1084–1098.

Sidwell, G. M., R. L. Wilson, and M. E. Hourihan. 1971. "Production in some pure breeds of sheep and their crosses, IV: Effect of crossbreeding on wool production." *J. Animal Sci.*, 32:1099–1102.

Simkiss, K. 1962. "Viviparity and avian reproduction." *J. British Ornithologists' Union*, 104(1): 216–219.

Simons, N. 1962. *Wildlife in Kenya (Between the Sunlight and Thunder)*. London: Collins.

Simoons, F. J. 1961. *Eat Not This Flesh: Food Avoidances in the Old World*. Madison: Univ. of Wisconsin Press.

Simoons, F. J., and E. S. Simoons. 1968. *A Ceremonial Ox of India: The Mithan in Nature, Culture, and History*. Madison: Univ. of Wisconsin Press.

Singh, B. P., W. E. Rempel, D. Reimer, H. H. Hanke, K. P. Miller, and A. B. Solmela. 1967. "Evaluation of breeds of sheep on the basis of crossbred lamb performance." *J. Animal Sci.*, 26:261–266.

Sisson, S., and J. D. Grossman. 1938, 1953. *The Anatomy of Domestic Animals*. Philadelphia, Pa.: Saunders.

Skinner, J. D. 1967. "An appraisal of the eland as a farm animal in Africa." *Anim. Breed. Abstr.*, 35:176–186.

Skinner, J. D. 1970. "An appraisal of the eland as a farm animal in Africa." *Trop. Anim. Hlth. Prod.*, 2:151.

Slade, L. M., and D. W. Robinson. 1970. "Nitrogen metabolism in nonruminant herbivores, II: Comparative aspects of protein digestion." *J. Animal Sci.*, 30: 761–763.

Sluckin, W. 1965. *Imprinting and Early Learning.* Chicago, Ill.: Aldine.

Smith, A. M., G. L. Holck, and H. B. Spafford. 1966. "Calcium, phosphorous and vitamin D." *J. Dairy Sci.*, 49:239–243.

Smith, C. 1964. "The use of specialised sire and dam lines in selection for meat production." *Animal Prod.*, 6:337–344.

Smith, Harold DeWitt. 1944. "Textile fibers: An engineering approach to their properties and utilization." Paper presented to the 47th Annual Meeting of the American Society for Testing and Materials.

Smith, N. E. 1971. "Feed efficiency in intensive milk production." 10th Annual Dairy Cattle Day, University of Calif., p. 40.

Snider, R. S. 1958. "The cerebellum." *Scientific American,* 199(2):84. Available as Offprint no. 38 from W. H. Freeman and Company.

Solberg, M. 1968. "Factors affecting fresh meat color." *Proc. Meat Industry Res. Conf., AMSA and AMIF, Chicago, Ill.*, pp. 32–40.

Spaeth, C. W., Z. L. Carpenter, G. T. King, and M. Shelton. 1967. "Effects of forced exercise on lamb carcass merit." *J. Animal Sci.*, 26(4): 901. (Abstr.).

Spector, W. S. 1956. *Handbook of Biological Data.* Philadelphia, Pa.: Saunders.

Spinage, C. A. 1962. *Animals of East Africa.* London: Collins.

Srb, A. M., R. D. Owen, and R. S. Edgar. 1965. *General Genetics.* 2d ed. San Francisco, Ca.: W. H. Freeman and Co.

Stabenfeldt, G. H., J. P. Hughes, and J. W. Evans. 1972. "Ovarian activity during the estrous cycle of the mare." *Endocrinol.*, 90:1379.

Stadleman, W. J., and C. G. Haugh. 1971. "Poultry products." In *ASHRAE Guide and Data Book Applications* (New York: American Society of Heating, Refrigerating, and Air-Conditioning Engineers), pp. 311–322.

Stallcup, O. T., and G. V. Davis. 1965. "Assessing the feeding value of forages by direct and indirect methods." *Arkansas Agric. Exper. Stat. Bull.*, no. 704.

Stanislaus County DHIA. 1970. *Annual Report,* no. 48. Modesto: Calif. Agric. Ext. Serv.

Steel, W. S. 1968. "Technology of wildlife management and game cropping in the Luangwa Valley, Zambia." *E. Afr. Agric. For. J.*, 33:226.

Stewart, G. F. 1967. "Fifty years of research in egg quality." *Commonwealth Sci. and Ind. Res. Org. (Australia) Quarterly.*, 27:73–82.

Stewart, G. F., and J. C. Abbott. 1961. *Marketing Eggs and Poultry.* Rome, Italy: FAO.

Stoddart, L. A., and A. D. Smith. 1954. *Range Movement.* 2d ed. New York: McGraw-Hill.

Stouffer, J. R., and G. H. Wellington. 1960. "Ultrasonics for evaluation of live animal and carcass composition." *Proc. 12th Res. Conf., AMIF, Chicago, Ill.*, pp. 81–87.

Stroock, S. I. 1937. *Llamas and Llamaland.* New York: Stroock.

Sturkie, P. D. 1965. *Avian Physiology.* Ithaca, N.Y.: Comstock.

Sud, S. C., H. A. Tucker, and J. Meites. 1968. "Estrogen-progesterone requirements for udder development in ovariectomized heifers." *J. Dairy Sci.*, 51:210.

Swanson, E. W. 1967. "Optimum growth patterns for dairy cattle." *J. Dairy Sci.*, 50:244.

Swenson, M. J., ed. 1970. *Duke's Physiology of Domestic Animals.* Ithaca, N.Y.: Comstock.

Swerdlow, M., and G. S. Seeman. 1948. "A method for the electron microscopy of wool." *Textile Res. J.*, 18:536–556.

Swift, R. W. 1957. "The nutritive evaluation of forages." *Pennsylvania Agric. Exper. Stat. Bull.*, no. 615.

Swift, R. W., and C. E. French. 1954. *Energy Metabolism and Nutrition.* Washington, D.C.: Scarecrow Press.

Symposium on Fibrous Proteins. 1967. Australia.

Szczesniak, A. S., and K. W. Torgeson. 1965. "Methods of meat texture measurement viewed from the background of factors affecting tenderness." *Advances Food Res.*, 14:33–165.

Taiganides, E. P., and R. K. White. 1968. *Origin, Identification, Concentration, and Control of Noxious Gases in Animal Confinement Production Units.* Columbus: Ohio State University Research Foundation.

Taylor, C. R. 1968a. "Minimum water requirements on some East African bovids." In Crawford (1968a), p. 195.

Taylor, C. R. 1968b. "Hygroscopic food: a source of water for desert animals." *Nature,* 219:181.

Taylor, C. R. 1969. "Metabolism, respiratory changes and water balance of the antelope and the eland." *Amer. J. Physiol.,* 217:907.

Taylor, C. R., and C. P. Lyman. 1967. "A comparative study of the environmental physiology of an East African antelope, eland and Hereford steer." *Physiol. Zool.,* 40:280.

Teal, J. J., Jr. 1970. "Domesticating the wild and wooly musk ox." *Nat. Geogr. Mag.* 137, no. 6: 862–879.

Tener, J. S. 1965. *Musk Oxen in Canada: A Biological and Taxonomic Review.* Ottawa, Canada: Queen's Printer.

Termohlen, W. D. 1968. "Past history and future developments." *Poultry Sci.,* 47:6–22.

Terrill, C. E. 1949. "The relation of face covering to lamb and wool production in range Rambouillet ewes." *J. Animal Sci.,* 8:353.

Terrill, C. E. 1958. "Fifty years of progress in sheep breeding." *J. Animal Sci.,* 17:944.

Terrill, C. E. 1962. *Heritability Estimates of Farm Animals, Part II: Sheep in Growth.* Washington, D.C.: Federation of American Societies for Experimental Biology.

Terrill, C. E. 1968. "The artificial insemination of farm animals." In Perry (1968), pp. 215–243.

Terrill, C. E. 1970. "The exotic breeds of sheep and the possibilities of their strengthening the American sheep industry." Paper presented at 62nd Annual Meeting of ASAS.

Terrill, C. E., and J. A. Stoehr. 1942. "The importance of body weight in selection of range ewes." *J. Animal Sci.,* 1:221–228.

Thompson, G. F. 1901. *Information Concerning the Angora Goat.* Bull. no. 27. Washington, D.C.: USDA.

Tinbergen, N. 1969. *The Study of Instinct.* 2d ed. London: Oxford Univ. Press.

Topel, D. G., ed. 1968. *The Pork Industry: Problems and Progress.* Ames: Iowa State University Press.

Touchberry, R. W. 1970. "A comparison of the merits of purebred and crossbread dairy cattle resulting from twenty years (four generations) of crossbreeding." *Proc. 19th Ann. National Breeders' Roundtable, Kansas City, Mo., May 1970.*

Towne, C. W., and E. N. Wentworth. 1950. *Pigs from cave to cornbelt.* Norman: Univ. of Oklahoma Press.

Treus, V., and D. V. Krevchenko. 1968. "Methods of rearing and economic utilisation of eland in the Askanya-Nova Zoological Park." In Crawford, (1968a), pp. 395–411.

Trevelyan, G. M. 1944. *English Social History.* London: Longman's Green.

Tuckey, S. L., and D. B. Emmons. 1967. *Cottage Cheese and Other Cultured Milk Products.* New York: Chas. Pfizer.

Turkington, R. W., and Y. J. Topper. 1966. "Stimulation of casein synthesis and histological development of mammary gland by human placental lactogen in vitro." *Endocrinol.,* 79:175.

Turnbull, P., and C. A. Reed. "Faunal study of Palegawra, a Zarzian site from the late Pleistocene of northern Iraq." *Fieldiana,* in press.

Turner, A. B., D. S. H. Smith, and A. M. Mackie. 1971. "Characterization of the principal steroidal saponins of the starfish *Marthasterias glacialis:* Structures of the aglycones." *Nature,* 233: 209–210.

Turner, C. D., and J. T. Bagnara. 1971. *General Endocrinology.* Philadelphia, Pa.: Saunders.

Turner, C. W. 1952. *The Mammary Gland, I: The Anatomy of the Udder of Cattle and Domestic Animals.* Columbia, Mo.: Lucas.

Turner, C. W. 1969. *Harvesting Your Milk Crop.* Chicago, Ill.: Bobson.

Turner, H. N., and S. S. Y. Young. 1969. *Quantitative Genetics in Sheep Breeding.* Ithaca, N.Y.: Cornell Univ. Press.

Turton, J. D. 1969. "Recent research in cattle breeding and production in the U.S.S.R.: A review." *Animal Breeding Abst.,* 37:347.

Tyrrell, H. F., and W. P. Flatt. 1971. "Energetics of body tissue mobilization." *J. Dairy Sci.,* 54: 548.

Ucko, P. J., and G. W. Dimbleby, eds. 1969. *The Domestication and Exploitation of Plants and Animals.* London: Duckworth.

Underwood, E. J. *Trace Elements in Human and Animal Nutrition.* 3d ed. New York: Academic Press.

United Nations. 1958. *The Future Growth of World Populations.* Population Studies no. 28. New York: UN.

United Nations. 1964. *Provisional Report on World Population Prospects, As Assessed in 1963.* New York: UN.

United Nations. 1965. *World Population Prospects*

up to the Year 2000. Population Commission Document no. E/CN.9/186. New York: UN.

United Nations. 1969. "New findings on population trends." *Population Newsletter,* no. 7, pp. 1–4. New York: UN.

Upadhaya, R. M. 1969. "Chauries and their place in Nepalese economy." *Zootechnica Veterinaria,* 24:281–283.

USDA. 1949. *Comparison of Core Tests and Visual Estimates of Shrinkage with Actual Mill Scouring Results on 96 lots of Wool.* Washington, D.C.: USDA.

USDA. 1956a. *Slaughter Vealers and Calves.* Effective October 1, 1956. Washington, D.C.: USDA.

USDA. 1956b. *Sow Carcasses.* Effective September 1, 1956. Washington, D.C.: USDA.

USDA. 1956c. *Veal and Calf Carcasses. Effective* October 1, 1956. Washington, D.C.: USDA.

USDA. 1960a and 1969e. *Lamb, Yearling Mutton, and Mutton Carcasses.* Quality, effective March 1, 1960. Yield, effective March 1, 1969. Washington, D.C.: USDA.

USDA. 1960b and 1969f. *Slaughter Lambs and Sheep.* Quality, effective June 18, 1960. Yield, effective March 1, 1969. Washington, D.C.: USDA.

USDA. 1963. *Composition of Foods.* Agric. Handbook no. 8. Washington, D.C.: USDA.

USDA. 1964a. *Feeder Cattle.* Effective September 25, 1964. Washington, D.C.: USDA.

USDA. 1964b. *A Graphic Summary of World Agriculture.* Misc. Pub. no. 705, rev. ed. Washington, D.C.: USDA.

USDA. 1965a. *Carcass Beef (Steer, Heifer, Cow, Bull, and Stag).* Effective June 1, 1965. Washington, D.C.: USDA.

USDA. 1965b. *Losses in Agriculture.* Agric. Handbook no. 291. Washington, D.C.: USDA.

USDA. 1965c. *U.S. Food Consumption.* Statistical Bulletin no. 364. Washington, D.C.: USDA.

USDA. 1966a. *Changes in Farm Productivity and Efficiency: A Summary Report.* Statistical Bull. no. 233. Washington, D.C.: USDA.

USDA. 1966b. *Official Standards of the United States for Grades of Wool.* Consumer and Marketing Service Announcement no. 135 (reprinted with amendments, April 1966). Washington, D.C.: USDA.

USDA. 1966c. *Slaughter Cattle (Steers, Heifers, Cows, Bulls, and Stags).* Effective April 23, 1966. Washington, D.C.: USDA.

USDA. 1968a. *Barrow and Gilt Carcasses.* Effective April 1, 1968. Washington, D.C.: USDA.

USDA. 1968b. *Recommendations for Uniform Sheep Selection Programs.* Washington, D.C.: USDA.

USDA. 1968c. *Slaughter Barrows and Gilts.* Effective July 1, 1968. Washington, D.C.: USDA.

USDA. 1968d. *Yield Grades for Beef.* Marketing Bull. no. 45. Washington, D.C.: USDA.

USDA. 1969a. *Dairy Situation.* ERS DS, no. 327. Washington, D.C.: USDA.

USDA. 1969b. *Egg Grading Manual.* Agriculture handbook no. 75. Washington, D.C.: USDA.

USDA. 1969c. *Feeder Pigs.* Effective April 1, 1969. Washington, D.C.: USDA.

USDA. 1969d. *Food Consumption, Prices, and Expenditures.* ERS Supp. Agric. Econ. Rep. no. 138. Washington, D.C.: USDA.

USDA. 1969g. *Slaughter Sows.* Effective September 1, 1969. Washington, D.C.: USDA.

USDA. 1969h. *Summary of Major Provisions of Federal Milk Marketing Orders.* Washington, D.C.: USDA.

USDA. 1970a. *Foreign Agric. Circular,* May 1970. Washington, D.C.: USDA.

USDA. 1970b. *Grease Mohair Standards—Proposed Standards for Grade.* CMS 7CFR, part 32. Washington, D.C.: USDA.

USDA. 1970c. *Wool Statistics and Related Data.* Statistical bull. no. 455. Washington, D.C.: USDA.

USDA. 1970d. *Yield Grades for Lamb.* Marketing Bull. no. 52. Washington, D.C.: USDA.

USDA. 1971a. *Dairy Herd Improvement Letter,* 47(5). Washington, D.C.: USDA.

USDA. 1971b. Foreign Agric. Circular, Jan. 1971. Washington, D.C.: USDA.

USDA. 1971c. "Inspection of poultry and poultry products." *Federal Register,* 36, no. 103:9716–9754.

Vander, A. J., J. H. Sherman, and D. S. Luciano. 1970. *Human Physiology: The Mechanism of Function.* New York: McGraw-Hill.

Van Es, L., and J. F. Olney. 1940. "An enquiry into the influence of environment on the incidence of poultry disease." *Nebraska Agric. Exper. Stat. Res. Bull.,* no. 118.

Van Soest, P. J. 1967. "Development of a comprehensive system of feed analysis and its application to forages." *J. Anim. Sci.,* 26:119–128.

Van Tienhoven, A., 1968. *Reproductive Physiology of Vertebrates.* Philadelphia, Pa.: Saunders.

Varo, M. 1965. "Some coefficients of heritability in horses." *Annls. Agric. Fenn.*, 4:223.

Veisseyre, R. 1969. "Revue lait." Cited in *Dairy Sci. Abst.*, 31, 3737.

Verhave, T. 1970. "The inspector general is a bird." In N. Adler, ed., *Readings in Experimental Psychology Today* (Del Mar, Ca.: CRM Books), pp. 91–96.

Verley, F. A., and R. W. Touchberry. 1961. "Effects of crossbreeding on reproductive performance of dairy cattle." *J. Dairy Sci.*, 44:2058.

Vesey-Fitzgerald, D. F. 1965. "The utilisation of natural pastures by wild animals in the Rukwa Valley, Tanganyika." *E. Afr. Wildl. J.*, 4:38.

Villarroel, L. J. 1966. "The production and industry of alpaca, llama and vicuna in Peru." *Proceedings of International Symposium on Technical Economic Problems of the Production of Sheep, Goats, Auchenids, and Fur Animals, Milan, Italy.*

Von Bergen, Werner, ed. 1963, 1969, 1970. *Wool Handbook.* 3d ed., 3 vols. New York: Interscience.

Von Neuhaus, U. 1959. "Milch and milchgewinnung von Pferdestuten." *Zeitschr. für Tierzuchlung und Zuchtingsbiol.*, 73:370–392.

Wallace, L. R. 1948. "The growth of lambs before and after birth in relation to the level of nutrition." *J. Agric. Sci., Camb.*, 38:367–401.

Walsh, E. G. 1964. *Physiology of the Nervous System.* London: Longmans, Green.

Walton, A., and J. Hammond. 1938. "The maternal effects on growth and conformation in Shire horse–Shetland pony crosses." *Proc. Roy. Soc. London, B*, 125:311.

Wanderstock, J. J. 1968. "Food analogs." *Cornell Hotel and Restaurant Admin. Quarterly*, 9(2): 29–33.

Warwick, E. J. 1958. "Fifty years of progress in breeding beef cattle." *J. Animal Sci.*, 17:922–943.

Warwick, E. J. 1968. "Crossbreeding and linecrossing beef cattle: Experimental results." *World Rev. of Animal Prod.*, 4, no. 19–20:34–45.

Watson, J. B. 1914. *Behavior: An Introduction to Comparative Psychology.* New York: Henry Holt.

Watson, J. D. 1970. *Molecular Biology of the Gene.* 2d ed. New York: Benjamin.

Webb, B. H., and A. H. Johnson. 1965. *Fundamentals of Dairy Chemistry.* Westport, Conn.: Avi.

Webb, B. H., and E. O. Whittier. 1970. *Byproducts from Milk.* 2d ed. Westport, Conn.: Avi.

Webster, C. C., and P. N. Wilson. 1966. *Agriculture in the Tropics.* London: Longmans, Green.

Weir, C. E. 1960. "Palatability characteristics of meat." In AMIF (1960), pp. 212–221.

Weir, W. C., and R. Albough. 1954. "California sheep production." *Calif. Agric. Exper. Stat. Manual*, no. 16.

Weiss, G. M., D. G. Topel, R. C. Ewan, R. E. Rust, and L. L. Christian. 1971. "Growth comparison of a muscular and fat strain of swine, I: Relationship between muscle quality and quantity, plasma lactate, and 17-hydroxycorticosteroids. *J. Animal Sci.*, 32:1119.

Wentworth, E. N. 1948. *America's Sheep Trails.* Ames: Iowa State Univ. Press.

White, A. and others. 1959. *Principles of Biochemistry.* 2d ed. New York: McGraw-Hill.

Wiersma, F., and G. H. Stott. 1965. "Micro-climate modification for hot weather stress relief in dairy cattle." *Trans. Am. Soc. Agr. Engrs.*, no. 65–404.

Wilcox, C. J., J. A. Curl, J. Roman, A. H. Spurlock, and R. B. Becker. 1966. "Life span and livability of crossbreed dairy cattle." *J. Dairy Sci.*, 49:991.

Wilcox, C. J., S. N. Gaunt, and B. R. Farthing. 1971. *Genetic Interrelationships of Milk Composition and Yield.* Southern Coop. Series Bull. no. 155. Florida Agric. Exper. Stat.

Wilde, P. F., and R. M. C. Dawson. 1966. "The biohydrogenation of a linolenic acid and oleic acid by rumen micro-organisms." *Biochem. J.*, 98:469.

Wilhelm, L. A. 1939. "Egg quality: A literative review." *U.S. Egg and Poultry Mag.*, 45:565–575, 588–594, 619–624, 675–679, 687–694.

Wilkinson, P. 1971. "The domestication of the musk-ox." *Polar Rec.*, 15:683–690.

Williams, C. M. 1967. "Livestock production in cold climates." In Shaw (1967), pp. 221–232.

Williamson, G., and W. J. A. Payne. 1965. *An Introduction to Animal Husbandry in the Tropics.* London: Longmans, Green.

Wilson, D. 1965. *Wool Pools—Organization, Prac-*

tices, Services, and Problems. General Rept. no. 127, Farmer Cooperative Service. Washington, D.C.: USDA.

Wilson and Company, in cooperation with the USDA and the University of Illinois. 1943. *Yields from Different Grades and Weights of Steer Carcasses.* Chicago, Ill.: Wilson and Co.

Winchester, C. F., R. E. Davies, and R. L. Hiner. 1967. "Malnutrition of young cattle: Effect on feed utilization, eventual body size, and meat quality." *USDA Tech. Bull.* no. 1374.

Winn, P. N., Jr., and E. F. Godfrey. 1966. "The effect of humidity on growth and feed conversion of broiler chickens." *Proc. 4th Biometeor. Congr. Rutgers, New Jersey.*

Wood-Gush, D. G. M. 1959. "A history of the domestic chicken from antiquity to the 12th century." *Poultry Sci.,* 39:321–326.

Woods, G. T., A. H. Jensen, T. H. Berry, and H. E. Rhodes. 1962. "Production of primary specific pathogen-free pigs. I: Birth to eight weeks of age." *Illinois Vet.,* 5:27.

Woodwell, George M. "The energy cycle of the biosphere." *Scientific American,* 223(3):64–74. Available as Offprint no. 1190 from W. H. Freeman and Co.

Wool Science Review, vols. 1–31. London: International Wool Secretariat.

Wright, H. E., Jr. 1968. "Natural environment of early food production north of Mesopotamia." *Science,* 161:334–339.

Yazan, Y., and Y. Knorre. 1964. "Domesticating elk in a Russian national park." *Oryx,* 7:301–304.

Yeates, Neil T. M. 1965. *Modern Aspects of Animal Production.* Washington, D.C.: Butterworth.

Young, C. W., W. J. Tyler, A. E. Freeman, H. H. Voelker, L. D. McGilliard, and T. M. Ludwick. 1969. *Inbreeding Investigations with Dairy Cattle in the North Central Region of the United States.* N.C. Tech. Bull. no. 266. Minn. Agric. Exper. Stat.

Young, G. A., N. R. Underdahl, and R. Hintz. 1955. "Procurement of baby pigs by hysterectomy." *Am. J. Vet. Res.,* 16:123.

Young, S. P., and E. A. Goldman. 1964. *The Wolves of North America.* New York: Dover.

Zeuner, F. E. 1954. "Domestication of animals." In C. Singer, E. J. Holmyard, and A. R. Hall, eds., *A History of Technology* (Oxford, Eng.: Clarendon Press), I, 327–352.

Zeuner, F. E. 1963. *A History of Domesticated Animals.* London: Hutchinson.

Zhigunov, P. S., ed. 1968. *Reindeer Husbandry* 2d ed. (Tranl. from Russian). Washington, D.C.: USDI and NSF.

Zinn, D. W., R. M. Durham, and H. B. Hedrick. 1970. "Feedlot and carcass grade characteristics of steers and heifers as influenced by days on feed." *J. Animal Sci.,* 31:302.

Index